D1001938

Developments in Earth Surface Processes

Volume 19

Principles and Dynamics of the Critical Zone

Developments in Earth Surface Processes, 19

Series Editor – J.F. Shroder Jr

For previous volumes refer
http://www.sciencedirect.com/science/bookseries/09282025

Developments in Earth Surface Processes

Volume 19

Principles and Dynamics of the Critical Zone

Edited by

John R. Giardino

Texas A&M University,
College Station,
TX, USA

Chris Houser

Texas A&M University,
College Station,
TX, USA

ELSEVIER

AMSTERDAM • BOSTON • HEIDELBERG • LONDON
NEW YORK • OXFORD PARIS • SAN DIEGO
SAN FRANCISCO • SINGAPORE • SYDNEY • TOKYO

Elsevier
Radarweg 29, PO Box 211, 1000 AE Amsterdam, Netherlands
The Boulevard, Langford Lane, Kidlington, Oxford OX5 1GB, UK
225 Wyman Street, Waltham, MA 02451, USA

British Library Cataloguing-in-Publication Data
A catalogue record for this book is available from the British Library

Library of Congress Cataloging-in-Publication Data
A catalog record for this book is available from the Library of Congress

ISBN: 978-0-444-63369-9
ISSN: 0928-2025

For information on all Elsevier publications
visit our website at http://store.elsevier.com/

This book is dedicated to my daughter,
Anne-Marie Giardino-Wheatley.

Rick

This book is dedicated to my family,
Tina, Jonathan, and Allie.

Chris

Contents

List of Contributors

Suzanne Anderson, Department of Geography, University of Colorado at Boulder, Boulder, Colorado, USA

Anthony Aufdendkampe, Stroud Water Research Center, West Grove, Pennsylvania, USA

Nazgol Bagheri, Department of Political Science and Geography, University of Texas at San Antonio, San Antonio, Texas, USA

Roger Bales, Sierra Nevada Research Institute, University of California, Merced, California, USA

Steve Banwart, Kroto Research Institute, North Campus, University of Sheffield, Broad Lane, Sheffield, UK

Patrick Barrineau, Department of Geography, Texas A&M University, College Station, Texas, USA

Greg Barron-Gafford, School of Geography and Development, Biosphere 2, University of Arizona, Oracle, Arizona, USA

Michael P. Bishop, Department of Geography; Center for Geospatial Sciences, Applications & Technology, Texas A&M University, College Station, Texas, USA

Susan Brantley, Earth and Environmental Systems Institute, Pennsylvania State University, University Park, Pennsylvania, USA

Aniela Chamorro, High Alpine and Arctic Research Program, Department of Geology and Geophysics, Texas A&M University, College Station, Texas, USA

Jon Chorover, Department of Soil Water and Environmental Science, University of Arizona, Tucson, Arizona, USA

Lou Derry, Earth and Atmospheric Sciences, Cornell University, Cornell, New York, USA

William Dietrich, Earth and Planetary Science, University of California, Berkeley, California, USA

John C. Dixon, Department of Geosciences, University of Arkansas, Fayetteville, Arkansas, USA

Iliyana D. Dobreva, Department of Geography; Center for Geospatial Sciences, Applications & Technology, Texas A&M University, College Station, Texas, USA

Trenton W. Ford, Department of Geography and Environmental Resources, Southern Illinois University, Illinois, USA

Kevin R. Gamache, Water Management and Hydrological Sciences Graduate Program; High Alpine and Arctic Research Program, Department of Geology and Geophysics; Office of the Vice President for Research, Texas A&M University, College Station, Texas, USA

John R. Giardino, High Alpine and Arctic Research Program, Department of Geology and Geophysics; Water Management and Hydrological Science Graduate Program, Texas A&M University, College Station, Texas, USA

Raquel Granados-Aguilar, High Alpine and Arctic Research Program, Department of Geology and Geophysics, Texas A&M University, College Station, Texas, USA

Amalia Gutiérrez, Laboratorio Nacional de Materiales y Modelos Estructurales, Universidad de Costa Rica, Ciudad de la Investigación, Costa Rica

Brianna Hammond, Department of Geography, Texas A&M University, College Station, Texas, USA

Chris Houser, Department of Geography; Department of Geology and Geophysics, Office of the Dean of Geosciences; Center for Geospatial Sciences, Applications & Technology, Texas A&M University, College Station, Texas, USA

Praveen Kumar, Colonel Harry F. and Frankie M. Lovell Endowed Professor of Civil and Environmental Engineering, Civil and Environmental Engineering, University of Illinois at Urbana-Champaign, Urbana, Illinois, USA

Kathleen Lohse, Soil and Watershed Biogeochemistry, Biological Sciences, Idaho State University, Pocatello, Idaho, USA

Sara Mana, Department of Earth and Environmental Sciences, University of Iowa, Iowa City, Iowa, USA

Eric V. McDonald, Desert Research Institute, Division of Earth and Ecosystem Sciences, Reno, Nevada, USA

Bill McDowell, New Hampshire Water Resources Research Center, Department of Natural Resources & the Environment, University of New Hampshire, Durham, New Hampshire, USA

Kevin McGuire, Forest Resources and Environmental Conservation, Virginia Water Resources Research Center, Virginia Tech, Blacksburg, Virginia, USA

Georgianne Moore, Ecosystem Science and Management, Texas A&M University, Texas, USA

Patrick Pease, Department of Geography, University of Northern Iowa, Cedar Falls, Iowa, USA

Julia Perdrial, Department of Geology, University of Vermont, Burlington, Vermont, USA

Gregory A. Pope, Department of Earth and Environmental Studies, Montclair State University, Montclair, New Jersey, USA

Amy E. Price, Department of Geology and Geophysics, Texas A&M University, College Station, Texas, USA

Steven M. Quiring, Department of Geography, Climate Science Lab, Texas A&M University, Texas, USA

Netra R. Regmi, Department of Soil, Water and Environmental Sciences, University of Arizona, Tucson, Arizona, USA

Dan Richter, Soils and Forest Ecology, Duke University, Durham, North Carolina, USA

Taylor Rowley, High Alpine and Arctic Research Program; Water Management and Hydrological Science Graduate Program, Texas A&M University, College Station, Texas, USA

Paulo Ruiz, Laboratorio Nacional de Materiales y Modelos Estructurales, Universidad de Costa Rica, Ciudad de la Investigación, Costa Rica

Vatche Tchakerian, Department of Geography and Geology & Geophysics, Texas A&M University, College Station, Texas, USA

Aaron Thompson, Department of Crop and Soil Science, University of Georgia, Athens, Georgia, USA

Sarah Trimble, Department of Geography, Texas A&M University, College Station, Texas, USA

Peter Troch, Hydrology and Water Resources, Biosphere 2, University of Arizona, Oracle, Arizona, USA

John D. Vitek, High Alpine and Arctic Research Program, Department of Geology and Geophysics; Water Management and Hydrological Science Graduate Program, Texas A&M University, College Station, Texas, USA

Quanrong Wang, School of Environmental Studies, China University of Geosciences, Wuhan, Hubei, China

Phillipe Wernette, Department of Geography, Texas A&M University, College Station, Texas, USA

Bradley Weymer, Department of Geology and Geophysics, Texas A&M University, College Station, Texas, USA

Timothy White, Earth and Mineral Sciences, Penn State University, Pennsylvania State University, State College, Pennsylvania, USA

Ellen Wohl, Department of Geosciences, Colorado State University, Fort Collins, Colorado, USA

Shanshui Yuan, Department of Geography, Climate Science Lab, Texas A&M University, Texas, USA

Hongbin Zhan, Department of Geology and Geophysics, Texas A&M University, College Station, Texas, USA

Foreword

The Critical Zone as originally visualized in 1998 as a way to integrate the research of the four scientific spheres (lithosphere, hydrosphere, biosphere, and atmosphere) at the surface of Earth and to study the linkages, feedbacks, and record of processes. In 2001, the National Research Council recommended that a high-priority research opportunity be established. As a concept, it was intended to be both *very* specific as to its context (i.e., the surface of Earth), but *very* broad in scope as to its potential applications. Exploration of the Deep Time variability of Critical Zones was explicitly anticipated. The Critical Zone concept represents the *spirit of system science.* Rather than closeting studies by a variety of disciplines into their respective pigeonholes the CZ perspective provides the symbiotic framework from which the tendrils of improved understanding can radiate outward to new disciplines and/or feedback into the component disciplines.

The establishment of the NSF-sponsored Critical Zone Observatories in 2005 produced a wealth of information on short-term, near-surfaces processes. This book is timely, in that it captures the spectrum of research thrusts, broad range of environments, and the myriad of chemical, physical, and biological processes emanating from a large community of scientists now at work. It is clear from the vast accomplishments to date that the community has a solid base from which to now actively tackle the fourth dimension (change of CZ through time): to investigate and unravel the record of the *paleo-critical zones* (PCZ) back to early Earth. And, looking forward into the Anthropocene (it may not be all bad) to predict future changes in the inhabitable zone on Earth. The concepts of the critical zone studies should soon be applied on Mars as our knowledge base expands and allow us to design increasingly insightful experiments. The water and hydrocarbon bearing satellites (e.g., Europa, Enceladus, Titan) of Jupiter, Saturn, and other gas-giant planets could also move under the Critical Zone umbrella in the foreseeable future.

Gail M. Ashley

Quaternary Studies Program, Department of Earth and Planetary Sciences, Rutgers University, Piscataway, NJ, USA

Chapter 1

Introduction to the Critical Zone

John R. Giardino* and Chris Houser**

*High Alpine and Arctic Research Program, Department of Geology and Geophysics and the Water Management and Hydrological Science Graduate Program, Texas A&M University, College Station, Texas, USA; **Department of Geography, Department of Geology and Geophysics, Office of the Dean of Geosciences, Texas A&M University, College Station, Texas, USA

It ain't what they call you, it's what you answer to.

W.C. Fields

1.1 INTRODUCTION

Pick up a newspaper or magazine. Turn on the television or radio and watch or listen to the news or typical talk show. All of these will have articles or programs that focus on global change and human-caused actions that focus on drought, floods, earthquakes, oil spills, hurricanes, tsunamis, forest fires, coastal erosion, landslides, avalanches, and pollution of air and water resources. The list can go on and on. Partly as a result of the impact of these events on humans and the environment, the Nation Research Council's (NRC) Committee on Basic Research Opportunities in the Earth Sciences, and the National Science Foundation (NSF) created the Critical Zone Observatory (CZO) program (NRC, 2001). Whereas the NRC conceptualized the term, the funding of the program fell on the shoulders of NSF.

In the *Principles and Dynamics in the Critical Zone*, we have assembled a group of authors to provide a broad-ranging approach to Critical Zone research that extends beyond the current CZOs. Each chapter has been written with the perspective of integrating the viewpoints from specific fields into an interdisciplinary view of the Critical Zone. Each chapter provides a view of a specific discipline or from a specific environment and the contributions it brings to Critical Zone research. As a first step in reading this volume, we begin by providing a succinct overview of the Critical Zone.

Developments in Earth Surface Processes, Vol. 19. http://dx.doi.org/10.1016/B978-0-444-63369-9.00001-X

1.2 BRIEF HISTORY AND BACKGROUND OF THE CRITICAL ZONE OBSERVATION NETWORK AND CRITICAL ZONE OBSERVATORIES

In 2001, a panel of the US National Research Council (NRC, 2001) recommended an integrated study of the *Critical Zone* as one of the most compelling research areas in Earth sciences in the twenty-first century. They (NRC, 2001, p. 2) went on to define the Critical Zone as "... *the heterogeneous, near surface environment in which complex interactions involving rock, soil, water, air and living organisms regulate the natural habitat and determine availability of life sustaining resources*." From this original definition, many, now loosely worded definitions have been crafted to define the limits of this zone as ranging from the top of the canopy layer down to the bottom of the aquifer (Fig. 1.1).

Fortunately, today these various refinements and rewordings of the definition of the Critical Zone have brought additional clarity. For example, groundwater has been constrained from all groundwater to "*freely circulating fresh groundwater*," instead of groundwater *sensu lato*. Aquifers that contain deep connate brines and confined aquifers have also been excluded (White, 2012). In reading the original definition, it is unclear what environments were to be included, and in fact, if some had not been excluded. One was left to assume that any environment present on the surface of Earth was included, but that was not clear. White (2012) has provided clarity to the definition, so that it encompasses polar/arctic,

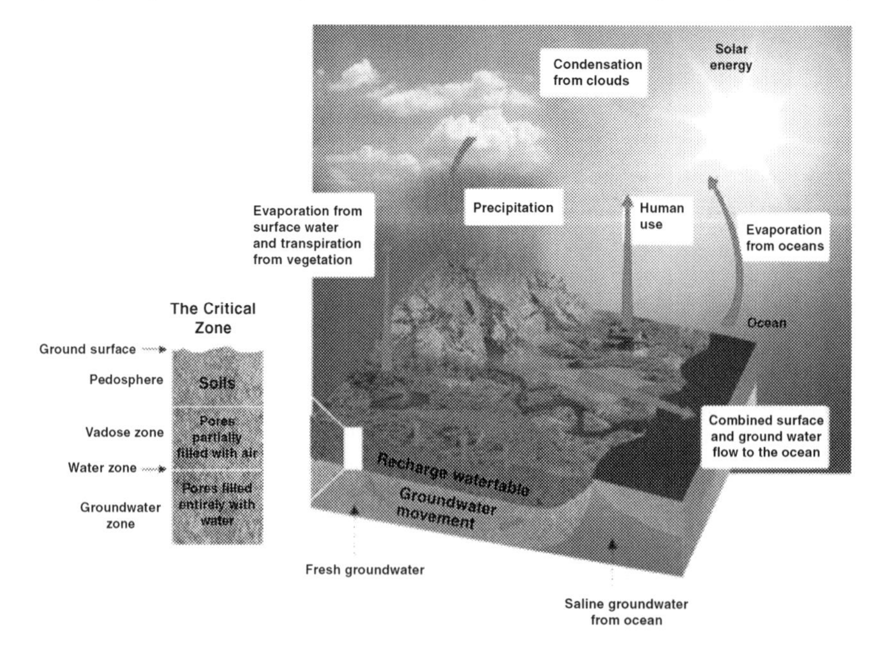

FIGURE 1.1 The diagram illustrates the extent of the Critical Zone from the top of the canopy to the bottom of the aquifer. The pathways, which act as linkages for flows of energy and mass between the various subsystems of Earth, are shown as gray arrows [purple arrows in the web version]. *(Modified from NRC (2001, p. 36).)*

alpine, and desert environments, but not coastal. Whereas still limited, we think this clarification of the broader definition of the term Critical Zone has a more meaningful application than originally penned.

The NRC was not the creator of the "Critical Zone" term. It has been around for a long time. Tsakalotos (1909) first introduced the term into the chemical literature to describe the binary mixture of two fluids and geologists have long used the term to refer to the complex geology of the Bushveld Complex in South Africa (Cameron, 1963). The precursor application of the term Critical Zone to this relatively thin zone of Earth was first suggested by Gail Ashley (1998). She noted that the term applied where the soil connects the vegetation canopy to the soil; the soil connects to the weathered materials, and the weathered materials connect to bedrock, and bedrock provides the connection to the aquifer.

Critical Zone is a term that brings attention to both the scientific community and the layperson, of the important, critical role of what this relatively thin zone plays in the existence of life on Earth. Latour (2014) provided an interesting view of the Critical Zone from a geopolitical point-of-view. He argued that the important contributions of CZOs and the data they produce are providing various types of consistent observations from selected locations on Earth. More importantly, the Critical Zone concept deconstructs the planet into much smaller components of the "*living planet.*" Using the Critical Zone concept that focuses on an individual location brings understanding to the public of the linkages between a single, garden parcel to a total watershed to the complete planet. He pointed out how important the concept of scale is for humans to understand and accept responsibility for stewardship of Earth. This idea has been pushed by environmental groups who have coined the phrase "*Think globally – Act locally.*"

The creation of the CZOs is not the first time NSF has been involved with major contributions that benefit society. In fact, several authors have suggested that it was the contribution of scientists and engineers that helped win World War II that ultimately lead to the establishment of NSF (Mazuzan, 1988; Bronk, 1975; Kleinman, 1995). Today NSF is the only federal agency solely dedicated to the support of fundamental research and education. Over the past few years, NSF has shifted their focus from funding strictly fundamental research to requiring researchers to expand fundamental research to include contributions to society and outreach. The CZOs is one outcome of this new, important focus at NSF.

It was not until 2003 that funding for Critical Zone research appeared. The initial step in creating CZOs began with the funding of *The Weathering System Science Workshop*. The main focus was on the development of weathering-system science. The initial idea was that this new field should embrace the fields of chemistry, biology, physics, and geology (Brantley et al., 2006). During the following year, The Weathering System Science Consortium (WSSC) was formed, and it was morphed in 2006 to what has been referred to as the Critical Zone Exploration Network (CZEN).

NSF issued a request for proposal in 2005 to commence the establishment of Critical Zone research. Soon after this, NSF sponsored the *Frontiers in Exploration of the Critical Zone Workshop* to help establish goals and agenda for future directions and research. Attendees at the workshop suggested the need for creating an initiative to study the Critical Zone based on an international focus. This eventually led to the establishment of CZOs in various parts of the planet. In addition to the NSF-funded CZOs, the Soil Transformations in European Catchment (SoilTrEC), which is a consortium of European Union members, also established Critical Zone centers.

In 2006, NSF solicited proposals to establish CZOs; the solicitation was through the Division of Earth Sciences. From the first solicitation, three locations (Southern Sierra (California), Boulder Creek (Colorado), and Susquehanna Shale Hills (Pennsylvania)) were selected and established in 2007. Three more CZOs were launched in 2009; the three CZOs are Jemez River Basin – Santa Catalina Mountains (Arizona/New Mexico), Christina River Basin (Delaware/Pennsylvania), and Luquillo Mountains (Puerto Rico). In 2013, four additional CZOs were established. These four CZOs are the Calhoun Forest (S. Carolina), Intensively Managed Landscape (Minnesota, Illinois, Iowa), Reynolds Creek (Idaho), and Eel River (northern California) (Table 1.1). Thus, the number of

TABLE 1.1 Locations of Critical Zone Sites Located Around the World

	CZO	Location
1	Southern Sierra CZO	Southern Sierra Nevada, California
2	Boulder Creek CZO	Colorado Front Range, Rocky Mountains
3	Susquehanna Shale Hills CZO	Central Pennsylvania
4	Jemez River Basin CZO	New Mexico
5	Christina River Basin CZO	South-eastern Pennsylvania and Northern Delaware
6	Luquillo	Luquillo, Puerto Rico
7	Calhoun LTSE	South Carolina
8	Intensively Managed Landscape	Minnesota, Illinois, Iowa
9	Reynolds Creek Watershed	Southwest Idaho
10	Eel River	California
11	Clear Creek	Iowa
12	Adirondack Mountains	Southwestern Adirondacks
13	AGRHYS	Brittany
14	AMMA-CATCH	S-N ecoclimatic gradient in West Africa

TABLE 1.1 Locations of Critical Zone Sites Located Around the World *(cont.)*

	CZO	Location
15	Damma Glacier	Canton URI, Switzerland
16	Bonanza Creek	Alaska
17	Central Great Plains	Colorado
18	DRAIX_BLEONE	6.3′ E-44.1′ N, French South Alps
19	Riviere des Pluies Erorun	Reunion Island, Indian Ocean
20	French Karst Observatory	Langiedoc, Jura, Provence, Pyrenees, Paris Basin, Aquitanien Basin
21	Fuchsenbigl	East Austria
22	Galapagos CZO	Santa Cruz Island, Galapagos
23	Guadeloupe	Guadeloupe, French West Indies
24	Hawaii	Hawaii
25	Hoffman Creek site	Oregon
26	Hubbard Brook Experimental Forest	New Hampshire
27	HYBAM: Hydrological and Geochemical observatory of the Amazon Basin	Amazon Drainage Basin
28	Illinois River Basin	Illinois
29	Kindla	Kindla, Bergslagen
30	Kouliaris River Basin	
31	Lowlands CZO	Netherlands
32	Lysina	Slavkov Forest
33	Marcellus Shale	Pennsylvania
34	Merced River Chronosequence	California
35	Montousse	Gascogne
36	MSEC: Management of Soil Erosion Consortium	SE Asia (three sites)
37	MSEC Dong Cao long-term monitoring catchments	20′57′40″ N-105′29′10″ E
38	MSEC Houay Panoi long-term monitoring catchments	19′51′10″ N-102′10′45″ E
39	Mule Hole (Bandipur National Park)	Southern India (Mule Hole:11′72′N 76′42E
40	Muskingum Watershed	Ohio
41	Na Zelenem	Western Bohemia
42	NC2	New Caledonia

(Continued)

TABLE 1.1 Locations of Critical Zone Sites Located Around the World *(cont.)*

	CZO	Location
43	NevCAN Sheep Range and Snake Range Transects (NevCAN)	Southern and East Central Nevada
44	North Ogilvie Mountains	Yukon Territory
45	North Eastern Soil Monitoring Cooperative	North eastern Soil Monitoring Cooperative
46	Nsimi	Cameroon
47	OBSERA	Guadeloupe (Lesser Antilles)
48	OHM-CV	Cervennes-Vivarais (four sites)
49	OMERE	Brie, Paris Basin
50	ORACLE	Languedoc and Cap Bon (two sites)
51	Panola Mountain	Atlanta
52	Pluhuv Bor	Slavkov Forest
53	Plynlimon	Mid Wales
54	Red Soil Site	Yingtan Jiangxi Province
55	Santa Catalina Mountains CZO	Saguaro National Park
56	SEQ periurban supersite	South East Queensland
57	Stengbach	Vosges Mountains
58	Tenderfoot Creek Experimental Forest	Montana Southwest Alberta and Wyoming
59	The Prairie Pothole Region CZO	South Central North Dakota
60	The Rogers Glen (Shale Hills CZO) Satellite Site	Chadwicks, New York
61	Trindle Road Appalachian Trail Diabase	Pennsylvania
62	TUM CZO	Bavaria
63	Beacon Farm 1	Rakaia River Catchment, Canterbury Plains
64	Omere Site	Tunisia

The numbers in the table correspond to the Critical Zone sites shown on the map in Fig. 1.2.

CZOs funded by NSF today is 10. In 2014, NSF established a National Critical Zone Office with Lou Derry serving as the first director and Tim White serving as the first coordinator. To date to carry out the directive of the Critical Zone mission, NSF has created 10 CZOs and a national office (Fig. 1.2).

The 10 NSF CZOs, each operate as environmental laboratories and focus on specific problems. A detailed discussion of the 10 NSF-funded CZOs is provided by White in Chapter 2 of this volume.

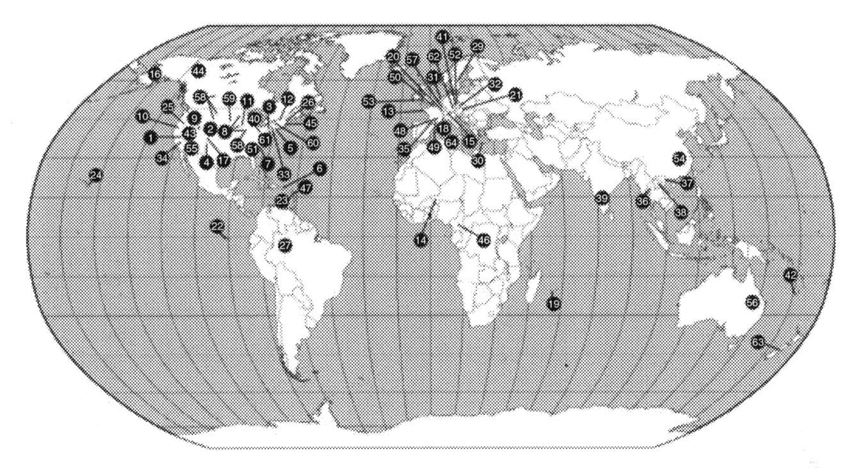

FIGURE 1.2 **Locations of Critical Zone sites located around the world.** The number in the marker corresponds to the Critical Zone site listed in the table. *(Map compiled from Banwart et al. (2013) and NRC (2001).)*

1.3 DEVELOPMENT OF THE GLOBAL CRITICAL ZONE NETWORK

The number of CZOs and the spatial distribution were increased in 2009 when the European Commission funded the development of an international network of observatories, which are located in Europe, China, and the United States. A major requirement of the new program was to work with the US- established CZOs (Banwart et al., 2013). Soon after, France, Germany, and China founded additional CZOs. Today the number of CZOs around the world stands around 64 (Fig. 1.2 and Table 1.1). Participants at 2013 Critical Zone Meeting in Delaware, USA, originally compiled the table (Banwart, 2015). Additional sites in France (Critex programme), Germany (TERENO), and China will be coming on board in the coming years. In addition, several locations in Australia are being developed as future CZOs (Banwart, 2015). We have added additional sites to the table produce by Banwart et al. (2013).

The raison d'être of the Critical Zone network is driven by basic principles of science. All the research at each location is focused on asking fundamental questions and collecting and building data banks. The modus operandi of the CZOs can be summed up as seven tasks: (1) hypothesis testing; (2) process understanding across temporal and spatial scales; (3) development of mathematical models; (4) utilization of multiple sensor and sampling methods; (5) installation and use of high-density instrument arrays; (6) undertaking time series/real-time measurements of coupled process dynamics; and (7) combination of large data sets with numerical simulations (Banwart et al., 2011, 2013).

Banwart et al. (2013) provided an excellent summary of progress being made by the CZO Network, as well as the short- and long-term goals. The scientific foundation of the Observatories was strengthened with the articulation of six big science questions being formulated. Banwart et al. (2012, p. 20) list the six big science questions as:

Short-Term Processes and Impacts

- What controls the resilience, response, and recovery of the CZ and its integrated geophysical–geochemical–ecological functions to perturbations such as climate and land use changes, and how can this be quantified by observations and predicted by mathematical modeling of the interconnected physical, chemical, and biological processes and their interactions?
- How can sensing technology, e-infrastructure, and modeling be integrated for simulation and forecasting of essential terrestrial variables for water supplies, food production, biodiversity, and other major benefits?
- How can theory, data, and mathematical models from the natural- and social-sciences, engineering, and technology be integrated to simulate, value, and manage Critical Zone goods and services and their benefits to people?

Long-Term Processes and Impacts

- How has geological evolution and paleobiology of the CZ established ecosystem functions and the foundations of CZ sustainability?
- How do molecular-scale interactions between CZ processes dictate the linkages in flows and transformations of energy, material, and genetic information across the vertical extent of aboveground vegetation, soils, aquatic systems, and regolith – and influence the development of watersheds and aquifers as integrated ecological–geophysical units?
- How can theory and data be combined from molecular to global scales in order to interpret past transformations of Earth's surface and forecast CZ evolution and its planetary impact?

To solve these six big questions, it requires integrative research that spans traditionally siloed disciplines and the long-term studies that are hierarchically structured in both space and time. Each CZO works on shared CZO goals, but also focuses on aspects of Critical Zone science that fits the strengths of its investigators and its environmental setting. Each CZO consists of field sites within a watershed. The sites are instrumented for a variety of ecological, geomorphological, pedological, hydrological, and geochemical measurements, as well as sampled for soil, canopy, and bedrock materials, but unfortunately research at the CZOs is not always integrative in space or in time.

1.4 WATER, THE TRUE THREAD OF THE CRITICAL ZONE

As the Critical Zone was being defined, three recommendations associated with the Critical Zone were tenured by the NRC Committee: make soil science stronger, continue to build hydrological sciences, and strengthen study of coastal zone processes (NRC, 2001, p. 6). From our perspective, it is interesting to note that from these recommendations somehow soil, and not water, became the thread that ties the various systems of the Critical Zone together. We understand and appreciate that focus on world hunger and increasing rates of soil erosion along with The Millennium Ecosystem Assessment (MEA, 2005) and the EU Thematic Strategy for Soil Protection (European Commission, 2006) provided a strong foundation for soil being viewed as the link between soil erosion, food production, and worldwide human welfare. We also agree that soil plays an important role, but we do not think it is the link between all the systems. It is easy to understand that the soil profile does connect the vegetation canopy to the soil; the soil connects to the weathered materials and the weathered materials connect to bedrock, and bedrock provides the connection to the aquifer. These pathways facilitate flows (i.e., energy, mass, biogeochemical) from the top to the bottom and vice versa. But, none of these energy and material flows would be possible without water.

Thus, we strongly argue that the link between all these components or systems is water, rather than soil. Water is the true thread that stitches these systems together. It is water that travels from the atmosphere, to and through the biosphere, and into the lithosphere where it is stored as surface water, soil water, and groundwater. Water is a key component in chemical weathering processes, a sustainer of life, and is responsible for floods or absent to cause drought. Although the NSF never mentioned the above concern, we believe that they (NRC, 2005) also saw the "*thread problem*" when they recommended that focus needed to be brought to water as the unifying theme in the study of complex environmental systems.

As Lin (2010) has pointed out, the Critical Zone term has been met with both enthusiasm as well as skepticism. These two visions have led to numerous perceptions, both right and wrong of the concept. Lin (2010) presented an excellent summary and discussion on the mixed messages of the Critical Zone concept. He listed four issues: (1) researchers consider the Critical Zone (nearly) synonymous with the term soils; (2) some researchers equate the Critical Zone with the term regolith; (3) some researchers question the effectiveness of the Critical Zone concept because (they think soil is the Critical Zone) the lower boundary is so highly inconsistent and ill-defined; and (4) some believe the Critical Zone concept is useful because it is intrinsically process-oriented and serves as a uniting concept accommodating the various components and systems of Earth. If no clear definition exists of the Critical Zone concept, it is hard to imagine that Critical Zone research is sufficiently integrated and directed to solve the six big questions outlined by Banwart et al. (2012).

We do not want the reader to develop the opinion that we have unenthusiastic feelings about the Critical Zone concept. This is not so. We are very strong supporters of the Critical Zone concept, and we applaud the foresight of creating a 4D approach to studying the processes that form and shape the surface and near-surface features and resources of Earth. This is critical since the Critical Zone concept calls for a focus on the zone where humans live and work. For life, as we know it, it is the most important and critical area for the human species. Human activities are significantly influencing the environment of Earth in many ways, in addition to greenhouse gas emissions and climate change.

Anthropogenic changes to the land–surface, oceans, coasts, and atmosphere; biological diversity; the water cycle and biogeochemical cycles of Earth are clearly discernable beyond natural variability. This impact is on a par with some of the greatest forces of nature in spatial extent and magnitude. Unfortunately, many are accelerating. In spite of wholesale neglect and outright denial, global change is real and appears to be happening now.

Earth-System dynamics are characterized by critical thresholds and abrupt changes. Human activities could inadvertently trigger such changes with severe consequences for the environment and inhabitants of the planet. The Earth System has operated in different states over the last half million years, with abrupt transitions (a decade or less) sometimes occurring between them. Human activities have the potential to switch the Earth System to alternative modes of operation that may prove irreversible and less hospitable to humans and other life. The probability of a human-driven abrupt change in the environment of Earth has yet to be quantified but is not negligible. In terms of some key environmental parameters, the Earth System has moved well outside the range of the natural variability exhibited over the last half million years, at least. The nature of changes now occurring simultaneously in the Earth System, their magnitudes and rates of change are unprecedented. Earth is dynamic.

Global change cannot be understood in terms of a simple cause-and-effect paradigm. Human-driven changes cause multiple effects that cascade through the Earth System in complex ways. These effects interact with each other and with local- and regional-scale changes in multidimensional patterns that are difficult to understand and even more difficult to predict. It is in this arena that the CZOs will play a significant role. The CZOs are intended to identify flow pathways, and energy transformations at various scales. Through this research they will be creating a continuum of change that will provide keys to understanding the linkages between the components in Earth Systems. Many of the CZOs are employing a hind-casting approach to accomplish a forecasting approach to their research. This type of research will provide the data to predict and understand future change and global impacts.

1.5 THE FASHION OF CRITICAL ZONE RESEARCH

Critical Zone is a fashionable term. Much of what is being undertaken at the various CZOs is proven science. What the term Critical Zone is driving is a new awareness of interdisciplinary study of single locations. The CZO fosters conventional studies, governance, and coordination of data on a global scale. Banwart et al. (2013) pointed out that the real contributions from the CZO network will be the understanding of the relationship between the strength of a response, the time of recovery, and the overall resilience, to environmental perturbations of our planet. A 4D architecture approach will be required to achieve this goal. Pursuit of that goal will result in the development of new sensing technologies, as well as new spatial statistics and modeling.

After studying all the research being produced by the various CZOs, we were both elated and depressed. Elated to see that researchers from various disciplines are focusing on specific locations, creating a warehouse of data, and beginning to engage in what we would describe as true interdisciplinary work. We were depressed to see that it appears that a lot of the research being produced at the various CZOs is still discipline-focused and published in discipline-specific journals. After we delved into all the Critical Zone literature, we came to the conclusion that much of what is being referred to as a new approach, the Critical Zone is, in fact, a new name for old science. We find ourselves in a sort of a dichotomy in accepting the new approach and concept of the Critical Zone. On the one hand, the concept of the Critical Zone is a fresh way of showing the connection and flow of mass and energy from the top of the canopy to the bottom of the aquifer. The fundamental precepts of the Critical Zone appear to be based on much of the conceptualization of General Systems Theory as developed by von Bertalanffy in the 1950s (von Bertalanffy, 1950, 1951, 1968). And, although the Critical Zone is a very useful concept, it is by no means a new or novel idea. This attitude also underlines the importance of the system perspective.

However, much positive exists about the Critical Zone concept. It is resulting in enormous amounts of data being collected at specific locations by a wide range of disciplines; it is developing new ways to collect and model these data; it is providing methods and data to combine data from various CZOs to model change of Earth in both longitudinal as well as latitudinal directions that is engaging researchers from various fields to once again visit the idea of interdisciplinary research and to view activity on Earth from a system's point of view. Probably one of the most important applied aspects of the global CZO network will be the development of reliable data that will lead to enlightened policy and management of the geoscience base of Earth. As geoscientists continue on their quest for a unifying principle, the Critical Zone approach might just be that step in the right direction.

The current popularity of the Critical Zone concept is very welcome, as it refocuses attention on the need to study the complex nature of the upper zone of Earth from an integrative, holistic approach. This fresh approach reminds us of the argument of Sperber's (1990) and Sherman's (1996, p. 87) comments on *"fashion change from the design and arts disciplines to explain one means of controlling the developments and directions of science ... changes in the goals, subjects, methods, philosophies, or practice of science can often be attributed to the emergence of an opinion (or fashion) leader, pointing toward a different path – setting out the new fashion. The fashion process relies upon fashion dudes to advance their disciplines."* We think we are seeing a large group of fashion dudes, stepping forward to configure a transformative science of the surface Earth; the Critical Zone concept is a perfect unifying principle for the geosciences and the means by which we are going to move towards integrative studies of the biophysical environment and the role of humans within that environment.

As we mentioned at the beginning of the introduction, *Principles and Dynamics of the Critical Zone* contains chapters written by various professionals in their respective fields. Our purpose in writing this book is to provide an all-encompassing, broad view of the Critical Zone. The chapters run the gamut from very general discussions on the paradigm of the Critical Zone to chapters that focus on more detailed aspects of the Critical Zone and include environments that have never been traditionally described from a Critical Zone perspective. We have tried to be somewhat systematic in the order of the Table of Contents. The reader will note that our coverage is not complete. Whereas we set out with the goal of providing a complete coverage of all aspects of the Critical Zone, unfortunately, a couple of authors who agreed to provide chapters never followed through and we had to supply a manuscript to the publisher. Although not able to accomplish what we set out to do, we offer this as an overview of the first approximation of the principles and dynamics of the Critical Zone.

REFERENCES

Ashley, G.M., 1998. Where are we headed? "Soft" rock research into the new millennium. Geological Society of America Abstract/Program, vol. 30. p. A-148.

Banwart, S. 2015. Personal communication.

Banwart, S., Bernasconi, S.M., Bloem, J., Blum, W., Brandao, M., Brantley, S., Chabaux, F., Duffy, C., Kram, P., Lair, G., Lundin, L., Nikolaidis, N., Novak, M., Panagos, P., Ragnarsdottir, K.V., Reynolds, B., Rousseva, S., de Ruiter, P., van Gaans, P., van Riemsdijk, W., White, T., Zhang, B., 2011. Soil processes and functions in Critical Zone Observatories: hypotheses and experimental design. Vadose Zone J. 10, 974–987.

Banwart, S., Chorover, J., Sparks, D., White, T., 2012. Sustaining Earth's Critical Zone. Report of the International Critical Zone Observatory Workshop. November 8–11, 2011, Delaware, USA.

Banwart, S.A., Chorover, J., Gaillardet, J., Sparks, D., White, T., Anderson, S., Aufdenkampe, A., Bernasconi, S., Brantley, S.L., Chadwick, O., Dietrich, W.E., Duffy, C., Goldhaber, M., Lehnert, K., Nikolaidis, N.P., Ragnarsdottir, K.V., 2013. Sustaining Earth's Critical Zone Basic Science and Interdisciplinary Solutions for Global Challenges. The University of Sheffield, Sheffield, United Kingdom.

Brantley, S.L., White, T.S., White, A.F., Sparks, D., Richter, D., Pregitzer, K., Derry, L., Chorover, J., Chadwick, O., April, R., Anderson, S., Amundson, R., 2006. Frontiers in exploration of the Critical Zone, report of a workshop sponsored by the National Science Foundation (NSF), Newark, DE, USA, 30, 1–30.

Bronk, D.W., 1975. The National Science Foundation: origins, hopes, and aspirations. Science 188, 409–414.

Cameron, E.N., 1963. Structure and rock sequence of the critical zone of the Eastern Bushveld Complex. Min. Soc. Am. Spec. Pap. 1, 93–107.

European Commission, 2006. Thematic Strategy for Soil Protection. COM (2006). Commission of European Communities, Brussels. p. 231.

Kleinman, D.L., 1995. Politics on the Endless Frontier: Postwar Research Policy in the United States. Duke University Press, Durham, NC.

Latour, B., 2014. Some advantages of the notion of "Critical Zone" for geopolitics. Procedia Earth and Planetary Science Geochemistry of the Earth's surface GES-10 Paris France, 10, 3–6.

Lin, H., 2010. Earth's Critical Zone and hydropedology: concepts, characteristics, and advances. Hydrol. Earth Syst. Sci. 14, 25–45.

Mazuzan, G.T., 1988. The National Science Foundation: A Brief History. National Science Foundation, Washington, DC.

MEA, 2005. Ecosystems and Human Well-Being Synthesis Millennium Ecosystem Assessment. Island Press, Washington, DC.

National Research Council Committee on Basic Research Opportunities in the Earth Sciences, 2001. Basic Research Opportunities in the Earth Sciences. National Academies Press, Washington, DC.

NRC, 2005. Review of the GAPP Science and Implementation Plan. Committee to Review the GAPP Science and Implementation Plan, National Research Council. Washington, DC.

Sherman, D., 1996. Fashion in geomorphology. In: Rhoads, B.L., Thorn, C.E. (Eds.), The Scientific Nature of Geomorphology: Proceedings of the 27th Binghamton Symposium in Geomorphology. John Wiley & Sons Ltd., Chichester, England. pp. 87–114.

Sperber, L., 1990. Fashions in Science: Opinion Leaders and Collective Behavior in the Social Sciences. University of Minnesota Press, Minneapolis, p. 303.

Tsakalotos, D.E., 1909. The inner friction of the critical zone. Z. Phys. Chem. – Stoch. Ve. 68, 32–38.

von Bertalanffy, L., 1950. An outline of general system theory. Br. J. Philos. Sci. 1, 114–129.

von Bertalanffy, L., 1951. General system theory – a new approach to unity of science (Symposium). Hum. Biol. 23, 303–361.

von Bertalanffy, L., 1968. General System Theory: Foundations, Development, Applications. George Braziller, New York.

White, T., 2012. Special focus: the US Critical Zone Observatories. Int. Innovation August, 108–127.

Chapter 2

The Role of Critical Zone Observatories in Critical Zone Science

Timothy White*, Susan Brantley**, Steve Banwart†, Jon Chorover‡, William Dietrich§, Lou Derry¶, Kathleen Lohse††, Suzanne Anderson‡‡, Anthony Aufdendkampe§§, Roger Bales***, Praveen Kumar†††, Dan Richter‡‡‡, and Bill McDowell§§§

*Earth and Mineral Sciences, Penn State University, Pennsylvania State University, State College, Pennsylvania, USA; **Earth and Environmental Systems Institute, Pennsylvania State University, University Park, Pennsylvania, USA; †Kroto Research Institute, North Campus, University of Sheffield, Broad Lane, Sheffield, UK; ‡Department of Soil Water and Environmental Science, University of Arizona, Tucson, Arizona, USA; §Earth and Planetary Science, University of California, Berkeley, California, USA; ¶Earth and Atmospheric Sciences, Cornell University, Cornell, New York, USA; ††Soil and Watershed Biogeochemistry, Biological Sciences, Idaho State University, Pocatello, Idaho, USA; ‡‡Department of Geography, University of Colorado at Boulder, Boulder, Colorado, USA; §§Stroud Water Research Center, West Grove, Pennsylvania, USA; ***Sierra Nevada Research Institute, University of California, Merced, California, USA; †††Colonel Harry F. and Frankie M. Lovell Endowed Professor of Civil and Environmental Engineering, Civil and Environmental Engineering, University of Illinois at Urbana-Champaign, Urbana, Illinois, USA; ‡‡‡Soils and Forest Ecology, Duke University, Durham, North Carolina, USA; §§§New Hampshire Water Resources Research Center, Department of Natural Resources & the Environment, University of New Hampshire, Durham, New Hampshire, USA

2.1 THE CRITICAL ZONE

The Critical Zone (CZ), a term first coined by the US National Research Council (2001), encompasses the thin outer veneer of Earth's surface extending from the top of the vegetation canopy down to the subsurface depths of fresh groundwater. Complex biogeochemical-physical processes combine in the CZ to transform rock and biomass into soil that in turn supports much of the terrestrial biosphere. Processes in the Critical Zone are represented by coupled physical, biological, and chemical processes (Fig. 2.1), and scientific expertise from an array of disciplines is needed to understand the zone and its processes: geology, soil science, biology, ecology, geochemistry, hydrology, geomorphology, atmospheric science, and many more. The zone sustains most aboveground terrestrial life including humanity. Yet, the science of coupled human-natural systems is far

Developments in Earth Surface Processes, Vol. 19. http://dx.doi.org/10.1016/B978-0-444-63369-9.00002-1

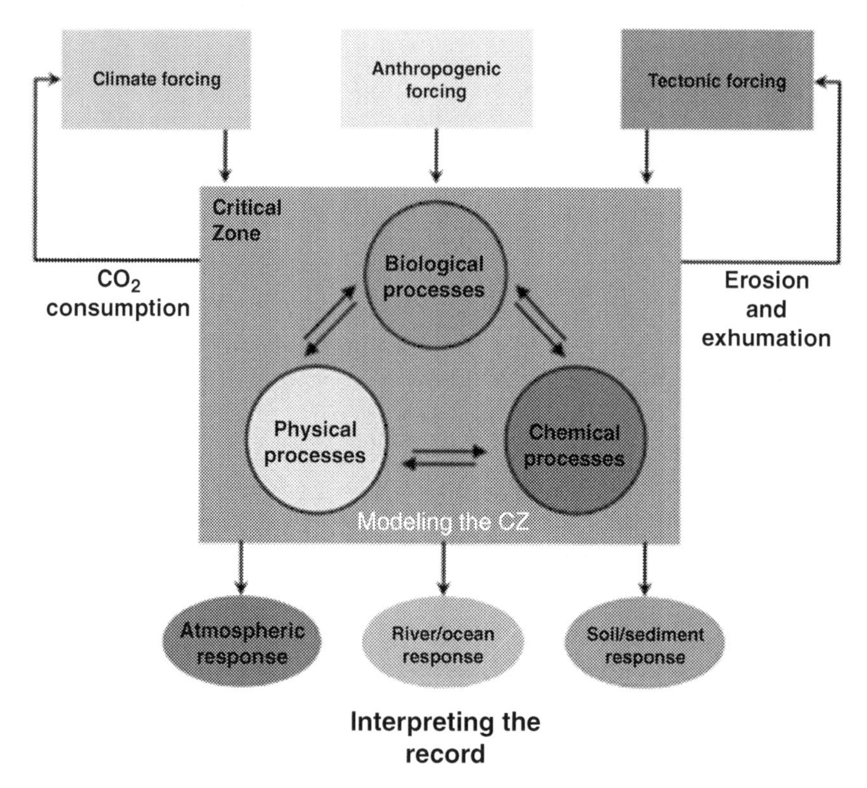

FIGURE 2.1 Physical, chemical, and biological processes in the Critical Zone (CZ) are subjected to climate, tectonic, and anthropogenic forcing that lead to responses in the atmosphere, biosphere, hydrosphere, lithosphere, and pedosphere. The challenge of CZ science is to interpret CZ processes over both short and long timescales: for example, CZ scientists attempt to understand sediment and soil records for comparison to changes in the CZ associated with ongoing climate and land-use change. *(From Brantley et al. (2007).)*

from developing a theory that could offer the potential to predict, or earthcast, the environment of the future (e.g., Godderis and Brantley, 2014), not to mention one that could yield the knowledge needed to slow or reverse environmental degradation in a sustainable fashion (National Research Council, 2010).

The structure and functioning of the CZ have evolved in response to climatic and tectonic perturbations throughout Earth's history with the processes driving change more recently accelerated by human activities in the Anthropocene (e.g., Vitousek et al., 1996; Wilkinson, 2005; Steffen et al., 2007; see Fig. 2.2). The degraded state of Earth's surface has been well documented, for example, in the United Nations Environment Programme's (UNEP) Millennium Ecosystem Assessment report (2005), http://www.unep.org/maweb/en/index.aspx, and UNEP (2005), One Planet Many People: Atlas of Our Changing Environment (2005); the Intergovernmental Panel on Climate Change Fifth Assessment Report, http://www.ipcc.ch/; Smith (2012), Penguin State of the World Atlas: Ninth Edition (2012); and Hoekstra et al. (2010), The Atlas of Global Conservation: Changes, Challenges, and Opportunities to Make a Difference.

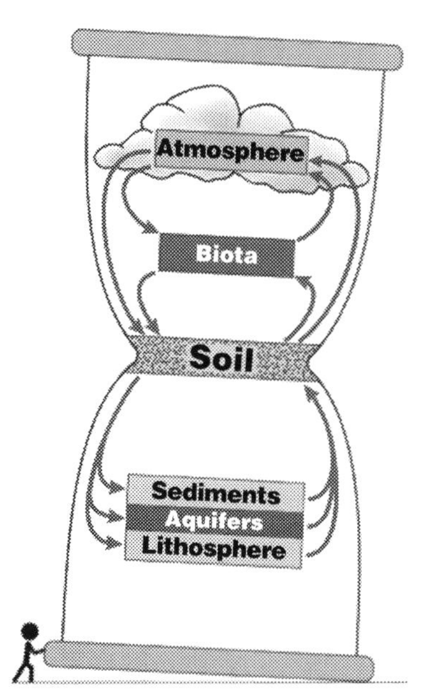

FIGURE 2.2 The Critical Zone Observatory mission is to learn to: measure the fluxes occurring today, read the geological record of the cumulative effect of these fluxes through time, and develop quantitative models of CZ evolution. By measuring what is happening today and reading what happened yesterday, we will learn to project what will happen tomorrow – including humanity's role. *(From Godderis and Brantley (2014).)*

Hooke et al. (2012) summed up some of the topics and tone of these reports, concluding that humans have modified more than half of Earth's land surface, that the current rate of land transformation is unsustainable, and that "changes that human activities have wrought in Earth's life support system have worried many people." To many scientists and citizens, these threats to this essential component, that is, the CZ, of our life-support system, have reached an acute level, yet the science of understanding and managing these threats mostly still remains embedded within individual disciplines, and the science has largely remained qualitative – never has a more important time occurred for an international interdisciplinary approach to accelerate our understanding of processes in the CZ and how to intervene positively to mitigate threats and ensure CZ function. The immediate challenge is to develop a robust predictive ability for how the CZ attributes, processes, and outputs will respond to projected climate and land-use changes to guide societal adaptations toward a more sustainable future. This predictive ability must be founded on a sufficiently broad knowledge of the CZ system and CZ processes to describe the interactions of the varied climatic, ecologic and geologic factors that distinguish different geographic regions – a primary focus of the scientific and educational efforts at Critical Zone Observatories (CZOs). The aim of the US CZO program is largely focused on developing methods to quantitatively project the dynamics of Earth surface processes – from the past, through today into the future.

The key to CZ science is to use observatories as time telescopes that allow focus, not only on the processes and fluxes operating today, but to compare these to the record of the processes in the rock and soil and sediment record – then to use quantitative models parameterized from these observations across scales of space and time to project the future using various scenarios of human behavior. One example of forward-projecting CZ science is shown in papers investigating the record of change in a soil climosequence along the Mississippi River in the United States. (Williams et al., 2010) using climate models to drive soil-development models (Godderis et al., 2010) that are in turn used to project the future of the soils and water (Godderis et al., 2013). By implementing such an approach at CZOs where datasets are less sparse, we will learn how mass and energy fluxes interact with biota and lithology over geological timescales, transforming bedrock into soils. We will also learn how the same, coupled processes enact feedbacks between the CZ, changing climate and land use over timescales of human decision-making. Furthermore, while the example focuses on weathering, many other CZ processes can be similarly addressed.

2.2 CRITICAL ZONE OBSERVATORIES (CZOs)

CZOs are natural watershed laboratories for investigating Earth-surface processes mediated by fresh water. Research at the CZO scale seeks to understand these little-known coupled processes through monitoring of streams, climate/weather, and groundwater. CZOs are instrumented for hydrogeochemical measurements and are sampled for soil, canopy, and bedrock materials. CZOs involve teams of cross-disciplinary scientists whose research is motivated and implemented by both field and theoretical approaches, and include substantial education and outreach.

The interdisciplinary and integrative science of the CZOs is a relatively new scientific endeavor formalized through the US NSF funding of the CZO program. The US CZOs were chosen through a rigorous NSF peer review process, based on standard NSF criteria. A CZO National Office (CZONO) is now in place to guide network-level activities. Details of the US CZO program and each of the 10 observatories (described later) may be found at: http://criticalzone.org/.

The CZOs – whether funded by the US NSF, the European Commission or similar entities in France, Germany, Australia, China, or other nations – uniquely address the challenge of understanding terrestrial life's support system. Of all the environmental observatories and networks, the CZOs are the only ones to tightly integrate ecological and geological sciences to combine with computational simulation, and to project from the deep geologic past to that of human life spans. As such, CZOs represent a unique opportunity to transform our understanding of coupled surface Earth processes and to address quantitatively the impacts of climate and land use change and the value of Critical Zone functions and services. Indeed, a fundamental concept applied from ecological economics is that the CZ embodies natural capital as a means of production to support flows of materials, energy, genetic information, and human population over time.

Although all the observatories worldwide are not called CZOs, many of the scientists worldwide use the energizing framework and nomenclature of "Critical Zone science" in their strategies of national science. This is because CZOs provide essential data sets and a coordinated community of researchers that integrate hydrological, ecological, geochemical, and geomorphic-process science from grain to watershed scales and, perhaps as importantly, from deep time to human timescales. Furthermore, scientists in each country have found the paradigm of CZ science to be compelling in addressing problems and attracting students to the field. Nonetheless, each national program has its own approach and strategy. For example, European CZOs are integrated with social sciences related to ecological economics and management science in order to translate natural science advances into European Union policy. CZOs are the lenses through which understanding will be gained of the complexity of interactions between the lithosphere, the hydrosphere, the biosphere, the atmosphere, and the pedosphere through time.

The NSF CZO program began in 2007 with support of the Susquehanna-Shale Hills Observatory in Pennsylvania, the Southern Sierra Observatory in California, and the Boulder Creek Observatory in Colorado. In 2009, three additional observatories were added to the program: Luquillo Mountains Observatory in Puerto Rico; Christina River Basin CZO in Delaware and Pennsylvania; and the Jemez River Basin/Santa Catalina Mountains CZO in Arizona and New Mexico. Most recently, four new observatories were selected for funding: Eel River CZO in northern California; Reynolds Creek CZO in Idaho; the Intensively Managed Landscape CZO in Illinois, Iowa and Minnesota, and the Calhoun Forest CZO in northern South Carolina (Fig. 2.3). The following descriptions of each US CZO are chronologically organized based on the timing of funding from the NSF and hence their formal designation as a US CZO.

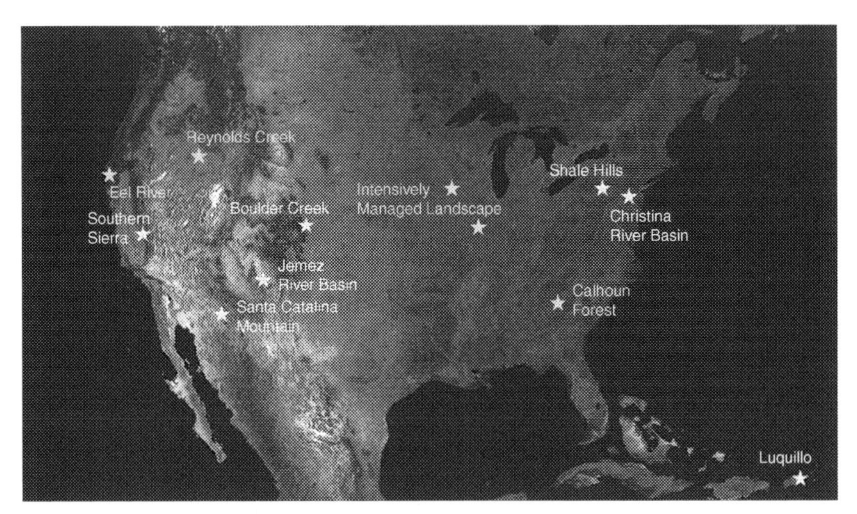

FIGURE 2.3 The United States CZO network consists of seven sites developed since 2007 (two linked as one CZO in NM and AZ) shown in white, with an additional four sites recently designated in 2014 (light gray [yellow in the web version]).

2.2.1 The First Observatories (2007)

2.2.1.1 *Boulder Creek (BcCZO)*

The BcCZO is situated in the Colorado Front Range, one of the Laramide ranges. Boulder Creek flows east, from the Continental Divide to the eastern Great Plains (Fig. 2.4), crossing landscapes shaped by glaciers and permafrost, fluvial canyon incision, and exhumation of the former Cretaceous seaway sedimentary rocks on the Great Plains (Murphy, 2006).

The BcCZO team focuses on three catchments within the greater Boulder Creek watershed (Fig. 2.5). Betasso, at an elevation of 1810–2024 m, is underlain by Precambrian granodiorite (GD) in lower Boulder Canyon. Its steep ephemeral stream cascades into Boulder Canyon from a low relief upland. Grasses and Ponderosa pine dominate vegetation. Mean annual precipitation is ~400 mm, of which 60% falls as snow. Gordon Gulch (2440–2740 m) is on the low relief Rocky Mountain surface, underlain by Precambrian biotitic gneiss. Long evolution, notably including past periglacial conditions (Leopold et al., 2014), shaped this landscape. Different aspect slopes present contrasts in regolith and vegetation: thinner, less-weathered regolith with grasses and Ponderosa pine exists on south-facing slopes, whereas thicker, more weathered regolith with dense Lodge pole pine exists on north-facing slopes (Befus et al., 2011; Anderson et al., 2014). Mean annual precipitation here is ~500 mm, of which 70% falls as snow. Green Lakes valley (3560–4020 m) is a glaciated alpine watershed on Precambrian biotitic gneiss and GD. As part of the Niwot Ridge

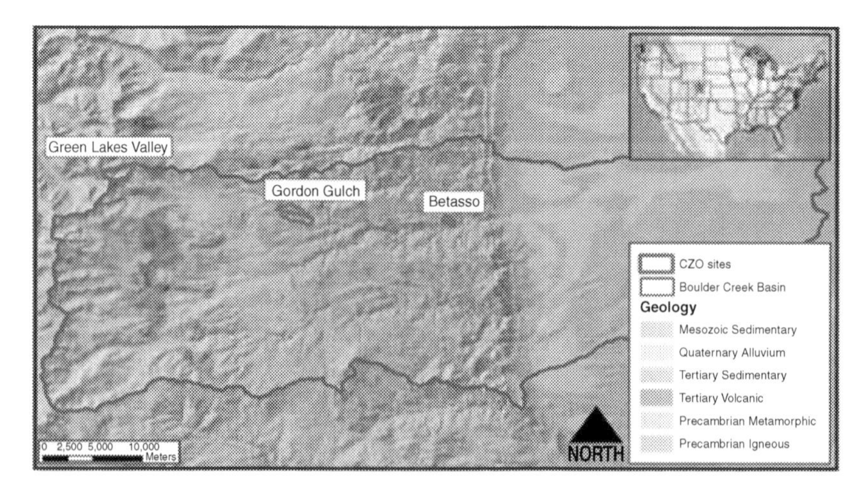

FIGURE 2.4 Map of Boulder Creek watershed (light gray outline [blue outline in the web version]); inset shows location in the USA. The watershed crosses crystalline rocks from the alpine Continental Divide (~4100 m) on the west, and through the forested Rocky Mountain surface (~2000–2700 m). The eastern half of the watershed drains grasslands of the Great Plains (~1480 m) underlain by Mesozoic sedimentary rocks (chiefly shale). Three instrumented subcatchments (dark gray outline [red outline in the web version]) represent different vegetation, climate, and erosional histories.

FIGURE 2.5 Vegetation, topography, and climate vary in each instrumented catchment in BcCZO. Green Lakes valley is above tree line, and its precipitation is dominated by snow. Gordon Gulch is below the glacier limit on the rolling terrain of the Rocky Mountain surface. It is forested, and has strongly contrasting north and south-facing slopes. Betasso stretches from the bottom of the bedrock-dominated Boulder Canyon, up to the gentler Rocky Mountain surface in a forested area of ephemeral streams and ephemeral snow.

LTER, monitoring data extends back several decades (e.g., Caine, 2010). Mean annual precipitation is ~1200 mm, of which 85% falls as snow.

Research infrastructure in the CZO includes meteorological stations at Betasso and Gordon Gulch, soil-temperature and soil-moisture probes at multiple depths within soil profiles at all locations, arrays of soil water samplers (zero tension, ceramic and fused quartz suction lysimeters) at Gordon Gulch, automated, snow-depth sensors at all locations, time-lapse cameras at Gordon Gulch and Green Lakes Valley, stream gauges at all locations, automated, stream-water samplers at Gordon Gulch, manual, surface-water and snow-sampling program at all locations, and groundwater wells at Betasso and Gordon Gulch. The BcCZO team engages substantially in community outreach and education, including a partnership with the University of Colorado's Science Discovery to bring CZ science to K-12 students and teachers, summer classes entitled "Go with the flow" and "Fire and ice," middle school "Science Explorers" workshops, a high-school experience with mountain research, and a field course for professional development of Colorado teachers (http://www.colorado.edu/sciencedisovery/).

The BcCZO science team aims to understand how Critical Zone architecture evolves over time (Anderson et al., 2012), how it conditions hydrologic and biogeochemical response and ecosystem structure (Eilers et al., 2012; Gabor et al., 2014; Hinckley et al., 2014b), and how it will respond to future changes in climate (Fig. 2.6). These goals are addressed by documenting CZ architecture, denudation processes and rates; studying weathering-front advance and hydro-biogeochemical coupling; and through modeling these systems.

Key research highlights to date include understanding of the: (1) Critical Zone role in exhumation of the Plains (Wobus et al., 2010). Cosmogenic radionuclide

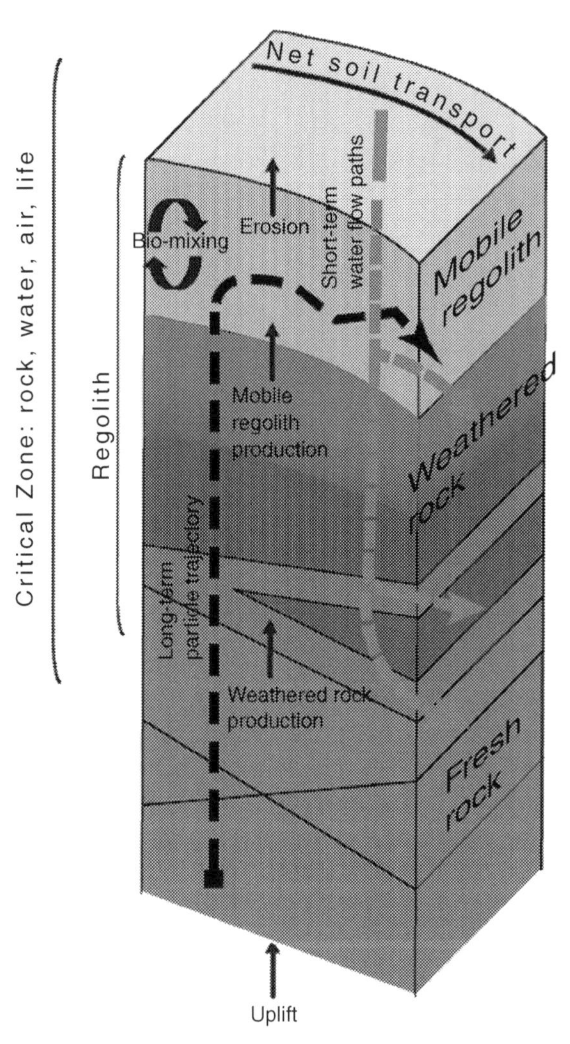

FIGURE 2.6 A schematic diagram of Critical Zone architecture displaying typical layering of mobile regolith, weathered rock and fresh rock, and the flow paths of water (chiefly downward), and trajectories of parcels of rock (vertically up, until it reaches the bio-mixing zone of the mobile regolith layer). *(Figure modified from Anderson et al. (2007).)*

ages of strath terrace sediment covers on the Plains suggest that terrace planation occurs during glacial episodes, and river incision during deep interglacial episodes (Dühnforth et al., 2012). This further indicates that change in sediment delivery from hill slopes (Anderson et al., 2013) is a key driver of river down cutting, which itself incites transient adjustment of hill slopes. (2) Slope aspect provides insight into controls on weathering front advance and slope processes (Anderson et al., 2014). At the rain–snow transition, aspect strongly controls snowpack presence, which controls water delivery (Hinckley et al., 2014a). The thickness of weathered rock varies with slope aspect as well (Befus et al., 2011), which could reflect these differences in water flow path or some other process difference over the order of 10^5 years residence time in the weathered rock layers. (3) Multiple roles of trees. In addition to sediment transport by tree fall, trees also stress rock and transport regolith, simply by growing and dying. Over the lifetime of a tree (on the order of 100 years), rock and regolith are dilated and collapse; roots act as slow explosions, breaking and prying apart rock (Hoffman and Anderson, 2014). (4) Hydraulic conductivity and porosity structure of the Critical Zone controls water flow, with feedbacks on the evolution of structure within ecosystems and rock porosity (Gabor et al., 2014). For example, the decline in porosity with depth leads to the potential for threshold-like behavior in water delivery to rocks and for strong lateral flow even in the vadose zone (Langston et al., 2011).

2.2.1.2 Southern Sierra

The Southern Sierra Critical Zone Observatory (SSCZO) was established in 2007 as a community platform for research on Critical Zone processes, and is based on a strategic partnership between the University of California and the Pacific Southwest Research Station (PSW) of the US Forest Service. The SSCZO is co-located with PSW's Kings River Experimental Watersheds (KREW), a watershed-level, integrated ecosystem project established in 2002 for long-term research to inform forest management. The SSCZO is built on a transect of instrumented sites at elevations from 400 m to 2700 m, anchored by the oldest site in a productive mixed-conifer forest at the rain–snow transition (1750–2100 m) (Fig. 2.7). The main SSCZO site includes three headwater catchments with a dominant southwest aspect (Fig. 2.8).

Soils within the Providence watersheds developed from residuum and colluvium of granite, GD, and quartz diorite parent material. Soils are weakly developed as a result of the parent material's resistance to chemical weathering and cool temperatures. Upper-elevation soils are at the lower extent of late Pleistocene glaciations. Shaver and Gerle–Cagwin soil families dominate the watershed. Soils are gravely sand to loamy sand, with a sand fraction of about 0.75. Soils are shallow (<50 cm) in parts of the watershed, with low tree density and many rock outcrops. Soils in more gently sloping terrain with linear or convex hill slopes are moderately deep; and landforms with the deepest soils (>150 cm) support a high tree density.

The area has a high forest density, with canopy closures up to 90%. PSW plans to thin and/or carry out controlled burns in two of the three-headwater catchments as part of the KREW study, to inform forest managers about impacts of thinning on ecosystem services.

The southern Sierra Nevada is a Mediterranean climate, and experiences relatively wet winters and dry summers. Annual average temperatures are about 8°C at the bottom and 1–2°C cooler at the top of the SSCZO catchments, differences that are driven largely by daytime temperatures and cold-air drainage at night. Daytime winds are upslope and nighttime winds downslope, with wind speeds generally under 1–2 m/s. Precipitation averages about 120 cm per year, and is about 20–60% snow. Photosynthesis persists through the winter, and

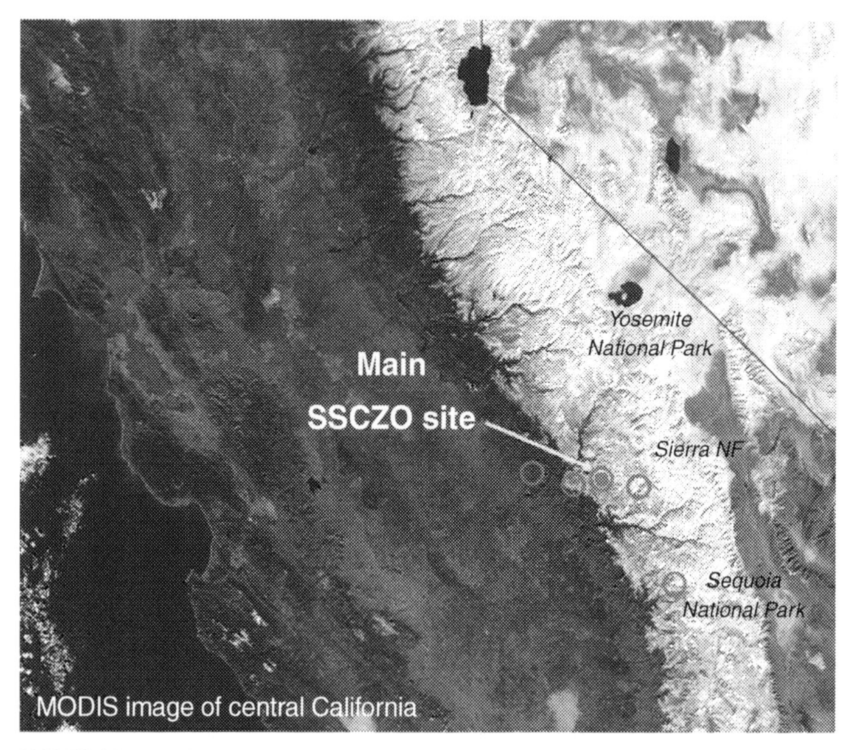

FIGURE 2.7 The Southern Sierra CZO spans an elevation gradient from 400 m to 2700 m. Oak–pine woodlands dominate the lowest elevations, middle elevation sites are in mixed conifer forest, and the highest elevations lie in subalpine forest.

soils and regolith store enough water for photosynthesis to occur all summer. As soils dry out, trees apparently extract water from the deeper soils. Annual runoff is about 15–30% of precipitation in dry years, increasing to 30–50% in wet years. The ground is snow covered for 4–5 months each year, and may experience multiple melt events during the winter and spring (Fig. 2.9).

Two meteorological stations exist, a 60-m flux tower, a 60-node wireless embedded sensor network, 215 EC-TM sensors for volumetric water content, over 110 MPS sensors for matric potential, 60 snow-depth sensors, meadow piezometers and wells, sap-flow sensors, stream gauges and water-quality measurements.

Level 2 data (cleaned, calibrated) from core field measurements are made available by water year. These include precipitation, energy balance, snow, stream flow, soil moisture, sap flow, temperatures, stream geochemistry, soil chemistry, flux-tower data, meadow water levels, vegetation, and various other characterization data sets. Raw data are posted as it is collected. Current-year level-1 data are available by request. Investigator-specific data are available as per the NSF data policy. The area has multiple LIDAR coverage, and a variety of other data sets are available through PSW scientists. It is planned to locate NEON flux towers within the current elevation transect of four SSCZO towers.

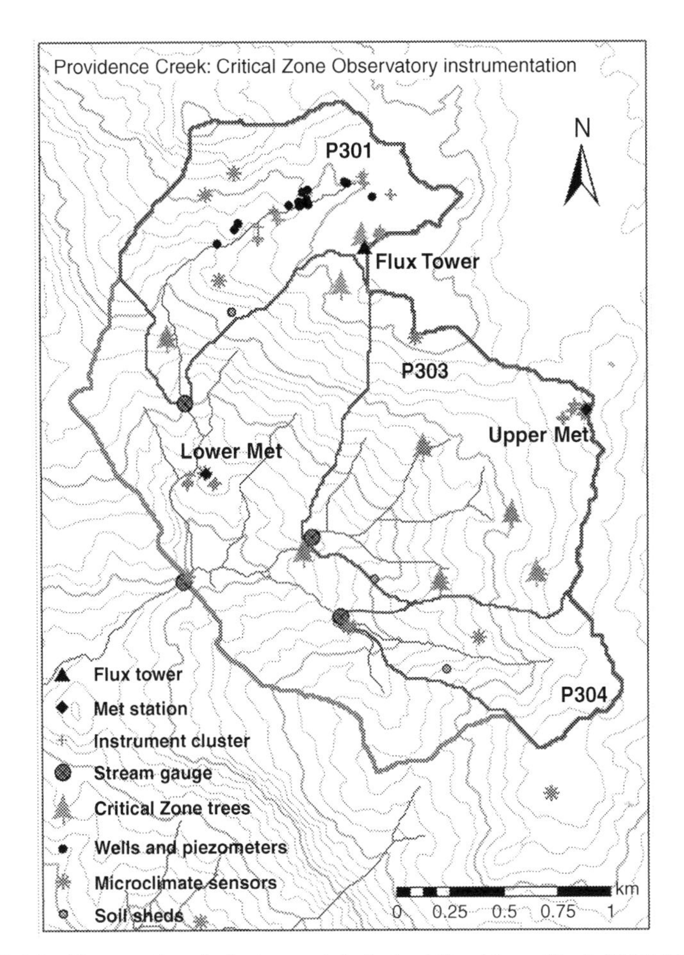

FIGURE 2.8 The most heavily instrumented site is at Providence Creek (1700–2000 m in elevation). This site is also part of the Forest Service KREW (Kings River Experimental Watershed) project. The map shows instrumentation at the site by the Southern Sierra CZO and KREW.

The conceptual science model for the SSCZO is built around bi-directional links between landscape/climate variability and water/material fluxes across the rain–snow transition. Ongoing research focuses on water balance, nutrient cycling and weathering across the rain–snow transition; soil moisture is an integrating variable. Science questions currently being addressed include: (1) How landscape variability controls? How soil moisture, evapotranspiration, and stream flow respond to snowmelt and rainfall? (2) How soil moisture is linked to topographic variability, soil formation and weathering? (3) What physiological mechanisms are controlling how vegetation distribution and function vary with climate? (4) How vegetation attributes influence cycling of water, energy, and CO_2? (5) What links occur between soil heterogeneity, water fluxes and nutrient availability?

FIGURE 2.9 **The road into the P301 subcatchment climbs a rise to 2000 m.** Snow can persist into June, or melt out in March at this site. Daily photos taken near this point at the site (at Critical Zone Tree-1) show intermittent snow cover with multiple melt events in some winters.

Ecosystem types. The mid-elevation Providence catchments are largely Sierra mixed-conifer forest, with some mixed chaparral and rock outcrops. Dominant trees are white fir, ponderosa pine, Jeffrey pine, sugar pine, incense cedar and black oak. Several species of shrubs are also present. Of the three perennial streams, one borders meadow over 90% of its length, one has no meadow, and the third is intermediate. The lowest elevation site (400 m) is located at the San Joaquin Experimental Range. This site has oak-pine woodlands and annual grasses. The mixed-conifer forest of the lower mid-elevation site (1160 m) is dominated by ponderosa pine and black oak. Subalpine Lodge pole pine forests cover the highest elevation site (2700 m).

Research highlights. The distributed snow and soil moisture measurements show a close coupling between snowmelt and soil drying in spring/summer, with systematic variability across elevation, aspect and canopy cover. Evapotranspiration (ET) decreased proportionally as soils dried, going from about 1 mm/d to 0.1 mm/d over the summer. However, about half of the ET occurred after snow melted. Runoff increased with elevation, corresponding to decreasing temperature, more precipitation falling as snow, decreasing vegetation density and coarser soils. Nutrient hotspots in soils are important for nitrogen cycling, and do not necessarily correspond to preferential flow paths for water. Annual erosion rates measured in headwater catchments are only 1% of long-term rates, with head cut and bank erosion dominating. Sensor networks that are wireless and embedded statistically sample the landscape variability and have proven to be an economical, scalable approach to measurement design; this approach is being replicated at other locations in the Sierra Nevada.

2.2.1.3 *Shale Hills*

The focus of the SSHCZO is the Shale Hills watershed (red square on Fig. 2.10), an 8 hectare catchment which lies within the Valley and Ridge physiographic province of the central Appalachian Mountains in Huntingdon County, Pennsylvania (40°39′52.39″N 77°54′24.23″W) (Lin, 2006). It is a first order, V-shaped basin characterized by relatively steep slopes (25–35%) and narrow ridges. The stream is a tributary of Shavers Creek that eventually debauches into the Juniata River, a part of the Susquehanna River Basin. Since 2013, the focus of SSHCZO has grown to include sites within the larger Shavers Creek watershed (Fig. 2.10), including a small, forested satellite catchment located on sandstone

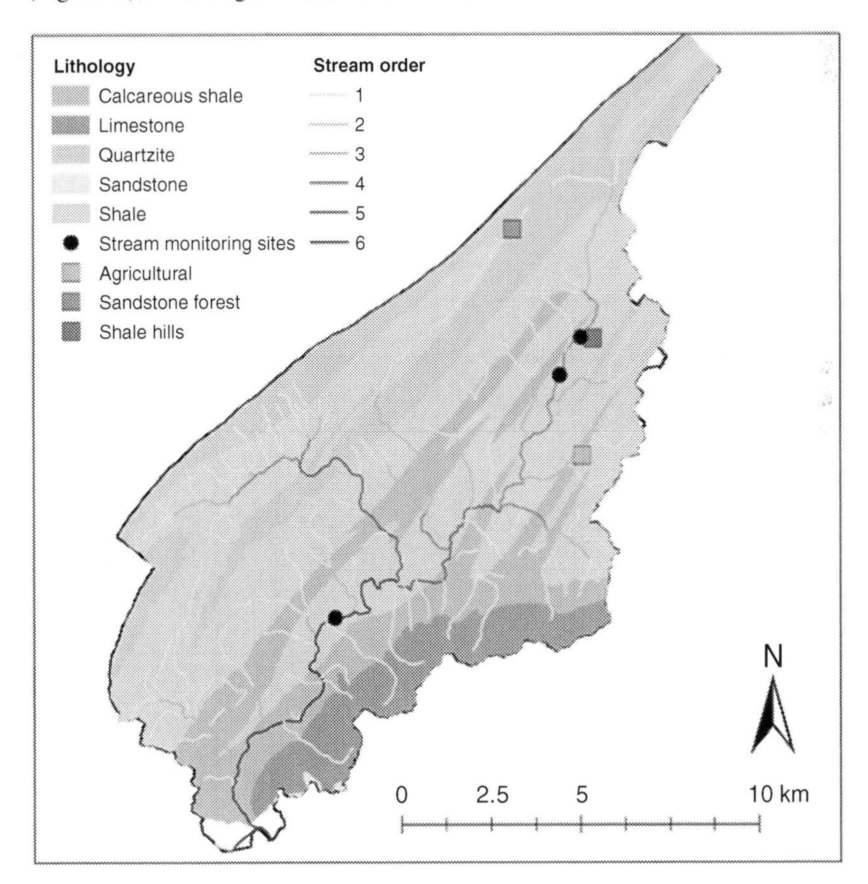

FIGURE 2.10 Geological map of the Shavers Creek Watershed in central PA. Shale Hills (darkest gray square [red square in the web version]) is the original focus of the CZO. New stream monitoring sites (black circles) have been established on the main stem of the Creek, and two new subcatchments (two lighter gray squares) are being implemented as shown.

(blue symbol). The CZO will eventually include a similarly sized agricultural catchment on calcareous shale (green symbol). Study of the small catchments is a step toward scaling up to the entire Shavers Creek watershed. Outreach to watersheds in northern PA that are impacted by shale-gas development is also ongoing.

The Shale Hills catchment is oriented in an east-west direction such that the major side slopes are north- and south-facing. Elevation ranges from 256 m at the outlet to 310 m at the highest ridge. The relatively uniform side slopes are periodically interrupted by seven distinct topographic depressions (swales). The catchment is underlain by shale of the Silurian Rose Hill Formation, whereas Shavers Creek watershed includes sandstone, calcareous shale, and limestone. The folded and faulted sedimentary strata were uplifted during the Alleghenian Orogeny ~300 million years ago. Pleistocene permafrost and associated solifluction occurred most recently during and shortly after the last glacial maximum when the area experienced periglacial conditions.

Shallow to moderately deep, gently-dipping to steep, well-drained residual shale soils occur on the ridge tops of Shale Hills, whereas along slopes and in the valley bottom, soils have formed on a colluvial and alluvial mantle of shale chips (Jin et al., 2010; Lin et al., 2006). Soils are typically saturated along the stream where they exhibit redoximorphic features as a result of seasonal soil saturation (Lin et al., 2010). A 3–5 cm organic layer of decaying leaf litter overlies all soils in the watershed. Typical surface soil textures are silt loam, with the percentage of channery shale increasing with depth. Effective rooting depth (depth to bedrock) ranges from 15 cm on ridge tops to 165 cm, especially near valley floor. Soil structure is moderately developed throughout the basin. Small roots have been observed to penetrate into fractured shale beneath the augerable soil.

SSHCZO is characterized by a humid continental climate. Temperatures average 9.5°C, with large seasonal variations: January temperature is –5.4°C, July is 19.0°C. The highest temperature recorded is 33.5°C (April 27, 2009) lowest –24.8°C (January 17, 2009). Annual average relative humidity is 70.2%. Atmospheric deposition in PA is characterized by acidic (pH~4) precipitation. The Shale Hills water balance for 2009–2010 is summarized as follows:

	2009	**2010**
Precipitation (mm)	1028	958
Interception (mm)	284	276
Evapotranspiration (mm)	594	586
Recharge (mm)	319	306
Runoff (mm)	509	364
Runoff ratio (%)	49.51	38.00

The Shale Hills forest ecosystem is dominated by oak, hickory, and pine species. Hemlock, red maple, white oak and white pine line the deep, moist soils of the stream banks, whereas on the drier, shallower north- and south-facing slopes, red oak, chestnut oak, pignut hickory and mockernut hickory are dominant, with Virginia pine appearing on the southern ridge (Naithani et al., 2013). Understory woody species include plants in the Ericaceous family, service berry, hawthorn, raspberry/blackberry, sugar maple saplings, and witch hazel. Forest surveys at the new sites are ongoing.

Historically, the region was logged for charcoal to support a nineteenth and twentieth century iron industry. Today, Shale Hills and the sandstone forest site experience low human impact. Shale Hills is used for recreation and education and the catchment is managed for timber with set-asides for research within the Penn State Forest. The sandstone site is managed by Rothrock State Forest, PA Bureau of Forestry. The lower Shavers Creek watershed is characterized by residential and agricultural land use (row crops and pasture for dairy farms). Research is now being extended into the sandstone catchment, as shown on the map earlier. This new catchment is largely pristine.

The Shale Hills watershed has a comprehensive base of instrumentation for characterization of water, energy, stable isotopes and geochemical conditions. This includes a dense network of sites for soil-moisture observation at multiple depths (120), a network of shallow observation wells (24 wells), soil lysimeters at multiple depths (+80), a cosmic-ray sensor for soil moisture, a research weather station, including eddy flux measurements for latent and sensible heat flux, CO_2, and water vapor, radiation, barometric pressure, temperature, relative humidity, wind speed/direction, snow-depth sensors, leaf-wetness sensors, and a load-cell precipitation gauge. A laser precipitation monitor (LPM: rain, sleet, hail, snow, etc.) was installed in 2008, as were automated water samplers (daily) for precipitation, groundwater, and stream water. Sap flow is measured as a function of tree species. Several hill slope transects have been investigated for collection of soil pore water and gas. Stream, groundwater, and precipitation samples have also been collected and analyzed over extended periods. A wireless sensor network is being deployed to allow near real-time observations of soil moisture, groundwater level, ground temperature, and electrical conductance. Data collected from sensors at Shale Hills are compiled in an online database for use by anyone.

As the new catchments are being implemented, we are decreasing the frequency and spatial density of measurements in the Shale Hills catchment itself. To extend observations to the sandstone and agricultural catchments has required development of a paradigm of fewer measurements per catchment. The new plan targets observations along catenae where soil pits are dug and instruments are deployed.

Three airborne LiDAR flights were flown over the Shavers Creek watershed. The most recent flight (0.5-m resolution) was used to evaluate micro-topography and tree species. Depth to bedrock in Shale Hills has been surveyed. Ground-based LiDAR and total-station surveys were completed for all instrument elevations. Trees >20.3 cm (8 in.) were surveyed for species, biomass, and crown height. Leaf area index (LAI), greenness index, distribution and CO_2 flux have been measured. Borehole logging was completed at three locations to 17 m.

The SSHCZO team is working to quantify the rates of formation and evolution of the structure and function of the Critical Zone, focusing on the fluxes of water, energy, solutes, and sediments (WESS) as a function of the geochemical, hydrological, biological, and geomorphological processes operating in a temperate, forested landscape on sedimentary rock. Our interdisciplinary team works collaboratively in one observatory to advance methods for characterizing regolith and WESS fluxes to provide a theoretical basis for predicting the distribution and character of regolith as well as impacts of regolith on fluid pathways and rates. The modeling effort centers around the Penn State Integrated Hydrological Model (PIHM) (Duffy et al., 2014). Some research highlights are as follows. In hydroclimatology, the instrument array enabled an investigation of explicit coupling and feedback for sub-surface–land surface–atmosphere interaction using fully coupled models (Shi et al., 2013). Biogeochemical measurements of soil N_2 and CO_2, total soil carbon (SC) and dissolved organic carbon are helping to predict how soil development affects SC storage and transport. In ecology, water uptake and isotopic composition of trees are being tracked to understand tree-water use. Stable isotope measurement of deuterium and $\delta^{18}O$ has been used to investigate water fluxes (Jin et al., 2011; Thomas et al., 2013). In geomorphology, rates of regolith formation, the role of tree-throw in the downslope transport of material, and the residence time of material in regolith are being measured (Ma et al., 2010; West et al., 2013). Field-scale and lab-scale tracer tests are being conducted to elucidate preferential pathways and dual-domain solute transport behavior. In hydropedology, ongoing studies target how landscape water influences soil genesis, evolution, variability, and function (Lin et al., 2010). Finally, studies of the physical, biological and hydrogeochemical processes that operate within this catchment though quantification of long-term physical–chemical weathering fluxes are being measured (Brantley et al., 2013).

2.2.2 Expansion to Six Observatories (2009)

2.2.2.1 Christina River Basin Critical Zone Observatory (CRB-CZO)

The Christina River Basin Critical Zone Observatory (CRB-CZO) was established in 2009 as a community platform for research on Critical Zone processes, and is based on a strategic partnership between the University of Delaware (UD), the Stroud Water Research Center (SWRC), the University of Minnesota and numerous other organizations. The 6th order Christina River Basin (1440 km^2) and its four subwatersheds – White Clay Creek (277 km^2), Red Clay Creek (140 km^2), Brandywine Creek (842 km^2), and the tidal Christina River (202 km^2) – flow from Pennsylvania, Delaware, and Maryland into the Delaware Estuary and Bay (Fig. 2.11).

The Christina River Basin transitions from Piedmont into Atlantic Coastal Plain, the two most populated physiographic provinces in the United States.

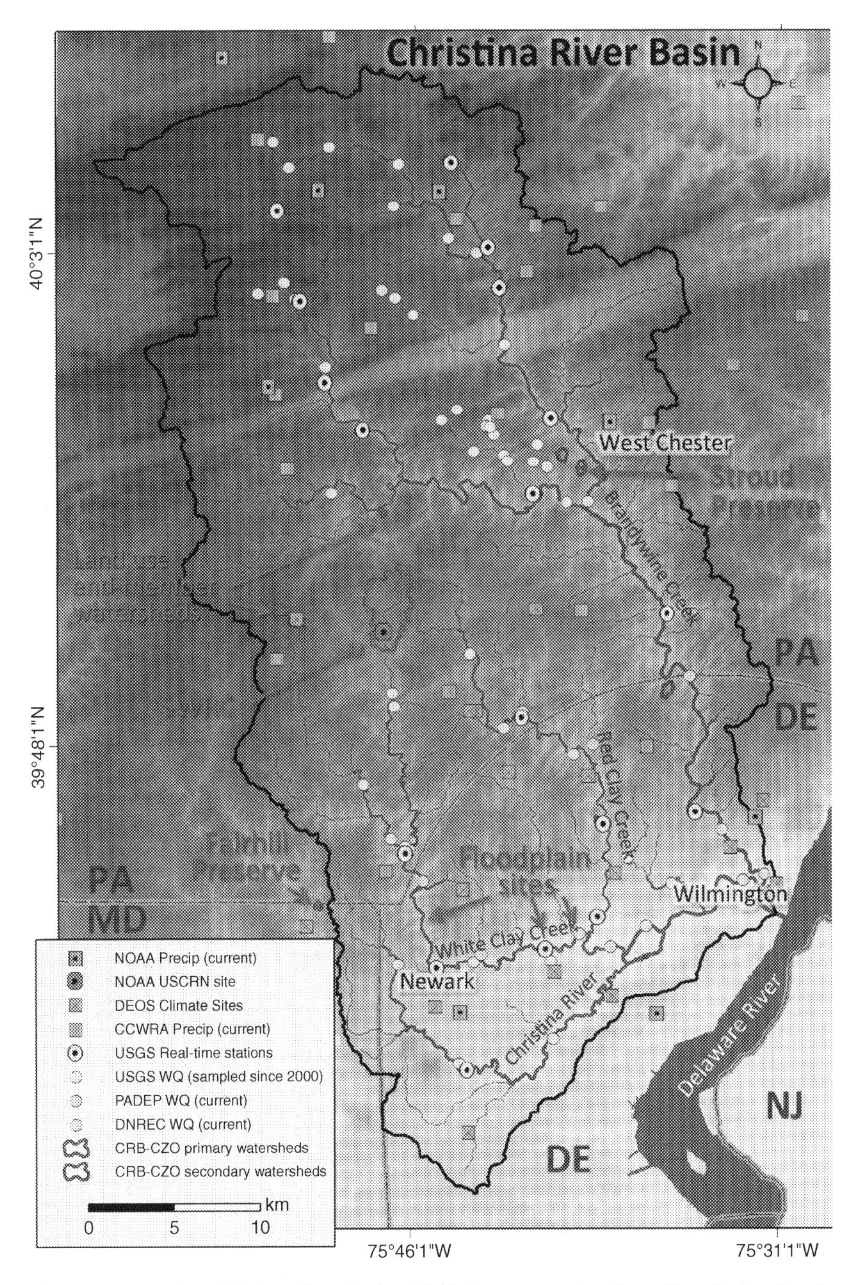

FIGURE 2.11 The Christina River Basin (CRB) is an exceptionally well-studied and well-protected 1400 km^2 watershed that traverses the rural historic colonial corridor between Philadelphia and Baltimore. The CRB thus offers innumerable opportunities to research human impacts to Critical Zone evolution, structure and function through paired and nested plot and watershed approaches.

FIGURE 2.12 Mixed forest and agricultural land uses have dominated the Christina River Basin since colonial times. Although suburban development is very active within the CRB, and urban/industrial land uses are dominant in the lower Christina, most of the core landscapes are preserved by conservation easements and actively protected by a larger number of conservation organizations.

The human footprint within the region spans centuries and current land covers include mature forest, agriculture, suburbia, urban, commercial and industrial (Fig. 2.12), providing an ideal natural laboratory to study the gradient of human impacts on Critical Zone processes. A diverse meta-sedimentary lithology – ranging from micaceous schist and gneiss to quartzite to marble – is overlain by deep, unglaciatied soils of diverse chemical and physical characteristics – from Entisols to Ultisols to Histosols. Nearly all stream valleys are filled with 1–3 m of sediment eroded and deposited since colonial times by intensive deforestation, agriculture and mill damming. White Clay Creek is the only entire watershed designated within the Wild and Scenic Rivers System.

Unique aspects of CRB-CZO include the location in the Piedmont and Atlantic Coastal Plain physiographic provinces, including satellite coastal sites along the Delaware Estuary, the diversity of human land use for centuries, the role as a drinking water source to a million plus people, and its history as being exceptionally well studied by US Geological Survey (USGS), SWRC, environmental protection agency (EPA), and numerous local agencies.

The Christina River Basin and its four sub-basins may be one of the best-studied watersheds of its size in the nation. Studies include seminal fluvial geomorphology of the Brandywine Creek by Luna Leopold, >100 peer-reviewed publications by the Stroud Water Research Center, and extensive long-term monitoring efforts by EPA, USGS and others maintaining quality water supplies to one million people and developing total maximum daily load regulations. Within the CRB exist 19 USGS stream/river gauging stations (6 in DE, 13 in PA) and 5 of the PA stations continuously monitor water quality properties (i.e., turbidity, pH, conductivity, dissolved oxygen). SWRC has historically maintained continuous discharge and other datasets at three stations and is expanding that number for the CZO and other programs. Non-continuous data collected by USGS and SWRC are available for 141 stations. Five USGS stations and a NOAA Climate Reference Network station have continuously recorded weather data.

The holistic study of the entire 1440 km^2 Christina River Basin has the overall goal to integrate the feedbacks among the water cycle, mineral cycle and carbon cycle within contrasting land uses as materials are transported and transformed across geophysical boundaries from "saprolite to sea" that traditionally separate scientific disciplines; that is, bedrock → saprolite → topsoil → aquifers → riparian floodplains → wetlands → river networks → salt marshes → estuaries → sea. Such a study requires a tightly integrated team, sophisticated field sensors and samplers to capture hot spots and hot moments at each geophysical boundary from source to sink, mechanistic models to test the bleeding edge of theory and the coupling of processes with targeted high-resolution data, and a field area that includes both actively weathering and eroding headwaters and actively accumulating depositional zones. Overall hypotheses (Fig. 2.13) are that: (1) Hydrological, chemical, and biological processes that produce and mix mineral surfaces and organic carbon are rate limiting to watershed-scale chemical weathering, soil production and carbon sequestration; and, (2) Humans accelerate rates of carbon-mineral mixing, altering Critical Zone function and resulting in anthropogenic carbon sequestration significant to local, regional and global budgets.

Research highlights from the CRB-CZO include: (1) intensive study sites for the impacts of three land-use end-members (mature forest, row crop agriculture, landfill) on hydrological, pedological, and geomorphological processes; (2) long-term continuous datasets as a basis for hydrological and material transport models, including implementation of the PIHM; (3) intensive geochemical characterization of mineral weathering, carbon-mineral complex formation and carbon stabilization in soils and sediments of different origin and land use; (4) source-to-sink sediment tracing study using radio-isotopes and other geochemical "fingerprints"; and (5) *in-situ* sensors. The CRB-CZO is developing and deploying an advanced sensor network for real-time observations of hydrological and biogeochemical processes. These sensors range from meteorological and hydrological sensors that can be widely deployed at low cost to advanced field instruments such as submersible UV-Vis spectrometry. The wireless

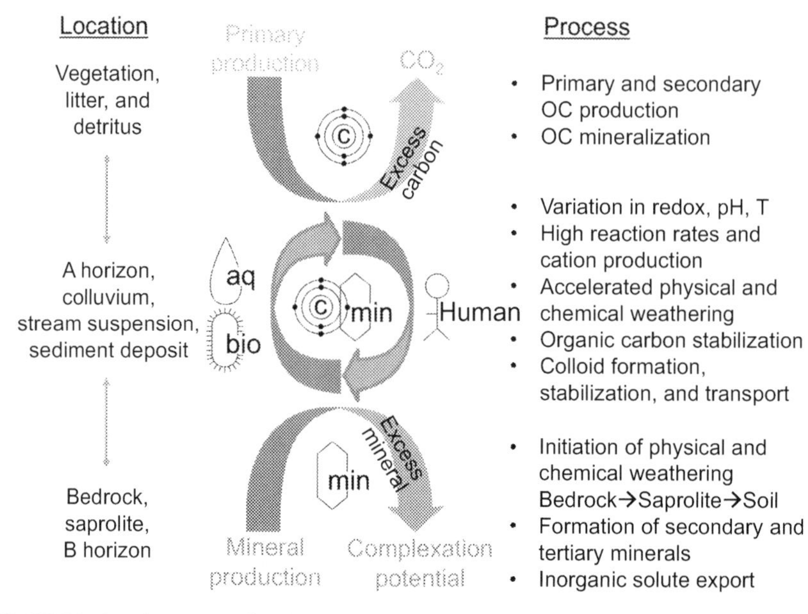

FIGURE 2.13 The conceptual model driving much of CRB-CZO research is that hydrological, chemical, biological, and human processes that produce and mix mineral surfaces and organic carbon are rate limiting to watershed-scale chemical weathering, soil production and carbon sequestration. These processes occur at a range of spatial and temporal scales and across a diversity of landscape positions and environmental settings. Therefore the CRB-CZO maintains that a whole watershed, saprolite-to-sea integration is required to fully understand the feedbacks among these processes.

sensor network is based on the open source electronics "Arduino" platform with ZigBee-based radio networking. With over half a million users worldwide, this approach is robust, easy-to-use and low-cost, which allows SWRC to invest resources on widespread deployment of high quality sensors rather than on data communication infrastructure.

2.2.2.2 Luquillo CZO (LCZO)

The LCZO is located in the Luquillo Mountains of northeastern Puerto Rico in one of the wettest zones of the greater Caribbean Sea region (Fig. 2.14). Over a distance of 10–20 km, the mountains rise from sea level to an elevation of 1075 m while precipitation increases from <1000 mm/year to >5000 mm/year. An overview of the geology, biology, and biogeochemistry of the Luquillo Mountains is found in McDowell et al. (2012).

The two main study watersheds in the watershed, the Río Mameyes and the Río Blanco, have similar climates and environmental histories but differing lithology (Fig. 2.14). The Mameyes watershed (Fig. 2.15) is primarily underlain by volcaniclastic (VC) bedrock that weathers to produce clays and boulders. The Río Blanco watershed is underlain by GD that weathers into a saprolite comprised of sand and large core stones. These differences in weathering patterns

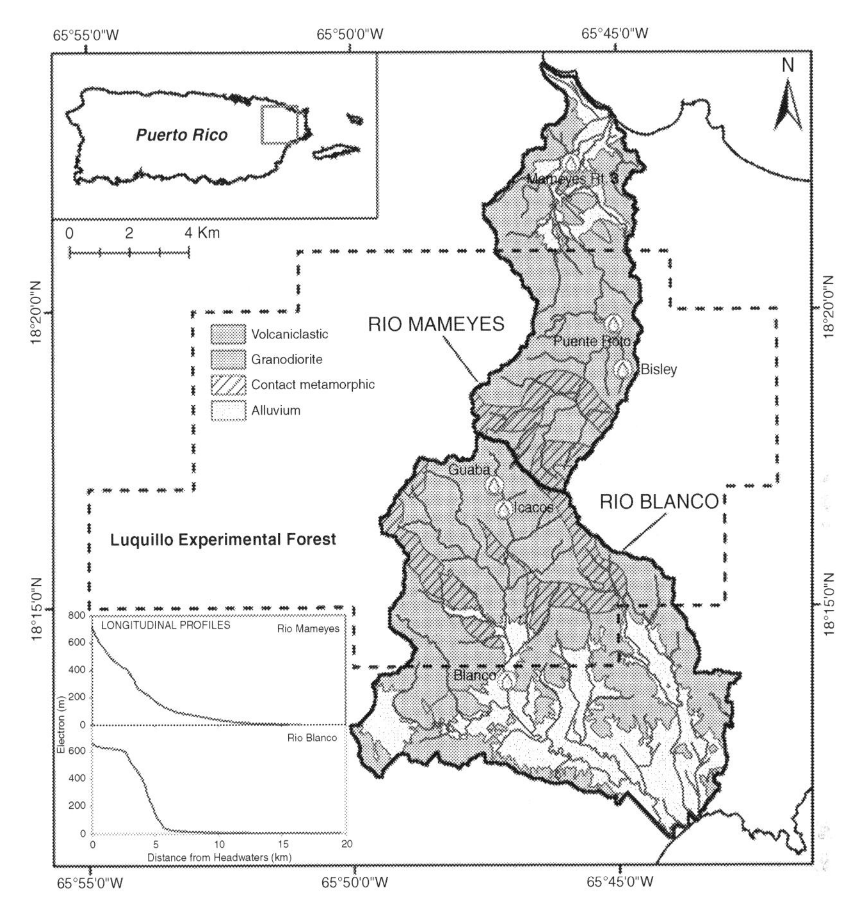

FIGURE 2.14 Location of the LCZO and its primary study watersheds, the Rio Mameyes and Rio Blanco basins. The outline of the Luquillo Experimental Forest is shown as a dashed line. Stream gauges are labeled. Longitudinal profiles of the two basins are modified from Pike et al. (2010).

have a profound influence on landslide frequency, chemical denudation, and the morphology of streams and hill slopes.

The Observatory has a subtropical, humid, maritime climate that is influenced by both orographic and global-scale synoptic weather systems. Rainfall occurs throughout the year and exceeds 5000 mm/year at the highest elevations. Rainfall events at mid-elevations are generally small (median daily rainfall 3 mm/day) but numerous (267 rain days per year) and of relatively low intensity (<5 mm/h). Nevertheless, individual storms greater than 125 mm/day occur annually, and daily rainfalls greater than 600 mm have been recorded (McDowell et al., 2012).

Six Holdridge life zones occur in the Luquillo Mountains and surrounding areas: lower montane wet, lower montane rain, subtropical moist, wet, rain, and dry (Ewel and Whitmore, 1973). Luquillo forests are classified into four main types: "tabonuco forest," so named for the dominant tree *Dacryodes excelsa*,

FIGURE 2.15 Rio Mameyes at the La Mina waterfall. Waterfalls are important geomorphic breaks in the river systems of the Luquillo Mountains, which structure the aquatic biota above and below the break (Covich et al., 2009). *(W. McDowell.)*

which covers lower elevations up to about 600 m and harbors about 168 tree species. The "Colorado forest," named for *Cyrilla racemiflora* trees, occurs above the tabonuco forest and extends to about 900 m. This forest type is a montane cloud forest that has a canopy of about 15 m and harbors 53 tree species. At this same elevation, and in especially steep and wet areas, is the third forest type, the "palm forest," which is dominated by *Prestoea acuminata*. On the highest peaks are the "elfin forest," a dense, short-statured cloud forest with abundant epiphytes and a canopy that can be less than 3-m tall.

Two striking features of the region's disturbance regime are the long history of natural hurricanes and the recent history of human disturbance and recovery. In addition to hurricanes, other natural disturbances include landslides, tree falls, floods, and droughts. Human disturbances are most common at lower elevations and include historic clearing for pasture, crops, and coffee plantations; logging; road construction, and water diversions.

Researchers have access to 11 stream gages, 2 walk-up canopy towers, 4 meteorological stations, 3 deep observation wells, 20 shallow riparian wells, lysimeter nests, an extensive GIS system and numerous long-term vegetation plots at the site. Because the Luquillo Mountains have been a center for research on tropical forests for over a century, many long-term environmental data sets exist. Three signature data sets include (1) micrometeorology and hydrology: hourly and daily measurement of radiation, air pressure, temperature, relative humidity, precipitation, wind speed, and wind direction (four stations maintained by USFS-IITF). The world's longest known record of weekly rainfall and through fall, and associated chemistry is maintained at the site and available online. Eight stream gages are maintained by the USGS, and the USFS-IITF maintains three stream gages in the Bisley Watershed. (2) Geochemistry and biogeochemistry: the LCZO is developing an extensive data set of Luquillo soils and bedrock geochemistry. The chemistry of weekly rainfall, through fall, and stream flow is maintained and available from the LCZO and Luquillo LTER web pages. Additional data are available on plant species composition, allometry, and chemistry. (3) Spatial data sets: a 10-m DEM and associated spatial data sets are also available for the upper Luquillo Mountains.

The role of hot spots and hot moments in tropical landscape evolution and Critical Zone function is the overarching focus of the LCZO. The infrastructure, sampling strategy, and data management system are designed to address this question in watersheds underlain by GD and VC bedrock in the natural laboratory of the Luquillo Mountains, Puerto Rico. LCZO research is centered on four interrelated focal areas: (1) the importance of knick points and different landscape positions as hot spots for weathering, soil development, and biogeochemical cycling; (2) the role of hot spots and hot moments in redox cycling, and their effects on weathering, as well as the storage and loss of carbon and nutrients in soils over a range of spatial and temporal scales; (3) the role of hot moments in the transport of sediment, C, and nutrients in stream flow, and hot spots that determine the distribution of material across the landscape; (4) scaling up hot spots and hot moments in time and space using climate and hydrologic modeling, and assessing the role of rain, cloud water, and dust inputs on landscape evolution and Critical Zone function. Overall, the research conducted by LCZO will provide a well-integrated assessment of Critical Zone properties and processes that scale from microsites to catenae, watersheds, landscapes, and the region; and from minutes to hours, days, months, and years. Because of the long-term ecological studies that are also conducted at the site, the LCZO is ideally suited to examine interactions between Critical Zone processes and the biology of tropical forests.

2.2.2.3 Santa Catalina Mountains – Jemez River Basin CZO

The Santa Catalina Mountains – Jemez River Basin Critical Zone Observatory (SCM-JRB CZO) was established in 2009 as part of the National CZO Program funded by NSF. It comprises an elevation gradient that begins with a granite-schist "sky island" mountain range rising out of Sonoran desert scrub

FIGURE 2.16 The Santa Catalina Mountains – Jemez River Basin CZO comprises two geographical locations (in Southern Arizona and Northern New Mexico) to capture a climatic parameter space that encompasses much of the water-limited Southwestern United States.

in southern Arizona (SCM) and extends into a rhyolite montane caldera in northern New Mexico (JRB) (Figs 2.16 and 2.17). Inclusion of both sites in the observatory design enables researchers to explore a wider range of CZ-forcing parameter space than would be possible with either site alone. The CZO includes lithological, climatic, and disturbance variation characteristic of much of the southwestern United States (Chorover et al., 2011).

Precambrian- and Paleogene-aged granites and GD, in combination with Paleozoic-aged metamorphic rocks such as schist and quartzite, dominate SCM bedrock. The terrain is steep and rugged. Soils are shallow at low elevation (<25–50 cm depending on landscape position) where weathering depth is limited by hot, dry climate conditions, and deeper (ca. 50–100 cm) at higher elevation where cool, wet conditions prevail. Schist soils are more deeply weathered, finer in texture, and contain more organic matter than granite soils (Lybrand et al., 2011; Pelletier et al., 2013).

Silica-rich extrusive volcanic rocks dominate JRB bedrock: rhyolite tuff, rhyolite, andesite, dacite, and silica-rich volcanic ash. Instrumented catchments in JRB are located primarily on rhyolite tuff, which facilitates formation of deep soils (70–>200 cm depending on landscape position). The thick soil cover corresponds with a relatively diffuse landscape structure. Tuff-derived soils exhibit substantial clay accumulation in the subsurface. Upper soil horizons contain appreciable volcanic glass and kaolin, whereas subsurface horizons have less glass and more smectite (Vazquez-Ortega et al., 2014).

Instrumented zero order basins (ZOBs) are in Upper Sonoran Desert, Desert Woodland-Grassland, Ponderosa Pine, and Mixed Conifer forest ecosystems that have been managed for multiple use. Both SCM and JRB have been subjected to regular wildfires, including two in 2011 and 2013 in the Valles Caldera

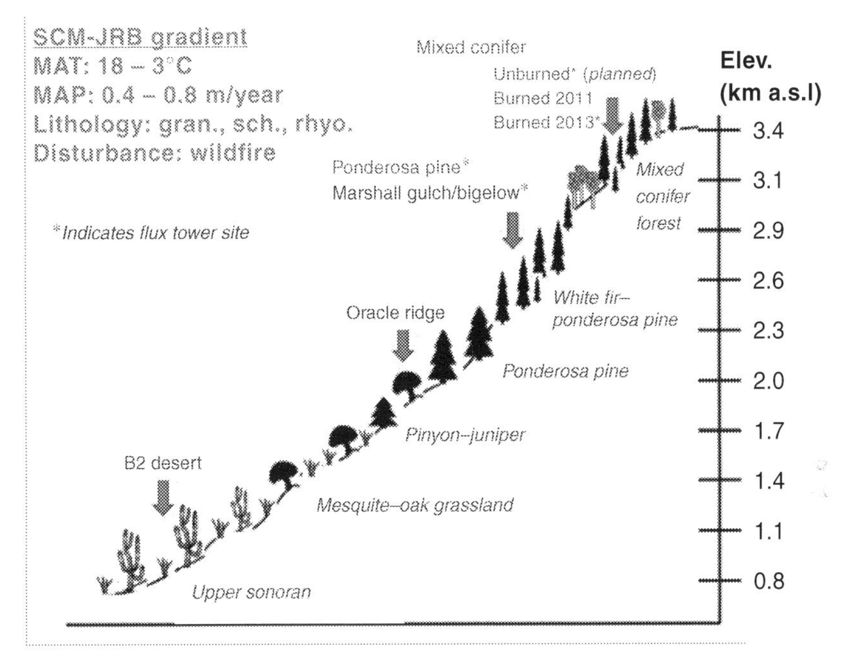

FIGURE 2.17 Ecosystems colonizing the SCM-JRB CZO extend from Sonoran Desert scrub to mixed conifer forest. Mean annual temperature decreases and mean annual precipitation increases with increasing vegetation from 800 m to 3400 m above sea level. Forests at higher elevations are seasonally-snow covered.

National Preserve (VCNP, JRB) that provide a chronosequence of mixed conifer forest disturbance. Pre- and postfire, LiDAR studies combined with sediment flux, topographic incision analysis and ^{10}Be cosmogenic radionuclide data indicate that ca. 99% of long-term denudation occurs during the brief pulses of erosion that immediately follow wildfire (Orem and Pelletier, 2015).

Precipitation and temperature throughout the southwestern United States are highly dependent on elevation, and SCM-JRB climate covers much of that encountered in the southwestern United States. Most of the water-sustained urban populations in semi-arid basins of the southwestern United States. (e.g., Tucson, Phoenix, Albuquerque, El Paso) originates in snowpack delivered to montane sites (Harpold et al., 2012). As a result, the dynamics of snowpack accumulation and ablation, including the impacts of wildfire disturbance, are of significant societal concern (Harpold et al., 2014a, 2014b). The SCM-JRB CZO seeks to better resolve the water, carbon, and weathering dynamics of high elevation receiving catchments, including their resilience to climatic variation and associated disturbances such as wildfire.

ZOBs nested within catchments are an integrating unit of study in the SCM-JRB CZO (Chorover et al., 2011; Perdrial et al., 2012). For example, as shown in Fig. 2.18, two instrumented ZOBs are situated within the Marshall Gulch

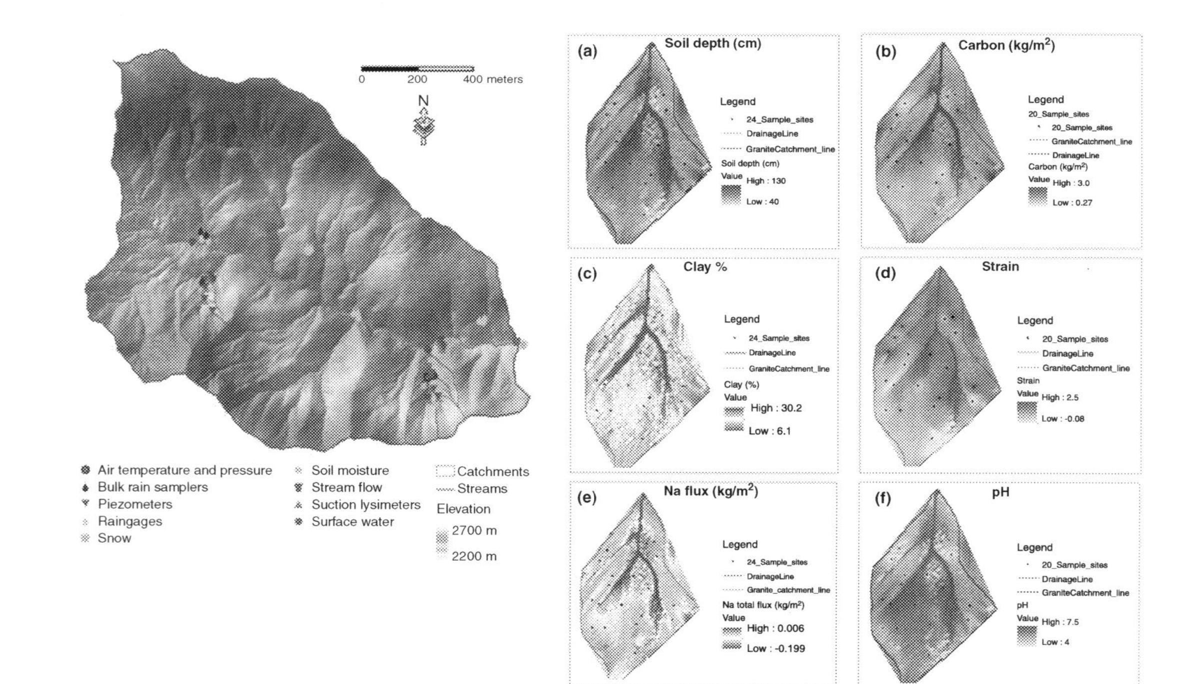

FIGURE 2.18 The SCM-JRB CZO has instrumented zero order basins (ZOBs) nested within catchments. For example, the Marshall Gulch Catchment (1.54 km^2, in the SCM, left side of figure above), located southeast of Mount Lemmon peak (32°25'45"N, 110°46'0"W)) contains a weir for integrated response and instrumented ZOBs such as the "granite ZOB" shown above right (Holleran et al., 2014).

Catchment in the SCM: the western ZOB is underlain by granite whereas the eastern ZOB is underlain by schist. The mean transit time of water in this catchment shows strong dependence on initial conditions (Heidbuchel et al., 2012). Digital maps of soil properties in the granite ZOB of Marshall Gulch are shown in (a) to (e) at right (Holleran et al., 2014).

At each location (JRB and SCM), field equipment deployed in ZOBs provide continuous or periodic measures of water, carbon, and energy stores and fluxes across the CZ. Instrumentation includes flux towers for measuring eddy covariance, sap-flow sensors, phenocams, weather stations, rain gauges, rainwater samplers, stream flow flumes, snow-depth sensors, snow-melt lysimeters, shallow groundwater piezometers, soil–moisture and soil–temperature probes, soil–water tensiometers, and soil–water solution samplers.

The SCM-JRB CZO focus is on understanding how variability in climate, lithology, and disturbance influence CZ structure and function over both short (e.g., hydrologic event) and long (e.g., landscape evolution) time scales. We are addressing this issue using a theoretical framework that quantifies system inputs in terms of effective energy and mass transfer (EEMT, MJ/m year, Rasmussen et al., 2011, see Fig. 2.19). Science questions being addressed in the current phase of CZO research include: (1) How do long-term drivers of CZ structure and function (EEMT and tectonics) interact with parent material to control current CZ structure and response to perturbation? (2) How is long-term CZ evolution affected by ecosystem process controls? (3) What is the impact of CZ structure on buffering climate- and disturbance-driven variability in water, soil, and vegetation resources, and how does this translate into changes in CZ services? Highlights of SCM-JRB research so far include the recognition that isotope hydrology, trace element geochemistry, pedogenic studies, and landscape evolution modeling show that weathering trajectories at the pedon and

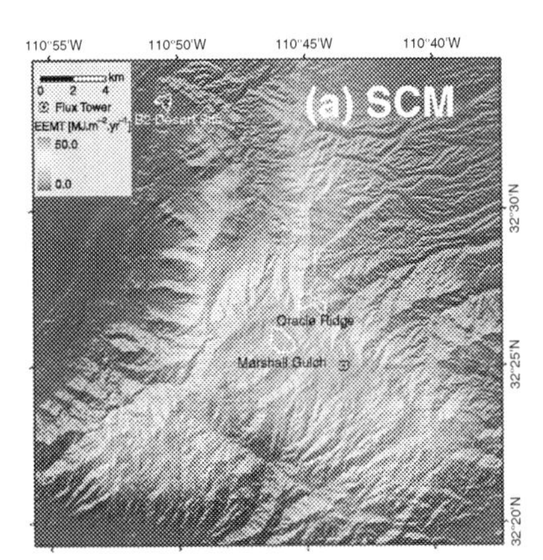

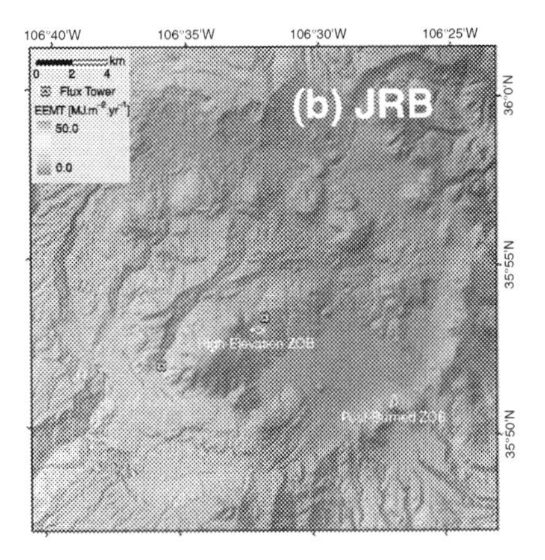

FIGURE 2.19 EEMT maps of the SCM (a) and JRB (b) surfaces and associated instrumented sites enable visualization of how climatic forcing varies with elevation (as manifest in water and carbon throughputs to the CZ subsurface).

ZOB scale exhibit strong dependence on landscape position and lithology in similar water/energy (EEMT) regimes. In addition, CZ evolution occurs episodically; landscape-disturbance events, such as wildfire, disproportionately affect long-term rates.

2.2.3 Creating a CZO Network and a National Office (2014)

2.2.3.1 Calhoun CZO

The Calhoun CZO is located in north central South Carolina, in the Southern Piedmont physiographic province that extends from Virginia to Alabama (Fig. 2.20). The new Calhoun CZO leverages more than 60 years of USDA Forest Service and Duke University research on land and water degradation and soil change based at the Forest Service's Calhoun Experimental Forest in the Sumter National Forest (Richter et al., 2014; Fig. 2.21). Calhoun investigators include researchers and educators from Duke University, University of Georgia, Georgia Tech, University of Kansas, Mississippi State University, Roanoke College, as well as the USDA Forest Service.

The Calhoun CZO aims to improve understanding of the dynamics and evolution of biota, landforms, soils, saprolites, hill slope hydrology, stream channels, and sediments that comprise the CZ with belowground systems that are ancient (>10^6 years), deep (~30 m on interfluves), and of advanced-weathering stage, commonly with no weatherable primary minerals for many meters below

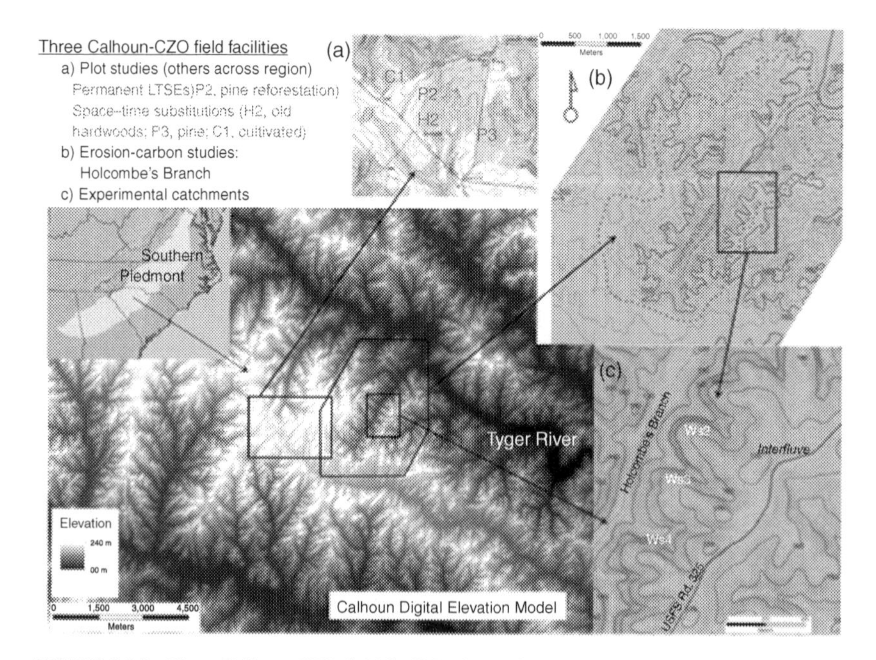

FIGURE 2.20 Three Calhoun CZO field facilities located on topographic and DEM maps.

FIGURE 2.21 A circa 1948 photograph of the USDA Forest Service's Calhoun Experimental Forest showing the intensity of soil erosion and land degradation that had occurred at this site prior to reforestation to the present state.

the ground surface (Markewitz et al., 1998; Bacon et al., 2012). These attributes suggest a portion of the CZ that is highly vulnerable to human alteration, and indeed, much of the Southern Piedmont including the Calhoun Experimental Forest has a history that involves some of the most serious agricultural land and water degradation in the nation. By the mid-twentieth century, nearly 18 cm of soil over more than 10 million ha were estimated to have been lost to erosion, rivers carried enormous sediment yields, cultivation-based crops were no longer viable, and large numbers of farmers had abandoned the land.

By the late twentieth century, the eroded and commonly abandoned Piedmont farmland had been extensively reforested (Fig. 2.22), motivating many to adopt the perspective that the degraded land had been restored in a matter of decades by a process known as "old-field succession." The Calhoun CZO research team has a more critical perspective, and is guided by a hypothesis that the impressive reforestation masks fundamental alterations in CZ hydrology, geomorphology, biology, and biogeochemistry and that post-disturbance CZ evolution may not so much recover as restabilize in altered states. Given all this, the Calhoun CZO provides an important opportunity for meeting the growing need to understand CZs "in the face of land use change ... to inform strategies for sustaining a wide range of human activities" (from NSF's CZO Program Solicitation, NSF 12-575).

The Calhoun CZO team seeks to understand how Earth's CZ responds to severe soil erosion and land degradation. Re-instrumented catchments are used to measure and model changes in ecohydrology and biogeochemistry of

FIGURE 2.22 A circa 2006 photograph from the Calhoun Experimental Forest showing the current reforested state of the Piedmont region.

interfluves, hill slopes, and terraces that were historically subject to accelerated erosion and deposition. Calhoun research focuses on how land use stresses networks and processes that connect the CZ's surface and subsurface subsystems. Calhoun researchers use historic and contemporary data of vegetation, soil, catchments, and sensor networks to help hind cast and forecast CZ dynamics and evolution across temporal and spatial scales.

A key objective of the Calhoun CZO is to help integrate contemporary and historic land use into CZ science since over most of Earth's surface CZs are affected by natural and human forcings. Calhoun CZ science aims to be transformational in exploring the CZ's lower boundary conditions and processes, and the CZ's evolution following land degradation. The Calhoun CZO is organized by research questions motivated by the concept that the CZ is an integrated system from the atmosphere and upper plant-canopy boundary layer to the water in the deepest aquifers and that human forcings typically accelerate CZ processes associated with vegetation, atmosphere, and surface hydrology and soils, thus stressing temporal and spatial networks that connect surface and deep subsurface components of the CZ.

The questions that organize Calhoun CZO research build directly on past research at the site and span multiple scales of time and space: (1) Do land-use change, land degradation, and erosion decouple upper and lower CZ systems by destroying macro porosity networks that are conduits of gas and water

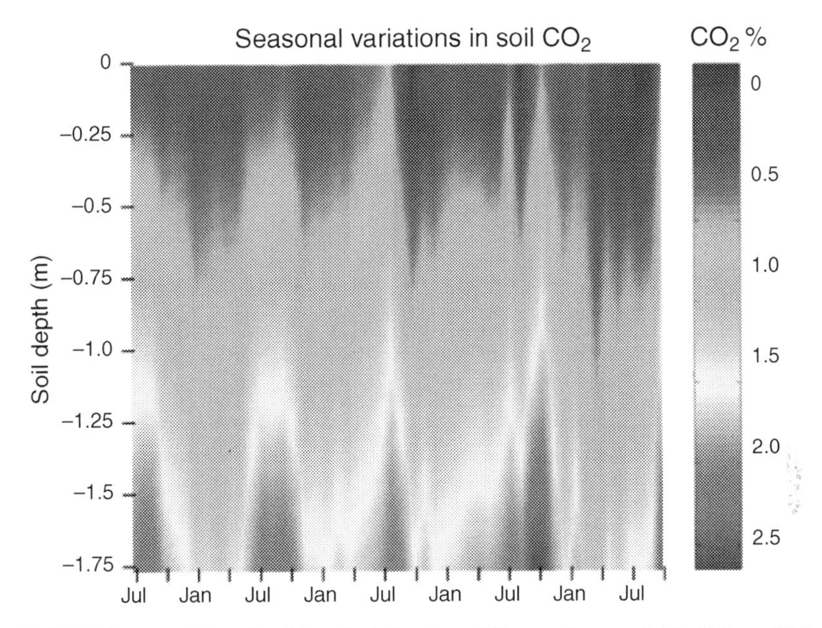

FIGURE 2.23 Deep CO_2, soil solid, and solution data will be greatly expanded in Calhoun CZO studies (Oh et al., 2004; Richter and Markewitz, 2001).

exchange? How rapidly can re-forestation recover CZ porosity and re-network the CZ into an integrated system? (2) How has the legacy of severe erosion re-distributed and altered organic carbon dynamics on both eroded uplands and in anoxic alluvium filled with historic sediment (Billings et al., 2010)? (3) How do human-forced changes in the CZ interact with human livelihoods, adaptation, and governance? (4) Can human-forced CZs enter new steady states, complete with positive feedbacks and attractors that resist recovery?

Re- and up-instrumented catchments, first instrumented by hydrologists in the late 1940s, will allow investigators to measure, experiment, and model eco-hydrologic and biogeochemical dynamics from interfluves to a variety of hill slopes to toe slopes and terraces, all across multiple temporal and spatial scales (Fig. 2.23). Sensors connected by wireless networks and samplers of gases and water will be co-located along transects and in depth-dependent arrangements to examine the recovery of integration in the degraded Critical Zone systems. Interdisciplinary models are being coupled with radiocarbon and stable isotope analyses to hind cast and forecast system hydrology, soil properties and processes, and overall CZ biogeochemistry.

The Calhoun CZO is a center for research and education. The CZO is linked to a Duke University IGERT – training center on intelligent sensor networks; annual CZO science meetings will have oral and poster presentations, discussions, and training sessions that involve scientists and students as fundamental members of the CZO team. We plan a range of outreach efforts to local, regional, and national publics with multi-media science, history, and community-based components, using field days, hardcover books, op-eds, Facebook, and Twitter. An important focus of the CZO is on undergraduate research and education, and we are developing a set of classroom-tested, web-based laboratory, and classroom activities for undergraduates and advanced high school students based, in part, on real-time and historic data from the Calhoun CZO.

2.2.3.2 Eel River

The Eel and Russian rivers are the two large river systems of coastal Northern California (Fig. 2.24), with the Eel draining northward through mostly timber and grazing lands (but with increasing irrigated agricultural use), and the Russian draining to the south through increasing vineyard and housing development towards Santa Rosa, the largest city in the basin. Although more developed than the Eel, only about 13% of the Russian watershed is in agricultural lands. The Eel has two dams in the main stem headwaters where the Potter Valley diversion reroutes Eel flow into the Russian River. The Russia River has two large dams, one in the headwaters (that receives flow from the Eel), and one on a major tributary, Dry Creek, farther down the river. But like other river systems in active agricultural lands, the Russian has some 500 small dams along the tributaries. Salmon (including the endangered Coho Salmon) spawn and migrate in both the Eel and Russian; their populations are in great decline, which is attributed to many factors, including reduced summer base flows and elevated temperatures (Katz et al., 2012).

The Eel and Russian have competed over geologic time for drainage area at their mutual headwaters as tectonic pulses swept through the region (Lock et al., 2006). The underlying geology of this region records the complex accreted terrain of the North American plate margin. Three dominant rock types exist in the Franciscan terrain of northern coastal California: the Coastal Belt (argillite, sandstones and conglomerates), the Central Belt (mélange) and the Eastern Belt (metasedimentary and metavolcanic rocks) (Fig. 2.24), with the Central Belt swept by large, active earthflows in the mechanically weak rocks (e.g., Mackey et al., 2011; Mackey and Roering, 2011; Booth et al., 2013). Most of the Eel River watershed emerged above sea level in just the past 4 million years (Lock et al., 2006) and uplift rates increase from south to north from about 0.4 mm/year to greater than 4 mm/year (Merritts and Bull, 1989). Despite high uplift and erosion rates (Fuller et al., 2009; Willenbring et al., 2013), a deep and well-developed, hydrologically active weathered bedrock zone has formed on Coastal Belt rocks (Salve et al., 2012) (no studies have been done yet on the Central and Eastern Belts).

A striking pattern in the Eel watershed is the correspondence between the three geologic units and the dominant vegetation (Fig. 2.24). The Coastal and Eastern Belts support conifer forests, whereas hardwood trees and herbaceous vegetation predominate in the mélange of the Central Belt. The vegetation distribution does not follow elevation or precipitation patterns. We hypothesize that the approximate correspondence of vegetation type with bedrock results from higher available moisture at higher water potential (less negative pressure) in

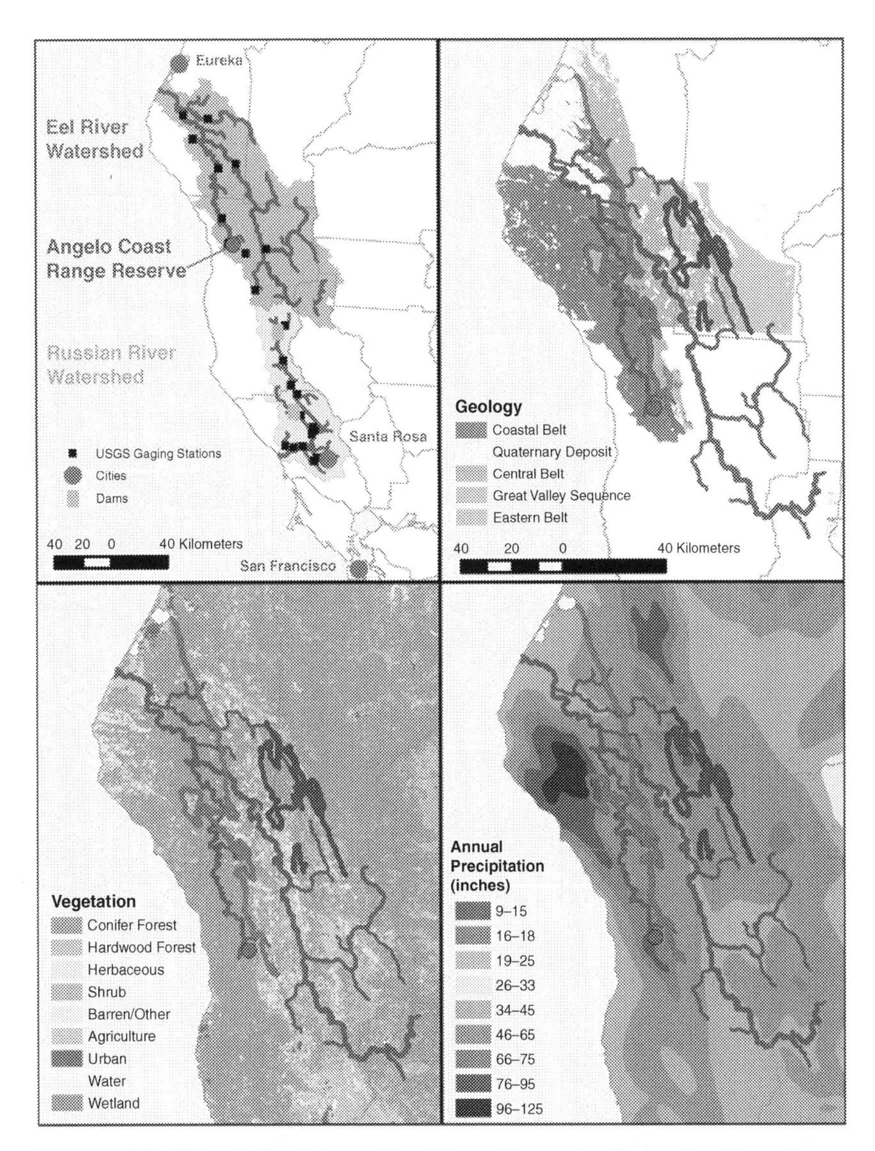

FIGURE 2.24 Watershed and Angelo Coast Range Reserve locations and patterns of mean annual precipitation, dominant vegetation, and geology in Northern California. Pink line is the western boundary between the Coastal and Central Belt bedrock in all maps. Purple line denotes eastern boundary of Central Belt with Eastern belt. Note their approximate correspondence with conifer and hardwood to hardwood/herbaceous boundary. *(Precipitation data and the vegetation map are from the Cal-Atlas geospatial clearinghouse (precipitation mean from 1900 to 1960). Geology map is from the California Geological Survey (data are lacking to the east and south).*

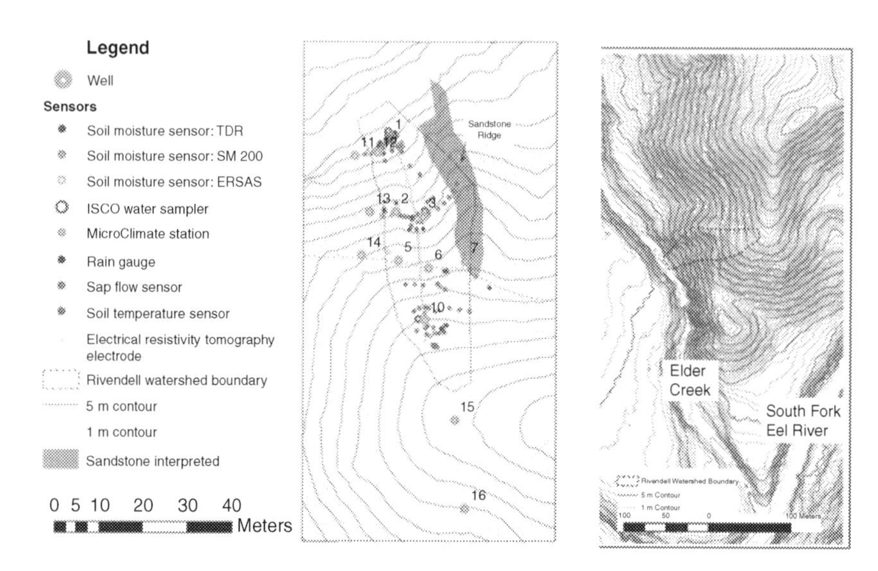

FIGURE 2.25 Location of Rivendell site on Elder Creek (right panel) and instrumentation field at the site (left panel). The instruments include time-domain, reflectometry probes (TDR), and Electrical Resistance Sensory Array System (ERSAS) (Salve, 2011) for relative rock moisture profiles. Contour interval is 1 m in both cases. *(Modified from Salve et al. (2012).)*

the fractured bedrock of the Coastal and Eastern belts than in the fine-grained mélange. Rainfall in this region is strongly seasonal, with summer months hot and dry, hence, deep sources of moisture may be critical for plants.

The Eel River observatory is based mainly in the Eel River watershed (Fig. 2.24), and is composed of four nested components, increasing in spatial scale from an intensively instrumented hill slope to a region encompassing both the Eel and Russian River watersheds (Figs 2.24 and 2.25):

1. Rivendell, a 4000 m^2 sub-basin of Elder Creek located in the Angelo Coast Range Reserve, where hill slope-scale intensive field investigations have been underway since 2007 (Fig. 2.25).
2. The Angelo Coast Range Reserve, a University of California Berkeley-administered research reserve in the University of California Natural Reserve System, protecting 31 km^2 of steep forested terrain, 5 km of the upper South Fork Eel River and the entire watersheds of Elder Creek and several other tributaries.
3. The entire Eel River watershed (9546 km^2).
4. Rivers of the California North Coast, focusing on Eel and Russian River systems (13,800 km^2).

2.2.3.2.1 Rivendell

In 2007, researchers at the University of California at Berkeley (UC Berkeley) began an intensive monitoring program of the Critical Zone on a small (~4000 m^2) steep sub-basin, nicknamed "Rivendell," on a north-facing hill slope adjacent to

Elder Creek (Salve et al., 2012). Elder Creek is a tributary of the South Fork Eel River in Northern California within the Angelo Coast Range Reserve (Fig. 2.25).

The ~30° hill slope at the Rivendell installation site is underlain primarily mudstone (argillite). Cosmogenic dating of sediments (Fuller et al., 2009), direct measurement of bedrock erosion (Stock et al., 2005), and modeling (Seidl and Dietrich, 1992; Sklar and Dietrich, 2006) have documented active channel incision driving hill slope processes, with a local pace of about 0.2–0.4 mm/year. A wireless radio network, powered by treetop solar panels, supports a dense sensor network for environmental monitoring. A network of 12 wells as deep as 30 m have been drilled through weathered bedrock and into fresh bedrock. The installation includes extensive soil and rock moisture monitoring devices, sap-flow sensors on 30 trees, and 4 meteorological stations (Fig. 2.25). The entire system, some 750 sensors, together with a USGS stream gauge, record data at <30-min frequency. These data are transmitted every 4 h to UC Berkeley and placed into a sensor database that displays them near real-time. Four automated ISCO samplers are used to collect water samples (daily or more frequently during runoff events) for water chemistry analysis. Repeated campaigns are carried out to sample water for isotope analysis. Weather stations are also located in the meadow across from Rivendell, at two other nearby meadows, and at Cahto Peak (at the headwaters of Elder Creek).

Figures 2.26 and 2.27 illustrate the Critical Zone profile and hydrologic processes. The Rivendell site crosses a hill slope that faces north where it borders Elder

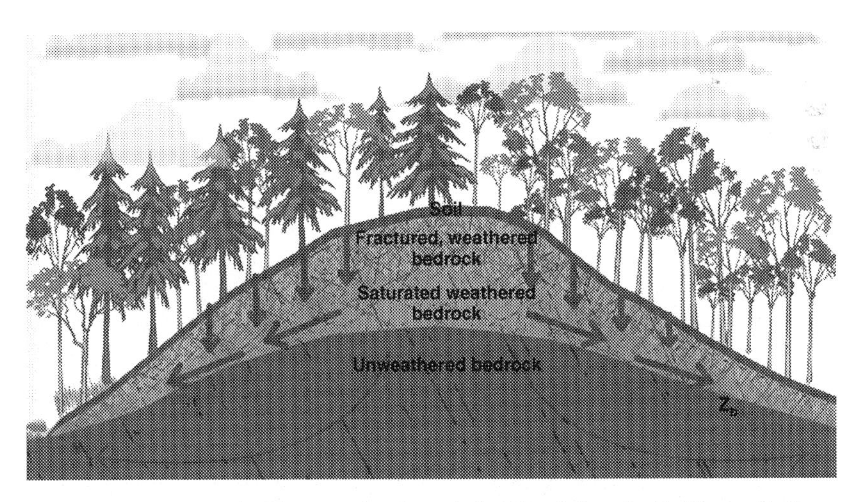

FIGURE 2.26 An idealized cross-section through the Rivendell study site. The conifer Douglas fir-dominated hill slope (left side) faces North whereas the evergreen leaf trees (e.g., Madrone, *A. menziesii*; Tan Oak, *Notholithocarpus densiflorus*; California bay, *U. californica*) dominate the south-facing slope. Soils on the underling argillite are thin (<50 cm), stony and highly conductive. No overland flow occurs. Instead excess precipitation passes through the fractured, weathered bedrock (vertical arrows) and perches on the boundary between the weathered and fresh bedrock (Z_b). This perched water forms a seasonally dynamic groundwater that runs laterally to channels (sloping arrows), providing both storm flow and, importantly, the summer base flow that sustains the river ecosystem. Rempe and Dietrich (2014) presented a simple model for the elevation profile of Z_b under hill slopes. *(Figure modified from Rempe and Dietrich (2014).)*

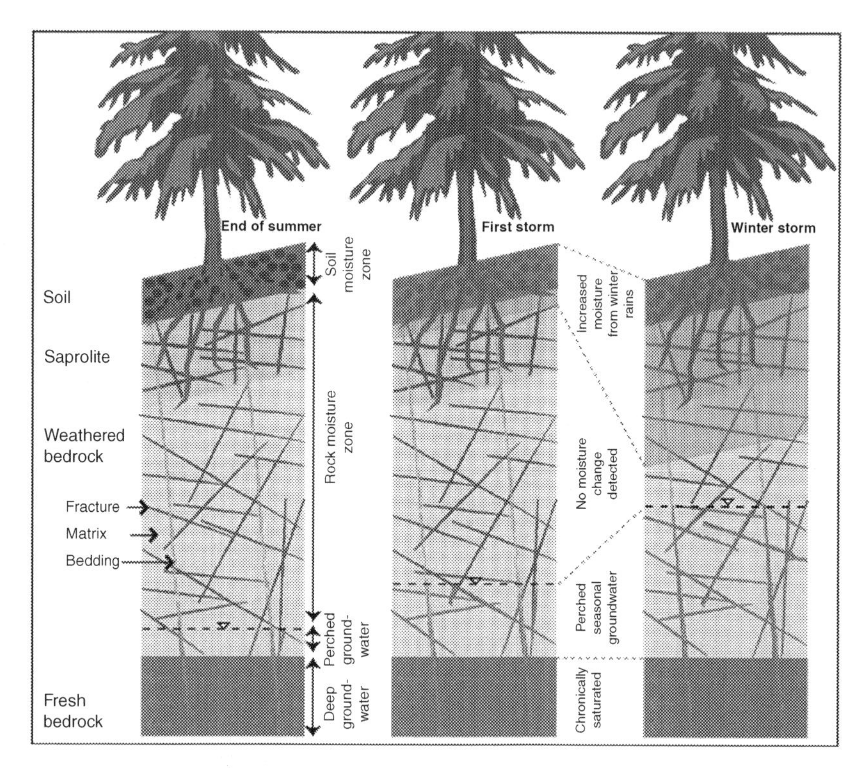

FIGURE 2.27 Illustration of how winter rains vertical structure and seasonal rapid injection of winter rain via fracture flow to a perched water table as soil and saprolite and weathered bedrock more slowly gain moisture (from Salve et al., 2012).

Creek and the ERCZO will extend monitoring to the south-facing slope where it borders what may be an ancient deep-seated landside. Douglas fir (*Pseudotsuga menziesii*), up to 60-m tall mixed with evergreen hardwoods such as interior live oak (*Quercus wislizeni*), madrone (*Arbutus menziesii*), and California bay (*Umbellularia californica*) dominate the north-facing slope, whereas madrone mixed with oak and bay dominate the south-facing slope. Shallow stony soils overlie vertically dipping argillite with interbeds of lenticular sandstone bodies, the largest of which defines and outcrops along the eastern divide of the north slope. The argillite intensely fractures upon weathering forming a soil-like material (saprolite) near the surface and fracture density decreases with depth. This weathered bedrock layer is bounded by fresh bedrock (Z_b) (Fig. 2.26).

All excess precipitation travels vertically through the weathered bedrock, perches on the dense fresh bedrock and then flows to adjacent streams. Figure 2.27 illustrates a proposed seasonal cycle of rain and subsurface moisture change and runoff. First rains at end of summer (September and October) advance moisture into the soil and a limited distance into the weathered bedrock. Some of that rain water flows along fractures, reaching the groundwater causing minor water table rise. Subsequent rain in the winter advances a wetting front through the weathered bedrock zone and delivers water to a dynamic, fractured-controlled groundwater response system (Salve et al., 2012). Graduate student led research has been initiated on: (1) prediction of the Critical Zone development (Rempe and Dietrich, 2014); (2) dynamics of runoff and rock moisture availability to the forest canopy (Oshun et al., 2012); (3) chemical evolution of water and gasses through the Critical Zone (Kim et al., 2012, 2014); and (4) influence of vegetation on regional climate (Link et al., 2014).

2.2.3.2.2 Angelo Coast Range Reserve

The Angelo Reserve is one of forty natural reserves protected by the University of California Natural Reserve System for university-level teaching and research. Angelo is the field base for the Eel River Critical Zone Observatory and where meetings and workshops will be held. From 2004 to 2013, researchers and students from 50 different institutions, including 19 California colleges and universities, and 28 outside of California, with an average of 1571 user-days per year, used Angelo. Mapping, monitoring, and experimental field manipulations within the reserve over the past 25 years have documented the food-web ecology during summer low flow and biogeochemical dynamics along the upper South Fork Eel River and twelve of its tributaries, including Elder Creek (Power et al., 2008, 2009; Finlay et al., 2002, 2011; Kupferberg et al., 2011; Sabo and Power, 2002a,b; Suttle et al., 2004, 2007; Wootton et al., 1996). Among many findings, this work has identified a drainage area or network-based dependency in many ecosystem attributes, including total-dissolved nitrogen, algae abundance and taxonomic distribution, salmonid densities and energy sources, and aquatic insect emergence. Current research is revealing remarkable patterns of aquatic insect migration and reproduction that link main stem and tributary food webs in ways that support summer-rearing salmonids (Uno and Power, 2013). Other ongoing research probes environmental temperature, nutrient and flow thresholds that differentiate salmon-supporting ecosystems from those degraded by cyanobacteria blooms.

The Eel's summer base flow is entirely derived from slow drainage of perched ground water in the Critical Zone, which also influences stream temperature and nutrient loading. Continuous thermal records (15-min frequency) from iButton recorders deployed throughout the study basin are available for several years. Seven water level recorders in tributaries, established to investigate sediment transport (Scheingross et al., 2013) also show strong diurnal oscillations in water level, a signal that can be seen down the South Fork Eel to drainage areas of 642 km^2. Interestingly, tributaries of similar location and size differ in magnitude and timing of fluctuations. These data offer an opportunity to establish more directly the link between terrestrial vegetation and stream flow and temperature.

The Environmental Center lies at the south boundary of the Angelo Reserve, which is still on the electrical grid. This complex of buildings includes a large (30–50 person) meeting room supported by a small kitchen, two laboratories and a microscope room supporting chemical, biological, and earth science research, a screened lathe house for experimental work under ambient light and temperatures, a computer room, and a small office. Wireless Internet connections are available throughout the complex.

2.2.3.2.3 Eel River Watershed

By scaling up to the entire Eel River watershed (9546 km^2) researchers at the ERCZO are exploring how different rainfall, topography, vegetation, and geology regimes affect Critical Zone currencies and their ecosystem consequences. Collaborations among climate scientists, ecologists, and ocean scientists have documented and predicted the influence of algae and nutrients flushed from the Eel River into coastal marine ecosystems (Olhsson, 2013; Ng, 2012). The plume from the Eel appears to be associated at times with increased coastal primary production (as visible from satellite records). Salmon and lamprey migrations also connect the ocean to the river. The Eel River watershed is of sufficient size to apply climate modeling. Seasonal water balance changes are detectable in GRACE imagery. The Eel River has 10 USGS real-time stream gauges, which tap drainages of different mean precipitation, vegetation, and geology. The ERCZO maintains an additional former USGS stream gauge ("Branscomb") within the Angelo Reserve on the South Fork Eel.

2.2.3.2.4 California North Coast region, focusing on Eel and Russian Rivers

Twenty-one stream gauges are operated by the USGS on the Russian River, along with another 17 operated by the State Water Resources Control Board (SWRCB) and National Marine Fisheries Service (NMFS). This high level of monitoring reflects the importance of water in the Russian River, especially for agricultural and domestic use and as critical flow for the survival of salmonids. Great concern occurs about significant drawdown during water spraying in the spring (to protect vineyards from frost) and its effects on survival of the juveniles of the endangered and threatened salmon species (Deitch et al., 2008; NMFS, 2009). Rising stream temperature, perhaps due to reduced summer flows, is also a concern. Extreme low flows due to drought or water withdrawals that warm main stems above temperatures tolerable to salmonids, favor invasive warm-water fishes including pike minnow that threaten salmonids (Reese and Harvey, 2002), and in the Eel River, have been associated with localized toxic cyanobacteria blooms (Puschner et al., 2008).

Research Focus: Through intensive field monitoring in the Critical Zone, researchers at the ERCZO are following watershed currencies – water, solutes, gases, sediment, biota, energy, and momentum – through a subsurface physical environment and microbial ecosystem into the terrestrial ecosystem, up into the atmosphere, and out through diverse drainage channel networks in which aquatic ecosystems interact with these currencies, mediating the delivery of nutrients to coastal ecosystems at the river mouth. Nine Berkeley faculties from four departments and three different colleges lead the ERCZO research.

In the first five years of the ERCZO, the primary research questions are: (1) Does lithology control rock moisture availability to plants and therefore overall resilience of vegetation to climate change in seasonally dry environments? (2) How are solute and gas effluents from hill slopes influenced by biota in changing moisture regimes? (3) What controls the spatial extent of wetted channels in the channel networks of seasonally dry environments? (4) Will changes in Critical Zone currencies induced by climate or land use change lead to threshold-type switches in river and coastal ecosystems?

Synthesis modeling will incorporate ERCZO advances in understanding mechanisms at a finer scale, couple the different Critical Zone subsystems, and be used to explore long-term and large-scale consequences of the dynamics of the Critical Zone in the context of climate and land-use change, and water management policy.

2.2.3.3 Intensively Managed Landscapes CZO

The intensively managed landscapes (IML) CZO (Fig. 2.28) consists of two core sites: the 3,690 km^2 Upper Sangamon River Basin in Illinois and 270 km^2 Clear

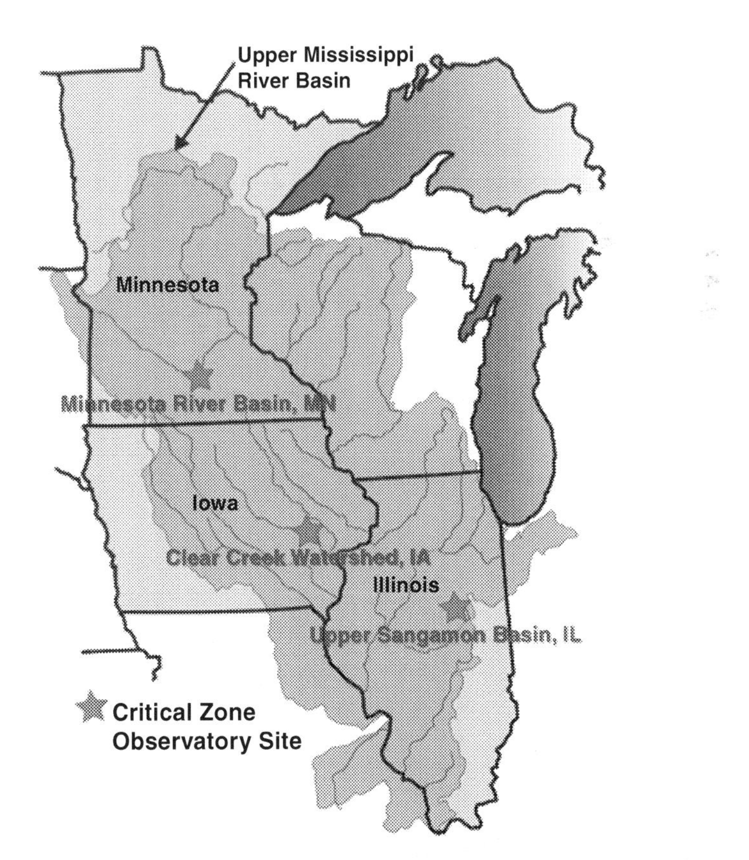

FIGURE 2.28 **Locations of the study sites in the Intensively Managed Landscapes CZO.**

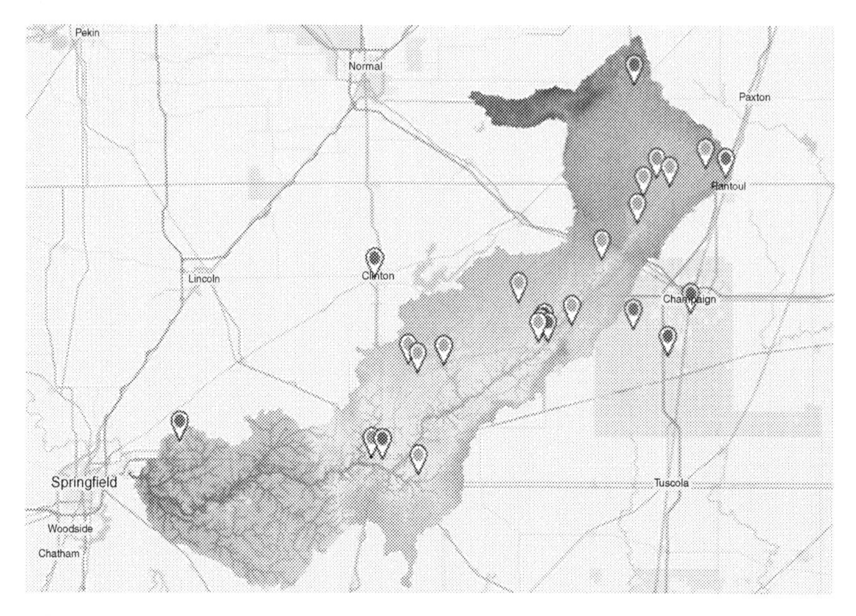

FIGURE 2.29 Monitoring locations of select variables in the Upper Sangamon River Basin, IL. The teardrop colors represent different variables or groups of variables monitored at this site.

Creek Watershed in Iowa, along with the 44,000 km^2 Minnesota River Basin in Minnesota as third participating site. These sites together represent a broad range of physiographic variations occurring throughout the glaciated Midwest.

The Upper Sangamon River Basin (USRB; Fig. 2.29) is a well-studied watershed, representative of low relief, glaciated regions in the Midwest (Keefer et al., 2010). The USRB is intensively managed (USDA, 2006), following conversion to agricultural production from native, tall grass prairie in the 1800s (Gates, 1934; Bogue, 1951). Subsequently, tile installation in poorly drained soils; construction of drainage ditch systems, and channelization of headwater reaches drained seasonal wetlands and altered the hydrology and biogeochemistry of the region (Alexander and Darmody, 1991; Fig. 2.30).

The Clear Creek, IA Watershed (CCW; Fig. 2.31) empties into the Iowa River in east-central Iowa. Clear Creek is representative of most Midwestern watersheds with its mollic soils, humid-continental climate with freeze-thaw cycles, and predominantly agricultural land use (Bettis, 1995; Abaci and Papanicolaou, 2009). The combination of intensive agriculture with a wet climate on highly erodible soils has dramatically decreased the time for material transport through the system, making Clear Creek a good example of the shift from transformer to transporter (Papanicolaou and Abaci, 2008; Wilson et al., 2009; Dermisis et al., 2010; Wilson et al., 2012).

The Minnesota River Basin (MRB) feeds to the Upper Mississippi River Basin and the main stem of the river is actively aggrading with many of its

FIGURE 2.30 Anthropogenic activities in the IML CZO include: (a) channelization and tiling, as well as (b) tillage, that have altered Critical Zone processes and services.

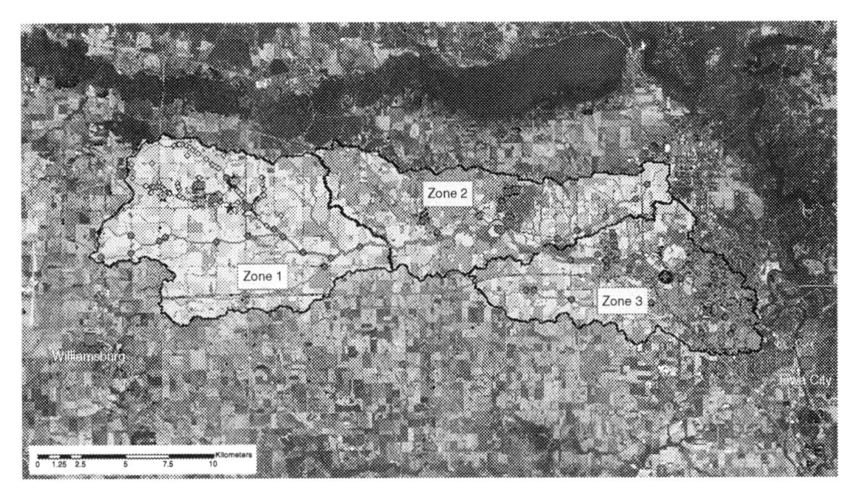

FIGURE 2.31 The established measurement locations in the Clear Creek, IA watershed.

tributaries rapidly incising (Wilcock, 2009; Gran et al., 2009). Following retreat of the Wisconsin ice sheet, glacial Lake Agassiz catastrophically drained through the proto-Minnesota River valley resulting in a nearly instantaneous, 70-m vertical incision of the main stem (Thorleifson, 1996; Belmont et al., 2011), which produced steep gradient knick points at the confluence of each tributary along the main stem.

The IML-CZO combines the breadth and depth of expertise in research and education pertaining to the sustainability of Critical Zone processes and services from eight academic institutions, namely the Universities of Illinois, Tennessee, Iowa, Northwestern, Pennsylvania State, Purdue, Minnesota, and Utah State. Additionally, the partnerships exist with researchers and scientists from several state and federal agencies in the Midwest including the Illinois State Water Survey, Illinois State Geological Survey, Iowa Geological Survey, and United States Geological Survey.

Agricultural intensification, artificial drainage, wetland loss, and urbanization in IMLs of the US Midwest have dramatically altered the Critical Zone extending from the top of the plant canopy to the depth of mineral weathering (e.g., Amundson et al., 2003) in the soil column. Less than one percent of the Midwest's original tall grass prairie has been left unmodified and although this extreme transformation (Fig. 2.1) has made the region a global agricultural leader, it has also had many other far-reaching, often deleterious effects on the Critical Zone (e.g., Mutel, 2008; Papanicolaou et al., 2011). To assess better the short- and long-term resilience of the key Critical Zone services in IMLs related to biodiversity, ecology, hydrology, geomorphology, and climate, an observational network has been established consisting of three sites in Illinois, Iowa, and Minnesota. These sites capture the geological diversity of the low relief, glaciated, and tile-drained landscape in the Midwest. This IML-CZO aims to understand better the present day dynamics of managed landscapes in the context of their long-term co-evolution with soils and biota. Ultimately, the IML-CZO will provide informed management strategies to reduce the vulnerability of the IMLs to present and emerging trends in human activities such as bioenergy crop expansion (Le et al., 2011) and climate change (Kumar, 2013), as well as help develop the next generation of scientists and practitioners to sustain the Critical Zone processes and services.

The central hypothesis of the IML-CZO is that the Critical Zone of IMLs has passed a tipping point (or threshold) and has gradually shifted from being a transformer of material flux with high nutrient, water, and sediment storage to being a transporter of material flux with low nutrient, water and sediment storage. Much of this shift can be attributed to human modification of the landscape. Anthropogenic management of the landscape has reduced storage and residence times, and the emergent patterns of connectivity, dynamics, and responses reflect this change in the residence times. Further, this shift in landscape conditions may affect the capacity of the Critical Zone to acclimate to future impacts, that is, its resilience, associated with ongoing human activity or with extreme weather events caused by climate change. As a result, this trajectory of rapid land-use change is unsustainable for maintaining and enhancing critical services. Conservation programs have helped reverse this course in some instances, but a comprehensive scientific underpinning of these strategies will be further guided by IML-CZO (e.g., Nikolaidis, 2011; Banwart, 2011; Papanicolaou et al., 2011; Banwart et al., 2012).

Through strategic measurements, analysis, and modeling, the IML-CZO is addressing the following questions centered on the hypothesis given earlier: (1) How do different time scales of geologic evolution and anthropogenic influence interact to determine the trajectory of critical structure and function? (2) How is the co-evolution of biota, consisting of both vegetation and microbes, and soil affected due to intensive management? (3) How have dynamic patterns of connectivity, which link across transition zones and heterogeneity, changed by anthropogenic impacts? (4) How do these changes affect residence times and aggregate fluxes of water, carbon, nutrients, and sediment?

Two overarching objectives have been developed from these questions through five key research thematic areas. The first objective is to evaluate changes in the Critical Zone of the Midwest by examining the following:

- Glacial legacy effects of surficial geologic materials as they influence the Critical Zone response to human impacts by establishing baseline conditions of soil resilience, as well as landscape and soil evolution trajectories and rates.
- The interplay between intensive cultivation, with increased commodity and bioenergy crop production, and changing weather patterns (i.e., prevalent freeze–thaw cycles, intense rain events, and droughts), as it alters soil organic matter storage and export with implications to soil aggregate structure evolution, landscape water holding capacity, microbial activity, runoff partitioning to surface and subsurface flow, soil strength, and erodibility.
- Anthropogenic alteration of the hydrological landscape with engineered drainage infrastructure through tiles, as it affects water fluxes and biogeochemical cycling in surface and subsurface domains through changes in residence times and biogeochemical transformations and transport.
- Channelization through drainage ditches and headwater extensions as it alters regimes of hydrologic transport, sediment production and storage, and changes drastically landscape morphodynamic processes.

The second objective is to capture the connectivity of modern hydrological, geomorphological, and biogeochemical processes along different flow paths by quantifying the stores and residence times of water, soil/sediment, SOM/nutrients, identifying provenance across scales, and assessing shifts in the system behavior through measures of thresholds, resilience, reversibility/irreversibility to change, and hysteresis using key state variables for all phases under forcings of climate and local anthropogenic activity.

Along with *in situ* and remote sensing measurements, analysis and modeling will be used to: (1) quantify the fluxes and transformations, interactions, thresholds, and dynamic feedbacks of water, nutrients, and sediment; and (2) assess how rapid land use changes have altered the vulnerability and resilience of these systems. These studies will range from event dynamics to longer time scales that are shaped by long-term dynamics. The CZO will take advantage of existing and new infrastructure to measure short-term dynamic fluxes of water, soil aggregates, crop residue, and soil organic matter during high-intensity rains in early summer (Elhakeem and Papanicolaou, 2009), and also during freeze-thaw cycles. Common or core measurements and associated protocols will be consistent with that of the other CZOs. Additionally, the IML-CZO will make specific measurements that are unique to the characteristic of the study sites.

Characterization of post-settlement alluvium (PSA), unique to IML-CZO, will provide the historical movement of sediment at the hill slope and subwatershed scales to determine time-incremented, sediment-delivery ratios. PSA will be investigated using GPR and coring coupled with fly ash and radiogenic nuclide (i.e., ^{137}Cs and ^{210}Pb) studies. Through experiments, IML-CZO will measure fluxes of water, sediment, and carbon, as well as enrichment ratios (i.e., measure of SOM enrichment in eroded soil relative to that in the soil column), the role of aggregates in stabilizing SOM, and decomposition. It will also measure net ecosystem exchange (NEE), water-vapor flux, radiation and heat energies, using eddy flux towers to understand the impact of climate change on vegetation (Drewry et al., 2010a,b) and subsequent impact of the Critical Zone processes. To quantify dynamic surface and groundwater interactions, tracking of water, solute/particle, and C, N, P between the landscape and stream channels will be conducted. Additionally, the impact of tile drainage networks on transport and transformation will be characterized through measurements of tile-drainage intensity, drainage coefficient, and the flashiness index, as well as the hysteresis in water, sediment, and nutrients. Finally, dynamic measurements will include storm-based evaluations of sediment production and transfer with field measurements of lag times, enrichment ratios, and sediment concentrations. Event-based sediment rating curves will be developed to determine the hysteresis in the system and will be complemented with sediment source fingerprinting.

2.2.3.4 Reynolds Creek

The Reynolds Creek (RC) CZO was established in 2013 and is located in the Reynolds Creek Experimental Watershed (RCEW) in southwestern Idaho. The RC CZO is based on a strategic partnership among the United State Department of Agriculture – Agricultural Research Service (USDA-ARS), Northwest Watershed Research Center (NWRC), Idaho State University and Boise State University. The RCEW has been administered by the USDA-ARS-NWRC for over 50 years. The RCEW is an ideal location for the establishment of a SC CZO for the following reasons: (1) the RCEW is an intermediate scale watershed (239 km^2), (2) it is physically diverse and has a wide range of climate conditions, (3) it supports a preexisting, long-term, spatially extensive data collection, and (4) it is the site of evaluation of land-management practice. These features are further expanded on in the following sections.

The RCEW encompasses a wide range of ecohydrological environments typical of the intermountain region of the western USA. An extensive description of the RCEW environment can be found in Seyfried et al. (2001a). The environmental variability is driven by the nearly 1000-m elevation range and variable geology. Precipitation in the RCEW is not strictly a function of elevation, but generally increases with elevation from less than 250 mm/year to greater than 1100 mm/year while mean annual temperature decreases about 5°C. Rain is the dominant form of precipitation in the RCEW, with snow dominating in the highest elevations. Corresponding vegetation types include sagebrush steppe in the lower elevations, transitioning to mountain sagebrush, western juniper, aspen, and coniferous forest (Fig. 2.32).

The existing scientific infrastructure is a key advantage for the RCEW site as a CZO. Most CZO sites require substantial funding to produce a

FIGURE 2.32 Vegetation types include sagebrush steppe in the lower elevations and transition to mountain sagebrush, western juniper, aspen, and coniferous forest.

hydro-meteorological network that falls short of that available currently at the RCEW. This network is critical because it forms the basis for understanding how the soil environment varies over time and space. Detailed descriptions of the RCEW and published data can be found in Hanson (2001); Hanson et al. (2001); Marks (2001); Seyfried et al. (2001a, 2001b, 2001c, 2001d, 2011); Slaughter et al. (2001); Nayak et al. (2010); Chauvin et al. (2011); Reba et al. (2011a, b) (see also ftp.nwrc.ars.gov for data).

The existing, publically available hydroclimatic data are long-term and spatially extensive. The long-term nature of the data research can be conducted in the context of the climate at different locations and how it is changing. An increase in temperature (1–2°C), reduction of snow, and temporal shift of stream flow with no change in total precipitation or soil-water storage has been documented at the RCEW (Nayak et al., 2010; Seyfried et al., 2011). The spatially extensive nature of the data is critical, given the now understood horizontal, as well as vertical, variability of the climate within the RCEW. These data are not collected on a regular grid, but spaced with higher density in areas of steeper environmental gradients and at specific special study sites.

Characterization of the soil environment, as opposed to the climate, is central to this CZO. The original network of soil-water and temperature-data collection, which extends back more than 30 years, has been dramatically expanded in the last 10 years to include robust, well-calibrated (Seyfried et al., 2005) soil

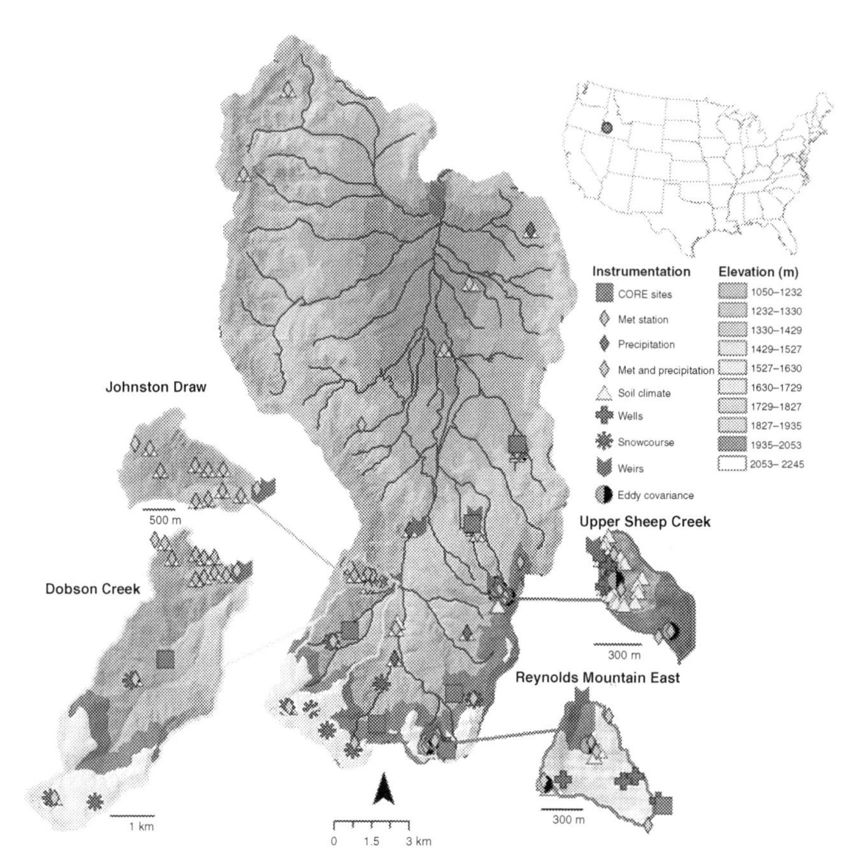

FIGURE 2.33 The RC CZO extends from ~1000 m in elevation to over 2200 m and a wide range of vegetation and climate. Instrumentation including CORE sites for eddy covariance (EC), current meteorological, precipitation, soil-climate stations, wells, snow courses and weirs are shown with different symbols.

water and temperature sensors. These kinds of data are necessary for confirming the accuracy of the SC models used to calculate SC dynamics across the landscape (e.g., RCEW). Note that data collection is spread throughout the RCEW and also concentrated in specific research sites. For example, data collected at Johnston Draw is intended to elucidate topographic influences (Fig. 2.33), while that at Upper Sheep Creek is focused on differential snow distribution. Other sites (not shown) are oriented toward grazing effects in the low elevations.

SC varies widely within the RCEW, both in amount and type. For example, at one, high-elevation site under aspen and affected by snow drifting, the depth-weighted average (150 cm) soil organic carbon (SOC) content is 20.3 g C/kg with no measureable SIC. (Soil pH is about 6.3 at all depths.) This is contrasted with a depth-weighted average SOC content of 5.0 g C/kg at a low elevation, much-drier site under sagebrush. At the low-elevation site, however, 39.8 g C/kg of SIC was measured, so that considerably more total SC is stored at the low-elevation site. Strong vertical gradients of SOC occur in both profiles, and no SIC was detected above 76 cm at the low-elevation site. This "flipping" of the predominant SC form with elevation, or more precisely, with soil environment, is evident in the detailed, watershed-specific soil survey that was conducted at the watershed (Stephenson, 1977).

Determination of present day C balance, in conjunction with the water balance is critical. The existing five EC instruments have been used for hydrological research to date (Marks et al., 2008; Reba et al., 2009, 2012; Flerchinger et al., 2010, 2012) but C flux has been monitored throughout, providing a potentially powerful starting point. The instruments will be redeployed to reflect a shift in research emphasis to include C fluxes at drier, SIC dominated sites.

Reynolds Creek has LiDAR (point clouds and processed bare earth rasters) for the entire experimental watershed, with a point-cloud density of ~5 pts/m^2. The dataset is adequate for estimation of aboveground biomass estimates for large shrubs and trees but is not entirely sufficient to resolve estimates of aboveground, sagebrush-steppe biomass estimates. Fire frequency and extent are increasing in the Western United States with changes in climate (Westerling et al., 2006) and vegetation (introduced species such as *Bromus tectorum* (cheatgrass)) (Allen et al., 2011). Use of prescribed fire as a management practice has emerged as a tool to control fuel loads (McIver et al., 2010); the NWRC has undertaken a prescribed fire management program in cooperation with the Bureau of Land Management (BLM). The NWRC selects sites, conducts experiments and monitors and evaluates fire effects, whereas the BLM assists in site selection and conducts the fires. The primary ecohydrological criteria for site selection is that precipitation be sufficient such that invasive species such as cheatgrass, and yellow star thistle do not expand as a result of the fire. To date, three fires (2002, 2004, and 2007) have been conducted in the RCEW. The next fire is scheduled for 2015. The temporal sequence of past fires and the ability to participate in the planning of future fires provides a rare opportunity for research into the effects of fire on SC.

In general, the management and ownership of the RCEW lands are typical of much of the western United States, which is to say that the RCEW is a "working" as opposed to a pristine, watershed. Most of the land in the watershed (77%) is owned by either the state or federal government, and, in this case, managed by the BLM. The remaining, privately held land is managed by local ranchers, primarily four families that live in, or adjacent to the watershed and derive their livelihood from cattle ranching. In addition to cattle grazing, a small part of the valley is used to raise hay and some timber harvesting occurs. The mission of the NWRC compliments the objectives of the CZO program: "To provide knowledge and technology for management of semi-arid rangeland watersheds; to quantitatively describe the hydrologic processes and interactive influences of climate, soils, vegetation, topography, and management on rangeland systems; to develop information, simulation models, and tools that can be used by action agencies and producers in determining optimum management strategies; and to maintain long-term databases for scientific applications." Much of the success of the unit has been through cooperative research with academic institutions. In fact, the RCEW is intended to provide a springboard for complimentary research. Accommodation, with Wi-Fi and rudimentary lab space is provided for visitors, which typically log about 100 visitor-hours each year.

RC CZO is focused on the quantification of SC (C) and the Critical Zone processes governing it. Most of the world's terrestrial C is found in the Critical Zone (Lal, 2004), where it is predominantly stored as SC and sensitive to climate change and land management. Despite its importance, SC remains a large source of uncertainty in both C cycling and global climate models (Jones et al., 2005; Friedlingstein et al., 2006; Cadule et al., 2010; Falloon et al., 2011; Hopkins et al., 2012). That uncertainty arises due to both an incomplete understanding of the processes dictating SC fate and the challenge of up-scaling commonly highly spatially and temporally heterogeneous soil processes to the landscape or global level (Todd-Brown et al., 2013).

Conceptual models of SOC dynamics are being revisited as flaws in traditional models are identified (Conant et al., 2011; Schmidt et al., 2011; Hopkins et al., 2012). Indeed, consensus is growing that rates of SC storage and release are not particularly sensitive to the chemical properties of the organic C (Marschner et al., 2008; Amelung et al., 2008; Kleber and Johnson, 2010; Conant et al., 2011; Schmidt et al., 2011). Instead, soil physical, chemical, and biological variables (e.g., soil moisture, temperature, structure, bacterial assemblage, root behavior, bio char), more strongly dictate SC fate (Torn et al., 1997; Jobbágy and Jackson, 2000; Davidson and Janssens, 2006; Sollins et al., 2007; Ekschmitt et al., 2008; Totsche et al., 2010; Conant et al., 2011).

Understanding and predicting SC and associated processes are further complicated by the fact that most studies are conducted at the plot scale, but processes operating at larger spatial and temporal scales such as fire and vegetation change may ultimately determine the impact of SC on the global budget (Westerling et al., 2006; Trumbore and Czimczik, 2008). For example, increasing burn frequency or area, a trend in much of the Western United States (Westerling et al., 2006), may return C to the atmosphere faster than it can accumulate, as observed in fire-prone Mediterranean and boreal regions (Harden et al., 2000; Trumbore and Czimczik, 2008). Moreover, a scaling challenge exists in distributing SC, a Critical Zone property that is highly heterogeneous in nature, across the landscape. To address this challenge, many environmental parameters have been used to describe SC distribution using statistical approaches (Arrouays et al., 1998; Jobbágy and Jackson, 2000; Kulmatiski et al., 2004; Garcia-Pausas et al., 2007; Hirmas et al., 2010; Kunkel et al., 2011), yet these approaches are limited because they often use surrogates for the soil environment (precipitation, topography, etc.) rather than actual soil environment variables (soil water content, temperature, or net water flux), and they are not necessarily transferable and grounded in process-based understanding.

The RC CZO is addressing the grand challenges of improving prediction of SC storage and flux from the pedon to landscape scale. The overarching hypothesis is that soil environmental variables (e.g., soil water content, soil temperature, net water flux) measured and modeled at the pedon and watershed scale will improve our understanding and prediction of SC storage, flux, and processes. Research priorities for the CZO include: (1) determining the relationship between measured SC storage and the soil environment at high spatial resolution across a broad, regionally significant environmental gradient; (2) measuring net C flux in conjunction with components of the SC cycle (soil respiration, litter decomposition, SC characteristics) at the pedon to landscape scale; (3) evaluating SC model performance in terms of: (a) SC distribution across the landscape, and (b) representation of critical C fluxes at the pedon to landscape scale; (4) being a community resource and magnet for global climate modeling and C cycle research.

2.2.3.5 CZO National Office

The growth in size and scope of the CZO program led to the recognition that a National Office (NO) was a primary need, and one was formalized in 2014. The goals of the CZONO include enhancing communication among CZO researchers and students, promoting common measurement and data protocols, providing a single point of contact for the CZO program for the public and other scientists, developing new educational and outreach initiatives for students at the graduate, undergraduate and K-12 levels, and enhancing interaction of the CZO program with a broad range of scientists, all within a framework of advancing Critical Zone science as a tool for sustainability. The NO will accomplish these goals by: establishing regular communication among CZO PIs, organizing two national meetings annually, offering graduate/young scientist's workshops, and developing electronic delivery of educational resources for Critical Zone science. The CZONO will lead the maintenance and further development of the criticalzone.org website, a key resource for the CZO and the broader scientific community. It will develop and maintain a visible presence in the scientific community at national and international meetings and via new media. The office will also seek to identify and raise funds for larger initiatives, including international summer schools in Critical Zone science and a national Critical Zone K-12 education strategy. An important function of the CZONO will be to integrate scientists and students currently outside the CZO community to take advantage of the CZO as an outstanding scientific resource. The office will act as a liaison between the CZO program and related US programs including CUAHSI, NEON, and LTER. It will act as a liaison to international programs such as SoilTrEC, the EU CZO program, and the French RBC (Network of River Basins), and other evolving CZO and CZ-related networks for example in China, Australia, and other European nations. Improved integration with other science programs that have overlapping goals will extend the Critical Zone network concept over a wider range of environments and processes than the US CZO program can achieve on its own.

2.3 COMMON SCIENCE QUESTIONS

The 10 US CZOs represent a wide range of geological, climatic, and land use settings that can provide an opportunity to develop a broad and general understanding of the evolution and function of the Critical Zone. By identifying shared or "common" research questions across these CZOs, an opportunity arises to advance new understanding in key issues. The work of many CZOs can accomplish what no one observatory can provide. This is due to the diversity of sites, as well as the diversity of the researchers who collectively bring the essential observational and theoretical skills and knowledge to these common problems.

Research at the CZOs involves the following shared conceptual framework: (1) The Critical Zone evolves a structure that influences the storage and flux of water, solutes, sediments, gases, biota and energy. (2) By mediating these stores and fluxes, the Critical Zone provides ecosystem services, and is thus critical to people.

As well, CZO research may be summarized by the following three general shared questions: (1) What controls Critical Zone properties and processes? (2) What will be the response of the Critical Zone structure, and its stores and fluxes, to climate and land use change? (3) How can improved understanding of the Critical Zone be used to enhance ecosystem resilience and sustainability, and restore ecosystem function?

Intensive field measurements at the observatories will provide the data to guide process understanding to develop models that explain Critical Zone evolution, to forecast possible future states, and to guide land-use decisions. All of the CZOs have modeling components, though a wide range of approaches is used.

2.4 COMMON MEASUREMENTS CONCEPTUAL FRAMEWORK AND GOALS

Although the network of CZOs has been developing to increasing levels of coordination and large efforts have been made to develop the cyber infrastructure for sharing data (e.g., Horsburgh et al., 2009; Niu et al., 2011), it is obvious that different measurement and modeling strategies are used at each site. In some cases this is appropriate and beneficial. Indeed, the CZOs must remain incubators of methodologies and innovation. However, some of the science questions require the use of identical measurement and modeling strategies across sites. This latter need is especially important as the CZO network expands globally.

Measurements at the US CZOs include a common set of variables that quantify CZ architecture and evolution; fluxes across the CZ boundaries; and, fluxes and changes in storage of the major CZ reservoirs at the catchment scale. The CZOs have recognized that many of the details of overarching science questions can best be addressed if a core set of variables is measured across the CZOs, and if those core measurements are made using the same or readily comparable methods. Thus, while each site develops and shares novel approaches to quantifying the CZ, all sites aim to make a set of cross-CZO comparable measurements. A key aspect of such "common" measurements is sampling that must be guided by site conditions and the science questions at hand, but to the extent possible should be carried out using materials and methods that are similar. The site-specific approach is based on the principal of "best technique and sampling design" for the individual CZO. The data collected are comparable with local, regional, and global monitoring efforts.

Despite this network-wide agreement, identification of essential Critical Zone variables – the data without which Critical Zone processes and functions cannot be properly understood and modeled – has so far proven to be challenging. Given the wide range of measurement activities each CZO is engaged in, the development of common sampling and measurement protocols is not a small task. Nonetheless, progress has already been made along these lines. A CZO Common Measurements working group coalesced during Fall 2012 and helped write and edit the so-called Common Measurements document available at: http://criticalzone.org/christina/publications/pub/chorover-et-al-2012-common-critical-zone-observatory-infrastructure-and-mea/. A CZO Graduate Research Group was encouraged to help determine "commonality" across the CZO network. That effort showed that all or most of the CZOs are engaged in twenty-five categories of specific instrumentation and campaign-style measurements to quantify the composition and fluxes across the land-atmosphere boundary through vegetation, regolith, and ground and stream-water through space and time. More recently, the CZO PIs concluded that the CZOs share the common view that the CZ evolves a structure that influences the storage and flux of water, solutes, sediments, gases, biota and energy; and, that by mediating these stores and fluxes, the CZ provides services that are critical to humanity. This review also concluded that science strategy at the 10 CZOs share three general questions: (1) What controls CZ properties and processes? (2) What will be the response of the CZ structure, and its stores and fluxes, to climate and land use change?; and, (3) How can improved understanding be used to enhance CZ resilience and sustainability, and restore CZ function?

Despite these efforts, the scale of measurement and methodology used across the CZOs to address societally relevant issues has not been standardized and the ability to initiate comparative studies remains an elusive goal. The list in Table 2.1 is drawn from the aforementioned Common Measurements document and includes a minimum set of processes that should be measured to adequately characterize the CZ – this important list provides a starting point from which future conversations can coalesce and evolve.

TABLE 2.1 Categories of Instrumentation and Measurements Made at the United States CZOs.

1. Land-atmosphere
   a. LiDAR datasets
   b. Eddy flux for momentum, heat, water vapor, CO_2
   c. Wind speed and direction sensors
   d. Solar radiation and temperature sensors
   e. Precipitation and through-fall samplers
   f. Wet and dry deposition samplers
2. Vegetation and associated microbiota
   a. Above- and below-ground vegetative and microbial composition
   b. Relations between ET and species composition and structure
   c. Soil/plant respiration, net ecosystem exchange
3. Soil (vadose zone)
   a. Solid phase (campaign sampling for spatial characterization)
   b. Elemental composition and mineralogy
   c. Texture and physical characterization
   d. Organic-matter content
   e. Stable and radiogenic isotope composition
   f. Fluid phase (sensors and samplers for time series)
   g. Soil moisture (sensors)
   h. Soil temperature (sensors)
   i. Soil-solution chemistry (samplers)
   j. Soil-gas chemistry (samplers/sensors)
   k. Rates of infiltration and groundwater flow
4. Saprolite and bedrock (saturated zone)
   a. Solid phase (campaign sampling for spatial characterization)
   b. Petrology and mineralogy
   c. Elemental-composition and organic-matter content
   d. Texture and other physical and architectural traits
   e. Fluid phase (sensors and samplers for time series)
   f. Potentiometric head and temperature (sensors)
   g. Groundwater chemistry (samplers/sensors)
   h. Gas chemistry (samplers/sensors)
5. Surface water
   a. Discrete and instantaneous discharge (flumes, weirs, with water quality sensors)
   b. Channel morphology
   c. Stream-water chemistry, dissolved and suspended (samplers/sensors)
   d. Sediment and biota (samplers/sensors)

2.5 INTERNATIONAL CZ PROGRAM OF RESEARCH AND EDUCATION

Since the initiation of conversations regarding an interdisciplinary effort to study the CZ, CZ researchers have recognized the need to engage colleagues globally. Beginning in 2007, the US NSF CZ International Scholars program funded 54 graduate and post-doctoral students to pursue CZ research in Europe. Additional students have been supported to attend training workshops in Crete and Iceland in that time frame. In Europe during 2007–2009, Critical Zone researchers were initially organized under the acronym SoilCritZone, a European Commission-funded project. That project facilitated a series of meetings and workshops that led to a report to policy makers on soil sustainability in Europe. Eventually a subset of these European researchers organized and developed a proposal to the European Commission that was funded, and the SoilTrEC project, the European counterpart to the US CZO program, began in January 2010. Back in the United States, by late 2009 the need for coordination of a variety of cross-CZO activities became necessary, and NSF funding was secured midway through 2010 to support such an effort. NSF support for US coordination of international collaborative activities, primarily between the US CZOs and the SoilTrEC project, was obtained at approximately the same time.

An international workshop was convened in November 2011 at the University of Delaware to develop an international CZ science agenda for the next decade. Eighty-nine scientists from 25 countries attended the meeting and debated and refined 6 key science questions and developed these into research hypotheses and framework experimental designs, in order to move a 10-year agenda forward (final workshop report available at: http://www.czen.org/files/czen/Sustaining-Earths-Critical-Zone_FINAL-290713.pdf). A common feature of the experimental design was the establishment of CZO networks along planetary-scale gradients of environmental change (Fig. 2.34), for example, gradients of climate and intensity of land use (Banwart et al., 2013).

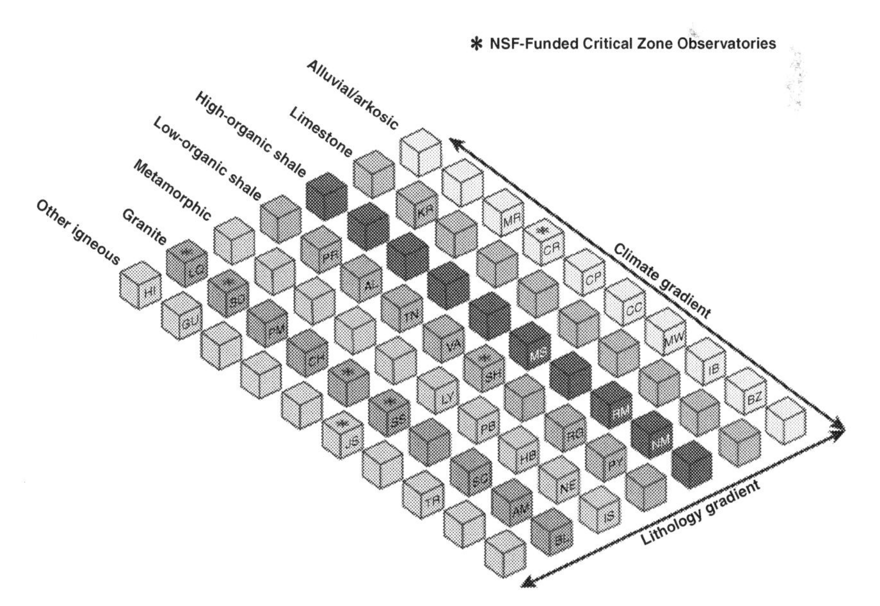

FIGURE 2.34 CZOs developed on different continents may still be understood within an organizational framework as shown schematically here. The sites are positioned along environmental gradients, in this case of lithology and climate. By understanding sites within these gradients we will evaluate the full range of Earth-surface conditions and processes.

Meeting participants recognized that international networks of CZOs offer enormous potential to globally integrate basic science with innovation in human adaptation to rapid and intensive environmental change. Achieving this vision requires a transformation in the ambition and integration of CZO science agendas worldwide. The CZO aim is to understand the resilience and vulnerabilities of the CZ, and to formulate interdisciplinary solutions to sustaining the CZ for future generations. To advance this global project requires continued development and implementation of the Delaware plan for a coordinated international program of CZO research and education.

2.6 CONCLUSION

The central idea of CZ science has captured the imagination of scientists worldwide: to learn to measure the panoply of processes occurring today in the CZ and to relate these to the history of these processes that is recorded in the soil and rock record. The idea of CZ science has also crossed agencies within the US federal government: ideas about the CZ are now driving some research within the Department of Energy's Terrestrial Ecosystem Science program, and in the US Geological Survey. The scientific community must learn to forecast CZ processes using both observation and computational simulation. New CZ knowledge will support quantitative models across scales of space and time to determine how the CZ has transformed in the Anthropocene, and to project how the CZ will continue to transform into the future. CZ science offers enormous potential to integrate basic knowledge of Earth's surface with sustainable adaptation to ongoing rapid and intensive land-use and climate change. CZOs are the environmental laboratories from which this knowledge will be gained.

REFERENCES

Abaci, O., Papanicolaou, A.N., 2009. Long-term effects of management practices on water-driven soil erosion in an intense agricultural sub-watershed: monitoring and modeling. Hydrol. Process. 23, 2818–2837.

Alexander, J.D., Darmody, R.G., 1991. Extent and organic matter content soils in Illinois soil associations and counties. Agronomy Special Report, University of Illinois.

Allen, E.B., Steers, R.J., Dickens, S.J., 2011. Impacts of fire and invasive species on desert soil ecology. Rangeland Ecol. Manag. 64, 450–462.

Amelung, W., Brodowski, S., Sandhage-Hofmann, A., Bol, R., 2008. Combining Biomarker with Stable Isotope Analyses for Assessing the Transformation and Turnover of Soil Organic Matter. Advances in Agronomy, 100, 155–250 (Chapter 6).

Amundson, R., Guo, Y., Gong, P., 2003. Soil diversity and land use in the United States. Ecosystems 6, 470–482.

Anderson, S.P., von Blanckenburg, F., White, A.F., 2007. Physical and chemical controls on the critical zone. Elements 3, 315–319.

Anderson, R.S., Anderson, S.P., Tucker, G.E., 2012. Landscape scale linkages in critical zone evolution. C. R. Geosci. 344, 586–596.

Anderson, R.S., Anderson, S.P., Tucker, G.E., 2013. Rock damage and regolith transport by frost: an example of climate modulation of critical zone geomorphology. Earth Surf. Process. Land. 38, 299–316.

Anderson, S.P., Hinckley, E.-L., Kelly, P., Langston, A., 2014. Variation in critical zone processes and architecture across slope aspects. Procedia Earth Planetary Science, 10, pp. 28–33.

Arrouays, D., Daroussin, J., Kicin, J.L., Hassika, P., 1998. Improving topsoil carbon storage prediction using a digital elevation model in temperate forest soils of France. Soil Sci. 163, 103–108.

Bacon, A.R., Richter, D.deB., Bierman, P.R., Rood, D.H., 2012. Coupling meteoric ^{10}Be with pedogenic losses of ^{9}Be to improve soil residence time estimates on an ancient North American interfluve. Geology 40, 847–850.

Banwart, S., 2011. Save our soils. Nature 474, 151–152.

Banwart, S.A., Chorover, J., Gaillardet, J., Sparks, D., White, T., et al., 2013, Sustaining Earth's Critical Zone – basic science and interdisciplinary solutions to global challenges. The University of Sheffield. Available from: www.czen.org.

Banwart, S.A., Menon, M., Bernasconi, S., Bloem, J., Blum, W., de Souza, D.M., Chabaux, F., Davíðsdóttir, B., Duffy, C., Lair, G.J., Kram, P., Lamacova, A., Lundin, L., Nikolaidis, N., Novak, M., Panagos, P., Ragnarsdóttir, K.V., Reynolds, B., Robinson, D., Rousseva, S., de Ruiter, P., van Gaans, P., Weng, L., White, T., Zhang, B., 2012. Soil processes and functions across an international network of Critical Zone Observatories: introduction to experimental methods and initial results. C. R. Geosci. Special Issue Erosion Weathering 344 (12), 758–772.

Befus, K.M., Sheehan, A.F., Leopold, M., Anderson, S.P., Anderson, R.S., 2011. Seismic constraints on critical zone architecture, Boulder Creek Watershed, Colorado. Vadose Zone J. 10, 915–927.

Belmont, P., Gran, K.B., Schottler, S.P., Wilcock, P.R., Day, S.S., Jennings, C., Lauer, J.W., Viparelli, E., Willenbring, J.K., Engstrom, D.R., Parker, G., 2011. Large shift in source of fine sediment in the Upper Mississippi River. Environ. Sci. Technol., 45.

Bettis, E.A., 1995. The Holocene stratigraphic record of entrenched stream systems in thick loess regions of the Mississippi River Basin. PhD Dissertation. The University of Iowa. Iowa City, IA.

Billings, S.A., Buddemeier, R.W., Richter, D.deB., Van Oost, K., Bohling, G., 2010. A simple method for estimating the influence of eroding soil profiles on atmospheric CO_2. Global Biogeochem. Cycles 24, GB3560.

Bogue, M.B., 1951. The swamp land act and wet land utilization in Illinois, 1850–1890. Agric. History 25, 169–180.

Booth, A.M., Roering, J.J., Rempel, A.W., 2013. Topographic signatures and a general transport law for deep-seated landslides in a landscape evolution model. J. Geophys. Res. (Earth Surf.) 118, 603–624.

Brantley, S.L., Goldhaber, M.B., Ragnarsdottir, V., 2007. Crossing disciplines and scales to understand the Critical Zone. Elements 3, 307–314.

Brantley, S.L., Holleran, M.E., Jin, L., Bazilevskaya, E., 2013. Probing deep weathering in the Shale Hills Critical Zone Observatory, Pennsylvania (U.S.A.): the hypothesis of nested chemical reaction fronts in the subsurface. Earth Surf. Proc. Land. 38, 18.

Cadule, P., Friedlingstein, P., Bopp, L., Sitch, S., Jones, C.D., Ciais, P., Piao, S.L., Peylin, P., 2010. Benchmarking coupled climate-carbon models against long-term atmospheric CO_2 measurements. Global Biogeochem. Cycles 24, GB2016.

Caine, N., 2010. Recent hydrologic change in a Colorado alpine basin: an indicator of permafrost thaw. Ann. Glaciol. 51 (56), 130–134.

Chauvin, G.M., Flerchinger, G.N., Link, T.E., Marks, D., Winstral, A.H., Seyfried, M.S., 2011. Long-term water balance and conceptual model of a semi-arid mountainous catchment. J. Hydrol. 400, 133–143.

Chorover, J., Troch, P.A., Rasmussen, C., Brooks, P.D., Pelletier, J.D., Breshears, D.D., Huxman, T.E., Kurc, S.A., Lohse, K.A., McIntosh, J.C., Meixner, T., Schaap, M.G., Litvak, M.E., Perdrial, J., Harpold, A., Durcik, M., 2011. How water, carbon, and energy drive critical zone evolution: the Jemez-Santa Catalina Critical Zone Observatory. Vadose Zone J. 10, 884–899.

Conant, R.T., Ryan, M.G., Agren, G.I., Birge, H.E., Davidson, E.A., Eliasson, P.E., Evans, S.E., Frey, S.D., Giardina, C.P., Hopkins, F.M., Hyvonen, R., Kirschbaum, M.U.F., Lavallee, J.M., Leifeld, J., Parton, W.J., Steinweg, J.M., Wallenstein, M.D., Wetterstedt, J.A.M., Bradford, M.A., 2011. Temperature and soil organic matter decomposition rates – synthesis of current knowledge and a way forward. Global Change Biol. 17, 3392–3404.

Covich, A.P., Crowl, T.A., Hein, C.L., Townsend, M.J., McDowell, W.H., 2009. Predator-prey interactions in river networks: comparing shrimp spatial refugia in two drainage basins. Freshwater Biol. 54, 450–465.

Davidson, E.A., Janssens, I.A., 2006. Temperature sensitivity of soil carbon decomposition and feedbacks to climate change. Nature 440, 165–173.

Deitch, M.J., Kondolf, G.M., Merenlender, A.M., 2008. Hydrologic impacts of small-scale instream diversions for frost and heat protection in the California Wine Country. River Res. Appl. 25, 118–134.

Dermisis, D., Abaci, O., Papanicolaou, A.N., Wilson, C.G., 2010. Evaluating the effects of grassed waterways in southeastern Iowa. Soil Use Manage. 26, 183–192.

Drewry, D.T., Kumar, P., Long, S., Bernachi, C., Liang, X.Z., Sivapalan, M., 2010a. Ecohydrological Responses of Dense Canopies to Environmental Variability Part 1: Interplay Between Vertical Structure and Photosynthetic Pathway. J. Geophys. Res. Biogeosci. 115, G04022.

Drewry, D.T., Kumar, P., Long, S., Bernachi, C., Liang, X.Z., Sivapalan, M., 2010b. Ecohydrological responses of dense canopies to environmental variability part 2: role of acclimation under elevated CO_2. J. Geophys. Res. Biogeosci. 115, G04023.

Duffy, C., Shi, Y., Davis, K., Slingerland, R., Li, L., Sullivan, P.L., Godderis, Y., Brantley, S.L., 2014. Designing a system of models to understand the critical zone. Procedia Earth Planetary Sci. 10, 7–15.

Dühnforth, M., Anderson, R.S., Ward, D.J., Blum, A.E., 2012. Unsteady late Pleistocene incision of streams bounding the Colorado Front Range from measurements of meteoric and *in situ* ^{10}Be. J. Geophys. Res. Earth Surf. 117, 20.

Eilers, K., Debenport, S., Anderson, S.P., Fierer, N., 2012. Digging deeper to find unique microbial communities: the strong effect of depth on the structure of bacterial and archaeal communities. Soil Biol. Biochem. 50, 58–65.

Ekschmitt, K., Kandeler, E., Poll, C., Brune, A., Buscot, F., Friedrich, M., Gleixner, G., Hartmann, A., Kästner, M., Marhan, S., Miltner, A., Scheu, S., Wolters, V., 2008. Soil-carbon preservation through habitat constraints and biological limitations on decomposer activity. J. Plant Nutr. Soil Sc. 171, 27–35.

Elhakeem, M., Papanicolaou, A.N., 2009. Estimation of the runoff curve number via direct rainfall simulator measurements in the State of Iowa, USA. Water Resour. Manage. 23, 2455–2473.

Ewel, J.J., Whitmore, J.L., 1973. The Ecological Life Zones of Puerto Rico and the U.S. Virgin Islands. U.S.D.A. Forest Service Research Paper ITF-18. Institute of Tropical Forestry, Rio Piedras, P.R.

Falloon, P., Jones, C.D., Ades, M., Paul, K., 2011. Direct soil moisture controls of future global soil carbon changes: an important source of uncertainty. Global Biogeochem. Cycles 25, GB3010.

Finlay, J.C., Khandwala, S., Power, M.E., 2002. Spatial scales of energy flow in food webs of the South Fork Eel River. Ecology 83, 1845–1859.

Finlay, J.C., Hood, J.M., Limm, M., Power, M.E., Schade, J.D., Welter, J.R., 2011. Light mediated thresholds in ecosystem nutrient stoichiometry in a river network. Ecology 92, 140–150.

Flerchinger, G.N., Marks, D., Reba, M.L., Yu, Q., Seyfried, M.S., 2010. Surface fluxes and water balance of spatially varying vegetation within a small mountainous headwater catchment. Hydrol. Earth Syst. Sci. 14, 965–978.

Flerchinger, G.N., Reba, M.L., Marks, D., 2012. Measurement of surface energy fluxes from two rangeland sites and comparison with a multilayer canopy model. J. Hydrometeorol. 13, 1038–1051.

Friedlingstein, P., Cox, P., Betts, R., Bopp, L., Von Bloh, W., Brovkin, V., Cadule, P., Doney, S., Eby, M., Fung, I., Bala, G., John, J., Jones, C., Joos, F., Kato, T., Kawamiya, M., Knorr, W., Lindsay, K., Matthews, H.D., Raddatz, T., Rayner, P., Reick, C., Roeckner, E., Schnitzler, K.G., Schnur, R., Strassmann, K., Weaver, A.J., Yoshikawa, C., Zeng, N., 2006. Climate-carbon cycle feedback analysis: results from the C(4)MIP model intercomparison. J. Climate 19, 3337–3353.

Fuller, T.K., Perg, L.A., Willenbring, J.K., Lepper, K., 2009. Field evidence for climate-driven changes in sediment supply leading to strath terrace formation. Geology 37, 467–470.

Gabor, R.S., Eilers, K.G., McKnight, D.M., Fierer, N., Anderson, S.P., 2014. From the litter layer to the saprolite: chemical changes in water-soluble soil organic matter and their correlation to microbial community composition. Soil Biol. Biochem. 68, 166–176.

Garcia-Pausas, J., Casals, P., Camarero, L., Huguet, C., Sebastià, M.T., Thompson, R., Romanyà, J., 2007. Soil organic carbon storage in mountain grasslands of the Pyrenees: effects of climate and topography. Biogeochemistry 82, 279–289.

Gates, P.W., 1934. Illinois Central Railroad and its Colonization Work. Harvard University Press, Harvard.

Godderis, Y., Brantley, S., Francois, L., Schott, J., Pollard, D., Deque, M., Dury, M., 2013. Rates of consumption of atmospheric CO_2 through the weathering of loess during the next 100 yr of climate change. Biogeosciences 10, 135–148.

Godderis, Y., Brantley, S.L., 2014. Earthcasting the future Critical Zone. Elementadoi: 10.12952/journal.elementa.000019.

Godderis, Y., Williams, J., Schott, J., Pollard, D., Brantley, S., 2010. Time evolution of the mineralogical composition of Mississippi Valley loess over the last 10 kyr: Climate and Geochemical Modeling. Geochim. Cosmochim. Acta 74 (22), 6357–6374.

Gran, K.B., Belmont, P., Day, S.S., Jennings, C., Johnson, A., Perg L., Wilcock, P.R., 2009. Geomorphic evolution of the Le Sueur River, Minnesota, USA, and implications for current sediment loading. In: James, L.A., Rathburn, S.L., Whittecar, G.R. (Eds.), Management and Restoration of Fluvial Systems with Broad Historical Changes and Human Impacts: Geological Society of America Special Paper 451. Geological Society of America, Boulder, CO pp. 119–130.

Hanson, C.L., 2001. Long-term precipitation database, Reynolds Creek Experimental Watershed, Idaho, United States. Water Resour. Res. 37, 2831–2834.

Hanson, C.L., Marks, D., Van Vactor, S.S., 2001. Long-term climate database, Reynolds Creek Experimental Watershed, Idaho, United Sates. Water Resour. Res. 37, 2839–2841.

Harden, J.W., Trumbore, S.E., Stocks, B.J., Hirsch, A., Gower, S.T., O'Neill, K.P., Kasischke, E.S., 2000. The role of fire in the boreal carbon budget. Global Change Biol. 6, 174–184.

Harpold, A., Brooks, P., Rajagopal, S., Heidbuchel, I., Jardine, A., Stielstra, C., 2012. Changes in snowpack accumulation and ablation in the intermountain west. Water Resour. Res. 48, doi: 10.1029/2012WR011949.

Harpold, A.A., Biederman, J.A., Condon, K., Merino, M., Korgaonkar, Y., Nan, T.C., Sloat, L.L., Ross, M., Brooks, P.D., 2014a. Changes in snow accumulation and ablation following the Las Conchas Forest Fire, New Mexico, USA. Ecohydrology 7, 440–452.

Harpold, A.A., Guo, Q., Molotch, N., Brooks, P.D., Bales, R., Fernandez-Diaz, J.C., Musselman, K.N., Swetnam, T.L., Kirchner, P., Meadows, M.W., Flanagan, J., Lucas, R., 2014b. LiDAR-derived snowpack data sets from mixed conifer forests across the Western United States. Water Resour. Res. 50, 2749–2755.

Heidbuchel, I., Troch, P.A., Lyon, S.W., Weiler, M., 2012. The master transit time distribution of variable flow systems. Water Resour. Res. 48, W06520.

Hinckley, E.-L., Ebel, B.A., Barnes, R.T., Anderson, R.S., Williams, M.W., Anderson, S.P., 2014a. Aspect control of water movement on hillslopes near the rain–snow transition of the Colorado Front Range, U.S.A. Hydrol. Process. 28, 74–85.

Hinckley, E-L., Barnes, R.T., Anderson, S.P., Williams, M.W., Bernasconi, S., 2014b. Nitrogen retention and transport differ by hillslope aspect at the rain–snow transition of the Colorado Front Range. J. Geophys. Res. Biogeosci. 119 (7), 1281–1296.

Hirmas, D.R., Amrhein, C., Graham, R.C., 2010. Spatial and process-based modeling of soil inorganic carbon storage in an arid piedmont. Geoderma 154, 486–494.

Hoekstra, J., Molnar, J., Jennings, M., Revenga, C., Spalding, M., Boucher, T., Robertson, J., Hiebel, T., Ellison, K., 2010. In: Molnar, J. (Ed.), The Atlas of Global Conservation: Changes, Challenges, and Opportunities to Make a Difference. California Press, Berkeley, p. 234.

Hoffman, B., Anderson, R.S., 2014. Tree root mounds and their role in transporting soil on forested landscapes. Earth Surf. Proc. Land. 39 (6), 711–722.

Holleran, M.E., Levi, M.R., Rasmussen, C., 2014. Quantifying soil and critical zone variability in a forested catchment through digital soil mapping. Soil 1, 1–49.

Hooke, et al. 2012. Land transformations by humans: a review. GSA Today 22 (12), 4–10.

Hopkins, F.M., Torn, M.S., Trumbore, S.E., 2012. Warming accelerates decomposition of decades-old carbon in forest soils. Proc. Natl. Acad. Sci. 109, 1753–1761.

Horsburgh, J., Tarboton, D., Piasecki, M., Maidment, D., Zaslavsky, I., Valentine, D., Whitenack, T., 2009. An integrated system for publishing environmental observations data. Environ. Model. Softw. 24, 879–888.

IPCC, 2014. Intergovernmental Panel on Climate Change Fifth Assessment Report, http://www.ipcc.ch/.

Jin, L., Ravella, R., Ketchum, B., Bierman, P.R., Heaney, P., White, T., Brantley, S.L., 2010. Mineral weathering and elemental transport during hillslope evolution at the Susquehanna/Shale Hills Critical Zone Observatory. Geochim. Cosmochim. Ac 74, 3669–3691.

Jin, L., Andrews, D.M., Holmes, G.H., Lin, H., Brantley, S.L., 2011. Opening the "black box": water chemistry reveals hydrological controls on weathering in the Susquehanna Shale Hills Critical Zone Observatory. Vadose Zone J. 10, 928–942.

Jobbágy, E.G., Jackson, R.B., 2000. The vertical distribution of soil organic carbon and its relation to climate and vegetation. Ecol. Appl. 10, 423–436.

Jones, C., McConnell, C., Coleman, K., Cox, P., Falloon, P., Jenkinson, D., Powlson, D., 2005. Global climate change and soil carbon stocks; predictions from two contrasting models for the turnover of organic carbon in soil. Glob. Change Biol. 11, 154–166.

Katz, J., Moyle, P.B., Quiñones, R.M., Israel, J., Purdy, S., 2012. Impending extinction of salmon, steelhead, and trout (Salmonidae) in California. Environ. Biol. Fish., 1–18.

Keefer, L.L., Bauer, E., Markus, M., 2010. Hydrologic and Nutrient Monitoring of the Lake Decatur Watershed: Final Report 1992–2008. Illinois State Water Survey Contract Report 2010-07, Champaign, IL, 82 pp.

Kim, H., Bishop, J.K.B., Wood, T.J., Fung, I.Y., 2012. Autonomous water sampling for long-term monitoring of trace metals in remote environments. Environ. Sci. Technol. 46 (20), 11220–11226.

Kim, H., Bishop, J.K.B., Dietrich, W.E., Fung, I.Y., 2014. Process dominance shift in solute chemistry as revealed by long-term high-frequency water chemistry observations of groundwater flowing through weathered argillite underlying a steep forested hillslope. Geochim. Cosmochim. Ac. 140, 1–19.

Kleber, M.J., Johnson, M.G., 2010. Advances in understanding the molecular structure of soil organic matter: implications for interactions in the environment. Adv. Agron. 106, 77–142.

Kulmatiski, A., Vogt, D.J., Siccama, T.G., Tilley, J.P., Kolesinskas, K., Wickwire, T.W., Larson, B.C., 2004. Landscape determinants of soil carbon and nitrogen storage in southern New England. Soil Sci. Soc. Am. J. 68, 2014–2022.

Kumar, P., 2013. Hydrology: seasonal rain changes. Nat. Clim. Change 3, 783–784.

Kunkel, M.L., Flores, A.N., Smith, T.J., McNamara, J.P., Benner, S.G., 2011. A simplified approach for estimating soil carbon and nitrogen stocks in semi-arid complex terrain. Geoderma 165, 1–11.

Kupferberg, S.J., Lind, A.J., Thill, V., Yarnell, S.M., 2011. Water velocity tolerance in tadpoles of the foothill yellow-legged frog (*Rana boylii*): swimming performance, growth, and survival. Copeia 1, 141–152.

Lal, R., 2004. Soil carbon sequestration impacts on global climate change and food scarcity. Science 304, 1623–1628.

Langston, A.L., Tucker, G.E., Anderson, R.S., Anderson, S.P., 2011. Exploring links between vadose zone hydrology and chemical weathering in the Boulder Creek Critical Zone Observatory. Appl. Geochem. 26, S70–S71.

Le, P., Kumar, P., Drewry, D., 2011. Implications for the hydrologic cycle under climate change due to the expansion of bioenergy crops in the Midwestern United States. Proc. Natl. Acad. Sci. 108 (37), 15085–15090.

Leopold, M., Völkel, J., Dethier, D.P., Williams, M.W., 2014. Changing mountain permafrost from the 1970s to today – comparing two examples from Niwot Ridge, Colorado Front Range, USA. Zeitschrift für Geomorphologie Suppl. Issue 58 (1), 137–157.

Lin, H.S., 2006. Temporal stability of soil moisture spatial pattern and subsurface preferential flow pathways in the Shale Hills catchment. Vadose Zone J. 5, 317–340.

Lin, H., Kogelmann, W.J., Walker, C., Bruns, M.A., 2006. Soil moisture patterns in a forested catchment: a hydropedological perspective. Geoderma 131, 345–368.

Lin, H.S., Fluhler, H., Otten, W., Vogel, H.J., 2010. Soil architecture and preferential flow across scales. J. Hydrol. 393, 1–2.

Link, P., Simonin, K., Maness, H., Oshun, J., Dawson, T., Fung, I., 2014. Species differences in the seasonality of evergreen tree transpiration in a Mediterranean climate: analysis of multiyear, half-hourly sap flow observations. Water Resour. Res. 50 (3), 1869–1894.

Lock, J., Kelsey, H., Furlong, K., Woolace, A., 2006. Late Neogene and Quaternary landscape evolution of the northern California Coast Ranges: evidence for Mendocino triple junction tectonics. Geol. Soc. Am. Bull. 118 (9–10), 1232–1246.

Lybrand, R., Rasmussen, C., Jardine, A., Troch, P., Chorover, J., 2011. The effects of climate and landscape position on chemical denudation and mineral transformation in the Santa Catalina mountain critical zone observatory. Appl. Geochem. 26, S80–S84.

Ma, L., Chabaux, F., Pelt, E., Blaes, E., Jin, L., Brantley, S.L., 2010. Regolith production rates calculated with uranium-series isotopes at the Susquehanna/Shale Hills Critical Zone Observatory. Earth Planet. Sc. Lett. 297, 211–225.

Mackey, B.H., Roering, J.J., 2011. Sediment yield, spatial characteristics, and the long-term evolution of active earthflows determined from airborne LiDAR and historical aerial photographs, Eel River, California. Geol. Soc. Am. Bull. 123, 1560–1576.

Mackey, B.H., Roering, J.J., Lamb, M.P., 2011. Landslide-dammed paleolake perturbs marine sedimentation and drives genetic change in anadromous fish. Proc. Natl. Acad. Sci. 108 (47), 18905–18909.

Markewitz, D., Richter, D.D., Allen, H.L., Urrego, J.B., 1998. Three decades of observed soil acidification in the Calhoun Experimental Forest: has acid rain made a difference? Soil Sci. Soc. Am. 62, 1428–1439.

Marks, D., 2001. Introduction to special section: Reynolds Creek Experimental Watershed. Water Resour. Res. 37, 2817.

Marks, D.G., Reba, M., Pomeroy, J., Link, T., Winstral, A., Flerchinger, G., Elder, K., 2008. Comparing simulated and measured sensible and latent heat fluxes over snow under a pine canopy to improve an energy balance snowmelt model. J. Hydrometeorol. 9, 1506–1522.

Marschner, B., Brodowski, S., Dreves, A., Gleixner, G., Gude, A., Grootes, P.M., Hamer, U., Heim, A., Jandl, G., Ji, R., Kaiser, K., Kalbitz, K., Kramer, C., Leinweber, P., Rethemeyer, J., Schäffer, A., Schmidt, M.W.I., Schwark, L., Wiesenberg, G.L.B., 2008. How relevant is recalcitrance for the stabilization of organic matter in soils? J. Plant Nutr. Soil Sci. 171, 91–110.

McDowell, W.H., Scatena, F.N., Waide, R.B., Brokaw, N.V., Camilo, G.R., Covich, A.P., Crowl, T.A., Gonzalez, G., Greathouse, E.A., Klawinski, P., Lodge, D.J., Lugo, A.E., Pringle, C.M., Richardson, B.A., Richardson, M.J., Schaefer, D.A., Silver, W.L., Thompson, J., Vogt, D., Vogt, K., Willig, M., Woolbright, L., Zou, X., Zimmerman, J., 2012. Geographic and ecological setting. In: Brokaw, N., Crowl, T.A., Lugo, A.E., McDowell, W.H., Scatena, F.N., Waide, R.B., Willig, M.W. (Eds.), A Caribbean Forest Tapestry: The Multidimensional Nature of Disturbance and Response. Oxford University Press, New York.

McIver, J., Brunson, M., Bunting, S., Chambers, J., Devo, N., Doescher, P., Grace, J., Johnson, D., Knick, S., Miller, R., Pellant, M., Pierson, F., Pyke, D., Rollins, K., Roundy, B., Schupp, E., Tausch, R., Turner, D., 2010. The Sagebrush Steppe Treatment Evaluation Project (SageSTEP): A Test of State-and-Transition Theory. General Technical Report RMRS-GTR-237. U.S. Department of Agriculture, Forest Service, Rocky Mountain Research Station, Fort Collins, CO.

Merritts, D., Bull, W., 1989. Interpreting Quaternary uplift rates at the Mendocino triple junction, northern California, from uplifted marine terraces. Geology 17, 1020–1024.

Murphy, S.F., 2006. State of the watershed: water quality of Boulder Creek, Colorado. U.S. Geological Survey Circ. 1284, 34.

Mutel, C., 2008. The Emerald Horizon: Restoring Iowa's Greatly Changed Landscape. The University of Iowa Press, Iowa City, IA, p. 328.

Naithani, K.J., Gaines, K., Baldwin, D., Lin, H.S., Eissenstat, D.M., 2013. Spatial distribution of tree species governs the spatio-temporal interaction of leaf area index and soil moisture across a landscape. PloS ONE 8, 12.

National Research Council, 2010. Landscapes on the Edge: New Horizons for Research on Earth's Surface. The National Academy Press, Washington, DC, p. 5.

Nayak, A., Marks, D., Chandler, D.G., Seyfried, M., 2010. Long-term snow, climate, and streamflow trends at the Reynolds Creek Experimental Watershed, Owyhee Mountains, Idaho, United States. Water Resour. Res. 46, W06519.

Ng, C.M., 2012. The transport of chemicals and biota into coastal rivers and marine ecosystems. PhD dissertation. U.C. Berkeley.

Nikolaidis, N.P., 2011. Human impacts on soils: tipping points and knowledge gaps. Appl. Geochem. 26, S230–S233.

Niu, X., Lehnert, K., Williams, J., Brantley, S., 2011. Advancing management and access of critical zone geochemical data. Appl. Geochem. 26, S108–S111.

NMFS, 2009. Draft frost protection threat assessment for threatened and endangered salmonids in the Russian River watershed, Testimony prepared for the State Water Resources Control Board public workshop on frost protection on November 18, 2009. Sacramento, California.

Oh, N.-H., Richter, Jr., D.D., 2004. Soil acidification induced by elevated atmospheric CO_2. Global Change Biol. 10 (11), 1936–1946.

Olhsson, E., 2013. Coupling terrestrial ecosystems and watershed hydrology to coastal ocean productivity for the Eel River Basin, University of California, Berkeley, PhD dissertation (in prep).

Orem, C.A., Pelletier, J.D., 2015. The predominance of post-wildfire denudation in the long- term evolution of forested, mountainous landscapes. Proc. Natl. Acad. Sci. USA (in press).

Oshun, J., Rempe, D.M., Link, P., Simonin, K.A., Dietrich, W., Dawson, T.E., Fung, I., 2012. A look deep inside a hillslope reveals a structured heterogeneity of isotopic reservoirs and distinct water use strategies for adjacent trees, Abstract B33A-0498, presented at 2012 Fall Meeting, AGU.

Papanicolaou, A.N., Abaci, O., 2008. Upland erosion modeling in a semi-humid environment via the Water Erosion Prediction Project model. J. Irrig. Drain. Eng. 134, 796–806.

Papanicolaou, A.N., Dermisis, D., Wacha, K., Abban, B., Wilson, C., 2011. Agricultural soil erosion and soil organic carbon (SOC) dynamics in the U.S. Midwest: bridging the knowledge gap across scales. The Geological Society of America (GSA) Meeting, Minneapolis, Minnesota, USA, October 9–12, 2011.

Pelletier, J.D., Barron-Gafford, G.A., Breshears, D.D., Brooks, P.D., Chorover, J., Durcik, M., Harman, C.J., Huxman, T.E., Lohse, K.A., Lybrand, R., Meixner, T., McIntosh, J.C., Papuga, S.A., Rasmussen, C., Schaap, M., Swetnam, T.L., Troch, P.A., 2013. Coevolution of nonlinear trends in vegetation, soils, and topography with elevation and slope aspect: a case study in the sky islands of southern Arizona. J. Geophys. Res. Earth 118, 741–758.

Perdrial, J.N., Perdrial, N., Harpold, A., Gao, X.D., Gabor, R., LaSharr, K., Chorover, J., 2012. Impacts of sampling dissolved organic matter with passive capillary wicks versus aqueous soil extraction. Soil Sci. Soc. Am. J. 76, 2019–2030.

Pike, A.S., Scatena, F.N., Wohl, E.E., 2010. Lithological and fluvial controls on the geomorphology of tropical montane stream channels in Puerto Rico. Earth Surf. Process. Land. 35 (12), 1402–1417.

Power, M.E., Parker, M.S., Dietrich, W.E., 2008. Seasonal reassembly of a river food web: floods, droughts, and impacts of fish. Ecol. Monogr. 78, 263–282.

Power, M.E., Lowe, R., Furey, P.C., Welter, J., Limm, M., Finlay, J.C., Bode, C., Chang, S., Goodrich, M., Sculley, J., 2009. Algal mats and insect emergence in rivers under Mediterranean climates: towards photogrammetric surveillance. Freshwater Biol. 54, 2101–2115.

Puschner, B., Hoff, B., Tor, E.R., 2008. Diagnosis of anatoxin-a poisoning in dogs from North America. J. Vet. Diagn. Invest. 20, 89.

Rasmussen, C., Troch, P.A., Chorover, J., Brooks, P., Pelletier, J., Huxman, T.E., 2011. An open system framework for integrating critical zone structure and function. Biogeochemistry 102, 15–29.

Reba, M., Link, T., Marks, D., Pomeroy, J., 2009. An assessment of techniques to improve estimates of turbulent fluxes over snow by eddy covariance in mountain environments. Water Resour. Res. 45, W00D38.

Reba, M., Marks, D., Winstral, A., Link, T., Kumar, M., 2011a. Sensitivity of the snowcover in a mountain basin to variations in climate. Hydrol. Process. 25, 3312–3321.

Reba, M.L., Marks, D., Seyfried, M., Winstral, A., Kumar, M., Flerchinger, G., 2011b. A long-term data set for hydrologic modeling in a snow-dominated mountain catchment. Water Resour. Res. 47.

Reba, M.L., Marks, D., Link, T.E., Pomeroy, J., Winstral, A., 2012. Sensitivity of model parameterizations for simulated latent heat flux at the snow surface for complex mountain sites. Hydrol. Process. 28 (3), 868–881.

Reese, C.D., Harvey, B.C., 2002. Temperature-dependent interactions between juvenile Steelhead and Sacramento Pikeminnow in Laboratory streams. Trans. Am. Fish. Soc. 131 (4), 599–606.

Rempe, D.M., Dietrich, W.E., 2014. A bottom-up control on fresh-bedrock topography under landscapes. Proc. Natl. Acad. Sci. 111 (18), 6576–6581.

Richter, D.D., Markewitz, D., 2001. Understanding Soil Change: Soil Sustainability over Millennia, Centuries, and Decades. Cambridge University Press, UK, p. 255.

Richter, D.deB., Bacon, A.R., Billings, S.A., Binkley, D., Buford, M., Callaham, M.A., Curry, A.E., Fimmen, R.L., Grandy, A.S., Heine, P.R., Hofmockel, M., Jackson, J.A., Lemaster, E., Li, J., Markewitz, D., Mobley, M.L., Morrison, M.W., Strickland, M., Waldrop, T., Wells, C.G., 2014. Evolution of a half-century of soil, ecosystem, and critical zone research at the Calhoun Experimental Forest. In: Research for the Long Term. Springer-Verlag, New York.

Sabo, J.L., Power, M.E., 2002a. River-watershed exchange: effects of riverine subsidies on riparian lizards and their terrestrial prey. Ecology 83, 1860–1869.

Sabo, J.L., Power, M.E., 2002b. Numerical response of lizards to aquatic insects and short-term consequences for terrestrial prey. Ecology 83 (11), 3023–3036.

Salve, R., 2011. A sensor array system for profiling moisture in unsaturated rock and soil. Hydrol. Process. 25 (18), 2907–2915.

Salve, R., Rempe, D.M., Dietrich, W.E., 2012. Rain, rock moisture dynamics, and the rapid response of perched groundwater in weathered, fractured argillite underlying a steep hillslope. Water Resour. Res. 48, W11528.

Scheingross, J.S., Winchell, E.W., Lamb, M.P., Dietrich, W.E., 2013. Influence of bed patchiness, slope, grain hiding, and form drag on gravel mobilization in very steep streams. J. Geophys. Res. Earth Surf. 118, 982–1001.

Schmidt, M.W.I., Torn, M.S., Abiven, S., Dittmar, T., Guggenberger, G., Janssens, I.A., Kleber, M., Kögel-Knabner, I., Lehmann, J., Manning, D.A.C., Nannipieri, P., Rasse, D.P., Weiner, S., Trumbore, S.E., 2011. Persistence of soil organic matter as an ecosystem property. Nature 478, 49–56.

Seidl, M.A., Dietrich, W.E., 1992. The problem of channel erosion into bedrock. In: Schmidt, K.H., de Ploey, J. (Eds.), Functional Geomorphology: Landform Analysis and Models. Catena Supplement, vol. 23, pp. 101–124.

Seyfried, M.S., Flerchinger, G.N., Murdock, M.D., Hanson, C.L., 2001a. Long-term soil temperature database, Reynolds Creek Experimental Watershed, Idaho, United States. Water Resour. Res. 37, 2843–2846.

Seyfried, M.S., Hanson, C.L., Murdock, M.D., Van Vactor, S., 2001b. Long-term lysimeter database, Reynolds Creek Experimental Watershed, Idaho, United States. Water Resour. Res. 37, 2853–2856.

Seyfried, M.S., Harris, R., Marks, D., Jacob, B., 2001c. Geographic database, Reynolds Creek Experimental Watershed. Idaho, United States. Water Resour. Res. 37, 2825–2829.

Seyfried, M.S., Murdock, M.D., Hanson, C.L., Flerchinger, G.N., Van Vactor, S., 2001d. Long-term soil water content database, Reynolds Creek Experimental Watershed, Idaho, United States. Water Resour. Res. 37, 2847–2851.

Seyfried, M.S., Grant, L.E., Du, E., Humes, K., 2005. Dielectric loss and calibration of the hydra probe soil water sensor. Vadose Zone J. 4, 1070–1079.

Seyfried, M., Chandler, D., Marks, D., 2011. Long-term soil water trends across a 1000-m elevation gradient. Vadose Zone J. 10, 1275–1285.

Shi, Y., Davis, K.J., Duffy, C.J., Yu, X., 2013. Development of a coupled land surface hydrologic model and evaluation at a critical zone observatory. J. Hydrometeorol. 14, 1401–1420.

Sklar, L.S., Dietrich, W.E., 2006. The role of sediment in controlling steady-state bedrock channel slope: implications of the saltation – abrasion incision model. Geomorphology 82 (1), 58–83.

Slaughter, C.W., Marks, D., Flerchinger, G.N., Van Vactor, S.S., Burgess, M., 2001. Thirty-five years of research data collection at the Reynolds Creek Experimental Watershed, Idaho, United States. Water Resour. Res. 37, 2819–2823.

Smith, D., 2012. The Penguin State of the World Atlas, ninth ed. Penguin Books, New York, 144 pp.

Sollins, P., Swanston, C., Kramer, M., 2007. Stabilization and destabilization of soil organic matter – a new focus. Biogeochemistry 85, 1–7.

Steffen, W., Crutzen, P., McNeill, J., 2007. The Anthropocene: are humans now overwhelming the great forces of nature. AMBIO J. Hum. Environ. 36 (8), 614–621.

Stephenson, G., 1977. Soil, geology, vegetation inventories, Reynolds Creek Experimental Watershed, Idaho. Micellaneous Series No., p. 73.

Stock, J.D., Montgomery, D.R., Collins, B.R., Sklar, L., Dietrich, W.E., 2005. Field measurements of incision rates following bedrock exposure: implications for process controls on the long-profiles of valleys cut by rivers and debris flows. Geol. Soc. Am. 117 (11–12), 174–194.

Suttle, K.B., Power, M.E., Levine, J.A., McNeely, F.C., 2004. How fine sediment in river beds impairs growth and survival of juvenile salmonids. Ecol. Appl. 14, 969–974.

Suttle, K.B., Thomsen, M.A., Power, M.E., 2007. Species interactions reverse grassland responses to changing climate. Science 315, 640–642.

Thomas E., Lin, H., Duffy, C., Sullivan, P., Holmes, G.H., Brantley, S.L., Jin, L., 2013. Spatiotemporal patterns of water stable isotope compositions at the Shale Hills Critical Zone: Linkages to subsurface hydrologic processes. Vadose Zone J., Vol. 12, No. 4.

Thorleifson, L.H., 1996. Review of Lake Agassiz history. In: Teller, J.T., Thorleifson, L.H., Matile, G., Brisbin, W.C. (Eds.), Sedimentology, Geomorphology and History of the Central Lake Agassiz Basin: Geological Association of Canada/Mineralogical Association of Canada Annual Meeting, Winnipeg, Manitoba, May 27–29, 1996, Field Trip Guidebook B2., Geological Association of Canada, St. John's, NL pp. 55–84.

Todd-Brown, K.E.O., Randerson, J.T., Post, W.M., Hoffman, F.M., Tarnocai, C., Schuur, E.A.G., Allison, S.D., 2013. Causes of variation in soil carbon simulations from CMIP5 Earth system models and comparison with observations. Biogeosciences 10, 1717–1736.

Torn, M.S., Trumbore, S.E., Chadwick, O.A., Vitousek, P.M., Hendricks, D.M., 1997. Mineral control of soil organic carbon storage and turnover. Nature 389, 170–173.

Totsche, K.U., Rennert, T., Gerzabek, M.H., Kögel-Knabner, I., Smalla, K., Spiteller, M., Vogel, H.J., 2010. Biogeochemical interfaces in soil: the interdisciplinary challenge for soil science. J. Plant Nutr. Soil Sc. 173, 88–99.

Trumbore, S.E., Czimczik, C.I., 2008. An uncertain future for soil carbon. Science 321, 1445.

United Nations Environment Programme's Millennium Ecosystem Assessment Report, 2005, http://www.unep.org/maweb/en/index.aspx.

Uno, H., Power, M., 2013. Unusual mayfly life cycle connects mainstem and tributary food webs, Abstract, Society of Freshwater Sciences, May 2013 meeting, Jacksonville, FL.

UNEP, 2005. One Planet Many People: Atlas of Our Changing Environment. Division of Early Warning and Assessment, UNEP, Nairobi, Kenya, 322 pp.

USDA, 2006. Land resource regions and major land resource areas of the United States, the Caribbean, and the pacific basin. In: Agriculture Handbook 296. USDA, Washington, DC. http://soils.Usda.gov/survey/geography/mlra/index.Html.

Vazquez-Ortega, A., Perdrial, J., Harpold, A., Zapata-Rios, X., Rasmussen, C., McIntosh, J.C., Schaap, M., Pelletier, J.D., Amistadi, M.K., Chorover, J., 2014. Rare earth elements as reactive tracers of biogeochemical weathering in the Jemez River Basin Critical Zone Observatory. Chem. Geol., 391 (6), 19–32.

Vitousek, P., Mooney, H., Lubchenko, J., Melillo, J., 1996. Human domination of Earth's ecosystems. Science 277 (5325), 494–499.

West, N., Kirby, E., Bierman, P.R., Slingerland, R., Ma, L., Rood, D., Brantley, S.L., 2013. Regolith production and transport at the Susquehanna Shale Hills Critical Zone Observatory, Part 2: insights from meteoric ^{10}Be. J. Geophys. Res. Earth Surf. 118, 20.

Westerling, A.L., Hidalgo, H.G., Cayan, D.R., Swetnam, T.W., 2006. Warming and earlier spring increase Western U.S. forest wildfire activity. Science 313, 940–943.

Wilcock, P., 2009. Synthesis Report for Minnesota River Sediment Colloquium. Report for MPCA. http://www.lakepepinlegacyalliance.org/SedSynth_FinalDraft-formatted.pdf.

Wilkinson, B., 2005. Humans as geologic agents: a deep-time perspective. Geology 33 (3), 161–164.

Willenbring, J.K., Gasparini, N.M., Crosby, B.T., Brocard, G., 2013. What does a mean mean? The temporal evolution of detrital cosmogenic denudation rates in a transient landscape. Geology 41 (12), 1215–1218.

Williams, J., Bandstra, J., Pollard, D., Brantley, S., 2010. The temperature dependence of feldspar dissolution determined using a coupled weathering-climate model for Holocene-aged loess soils. Geoderma 156 (1–2), 11–19.

Wilson, C.G., Papanicolaou, A.N., Abaci, O., 2009. SOM dynamics and erosion in an agricultural test field of the Clear Creek, IA watershed. Hydrol. Earth Syst. Sci. Discuss. 6, 1581–1619.

Wilson, C.G., Papanicolaou, A.N., Denn, K.D., 2012. Quantifying and partitioning fine sediment loads in an intensively agricultural headwater system. J. Soils Sediments 12, 966–981.

Wobus, C.W., Tucker, G.E., Anderson, R.S., 2010. Does climate change create distinctive patterns of landscape incision? J. Geophys. Res. 115, 12.

Wootton, J.T., Parker, M.S., Power, M.E., 1996. The effect of disturbance on river food webs. Science 273, 1558–1560.

Chapter 3

Climate of the Critical Zone

Steven M. Quiring*, Trenton W. Ford**, and Shanshui Yuan*
*Department of Geography, Climate Science Lab, Texas A&M University, Texas, USA;
**Department of Geography and Environmental Resources, Southern Illinois University, Illinois, USA

3.1 INTRODUCTION

As noted in Chapter 1 of this book, the Critical Zone (CZ) extends from the top of the vegetative canopy through to the bottom of weathered bedrock. The CZ is the medium in which many important interactions and exchanges, including water, energy, momentum and matter, take place between the land surface and the atmosphere. These exchanges are controlled by soil, vegetation, and atmospheric processes. Therefore, the CZ and the associated interactions between the land and atmosphere play an important role in Earth's climate system.

3.1.1 How Does Climate Influence the Critical Zone?

Climate variability, both in space and in time, has a significant impact on nearly every aspect of the CZ. Many of the National Science Foundation's Critical Zone Observatories are examining different aspects of how climate influences the structure and function of the CZ (Anderson et al., 2013; Chorover et al., 2011; Mahmood and Vivoni, 2014; Troch et al., 2013). For example, land-cover patterns (excluding human influence) are primarily determined by gradients and variability in climate (Chorover et al., 2011; Clifford et al., 2013; Troch et al., 2013). Spatial patterns in vegetation across large landscapes are predominantly controlled by climate (Sarr et al., 2005). Climate variability and climate change have been shown to cause changes in vegetation distributions, including tree-line extent in high-altitude and high-latitude regions (Holtmeier and Broll, 2005). Climate change and climate variability have also been connected to the location and size of terminal glacial moraines from the last glacial maximum (Anderson et al., 2014). Presently, climate change and climate variability influence the spatial extent of high-altitude glaciers in tropical (Vuille et al., 2008) and mid-latitude environments (Scherler et al., 2011).

In the subsurface, climate has been shown to impact the chemistry of water in the unsaturated (Gurdak et al., 2007) and saturated zones (Anderson et al., 2007). Changes in the lithosphere have also been attributed to long-term climate variability. For example, temperature variability due to aspect differences

Developments in Earth Surface Processes, Vol. 19. http://dx.doi.org/10.1016/B978-0-444-63369-9.00003-3

has been connected to changes in transport and damage efficiencies of mobile regolith on bedrock (Anderson et al., 2013). Concurrently, the chemical composition of the lithosphere is also impacted by climate, as sodium-weathering rates in regolith have been shown to strongly correlate with precipitation variability (Rasmussen et al., 2011). Overall the spatial and temporal variability of climate, combined with long- and short-term climate change, can impact nearly all aspects of the CZ, from the landscape to the bedrock.

3.1.2 How Does the Critical Zone Influence Climate?

Many components of the CZ influence the local and regional climate, including vegetation, land use/land cover, albedo, snow cover, and soil moisture (SM) (Pielke, 2001; McPherson et al., 2004; Mahmood et al., 2014). Soils and vegetation are two fundamental factors that influence the climate, especially land–atmosphere interactions. The persistence of SM (i.e., SM memory) is controlled by soil characteristics. For example, soil texture, soil structure and the amount of organic matter in the soil influence the movement of water in the soil (infiltration and percolation) and determine the water-holding capacity (Seneviratne et al., 2010). Seneviratne and Koster (2011) found that SM memory in general circulation models (GCMs) is influenced by water-holding capacity; and soils with a larger water-holding capacity tended to have longer memory.

Land cover and vegetation dynamics are another critical factor that influences the climate. Dirmeyer et al. (2006) found that transpiration is the primary way that SM influences evapotranspiration. Therefore, plants play an important role in modulating land–atmosphere interactions. For example, during the 2003 European heat wave, temperature anomalies were greater in locations with crops and pastures than in forested locations (Zaitchik et al., 2006). Nobre et al. (1991) used GCMs to examine the impacts of Amazonian deforestation and found that deforestation increased mean annual temperature and decreased evapotranspiration and precipitation.

3.1.3 Focus of This Chapter

This chapter examines how the CZ influences the local and regional climate through CZ-climate (land–atmosphere) interactions and how the coupling between the land and atmosphere varies over time and space. CZ-climate interactions are an active field of research and so it would be impossible to provide an exhaustive review of all aspects of the relationships between the CZ and the climate. We have chosen to focus on SM because it is an important component of water balance of the CZ and it is a key parameter that influences land–atmosphere interactions in the climate system by modifying energy and water fluxes in the boundary layer (Eltahir, 1998; Legates et al., 2010). SM plays an integrative role because it directly influences atmospheric, geomorphic, hydrologic, and biologic processes (Legates et al., 2010). Therefore, SM directly or indirectly drives the relationships between the climate and the CZ.

3.2 SOIL MOISTURE

Soil moisture represents the amount of water available to plants as well as the relative state of water supply (from precipitation or irrigation) versus water demand (from evapotranspiration). SM in the root zone of vegetated regions has a significant influence on evapotranspiration rates (McPherson, 2007; Alfieri et al., 2008) and latent and sensible heat exchange (Dirmeyer et al., 2000; Basara and Crawford, 2002). Wet soils increase evaporation, decrease surface temperature, and enhance convective available potential energy (CAPE) (Eltahir, 1998; Pal and Eltahir, 2001). Dry soils can induce and amplify warm and dry conditions, especially during the summer, by reducing local evaporation and modifying patterns of moisture convergence/divergence and atmospheric circulation (Namias, 1991). Therefore SM can have a strong impact on the surface Bowen ratio (Quintanar et al., 2008), CAPE (Pal and Eltahir, 2001), development of clouds (Findell and Eltahir, 2003), near-surface air temperature (Mahmood et al., 2006), and the persistence of heat waves (Lorenz et al., 2010), precipitation (Koster et al., 2003; Pan et al., 1995), planetary boundary layer (Quintanar et al., 2008), and atmospheric circulation and the surface wind field (Arrigo and Salvucci, 2005; Quintanar et al., 2009; Taylor et al., 2007). These changes can, in turn, often intensify and extend the anomalous SM conditions.

The time rate of change of SM is determined by the basic equation that governs the hydrologic cycle:

$$\frac{\mathrm{d}S}{\mathrm{d}t} = P - E - R_{\mathrm{s}} - R_{\mathrm{g}} \tag{3.1}$$

where $\mathrm{d}S/\mathrm{d}t$ is the change of water content in the soil, P is the precipitation, E is the evapotranspiration (including bare soil evaporation, transpiration, interception evaporation, snow sublimation, and evaporation from open water), R_{s} is the surface runoff, and R_{g} is the drainage or groundwater runoff (Fig. 3.1). This water balance is the embodiment of the hydrologic cycle and is governed by conservation of mass (Entekhabi et al., 1992; Seneviratne et al., 2010).

SM is persistent (Delworth and Manabe, 1988, 1993; Robock et al., 2000; Mostovoy and Anantharaj, 2008). The degree of persistence varies by region, but has been shown to range from 1 to 3 months in the central United States. This is commonly referred to as SM memory. Therefore, SM can influence atmospheric conditions over land on monthly to seasonal timescales (Liu, 2003; Karl, 1986; Wang and Kumar, 1998) and they are considered to be important sources of predictability of seasonal climate (Koster et al., 2003, 2004; Koster and Suarez, 2001). Once SM levels that are drier (or wetter) than normal become established, these conditions tend to persist, especially in the deeper layers of the soil, for several months (Entin et al., 2000; Pielke et al., 1999; Vinnikov et al., 1996). This multimonth memory is important because it provides the basis for seasonal climate forecasts. Implementation of SM in seasonal climate predictions has been shown to increase the accuracy

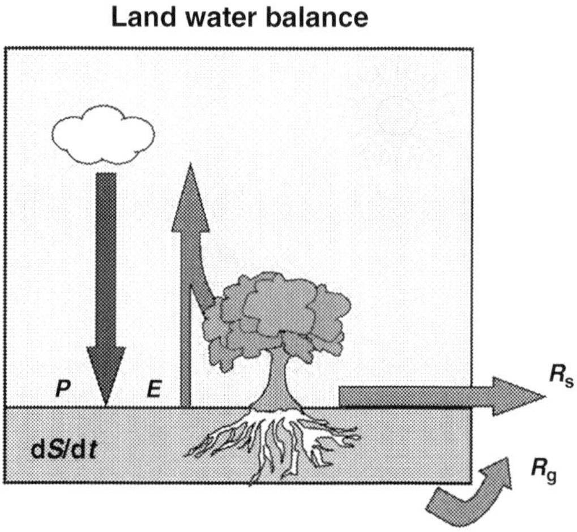

FIGURE 3.1 Depiction of the components of the terrestrial water balance. P is precipitation, E is evaporation, R_s is the surface runoff, R_g is the drainage or groundwater runoff, and dS/dt is the change of water content in the soil. *(Figure from Seneviratne et al. (2010).)*

(McPherson et al., 2004; Meng and Quiring, 2010; Ford and Quiring, 2014). SM has an impact on evapotranspiration (Section 3.4), temperature (Section 3.5), and precipitation (Section 3.6) at seasonal scales and this chapter will examine these influences.

3.3 ANTHROPOGENIC INFLUENCE

Although the primary focus of this chapter is SM –atmosphere interactions, the role of anthropogenic activity in modifying these dynamic relationships needs to be acknowledged. Humans are dependent on ecosystem services in the CZ, particularly soil fertility and water holding capacity, for their roles in food and fiber production (Palm et al., 2007; Janzen et al., 2011), the hydrologic cycle (Loescher et al., 2007; Ines and Mohanty, 2008, 2009), the carbon cycle (Amundson, 2001), and the climate system (Lal, 2004; Brevik, 2012). However, human activities also have a major influence on SM, both directly and indirectly. Irrigation directly changes soil water content and also has a significant influence on evapotranspiration. For example, conversion of grassland to irrigated corn in Nebraska has resulted in an additional 50 million cubic meters of evapotranspiration per year (Mahmood and Hubbard, 2002). These estimates are based on model simulations from a single representative county in Nebraska. It has been demonstrated that large-scale irrigation influences the local and regional climate by increasing the amount of water vapor in the lower atmosphere (i.e., increased latent heat and decreased sensible heat), increasing the dew point temperature, decreasing the maximum daily air temperatures, and increasing the minimum air temperatures (Mahmood et al., 2004). These changes are concentrated during the growing season and are controlled to an extent by the phenologic growth stage of the crop. For example, the maximum evapotranspiration for corn occurs during the tasseling and reproductive growth stages (Quiring, 2004). Because crops can have a different phenology than natural vegetation, this means that the strength and timing of energy and water fluxes can be modified by agricultural practices. Human modifications of the land surface tend to have the greatest impact during the growing season (Mahmood et al., 2014). Therefore, irrigated agriculture in the US cornbelt has the greatest impact on the local and regional climate during May–September. In other regions the strength and timing of the impact will vary based on the growing season and type of crop. Many of these changes occur during the warm season when most precipitation is convective and therefore the atmosphere is much more sensitive to local changes in land surface conditions and the resulting impact on energy and water fluxes. Although irrigation has a major influence on SM, any changes in land use/land cover or cropping patterns can impact SM as well as the local and regional climate.

Human activities have the potential to influence soil characteristics (fertility, organic matter, bulk density, soil water content, etc.) and land–atmosphere interactions in the CZ. Agricultural regions have been greatly influenced by anthropogenic activities. For example, soil compaction and erosion reduce water-holding capacity, altering infiltration and runoff (Das Gupta et al., 2006a,b). In turn, soil organic matter is lost and microbial activities are reduced to levels that influence soil fertility, which may alter rates of carbon sequestration belowground. Similarly, reduced fertility affects crop and forest yields (den Biggelaar et al., 2004a,b; Pimentel, 2006; Montgomery, 2007), which can also impact carbon sequestration (Van Oost et al., 2012).

One way to conceptualize the relationship between water in the CZ and human activities (e.g., land-use practices), including the feedbacks to hydroclimate dynamics, is using a socio-hydrologic system (SHS) framework to examine the complex interactions. This is an adaptation of Ostrom's socio-ecologic system (SES) framework (Ostrom, 2007), but reworks it to focus on how water interacts with society; the "hydrosocial" cycle (Bakker, 2003; Swyngedouw, 2004, 2009; Budds, 2008, 2009). In this context, SM is a critical hydrologic pathway that links the SHS with land–atmosphere interactions (Reenberg, 2009). Human activities, such as land management, modify biogeochemical processes, biodiversity, and flows of energy and water. Moreover, these activities affect changes in atmospheric composition and hydro-climate dynamics.

Our conceptual model is presented in Fig. 3.2. It focuses on SM as a critical pathway in the hydrologic dynamics of coupled climate-human systems. The outcome (O) in the SHS framework is SM, a critical variable that we can measure to understand land–atmosphere interactions. This conceptual model requires further exploration, theoretically and empirically. For example, the spatial or temporal scales of these interactions, feedbacks, and pathways are not

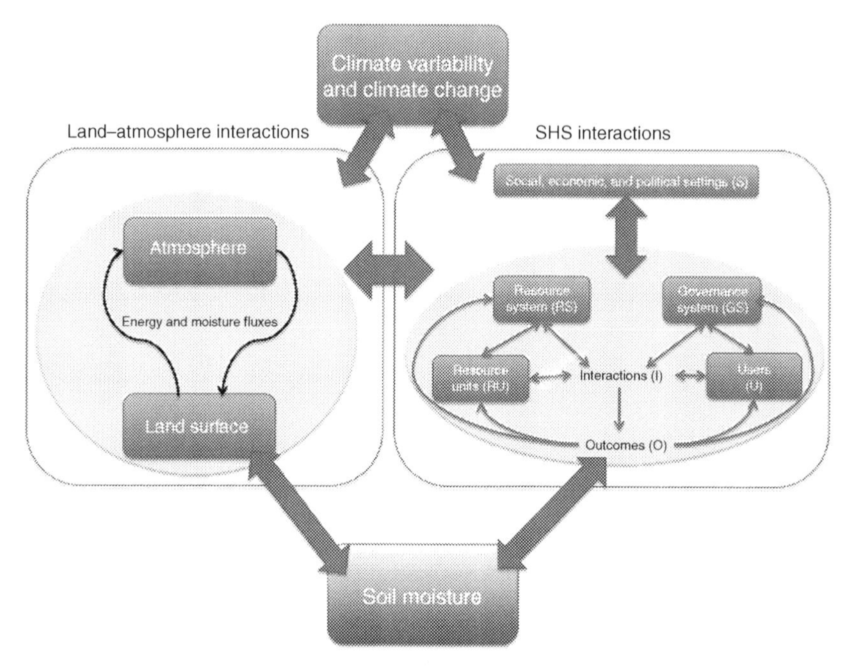

FIGURE 3.2 Conceptual model of the SHS developed by Wendy Jepson and Steven Quiring. SHS conceptual model inspired by Ostrom (2007) SES.

well understood. However, this conceptual model is helpful for understanding soil water dynamics in the CZ because it may help to reveal complex patterns and processes of environmental change and resource use not evident when these systems are studied separately. Human adaptation to changes in water availability, and SM, in particular, represent key factors for agricultural production. SM is an important factor influencing how land users perceive and respond to changes in the provisioning and uptake of water for crops. The focus on SM–land/water manager interaction, therefore, may provide important insights on human decisions about water and land use that allow us to explore how these decisions, collectively, feed back into the land–atmosphere system.

3.4 SOIL MOISTURE–EVAPOTRANSPIRATION COUPLING

Root-zone SM in vegetated regions has a significant influence on evapotranspiration rates (McPherson, 2007; Alfieri et al., 2008) and latent and sensible heat exchange (Dirmeyer et al., 2000; Basara and Crawford, 2002). Previous studies have determined that the nature of the soil moisture–evaporative fraction (SM–EF) relationship is dependent on whether the situation is SM -limited or energy-limited. EF is the ratio of latent heat to total surface energy flux. Dirmeyer et al. (2000) demonstrated that SM–EF relationships in land-surface models are characterized by two disparate regimes. When SM is between the

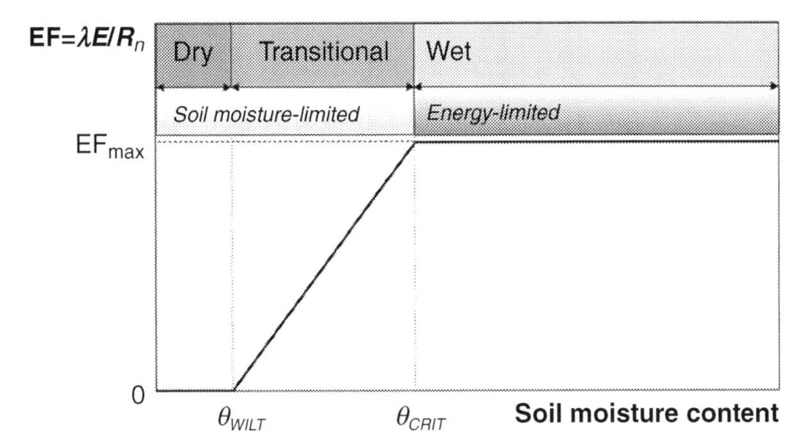

FIGURE 3.3 Depiction of the relationship between soil water content and EF (ratio of latent heat to sensible heat) that shows the two modes of SM–evaporation relationships: SM-limited and energy-limited. *(Figure from Seneviratne et al. (2010).)*

wilting point and critical value (typically 80% of field capacity), evaporation is strongly controlled by soil-water content and a linear relationship exists between SM and EF. When SM is not limited (SM > critical value), evaporation is controlled by available energy. Under this regime, evaporation is essentially decoupled from SM (i.e., uncorrelated with SM). Dirmeyer et al. (2000) evaluated three different land surface models and found that these two modes of behavior were present in all of them. Seneviratne et al. (2010) also defined two evapotranspiration regimes, SM-limited and energy-limited (Fig. 3.3). When SM is abundant (larger than a critical value), the main factor controlling evapotranspiration is energy, and when SM is limited (less than wilting point), there is no evapotranspiration. Under these two conditions, SM does not strongly influence evapotranspiration. SM-EF interactions in the SM-limited regime are typically linear and strongest when SM is between the wilting point and a critical value (Seneviratne et al., 2010).

3.4.1 Temporal Variations in Soil Moisture–Evapotranspiration Coupling

Previous studies have investigated the coupling between SM and surface heat fluxes, both with land-surface models and observations. Dirmeyer et al. (2000) used three land surface schemes to examine the relationship between the partitioning of latent and sensible heat and SM at sites with different land-cover. They demonstrated that surface energy partitioning was strongly controlled by SM in all three schemes, although the form of the relationship varied somewhat between schemes. Basara and Crawford (2002) used a correlation analysis to evaluate the relationship between observed SM and latent and sensible heat flux in central Oklahoma. They found strong, linear correlations between SM and surface energy flux, representing strong land–atmosphere coupling on days that they identified as ideal (i.e., clear skies, low winds, and no recent precipitation). Koster et al. (2009) employed an atmospheric general circulation model (AGCM) to characterize hydroclimatologic regimes around the globe. They used the relationship between temperature and precipitation as a proxy for SM–evaporation coupling and found that SM most strongly influences evaporation in semiarid areas where decreases in precipitation were associated with increases in air temperature. Brimelow et al. (2011) proposed that reductions in latent heat flux due to dry soils resulted in a deeper, warmer boundary layer and less convective storms during the summer of 2002 in Alberta, Canada. These studies suggest that SM can strongly influence the partitioning of latent and sensible heat flux, particularly under dry conditions when moisture is a limiting factor to evapotranspiration.

Because *in situ* observations of SM and surface heat fluxes are not readily available, past efforts to study land–atmosphere interactions have predominantly relied on land surface models. Recently, Ford et al. (2014) addressed this knowledge gap by evaluating surface-atmosphere relationships using both *in situ* observations and model-derived data. They examined the nature of the relationship between SM and surface energy fluxes using data from four stations in the Southern Great Plains (GP) of the United States. They evaluated how the relationship between SM and surface energy partitioning varies by location and by season. They examined why these variations occur, and they compared the model-derived and observed SM-energy flux relationships. Ford et al. (2014) examined the SM–EF relationship during May–October between 2004 and 2008. A great deal of variability occurred in the observed SM–EF relationships during the twenty growing seasons (four sites, five years) that they examined and they found evidence of both linear and nonlinear interactions. The relationship between SM–EF appears to be constrained by SM, as EF responds most strongly to changes in SM when SM is below the 30th percentile (Fig. 3.4a). However, when SM is greater than the 30th percentile little to no relationship appears between SM and EF. This pattern reflects, to some extent, the nonlinear relationships between SM and EF that were reported in Dirmeyer et al. (2000) and Santanello et al. (2007). Figure 3.4 shows the observed relationship between SM and EF in Ashton, Kansas during the 2006 growing season. The best-fit lines for the SM–EF relationships are based on 184 daily data points. They used the 30th percentile as the critical value. The relationship between SM–EF based on the *in situ* observations is statistically significant when SM is less than the 30th percentile, compared to an insignificant relationship when SM is greater than the 30th percentile. The model-simulated SM–EF relationship differs from that based on the *in situ* observations. The model-simulated relationship between SM–EF is only weakly linear when SM is less than the 30th percentile and the slope of the best-fit line does not change much when SM is greater than the 30th percentile. In fact, the model SM–EF relationship is

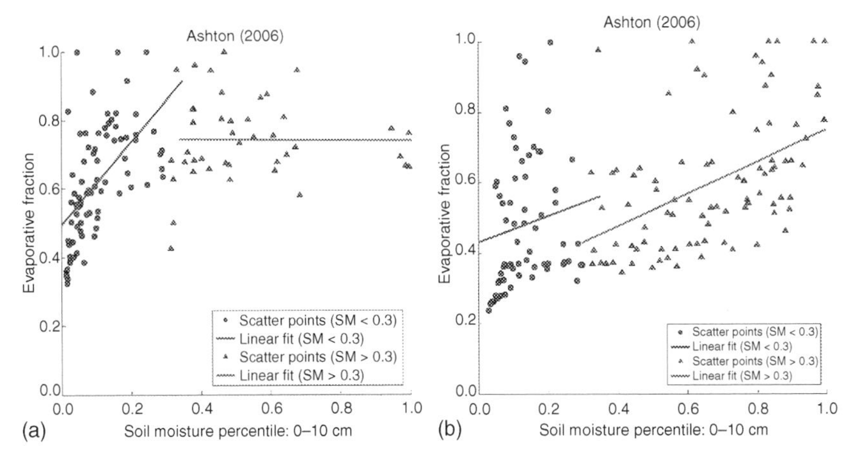

FIGURE 3.4 SM **percentiles (0–10cm) versus EF in Ashton, Kansas based on (a)** ***in situ*** **observations and (b) variable infiltration capacity model simulations.** Each point represents a daily average; all days during the 2006 growing season are shown. The light gray [blue in the web version] (dark gray [red in the web version]) line is the best-fit line between SM and EF when SM is <30th (>30th) percentile. Each graph contains 184 days (data points). *(Figure from Ford et al. (2014).)*

statistically significant for both SM conditions. The SM–EF relationship based on the observations reflects the dual regime behavior, where EF is a strong function of SM when SM is below the critical value; however the model-simulated SM–EF does not show the same response.

There is noticeable interannual variability in the relationship between SM–EF in the central United States. To better document this variability, the linear relationship between observed SM–EF for all four stations is separated by year (Fig. 3.5). Each line in Fig. 3.5 shows the linear fit between daily SM and EF, and fits are assessed separately for days when SM is less than and greater than the 30th percentile (our critical value). Substantial year-to-year variability occurs in the slope of the linear fit, especially at Cordell and El Reno. Figure 3.6 shows the same information as Fig. 3.5, except the linear fits are based on model-derived SM–EF data. The model-derived SM–EF relationships at Cordell and Coldwater are less variable than the model-derived SM–EF relationships at Ashton and El Reno. In general, the SM–EF relationships in both the observed and model-simulated datasets are similar to those reported in previous studies (e.g., Santanello et al., 2007). During most years, EF responds most strongly (linearly) to SM variations below the 30th percentile.

During the majority of growing seasons EF is a linear function of SM, when SM is below a critical value. In both the observations and model, the critical value for SM corresponds approximately with the 30th percentile. When SM is wetter than this, EF tends to function independently of SM conditions (Seneviratne et al., 2010). Ford et al. (2014) confirm that these are present in both the observed and model-simulated SM–EF data at four sites in the US GP.

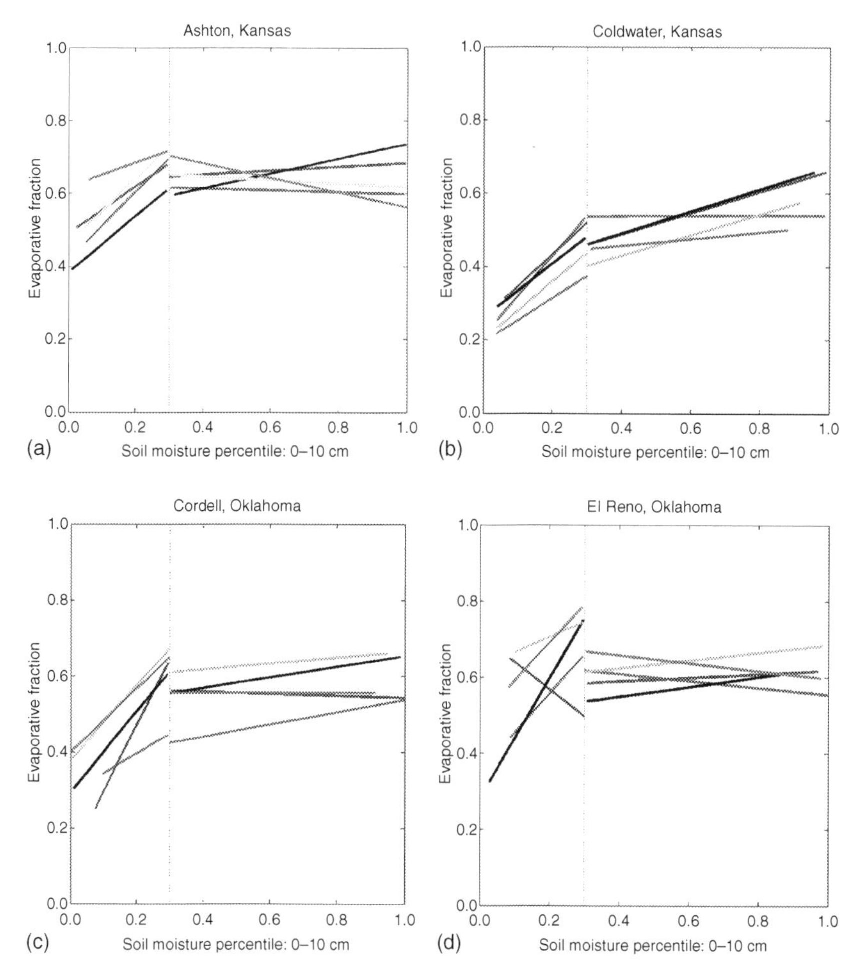

FIGURE 3.5 Observed SM (0–10 cm) and EF from ARM *in situ* measurements in Ashton, Kansas (upper left), Coldwater, Kansas (upper right), Cordell, Oklahoma (lower left), and El Reno, Oklahoma (lower right). The lines represent the linear fit between daily SM and EF. The fit is assessed separately for days when SM is less than the 30th percentile, and days when SM is greater than the 30th percentile, separated by the vertical black line. Fit lines are color coded by year. Each year contains 184 days, from which the two regimes are fitted. *(Figure from Ford et al. (2014).)*

Despite confirming the dual regime behavior, Ford et al. (2014) also found that considerable interannual variability exists in the observed and model-simulated SM–EF relationships. For example, the SM–EF relationship in 2007 and 2008 in El Reno does not fit the dual regime form. A more detailed analysis showed that net radiation has a strong influence on the SM–EF relationship (results not shown). In the observations, EF varies as a function of SM when SM is less than the critical value and net radiation is above normal. Basara and Crawford (2002) found a similar relationship between SM and latent heat during

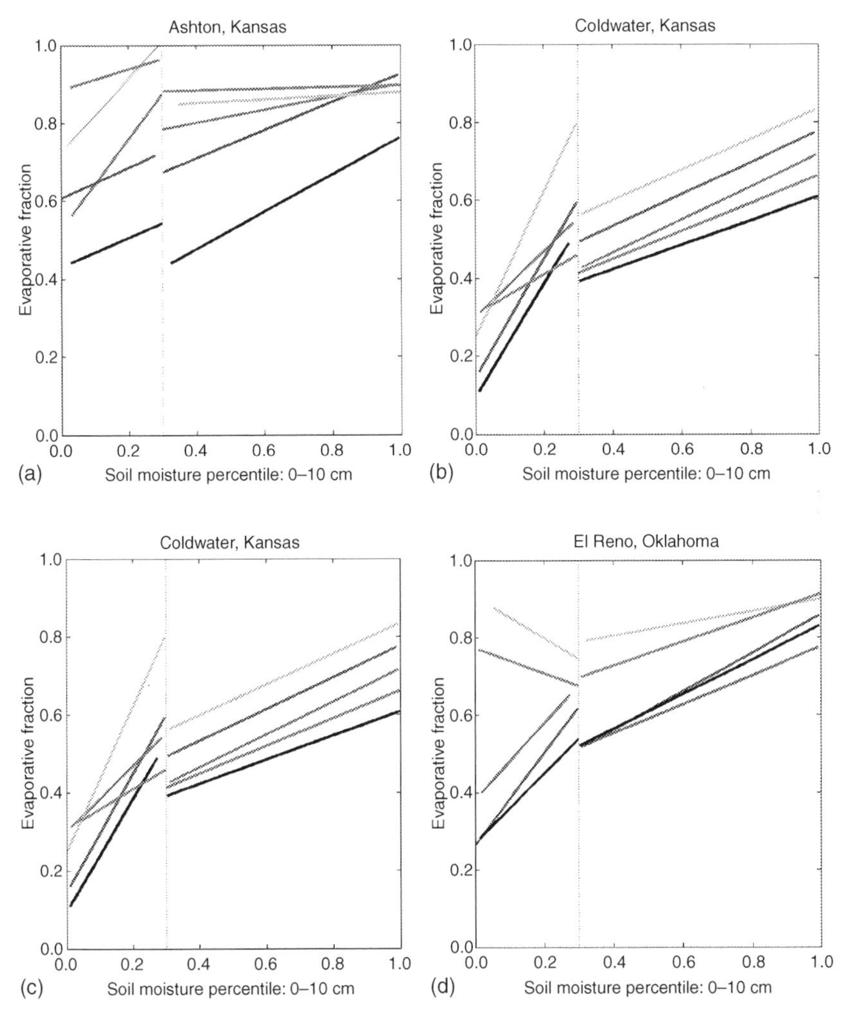

FIGURE 3.6 Model-derived SM (0–10 cm) and EF from the variable infiltration capacity model in Ashton, Kansas (upper left), Coldwater, Kansas (upper right), Cordell, Oklahoma (lower left), and El Reno, Oklahoma (lower right). The lines represent the linear fit between daily SM and EF. The fit is assessed separately for days when SM is less than the 30th percentile, and days when SM is greater than the 30th percentile, separated by the black, dotted line. Fit lines are color coded by year. Each year contains 184 days, from which the two regimes are fitted. *(Figure from Ford et al. (2014).)*

the growing season in Oklahoma. Their results showed that latent heat varied as a strong, linear function of SM during "ideal days", characterized by clear skies, low near-surface wind speed and negligible precipitation during preceding days. In accordance with our findings, EF variability is not only constrained by soil water, but also by net radiation and general atmospheric conditions (cloudiness, presence/absence of precipitation).

The ability of SM to impact near-surface atmospheric conditions depends on SM being the primary driver for partitioning latent and sensible heat (Wei and Dirmeyer, 2012). Koster et al. (2009) suggested that SM can impact near-surface temperatures more strongly in the south-central United States than in other regions because of the tight coupling between SM and EF and sufficient intra-annual SM variability. Koster et al. (2009) demonstrated that because the GP is a semiarid region, it is typically characterized by a moisture-limited evaporative regime, which occasionally becomes energy-limited. The differences in SM–EF interactions between 2006 (moisture-limited) and 2008 (energy-limited) in the results of Ford et al. (2014) corroborate those of Koster et al. (2009). In addition, the findings of Ford et al. (2014) suggest that the SM–EF relationship cannot be determined solely based on SM conditions. SM–EF interactions are a function of both SM and net radiation and both of these variables are necessary to characterize land–atmosphere interactions, particularly in semiarid environments such as the US GP. The impact that SM has on atmospheric temperature, humidity, and subsequent precipitation is dependent on the strength of the relationship between SM and EF (Findell et al., 2011). Therefore, the findings of Ford et al. (2014) are relevant for land–atmosphere interactions in other regions around the world because they have shown that SM is strongly, linearly related to EF under certain conditions; and that these conditions are dependent on both SM and the amount of available energy (net radiation).

3.4.2 Spatial Variations in Soil Moisture–Evapotranspiration Coupling

As discussed earlier, SM has a strong impact on EF (evapotranspiration) only when SM is a limiting factor. This suggests that the coupling between SM and evapotranspiration will be stronger in some regions of the world than others. Previous studies used both models (Budyko et al., 1965; Betts et al., 1996; Koster et al., 2004; Guo et al., 2006; Comer and Best, 2012; Dirmeyer et al., 2013) and observations (Baldocchi et al., 2001; Betts et al., 1996; Teuling et al., 2009; Hartmann et al., 2013) to investigate spatial variations in coupling strength between SM and evapotranspiration.

Because the manner in which evapotranspiration is represented differs among land surface models, the Global Land–Atmosphere Coupling Experiment (GLACE) (Koster et al., 2004; Guo et al., 2006) assembled 12 models to identify where "hot spots" of SM–evapotranspiration coupling are located. Koster et al. (2004) found that hot spots were located in transitional zones between dry and wet climates (e.g., the Southern GP, Sahel and Northern India). Dirmeyer et al. (2012) used the European Centre for Medium-Range Weather Forecasts operational model to investigate changes of SM–evapotranspiration coupling and found an increase in the responsiveness of surface evapotranspiration to SM variations induced by warming climate. This suggests that these "hot spots" of SM–evapotranspiration coupling may shift due to climate change (Seneviratne et al., 2006; Koster et al., 2011; Guo and Dirmeyer, 2013).

Although observational datasets of SM and evapotranspiration are rare due to the lack of *in situ* SM observations (Seneviratne et al., 2010), some studies have attempted to validate model-based results using observations. The FLUXNET project (Baldocchi et al., 2001) provides flux observations, and these observations have been used to confirm the location of regions with strong SM–evapotranspiration coupling in Europe and North America (Teuling et al., 2009).

Finally, both observations and model simulations point out long-term vegetation dynamics and human-induced land use changes can also influence the SM–evapotranspiration coupling (Zhang et al., 2001; Zaitchik et al., 2006).

3.5 SOIL MOISTURE–TEMPERATURE COUPLING

As documented previously, SM controls the amount of latent and sensible heat flux. Dry soils lead to a larger proportion of sensible heat versus latent heat, which corresponds to a decrease in evapotranspiration (Fig. 3.3). This results in an increase in near-surface air temperature and provides a direct connection between SM and near-surface air temperature. SM coupling is apparent throughout the entire temperature range; however, it is particularly relevant for extreme hot temperatures and heat waves (Diffenbaugh et al., 2007; Della-Marta et al., 2007; Fischer et al., 2007; Hirschi et al., 2011; Mueller and Seneviratne, 2012; Miralles et al., 2012). The onset of some extreme drought events has also been partially attributed to antecedent dry soils in several regions of the world (Fischer et al., 2007; Wang et al., 2011; Hoerling et al., 2013). Drought conditions further dry out soils due to high temperatures and low humidity (high atmospheric demand).

3.5.1 Regions of Strong Soil Moisture–Temperature Coupling

SM–temperature coupling has been shown to be important in many regions of the world, as moisture deficits correspond strongly with extreme temperatures in Europe (Hirschi et al., 2011; Teuling et al., 2010; Stéfanon et al., 2014), China (Meng and Shen, 2014), and the Southern United States (Mueller and Seneviratne, 2012). Figure 3.7, taken from Mueller and Seneviratne (2012), identifies regions of strongest SM–temperature coupling as those with the highest correlation between the number of hot days and preceding precipitation deficits. Figure 3.7a shows the geographic distribution of the most frequent hottest month, and Fig. 3.7b–d shows correlations of hot days with 3-month (6-month, 9-month) precipitation deficits as the standardized precipitation index (SPI). Their study showed that strong coupling between SM and temperature in the hottest extreme conditions are present in the majority of the globe. Both model and observations-based studies of SM–temperature coupling show global patterns of coupling

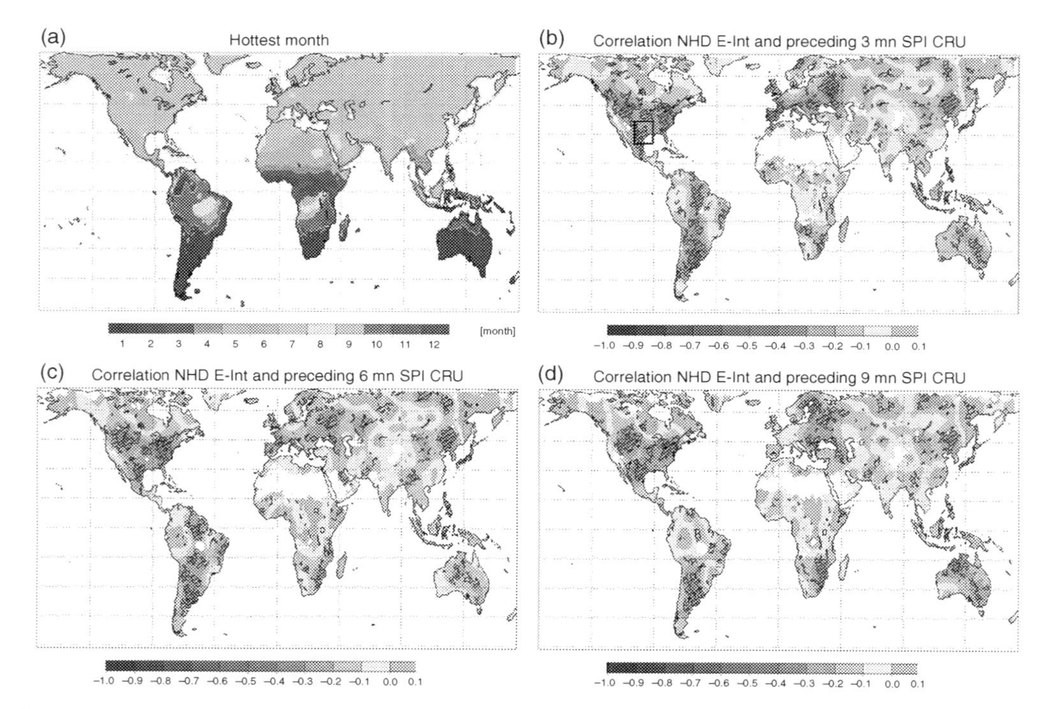

FIGURE 3.7 Relationship between the number of hot days in the hottest month of each year and preceding precipitation deficit based on the SPI. (a) Geographic distribution of the most frequent hottest month. (b) Correlation between number of hot days in the hottest month with the 3-month SPI, (c) 6-month SPI, and (d) 9-month SPI. Maps have been smoothed with a boxcar filter of width 10. Significance levels (hatched regions; 90%) are not smoothed. White areas indicate missing values. Datasets are ERA-Interim for number of hot days and CRU for SPI. *(Figure taken from Mueller and Seneviratne (2012).)*

strength. However, hot spots of SM–temperature coupling in most global climate model simulations are biased toward moisture-limited evapotranspiration regimes (Seneviratne et al., 2006) and therefore underestimate the interactions between SM and temperature in more humid climates.

3.5.2 Soil Moisture and Extreme Heat

Extreme heat has been linked to increased human health risks (Haines et al., 2006; Vanos and Cakmak, 2014), emergency response calls (Dolney and Sheridan, 2006; Schaffer et al., 2012) and heat-related mortality (Semenza et al., 1996; Vanos et al., 2014). Forecasting extreme heat events is crucial for public-health preparation, particularly in urban regions where vulnerability to extremely warm temperatures has consistently increased (Vanos et al., 2012). Studies have found that the frequency of extreme hot days increased during the latter half of the twentieth century (Della-Marta et al., 2007; Diffenbaugh et al., 2007). Therefore, accurate and timely forecasts of extreme warm temperatures are important for informing planning and mitigation activities (Quesada et al., 2012).

Most of the previous studies linking extreme temperatures with land surface conditions have typically used modeled or assimilated SM data from climate models (Fischer et al., 2007; Jaeger and Seneviratne, 2011), and satellite products (Miralles et al., 2012), or a proxy for SM such as the SPI (Hirschi et al., 2011). Although SPI has been shown to correspond with SM on timescales less than 3 months (Sims et al., 2002), it is better to use *in situ* SM observations to assess the SM–extreme temperature coupling.

Ford and Quiring (2014) examined the statistical relationship between monthly extreme temperatures and observed SM in Oklahoma, USA using quantile regression (a methodology previously used in studies using precipitation deficits as proxy for SM (e.g., Hirschi et al., 2011; Mueller and Seneviratne, 2012), but not with *in situ* SM measurements). They also assessed the potential of SM for predicting the probability of extreme heat events using SM from the previous month. Ford and Quiring (2014) used SM and air temperature data from 56 sites in Oklahoma and quantile regression to determine the relationship between SM anomalies and percent hot days. Percent hot days (%HD) are a commonly used heatwave index (Frich et al., 2002; Hirschi et al., 2011), and they represent the percent of the total days in a month during which the maximum temperature is greater than the 90th percentile of the maximum temperature distribution.

The regression results show that SM most strongly impacts extreme heat events at the high end of the conditional %HD distribution (Fig. 3.8). Regression slopes between SM and %HD are steepest in the 95th, suggesting that SM anomalies have the largest impact on the most extreme heat events. These findings are similar to those of Hirschi et al. (2011), Mueller and Seneviratne (2012), and Meng and Shen (2014); however, Ford and Quiring (2014) used *in situ* SM observations instead of precipitation-based proxies or modeled/

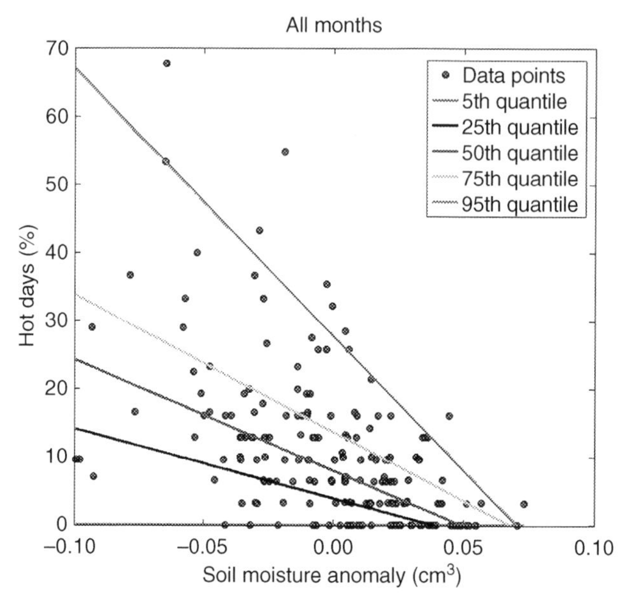

FIGURE 3.8 Monthly percent hot days (%HD) versus SM anomalies in Oklahoma based on March–November (1998–2013). The regression lines represent the fit at the 95th, 75th, 50th, 25th, and 5th quantiles. *(Figure from Ford and Quiring (2014).)*

assimilated SM data. The results of the Ford and Quiring (2014) study highlight the strong, conditional relationship between SM conditions and extreme heat events in Oklahoma.

Ford and Quiring (2014) also assessed the predictability of extreme heat events in Oklahoma using SM from the previous month. Table 3.1 shows the accuracy of the quantile regression model for predicting below normal, normal, or above normal %HD using SM anomalies using the Heidke Skill Score (HSS).

TABLE 3.1 Skill of the Percent Hot Days (%HD) Forecast Based on the Quantile Regression Model[a]

	Below normal	Normal	Above normal	
Sensitivity	0.34	0.63	0.46	
Specificity	0.86	0.56	0.79	
Season	Spring	Summer	Autumn	Overall
Heidke skill score	28.13	15.63	21.88	22.83

[a]*Shown are model sensitivity and specificity by tercile prediction and the Heidke skill scores.*

The HSS is used by the NOAA Climate Prediction Center to validate probabilistic temperature and precipitation forecasts. It compares how often the predicted category (below, normal, and above) matches the observed category, standardized by the number of correct predictions one would expect by chance alone. The HSS values range from −50 to 100, representing the worst and best possible models. Ford and Quiring (2014) found that the quantile regression model had an HSS score of 22.83. This suggests that the model has more skill than a random forecast, which has an HSS of 0. Seasonal HSS scores, also shown in Table 3.1, show some intra-annual variability, ranging from 15.6 (summer) to 28.1 (spring). Given the relatively small sample size ($n=48$ for each season), the difference in HSS between spring and summer equates to just four additional correct forecasts in spring. Although the HSS scores for spring are higher than those for summer, a larger sample size is necessary to conclude that %HD can be more accurately forecasted in the spring than the summer. Overall, model sensitivity for below and above normal predictions is less than 0.5. This is lower than expected and suggests that the probability of a model prediction that the %HD will be near normal is less than 50%. However, the model specificity is much higher. This suggests that the certainty of model predictions that %HD will not be below normal is 86%, and the certainty of model predictions that %HD will not be above normal is 79%. Overall, the quantile regression model using only the SM anomaly in the previous month provides reasonably accurate forecasts of whether %HD in a given month will be below normal, normal, or above normal. Using similar forecasting methodology, the NOAA CPC monthly mean temperature forecasts (http://www.cpc.ncep.noaa.gov/products/predictions/30day/) achieve HSS values between 10 and 20 at a two-week lead time. Although the predictor and predicted variables are different for the CPC forecasts, our %HD forecast skill is comparable.

The strong coupling between SM and temperature suggests that SM conditions can be used to predict the probability of extreme temperatures and heat waves at a one-to-three month lead-time with reasonable skill. SM–temperature coupling in this way is useful for public health and water resource management through extreme temperature forecasts.

3.6 SOIL MOISTURE–PRECIPITATION COUPLING

As shown in previous sections, SM has a significant influence on evapotranspiration and latent and sensible heat exchange. SM in the CZ can also influence precipitation on seasonal scales. Inclusion of SM in seasonal climate predictions has been shown to increase seasonal precipitation predictive accuracy in some cases (McPherson et al., 2004; Meng and Quiring, 2010; Koster et al., 2011; Guo et al., 2012). Meng and Quiring (2010) found that spring SM influenced summer precipitation in the Community Atmosphere Model (CAM 3). SM is thought to influence precipitation and temperature on seasonal scales through soil-moisture memory (Wu and Dickinson, 2004), such that anomalously low

SM reduces the amount of moisture available for precipitation recycling. Under these conditions, locally sourced precipitation is less likely than when SM is normal or wetter than normal (Dirmeyer et al., 2009).

SM–precipitation interactions have been a major avenue of hydroclimatic research for decades. Schubert et al. (2004) investigated the causes of droughts in the US GP using ensembles of long-term GCM forced with observed sea surface temperatures (SST) and found that approximately two-thirds of the low-frequency rainfall variance can be explained by land–atmosphere interactions (e.g., SM) and the remaining variance can be attributed to sea-surface temperature anomalies. In contrast the observational study of Findell and Eltahir (1997) only attributed ~16% of the variance in summer precipitation to spring SM conditions. Some have suggested that this discrepancy is due to the inability of some models to accurately represent land surface processes (Ruiz-Barradas and Nigam, 2006).

Terrestrial precipitation is comprised of water vapor that originates from both oceanic and terrestrial sources. Precipitation that originates from local evapotranspiration is referred to as recycled precipitation. The mean precipitation recycling ratio in the US GP varies from 12% to 21% (Dominguez et al., 2006; Raddatz, 2000; Trenberth, 1999). This variation primarily arises from differences in the methods used to calculate the recycling ratio. In particular, the recycling ratio is strongly controlled by the area. The recycling ratio systematically increases as the area used to calculate it increases because a greater chance occurs of a water molecule evaporated from the surface returning as precipitation within that area (Dominguez et al., 2006). Dominguez et al. (2006) found that the mean precipitation recycling ratio in the GP was 14%, but it varied from 11% to 19% between 1979 and 2000.

The amount of recycled precipitation (blue line) is not constant; it increases during dry summers and decreases during wet summers (Fig. 3.9). Summer moisture availability (red line) is strongly correlated with recycled precipitation in the GP. However, because there is less precipitation during drought years, the precipitation recycling ratio actually increases, demonstrating that local moisture recycling accounts for a greater percentage of rainfall in dry years (not shown). Although remote moisture sources are still, by far, of greater importance, local precipitation recycling can contribute to the intensity and duration of droughts. Hence, understanding local land–atmosphere interactions (e.g., SM) is key for understanding and predicting the occurrence of droughts.

3.6.1 Spatial and Temporal Variations in Soil Moisture–Precipitation Coupling

The strength of the coupling between SM and precipitation varies from place to place and from time to time. Koster et al. (2004) identified regions of strong land–atmosphere coupling using 16 GCMs. They found that locations with strong SM–precipitation coupling (e.g., hot spots) are generally located in transition zones between wet and dry climates such as those occurring in the central

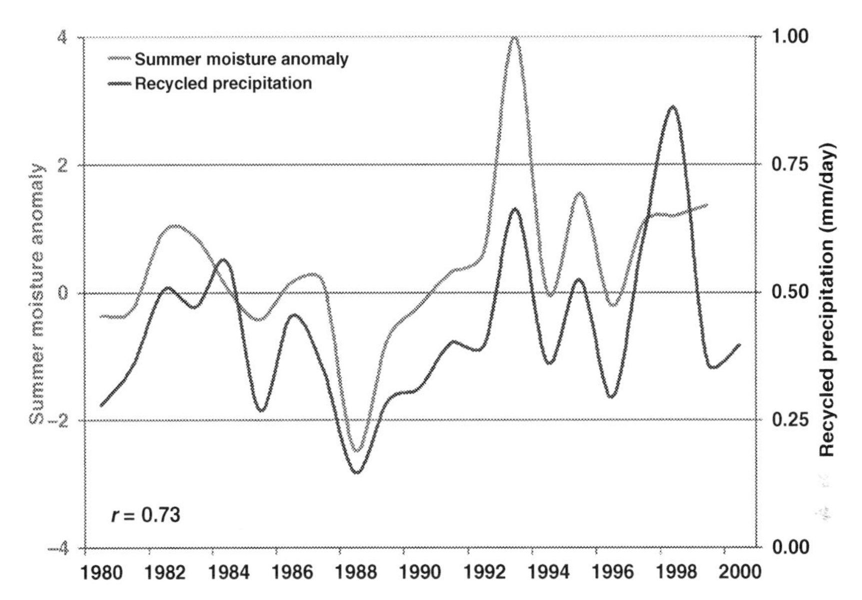

FIGURE 3.9 **Summer moisture anomaly (Palmer's Z-index) (dark gray line [red line in the web version]) and recycled precipitation (light gray line [blue line in the web version]) in the northern US GP (1979–2000).**

GP of North America, the Sahel, equatorial Africa, and India. These are regions where potential evapotranspiration is consistently high, whereas actual evapotranspiration is sensitive to SM availability. Zhang et al. (2008) used precipitation and SM from the global land data assimilation system (GLDAS) to assess the land–atmosphere coupling and also found that hot spots are primarily located in arid to semiarid transition zones. Both of these studies identified the US GP as a region of strong land–atmosphere coupling. However, Meng and Quiring (2010) demonstrated that significant intra-regional variability occurs in the strength of these land–atmosphere interactions within the GP and this variability has also been documented in other studies (Guo et al., 2006; Koster et al., 2006). More research is needed to identify how and why there is intra-regional variability in the strength of land–atmosphere interactions in the GP.

Coupling strength also varies by season and generally is strongest during summer because this is when convective precipitation dominates (Zhu et al., 2005). Using observed SM and rainfall data from Illinois, Findell and Eltahir (1997) found that SM had a significant correlation with subsequent precipitation in the summer and little or no correlation during the rest of year. Kim and Wang (2007) investigated the impact of initial SM anomalies on subsequent precipitation in the coupled land–atmosphere model CAM3 and also found that SM–precipitation feedback is significantly reduced after September due to the decrease in local convective precipitation. Therefore, significant feedback between SM and precipitation in summer is due to the fact that SM anomalies primarily influence convective precipitation, which is only dominant in summer. In addition, the strength of land–atmosphere interactions also varies on an interannual basis as documented by Meng and Quiring (2010). These interannual variations are caused by remote forcings such as tropical Pacific SST. For instance, Hu and Feng (2004) demonstrated that winter precipitation can affect summer rainfall in southwestern United States when the persistence of SST anomalies is weakened. However, this land–atmosphere coupling weakens when North Pacific SST anomalies are persistent. Therefore it is also inappropriate to assume that SM will have a consistent influence on precipitation on a year-to-year basis. Both the spatial and temporal variations in the strength of land–atmosphere coupling pose significant challenges for understanding how SM influences the climate.

The nature of the SM–precipitation feedback also depends on the timing of initial SM anomalies. Pal and Eltahir (2001) found the greatest SM–rainfall sensitivity during June and July in NCAR regional climate model (RegCM). Oglesby (1991) imposed desert-like SM anomalies on May 1st and March 1st in NCAR community climate model and found that the initial SM anomalies on May 1st persisted through summer whereas those imposed on March 1st did not. Kim and Wang (2007) investigated how summer precipitation responded to initial SM anomalies imposed in spring and summer months and found a significant difference in terms of magnitude. Summer precipitation response to dry SM anomalies initialized on April 1st was more significant to wet SM anomalies applied on the same date. This was reversed when SM anomalies were imposed on August 1st.

3.6.2 Positive and Negative Feedback

Many previous studies have found evidence of a positive SM–precipitation feedback in which anomalously wet soils lead to elevated latent heat flux, increased locally sourced moisture into the low-level atmosphere, destabilization of the lower atmosphere, and a higher likelihood of convective initiation (Pielke, 2001; Findell and Eltahir, 2003; Pal and Eltahir, 2001; Koster et al., 2004; Ferguson and Wood, 2011). For example, Fig. 3.10 shows "hot spots" of strong positive SM–precipitation coupling identified by ensemble global climate model simulations. The Southern GP, Sahel region of Africa and Northern India are identified as semiarid climate regions in which SM has a strong impact on the occurrence of subsequent precipitation. Many studies using global climate models and land surface models show strong, positive feedbacks in many regions of the world, meaning that wet soils increase the likelihood of precipitation and dry soils decrease the likelihood of precipitation.

In contrast, other studies have found that anomalously dry soils can impact convective initiation more strongly than wet soils through increased sensible heat flux and a corresponding decreased lower atmospheric stability and faster planetary boundary layer (PBL) growth (Taylor et al., 2007, 2012; Santanello

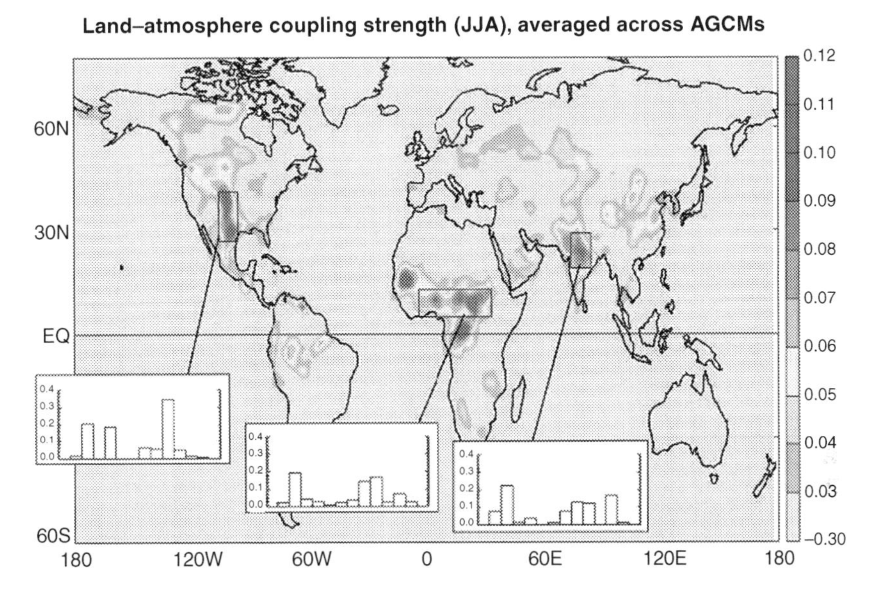

FIGURE 3.10 Depiction of land–atmosphere coupling strength, derived from global climate model ensemble averages. The three regions delineated (GP, Sahel, Northern India) are "hot spots" of strong, positive land–atmosphere feedback. *(From Koster et al. (2004).)*

et al., 2011; Taylor et al., 2012). Taylor and Ellis (2006) found that strong horizontal gradients in observed sensible heat flux, forced by SM gradients, can generate sea breeze-like circulations in the lower atmosphere. These mesoscale circulations can provide the focus for moist convection, which they show generates precipitation preferentially over dry soils adjacent to wet soils (Taylor and Ellis, 2006). Garcia-Carreras et al. (2011) similarly demonstrated that initiation of mesoscale convection occurs preferentially on the warm (dry) side of dry-wet soil boundaries. Figure 3.11, from Taylor et al. (2012), shows a dominant preference for afternoon convective precipitation to fall over dry soils. Their findings suggest that SM–precipitation coupling is strongest when soils are drier than normal (negative feedback). However, Roundy et al. (2013) found that high SM more strongly destabilizes the atmospheric profile in the PBL than low SM. Similarly Myoung and Nielsen-Gammon (2010) found that anomalously high SM values were negatively correlated with convective inhibition in the North American Southern GP.

The lack of agreement in the sign of SM–precipitation feedback suggests that different mechanisms may be at work in different regions (or in the same region at different times). Anomalously dry soils (negative feedback), for example, increase sensible heat flux and surface heating, which destabilizes the atmospheric profile in the PBL and creates an area of local convergence, possibly leading to convection and precipitation over the relatively dry soils. For

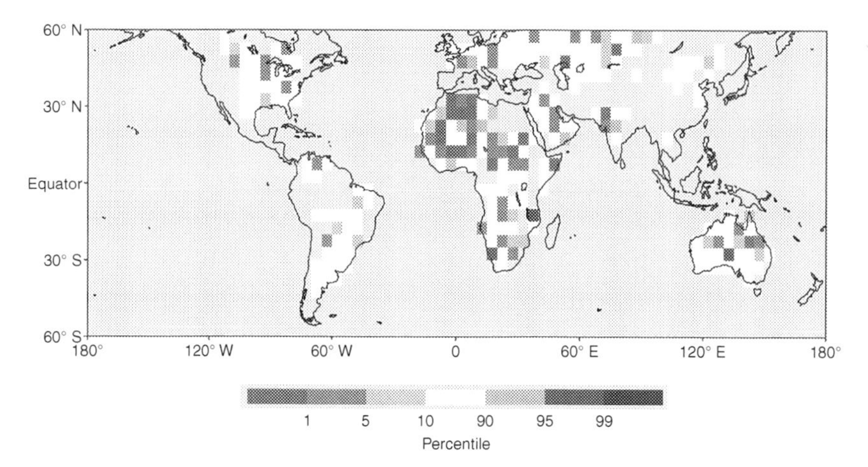

FIGURE 3.11 Global preference for convective precipitation over relatively dry (dark gray [red in the web version]) or relatively wet (light gray [blue in the web version]) soils. SM is taken from AMSR-E satellite retrievals while precipitation is identified with the CMORPH product. *(From Taylor et al. (2012).)*

wet soil (positive) feedback, increased moisture flux into the atmosphere from evapotranspiration, when locally sourced by high SM, can increase the likelihood of convective precipitation.

Ford et al. (2015) used *in situ* SM data to determine if afternoon precipitation occurs preferentially over drier or wetter soils in Oklahoma, USA. They separated event days based on synoptic conditions and the presence of the GP low-level jet (LLJ) to isolate conditions representative of large-scale atmospheric forcing from those without. They examined convective precipitation events during the warm season (May–September), 2003–2012. Figure 3.12, from Ford et al. (2015), shows that afternoon precipitation events occurring during the presence (absence) of the LLJ fell preferentially over drier (wetter) than normal soils. The results do not show unequivocal evidence for strong SM–precipitation coupling in Oklahoma, but instead suggest that convective precipitation is more likely to fall over drier or wetter than normal soils depending on the synoptic conditions and atmospheric dynamics. More specifically, convective events that occur with the LLJ present, tend to occur more frequently over drier than normal soils, whereas those without an LLJ occur over wetter than normal soils. These preferences were statistically significant for all event classes (at 95% confidence level). SM in the morning of convective events had statistically significant correlations with both near-surface atmospheric humidity and temperature (at noon), such that humidity (temperature) increased (decreased) as SM increased.

3.6.3 Confounding Factors

Several factors obscure the nature of the relationship between SM and convective precipitation. One issue is the different mechanisms by which SM can

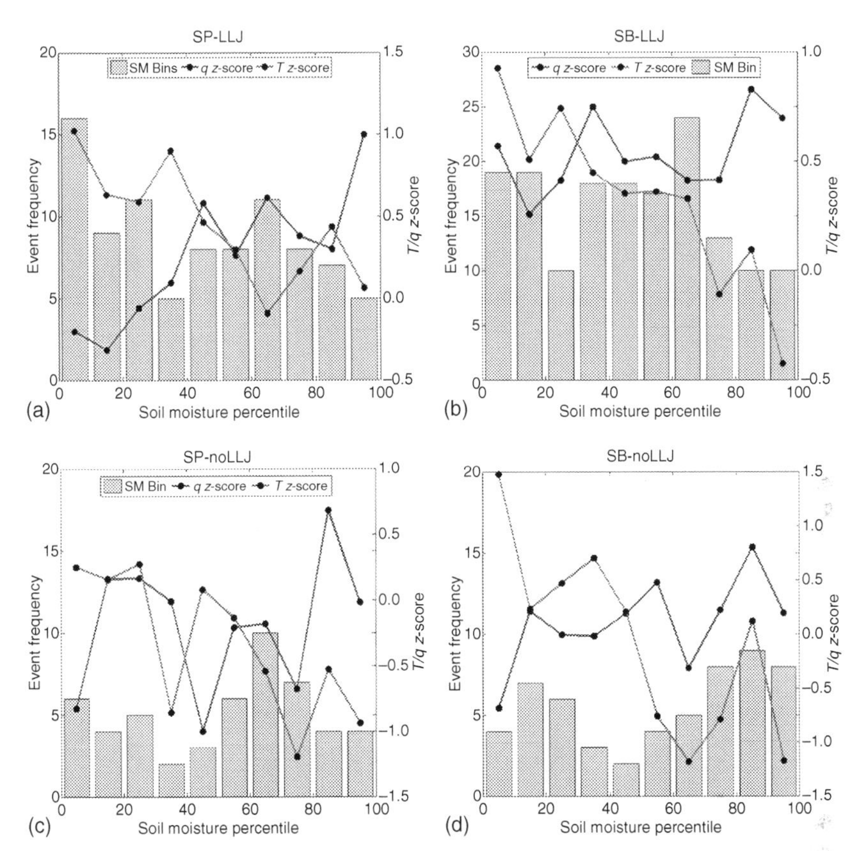

FIGURE 3.12 Histograms show the frequency of convective precipitation events in Oklahoma that occurred during (a) synoptically prime conditions with a LLJ, (b) synoptically benign conditions with a LLJ, (c) synoptically prime conditions with no LLJ, and (d) synoptically benign conditions with no LLJ. SM is binned into 10 groups based on percentiles. Near surface humidity (q, light gray line [blue line in the web version]) and temperature (T, dark gray line [red line in the web version]) z-scores are averaged for each percentile bin and plotted in each panel. *(Source: Ford et al. (2015).)*

influence convection. Taylor et al. (2012) found that convective precipitation fell preferentially over dry soils in the Sahel region of Africa. Their explanation was that anomalously dry soils increase sensible heat flux and surface heating, which destabilizes the atmospheric profile in the PBL and creates an area of local convergence, possibly leading to convection and precipitation over the relatively dry soils. In contrast, Kang and Bryan (2011) found that greater cooling and moistening in the PBL over relatively a moist surface is consistent with earlier convective initiation than over relatively dry surface patches. Brimelow et al. (2011) compared convection-related indices with observations of normalized difference vegetation index, a proxy for evapotranspiration. They found that reduced latent heat flux resulting from (inferred) low SM and stressed vegetation were largely responsible for a deeper, drier boundary layer. In this case, anomalously dry soils were connected to decreased CAPE, increased lifting condensation level, and increased convective inhibition. Increased moisture flux into the atmosphere from evapotranspiration, when locally sourced by high SM, is thought to increase the likelihood of convective precipitation. However, increased surface heating and corresponding local areas of convergence through low SM patches is also thought to increase the likelihood of convective precipitation. Another issue with quantifying SM–precipitation coupling is the indirect connection between the two variables. In reality, SM does not directly force precipitation, but instead moderates latent and sensible heat flux in the near surface atmosphere (Pielke, 2001; Gu et al., 2006; Wei and Dirmeyer, 2012). The corresponding energy and moisture flux in the near-surface atmosphere can influence convective initiation and precipitation (Taylor et al., 2007; Jones and Brunsell, 2009; Matyas and Carleton, 2010).

Despite the overwhelming evidence of SM–precipitation coupling in modeling studies, few investigations have been able to conclusively show SM impact on subsequent precipitation using observations of both SM and precipitation. This is partly due to the lack of spatially and temporally extensive *in situ* SM observations (Robock et al., 2000; Seneviratne et al., 2010; Ford and Quiring, 2014) as well as the autocorrelation of precipitation, which can mask any signal of SM forcing (Wei et al., 2008).

3.6.4 Summary of Soil Moisture–Precipitation Relationships

Past studies have shown evidence of SM feedback to precipitation from wet soils (Brimelow et al., 2011) and dry soils (Westra et al., 2012). Mechanistically, wet soils partition incoming energy into a higher ratio of latent heat over sensible heat, which increases moist static energy in the near-surface atmosphere (Pal and Eltahir, 2001). Higher atmospheric humidity near the surface lowers the lifting condensation level, and increases CAPE (Taylor and Lebel, 1998). Drier than normal soils lead to increases in sensible heat flux and rapid growth of the PBL leading to entrainment of drier, warmer air (Ek and Holtslag, 2004). Both of these mechanisms lead to increased air temperatures and decreased atmospheric humidity near the surface over dry soils. Despite less moisture flux from drier soils to the atmosphere, convection may be favored over dry soils due to rapid PBL growth, particularly if the PBL reaches the lifting condensation level (Santanello et al., 2011). The results of Ford et al. (2015) are consistent with these mechanisms for SM feedback over wet and dry soils; events over wetter than normal soils correspond with increased humidity and stronger CAPE values and events over drier than normal soils coincided with increased air temperature and decreased atmospheric humidity. Interestingly, Ford et al. (2015) found that there is no statistically significant relationship between SM and precipitation if all of the convective events are combined into a single class. This agrees with Taylor et al. (2012) who did not find any statistically significant positive or negative SM–precipitation feedbacks in the US GP. This suggests that in some regions, such as the US GP, both positive and negative SM–precipitation feedbacks can occur, depending on the large-scale atmospheric conditions.

3.7 CONCLUSIONS

As illustrated in this chapter, the CZ is the medium in which many important interactions and exchanges, especially related to water and energy, take place between the land surface and the atmosphere. Therefore, the CZ and the associated interactions between the land and atmosphere play an important role in Earth's climate system. This chapter has demonstrated that the CZ influences the local and regional climate through land–atmosphere interactions. Specifically, SM can be used to account for climate variability on seasonal to interannual timescales. Land surface moisture availability impacts energy partitioning and moisture flux into the atmosphere, which modulate atmospheric temperature and humidity. Studying land–atmosphere interactions, especially those modulated by the CZ, will enhance our ability to understand and predict climate variability at local to global scales. For example, severe drought conditions in previous seasons can impact moisture availability in subsequent seasons.

Under changing climate conditions, the hydrologic cycle is expected to be more vigorous. Therefore, precipitation recycling from the land surface could intensify, making SM–precipitation coupling vital for accurate precipitation forecasts. However, one of the biggest challenges facing the hydroclimate community is the lack of *in situ* measurements of fluxes of energy, water (including SM), and carbon in the CZ. Better monitoring and more comprehensive measurements of the dynamics in the CZ will allow scientists to address uncertainties in global climate models and satellite retrievals and develop more timely and accurate seasonal climate forecasts. An improved understanding of land–atmosphere interactions and the dynamics of the CZ will enhance our ability to understand and mitigate the effects of future climate variability and climate change.

REFERENCES

Alfieri, L., Claps, P., D'Odorico, P., Laio, F., Over, T.M., 2008. An analysis of the soil moisture feedback on convective and stratiform precipitation. J. Hydrometeorol. 9 (2), 280–291.

Amundson, R., 2001. The carbon budget in soils. Annu. Rev. Earth Pl. Sc. 29, 535–562.

Anderson, S.P., von Blanckenburg, F., White, A.F., 2007. Physical and chemical controls on the critical zone. Elements 3, 315–319.

Anderson, R.S., Anderson, S.P., Tucker, G.E., 2013. Rock damage and regolith transport by frost: an example of climate modulation of the geomorphology of the critical zone. Earth Surf. Process. Land. 38, 299–316.

Anderson, L., Roe, G., Anderson, R., 2014. The effects of interannual climate variability on the moraine record. Geology 42 (1), 55–58.

Arrigo, J.A.S., Salvucci, G.D., 2005. Investigation hydrologic scaling: observed effects of heterogeneity and nonlocal processes across hillslope, watershed, and regional scales. Water Resour. Res. 41, W11417.

Bakker, K., 2003. Archipelagos and networks: urbanization and water privatization in the South. Geogr. J. 169 (4), 328–341.

Baldocchi, D., 2001. FLUXNET: a new tool to study the temporal and spatial variability of ecosystem–scale carbon dioxide, water vapor, and energy flux densities. Bull. Am. Meteorol. Soc. 82 (11), 2415–2434.

Basara, J.B., Crawford, K.C., 2002. Linear relationships between root-zone soil moisture and atmospheric processes in the planetary boundary layer. J. Geophys. Res. Atmos. 107 (D15), ACL 10-11-ACL 10-18.

Betts, A.K., Ball, J.H., Beljaars, A.C.M., Miller, M.J., Viterbo, P.A., 1996. The land surface-atmosphere interaction: a review based on observational and global modeling perspectives. J. Geophys. Res. Atmos. 101 (D3), 7209–7225.

Brevik, E.C., 2012. Soils and climate change: gas fluxes and soil processes. Soil Horizons 53 (4), 12–23.

Brimelow, J.C., Hanesiak, J.M., Burrows, W.R., 2011. Impacts of land–atmosphere feedbacks on deep, moist convection on the Canadian Prairies. Earth Interact. 15 (31), 1–29.

Budds, J., 2008. Whose scarcity? The hydrosocial cycle and the changing waterscape of La Ligua River Basin, Chile. In: Goodman, Michael K., Boykoff, Maxwell T., Evered, Kyle T. (Eds.), Contentious Geographies: Environmental Knowledge, Meaning, Scale. Ashgate Studies in Environmental Policy and Practice. Ashgate, Farnham, Surrey, UK, pp. 59–68.

Budds, J., 2009. Contested H_2O: science, policy and politics in water resources management in Chile. Geoforum 40, 418–430.

Budyko, M.I., et al. 1965. Climate and waters. Soviet Geogr. 6 (5–6), 298–334.

Chorover, J., Troch, P.A., Rasmussen, C., Brooks, P., Pelletier, J., Breshears, D.D., Huxman, T., Lohse, K., McIntosh, J., Meixner, T., Papuga, S., Schaap, M., Litvak, M., Perdrial, J., Harpold, A., Durcik, M., 2011. How water, carbon, and energy drive critical zone evolution: the Jemez-Santa Catalina Critical Zone Observatory. Vadose Zone J. 10 (3), 884–899.

Clifford, M.J., Royer, P.D., Cobb, N.S., Breshears, D.D., Ford, P.L., 2013. Precipitation thresholds and drought-induced tree die-off: insights from patterns of *Pinus edulis* mortality along an environmental stress gradient. New Phytol. 200 (2), 413–421.

Comer, R.E., Best, M.J., 2012. Revisiting GLACE: understanding the role of the land surface in land–atmosphere coupling. J. Hydrometeorol. 13 (6), 1704–1718.

Das Gupta, S., Mohanty, B.P., Kohne, J.M., 2006a. Soil hydraulic conductivities and their spatial and temporal variations in a vertisol. Soil Sci. Soc. Am. J. 70, 1872–1881.

Das Gupta, S., Mohanty, B.P., Kohne, J.M., 2006b. Impacts of juniper vegetation and karst geology on subsurface flow processes in the Edwards Plateau, Texas. Vadose Zone J. 5, 1076–1085.

Della-Marta, P.M., Haylock, M.R., Luterbacher, J., Wanner, H., 2007. Doubled length of western European summer heat waves since 1880. J. Geophys. Res. 112 , D15103.

Delworth, T.L., Manabe, S., 1988. The influence of potential evaporation on the variabilities of simulated soil wetness and climate. J. Climate 1, 523–547.

Delworth, T.L., Manabe, S., 1993. Climate variability and land-surface processes. Adv. Water Resour. 16 (1), 3–20.

den Biggelaar, C., Lal, R., Wiebe, K., Breneman, V., 2004a. The global impact of soil erosion on productivity. I. Absolute and relative erosion-induced yield losses. Adv. Agron. 81, 1–48.

den Biggelaar, C., Lal, R., Wiebe, K., Eswaran, H., Breneman, V., Reich, P., 2004b. The global impact of soil erosion on productivity. II. Effects on crop yields and production over time. Adv. Agron. 81, 49–95.

Diffenbaugh, N.S., Pal, J.S., Giorgi, F., Gao, X., 2007. Heat stress intensification in the Mediterranean climate change hotspot. Geophys. Res. Lett. 34, L11706.

Dirmeyer, P.A., Zeng, F.J., Ducharne, A., Morrill, J.C., Koster, R.D., 2000. The sensitivity of surface fluxes to soil water content in three land surface schemes. J. Hydrometeorol. 1 (2), 121–134.

Dirmeyer, P.A., Gao, X., Zhao, M., Guo, Z., Oki, T., Hanasaki, N., 2006. GSWP-2: multimodel analysis and implications for our perception of the land surface. Bull. Am. Meteorol. Soc. 87, 1381–1397.

Dirmeyer, P.A., Schlosser, C.A., Brubaker, K.L., 2009. Precipitation, recycling, and land memory: an integrated analysis. J. Hydrometeorol. 10, 278–288.

Dirmeyer, P., et al. 2012. Simulating the diurnal cycle of rainfall in global climate models: resolution versus parameterization. Clim. Dyn. 39, 399–418.

Dirmeyer, P.A., Jin, Y., Singh, B., Yan, X., 2013. Evolving land–atmosphere interactions over North America from CMIP5 simulations. J. Climate 26 (19), 7313–7327.

Dolney, T.J., Sheridan, S.C., 2006. The relationship between extreme heat and ambulance response calls for the city of Toronto, Ontario, Canada. Environ. Res. 101, 94–103.

Dominguez, F., Kumar, P., Liang, X.Z., Ting, M.F., 2006. Impact of atmospheric moisture storage on precipitation recycling. J. Climate 19 (8), 1513–1530.

Ek, M.B., Holtslag, A.A.M., 2004. Influence of soil moisture on boundary layer cloud development. J. Hydrometeorol. 5, 86–99.

Eltahir, E.A.B., 1998. A soil moisture rainfall feedback mechanism: 1. Theory and observations. Water Resour. Res. 34 (4), 765–776.

Entekhabi, D., Rodriguez-Iturbe, I., Bras, R.L., 1992. Variability in large-scale water balance with land surface–atmosphere interaction. J. Climate 5 (8), 798–813.

Entin, J.K., Robock, A., Vinnikov, K.Y., Hollinger, S.E., Liu, S.X., Namkhai, A., 2000. Temporal and spatial scales of observed soil moisture variations in the extratropics. J. Geophys. Res. 105 (D9), 11865–11877.

Ferguson, C.R., Wood, E.F., 2011. Observed land–atmosphere coupling from satellite remote sensing and reanalysis. J. Hydrometeorol. 12, 1221–1254.

Findell, K.L., Eltahir, E.A.B., 1997. An analysis of the soil moisture-rainfall feedback, based on direct observations from Illinois. Water Res. Res. 33, 725–735.

Findell, K.L., Eltahir, E.A.B., 2003. Atmospheric controls on soil moisture-boundary layer interactions: three-dimensional wind effects. J. Geophys. Res. 108, 8385.

Findell, K.L., Gentine, P., Lintner, B.R., Kerr, C., 2011. Probability of afternoon precipitation in eastern United States and Mexico enhanced by high evaporation. Nat. Geosci. 4 (7), 434–439.

Fischer, E.M., Seneviratne, S.I., Vidale, P.L., Luthi, D., Schar, C., 2007. Soil moisture–atmosphere interactions during the 2003 European summer heat wave. J. Climate 20, 5081–5099.

Ford, T.W., Quiring, S.M., 2014. *In situ* soil moisture coupled with extreme temperatures: a study based on the Oklahoma Mesonet. Geophys. Res. Lett., 41.

Ford, T.W., Wulff, C.O., Quiring, S.M., 2014. Assessment of observed and model-derived soil moisture–evaporative fraction relationships over the United States. South. Great Plains 119 (11), 6279–6291.

Ford, T.W., Rapp, A.D., Quiring, S.M., 2015. Does afternoon precipitation occur preferentially over dry or wet soils in Oklahoma? J. Hydrometeor. 16 (2), 874–888.

Frich, P., Alexander, L.V., Della-Marta, P., Gleason, B., Haylock, M., Tank, A.M.G.K., Peterson, T., 2002. Observed coherent changes in climatic extremes during the second half of the twentieth century. Clim. Res. 19, 193–212.

Garcia-Carreras, L., Parker, D.J., Marsham, J.H., 2011. What is the mechanism for the modification of convective cloud distributions by land surface–induced flows? J. Atmos. Sci. 68, 619–634.

Gu, L., Meyers, T., Pallardy, S.G., Hanson, P.J., Yang, B., Heuer, M., Hosman, K.P., Riggs, J.S., Sluss, D., Wullschleger, S.D., 2006. Direct and indirect effects of atmospheric conditions and soil moisture on surface energy partitioning revealed by a prolonged drought at a temperate forest site. J. Geophys. Res. Atmos. 11, D16102.

Guo, Z.C., Dirmeyer, P.A., Koster, R.D., Bonan, G., Chan, E., Cox, P., Gordon, C.T., Kanae, S., Kowalczyk, E., Lawrence, D., Liu, P., Lu, C.H., Malyshev, S., McAvaney, B., McGregor, J.L., Mitchell, K., Mocko, D., Oki, T., Oleson, K.W., Pitman, A., Sud, Y.C., Taylor, C.M., Verseghy, D., Vasic, R., Xue, Y.K., Yamada, T., 2006. GLACE: the global land-atmosphere coupling experiment. Part II: analysis. J. Hydrometeol. 7 (4), 611–625.

Guo, Z., Dirmeyer, P.A., DelSole, T., Koster, R.D., 2012. Rebound in atmospheric predictability and the role of the land surface. J. Climate 25 (13), 4744–4749.

Guo, Z., Dirmeyer, P.A., 2013. Interannual variability of land–atmosphere coupling strength. J. Hydrometeorol. 14, 1636–1646.

Gurdak, J.J., Hanson, R.T., McMahon, P.B., Bruce, B.W., McCray, J.E., Thyne, G.D., Reedy, R.C., 2007. Climate variability controls on unsaturated water and chemical movement, High Plains Aquifer, USA. Vadose Zone J. 6, 533–547.

Haines, A., Kovats, R.S., Campbell-Lendrum, D., Corvalan, C., 2006. Climate change and human health: impacts, vulnerability and public health. Public Health 120 (7), 585–596.

Hartmann, D. L., 2013. Observations: atmosphere and surface. In: Climate Change 2013: The Physical Science Basis. Contribution of Working Group I to the Fifth Assessment Report of the Intergovernmental Panel on Climate Change, Cambridge University Press Cambridge, United Kingdom and New York, NY, USA.

Hirschi, M., Seneviratne, S.I., Alexandrov, V., Boberg, F., Boroneant, C., Christensen, O.B., Formayer, H., Orlowsky, B., Stepanek, P., 2011. Observational evidence from soil-moisture impact on hot extremes in southeastern Europe. Nat. Geosci. 4, 17–21.

Hoerling, M., Kumar, A., Dole, R., Nielsen-Gammon, J.W., Eischeid, J., Perlwitz, J., Quan, X., Zhang, T., Pegion, P., Chen, M., 2013. Anatomy of an extreme event. J. Climate 26, 2811–2832.

Holtmeier, F., Broll, G., 2005. Sensitivity and response of northern hemisphere altitudinal and polar treelines to environmental change at landscape and local scales. Global Ecol. Biogeogr. 14, 395–410.

Hu, Q., Feng, S., 2004. Why has the land memory changed? J. Clim. 17, 3236–3243.

Ines, A.V.M., Mohanty, B.P., 2008. Parameter conditioning with a noisy Monte Carlo genetic algorithm for estimating effective soil hydraulic properties from space. Water Resour. Res. 44, W08441.

Ines, A.V.M., Mohanty, B.P., 2009. Near-surface soil moisture assimilation to quantify effective soil hydraulic properties using genetic algorithm. 2. With air-borne remote sensing during SGP97 and SMEX02. Water Resour. Res. 45.

Jaeger, E.B., Seneviratne, S.I., 2011. Impact of soil moisture–atmosphere coupling on European climate extremes and trends in a regional climate model. Clim. Dyn. 36, 1919–1939.

Janzen, H.H., Fixen, P.E., Franzluebbers, A.J., Hattey, J., Izaurralde, R.C., Ketterings, Q.M., Lobb, D.A., Schlesinger, W.H., 2011. Global prospects rooted in soil science. Soil Sci. Soc. Am. J. 75, 1–8.

Jones, A.R., Brunsell, N.A., 2009. Energy balance partitioning and net radiation controls on soil moisture–precipitation feedbacks. Earth Interact. 13, 1–25.

Kang, S.-L., Bryan, G.H., 2011. A large-eddy simulation study of moist convection initiation over heterogeneous surface fluxes. Monthly Weather Rev. 139, 2901–2917.

Karl, T.R., 1986. The sensitivity of the Palmer drought severity index and Palmer's Z-index to their calibration coefficients including potential evapotranspiration. J. Clim. Appl. Meteorol. 25, 77–86.
Kim, Y., Wang, G., 2007. Impact of initial soil moisture anomalies on subsequent precipitation over North America in the Coupled Land–Atmosphere Model CAM3–CLM3. J. Hydrometeorol. 8, 513–533.
Koster, R.D., Suarez, M.J., 2001. Soil moisture memory in climate models. J. Hydrometeorol. 2, 558–570.
Koster, R.D., Suarez, M.J., Higgins, R.W., Van den Dool, H.M., 2003. Observational evidence that soil moisture variations affect precipitation. Geophys. Res. Lett. 30 (5), 1241.
Koster, R.D., Dirmeyer, P.A., Guo, Z., Bonan, G.B., Chan, E., Cox, P., Gordon, C.T., Kanae, S., Kowalczyk, E., Lawrence, D., Liu, P., Lu, C.H., Malyshev, S., McAvaney, B., Oleson, K., Pitman, A.J., Sud, Y.C., Taylor, C.M., Verseghy, D., Vasic, R., Xue, Y., Yamada, T., 2004. Regions of strong coupling between soil moisture and precipitation. Science 305, 1138–1140.
Koster, R.D., Guo, Z.C., Dirmeyer, P.A., Bonan, G., Chan, E., Cox, P., Davies, H., Gordon, C.T., Kanae, S., Kowalczyk, E., Lawrence, D., Liu, P., Lu, C.H., Malyshev, S., McAvaney, B., Mitchell, K., Mocko, D., Oki, T., Oleson, K.W., Pitman, A., Sud, Y.C., Taylor, C.M., Verseghy, D., Vasic, R., Xue, Y.K., Yamada, T., 2006. GLACE: The global land–atmosphere coupling experiment. Part I: overview. J. Hydrometeorol. 7 (4), 590–610.
Koster, R.D., Guo, Z., Yang, R., Dirmeyer, P.A., Mitchell, K., Puma, M.J., 2009. On the nature of soil moisture in land surface models. J. Climate 22 (16), 4322–4335.
Koster, R.D., Mahanama, S.P.P., Yamada, T.J., Balsamo, G., Berg, A.A., Boisserie, M., Dirmeyer, P.A., Doblas-Reyes, F.J., Drewitt, G., Gordon, C.T., Guo, Z., Jeong, J.H., Lee, W.S., Li, Z., Luo, L., Malyshev, S., Merryfield, W.J., Seneviratne, S.I., Stanelle, T., van den Hurk, B.J.J.M., Vitart, F., Wood, E.F., 2011. The second phase of the global land–atmosphere coupling experiment: soil moisture contributions to subseasonal forecast skill. J. Hydrometeorol. 12, 805–822.
Lal, R., 2004. Soil carbon sequestration impacts on global climate change and food security. Science 304 (5677), 1623–1627.
Legates, D.R., Mahmood, R., Levia, D.F., DeLiberty, T.L., Quiring, S.M., Houser, C., Nelson, F.E., 2010. Soil moisture: a central and unifying theme in physical geography. Prog. Phys. Geogr. 35 (1), 65–86.
Liu, Y., 2003. Spatial patterns of soil moisture connected to monthly-seasonal precipitation variability in a monsoon region. J. Geophys. Res. Atmos. 108, 8856.
Loescher, H.W., Jacobs, J.M., Wendroth, O., Robinson, D.A., Poulos, G.S., Mcguire, K., Reed, P., Mohanty, B.P., Shanley, J.B., Krajewski, W., 2007. Enhancing water cycle measurements for future hydrologic research. Bull. Am. Meteorol. Soc. 88 (5), 669–676.
Lorenz, R., Jaeger, E.B., Seneviratne, S.I., 2010. Persistence of heat waves and its link to soil moisture memory. Geophys. Res. Lett. 37, L09703.
Mahmood, R., Hubbard, K.G., 2002. Anthropogenic land-use change in the North American tall grass-short grass transition and modification of near-surface hydrologic cycle. Climate Res. 21, 83–90.
Mahmood, T.H., Vivoni, E.R., 2014. Forest ecohydrological response to bimodal precipitation during contrasting winter to summer transitions. Ecohydrology 7 (3), 998–1013.
Mahmood, R., Hubbard, K.G., Carlson, C., 2004. Modification of growing-season surface temperature records in the northern Great Plains due to land-use transformation: verification of modeling results and implication for global climate change. Int. J. Climatol. 24 (3), 311–327.
Mahmood, R., Foster, S.A., Keeling, T., Hubbard, K.G., Carlson, C., Leeper, R., 2006. Impacts of irrigation on 20th century temperature in the northern Great Plains. Global Planet. Change 54, 1–18.

Mahmood, R., Pielke, R.A., Hubbard, K.G., Niyogi, D., Dirmeyer, P.A., McAlpine, C., Carleton, A.M., Hale, R., Gameda, S., Beltran-Przekurat, A., Baker, B., McNider, R., Legates, D.R., Shepherd, M., Du, J., Blanken, P.D., Frauenfeld, O.W., Nair, U.S., Fall, S., 2014. Land cover changes and their biogeophysical effects on climate. Int. J. Climatol. 34 (4), 929–953.

Matyas, C., Carleton, A., 2010. Surface radar-derived convective rainfall associations with Midwest US land surface conditions in summer seasons 1999 and 2000. Theor. Appl. Climatol. 99, 315–330.

McPherson, R.A., 2007. A review of vegetation–atmosphere interactions and their influences on mesoscale phenomena. Prog. Phys. Geogr. 31 (3), 261–285.

McPherson, R.A., Stensrud, D.J., Crawford, K.C., 2004. The impact of Oklahoma's winter wheat belt on the mesoscale environment. Mon. Weather Rev. 132, 405–421.

Meng, L., Quiring, S.M., 2010. Observational relationship of sea surface temperatures and precedent soil moisture with summer precipitation in the U.S. Great Plains. Int. J. Climatol. 30 (6), 884–893.

Meng, L., Shen, Y., 2014. On the relationship of soil moisture and extreme temperatures in East China. Earth Interact. 18, 1–20.

Miralles, D.G., van den Berg, M.J., Teuling, A.J., de Jeu, R.A.M., 2012. Soil moisture–temperature coupling: a multiscale observational analysis. Geophys. Res. Lett. 39, L21707.

Montgomery, D.R., 2007. Soil erosion and agricultural sustainability. Proc. Natl. Acad. Sci. USA 104, 13268–13272.

Mostovoy, G.V., Anantharaj, V.G., 2008. Observed and simulated soil moisture variability over the Lower Mississippi Delta Region. J. Hydrometeorol. 9, 1125–1150.

Mueller, B., Seneviratne, S.I., 2012. Hot days induced by precipitation deficits at the global scale. Proc. Natl. Acad. Sci. 109 (31), 12398–12403.

Myoung, B., Nielsen-Gammon, J.W., 2010. The convective instability pathway to warm season drought in Texas. Part I: the role of convective inhibition and its modulation by soil moisture. J. Clim. 23, 4461–4473.

Namias, J., 1991. Spring and summer 1988 drought over the contiguous United States – causes and prediction. J. Climate 4, 54–65.

Nobre, C.A., Sellers, P.J., Shukla, J., 1991. Amazonian deforestation and regional climate change. J. Climate 4, 957–988.

Oglesby, R.J., 1991. Springtime soil moisture, natural climatic variability, and North American drought as simulated by the NCAR Community Climate Model 1. J. Clim. 4, 890–897.

Ostrom, E., 2007. A general framework for analyzing sustainability of social-ecological systems. Science 325, 419.

Pal, J.S., Eltahir, E.A.B., 2001. Pathways relating soil moisture conditions to future summer rainfall within a model of the land–atmosphere system. J. Climate 14 (6), 1227–1242.

Palm, C., Sanchez, P., Ahamed, S., Awiti, A., 2007. Soils: a contemporary perspective. Annu. Rev. Environ. Resour. 32, 99–129.

Pan, Z., Segal, M., Turner, R., Takle, E., 1995. Model simulation of impacts of transient surface wetness on summer rainfall in the United States during drought and flood years. Mon. Weather Rev. 123, 1575–1581.

Pielke, R.A., 2001. Influence of the spatial distribution of vegetation and soils on the prediction of cumulus convective rainfall. Rev. Geophys. 39, 151–177.

Pielke, R.A., Liston, G.E., Eastman, J.L., Lu, L.X., Coughenour, M., 1999. Seasonal weather prediction as an initial value problem. J. Geophys. Res. 104 (D16), 19463–19479.

Pimentel, D., 2006. Soil erosion: a flood and environmental threat. Environ. Dev. Sustain. 8, 119–137.

Quesada, B., Vautard, R., Yiou, P., Hirschi, M., Seneviratne, S.I., 2012. Asymmetric European summer heat predictability from wet and dry southern winters and springs. Nature 2, 736–741.

Quintanar, A.I., Mahmood, R., Loughrin, J., Lovanh, N.C., 2008. A coupled MM5-NOAH land surface model-based assessment of sensitivity of planetary boundary layer variables to anomalous soil moisture conditions. Phys. Geogr. 29, 54–78.

Quintanar, A.I., Mahmood, R., Motley, M.V., Yan, J., Loughrin, J., Lovanh, N., 2009. Simulation of boundary layer trajectory dispersion sensitivity to soil moisture conditions: MM% and Noah-based investigation. Atmos. Environ. 43, 3774–3785.

Quiring, S.M., 2004. Developing a real-time agricultural drought monitoring system for Delaware. Publ. Climatol. 57 (1), 104.

Raddatz, R.L., 2000. Summer rainfall recycling for an agricultural region of the Canadian prairies. Can. J. Soil Sci. 80, 367–373.

Rasmussen, C., Brantley, S., deB Richter, D., Blum, A., Dixon, J., White, A.F., 2011. Strong climate and tectonic control on plagioclase weathering in granitic terrain. Earth Planet. Sci. Lett. 301, 521–530.

Reenberg, A., 2009. Land system science: handling complex series of natural and socio-economic processes. J. Land Use Sci. 4, 1–4.

Robock, A., Vinnikov, K.Y., Srinivasan, G., Entin, J.K., Hollinger, S.E., Speranskaya, N.A., Liu, S., Namkhai, A., 2000. The Global Soil Moisture Data Bank. Bull. Am. Meteorol. Soc. 81, 1281–1299.

Ruiz-Barradas, A., Nigam, S., 2006. Great plains hydroclimatic variability: the view from North American regional reanalysis. J. Climate 19 (3004-3010), .

Roundy, J.K., Ferguson, C.R., Wood, E.F., 2013. Temporal variability of land–atmosphere coupling and its implications for drought over the Southeast United States. J. Hydrometeorol. 14, 622–635.

Santanello, J.A., Friedl, M.A., Ek, M.B., 2007. Convective planetary boundary layer interactions with the land surface at diurnal time scales: diagnostics and feedbacks. J. Hydrometeorol. 8 (5), 1082–1097.

Santanello, J.A., Peters-Lidard, C.D., Kumar, S.V., 2011. Diagnosing the sensitivity of local land–atmosphere coupling via the soil moisture–boundary layer interaction. J. Hydrometeorol. 12, 766–786.

Sarr, D.A., Hibbs, D.E., Huston, M.A., 2005. A hierarchical perspective of plant diversity. Quarter. Rev. Biol. 80, 187–212.

Scherler, D., Bookhagen, B., Strecker, M.R., 2011. Spatially variable response of Himalayan glaciers to climate change affected by debris cover. Nat. Geosci. 4, 156–159.

Schaffer, A., Muscatello, D., Broome, R., Corbett, S., Smith, W., 2012. Emergency department visits, ambulance calls, and mortality associated with an exceptional heat wave in Sydney, Australia, 2011: a time-series analysis. Environ. Health 11, 1–8.

Schubert, S.D., Suarez, M.J., Pegion, P.J., Koster, R.D., Bacmeister, J.T., 2004. Causes of long-term drought in the U.S. Great Plains. J. Clim. 17, 485–503.

Seneviratne, S.I., Luthi, D., Litschi, M., Schar, C., 2006. Land–atmosphere coupling and climate change in Europe. Nature 443, 205–209.

Seneviratne, S.I., Corti, T., Davin, E.L., Hirschi, M., Jaeger, E.B., Lehner, I., Orlowsky, B., Teuling, A.J., 2010. Investigating soil moisture–climate interactions in a changing climate: a review. Earth Sci. Rev. 99 (3–4), 125–161.

Seneviratne, S.I., Koster, R.D., 2011. A revised framework for analyzing soil moisture memory in climate data: derivation and interpretation. J. Hydrometeorol. 13, 404–412.

Semenza, J.C., Rubin, C.H., Falter, K.H., Selanikio, J.D., Flanders, W.D., Howe, H.L., Wilhelm, J.L., 1996. Heat-related deaths during the July 1995 heat wave in Chicago. New Engl. J. Med. 335, 84–90.

Sims, A.P., Niyogi, D.D.S., Raman, S., 2002. Adopting drought indices for estimating soil moisture: a North Carolina case study. Geophys. Res. Lett. 29 (8), 1183.

Stéfanon, M., Drobinski, P., D'Andrea, F., Lebeaupin-Brossier, C., Bastin, S., 2014. Soil moisture–temperature feedbacks at meso-scale during summer heat waves over Western Europe. Clim. Dyn. 42, 1309–1324.

Swyngedouw, E., 2004. Social Power and the Urbanization of Water: Flows of Power. Oxford University Press, USA.

Swyngedouw, E., 2009. The political economy and political ecology of the hydro-social cycle. J. Contem. Water Res. Educ. 142, 56–60.

Taylor, C.M., Ellis, R.J., 2006. Satellite detection of soil moisture impacts on convection at the mesoscale. Geophys. Res. Lett. 33, L03404.

Taylor, C.M., Parker, D.J., Harris, P.P., 2007. An observational case study of mesoscale atmospheric circulations induced by soil moisture. Geophys. Res. Lett. 34, L15801.

Taylor, C.M., Lebel, T., 1998. Observational evidence of persistent convective-scale rainfall patterns. Monthly Weather Rev. 126, 1597–1607.

Taylor, C.M., de Jeu, R.A.M., Guichard, F., Harris, P.P., Dorigo, W.A., 2012. Afternoon rain more likely over drier soils. Nature 489, 423–426.

Teuling, A.J., et al. 2009. A regional perspective on trends in continental evaporation. Geophys. Res. Lett. 36 (2), L02404.

Teuling, A.J., et al. 2010. Contrasting response of European forest and grassland energy exchange to heatwaves. Nat. Geosci. 3, 722–727.

Trenberth, K.E., 1999. Atmospheric moisture recycling: role of advection and local evaporation. J. Climate 12, 1368–1381.

Troch, P.A., Carrillo, G., Sivapalan, M., Wagener, T., Sawicz, K., 2013. Climate-vegetation-soil interactions and long-term hydrologic partitioning: signatures of catchment co-evolution. Hydrol. Earth Syst. Sci. 17, 2209–2217.

Van Oost, K., Verstraeten, G., Doetterl, S., Notebaert, B., Wiaux, F., Broothaerts, N., Six, J., 2012. Legacy of human-induced C erosion and burial on soil-atmosphere C exchange. Proc. Natl. Acad. Sci. USA 109 (47), 19492–19497.

Vanos, J.K., Cakmak, S., 2014. Changing air mass frequencies in Canada: potential links and implications for human health. Int. J. Biometeorol. 58, 121–135.

Vanos, J.K., Warland, J.S., Gillespie, T.J., Slater, G.A., Brown, R.D., Kenny, N.A., 2012. Human energy budget modeling in urban parks in Toronto and applications to emergency heat stress preparedness. J. Appl. Meteorol. Climatol. 51 (9), 1639–1653.

Vanos, J.K., Hebbern, C., Cakmak, S., 2014. Risk assessment for cardiovascular and respiratory mortality due to air pollution and synoptic meteorology in 10 Canadian cities. Environ. Pollut. 185, 322–332.

Vinnikov, K.Y., Robock, A., Speranskaya, N.A., Schlosser, A., 1996. Scales of temporal and spatial variability of midlatitude soil moisture. J. Geophys. Res. 101 (D3), 7163–7174.

Vuille, M., Francou, B., Wagnon, P., Juen, I., Kaser, G., Mark, B.G., Bradley, R.S., 2008. Climate change and tropical Andean glaciers: past, present and future. Earth Sci. Rev. 89, 79–96.

Wang, W., Kumar, A., 1998. A GCM assessment of atmospheric seasonal predictability associated with soil moisture anomalies over North America. J. Geophys. Res. 103, 28637–28646.

Wang, A., Lettenmaier, D.P., Sheffield, J., 2011. Soil moisture drought in China, 1950–2006. J. Climate 24, 3257–3271.

Wei, J., Dickinson, R.E., Chen, H., 2008. A negative soil moisture–precipitation relationship and its causes. J. Hydrometeorol. 9, 1364–1376.
Wei, J., Dirmeyer, P.A., 2012. Dissecting soil moisture–precipitation coupling. Geophys. Res. Lett. 39 (19), L19711.
Westra, D., Steeneveld, G.J., Holtslag, A.A.M., 2012. Some observational evidence for dry soils supporting enhanced relative humidity at the convective boundary layer top. J. Hydrometeorol. 13, 1347–1358.
Wu, W., Dickinson, R.E., 2004. Time scales of layered soil moisture memory in the context of land–atmosphere interaction. J. Clim. 17, 2752–2764.
Zaitchik, B.F., Macalady, A.K., Bonneau, L.R., Smith, R.B., 2006. Europe's 2003 heat wave: a satellite view of impacts and land–atmosphere feedbacks. Int. J. Climatol. 26 (6), 743–769.
Zhang, L., Dawes, W.R., Walker, G.R., 2001. Response of mean annual evapotranspiration to vegetation changes at catchment scale. Water Res. Res. 37, 701–708.
Zhang, J., Wang, W.-C., Wei, J., 2008. Assessing land-atmosphere coupling using soil moisture from the Global Land Data Assimilation System and observational precipitation. J. Geophys. Res. Atmos. 113, D17119.
Zhu, C.M., Lettenmaier, D.P., Cavazos, T., 2005. Role of antecedent land surface conditions on North American monsoon rainfall variability. J. Climate 18 (16), 3104–3121.

Chapter 4

Regolith and Weathering (Rock Decay) in the Critical Zone

Gregory A. Pope
Department of Earth and Environmental Studies, Montclair State University, Montclair, New Jersey, USA

4.1 INTRODUCTION

Weathering and the Critical Zone have been inextricably linked, as both nested process domains in Earth history, and as much more recent research priorities among environmental scientists. The United States National Research Council's (USNRC) (2001) report defined the Critical Zone as the "heterogeneous, near-surface environment in which complex interactions involving rock, soil, water, air and living organisms regulate the natural habitat and determine availability of life-sustaining resources." This is commonly identified as "the fragile skin of the planet defined from the outer extent of vegetation down to the lower limits of groundwater" (Brantley et al., 2007, p. 307). Shortly following the USNRC report, a collaborating body of Earth scientists initiated what was then called the Weathering Systems Science Consortium (WSSC) (Anderson et al., 2004), intent on studying "Earth's weathering engine" in the context of the Critical Zone. The WSSC evolved into a more-encompassing Critical Zone Exploration Network (CZEN) as the collaboration involved more ecologists, hydrologists, and pedologists less intent on examining the weathering engine. Still CZEN has been and is a significant locus of weathering research. Not to be ignored, however, is the longstanding pursuit of regolith science (Scott and Pain, 2008; Taylor and Eggleton, 2001; Cremeens et al., 1994; Ollier and Pain, 1996), more or less parallel to that of the Critical Zone although not named as such. Ollier and Pain (1996, p. vii) outlined in the preface of their book goals for "regolith studies" that are nearly exactly those espoused by CZEN for the same environment. In the effort to draw together disparate research approaches, this chapter on weathering systems will draw on a variety of sources and scientists who study weathering in the Critical Zone/regolith: geochemists, geomorphologists, soil scientists, hydrologists, petrologists, mineralogists, and others. Viles' (2013) discussion of weathering system synergies elucidates the environmental as well as disciplinary entanglements, pointing out disparities in questions

Developments in Earth Surface Processes, Vol. 19. http://dx.doi.org/10.1016/B978-0-444-63369-9.00004-5

asked and methodologies used, despite the convergence on a congruent topic of investigation.

Yatsu's (1988, p. 2) definition for "*weathering*" remains the simplest and most flexible: "Weathering is the alteration of rock or minerals *in situ*, at or near the surface of the earth and under the conditions which prevail there." Indeed, rocks and minerals do thermodynamically adjust within the active Critical Zone envelope. Yet, the term "weathering" implies a predominance of atmospheric control over processes, when in actuality the inherent properties of the rock and mineral may be foremost (Hall et al., 2012). Neither is the atmosphere itself the sole nor necessarily the dominant weathering agent. Hall et al. (2012) forwarded a replacement term, "*rock decay*." The same term was used by early geomorphologists (Merrill, 1906; Chalmers, 1898), and Ollier (1963) referred to chemical weathering as one of several types of "rock decay." Hutton (1788), in his famous paper, never used the term, "weathering," instead used "decay" (although the cause of which he generally ascribed to "sun and atmosphere"). The term, "decay," appears commonly in the stone conservation literature (e.g., Prikryl and Smith, 2007). A growing body of "weathering" scientists is following this example (cf. Dorn et al., 2013a).

Even so, "rock decay" is not quite sufficient to cover the realm of weathering of all earth surface materials. Materials need not be rock to decay. Unconsolidated sediments can decay *in situ*, in temporary surface storage prior to burial and cementation. Direct byproducts of surface weathering may also weather. For instance, neoformed duricrusts precipitated in the soil (Nash and McLaren, 2007) may undergo decay when environmental conditions modify, in which the precipitate is no longer stable in its present form. For illustration, Eren and Hatipoglu-Bagcit (2010) reported karst-like solution features in exposed hardpans of caliche in Turkey. Calcium carbonate precipitated in the soil in dry conditions, but the now-exhumed hardpan was exposed to solution from rainwater. Twidale (1987) explained karst-like sinkholes in laterized quartz and clay in tropical Australia that resulted from Holocene drying as well as biogenic decay agents.

Regolith is the "unconsolidated or secondarily cemented cover that overlies more coherent bedrock... formed by weathering, erosion, transport and/or deposition of older material. Everything from fresh rock to fresh air" (Scott and Pain, 2008, p. 427). Regolith therefore coincides with all of the Critical Zone apart from the subaerial vegetation, and happens to be the natural focus of rock and mineral decay. If the regolith develops from *in situ* bedrock, the term *saprolite* is sometimes used for decayed bedrock that retains the original rock structure and fabric. *Saprock* is the slightly weathered transitional form between saprolite and rock, whereas some authors use the term *saprolith* to cover the combination of saprolite and saprock (Scott and Pain, 2008). Finally, *soil* is the uppermost segment of the regolith generally (but not always) with incorporated decomposed biotic components as well as genetic horizon organization. Soil is at the direct interface of Earth materials at the atmosphere, surface water, and biosphere, and covered in this text by Dixon (Chapter 5). Figure 4.1 diagrams the profile of the regolith, otherwise known as the terrestrial portion of the Critical Zone, whereas Fig. 4.2 illustrates a real-world example, complete with its vegetative cover.

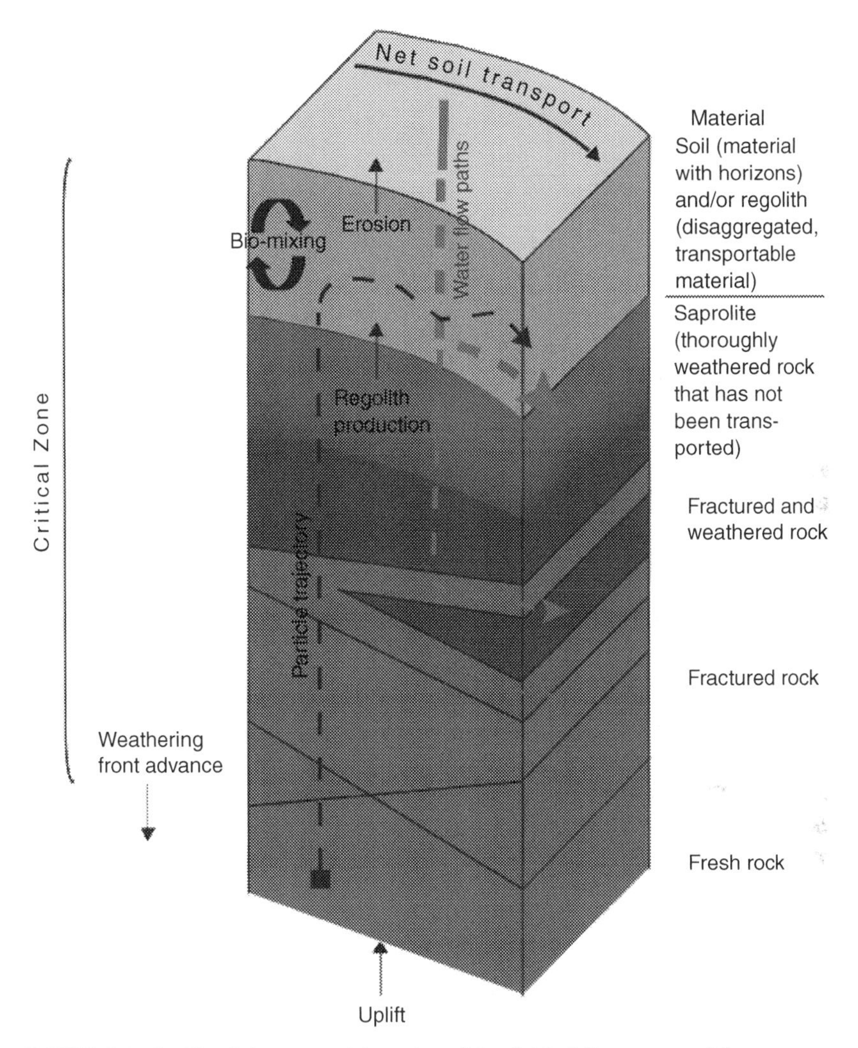

FIGURE 4.1 Profile of the terrestrial portion of the Critical Zone, or regolith, as conceptualized by Anderson et al. (2007). Penetration of the weathering front progresses from surface downward. Rock particles liberated by weathering have a net movement upward, then at the surface mixed and moved laterally and downslope. *(After Anderson et al. (2007).)*

4.2 WEATHERING RELEVANCE TO OTHER CRITICAL ZONE PROCESSES

Although a number of biological, atmospheric, and hydrologic processes operate and interact in the Critical Zone, the interfaces with the lithosphere illustrate the importance of weathering in the pedosphere and regolith, connected within the Critical Zone. The breakdown of rocks and minerals and the mobilization

FIGURE 4.2 **A regolith profile, in decayed schist and gneiss, in a road cut near Evergreen, Colorado, USA.** Relative scale is indicated by the nearby trees. Less resistant rock accounts for the concavities in the profile, while more resistant corestones and quartz veins stand out. Iron oxide staining is seen primarily along fracture zones and around corestones. The irregularity of the decay profile is evident, with highly decayed portions against less decayed portions incongruently, as identified by Phillips (2001b). *(Photo by author.)*

of elements are relevant to pedogenesis, water quality, atmosphere and climate, mineral resources, and life itself (Buss et al., 2013).

4.2.1 Pedogenesis

Soil is the literal foundation for terrestrial life, providing a physical platform, a water–air–nutrient reservoir, and a primary means to recycle both inorganic and organic elements for future use. Decay of organic material is beyond the focus of this chapter, although it is important to consider that almost all soil types have an organic component, and some byproducts of organic decay become rock/mineral decay agents. The processes of pedogenesis are well covered in Birkeland (1999), Schaetzl and Anderson (2005), and Phillips (1993, 2001a), and couched in the factorial controls on pedogenesis first expressed by Jenny (1941) and Simonson (1959). For the breakdown of rock and mineral, soil forming factors are essentially identical to those of weathering (Pope et al., 1995): the nature of the parent materials, influx of new material and efflux of byproducts, the atmospheric conditions over the timespan of the process (including temperature and moisture availability), the presence and type of biotic activity, the availability of abiotic and biotic weathering agents, and time. Pedogenesis in the Critical Zone is discussed in greater detail by Dixon (Chapter 5).

4.2.2 Water Quality

The transfer from atmospheric water to ground- and soil-water, and surface water takes place within the soil and regolith. Although diagenesis and hydrothermal processes below the Critical Zone impart dissolved solutes to groundwater, Critical Zone water also plays a role. With the assistance of water, rock-decay processes add dissolved solutes and suspended particulates deriving originally from rock. The net flow of water and solutes in the Critical Zone depends on the hydraulic gradient, downward under the influence of gravity (flushing components into the groundwater), or upward with strong vegetation uptake or evaporation (in which precipitates such as calcium and salts become part of the soil).

A segment of Critical Zone studies involves rock-decay processes and surface- and groundwaters interacting with polluted environments, such as exposed quarries, mine spoil, urban landfill, and agricultural soils. For instance, the problem of acid mine drainage involves water in contact with iron sulfide minerals in newly exposed rock (as mine spoil or as bare rock). One byproduct is sulfuric acid, itself a potent weathering agent, and a detrimental additive to aquatic life unaccustomed to low pH (Raymond and Oh, 2009; Bond et al., 2000). In the example of urban and industrial contaminants, biogeochemical surface reactions, essentially weathering processes although not necessarily identified as such in this body of research, are a key point in the analysis and remediation of polluted soils (cf. Trindade et al., 2005; Urum et al., 2004).

4.2.3 Supporting Life, Conditioning Ecosystems

Although ongoing research continues to identify a growing number of extremophiles (Rothschild and Mancinelli, 2001), the "normal" terrestrial ecosystem relies on a Critical Zone system that regulates and supports life, including recycling dead organisms, storing and making available nutrients and water, and physically supporting and sheltering organisms. Weathering processes help to provide nutrient elements out of decomposed minerals, and to produce soils that are acceptable to organism regeneration. The soil itself, as a product of weathering, strikes a balance of being porous enough to allow transfers of water, air, and nutrients, but also substantial enough to literally support the weight of vegetation or remain intact when tunneled by burrowing animals.

Members of the ecosystem present a feedback of weathering to the Critical Zone. Biotic agents contribute a significant, if not dominant, share of weathering work to the Critical Zone, involving multiple factors. First, root respiration is responsible for high concentrations of CO_2 in soil air, which in turn combines with water to form carbonic acid (Brook et al., 1983). Second, byproducts of organic decay directly contribute organic acids and chelates, discussed later. Third, penetration of roots and burrowing animals provide pathways for weathering agents and increased exposed surface area. Fourth, among the least understood, are the many symbiotic microorganismal, chemical reactions that exist in the pedosphere (cf. Leake et al., 2008; Chorover et al., 2007). Finally, there are newly discovered impacts of organisms beyond the classifications enumerated earlier. One such phenomenon is the verified enhancement of dissolution of Ca and Mg silicates associated with ant colonies (Dorn, 2014). Through mechanisms not yet understood, although perhaps involving geophagy of olivine and plagioclase, the process combines CO_2 to form pedospheric magnesium- and calcium carbonates, while acting as a terrestrial carbon sink.

4.2.4 Regulating the Environment and Climate

One of the most significant hypotheses put forth in weathering system science has been the role of weathering and the sequestration of carbon, and thus regulation of greenhouse gasses and climate. Originally developed by Raymo (1989), following a supposition by Chamberlin (1899), the hypothesis states that weathering of fresh silicate rocks (high on Reiche's (1950) weathering potential, see Figure 4.3) will draw down atmospheric carbon dioxide. Further, episodes of increased orogeny on a planetary scale (such as the Alpine orogeny) expose more fresh silicate rock, thereby allowing more weathering and more CO_2 drawdown, enough to precipitate global cooling. Indeed, the onset of large-scale orogeny does coincide with global cooling over late Cenozoic time, as well as earlier periods of coupled orogeny and climate change (Raymo and Ruddiman, 1992; Raymo, 1989). Originally beset by difficulties dating and quantifying the degree of weathering, the hypothesis has better support in recent work, by Raymo and her colleagues (cf. Raymo et al., 1997) as well as independent researchers (cf. Dupré et al., 2003; Kump et al., 2000; Liu et al., 2011). Still, some researchers find conflicting evidence. Dixon et al. (2012), working in the xeric climate of California, found evidence to suggest the opposite, tectonic/weathering forcing that increased orogeny beyond a certain threshold may retard chemical weathering. Nevertheless, the premise is strong justification for ongoing research, with age control, quantified rates and degrees of weathering, but over different climatic and tectonic settings.

4.2.5 Providing Natural Mineral Resources

The byproducts of weathering are occasionally useful as natural mineral resources, and their surface presence enhances their economic availability when compared to deeper strip mining, quarrying, or tunnel mining. Materials such as gravel and aggregates, glass sand, clay, and aluminum, magnesium, and iron ores may occur at the surface as end-products of weathering (Ollier and Pain,

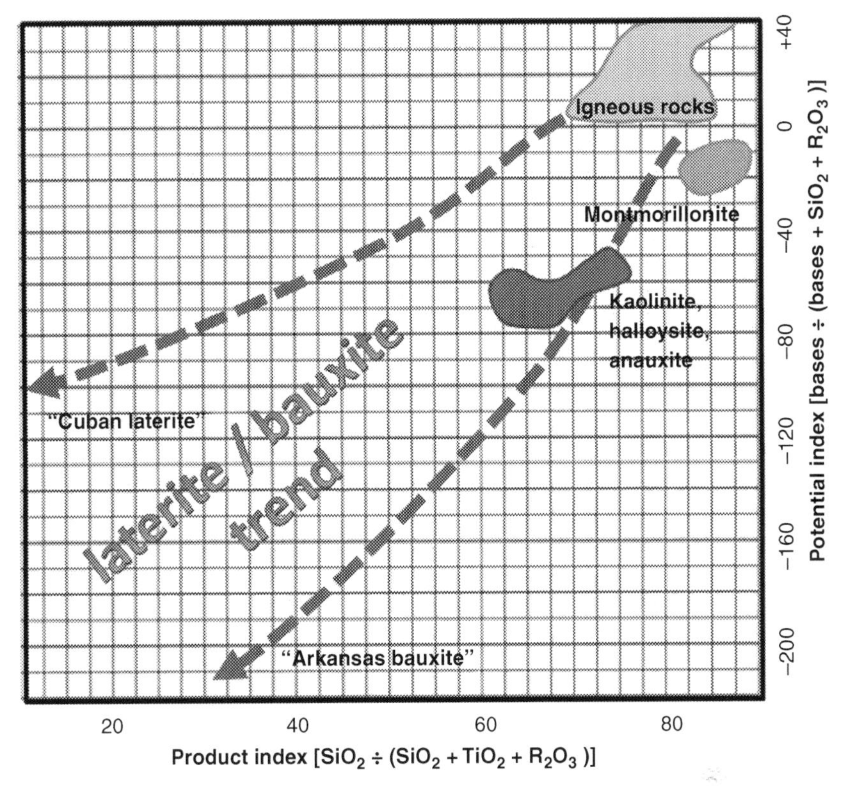

FIGURE 4.3 Reiche's (1950) weathering product versus weathering potential indexes, showing the trend from igneous rocks to end-stage bauxite and laterite. *(Reproduced from Pope (2013b).)*

1996; Taylor and Eggleton, 2001). Likewise, precious and strategic metals such as nickel, gold, silver, and copper tend to concentrate in the upper reaches of well-weathered regolith, or may occur in placer deposits downstream of regolith source areas (Ollier and Pain, 1996; Taylor and Eggleton, 2001). Australia has been the focus of elaborate exploration of regolith resources, including geophysical methods and remote sensing to identify regolith structure as well as chemistry (Butt et al., 2008; Pain, 2008). Wilford (2011a) employed airborne gamma-ray spectrometry to assess the presence of potassium, thorium, and uranium, as part of a weathering intensity index, resulting in a continent-scale mapping of weathering intensity for Australia.

Soil is of course a vital natural resource and byproduct of weathering. The roots of the science of pedology are found in initial attempts to understand soil and manage it. Soil is subject to both natural and human impacts. Soil management and impacts are beyond the scope of this chapter, but Richter (2007) provided an overview of the history and future human interfaces with soil resources.

4.3 TYPES OF WEATHERING (ROCK DECAY)

The traditional view of rock weathering in textbooks separates mechanical and chemical weathering processes. The former involves breakdown of the rock or mineral into smaller pieces without changing its chemical composition, whereas the latter involves breakdown of the rock or mineral by means of changes to the chemical compositions (adding or subtracting atoms, ions, or molecules, at the smallest scale). Many authors consider biological weathering to be a separate third type of weathering. Indeed, biological agents are important and in many cases dominant, although ultimately chemical or mechanical in nature. At the nano-scale of the mineral surface (Dorn et al., 2013b), the creation of smaller particles – by mechanical or chemical means – is almost semantic. Depetris et al. (2014) state that mechanical weathering is first to occur as the opening or exposure of surface area to subsequent chemical processes. Opening surface area to chemical alteration is accurate, although it may also be argued that the mechanical flaws in the rock or mineral that represents a locus of failure may well have been initiated by chemical attack. In reality, chemical and mechanical processes of decay, including the biological types, are synergistic, occurring in parallel and often with positive feedback between processes. Blair (1975) summarizes the relationship well: "It is frequently difficult and unnecessary to separate physical weathering from chemical weathering, for the two usually work together." A summary of the processes is provided here.

4.3.1 Normal Stress

Stress exists where pressure or force exceeds the ductile capacity of the rock or mineral. Normal stresses include compressive and tensile forces. The simplest example of compressive stress is that of the mass of a heavier object resting on a rock, which may induce fracture. Tectonic forces of compression also impart fractures, even at the crystal level; these fractures are in turn avenues for weathering by other agents.

A common subject of study in the rock decay/weathering literature is the pressure exerted by substances expanding within confined spaces in the rock or mineral, such as in cracks or voids. Ice is a common example, where the volume of frozen water increases over its liquid state. In confined cracks, pressure exerted by the ice widens the crack; the process may be repeated as more liquid water is allowed in and also freezes. It is argued (Bland and Rolls, 1998), however, that confining pressures may not be as much as anticipated. First, ice is capable of compressing in a ductile fashion. Second, cracks are almost always actually planes of weakness within the rock, capable of spreading the stress laterally instead of against the walls of the crack. Water pressure in advance of ice crystal growth can be more important (Bland and Rolls, 1998). Alternately, thermal stress (given later) may in fact account for mechanical fracture observed in cold (Hall and Hall, 1991) and hot (Viles, 2005) condition.

Salt-crystal growth is similar in principle to ice-crystal growth, although with less controversy. Salts are common in dry regions but also present near ocean coasts, in urban areas, and as long-transport aerosols (Goudie and Viles, 1997; McDowell et al., 1990), so more prevalent than would be expected. A substantial body of research on salt-induced decay derives from stone conservation and cultural heritage studies. Salts enter the rock in water solution, and then precipitate out as growing crystals as the water evaporates, so some duration of drying is required. Repeated wetting and drying is capable of inducing cycles of salt-crystal stress on a rock, more capable of damage to the rock. Not only at the scale of visible fractures or crystal/detrital grain boundaries, salt-crystal stress is also observed at the micro- and nano-scale, rupturing mineral weaknesses and lattices (Pope et al., 1995). Calcite crystal growth from calcium carbonate deposition in dry areas would present similar mechanical stress in soft or decayed rock (Boettinger and Southard, 1990).

The growth of plant roots has long been associated with mechanical weathering (cf. Ollier and Pain, 1996; Yatsu, 1988), and recent studies attribute the importance of both large and fine roots to mechanical weathering in the soil (Gabet and Mudd, 2010; Richter et al., 2007). Many introductory textbooks feature illustrations of boulders pried apart by tree roots, an image that would seem obvious to accept. Bland and Rolls (1998, p. 159) argued that the pressure of plant roots is insufficient to fracture rocks except in very weak rock, given that the radial pressure is only exerted onto two rock planes in a crack. This posits that plant roots simply take advantage of existing fractures initiated by other mechanical processes; the appearance of physical stress is simply coincidental. Root stress may be one of many processes where field-observed phenomena have yet to be rationalized by laboratory or numerical modeling. Plant roots do play a role in disturbing the soil, including wind-thrown root masses that may rip up weakened regolith rock (Gabet and Mudd, 2010; Phillips and Marion, 2006). Researchers appear to agree that microscale root hyphae are capable of mechanical breakup of minerals, attributed to lichens, algae, and fine roots of vascular plants (Richter et al., 2007; Duane, 2006; Chen et al., 2000; Lee and Parsons, 1999; Hall and Otte, 1990). Mechanical processes include both compression stress by root hydration as well as the tensile "ripping up" of attached particles by roots or hyphae.

Degrees of mineral expansion and contraction are possible with wetting and drying. Mineral hydration can expand the crystal dimensions, exerting stress similar to that of salt. The process of hydration is a chemical one, in which hydroxide molecules are incorporated into the mineral, thereby changing its chemical composition. However, the process also resembles a mechanical stress at the molecular scale. Hydroxide ions disrupt the mineral lattice, rendering it weaker in some cases, as in silica hydration (cf. Aomine and Wada, 1962), or expanding and splitting the crystals, as in salt, clay, or mica hydration (Velde and Meunier, 2008; Doehne, 2002). Hydration can take place in any environment where at least hygroscopic water is present. Dehydration works in the opposite fashion, removing hydroxide ions from hydrated minerals, which has the potential to collapse the mineral structure.

Salts, iron minerals, calcite, clays, and even amorphous glass (e.g., in volcanic ash) can undergo hydration and expansion. The reported mechanical stress of "fissuresols" (Villa et al., 1995) probably derives from clay-mineral expansion and calcite crystal growth (Dorn, 2011). Clay hydration is a good example of a chemo-mechanical process: Water bonds to the phyllosilicate crystal lattice, this in turn causes the lattice to expand, exerting physical pressure.

Apart from the above-mentioned root-ripping, tensile stresses are relatively less relevant to rock decay. One important exception is the process of pressure unloading, also sometimes known as sheeting, dilation, and exfoliation.[1] Blackwelder (1925) is credited with the first comprehensive description of pressure unloading. Where large rock masses, formerly buried under tonnes of crust, regolith, or glacial ice, are exposed to the atmosphere, confining pressures are relaxed, and the rock expands. Where it exceeds the plastic expansion limits, the rock fractures parallel to the exposed surface. Perpendicular fractures also develop because rock is incapable of stretching to its new expanded volume. Domed rock structures as well as cliff faces experience unloading stress. Prior to separation and falling, the widening fractures can channel water into the rock, along with biotic and other external materials, contributing to further weathering. The released rock slabs fall and become components of the detrital geomorphic system (Pope, 2013a).

4.3.2 Thermal Shock and Fatigue

The role of thermal shock, of rapid temperature excursions within the rock, has been a longstanding debate in the study of mechanical rock decay. Early researchers such as Blackwelder (1933) and Reiche (1950) dismissed the importance of thermal shock in producing rock fragments. However, with field data indicating the extremes of temperature change that exist on rock faces, Hall and Hall (1991) demonstrated the potential for tensile contraction forces during temperature excursions in dry polar rock surfaces. Viles (2005) presented comparable data from a hot desert situation. In fact, Sumner et al. (2004) found similar ranges of temperature excursions in deserts of South Africa and Antarctica, capable of thermal shock weathering. Repeated thermal excursions could result in eventual material fatigue (Hall, 1999).

A subset of thermal shock could include explosive detachment, caused by rapid expansion due to extreme heat. Numerous researchers note spalling and granular separation after fires (Dorn, 2003; Ollier, 1983a; Blackwelder, 1927), and the explosive expansion of water vapor just below the rock surface is a likely cause. The prevalence of natural fires in a number of ecosystems

1. "Exfoliation" is also used by some authors to describe a thin-layer (~ 1 cm) diminishing of rocks, often with a combination of mechanical and chemical processes. On corestones, exfoliation is sometimes referred to as "spheroidal weathering." The processes are not identical and the terminology is potentially confusing.

(Belcher, 2013) suggests that fire weathering may be more important than previously suspected. Even more extreme than fire, Knight and Grab (2014) make a good argument for the occurrence of lightning weathering, certainly qualifying as explosive, on exposed mountain ridges. They suggest that in some locations, lightning-induced weathering is a better explanation for shattered rock previously attributed to ice weathering processes.

4.3.3 Solution and Dissolution

Solution and dissolution involve the ability of water, often fortified with acids, to remove ions. Solution is the most straightforward process, able to disrupt molecules in one step by creating a solute ion and an aqueous byproduct. Solution of calcium carbonate in the presence of acid precipitation or acidic groundwater produces calcium ions and bicarbonate:

$$CaCO_{3(s)} + H_2O + CO_{2(aq)} \rightarrow Ca^{2+}{}_{(aq)} + 2HCO_3{}^-{}_{(aq)}$$

This process is relevant over a broad portion of the planet where carbonate rocks are present, and responsible in the extreme case for karst landscapes, but also for the common calcium load in surface waters.

Solution of silica, SiO_2, is also possible, although at much slower rates than solution of carbonate rocks. In the presence of water, silicate molecules dissociate into metastable monosilicic acid.

$$SiO_{2(s)} + nH_2O = Si(OH)_{4(aq)}$$

Solution is the process by which quartz, among the most resistant of minerals, is capable of decay. In its most extreme form, silica solution is responsible for silica karst observed in sandstone and quartzite rocks (Wray, 2013). Such landscapes have been otherwise stable enough, in relative tectonic quiescence, to be exposed to the long-term effects of solution. At much smaller scales, silica solution is probably responsible for overgrowths and glazing observed on quartz sand particles (Krinsley and Doornkamp, 2011).

Dissolution is a more complex process, but important to the silicate minerals other than quartz. Acid is involved, such as carbonic acid by atmospheric precipitation or by combination of water with soil, air, organic acids near the soil, and around sulfide deposits (including mine waste). For example, the feldspar mineral albite decays in weak acid to the aluminosilicate clay, kaolinite, releasing also quartz (often as a precipitated silicate cement or amorphous coating) and sodium ions in solution:

$$2NaAlSi_3O_8 + 2H^+ + H_2O = Al_2Si_2O_3(OH)_4 + 4SiO_2 + Na^{2+}$$

Hydrolysis, sometimes also known as incongruent dissolution due to its multistep and parallel decomposition sequence, is also relevant to the decomposition of silicate minerals. It may take place with acids or with water. For aluminosilicate minerals, the process results in end-product clays with ions of potassium, sodium, calcium, and quartz precipitate (dissociating from an initial silicic acid byproduct), as in this example for an alkali feldspar:

$$2(K, Na)AlSi_3O_8 + 11H_2O = Al_2Si_2O_3(OH)_4 + 4H_4SiO_4 + K^+ + Na^+ + 2OH^-$$

Not all silicate decomposition results in the formation of clay, as in this example of olivine, but does produce iron and magnesium ions (which may precipitate out), as well as precipitated quartz (again dissociating from a silicic acid byproduct):

$$(Fe, Mg)_2SiO_4 + 4H_2O = H_4SiO_4 + Fe^{2+} + Mg^{2+} + 4OH^-$$

Dolomite decomposition is an example of hydrolysis outside of the silicate minerals, producing calcium and magnesium ions as well as carbonic acid as a byproduct, which in turn take part in further solution reactions:

$$CaMg(CO_3)_2 + 2H_2O = Ca^{2+} + Mg^{2+} + 2OH^- + 2HCO_3^-$$

4.3.4 Rock and Mineral Decay by Organic Molecules

Organic chemicals that derive from living as well as decomposing dead organisms introduce unique and sometimes particularly intense decay agents into the near-surface regolith. The soil, particularly in the rhizosphere of root–soil interaction, is highly active chemically, including oxidation, solution, hydrolysis, and chelation reactions. Ollier and Pain (1996) pointed out the role of oxidation in plant uptake of iron and other nutrients, and the role of bacteria in silica depletion. Organic acids such as oxalic acid and humic acid can be effective in the decay of silicate minerals, such as this example of anorthite decomposition. (Scott and Pain, 2008, after Trencases, 1992):

$$CaAl_2Si_2O_8 + 8H^+ + [\text{organic ion}] = Ca^{2+} + 2(Al^{3+} + \text{organic anion}) + 4H_2O + 2SiO_2$$

The reaction releases Al as an organic compound, as well as calcium ions and a quartz precipitate.

Chelation is the process by which metals such as aluminum and iron are preferentially extracted by acids and organic molecules (such as ligands) derived from decomposing vegetation. Lichens, as pioneer organisms at the lithosphere/biosphere interface, incorporate chelation and oxalic acid in initial rock decay (Wilson and Jones, 1983; Schatz, 1963). Masiello et al. (2004) attributed chelation processes for the translocation of iron and aluminum with carbon in grassland soils.

4.4 FACTORS RELEVANT TO ROCK DECAY

The degree and rate of rock decay is influenced by three arch-factors: (1) time, (2) availability of rock and mineral decay agents, and (3) efficacy of rock and mineral decay agents, indicated in Table 4.1. Rock decay tends to be a slow process, but decay increases with time up to the point of weathering-agent saturation or

TABLE 4.1 Cascading Factors of Rock and Mineral Decay

<table>
<tr><td colspan="8">← More encompassing -More specific →</td></tr>
<tr><td rowspan="13">Time</td><td rowspan="13">Sources for decay agents</td><td rowspan="6">Environmental (exogenic) agents</td><td rowspan="13">Efficacy of decay agents</td><td rowspan="6">Vulnerability of the rock or mineral</td><td>Chemical composition</td><td>Molecular structure</td><td>Kinetic characteristics</td></tr>
<tr><td rowspan="5">Access to the rock or mineral</td><td rowspan="3">Crystal or grain structure</td><td>Macroporosity</td></tr>
<tr><td>Microporosity</td></tr>
<tr><td>Surface area</td></tr>
<tr><td rowspan="2">Depth</td><td>Surface erosion</td></tr>
<tr><td>Rate of decay front</td></tr>
<tr><td rowspan="7">Crustal (endogenic) agents</td><td rowspan="3">Power or energy of decay agent</td><td colspan="2" rowspan="3">Concentration and frequency</td><td>Recurrence factors</td></tr>
<tr><td>Fatigue factors</td></tr>
<tr><td>Dilution factors</td></tr>
<tr><td rowspan="4"></td><td colspan="2" rowspan="4">Environmental catalysts</td><td>Temperature</td></tr>
<tr><td>Water</td></tr>
<tr><td>Aeration</td></tr>
<tr><td>Biosphere</td></tr>
</table>

The factors listed are encompassing and representative, although not necessarily exhaustive, as additional factors may be found.
Derived from multiple sources, including Viles (2013, 2001), Anderson et al. (2007), Phillips (2005), Pope et al. (1995), Simonson (1959), and Jenny (1941).

end-product stability. With enough time, extensive rock decay can take place, anywhere. Time is therefore the determinant factor for all others. The weathering system also requires available decay agents, and materials to decay (assumed, not indicated in Table 4.1). Exogenic decay agents derive from the environment, whereas endogenic decay agents derive from the crust itself (such as gravity, tectonic stress, geothermal heat). Decay agents are in turn rendered more or less effective by intrinsic properties of the rock or mineral, and by factors in the environment that catalyze or diminish the power of the decay agent. Additional sub-factors become relevant specific to the time and location. The environment is clearly important, but equally so is the nature of the rock and mineral.

Reiche (1950) introduced a concept of "weathering potential," useful to distinguish different materials with respect to eventual thermodynamic equilibrium, therefore a chemical weathering attribute. Materials with a high weathering potential consist of freshly exposed igneous rock, most out of balance with the ambient surface conditions of the Critical Zone. Materials with low weathering potential are near the end-stage of weathering, close to thermodynamically stable at the Critical Zone, such as bauxite and laterite. Reiche illustrated the weathering potential index (leachable bases compared to silica) graphed against a weathering product index (silica loss compared to more stable titanium and aluminum oxide), reproduced in Fig. 4.3.

It should not be surprising that different factors dominate in different settings; no single factor is capable of distinguishing or determining the degree of decay. This applies spatially, across regions and landscapes or at smaller scales such as slope elements, and vertically through the regolith profile. Climatic factors provide one example. Studies by Dixon et al. (2009) and Brady et al. (1999) show the importance of climatic variables such as rainfall, especially when combined with biotic weathering agents, in sampling transects over local to regional scales. However, Pope (2013b), Pope et al. (1995), Viles (2013), and Anderson et al. (2007) provide evidence for factors such as soil and regolith structure, overburden erosion rates, and differences in weathering agents over-ride presumed dominance of enhanced decay due to high temperatures and water availability in the tropics. Likewise, Dixon and Thorn (2005) reported the dominance of chemical processes in subarctic and alpine environments ordinarily presumed to have greatly dampened chemical activity. Decay agents and factors vary with depth as well. Anderson et al. (2007) illustrated the terrestrial Critical Zone profile (the regolith, reproduced in Fig. 4.1) with chemical or mechanical processes dominant nearest bedrock, and chemical processes dominant in the upper saprolite, whereas a mix of mechanical (including transport), chemical, and biological processes impact the soil. Anderson et al. pointed out that this is a generalization, where variation in these trends may be the result of weathering potential and weathering agent efficacy. Further, it is important to remember that weathering processes are, as mentioned earlier, synergistic, and not strictly only chemical or only mechanical. At the depths of the Critical Zone, the transition to the diagenesis-influenced crust, mechanical stresses help to break up the massive rock formed in the crust, yet chemical processes are not absent and may still take advantage of existing petrographic weaknesses, which may in turn fail under physical stress.

Scale is relevant to rock decay factors. Rock decay initiates with mineral decay, the smallest of building blocks within the rock (whether they be interlocking crystals or cemented detrital particles). Mineral decay ultimately rests on the nanoscale separation of molecules, either by chemical or mechanical means. Factors influencing the processes at the nanoscale "boundary layer" (Dorn et al., 2013b) may be quite different from those of ambient atmospheric conditions at the ground surface (Pope et al., 1995). Present "weather" and "climate" of the location may well be poor proxies for the actual nanoscale boundary layer. That said, the nanoscale boundary layer is not divorced from its environmental envelope. Nested environmental realms exert influence (Table 4.1), if not control, on successively smaller envelopes, from planetary and regional to local to site-specific and finally to the mineral surface environments.

Figure 4.4a and b illustrates the weathering/mineral system, including global factors but focusing on the weathering boundary layer. Note that both sides of

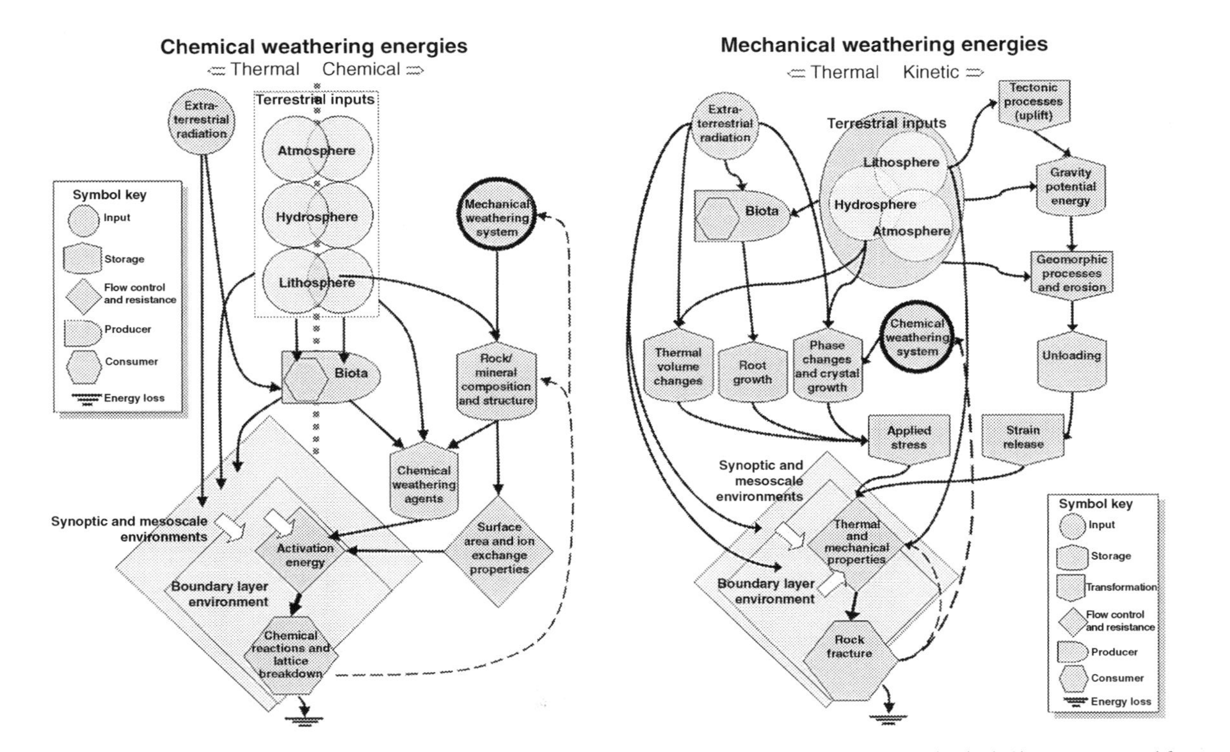

FIGURE 4.4 The weathering (rock decay) system, expressed in terms of cascading energy. Chemical (2a) and mechanical (2b) processes and factors are treated separately, but are connected by a feedback tunnel indicated in each diagram. Symbols follow the standard typology of systems.

the system, the mechanical and the chemical, are linked through a feedback "tunnel" indicated in each diagram. The system as illustrated is conceptual in energy contributed and used for weathering, borrowed in part from the ecosystem energetics of Odum (2000) and the factors illustrated in White et al. (1992). The combined lithosphere, hydrosphere, atmosphere, and biosphere (the terrestrial inputs) provide the chemical weathering agents, whereas the intensity and rates of these agents are modified by both external energy contributions and parent material properties (including mechanical processes). Likewise, mechanical weathering agents are sourced externally, and modified by external (such as geomorphic) and internal (rock/mineral) properties, as well as chemical weathering. As the process evolves, energy feedback (both positive and negative) moves into the system, as well as some net energy loss as the system progresses toward stability.

The concluding message is that it is possible to conceptually model decay agents and factors, but most researchers on the Critical Zone recognize that individual cases and locations are quite varied and often surprising. This is a reality that challenges our abilities to model the weathering system and to reconcile models with field observations (Dorn et al., 2013a).

4.5 ROCK DECAY IN THREE DIMENSIONS: THE "WEATHERING MANTLE"

Rock decay is readily apparent at the surface, with direct exposure to the elements, and at the near-surface, associated with pedogenesis. The Critical Zone is often much deeper (Fig. 4.1), and it is argued (Buss et al., 2013, for instance) that significant chemical decay is capable 10 m to more than 100 m below the surface. Terminology used in the weathering mantle varies; Table 4.2 illustrates the regolith zones and terminology used by several authors, suggesting congruence but some overlap. It should be noted that the terminology and classification is biased toward crystalline bedrock, and in humid-warm/hot environmental conditions (presently or presumed in the past). Figure 4.2 illustrates a typical regolith profile, in this case in weathered metamorphic rock.

In the deepest regolith, the influences of the surface (meteoric water, ambient surface temperatures) transition to those of the crust (geothermal heat, diagenetic processes), although in most cases, the Critical Zone derives minimal direct deep earth influence. Temperature ranges reach near-equilibrium at depths of 1 m or less (Anderson, 1998). The boundary between diagenetic and weathering processes is ill defined, and perhaps semantic. Ollier and Pain (1996, p. 72) rightly point out that decay occurs in both water-saturated (e.g., below the water table) and water-unsaturated zones; this entire thickness is encompassed in the defined "Critical Zone." Precipitated elements that would derive from hydrothermal interactions help distinguish hydrothermal alteration from surface-derived weathering (cf. Thuro and Scholz, 2003; May, 1994). Even so, the results are not straightforward, and sometimes controversial (Ollier, 1983b; Young and Dixon, 1983).

TABLE 4.2 Segments of the Regolith or Weathering Mantle as Described by Different Authors

Authors	Pavich et al. (1989)	Ackworth (1987)	Jones (1985)	Eggleton (2001) and Migoń (2013)	Anderson et al. (2007), see Fig. 4.1
Regions	*Appalachian Piedmont*	*Humid tropics*	*Humid tropics*	*"Idealized regolith"*	*Vertical Critical Zone*
←Increasing depth	Soil	A horizon	Quartz-rich soil	Pedolith	Soil and/or regolith
	Massive subsoil	B oxic horizon	Duricrust		
	Saprolite	Upper saprolite: sandy-clay/clayey-sand	Completely decomposed rock	Saprolite	Saprolite
		Mid-saprolite: massive secondary clays	Decomposed rock, original structure		
		Lower saprolite: altered rock	Weathered rock, some pseudomorphs	Saprock	
	Weathered rock	Transition bedrock, fractured	Slightly weathered rock		Fractured + weathered rock
					Fractured rock
			Weathering front	Weathering front	Weathering front
	Fresh rock	Fresh rock		Bedrock	Fresh rock

Deep regolith profiles are noted in tropical and subtropical locations, indeed one of the Critical Zone Observatories is established in the deeply weathered terrain of Luquillo, Puerto Rico, for the purposes of observing deep decay processes. Although conditions favoring intense and effective decay are present in tropical environments – abundant water, organic compounds, and high temperatures – time and stability are also relevant to developing deep profiles. Deep regolith profiles do occur outside the tropics, and although some argue that these can be relict of past tropical times over the long durations of regolith formation (by climate change or tectonic drift), deep weathering need not be tropical in origin (Pope, 2013b). Blair (1975), for instance, described deep grus in some places more than 50 m thick in Pikes Peak granite of the Colorado (USA) Front Range. In prestressed rock fabric, groundwater was responsible for chemical deterioration of biotite, lending to expansion and disaggregation. Although Chapin and Kelley (1997) presumed deep chemical weathering in the Pikes Peak granite to a warmer, wetter climate on the Eocene-aged erosion surface, Blair (1975) found the profile devoid of clay mineral formation characteristic of feldspar decay seen in the tropics. Bazilevskaya et al. (2013) summarized 27 studies in both temperate and tropical climates and found that although granitic terrain regoliths of similar age were slightly thicker, they were not statistically so, and regoliths of similar age on rocks of basaltic composition were thinner in tropical climates. (These results parallel the relative lack of variation in chemical denudation rates across climatic regions, noted by Saunders and Young (1983), and summarized and discussed in Pope et al. (1995).) Dosseto et al. (2011) showed that soil production rates (regolith in general) in granitic parent materials are "relatively insensitive to climate."

Regolith thickness can be thought of as a balance between regolith production by weathering at the saprolite/rock interface (the weathering front), and loss of material by erosional stripping at the surface. Ahnert (1987) modeled the production of regolith thickness determined by weathering rate and differentiation between chemical and mechanical weathering efficacy (Fig. 4.5), and precedes a similar conclusion by Gabet and Mudd (2009) by more than 20 years. In effect, regolith thickness influences the rate of weathering. The model predicts a "critical thickness" (C_c) at which rates of weathering peak, when there is at least some chemical weathering in the system (e.g., almost always). In systems of purely mechanical weathering (rare), weathering rates decrease continuously as disaggregation relieves the stress on solid rock, and further mechanical weathering is unlikely to have continuing impact. For systems with combined mechanical and chemical weathering (almost all) or only chemical

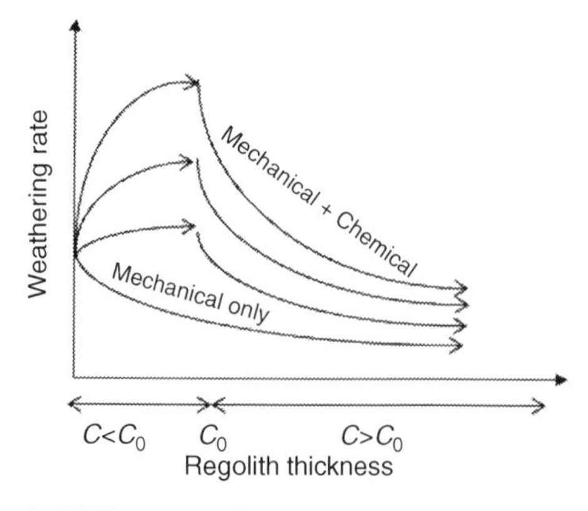

FIGURE 4.5 Ahnert's (1987) concept of regolith thickness as a function of weathering rate and type of weathering.

weathering (also rare), weathering rates increase from initial exposure up to the critical thickness point, owing to increasing porosity in the regolith and ability to maintain water. In a previous study, Ahnert (1976) described a zone of "optimal chemical weathering" somewhat below the surface which decreased as overburden cover increased (also observed in recent studies, such as Brantley and Lebedeva, 2011, and Bazilevskaya et al., 2013). The modeled relationship (see also Fig. 4.5) is as follows, for degree of weathering per unit time, W (an overall weathering rate), from the onset of weathering until the critical thickness has reached ($C < C_0$):

$$\text{for } C < C_c, \quad W = W_c \left(1.0 + k\frac{C}{C_c} - \frac{C^2}{C_c^2} \right)$$

where W_c is the chemical weathering rate, k is a coefficient, C is the regolith thickness, and C_c is the specific critical regolith thickness. Smaller values of regolith thickness C compared to the critical thickness C_c equate to lower unit weathering. Beyond (thicker than) the critical thickness ($C > C_0$), weathering rates decrease with progressively increasing regolith thickness:

$$\text{for } C > C_c, \quad W = W_c \ k \ e^{-k(C - C_c)}$$

This is a negative exponential function, in which the unit-weathering rate decreases with increasing regolith thickness. Beyond the critical thickness, it becomes more likely that weathering agents lose efficacy (for instance, become saturated) and also more likely that the regolith components approach thermodynamic equilibrium.

The regolith thickness model of Ahnert (1987), as well as Phillips' (2014) concept of convergent weathering intensity and weathering rate, could be applied to explain the "unexpected" regolith thickness differences observed by Bazilevskaya et al. (2013). In their study in the Virginia (USA) piedmont, granitic profiles were much thicker than those on diabase, which is the opposite of what was expected, based on mineralogical composition. Although potentially weathering is slower due to higher silica content, the granitic rock was deeply permeated through the loss of biotite, and was able to develop a critical depth thicker than that of nominally faster-weathering diabase. Migoń (2013) summarized a number of factors that are conducive to deep weathering mantles, reproduced in Table 4.3.

With long term decay of the regolith comes mass loss. Although Ahnert's (1987) model confines itself simply to weathering production, the impacts of surface material loss also determine regolith thickness. The concept of "weathering-limited" versus "transport-limited" landscapes comes into play. Transport-limited landscapes are those in which erosion processes are slower than weathering production. Tropical conditions of high moisture and temperature are commonly used to illustrate transport limitation. Weathering-limited is applied to locations where erosion exceeds weathering production. Textbooks cite

TABLE 4.3 Migoń's (2013) Summary of Factors of Enhancement or Limitation to Deep Weathering Mantles

Rock property factors	**Enhanced deep weathering**	**Limited deep weathering**
Mineralogy and rock chemistry	• Rocks with a large proportion of easily weatherable minerals • Rock with less Si, but more Fe, Ca, Na, and Mg	• Quartz-rich rocks • Rocks higher in Si, or rich in feldspars
Fabric	• Weak fabric, susceptible to granular disintegration • High primary or secondary porosity	• Interlocking crystal structure • Low primary porosity
Discontinuities	• High fracture density • Discontinuous, irregular joint network which slows water movement	• Massive rocks with widely spaced fractures • Wide and continuous joints, rapid through-flow
Geomorphic factors	**Enhanced deep weathering**	**Limited deep weathering**
Local relief	Mid- and foot- slope, water more available	Upper slopes and hilltops, less water
Regional relief	Low relief, limited surface erosion, retains weathered mantle	Higher relief, more erosion, lower residence time of mantle
Climatic factors	**Enhanced deep weathering**	**Limited deep weathering**
Precipitation	Higher precipitation, faster weathering	Less precipitation, slower weathering
Temperature	Higher temperature, faster chemical reactions	Lower temperature, slower chemical reactions; ground freezing shuts down deep weathering

deserts and polar areas as lacking sufficient weathering to keep pace with erosion. Pope (2013b) and Pope et al. (1995) pointed out that weathering efficacy is not the sole factor in weathering/erosion limitation. Factors that retard erosion, such as dense vegetation, lower slopes, and higher infiltration rates tend to be associated with many presumed "high weathering" climates. Conversely, slopes lacking vegetation and with higher rates of surface runoff would result in high erosion, regardless of weathering intensity.

Lastly, chemical denudation requires mention in regolith thickness. Weathering removes material by dissolved ions in ground- and soil-water. Volume is not necessarily lost: deeply decayed saprolite can be thought of as a "skeletal" remnant of former rock, with as much as one-third of the mass missing in the form of removed ions and translocated fines (Schaetzl and Anderson, 2005, p. 172). In other cases, mass loss results in a compaction of the surface, and sometimes lowering. Pavich (1989) attributed the undulating geomorphic surface of the Appalachian Piedmont to long-term decay and then settling of the resulting saprolite, as opposed to an old erosion surface (peneplain). This process may account for other deeply decayed and geomorphically old surfaces elsewhere (Phillips, 2002).

4.6 ROCK DECAY IN THE FOURTH DIMENSION: TIME AND RATES IN THE CRITICAL ZONE

Weathering rates have been studied from the scale of mineral lattices to soil profiles to entire landmasses. Although weathering varies by three-dimensional space, across geographic area and into the surface thickness, it also varies over time. Observing weathering over time introduces the fourth dimension, and the ability to classify rates of change done by weathering. Figure 4.6, based on the work of Wilford (2011b), illustrates the idealized evolution of regolith, with the concomitant losses of cations (useful for calculating weathering ratios), as well as changes in properties of the regolith such as permeability and water holding capacity, fertility, and biological activity.

In chemical decay processes, the relationship between reaction rate and temperature is demonstrated by the Arrhenius equation:

$$\ln k = \ln A - \ln\left(\frac{E_{\mathrm{a}}}{RT}\right)$$

where k is the rate constant, A is the molecular collision factor, E_{a} is the activation energy of the substance, R is the gas constant, and T is the temperature. In low temperature geochemical reactions such as rock decay, the nature of the rock composition factors into the activation energy, the efficacy of weathering agents would be reflected in the molecular collision factor, and temperature would become an important variable, the relation of rock decay to climate. Thus, rates would increase when ambient temperatures increase and/or decay agents increase. The equation can be solved for activation energy E_{a} when differing decay outcomes (such as mass loss, ΔM) can be compared at different temperatures (such as between locations, or the same location at different times):

$$E_{\mathrm{a}} = \frac{R\left(\ln(\Delta M_1/\Delta M_2)\right)}{T_1^{-1} - T_2^{-1}}$$

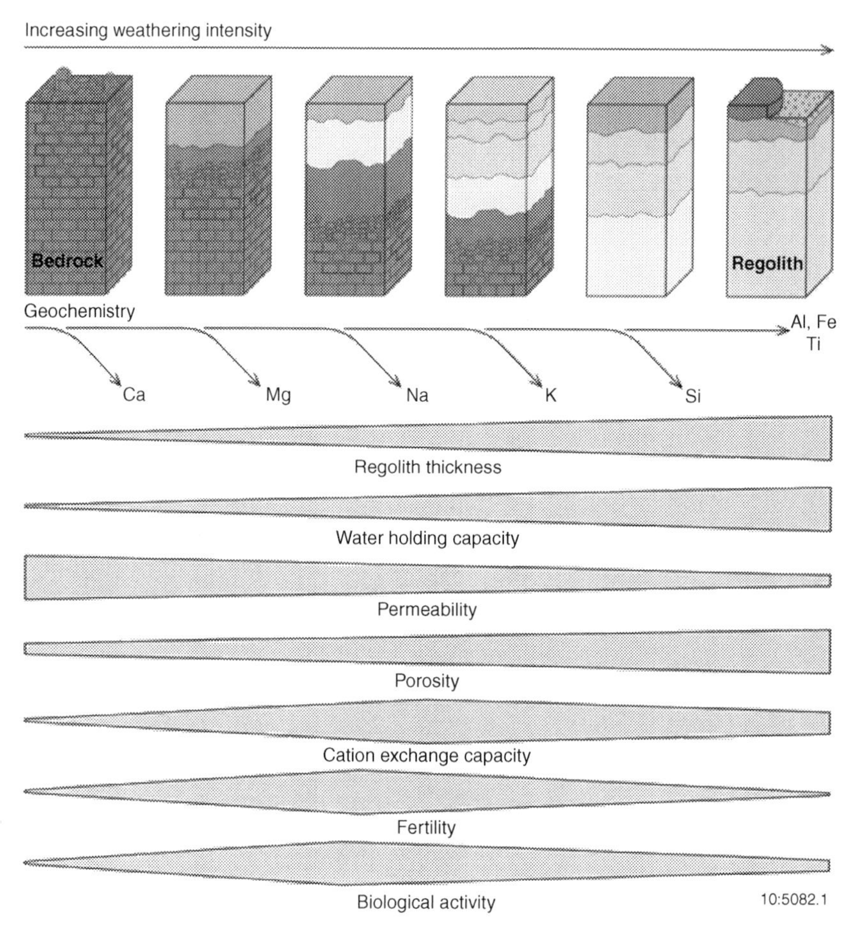

FIGURE 4.6 Evolution of an idealized regolith profile. *(From Wilford (2011b), weathering intensity map of the Australian continent. AusGeo News 101, March 2011.)*

In this case, it can be seen how the activation energy, the minimal energy needed for the reaction to take place, is lower when ambient temperatures are higher.

Contact time with water is important to chemical reactions in the regolith. The Critical Zone encompasses the vadose profile, in which the matrix is water-unsaturated except for specific saturation events or seasons. The amount of water in contact with soil or regolith particles ranges from adsorbed films on particles ("hygroscopic" water) to water filling small pores ("capillary" water) to saturated conditions in which pores are completely full, and varies considerably based on texture/particle size and composition (Schaetzl and Anderson, 2005). Saturated water is influenced by positive gravitational potential, whereas pore and adsorbed water would have negative potential (osmotic or matric, respectively), thus greater attraction to the particle (Brady and Weil, 2008). Adsorbed water is theoretically capable of chemical decay processes on mineral surfaces (Zilberbrand, 1999). However, without re-supply these thin film solutions acquire more negative potential, and would quickly saturate and lose effectiveness. Studies of mineral decay associated with adsorbed or hygroscopic films are mostly limited to the stone conservation literature (Camuffo, 1995; Larsen and Nielsen, 1990), with relatively little attention to natural environments, although the status of nanoscale observation is well developed (MacInnis and Brantley, 1993; Brantley et al., 1993; Dorn et al., 2013b).

In general, the residence time (t_R) of water in the soil and regolith is a function of both volume (V) and flow rate (Q) (Langmuir, 1997):

$$t_R = \frac{V}{Q}$$

Residence time for water in the soil increases with increasing depth in the soil and regolith. Within the soil profile, depending on texture and permeability, residence times of water are on the order of <1–4 months, whereas deeper into the regolith, residence times are measured in years to decades (Bleam, 2012; Matsutani et al., 1993; Langmuir, 1997; Stewart and McDonnell, 1991). Some mineral-water reactions take place nearly instantly, but reactions such as hydrolysis and recrystallization may evolve over many thousands of years (Langmuir, 1997). Thus, longer exposure to water results in greater potential for decay, up to the point of solute saturation in the water. As an illustration of contact time correlation to chemical activity, lake alkalinity has been shown to increase with time as the groundwater residence time increases in the soil and regolith (Wolock et al., 1989). Although chemical reaction rates would theoretically slow as solutions become saturated or decay products reach end-stage equilibrium, it appears that long-resident (25 years) regolith water does not show a rate decrease (Wolock et al., 1989).

At the mineral boundary layer, surface area and solution concentration are relevant to the decay rate. Lasaga (1994) modeled these relationships as follows, with a given mineral θ influenced by a solution i:

$$r = \frac{dc_i}{dt}\Big|_{diss} = \frac{A_\theta}{V}\beta_{i\theta}k_\theta$$

where dc_i/dt is the change in solution concentration as the dissolution reaction progresses, A_θ is the exposed area of the mineral, V is the volume of fluid in contact, $\beta_{i\theta}$ is the stoichiometric constant, and k_θ is the rate constant. Bricker et al. (2005) indicate that difficulties in estimating exposed surface area or "wetted surface area" (which also includes internal microfractures) would be cause for discrepancies in modeled versus field-observed dissolution rates. However, Pope (1995) demonstrated a means to quantify internal weathering, by image processing of backscatter electron micrographs that would be applicable to the model.

Individual minerals have different susceptibility to decay, so would have different rates of decay. A given rock (composed of several minerals) would decay initially as fast as its most susceptible minerals, although its remnants would comprise its least susceptible minerals. Lasaga et al. (1994) calculated the persistence of various minerals subjected to decay:

$$\frac{\mathrm{d}R}{\mathrm{d}t} = -k_i \overline{V}_i$$

where for a specific mineral grain i the change in the mineral grain radius over time is a product of the dissolution rate k and the molar volume $\overline{V}$ of the mineral grain. The results mirror those assumed by Goldich's (1938) sequence of mineral weathering. A single 1-mm crystal of quartz would persist 34 million years at a log rate of $-13.39\,\mathrm{mol/m^2s}$. Rock decay end-product kaolinite would persist six million years at a log decay rate of $-13.28\,\mathrm{mol/m^2s}$. At the other end of the spectrum, a 1-mm crystal of anorthite (calcium feldspar) would last just 112 years at a log decay rate of $-8.55\,\mathrm{mol/m^2s}$, and the pyroxene wollastonite (calcium silicate) persists 79 years at a log decay rate of $-8.00\,\mathrm{mol/m^2s}$. The minerals with low-residence-time and high-decay-susceptibility also corresponds high "weathering potential" (Reiche, 1950). It should be noted that these trends are not absolute. Exceptions are possible if minerals have different susceptibilities to decay agents, as in the case where olivine persists anomalously in microenvironments devoid of organic decay agents (Wasklewicz, 1994). For the most part, mineral persistence explains the prevalence of quartz in the terrestrial surface environment, and provides a means of measuring the maturity of sediments (Folk, 1951) and the relative age of soils by mineral depletion (Birkeland, 1999). Rates of weathering of individual minerals are covered well in earlier edited works such as Colman and Dethier (1986) and White and Brantley (1995).

Calculating the persistence of a rock, composed of many minerals, becomes more complicated, and the persistence of soils and regolith profiles much more complicated. Further generalizations are required. Although most researchers recognize the primacy of the nanoscale boundary layer as the first occurrence of decay, getting from there to rock, pedon, and regolith scale is not straightforward, with a "gray box" understanding of up-scale connections and factors. All the while, the holistic nature of combined processes and combined factors needs to be recognized. Decay outcomes of soil and regolith production (matched against negative feedbacks of erosion), and contribution to runoff solute loads, provide a sense of the combined processes. Recent studies calculated generalized regolith production rates from an assemblage of individual mineral or rock weathering rates. Ferrier et al. (2010) attempted this method at Rio Icacos, Puerto Rico, arriving at rates of <200 to >3000 moles/hectare/year. Cation and silica depletion across an assemblage of rock types was used by Hartmann and Moosdorf (2011) to estimate chemical flux across multiple watersheds in Japan, potentially extrapolated to global scales.

Rates of soil and regolith formation and residence time have been a perennial topic of research, but with limited ability to pin down definitively, until recently. Researchers relied on relative and semiquantitative dating techniques (such as soil development indexes, weathering rinds on cobbles, depletion of certain minerals, accumulation of caliche, cf. Chapter 8 in Birkeland, 1999). Paleosols and buried soils could sometimes be dated with radiocarbon, luminescence, or isotope decay in volcanic deposits on bracketing strata. However, the understanding of cosmogenic isotope pathways in the Critical Zone (Graly et al., 2010 for a recent study) led to revolutionary advances in discerning soil and regolith evolution. Pavich et al. pioneered the use of Be-10 to estimate ages of soil and regolith profiles, first in the Virginia piedmont, and later elsewhere in the Appalachian Mountains (Markewich et al., 1994; Pavich, 1989; Pavich et al., 1989, 1985), arriving at a regolith residence time approximating one million years. The methods have since been applied to many locales (cf. Heimsath et al., 1997; Riebe et al., 2003; Hewawasa et al., 2013). The residence time of a particular regolith particle increases as it progresses upward to the surface (see Fig. 4.1). Dosseto et al. (2011), using uranium-series isotopes, found soil residence times less than 10 ka at the lowest reaches of a regolith profile (~16 m), but exceeding 80 ka at the surface.

In the simplest sense, the advance of the rock decay front into fresh bedrock over time can be conceptualized as a crude linear rate, for instance, × meters per million years, presuming a profile of measurable depth and an exposure surface age. Hewawasa et al. (2013) conclude a weathering front advance rate of 6–14 mm/ka in slowly weathering crystalline rock in Sri Lanka, based on a regolith base ~10-m deep and ages constrained by cosmogenic nuclide accumulation in the soil as well as river dissolved loads. Rates in reality are most certainly not linear, instead are perhaps more nonequilibrium to a point, but most likely complex and chaotic (Phillips, 2001b), not unlike other geomorphic systems (Phillips, 2005, 2014).

4.7 CONCLUSIONS

Rock decay processes are at the center of the Critical Zone, the weathering engine that modifies the Earth's crust to adjust to surface atmospheric, hydrologic, and biotic conditions. It is for this reason that weathering research has reached tremendous output, not only to explain surface processes, but also to contribute toward better understanding of surface and near-surface hydrology, ecology, climate regulation, and applications in natural resources.

Weathering is a synergistic mechano-chemical process; individual processes such as stress release, pressure, and water and acid reactions provide positive feedback to each other, and in general work together. Weathering processes focus on a nanoscale boundary layer of mineral destruction, although nested environments, at increasing scales, are the sources of weathering agents and process modifying factors. But, although weathering takes place at a point location, the integration of weathering throughout the thickness of the regolith provides the key benefit of the Critical Zone concept: a holistic, three-dimensional porous body, leaky to the interior of the surface crust by groundwater, and to the subaerial exterior via organisms and the atmosphere. The formation of this regolith zone is found to be thousands if not millions of years of development. The rates of formation, and of depletion, provide a critical focus of ongoing studies across geographic areas.

4.7.1 A Note on Critical Zone Interdisciplinarity and Weathering System Science

The Weathering System Science Consortium, one of the first scientific collaborations inspired by the NRC's Critical Zone charge, is an ideal concept of converging ideas from different disciplines – geomorphology, hydrology, pedology, mineralogy, petrology, geochemistry – tackling a common research goal. It has been successful enough that the consortium broadened its name, and the CZEN now spreads into other sciences such as ecology and geohydrology. Yet, the Critical Zone movement has not reached all in the broad scope of weathering system science, nor has CZEN been able to recognize parallel (and in some cases long-term) efforts toward the same goals. Case in point, there are still "schools" of research that continue to talk past each other without much cross fertilization. This lack of cross-fertilization is best recognized in the keystone texts and seminal research papers. Powerful sources that they are, the works of Drever (2005) and Anderson and Anderson (2010), of Taylor and Eggleton (2001) and Scott and Pain (2008), and of Turkington et al. (2005), and of the contemporary researchers contributing or cited therein, seldom cite each other, although often working in the same types of environments. Partially this derives from approaching from different disciplinary cultures, attending only a comfortable availability of conferences and symposia, and some from differing research goals, such as resource development, environmental quality, or landscape evolution. CZEN attempts to bridge those gaps, and has gained the following, although some by happenstance and less by direct recruiting outside the immediate scholarly connections.

REFERENCES

Ackworth, R.I., 1987. The development of crystalline basement aquifers in a tropical environment. Quarter. J. Eng. Geol. 20, 265–272.

Ahnert, F., 1976. Brief description of a comprehensive three-dimensional model of landform development. Zeitschrift für Geomorphologie¸Supplementbände 25, 29–49.

Ahnert, F., 1987. Process-response models of denudation at different spatial scales. Catena Suppl. 10, 31–50.

Anderson, R.S., 1998. Near-surface thermal profiles in alpine bedrock: implications for frost-weathering of rock. Arctic Alpine Res. 30, 362–372.

Anderson, R.S., Anderson, S.P., 2010. Geomorphology: The Mechanics and Chemistry of Landscapes. Cambridge University Press, Cambridge, UK, 651 pp.

Anderson, S.P., Blum, J., Brantley, S.L., Chadwick, O., Chorover, J., Derry, L.A., Drever, J.I., Hering, J.G., Kirchner, J.W., Kump, L.R., Richter, D., White, A.F., 2004. Proposed initiative would study Earth's weathering engine. Eos Trans. AGU 85 (28), 265–272.

Anderson, S.P., von Blanckenburg, F., White, A., 2007. Physical and chemical controls on the critical zone. Elements 3, 315–319.

Aomine, S., Wada, K., 1962. Differential weathering of volcanic ash and pumice, resulting in formation of hydrated halloysite. Am. Mineral. 47, 1024–1048.

Bazilevskaya, E., Lebedeva, M., Pavich, M., Rother, M., Parkinson, D.Y., Cole, D., Brantley, S., 2013. Where fast weathering creates thin regolith and slow weathering creates thick regolith. Earth Surf. Process. Land. 38, 847–858.

Belcher, C.M. (Ed.), 2013. Fire Phenomena and the Earth System, an Interdisciplinary Guide to Fire Science. Wiley-Blackwell, Chichester, UK, p. 333.

Birkeland, P.W., 1999. Soils and Geomorphology. Oxford University Press, Oxford, 450 pp.

Blackwelder, E., 1925. Exfoliation as a phase of rock weathering. J. Geol. 33, 793–806.

Blackwelder, E., 1927. Fire as an agent in rock weathering. J. Geol. 35, 134–140.

Blackwelder, E., 1933. The insolation hypothesis of rock weathering. Am. J. Sci. 26, 97–113.

Blair, R.W., Jr., 1975. Weathering and geomorphology of the Pikes Peak granite in the Southern Rampart Range, El Paso County, Colorado. Unpublished PhD thesis. Colorado School of Mines, 115 pp.

Bland, W., Rolls, D., 1998. Weathering: An Introduction to the Scientific Principles. Arnold, London, 271 pp.

Bleam, W.F., 2012. Soil and Environmental Chemistry. Academic Press, San Diego, p. 478.

Boettinger, J.L., Southard, R.J., 1990. Micromorphology and mineralogy of a calcareous duripan formed in granitic residuum, Mojave Desert, California, USA. In: Douglas, L.A. (Ed.), Soil Micromorphology: A Basic and Applied Science. Proceedings of the 8th International Working Meeting of Soil Micromorphology, San Antonio, Texas, July 1988, Elsevier, Amsterdam, pp. 409–416.

Bond, P.L., Druschel, G.K., Banfield, J.F., 2000. Comparison of acid mine drainage microbial communities in physically and geochemically distinct ecosystems. Appl. Environ. Microbiol. 66 (11), 4962–4971.

Brady, N.C., Weil, R.W., 2008. The Nature and Properties of Soils, fourteenth ed. Pearson Prentice Hall, Upper Saddle River, New Jersey, 989 p.

Brady, P.V., Dorn, R.I., Brazel, A.J., Clark, J., Moore, R.B., Glidewell, T., 1999. Direct measurement of the combined effects of lichen, rainfall, and temperature on silicate weathering. Geochim. Cosmochim. Acta 63 (19/20), 3293–3300.

Brantley, S.L., Lebedeva, M., 2011. Learning to read the chemistry of regolith to understand the critical zone. Annu. Rev. Earth Planet. Sci. 39, 386–416.

Brantley, S.L., Blai, A.C., Cremeens, D.L., MacInnis, I., Darmody, R.G., 1993. Natural etching rates of feldspar and hornblende. Aquat. Sci. 55 (4), 262–272.

Brantley, S.L., Goldhaber, M.B., Ragnarsdottir, K.V., 2007. Crossing disciplines and scales to understand the critical zone. Elements 3, 307–314.

Bricker, O.P., Jones, B.F., Bowser, C.J., 2005. Mass-balance approach to interpreting weathering reactions in watershed systems. In: Drever, J.I. (Ed.), Surface and Ground Water, Weathering, and Soils: Treatise on Geochemistry, 5. Elsevier, Amsterdam, pp. 119–132.

Brook, G.A., Folkoff, M.E., Box, E.O., 1983. A world model of soil carbon dioxide. Earth Surf. Process. Land. 8 (1), 79–88.

Buss, H.L., Brantley, S.L., Scatena, F.N., Bazilievskaya, E.A., Blum, A., Schulz, M., Jiménez, R., White, A.F., Rother, G., 2013. Probing the deep critical zone beneath the Luquillo Experimental Forest, Puerto Rico. Earth Surf. Process. Land. 38, 1170–1186.

Butt, C.R.M., Scott, K.M., Cornelius, M., Robertson, D.M., 2008. Regoltih sampling for geochemical exploration. In: Scott, K.M., Pain, C.F. (Eds.), Regolith Science. Springer Science and CSIRO Publishing, Dordrecht and Collingwood, Australia, pp. 341–376.

Camuffo, D., 1995. Physical weathering of stones. Sci. Total Environ. 167 (1–3), 1–14.

Chalmers, R., 1898. The preglacial decay of rocks in eastern Canada. Am. J. Sci. 4 (5), 273–282.

Chamberlin, T.C., 1899. An attempt to frame a working hypothesis on the cause of glacial periods on an atmospheric basis. J. Geol. 7, 545–584, 667–685, 751–787.

Chapin, C.E., Kelley, S.A., 1997. The Rocky Mountain Erosion Surface in the Front Range of Colorado. Rocky Mountain Association of Geologists, Front Range Guidebook. Rocky Mountain Association of Geologists, Denver, pp. 101–114.

Chen, J., Blume, H.P., Beyer, L., 2000. Weathering of rocks induced by lichen colonization – a review. Catena 39, 121–146.

Chorover, J., Kretzschmar, R., Garcia-Pichel, F., Sparks, D.L., 2007. Soil biogeochemical processes within the critical zone. Elements 3 (5), 321–326.

Colman, S.M., Dethier, D.P. (Eds.), 1986. Rates of Chemical Weathering of Rocks and Minerals. Academic Press, Orlando, p. 620.

Cremeens, D.L., Brown, R.B., Huddleston, J.H. (Eds.), 1994. Whole Regolith Pedology. Soil Science Society of America Special Publication 34. Soil Science Society of America, Madison, USA, p. 136.

Depetris, P.J., Pasquini, A.I., Lecompte, K.L., 2014. Weathering and the Riverine Denudation of Continents. Springer, Dordrecht, p. 95.

Dixon, J.C., Thorn, C.E., 2005. Chemical weathering and landscape development in mid-latitude alpine environments. Geomorphology 67 (1–2), 127–145.

Dixon, J.L., Heismath, A.M., Amundson, R., 2009. The critical role of climate and saprolite weathering in landscape evolution. Earth Surf. Process. Land. 34, 1507–1521.

Dixon, J.L., Hartshorn, A.S., Heimsath, A.M., DiBiase, R.A., Whipple, K.X., 2012. Chemical weathering response to tectonic forcing: a soils perspective from the San Gabriel Mountains, California. Earth Planet. Sci. Lett. 323–324, 40–49.

Doehne, E., 2002. Salt weathering: a selective review. Natural stone, weathering phenomena, conservation strategies and case studies. Geol. Soc. Special Publ. 205, 51–64.

Dorn, R.I., 2003. Boulder weathering and erosion associated with a wildfire, Sierra Ancha Mountains, Arizona. Geomorphology 55, 155–171.

Dorn, R.I., 2011. Revisiting dirt cracking as a physical weathering process in warm deserts. Geomorphology 135, 129–142.

Dorn, R.I., 2014. Ants as a powerful biotic agent of olivine and plagioclase dissolution. Geology 42 (9), 771–774.

Dorn, R.I., Gordon, S.J., Allen, C.D., Cerveny, N., Dixon, J.C., Groom, K.M., Hall, K., Harrison, A., Mol, L., Paradise, T.R., Sumner, P., Thompson, T., Turkington, A.V., 2013a. The role of fieldwork in rock decay research: case studies from the fringe. Geomorphology 200, 59–74.

Dorn, R.I., Gordon, S.J., Krinsley, D., Langworthy, K., 2013b. Nanoscale: mineral weathering boundary. In: Shroder, J. (Editor-in-Chief), Pope, G.A. (Ed.), Treatise on Geomorphology, vol. 4. Weathering and Soils Geomorphology. Academic Press, San Diego, USA, pp. 44–69.

Dosseto, A., Buss, H., Suresh, P.O., 2011. The delicate balance between soil production and erosion, and its role on landscape evolution. Appl. Geochem. 26, S24–S27.

Drever, J.E. (Ed.), 2005. Surface and Ground Water, Weathering, and Soils. In: Holland H.D., Turekian K.K. (Executive Eds.), Treatise on Geochemistry, vol. 5. Elsevier, Amsterdam, 644 pp.

Duane, M., 2006. Coeval biochemical and biophysical weathering processes on Quaternary sandstone terraces south of Rabat (Temara), northwest Morocco. Earth Surf. Process. Land. 34, 1115–1128.

Dupré, B., Dessert, C., Oliva, P., Goddéris, Y., Viers, J., 2003. Rivers, chemical weathering and Earth's climate. C. R. Geosci. 335, 1141–1160.

Eggleton, R.A. (Ed.), 2001. The Regolith Glossary: Surficial Geology, Soils, and Landscapes. CRC LEME, Perth, Australia, p. 153.

Eren, M., Hatipoglu-Bagcit, Z., 2010. Karst surface features of the hard laminated crust (caliche hardpan) in the Mersin area, southern Turkey. Acta Carsol. 39 (1), 93–102.

Ferrier, K.L., Kirchner, J.W., Riebe, C.S., Finkel, R.C., 2010. Mineral-specific chemical weathering rates over millennial timescales: measurements at Rio Icacos, Puerto Rico. Chem. Geol. 277 (1–2), 101–114.

Folk, R.L., 1951. Stages of textural maturity in sedimentary rocks. J. Sediment. Petrol. 21, 127–130.

Gabet, E.J., Mudd, S.M., 2009. A theoretical model of coupling chemical weathering rates with denudation rates. Geology 37 (2), 151–154.

Gabet, E.J., Mudd, S.M., 2010. Bedrock erosion by root fracture and tree throw: a coupled biogeomorphic model to explore the humped soil production function and the persistence of hillslope soils. J. Geophys. Res. 115 (F04005), doi: 10.1029/2009JF001526.

Goldich, S.S., 1938. A study in rock weathering. J. Geol. 46, 17–58.

Goudie, A.S., Viles, H., 1997. Salt Weathering Hazards. Wiley, Chichester, 241 pp.

Graly, J.A., Bierman, P.R., Reusser, L.J., Pavich, M.J., 2010. Meteoric ^{10}Be in soil profiles – a global meta-analysis. Geochim. Cosmochim. Acta 74, 6814–6829.

Hall, K., 1999. The role of thermal stress fatigue in the breakdown of rock in cold regions. Geomorphology 31, 47–63.

Hall, K., Hall, A., 1991. Thermal gradients and rock weathering at low temperatures: some simulation data. Permafrost Periglac. Processes 2 (2), 102–112.

Hall, K., Otte, W., 1990. A note on biological weathering of nunataks of the Juneau Icefield, Alaska. Permafrost Periglac. Processes 1, 189–196.

Hall, K., Thorn, C., Sumner, P., 2012. On the persistence of 'weathering'. Geomorphology 149/150, 1–10.

Hartmann, J., Moosdorf, N., 2011. Chemical weathering rates of silicate-dominated lithological classes and associated liberation rates of phosphorus in the Japanese Archipelago – implications for global scale analysis. Chem. Geol. 287 (3/4), 125–157.

Heimsath, A.M., Dietrichs, W.E., Nishiizumi, K., Finkel, R.C., 1997. The soil production function and landscape equilibrium. Nature 388 (6640), 358–361.

Hewawasa, T., von Blanckenburg, F., Bouchez, J., Dixon, J.L., Schuessler, J.A., Maekeler, R., 2013. Slow advance of the weathering front during deep, supply-limited saprolite formation in the tropical Highlands of Sri Lanka. Geochim. Cosmochim. Acta 118, 202–230.

Hutton, J., 1788. Theory of the Earth; or an investigation of the laws observable in the composition, dissolution, and restoration of land upon the Globe. Trans. R. Soc. Edinb. 1 (2), 209–304.

Jenny, S.H., 1941. Factors of Soil Formation. McGraw-Hill, New York.

Jones, M.J., 1985. The weathered zone aquifers of the basement complex areas of Africa. Quarter. J. Eng. Geol. 18, 35–46.

Knight, J., Grab, S.W., 2014. Lightning as a geomorphic agent on mountain summits: evidence from South Africa. Geomorphology 204, 61–70.

Krinsley, D.H., Doornkamp, J.C., 2011. Atlas of Quartz Sand Surface Textures. Cambridge University Press, Cambridge, p. 102.

Kump, L.R., Brantley, S.L., Arthur, M.A., 2000. Chemical weathering, atmospheric CO_2, and climate. Annu. Rev. Earth Planet. Sci. 28, 611–668.

Langmuir, D., 1997. Aqueous Environmental Geochemistry. Prentice Hall, Upper Saddle River, New Jersey.

Larsen, E.S., Nielsen, C.B., 1990. Decay of bricks due to salt. Mater. Struct. 23 (1), 16–25.

Lasaga, A.C., Soler, J.M., Ganor, J., Burch, T.E., 1994. Chemical weathering rate laws and global geochemical cycles. Geochim. Cosmochim. Acta 58, 2361–2386.

Leake, J.R., Duran, A.L., Hardy, K.E., Johnson, I., Beerling, D.J., Banwart, S.A., Smits, M.M., 2008. Biological weathering in soil: the role of symbiotic root-associated fungi biosensing minerals and directing photosynthate-energy into grain-scale mineral weathering. Mineral. Mag. 72 (1), 85–89.

Lee, M.R., Parsons, I., 1999. Biomechanical and biochemical weathering of lichen-encrusted granite: a textural controls on organic–mineral interactions and deposition of silica-rich layers. Chem. Geol. 161 (4), 385–397.

Liu, Z., Dreybrodt, W., Liu, H., 2011. Atmospheric CO_2 sink: silicate weathering or carbonate weathering? Appl. Geochem. 26, S292–S294.

MacInnis, I.N., Brantley, S.L., 1993. Development of etch pit size distribution on dissolving minerals. Chem. Geol. 105, 31–49.

Markewich, H.W., Pavich, M.J., Lynn, W.C., Thomas, G.J., Johnson, R.C., Gerald, T.R., Phillips, G.G., 1994. Genesis and Residence Times of Soil and Weathering Profiles on Residual and Transported Parent Materials in the Pine Mountain Area of West-Central Georgia. United States Geological Survey Professional Paper 1589-E. United States Government Printing Office, Washington.

Masiello, C.A., Chadwick, O.A., Southon, J., Torn, M.S., Harden, J.W., 2004. Weathering controls on the mechanisms of carbon storage in grassland soils. Global Biogeochem. Cycles 18, GB4023.

Matsutani, J., Tanaka, T., Tsujimura, M., 1993. Residence times of soil, ground, and discharge waters in a mountainous headwater basis, Central Japan, traced by tritium. Traces in Hydrology. Proceedings of the Yokohama Symposium, July 1993. IAHS Publication no. 215, pp. 57–63.

May, F., 1994. Weathering or hydrothermal alteration? Examples from the Rhenish Massif, Germany. Extended Abstract of the V.M. Goldschmidt Conference, Held in Edinburgh, 29 August–2 September, 1994. Mineral. Mag. 58A, 577–578.

McDowell, W.H., Sánchez, C.G., Asbury, C.E., Ramos Pérez, C.R., 1990. Influence of sea salt aerosols and long range transport on precipitation chemistry at El Verde, Puerto Rico. Atmos. Environ. A Gen. Top. 24 (11), 2813–2821.

Merrill, G.P., 1906. A Treatise on Rocks, Rock Weathering, and Soils. MacMillan, New York, p. 400.

Migoń, P., 2013. Weathering mantles and long-term landform evolution. In: Shroder, J. (Editor-in-Chief), Pope, G.A. (Ed.), Treatise on Geomorphology, vol. 4, Weathering and Soils Geomorphology. Academic Press, San Diego, CA, pp. 127–144.

Nash, D.J., McLaren, S.J., 2007. Geochemical Sediments and Landscapes. Blackwell, Malden, Massachusetts, p. 465.

Odum, H.T., 2000. An energy hierarchy law for biogeochemical cycles. In: Brown, M.T. (Ed.), Emergy Synthesis: Theory and Application of the Emergy Methodology. The Center for Environmental Policy, Gainesville, Florida, USA, pp. 235–248.

Ollier, C.D., 1963. Insolation weathering: examples from central Australia. Am. J. Sci. 261, 376–381.

Ollier, C.D., 1983a. Fire and rock breakdown. Z. Geomorphol. 27, 363–374.

Ollier, C.D., 1983b. Weathering or hydrothermal alteration. Catena 10 (1–2), 57–59.

Ollier, C., Pain, C., 1996. Regolith, Soils, and Landforms. John Wiley and Sons, Chichester, UK, p. 316.

Pain, C.F., 2008. Regolith description and mapping. In: Scott, K.M., Pain, C.F. (Eds.), Regolith Science. Springer Science and CSIRO Publishing, Dordrecht and Collingwood, Australia, pp. 281–306.

Pavich, M.J., 1989. Regolith residence time and the concept of surface age of the Piedmont "Peneplain". Geomorphology 2 (1–3), 181–196.

Pavich, M.J., Brown, L., Valette-Silver, J., Nathalie, J., Klein, J., Middleton, R., 1985. ^{10}Be analysis of a quaternary weathering profile in the Virginia Piedmont. Geology 13, 39–41.

Pavich, M.J., Leo, G.W., Obermeier, S.F., Estabrook, J.R., 1989. Investigation of the Characteristics, Origin, and Residence Time of the upland residual mantle of the Piedmont of Fairfax County, Virginia. United State Geological Survey Professional Paper 1352. United States Government Printing Office, Washington, p. 69.

Phillips, J.D., 1993. Progressive and regressive pedogenesis and complex soil evolution. Quaternary Res. 40 (2), 169–176.

Phillips, J.D., 2001a. The relative importance of intrinsic and extrinsic factors of pedodiversity. Ann. Assoc. Am. Geogr. 91 (4), 609–621.

Phillips, J.D., 2001b. Inherited vs. acquired complexity in east Texas weathering profiles. Geomorphology 40, 1–14.

Phillips, J.D., 2002. Erosion, isostatic response, and the missing peneplains. Geomorphology 45, 225–241.

Phillips, J.D., 2005. Weathering instability and landscape evolution. Geomorphology 67, 255–272.

Phillips, J.D., 2014. Thresholds, mode switching, and emergent equilibrium in geomorphic systems. Earth Surf. Process. Land. 39, 71–79.

Phillips, J.D., Marion, D.A., 2006. Biomechanical effects of trees on soil and regolith: beyond treethrow. Ann. Assoc. Am. Geogr. 96 (2), 233–247.

Pope, G.A., 1995. Internal weathering in quartz grains. Phys. Geogr. 16 (4), 315–338.

Pope, G.A., 2013a. Weathering and sediment genesis. In: Shroder, J. (Editor-in-Chief), Pope, G.A. (Ed.), Treatise on Geomorphology, vol. 4, Weathering and Soils Geomorphology. Academic Press, San Diego, CA, pp. 284–293.

Pope, G.A., 2013b. Weathering in the tropics, and related extratropical processes. In: Shroder, J. (Editor-in-Chief), Pope, G.A. (Ed.), Treatise on Geomorphology, vol. 4, Weathering and Soils Geomorphology. Academic Press, San Diego, CA, pp. 179–196.

Pope, G.A., Dorn, R.I., Dixon, J.C., 1995. A new conceptual model for understanding geographical variations in weathering. Ann. Assoc. Am. Geogr. 85, 38–64.

Prikryl, R., Smith, B.J. (Eds.), 2007. Building Stone Decay, from Diagnosis to Conservation. Geological Society Special Publication no. 271. Geological Society of London, London, p. 344.

Raymo, M.E., 1989. Geochemical evidence supporting T.C. Chamberlin's theory of glaciation. Geology 19 (4), 344–347.

Raymo, M.E., Ruddiman, W.F., 1992. Tectonic forcing of late Cenozoic climate. Nature 359, 117–122.

Raymo, M.E., Oppo, D.W., Curry, W., 1997. The Mid-Pleistocene climate transition: a deep sea carbon isotopic perspective. Paleoceanography 12 (4), 546–559.

Raymond, P.A., Oh, N.-H., 2009. Long term changes of chemical weathering products in rivers heavily impacted from acid mine drainage: insights on the impact of coal mining on regional and global carbon and sulfur budgets. Earth Planet. Sci. Lett. 284 (1–2), 50–56.

Reiche, P., 1950. A Survey of Weathering Processes and Products. University of New Mexico Geological Seriesvol. 3University of New Mexico, Albuquerque, p. 95.

Richter, Jr., D. deB., 2007. Humanity's transformation of earth's soil: pedology's new frontier. Soil Sci. 172 (12), 957–967.

Richter, D.D., Oh, N.-H., Fimmen, R., Jackson, J., 2007. The rhizosphere and soil formation. In: Cardon, Z.G., Whitbeck, J.L. (Eds.), The Rhizosphere: An Ecological Perspective. Academic Press, Amsterdam, pp. 179–200.

Riebe, C.S., Kirchner, J.W., Finkel, R.C., 2003. Long-term rates of chemical weathering and physical erosion from cosmogenic nuclides and geochemical mass balance. Geochim. Cosmochim. Acta 67 (22), 4411–4427.

Rothschild, L.J., Mancinelli, R.L., 2001. Life in extreme environments. Nature 409, 1092–1101.

Saunders, I., Young, A., 1983. Rates of surface processes on slopes, slope retreat, and denudation. Earth Surf. Process. Land. 8 (5), 473–501.

Schaetzl, R., Anderson, S., 2005. Soils: Genesis and Geomorphology. Cambridge University Press, Cambridge, 817 pp.

Schatz, A., 1963. Chelation in nutrition, soil microorganisms and soil chelation: the pedogenic action of lichens and lichen acids. J. Agric. Food Chem. 11 (2), 112–118.

Scott, K.M., Pain, C.F. (Eds.), 2008. Regolith Science. Springer, Dordrecht, The Netherlands, p. 462.

Simonson, R., 1959. Outline of a generalized theory of soil genesis. Soil Sci. Soc. Am. Proc. 23, 152–156.

Stewart, M.K., McDonnell, J.J., 1991. Modeling base flow soil water residence times from deuterium concentrations. Water Resour. Res. 27 (10), 2681–2693.

Sumner, P.D., Nel, W., Hedding, D.W., 2004. Thermal attributes of rock weathering: zonal or azonal? A comparison of rock temperatures in different environment. Polar Geogr. 28 (2), 79–92.

Taylor, G., Eggleton, R.A., 2001. Regolith Geology and Geomorphology. John Wiley and Sons, Ltd, Chichester, UK, p. 375.

Thuro, K., Scholz, M., 2003. Deep weathering and alteration in granites – a product of coupled processes. In: GeoProc 2003, International Conference on Coupled T-H-M-C Processes in Geosystems. Fundamentals, Modelling, Experiments and Applications. Royal Institute of Technology (KTH), Stockholm, Sweden, October 13–15, 2003. http://www.geo.tum.de/people/thuro/pubs/2004_geoproc_granite.pdf.

Trencases, J.J., 1992. Chemical weathering. In: Butt, C.R.M., Zeegers, H. (Eds.), Regolith Exploration Geochemistry in Tropical and Subtropical Terrains. Elsevier, Amsterdam, pp. 25–40.

Trindade, P.V.O., Sobrai, L.G., Rizzo, A.C.L., Leite, S.G.F., Soriano, A.U., 2005. Bioremediation of a weathered and a recently oil-contaminated soils from Brazil: a comparison study. Chemosphere 58 (4), 515–522.

Turkington, A., Phillips, J., Campbell, S. (Eds.), 2005. Weathering and Landscape Evolution. Proceedings of the 35th Binghamton Symposium in Geomorphology, October 1–3, 2004. Elsevier, Amsterdam, 272 pp.

Twidale, C., 1987. Sinkholes (dolines) in laterised sediments, western Sturt Plateau, Northern Territory, Australia. Geomorphology 1 (1), 33–52.

United States National Research Council, 2001. Basic Research Opportunities in Earth Sciences. National Academic Press, Washington, DC, 168 pp.

Urum, K., Pekdemir, T., Copur, M., 2004. Surfactants treatment of crude oil contaminated soils. J. Colloid Interf. Sci. 276 (2), 456–464.

Velde, B., Meunier, A., 2008. The Origin of Clay Minerals in Soils and Weathered Rocks. Springer, Berlin, 406 pp.

Viles, H.A., 2001. Scale issues in weathering studies. Geomorphology 41, 63–72.

Viles, H.A., 2005. Microclimate and weathering in the central Namid Desert, Namibia. Geomorphology 67, 189–209.

Viles, H., 2013. Synergistic Weathering Processes. In: J. Shroder (Editor-in-Chief), Pope, G.A. (Ed.), Weathering and Soils Geomorphology, vol. 4, Treatise on Geomorphology,. Academic Press, San Diego, pp. 12–26.

Villa, N., Dorn, R.I., Clark, J., 1995. Fine material in rock fractures: aeolian dust or weathering? In: Tchakerian, V.P. (Ed.), Desert Aeolian Processes. Chapman and Hall, London, pp. 219–231.

Wasklewicz, T.A., 1994. Importance of environment on the order of mineral weathering in olivine basalts, Hawaii. Earth Surf. Process. Land. 19 (8), 715–734.

White, A.F., Brantley, S.L. (Eds.), 1995. Chemical Weathering of Silicate Minerals. Reviews in Mineralogy, vol. 31. Mineralogical Society of America, Chantilly, Virginia, USA, p. 583.

White, I.D., Mottershead, D.N., Harrison, S.J., 1992. Environmental Systems, second ed. Chapman and Hall, London, 616 pp.

Wilford, J., 2011a. A weathering intensity index for the Australian continent using airborne gamma-ray spectrometry and digital terrain analysis. Geoderma 1830184, 124–142.

Wilford, J., 2011b. Weathering intensity map of the Australian continent. AusGeo News 101, March 2011. http://www.ga.gov.au/ausgeonews/ausgeonews201103/weathering.jsp.

Wilson, M.J., Jones, D., 1983. Lichen weathering of minerals: implications for pedogenesis. Geol. Soc. London Special Publ. 11, 5–12.

Wolock, D.M., Hornberger, G.M., Beven, K.J., Campbell, W.G., 1989. The relationship of catchment topography and soil hydraulic characteristics to lake alkalinity in the northeastern United States. Water Resour. Res. 25, 829–837.

Wray, R.A.L., 2013. Solutional weathering and karstic landscapes on quartz sandstones and quartzite. In: Shroder, J. (Editor-in-Chief), Frumkin, A. (Ed.), Treatise on Geomorphology, vol. 6, Karst Geomorphology. Academic Press, San Diego, pp. 463–483.

Yatsu, E., 1988. The Nature of Weathering. Sozosha, Tokyo, p. 624.

Young, R.W., Dixon, J.C., 1983. Weathering and hydrothermal alteration: critique of Ollier's argument. Catena 10 (1–2), 439–440.

Zilberbrand, M., 1999. On equilibrium constants for aqueous geochemical reactions in water unsaturated soils and sediments. Aquat. Geochem. 5, 195–206.

Chapter 5

Soil Morphology in the Critical Zone: The Role of Climate, Geology, and Vegetation in Soil Formation in the Critical Zone

John C. Dixon
Department of Geosciences, University of Arkansas, Fayetteville, Arkansas, USA

5.1 INTRODUCTION

Soil lies at the core of the Critical Zone (Fig. 5.1) as it represents the "membrane" across which, or through which water and solutes, energy, gases, solids, and organisms are exchanged between the atmosphere, biosphere, hydrosphere, and lithosphere. Water is the force that drives the majority of these exchanges within the soil as well as in the Critical Zone as a whole (Lin, 2010). Soil is distinguished from regolith by two essential characteristics namely the presence of genetically related soil horizons and the presence of living and dead organic matter. Yet simultaneously, regolith is an essential component of the soil profile and the Critical Zone as a whole as it is either the lithified or unconsolidated parent material from which the soil is derived. Soil property variability within the Critical Zone is a function of the interaction of five fundamental factors: climate, organic activity, relief, parent material, and time (Jenny, 1941). This chapter focuses on the fundamental roles played by climate, vegetation, and geology (i.e., relief and parent-material) on the formation and associated properties of soil developed in the Critical Zone.

5.1.1 The Nature of Soil

Soil is a natural body comprised of solids (i.e., minerals and organic matter), liquid, and gases that occur at, or near, the land surface, and is characterized by one or both of the following: horizons, or layers, that are distinguishable from the initial formative material as a result of additions, losses, transfers, and transformations of energy and matter, and the ability to support rooted plants in a natural environment (Soil Survey Staff, 1994).

The upper limit of soil is the boundary between soil and air, shallow water, live plants, or plant materials that have not begun to decompose. Areas are not

Developments in Earth Surface Processes, Vol. 19. http://dx.doi.org/10.1016/B978-0-444-63369-9.00005-7

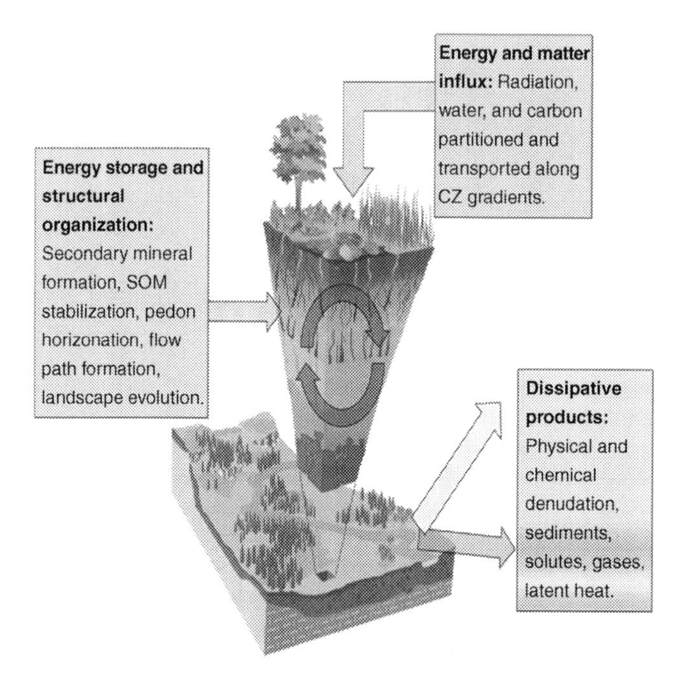

FIGURE 5.1 Conceptual model of the Critical Zone showing the prominent position of the soil.

considered to have soil if the surface is permanently covered with water too deep for the growth of rooted plants. The lower boundary that separates soil from the nonsoil underneath is more difficult to define. Soil consists of horizons near the surface of the Earth that, in contrast to the underlying parent material, have been altered by the interactions of climate, relief, and living organisms over time. Commonly, soil grades at its lower boundary to hard rock or to unconsolidated materials virtually devoid of animals, roots, or other indications of biological activity. For purposes of classification, the lower boundary of soil is arbitrarily set at 200 cm (Soil Survey Staff, 1994).

Soil can be viewed as the ultimate product of weathering of rocks and minerals at or near the surface of the Earth. It represents an equilibrium weathering product in which primary minerals are in metastable equilibrium with temperature, pressure, moisture, and gaseous conditions in Earth environments. As environmental conditions change however, they may be accompanied by responses in the weathering system. Weathering transformation of rock and regolith are examined in Chapter 4 by Pope.

5.1.2 The Soil Profile

Soils consist of a number of readily identified master horizons that reflect varying degrees of alteration of an original parent material (Fig. 5.2). Soils develop as a consequence of the biophysical and biogeochemical weathering of a parent

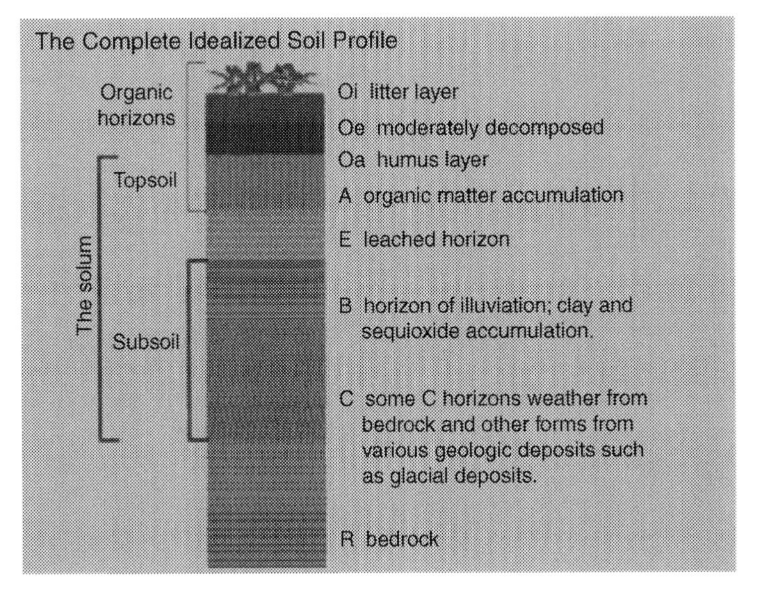

FIGURE 5.2 **Idealized soil profile showing the dominant master horizons present in well-developed soils of the middle latitudes.**

material. If the parent material is a lithified or crystallized rock material, the associated soil horizon is referred to as the "R" horizon, and defined as the consolidated bedrock underlying the soil (Birkeland, 1999). This horizon is by definition, one of fresh unaltered rock material in which weathering processes have yet to affect alteration, except for perhaps some oxidation. Above the "R" horizon is the "C" horizon. This horizon is frequently referred to as the regolith horizon and consists of material little affected by pedogenic processes, but displaying strong weathering. It lacks the properties of the overlying, truly pedogenic horizons. Most of the "C" horizons consist predominantly of mineral matter that may be like or unlike the material from which the overlying pedogenic horizons have been presumed to have formed (Soil Survey Staff, 1994). Where soil parent materials are unconsolidated they are most commonly designated as the "C" horizon. However their designation as a "D" horizon has been proposed by Tandarich et al. (1994), but this designation was never formally adopted. Despite this, some Critical Zone researchers do recognize a "D" horizon, as well as horizons transitional between "C" and "D" (Lin, 2010; Brantley et al., 2006, 2007).

Subsurface horizons that are so strongly impregnated with calcium carbonate that their morphology is determined by the carbonate are designated as "K" horizons (Gile et al., 1965). Calcium carbonate coats or engulfs nearly all primary grains in a continuous layer. The uppermost part of a strongly developed "K" horizon is laminated, brecciated, and/or pisolitic (Machette, 1985; Dixon and McLaren, 2009). Continuous layers of strongly cemented carbonate are referred to as hardpan calcrete (Dixon and McLaren, 2009).

Above the "C" or "K" horizon the first of the truly pedogenic soil horizons occur. The "B" horizon is characterized by the obliteration of all, or most of the original parent material structure, and shows one or more of the following characteristics:

1. The concentration of illuvial clay, iron, aluminum, organic matter carbonates, gypsum, or silica in combination or alone.
2. Concentration of residual sesquioxides.
3. Sesquioxide coatings that make the horizon conspicuously lower in color value, higher in chroma, or redder in hue without apparent iron illuviation.
4. Alteration that forms secondary silicate clays or oxides, or both, and that forms a granular, blocky, or prismatic structure if volume changes accompany changes in moisture content.
5. Brittleness or strong gleying (Soil Survey Staff, 1994).

Above the "B" horizon, and below the "A" horizon is the "E" horizon. This horizon is a mineral horizon distinguished by the loss of silicate clay, iron, aluminum, or some combination of these constituents leaving a concentration of sand and silt particles. This horizon exhibits obliteration of all or most of the original parent material structure.

The uppermost mineral horizon of a soil is designated as the "A" horizon. This is a horizon that has formed at the surface of Earth or below an organic matter-dominated horizon. It exhibits obliteration of all or much of the original parent-material structure and exhibits one or both of the following:

1. An accumulation of humified organic matter strongly mixed with the mineral fraction and not dominated by properties characteristic of "E" or "B" horizons.
2. Properties resulting from cultivation, pasturing, or similar kinds of disturbances (Soil Survey Staff, 1994).

Surface horizons dominated by organic matter are designated as "O" horizons. These horizons may possess a mineral matter component, but it is generally substantially less than 50% by weight. "O" horizons are susceptible to water logging, but are now typically drained. Some have never been saturated. The organic component consists of both living and dead floral and faunal components. Dead material is commonly in varying stages of decomposition (Soil Survey Staff, 1994).

5.2 MODELS OF SOIL FORMATION

Soil formation fundamentally involves the transformation of parent material to materials that are in some kind of equilibrium with the surface of Earth and near surface environmental conditions. Understanding how these transformations occur has been approached in a variety of different ways. This section presents some of the more common approaches that have been used to more fully understand how soils form and change over time, and how soil-forming factors interact to produce the variability in soil properties observed in the Critical Zone.

5.2.1 State Factor Model

One of the longest standing models of soil formation is that formulated by Jenny (1941). This fundamentally factorial model, fashioned after the pioneering work of Dokuchev (1883), identifies five state factors that interact to explain the natural variability in soil properties at different spatial scales. The five state factors are climate, organic activity, relief, parent material, and time.

Climate fundamentally drives rates of parent-material breakdown and biological processes through its control on temperature and moisture at various temporal and spatial scales. The biotic factor as originally defined by Jenny referred to the potential natural vegetation at a site. It referred only to biochemical influences of plants and specifically omitted the role of fauna in soil formation (Jenny, 1958). The relief, or topography factor relates to the configuration of the landscape and fundamentally influences the pattern of water movement across the landscape. In detail, it relates to elements such as slope angle, concavity and convexity, and aspect (Jenny, 1980). The parent material factor encompasses the physical and chemical characteristics of the material from which a soil is derived. It has profound influences on the fundamental physical and chemical properties of the resulting soil. The time factor refers to the length of time that a soil has been developing since the parent material was exposed to surface and near-surface environmental conditions. Fundamentally, it influences the degree of soil development, as expressed in the complexity of soil horizon development.

5.2.2 Process-Systems Model

By the end of the 1950s, models of soil development were beginning to shift from a functional-factorial approach to more of a process-oriented approach (Simonson, 1978; Schaetzl and Anderson, 2005). In the process-systems model, soil formation is envisioned to consist of two essential steps: (1) accumulation of parent material; and (2) differentiation of parent material into soil horizons (Yaalon, 1975). The focus of the process-systems model is on the second part of these steps. It identifies four process bundles: (1) additions, (2) removals, (3) transformations, and (4) translocations (Simonson, 1978). Additions refer to material added to the soil profile from outside sources and include materials delivered from the atmosphere and the biosphere. Similarly, removals refer to losses from the profile as a whole and refer essentially to dissolved materials transported to the groundwater and materials removed by erosion. Translocations refer to losses and additions between horizons, principally from higher to lower horizons, but also under specific conditions from lower to higher horizons. Transformations refer to alterations of inorganic and organic materials in the soil by weathering and decomposition processes. The four process bundles operate simultaneously in the soil profile. Their character and balance governs the ultimate nature of a soil.

5.2.3 Runge's Energy Model

Runge (1973) developed an energy-based model that is basically a hybrid of the state-factor model and the process-systems model. In this model two of Jenny's state factors: climate and relief become the dominant factors. Operating together they become an intensity factor, which Runge defined as the amount of water available for leaching (*w*). When combined, these two factors determine the potential for water to enter and percolate through the soil profile. Thus, the "*w*" factor acts as an organizing factor that utilizes gravitational energy to organize the soil profile. These ideas have more recently been refashioned into the concept of pedohydrology in which water becomes the chief influence on soil genesis, evolution, variability and function, and these pedogenic characteristics become the principal influence on patterns of water movement in space and time (Lin, 2010). Additionally, Runge's model combines the parent material and organisms factors into a single organic-matter factor (*o*). Runge's two factors (*w* and *o*) operating over time constitute the energy model of soil formation (Schaetzl and Anderson, 2005).

5.2.4 Dynamic-Rate Model

The dynamic-rate model stresses interactions between active and passive vectors of pedogenesis and their change over time. The model is an evolutional rather than developmental model and is based on the idea that soils evolve along interacting progressive and regressive pathways reflecting interactions between exogenous and endogenous vectors (Johnson et al., 1990). Soil evolution is a function of the interaction of dynamic vectors, which are fundamentally exogenous vectors and passive vectors, which are principally endogenous vectors. The dynamic vectors function very similarly to Runge's intensity factor.

The principal dynamic vector categories include energy fluxes, (insolation, heat transfers, entropy transfers, oxidation, and gravity), Mass fluxes (water, gases, and solids), frequency of wetting and drying, organic activity and pedoturbation. The principal passive vectors include parent material, soil chemical environment, water table environment, landscape stability, and pedogenic additions (Johnson et al., 1990).

Soil evolution is the product of interactions between the dynamic (**D**) and passive (**P**) vectors and the change in these vectors over time. Intimately associated with the dynamic-rate model are two pedogenic pathways: progressive and regressive pedogenesis (Johnson and Watson-Stegner, 1987). The dynamic-rate model assumes progressive pedogenesis during which there are progressive changes in parent material from a state of simplicity and disorder to a state of increasing order, physiochemical stability, and profile complexity. If pedogenesis subsequently shifts toward simplification, physicochemical instability, and/or shallower profiles, regressive pedogenesis dominates.

5.2.5 Effective Energy and Mass Transfer Models

Most recently, workers seeking to understand the structure and functioning of the Critical Zone have turned to integrating open-system and thermodynamic theory. This integrated approach investigates the Critical Zone (e.g., especially the soil component) as a system open to energy and mass fluxes forced by radiant, geochemical, and elevational gradients (Rasmussen et al., 2011). This approach uses rates of effective energy and mass transfer to quantify various flux-gradient relations. In particular, it permits the assessment of the relative importance of solar radiation, water, carbon, and physical/chemical denudation mass fluxes to the Critical Zone (soil) energy balance (Rasmussen et al., 2011). This approach is in many ways an extension of the Runge (1973) model.

The model predicts that structural organization within the Critical Zone, particularly soil horizonation, secondary clay mineral and colloid formation, preferential flow path development, vegetation distribution, and landscape evolution, evolves at the expense of increased entropy associated with dissipation of weathering products and gas and latent-heat fluxes (Charover et al., 2011).

5.3 GEOCHEMISTRY AND SOIL DEVELOPMENT

Soil formation fundamentally involves the modification of some parent material, either solid bedrock, or transported regolith, by the reaction of water and contained atmospherically and lithospherically derived gases, and the minerals present in the parent material. The "geochemical engine" operating in the soil to move materials around and evolve the soil profile consists of four fundamental components: additions, translocations, transformations, and removals (Simonson, 1978; Brantley et al., 2007; Dixon, 2013). Addition processes involve the contribution of salts, organic compounds, and mineralogenic dust from the atmosphere to the regolith and evolving soil where they are either rapidly transported to depth or accumulate.

Chemically, translocations involve the movement of chemical constituents down through the developing soil profile. Soluble salts are readily dissolved and transported vertically under the influence of gravity while more resistant chemical elements bound in aluminosilicate and other mineral groups may remain near the surface for extended periods of time until they can be released from the host minerals and transported by the process of chelation. Although translocations are fundamentally downward, under favorable circumstances they may occur vertically upward. In either case once the depth of water percolation or capillarity is reached, chemical constituents will be deposited. This results in the formation of a variety of salt-dominated soil horizons (e.g., calcic, sodic, gypsic, sesquioxide horizons) or surface crusts.

Transformations involve the alteration of primary rock-forming minerals and organic matter by a variety of chemical weathering processes (see Chapter 4 for a detailed discussion) including dissolution, oxidation, carbonation, hydration, and a variety of ligand reactions (Dixon, 2004). The result of these transformation reactions is the formation of new, secondary (principally clay) minerals as well as the release of soluble chemical components, which are either uptaken by vegetation, redeposited as chemical horizons, or removed completely from the soil profile. Transformations are responsible for the formation of clay-rich and salt depleted horizons. Their specific location within a soil profile is fundamentally determined by water table position (Anderson et al., 2007).

Removals collectively refer to those processes that completely remove geochemical constituents from the soil profile. They principally involve the leaching, or complete removal, of soluble materials from the soil profile as well as the removal of chemical constituents as a result of soil surface erosion (Brantley et al., 2007; Dixon, 2013).

5.3.1 Soils and Climate

In the majority of soil formation models, climate assumes premier position and continues to be viewed as a dominant factor in soil formation (Birkeland, 1999). The two most important components of climate are temperature and precipitation, as together they fundamentally control rates of chemical (and to some degree physical) weathering, rates of biomass production and decomposition, and the rates of translocation of solids and dissolved materials. Fundamentally, climate represents the overwhelmingly dominant direct and indirect source of energy in the soil system. The principal soil properties that correlate with climate include organic-matter content, clay content, clay mineralogy, iron mineralogy and color, and patterns of accumulation of calcium carbonate and soluble salts (Birkeland, 1984, 1999).

5.3.1.1 Climate and Organic Matter Content

Many different measures of organic matter contents in soils are to be found in the literature, including carbon content (Schlesinger, 1990), nitrogen content (Jenny, 1980), and C/N ratios (Post et al., 1985). A global ranking of soil organic matter content (kg/m^2 to 1-m depth) by soil order reveals the following sequence: Histosols $\gg$ Andisols > Inceptisols = Spodosols = Mollisols > Oxisols = Entisols = Ultisols > Alfisols = Vertisols > Aridisols (Eswaran et al., 1993). From quite contrasting geographic settings, Jenny (1961) identified general trends in soil carbon and nitrogen. He found that soil nitrogen increased logarithmically with increasing moisture and decreased exponentially with increasing temperature. For soils from India, he found that organic carbon increased with increasing precipitation. Global patterns of soil nitrogen abundance and C/N ratios closely follow world life zones. Grassland soil C/N ratios range from 10 to 15, whereas desert soil C/N ratios are similar or occasionally lower. This pattern is a reflection of low precipitation and infiltration in desert environments. Many forest and tundra soils display values of 10–20 and are frequently >20 as a result of the toughness of some leaves and needles and low mean annual temperatures (MATs). Low-latitude forests commonly display C/N ratios similar to desert and grassland soils as a result of abundant leaf fall and rapid rates of decomposition (Birkeland, 1999).

5.3.1.2 Climate and Clay Content

From first principles, it would be reasonable to expect that a linear relationship exists between clay content of soils and climate. One might expect that as temperature and precipitation increase, the clay content of soils would also increase. However, many factors complicate such a simple relationship. The relatively recent identification of aeolian dust as a substantial contributor to clay accumulation in all climates, parent-material lithology, and topography all play important roles. However, despite these complex influences, several studies exist that establish the fundamental nature of the climatic relationship. Jenny (1935) was one of the earliest to establish the nature of the climate–clay relationship. He showed that a linear relationship existed between moisture and an exponential relationship with temperature. Sherman (1952) working in Hawaii demonstrated an overall increase in clay abundance with increasing precipitation. Strakhov (1967) demonstrated progressive accumulation of weathering-derived clay with progressive increases in temperature and moisture.

Webb et al. (1986), in a climatic gradient from 64 to 200 mean annual precipitation (MAP) reported subsoil clay contents doubling. In a comprehensive N–S and E–W pair of climatic transects, Ruhe (1984a and b) reported contradictory trends in soil properties with increasing moisture and increasing temperature, respectively. These contradictory trends he ascribed principally to the effect of parent-material influences on clay formation.

5.3.1.3 Climate and Clay Mineralogy

Weathering in the Critical Zone is discussed in detail in Chapter 4 by Pope, so will only be explored here briefly. Climate strongly influences soil–clay mineralogy (Curtis, 1990). Primary rock-forming minerals are fundamentally unstable under Earth surface and near-surface conditions. Here moisture, temperature, gaseous, and pressure environments are markedly different from those that prevail where mineral formation takes place. Consequently, minerals transform into secondary products (i.e., generally clays) that are more kinetically stable under prevailing conditions. These transformations are driven by reactions of water, oxygen, and carbon dioxide with mineral surfaces. Rates of transformation are fundamentally controlled by temperature and moisture. In a classic study by White and Blum (1995) of 68 watersheds worldwide they concluded that chemical weathering rates are controlled by precipitation for any given temperature, but the strength of the precipitation correlation increases exponentially with increasing temperature. Accelerated chemical weathering occurs in warm, wet environments compared to cool, wet environments. This, however, does not mean that chemical weathering and, thus, soil formation is inhibited in cold climates (Dixon and Thorn, 2005). The nature of the secondary weathering products varies markedly with climate, principally in response to moisture availability. Generally speaking, warm, humid climates are dominated by kaolinite; cool, humid environments are dominated by smectite and montmorillonite; whereas, sepiolite and palygorskite dominate in arid environments (Birkeland, 1999).

Tardy et al. (1973) demonstrated a change from smectite/gypsum-dominated crest slopes and smectite/gypsum/halite-dominated toe slopes on soil catenas in arid environments; through kaolinite and Fe-hydroxides-dominated crest slopes and smectites and carbonates in toe slope soils in semiarid environments; to gibbsite and Fe-hydroxide-dominated crest slopes and gibbsite-dominated toe slopes in humid environments (Fig. 5.3). Even under constantly humid (wet) environmental conditions, specific clay minerals dominate, depending on the precise amounts of precipitation received (Sherman, 1952).

McFadden (1988) working in the southwestern United States, investigated the nature of clay–mineral assemblages along an arid-semiarid-xeric transect.

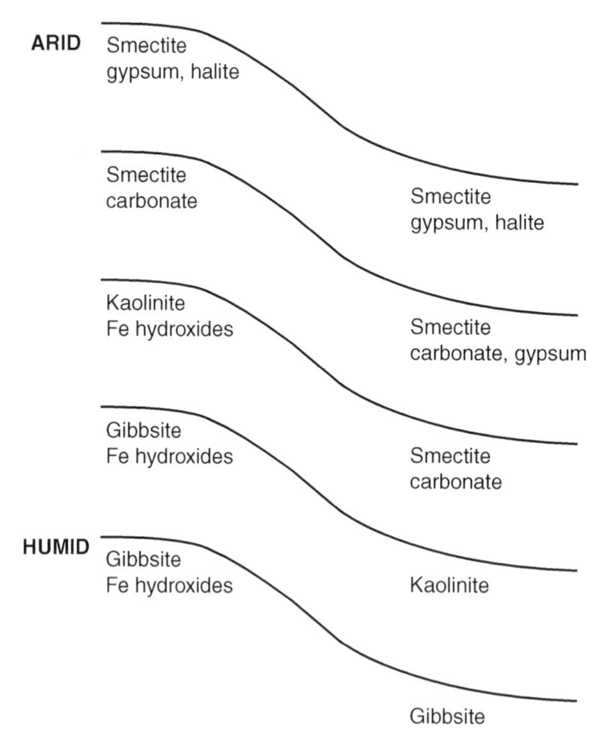

FIGURE 5.3 Schematic of the effects of climate on clay mineral formation and distribution on a generalized slope. *(After Tardy et al. (1973).)*

At the arid end of the transect, the dominant minerals are (i.e., in relative abundance) illite-smectite-palygorskite-kaolinite, whereas at the xeric end of the transect, the sequence is kaolinite-vermiculite- illite-smectite.

More recently, Egli et al. (2003) conducted a study of a soil climosequence in the Italian Alps, along a strong precipitation and temperature gradient. They found that weathering rates and the formation of soil smectites decreased both with increasing and decreasing elevation, and that this strongly nonlinear pattern was intimately related to pronounced podsolization at treeline where higher precipitation rates and the production of chelates promote smectite formation. This study amplifies the strong role vegetation plays in soil–clay mineral formation in conjunction with climate. Vegetation serves to bind soil particles and to reduce erosion, thus, increasing reaction time between soil solutes and mineral particles and completeness of mineral dissolution. In addition it provided H^+ for acid reactions while photosynthesis and subsequent degradation pump CO_2 through the soil, which facilitates additional chemical reactions (Curtis, 1990).

5.3.1.4 Soil Reddening

Soil color, especially soil redness is strongly correlated with climate (Birkeland, 1999). It appears that temperature exerts a stronger control than precipitation on redness as illustrated by the strong red colors of the southeastern United States, the humid tropics and hot deserts of the southwestern United States and Baja California (Birkeland, 1999). In contrast, soils are less red in cooler environments, such as those along the United States/Canadian border and Arctic and alpine environments (Birkeland, 1999). In a study of soil redness and iron mineralogy in a transect across the European Alps, Schwertmann et al. (1982) found that soil color reddens with increasing MAT and decreasing MAP. Iron mineralogy, as expressed by the ratio hematite:hematite + goethite, shows a strongly similar trend. A similar study by the authors in Brazil shows a similar trend with respect to mean annual air temperature (Kampf and Schwertmann, 1983).

5.3.1.5 Accumulation of Calcium Carbonate and Soluble Salts

With increasing aridity, the accumulation of calcium carbonate and other soluble salts begins to occur in soils. In general, carbonate begins to accumulate in soils where MAP is less than 50 cm, whereas gypsum accumulates in soils where MAP is less than 10 cm (Dan and Yaalon, 1982). MAT also influences the accumulation of the more soluble salts. In the mid-continent United States, the boundary between calcic and noncalcic soils is a MAP of 50 cm and a MAT of 5–6°C in the north. In southern Texas the boundary is at approximately 60 cm MAP and 22°C MAT. In the western United States, the climatic conditions associated with carbonate accumulation are radically different where deposition occurs under significantly lower precipitation regimes and leaching is substantially less effective. Here, carbonate accumulates under a MAP of 20 cm and a MAT of 9.5°C (Birkeland, 1999). Precipitation chemistry is also a significant component on patterns of carbonate accumulation.

Depth to the top of the carbonate layer varies directly with precipitation and is a reflection of the average depth of infiltration of moisture. In more arid environments, depths are generally shallower, whereas in wetter environments depths are greater. This pattern is complicated, however, by the fact that parent-material properties strongly influence the patterns of carbonate and salt accumulations (see Dixon and McLaren, 2009 and Dixon, 2009 for detailed discussions).

5.4 SOIL PROPERTIES AND GEOLOGY

For the purpose of the present discussion, geology is treated as a collective term for both the influence of topography and the influence of parent-material properties on soil development. Topography largely controls the pattern of water movement on slopes through slope form and angle (Ruhe, 1960, 1975; Conacher and Dalrymple, 1977; Schaetzl, 2013), the position of the water table, as well as aspect which integrates topography and climate (Cooper, 1960; Carter and Ciolkosz, 1991; Rech et al., 2001; Egli et al., 2007, 2010). Parent-material exerts strong controls on soil texture and structure, chemistry, color, all related essentially to patterns of water movement and associated transportation of solids and solutes as well as the influence it has on vegetation.

5.4.1 Soil Catenas

Soil catenas are associations of soils on a single slope. They display different characteristics depending on slope position, but they are all linked by the common association they share with respect to patterns of water movement on the slope and associated movement of solids, solutes, and gases (Schaetzl, 2013). The principal slope elements of a catena consist of summit, shoulder, backslope, footslope, and toeslope elements (Ruhe, 1960; Schaetzl, 2013) (Fig. 5.4).

Summit slopes are stable, with minimal erosion. Chemical weathering dominates over physical weathering. Soils are typically thick, as infiltration dominates, resulting in strong leaching and consequently more strongly developed soils than those on slope elements immediately downslope. In addition, they commonly have thick "A" horizons and accumulate organic matter (Ruhe, 1960; Schaetzl, 2013).

Shoulder slopes exhibit maximum convexity and, thereby, are dominated by runoff and erosion. Erosion may be so great as to completely outstrip soil formation (Walker and Ruhe, 1968). Soils occurring in this slope position are commonly thin, young, and least stable. They possess thin "A" horizons, lack organic matter accumulation and are commonly dry. They are unlikely to accumulate soil materials from higher on the slope (Ruhe, 1960; Schaetzl, 2013).

Backslopes are relatively steep, straight slopes lacking pronounced convexity or concavity. They are located between upslope erosional slopes and downslope depositional slopes and as such function principally as transportation

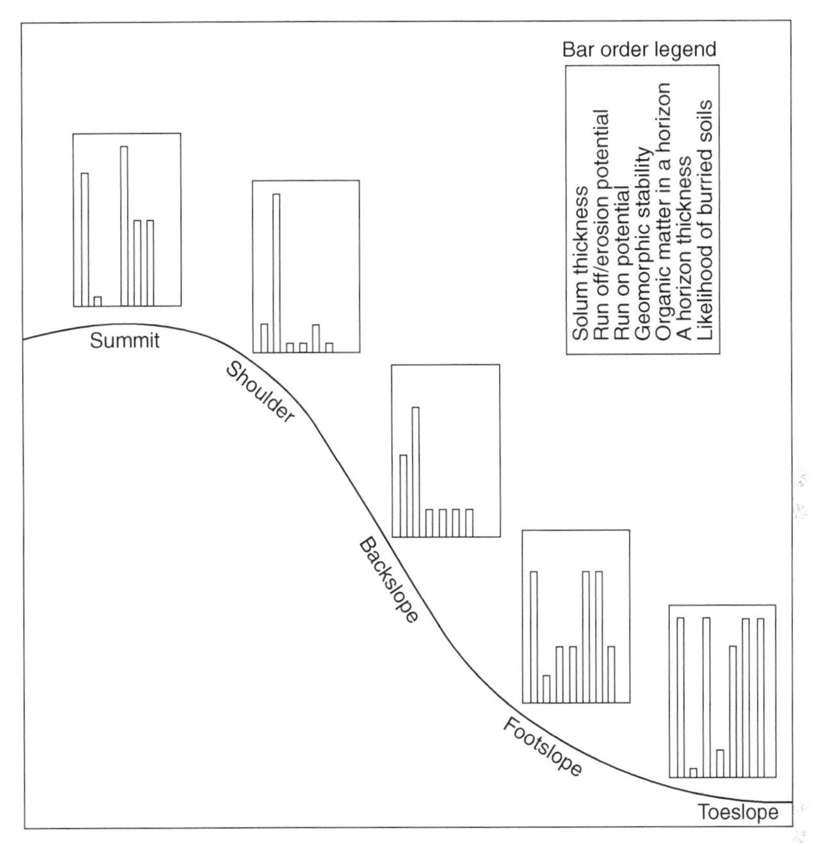

FIGURE 5.4 Effects of hillslope position on soil properties in a humid climate. *(After Schaetzl (2013).)*

slopes but are subject to potential erosion. The soils in these positions display considerable variability depending on the length of the slope segment but generally are thicker than those higher in the catena but slightly thinner than those in lower topographic positions. They typically display relatively thin "A" horizons and accumulate relatively small quantities of organic matter (Ruhe, 1960; Schaetzl, 2013).

Footslopes are the concave portions of a catena toward the bottom of the slope. This slope element is dominated by accumulation of materials carried from higher on the catena in suspension, solution and wash, both as surface wash and subsurface throughflow. Soils developed in this location are commonly thick because of the accumulation of materials from higher slope positions, as well as relative geomorphic stability, although potential for some erosion still exists. They display the development of thick "A" horizons and accumulation of organic matter. Soils developed in this location also possess the potential for the occurrence of buried soils (Ruhe, 1960; Schaetzl, 2013).

Toeslopes occur at the distal ends of footslopes and represent stable, constructional locations at the bottom of catenas. Soils developed in this location are commonly over-thickened as a result of both deposition of materials from higher on the slope, as well as deposition on the slope by fluvial processes. Commonly soils in this position possess very thick "A" horizons and contain multiple buried soils. Because of their inherently wet nature, toeslope soils typically exhibit high plant productivity and organic matter accumulation than at other locations down the catena (Ruhe, 1960; Schaetzl, 2013). These locations also can be sites of enhanced chemical weathering. Dixon (1986) noted enhanced clay formation from biotite weathering in these locations in catenas in the Front Range of the Rocky Mountains. The dominant secondary clays forming from biotite weathering were vermiculite and smectite.

The general structure and accompanying soil properties of catenas described above can be broadly applied across landscapes because of the fact that it expresses general patterns of water flow across and through hillslopes. However, in detail, the precise pattern of soil development and accompanying soil properties varies considerably depending on the prevailing climate. Dixon (2013) examined, in detail, the variability observed in catenas developed under a variety of humid, Arctic/alpine, and arid/semiarid climates.

5.4.2 Nine-Unit Slope Model

Whereas the catenary slope sequence outline can be generally applied at the individual hillslope scale, the nine-unit slope model was developed by Dalrymple et al. (1968) and Conacher and Dalrymple (1977) for the expressed purpose of recognizing that the catena concept should include soil differences that result from variations in drainage, as well as from differential transport of eroded solid and dissolved materials (Milne, 1935, 1936) (Fig. 5.5). In addition, the nine-unit slope model was developed to incorporate a consideration of soils developed in redeposited materials, hence the evolution of a pedo-geomorphic model. The model is particularly well suited to application in old, stable, deeply incised landscapes. It consists of nine geomorphic units, each with a distinctive set of associated soil properties. The nine units are:

Unit 1. Interfluve: Chemical weathering and *in situ* soil development dominate.

Unit 2. Seepage slope: Pedogenically, this unit is characterized by gleying above iron pans or other indurations. Reduced porosity and increased compaction is found in "E" compared to "B" horizons and in upper "B" compared to lower "B" horizon. Mottling, Mn concentrations, and concretions present.

Unit 3. Convex creep slope: Transportation of soil material from upslope by creep and surface and subsurface water movement. Better drained soils than in higher slope positions.

Unit 4. Fall face: Soils shallow to bedrock.

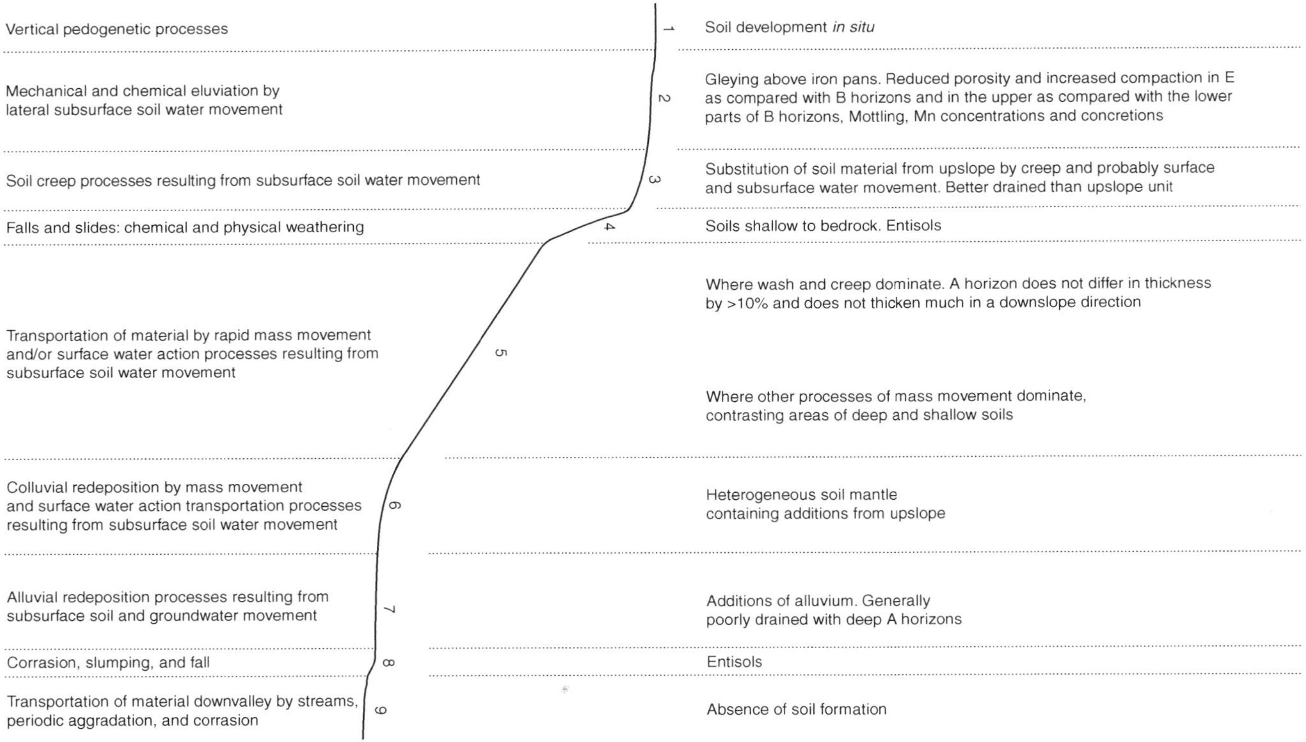

Unit	Geomorphic processes	Pedogenic processes
1	Vertical pedogenetic processes	Soil development *in situ*
2	Mechanical and chemical eluviation by lateral subsurface soil water movement	Gleying above iron pans. Reduced porosity and increased compaction in E as compared with B horizons and in the upper as compared with the lower parts of B horizons, Mottling, Mn concentrations and concretions
3	Soil creep processes resulting from subsurface soil water movement	Substitution of soil material from upslope by creep and probably surface and subsurface water movement. Better drained than upslope unit
4	Falls and slides: chemical and physical weathering	Soils shallow to bedrock. Entisols
5	Transportation of material by rapid mass movement and/or surface water action processes resulting from subsurface soil water movement	Where wash and creep dominate. A horizon does not differ in thickness by >10% and does not thicken much in a downslope direction Where other processes of mass movement dominate, contrasting areas of deep and shallow soils
6	Colluvial redeposition by mass movement and surface water action transportation processes resulting from subsurface soil water movement	Heterogeneous soil mantle containing additions from upslope
7	Alluvial redeposition processes resulting from subsurface soil and groundwater movement	Additions of alluvium. Generally poorly drained with deep A horizons
8	Corrasion, slumping, and fall	Entisols
9	Transportation of material downvalley by streams, periodic aggradation, and corrasion	Absence of soil formation

FIGURE 5.5 Nine-unit slope model showing relationships between hillslope components, dominant soil forming processes, and associated geomorphic processes. Dominant geomorphic processes are shown along the bottom of the figure and associated pedogenic processes are summarized at the top of the figure. *(Adapted from Dalrymple et al. (1968) and Conacher and Dalrymple (1977).)*

Unit 5. Transportational midslope: Where wash and creep dominate, "A" horizon thickness does not vary by >10% and does not thicken downslope. Where other mass wasting processes dominate, contrasting areas of thick and shallow soils occur.
Unit 6. Colluvial footslope: Heterogeneous soil mantle containing additions from upslope.
Unit 7. Alluvial toeslope: Addition of alluvium. Generally poorly drained soils with thick "A" horizons.
Unit 8. Stream channel wall: Pedogenesis is limited to entisol formation.
Unit 9. Stream channel: No pedogenesis.

5.4.3 Soils and Aspect

Soils developed on essentially north-facing slopes in the northern hemisphere or south-facing slopes in the southern hemisphere typically exhibit lower temperature regimes and greater moisture regimes than soils developed on south- and north-facing slopes, respectively. This contrast in thermal and moisture regimes has substantial impact on the properties exhibited by soils and also on the nature of the vegetation associated with such slope contrasts. The principal control being a climatic one in which effective moisture becomes the dominant factor (Cooper, 1960; Finney et al., 1962; Klemmedson, 1964; Macyk et al., 1978; Hunckler and Schaetzl, 1997; Rech et al., 2001).

Generally, soils developed on southerly-facing slopes experience more xeric conditions than those on north-facing slopes and consequently are less strongly developed. Soil horizons tend to be thinner, soil organic carbon tends to be lower, and because of less moisture, parent materials are less weathered (Fig. 5.6). Despite these generalities great variations also occur on specific hillslopes depending on the influence of the other soil-forming factors.

In a study of the effect of slope gradient and aspect on soils developed from sandstone in the Ridge and Valley physiographic province of the Appalachian Mountains in Pennsylvania, Carter and Ciolkosz (1991) examined the effect of

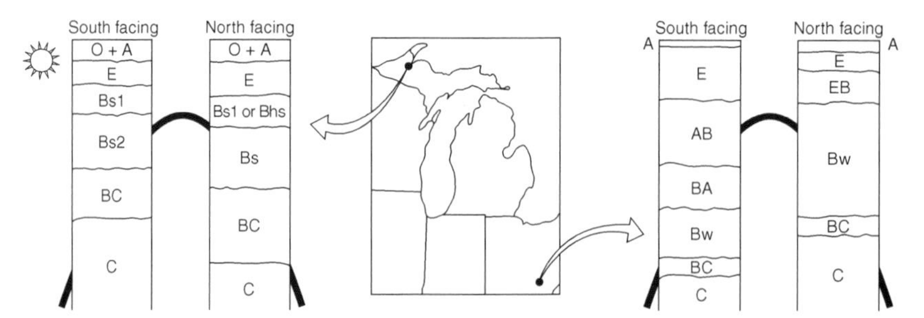

FIGURE 5.6 Two examples of the effect of aspect on soil profile complexity from Michigan and Ohio. Note the greater profile complexity, particularly as it relates to weathering and horizon thickness on north-facing slopes. *(After Schaetzl and Anderson (2005).)*

aspect on numerous soil properties. This study focused on soils developed on summit and shoulder slopes oriented in SW and NW directions. The authors found that with respect to aspect, solum depth, "B" horizon thickness, clay, Fe and Al indices showed a smaller decrease with increasing slope on the NW slopes than on the SW oriented slopes. In addition, "E" horizons were slightly thicker on the NW-facing slopes than the SW-facing slopes. The authors attributed these aspect differences principally to lower evaporation on the northerly oriented slopes and associated more effective eluviation and illuviation as a result of greater moisture availability.

Recent studies of soil formation in the Italian Alps have examined the influence of aspect on the weathering of clay minerals in soils and the patterns of soil organic-matter accumulation. In their study of clay mineral weathering, Egli et al. (2007) identified greater weathering intensities of clays as reflected in greater smectite abundances in soils developed on north-facing slopes than south-facing slope profiles. In these latter soils, smectite only occurred in surface horizons at the highest elevations. In addition, they noted that clay minerals in north-facing slope profiles underwent ionic substitutions in their octahedral sheets that led to reduction in their layer charge. A follow-up study (Egli et al., 2010) focused principally on the effects of exposure on organic matter and the effects of its degradation on soil chemistry. In this study, the authors found that there was a nonlinear relationship of soil organic carbon content to climate. The greatest amount of organic carbon was found to occur in soils developed at treeline. However, a strong aspect influence occurred as soils on north-facing slopes were found to contain greater amounts of organic matter, as well as a significantly lower degree of humification than south-facing slopes. Soils on north-facing slopes exhibited greater amounts of undecomposed and weakly degraded organic matter than soils on south-facing slopes. This difference was credited to lower temperatures and greater acidity and, thus, slower rates of decomposition. Overall, soils on north-facing slopes were more strongly leached than those on south-facing slopes as a result of easier transportation of fulvic acids and accompanying chelation of Fe and Al.

5.4.4 Soils and Parent Materials

Parent-material characteristics strongly influence the physical and chemical properties of soils. The principal parent-material properties that influence soil properties are those that control rock weathering: texture, porosity, drainage, and mineral composition (Ollier and Pain, 1996). Their influences are most strongly seen in such fundamental soil properties as color, structure, texture, clay mineralogy, and chemistry. Generally, the influence of parent material on soil properties is most strongly expressed in drier climates and in the early stages of soil development. In more humid environments and later stages of soil development, the influence of the other soil-forming factors tends to overshadow parent-material influences (Ollier and Pain, 1996; Birkeland, 1999).

Typically, light-colored parent materials give rise to light-colored soils, fine-textured parent materials give rise to fine-textured and strongly structured soils and coarse-textured parent materials give rise to coarse-textured and weakly-structured soils. Soil chemistry typically reflects parent-material chemistry, and the mineralogical composition of the parent material strongly controls the nature of the secondary clay species formed in soils (Ollier and Pain, 1996). A strong confounding factor to these general trends is the influence exerted by the addition of aeolian dust to soils, which has profound influences on all of these soil properties (Birkeland, 1999).

The influence of parent-material texture on soil development is especially well illustrated by the contrasting sequences of calcrete (i.e., calcic horizons) development in coarse- versus fine-grained parent materials in arid/semiarid environments (Gile et al., 1966, 1981; Machette, 1985; Dixon, 2013) (Fig. 5.7). Here the influence of contrasting patterns of water movement is particularly well seen (Holliday, 1994).

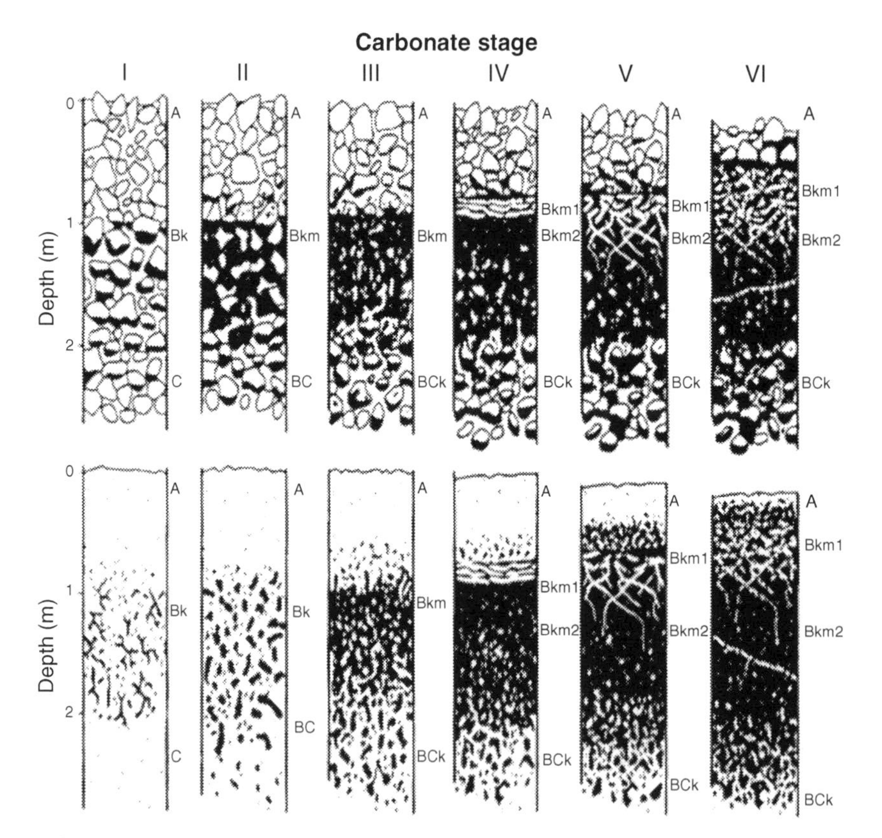

FIGURE 5.7 Morphological sequence of calcrete development in coarse-grained parent materials (top) and fine-grained parent materials (bottom). *(Adapted from Gile et al. (1966) and Machette (1985).)*

The influence of contrasting parent-material mineralogies has recently been well illustrated by a study of clay minerals formed in spodosols in the Italian Alps (Mirabella et al., 2002). Here all soil-forming factors except parent material are constant. Four spodosol profiles were investigated: two on till derived from granodiorite, one on slate-dominated till, and one on tonalite till. The profiles developed on the granodiorite-derived till displayed the presence of smectite derived from chlorite and interstratified vermiculite/mica in the "E" horizon and hydroxyl-interlayered vermiculite and smectite at depth. Soils developed on slate and tonalite-dominated tills lacked the "E" horizon smectite and vermiculite and exhibited hydroxyl-interlayered vermiculite/smectite through the entire depth of the profile.

Regionally, the soil order Andisols directly reflects the influence of their parent material volcanic ash. These soils develop dominantly on volcanic glass, a parent material possessing unique physical, chemical and mineralogical properties that can change with successive eruptive events (Dixon, 2002). Upon weathering, this parent material produces products that are predominantly noncrystalline aluminosilicates known as allophane, together with opaline silica and ferrihydrite. When humus is present, allophane–humus complexes are formed. Another secondary product commonly produced from the weathering of the volcanic ash is imogolite, which together with allophane tends to dominate subsoil horizons (Buol et al., 1989). Surface "A" horizons commonly display little evidence of weathering because of their high porosity. Under favorable drainage conditions, the subsurface horizons may develop into spodic horizons. Overall, however, these soils are weakly developed, are very freely drained, and possess low bulk densities; all direct reflections of the parent-material characteristics (Dixon, 2002).

5.5 SOIL PROPERTIES AND VEGETATION

Of all of the state factors identified by Jenny (1941) governing the formation of soils, general agreement appears to be that the role of the biotic (i.e., vegetation) factor is the most difficult to isolate (Birkeland, 1984; Schaetzl and Anderson, 2005). Jenny regarded the biotic factor to be dependent on other state factors and to principally control biochemical processes in soils, especially those related to nitrogen (Jenny, 1980). Initially the biotic factor was largely restricted to the role of plants with little consideration to the role of fauna (Schaetzl and Anderson, 2005). Vegetation patterns, although responding to climate, as well as to slope steepness and aspect, exert profound influences on moisture regulation, as well as influencing soil chemistry and mineralogy.

Plants function as geochemical pumps in the soil environment. They uptake biologically necessary nutrients from the soil in solution, use them in physiological processes, temporarily store them in plant tissue, and ultimately return them to the soil in litterfall, root decay, and decomposition of the plant upon its death (Amundson et al., 2007). Vegetation affects the availability and physical distribution of nutrients in unique ways. The cycling of nutrients, such as N, P, K, and Si by plants lead to upward migration within the soil and into the vegetation, commonly resulting in the surface enrichment of these elements in the soil surface (Lucas, 2001; Jobbágy and Jackson, 2004; Amundson et al., 2007). Plant-induced ascension of silicon has been proposed as the mechanism for the formation of kaolinite-rich horizons over Fe/Al-rich horizons in the warm tropics (Lucas, 2001). The release of Si from litterfall has also been proposed as the reason for the formation of Si-rich soil surface horizons (Amundson et al., 2007). Similarly, the formation of calcrete and its associated accumulation of Ca has been widely attributed to Ca pumping by vegetation (Klappa, 1979; Kosir, 2004). Plants also play a crucial role in weathering in the soil environment largely as a consequence of the huge amounts of CO_2 they produce and the effects of this on water-mineral reactions and release of chemical elements. In addition, plants produce chelates, which form soluble metal complexes that facilitate increased mineral solubility. Weathering in the Critical Zone is discussed in Chapter 4 by Pope.

At the regional scale, the role of vegetation in soil genesis is strongly expressed across the boundaries between the Great Plains/Eastern Temperate forests, and the Great Plains/North American Desert Ecoregion boundaries (United States Environmental Protection Agency, 2006). The Great Plains are dominated by mollisols with the influence of a predominantly grassland vegetation strongly imprinted on soil properties (Buol et al., 1989; Dixon, 2002). Mollisol formation is dominated by melanization, the darkening of soil by predominantly biological processes. These processes include the accumulation of organic materials from living and decaying grasses, soil mixing by organisms, eluviation and illuviation of organic materials and accompanying formation of dark-colored cutans, and the formation of dark-colored lignoproteins (Buol et al., 1989; Dixon, 2002).

Soils of the Eastern Temperate Forests are predominantly Alfisols and Ultisols (Brady and Weil, 2002). These soils are both characterized by the accumulation of clay in subsurface horizons, as well as the accumulation of organics and sesquioxides under favorable climatic regimes (Dixon, 2002). Decay of forest litter produces ligands, which play a crucial role in the transportation of sesquioxides by chelation (Buol et al., 1989). In addition, both of these soils experience considerable profile mixing as a result of treethrow.

The Western Desert Ecoregion is characterized by the widespread development of Aridisols. These soils are characterized, in part, by the accumulation of soluble salts and the notable absence of an organic "O" horizon (Southard, 2000). The source of salts, particularly calcium carbonate, is now widely ascribed to redistribution of atmospherically derived carbonate (Dixon and McLaren, 2009). However, plants play a crucial role in the concentration of carbonate within soil profiles. The occurrence of carbonate in the vicinity of plant roots is widely documented (Klappa, 1980; Phillips et al., 1987; McLaren,

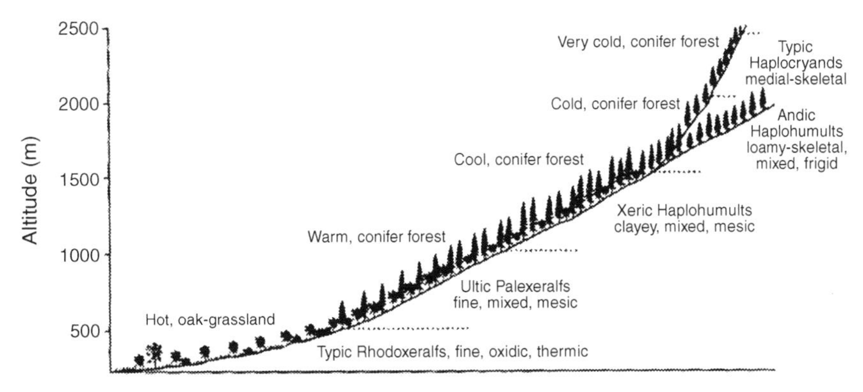

FIGURE 5.8 Relationship between soils and vegetation with increasing altitude in the Cascade Range of northern California. *(After Alexander et al. (1973).)*

1995; Dixon and McLaren, 2009). Mumm and Reith (2007) have demonstrated the complex role played by microorganisms and plants in biomediation of calcrete as biological processes modify parent-material chemistry.

Soil/vegetation relationships are perhaps most strongly demonstrated along topographic gradients. Alexander et al. (1973) investigated the relationship between soils and vegetation along a transect from a hot oak-grassland to a very cold coniferous forest (Fig. 5.8). The accompanying soils change from Alfisols under oak-grassland and warm coniferous forest, to Ultisols under cool coniferous forest and cold coniferous forest, to Andisols under very cold coniferous forest. Accompanying these soil changes is a progressive increase in organic-carbon content.

The relationship between vegetation distribution and soil properties has recently been investigated in the mountains of northern Sweden by Darmody et al. (2004). These investigators identified strong soil/vegetation associations largely influenced by soil moisture and nutrient availability, with strong elevation/climate influences. Spodosols dominate at lowest elevations and are strongly associated with birch and *Empetrum* heath vegetation. Between 600 m and 900 m asl, the dominant soils are Entisols and are associated with wet meadow and willow vegetation. These soils are characterized by high CEC, pH, Ca, Mg, K, and organic C and N. Soils developed under *Dryas octopetala* (650–900 m asl) display the highest pH, base saturation and Ca contents of any of the soils in the area. The soils are predominantly Entisols and Inceptisols, but at one location a Mollisol occurs. Soils in the alpine regions are the least well developed. Plant cover is minimal with a thin covering of cryptogams, at best. These soils display the lowest pH, base saturation, Ca, N and most of the other soil-fertility measures. The Al content of these soils is exceptionally high. Ca/Al ratios are exceptionally low, which may indicate Al phytotoxicity. The associated soils are dominantly Entisols, but where permafrost is present, gelisols are developed.

On a somewhat smaller scale, Dixon (1983) investigated the relationships between vegetation and soil chemistry and clay mineralogy of soils developed on Holocene-age soils in the Front Range of the Rocky Mountains. He found that under tundra- meadow- vegetation covers, clay mineralogy was dominated by hydrobiotite and vermiculite. Under higher life forms, such as spruce and willow, secondary minerals were dominated by smectite and interlayered smectite–vermiculite. Chemical trends in soil "B" horizons similarly reflected greater leaching under spruce and willow and less intense leaching under tundra meadow.

5.5.1 Pedogenesis and Time

Soil and regolith profile development and associated complexity are strongly time dependent (Birkeland, 1999; Anderson et al., 2007; Brantley et al., 2007). Birkeland (1999) pointed out that the time necessary for the production specific soil and weathering features varies. Generally speaking, however, soil properties associated with organic matter accumulation proceed rapidly whereas those associated with weathering generally proceed more slowly. Soil and regolith profiles generally thicken over time and horizon complexity increases. Increasing soil complexity is the result of progressive accumulation of materials resulting from additions, transformations and translocations as discussed previously. With time generally occurs a progressive buildup of soil profile form and constituents (Johnson and Watson-Stegner, 1987). This is referred to as progressive pedogenesis. Alternatively pedogenesis may be regressive: a situation in which soils no longer change or evolve in the same direction. Curves for the buildup of soil properties are initially typically steep but eventually the curves flatten out indicating less visible change with time. This flattening out is referred to soil steady state, which may be only transitory as environmental conditions change.

5.6 CONCLUSIONS

Soil constitutes a significant component of the Critical Zone. It represents the "membrane" across and through which water and solutes, energy, gases, solids, and organisms are exchanged with the atmosphere, biosphere, hydrosphere, and lithosphere. Water is the driving force of the majority of these exchanges within the soil, as well as the Critical Zone as a whole.

Effective water availability is strongly controlled by climate at all scales, by topographic form and aspect, and by vegetation type and abundance. Soil formation within the Critical Zone is a function of interactions between climate, geology, and vegetation. Whereas climate and geology are largely independent state factors, vegetation–controlled processes appear to be most appropriately regarded as dependent state factors. Whereas vegetation patterns and processes respond strongly to climatic and geologic processes, they clearly have profound influences on climate at multiple scales, as well as on geologic processes.

REFERENCES

Alexander, E.B., Mallory, J.I., Colwell, W.L., 1973. Soil-elevation relationships on a volcanic plateau in the southern Cascade Range, northern California, USA. Catena 20, 113–128.

Amundson, R., Richter, D.D., Humphreys, G.S., Jobbágy, E.G., Gaillardet, J., 2007. Coupling between biota and earth materials in the critical zone. Elements 3, 327–332.

Anderson, S.P., von Blanckenburg, F., White, A.F., 2007. Physical and chemical controls on the critical zone. Elements 3, 315–319.

Birkeland, P.W., 1984. Soils and Geomorphology. Oxford University Press, New York.

Birkeland, P.W., 1999. Soils and Geomorphology, third ed. Oxford University Press, New York.

Brady, N.C., Weil, R.R., 1999. The Nature and Properties of Soils, twelfth ed. Prentice Hall, New Jersey.

Brantley, S.L., Goldhaber, M.B., Ragnarsdottir, K.V., 2007. Crossing disciplines and scales to understand the critical zone. Elements 3, 307–314.

Brantley, S.L., White, T.S., White, A.F., Sparks, D., Richter, D., Pregitzer, K., Derry, L., Chorover, J., Chadwick, O., April, R., Anderson, S., Amundson, R., 2006. Frontiers in exploration of the critical zone. Report of a workshop sponsored by the National Science Foundation. Newark, Delaware, USA.

Buol, S.W., Hole, F.D., McCracken, R.J., 1989. Soil Genesis and Classification. Iowa State University Press, Ames, Iowa.

Carter, B.J., Ciolkosz, E.J., 1991. Slope gradient and aspect effects on soils developed from sandstones in Pennsylvania. Geoderma 49, 199–213.

Charover, J., et al. 2011. How water, carbon, and energy drive critical zone evolution: The Jemez-Santa Catalina critical zone observatory. Vadose Zone J. 10, 884–899.

Conacher, A.J., Dalrymple, J.B., 1977. The nine unit landsurface model: an approach to pedogeomorphic research. Geoderma 18, 1–154.

Cooper, A.W., 1960. An example of the role of microclimate in soil genesis. Soil Science 90, 109–120.

Curtis, C.D., 1990. Aspects of climatic influence on the clay mineralogy and geochemistry of soils, palaeosols and clastic sedimentary rocks. J. Geol. Soc. London 147, 351–357.

Dalrymple, J.B., Blong, R.R.J., Conacher, A.J., 1968. A hypothetical landsurface model. Z. Geomorphol. 12, 60–76.

Dan, J., Yaalon, D.H., 1982. Automorphic saline soils in Israel. Catena Suppl. 1, 103–115.

Darmody, R.G., Thorn, C.E., Schlyter, P., Dixon, J.C., 2004. Relationship of vegetation distribution to soil properties in Kärkevagge, Swedish Lapland. Arctic Antarct. Alp. Res. 36, 21–32.

Dixon, J.C., 1983. Chemical weathering of late quaternary cirque deposits, Colorado Front range. PhD Dissertation. University of Colorado, Boulder.

Dixon, J.C., 1986. Solute movement on hillslopes in the alpine environment. In: Abrahms, A.D. (Ed.), Hillslope Processes. Allen and Unwin, London, pp. 139–159.

Dixon, J.C., 2002. Weathering and soils. In: Orme, A.R. (Ed.), The Physical Geography of North America. Oxford University Press, Oxford, pp. 178–198.

Dixon, J.C., 2004. Weathering. In: Goudie, A. (Ed.), Encyclopedia of Geomorphology. Routledge, London, pp. 1108–1112.

Dixon, J.C., 2009. Aridic soils, patterned ground, and desert pavements. In: Parsons, A., Abrahams, A.D. (Eds.), Geomorphology of Desert Environments. Springer, London, pp. 101–122.

Dixon, J.C., 2013. Pedogenesis with respect to geomorphology. In: Pope, G.A. (Ed.), Treatise on Geomorphology, vol. 4: Weathering and Soils Geomorphology. Academic Press, San Diego, pp. 27–43.

Dixon, J.C., McLaren, S., 2009. Duricrusts. In: Parsons, A., Abrahams, A.D. (Eds.), Geomorphology of Arid Environments. Springer, London, pp. 123–152.
Dixon, J.C., Thorn, C.E., 2005. Chemical weathering and landscape development in mid-latitude alpine environments. Geomorphology, 127–145.
Dokuchev, V.V., 1883. Risskii Chernozem. Moscow.
Egli, M., Mirabella, A., Sartori, G., Fitze, P., 2003. Weathering rates as a function of climate: results from a climosequence of the Val Genova (Trentino, Italian Alps). Geoderma 111, 99–121.
Egli, M., Mirabella, A., Sartori, G., Giaccai, D., Zanelli, R., Plotze, M., 2007. Effect of slope aspect on transformation of clay minerals in Alpine soils. Clay Miner. 42, 373–398.
Egli, M., Sartori, G., Mirabella, A., Giaccai, D., 2010. The effects of exposure and climate on the weathering of late Pleistocene and Holocene Alpine soils. Geomorphology 114, 466–482.
Eswaran, H., Van Den Berg, E., Reich, P., 1993. Organic carbon in soils of the world. Soil Sci. Soc. Am. J. 57, 192–194.
Finney, H.R., Holowaychuk, N., Heddleson, M.R., 1962. The influence of microclimate on the morphology of certain soils of the Allegheny Plateau of Ohio. Soil Sci. Soc. Am. Proc. 33, 755–761.
Gile, L.H., Peterson, F.F., Grossman, R.B., 1965. The K horizon: a master soil horizon of carbonate accumulation. Soil Sci. 99, 74–82.
Gile, L.H., Hawley, J.W., Grossman, R.B., 1981. Soils and geomorphology in the basin and range area of southern New Mexico. Guidebook to the Desert Project. New Mexico Bureau of Mines and Mineral Resources. Memoir 39. Socorro New Mexico.
Gile, L.H., Peterson, F.F., Grossman, R.B., 1966. Morphological and genetic sequences of carbonate in desert soils. Soil Sci. 101, 347–360.
Holliday, V. T., 1994. The "State Factor" approach in geoarcheology. In: Factors of Soil Formation: A Fiftieth Anniversary Retrospective. Soil Science Society of America. Special Publication, vol. 33, pp. 65–86.
Hunckler, R.V., Schaetzl, R.J., 1997. Spodosol development as affected by geomorphic aspect, Baraga County, Michigan. Soil Sci. Soc. Am. J. 61, 1105–1115.
Jenny, H., 1935. The clay content of the soil as related to climatic factors, particularly temperature. Soil Sci. 40, 111–128.
Jenny, H., 1941. Factors of Soil Formation. McGraw Hill, New York.
Jenny, H., 1958. Role of the plant factor in the pedogenic functions. Ecology 39, 5–16.
Jenny, H., 1961. Derivation of state factor equations and ecosystems. Soil Sci. Soc. Am. Proc. 25, 385–388.
Jenny, H., 1980. The Soil Resource: Origin and Behavior. Ecological Studies, vol. 37. Springer-Verlag, New York.
Jobbágy, E.G., Jackson, R.B., 2004. The uplift of soil nutrients by plants: biogeochemical consequences across scales. Ecology 85, 2380–2389.
Johnson, D.L., Watson-Stegner, D., 1987. Evolution model of pedogenesis. Soil Sci. 143, 349–366.
Johnson, D.L., Keller, E.A., Rockwell, T.K., 1990. Dynamic pedogenesis: new views on some key soil concepts and a model for interpreting Quaternary soils. Quarter. Res. 33, 306–319.
Kampf, N., Schwertmann, U., 1983. Goethite and hematite in a climosequence in southern Brazil and their application in classification of kaolinite soils. Geoderma 29, 363–381.
Klappa, C.F., 1979. Calcified filaments in Quaternary calcretes: organo-mineral interactions in the subaerial vadose environment. J. Sediment. Petrol. 49, 955–968.
Klappa, C.F., 1980. Rhizoliths in terrestrial carbonates: classification, recognition, genesis and significance. Sedimentology 27, 613–629.
Klemmedson, J.O., 1964. Topofunction of soils and vegetation in a landscape. J. Am. Soc. Agron. Special Publ. 5, 176–189.

Kosir, A., 2004. Microcodium revisited: root calcification products of terrestrial plants on carbonate-rich substrates. J. Sediment. Res. 74, 845–857.

Lin, H., 2010. Earth's Critical Zone and hydropedology: concepts, characteristics, and advances. Hydrol. Earth Syst. Sci. 14, 25–45.

Lucas, Y., 2001. The role of plants in controlling rates and products of weathering: importance of biological pumping. Annu. Rev. Earth Planet. Sci. 29, 135–163.

Machette, M., 1985. Calcic soils of the southwest United States. Geol. Soc. Am. Special Pap. 203, 1–21.

Macyk, T.M., Pawluk, S., Lindsay, D., 1978. Relief and microclimate as related to soil properties. Can. J. Soil Sci. 58, 421–438.

McFadden, L.D., 1988. Climatic influences on rates and processes of soil development in Quaternary deposits of southern California. Geol. Soc. Am. Special Pap. 216, 153–177.

McLaren, S.J., 1995. Early diagenetic fabrics in the rhizosphere of late Pleistocene aeolian sediments. J. Geol. Soc. London 152, 173–181.

Milne, G., 1935. Some suggested units for classification and mapping, particularly for east African soils. Soil Res. 4, 183–198.

Milne, G., 1936. Normal erosion as a factor in soil profile development. Nature 138, 548.

Mirabella, A., Egli, M., Carnicelli, S., Sartori, G., 2002. Influence of parent material on clay minerals formation in Podsols of Trentino, Italy. Clay Miner. 37, 699–707.

Mumm, A.S., Reith, F., 2007. Biomediation of calcrete at the gold anomaly of the Barns prospect, Gawler Craton, South Australia. J. Geochem. Explor. 92, 13–33.

National Research Council, 2001. Basic Research Opportunities in Earth Science. National Academy Press, Washington, DC.

Ollier, C., Pain, C., 1996. Regolith, Soils and Landforms. John Wiley and Sons, Chichester.

Phillips, S.E., Milnes, A.R., Foster, R.C., 1987. Calcified filaments: an example of biological influences in the formation of calcrete in South Australia. Aust. J. Soil Res. 25, 405–428.

Post, W.M., pastor, J., Zinke, P.J., Stangenberger, A.G., 1985. Global pattern of soil nitrogen storage. Nature 317, 613–616.

Rasmussen, C., Troch, P.A., Chorover, J., Brooks, P., Pelletier, J., Huxman, T.E., 2011. An open system framework for integrating critical zone structure and function. Biogeochemistry 102, 15–29.

Rech, J.A., Reeves, R.W., Hendricks, D.M., 2001. The influence of slope aspect on soil weathering processes in the Springerville volcanic field, Arizona. Catena 43, 49–62.

Ruhe, R.V. 1960. Elements of the soil landscape. Transactions of the 7th International Congress of Soil Science. Madison, WI. pp. 165–170.

Ruhe, R.V., 1975. Climatic geomorphology and fully developed slopes. Catena 2, 309–320.

Ruhe, R.V., 1984a. Loess-derived soils of the Mississippi Valley region: II. Soil–climate system. Soil Sci. Soc. Am. J. 48, 864–867.

Ruhe, R.V., 1984b. Soil–climate system across the prairies in Midwestern U.S.A. Geoderma 34, 201–219.

Runge, E.C.A., 1973. Soil development sequences and energy models. Soil Sci. 115, 183–193.

Schaetzl, R.J., 2013. Catenas and soils. In: Pope, G.A. (Ed.), Treatise on Geomorphology, Weathering and soils geomorphology, vol. 4. pp. 145–158.

Schaetzl, R., Anderson, S., 2005. Soils: Genesis and Geomorphology. Cambridge University Press, Cambridge UK.

Schlesinger, W.H., 1990. Evidence from chronosequence studies for a low carbon-storage potential for soils. Nature 348, 232–234.

Schwertmann, U., Murad, E., Schulze, D.G., 1982. Is there Holocene reddening (hematite formation) in soils of axeric temperate areas? Geoderma 27, 209–223.

Sherman, G.D., 1952. The genesis and morphology of the alumina-rich lateritic clays. In: Problems of Clay and Laterite Genesis. American Institute of Mining and Metallurgical Engineers Symposium. St. Louis. pp. 154–161.

Simonson, R.W., 1978. A multiple-process model of soil genesis. In: Mahaney, W.C. (Ed.), Quaternary Soils. Geo-Abstracts, Norwich, UK, pp. 1–25.

Soil Survey Staff, 1994. Keys to Soil Taxonomy, sixth ed. United States Department of Agriculture. Pocahontas Press, Blacksburg.

Southard, R., 2000. Aridisols. In: Sumner, M.E. (Ed.), Handbook of Soil Science. CRC Press, Boca Raton, pp. E321–E338.

Strakhov, N.M., 1967. Principles of Lithogenesis. Oliver and Boyd, Edinburgh, UK.

Tandarich, J.P., Darmody, R.G., Follmer, L.R., 1994. The pedo-weathering profile: a paradigm for whole-regolith pedology from the glaciated midcontinental United States of America. Soil Sci. Soc. Am. Special Publ. 34, 97–117.

Tardy, Y., Bocquier, G., Paquet, H., Millot, G., 1973. Formation of clay from granite and its relation to climate and topography. Geoderma 10, 271–284.

United States Environmental Protection Agency, 2006. Ecoregions of the Continental United States. U.S. Government printing office, Washington, DC.

Walker, P.H., Ruhe, R.V., 1968. Hillslope models and soil formation: II Closed Systems. Trans. 9th Intl. Congr. Soil Sci. 4, 561–568.

Webb, T.H., Campbell, A.S., Fox, F.B., 1986. Effect of rainfall on pedogenesis in a climosequence of soils near Lake Pukaki, New Zealand. NZ J. Geol. Geophys. 29, 323–334.

White, A.F., Blum, A.E., 1995. Effects of climate on chemical weathering rates in watersheds. Geochim. Cosmochim. Acta 59, 1729–1747.

Yaalon, D.H., 1975. Conceptual models in pedogenesis: can the soil-forming functions be solved? Geoderma 14, 189–205.

Chapter 6

Soil Geochemistry in the Critical Zone: Influence on Atmosphere, Surface- and Groundwater Composition

Julia Perdrial*, Aaron Thompson**, and Jon Chorover†
*Department of Geology, University of Vermont, Burlington, Vermont, USA; **Department of Crop and Soil Science, University of Georgia, Athens, Georgia, USA; †Department of Soil Water and Environmental Science, University of Arizona, Tucson, Arizona, USA

6.1 INTRODUCTION

The capacity of the Critical Zone to support life derives, in large part, from the structured fabric of soil. This living porous medium is formed from chemical reactions occurring among minerals, microbes, natural organic matter (NOM), solutes, and mobile fluids that are the focus of the field of soil biogeochemistry. Soil, as defined here, is thus inclusive of all unconsolidated material above bedrock (e.g., the regolith profile including mobile colluvium and saprolite). This definition of soil is consistent with early notions of soil as "a thin layer over the Earth's surface" or "a medium for plant growth" (Bockheim et al., 2005; Churchman, 2010; Simonson, 1968) and its conceptualization as "a natural body independent of surface rocks and biota" (Coffey, 1909; Dokuchajev, 1883; Hilgard, 1882; Jenny, 1941; Shaler, 1892), as well as more recent concepts of soil as a landscape-scale expression of geomorphologic forces (Daniels and Hammer, 1992; Huggett, 1975).

This broad conceptualization of soil provides a framework for scaling biogeochemical approaches and fosters development of transdisciplinary research (Fig. 6.1). Soil is "the most complex biomaterial on the planet" (Young and Crawford, 2004), and so it is best understood by combining multiple disciplinary perspectives. In the landscape, soil operates as a biogeochemical reactor (Fig. 6.1), as discussed originally by Targulian and Sokolova (1996). This reactor is open to matter and energy fluxes, and subject to the constraints of elemental mass-balance (Chadwick et al., 1990; Crompton, 1960; Marshall and Haseman, 1942), and energy-driven (Minasny et al., 2008; Rasmussen, 2011; Rasmussen et al., 2005; Rasmussen and Tabor, 2007; Runge, 1973; Schaetzl and Schwenner, 2006; Volobuyev, 1964) concepts of pedogenesis. The soil reactor is

Developments in Earth Surface Processes, Vol. 19. http://dx.doi.org/10.1016/B978-0-444-63369-9.00006-9

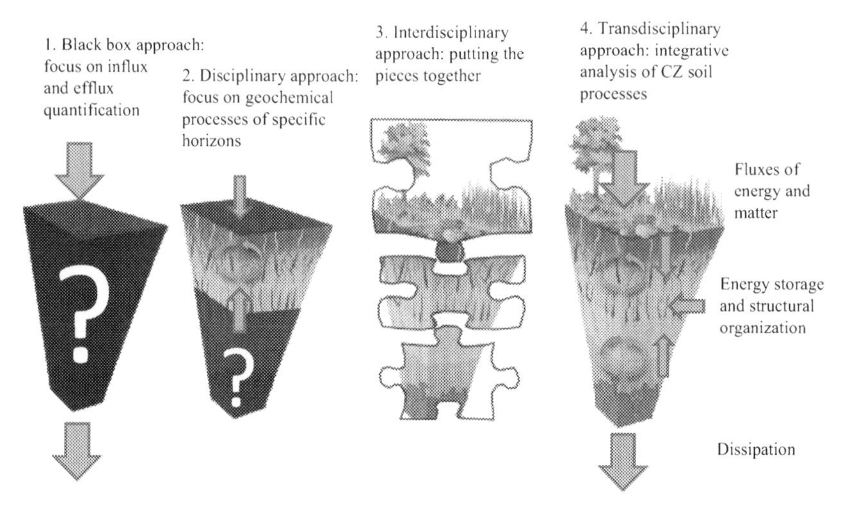

FIGURE 6.1 Different approaches used to study soil geochemical processes. The goal of integrative analysis of CZ processes is to combine all previous approaches in a synergistic effort *(after Chorover et al. (2007) and Rasmussen et al. (2011).)*

fed by material inputs (in gas, liquid, and solid form) from weathering rock and the atmosphere, including water and carbon influxes (see Section 6.2). Energetic inputs include heat and reduced carbon compounds fixed through photosynthesis. Reactants are reprocessed (see Section 6.3) within the soil by abiotic and biologically catalyzed reactions, with impacts on surface water, groundwater and atmospheric systems resulting from their receipt of the dissipative products of these reactions. Such dissipative products include those of, for example, respiration (CO_2), chemical and physical denudation (dissolved elements and mobilized particulates) and latent heat transfer (water vapor), that are transported through and beyond the CZ by volatilization, erosion, or water (stream/groundwater) fluxes (Banwart et al., 2011; Rasmussen et al., 2011).

Biogeochemical processes in the soil reactor occur at the molecular scale but influence (or reach) macroscopic to global scales. For example, soils constitute the largest terrestrial carbon (C) sink and accommodate roughly three times more C than the atmosphere (Houghton, 2007; Lal, 2005; Trumbore, 2006), but molecular-scale mechanisms of organo-mineral and organo-metal interaction are responsible for this C sequestration capacity (see Section 6.2.1). Another example is the soil's capacity for removing excess nutrients and contaminants from water, which is successfully relied on for on-site wastewater systems (i.e., septic tank leachfields, USEPA (2002)) and retardation of radioactive waste (Yasunari et al., 2011; Zachara et al., 2007). A suite of molecular-scale processes including sorption and transformation reactions (see Section 6.3) are responsible for providing this important service of "natural attenuation". Finally, it is molecular scale soil processes that provide ecosystems with life sustaining nutrients, including the fixation of nitrogen via microbially mediated redox reactions

(Sections 6.2 and 6.3.3), and the dissolution of P and other lithogenic macro- and micronutrients (e.g., Ca, Mg, K, Fe, Zn, etc.) via mineral weathering reactions (see Section 6.3.4). Chemical weathering in soils is the ultimate sink for atmospheric CO_2 in the geological carbon cycle, as it results in proton consumption that enables carbonate release to rivers and oceans where it is subsequently buried in marine sediments, a process of critical importance to climate sustainability (Gaillardet et al., 2011; Gislason et al., 2009).

6.2 MATERIAL FOR SOIL GEOCHEMICAL REACTIONS

Geochemical elements are fed to the soil reactor through multiple pathways that include biologic fixation, wet and dry deposition, and bedrock weathering. In this section we describe the processes leading to the introduction and stabilization of C, N and lithogenic elements. Many of the elements required for life derive from rock weathering (Ca, Mg, K, Fe, Zn, etc.), whereas others are present only at trace amounts in silicate rock (e.g., C, N) and therefore derive from biological fixation of atmospheric species (Fig. 6.2).

6.2.1 Carbon

Natural Organic Matter (NOM) enters the CZ via photosynthesis by autotrophic organisms (dominantly vascular plants) that transform atmospheric CO_2 into biomass and thereby provide a constant supply of biomolecular material in the form of decaying plant and microbial debris (Essington, 2002; Sparks, 2005). As a result, soils are significantly enriched in C relative to parent rock (Sposito, 2008). As a reduced substrate, NOM is the primary source of energy and carbon for

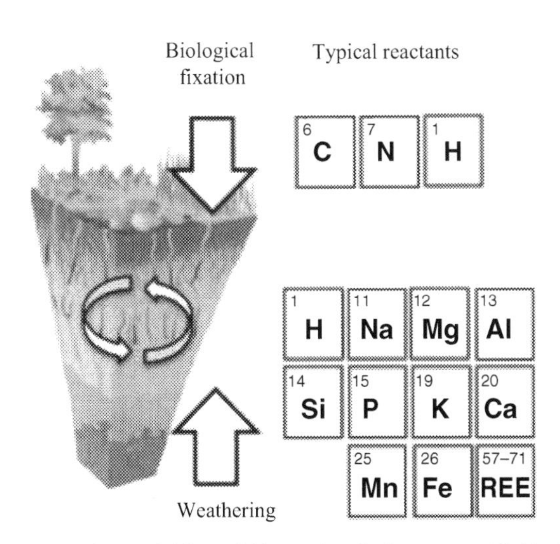

FIGURE 6.2 Sources of material for soil biogeochemical processes *(Critical Zone (CZ) sketch modified after Chorover et al. (2007) and Rasmussen et al. (2011).)*

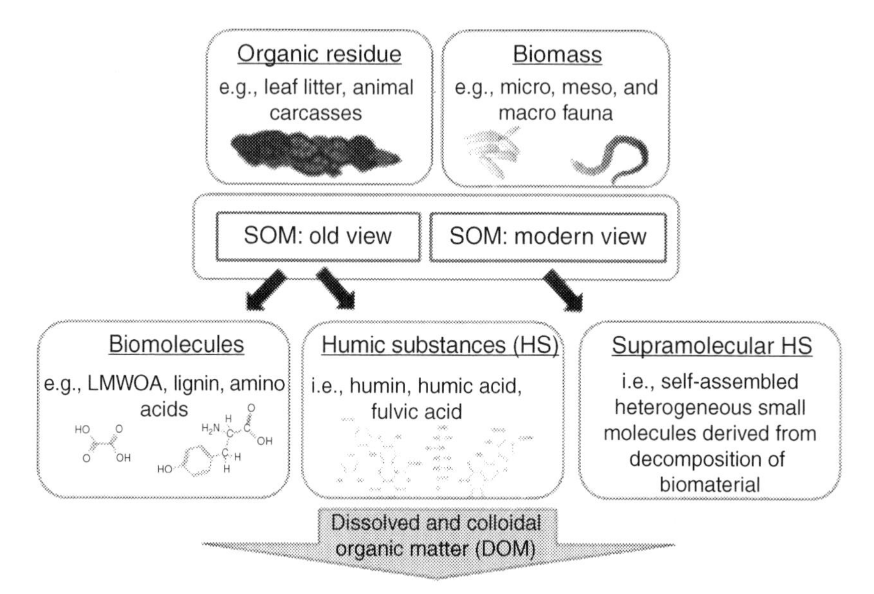

FIGURE 6.3 Schematic sketch of SOM showing the old view and new view of HS.

heterotrophic microorganisms, and the reactions fueled by NOM inputs underpin numerous foundational soil biogeochemical reactions as discussed in Section 6.3. Heterotrophic soil microbes oxidize this reduced C, producing soil partial pressures of CO_2 (P_{CO_2}) that are highly variable in space and time, and that can exceed atmospheric values by orders of magnitude with important implications for soil geochemistry (see Section 6.3). Different pools of C can be distinguished based on their origin and persistence. For example, plant debris, leaf litter, and animal carcasses are collectively termed *organic residues* and form one important pool of young, soil organic materials (Essington (2002), Fig. 6.3). A wealth of living organisms depend on this pool including macro and meso fauna as well as microorganisms, which themselves represent a second pool of organic material, termed *soil biomass* (Sparks, 2005). The third pool, broadly termed *soil organic matter* (SOM), is a result of the decomposition and rearrangement of materials from the first two pools and is more persistent in soils.

Traditionally, SOM was separated into two groups: SOM components that could be recognized as biochemical molecular classes were termed biomolecules (or nonhumic substances); whereas molecules with only remote resemblance of their precursor material were defined as humic substances (HS). The latter were long considered to represent high molar mass (>10,000 Da) polymerized degradation products of biomolecules. However, understanding of the molecular structure of SOM has advanced and more recent research suggests that SOM consists of smaller molecules, dominantly biomolecular fragments, that are aggregated to supramolecular structures (Kelleher and Simpson, 2006; Piccolo, 2002). In these structures, molecules are bound together via hydrophobic interaction, hydrogen bonding, and cation bridging that impart macromolecular characteristics of high molecular weight compounds (Chorover, 2011; Leenheer et al., 2001; Sutton and Sposito, 2005).

The modern view of SOM is central to understanding soil C dynamics since all previously mentioned bonding mechanisms are based on relatively weak interactions, rendering these SOM aggregates highly reactive and dynamic. Solution geochemistry exerts a strong control on these supramolecular entities where, for example, the presence of polyvalent cations increases the tendency for aggregation whereas the presence of low molecular weight organic acids (LMWOAs) has the opposite effect (Piccolo, 2002; Simpson, 2002).

Dissolved organic matter (DOM) represents that fraction of the waterborne organic matter (OM) pool that is, in most cases, defined operationally as passing through ca. 0.45 μm filtration. Although DOM represents only a small fraction of total SOM, it is the most mobile and influences most terrestrial and aquatic biogeochemical processes (Bolan et al., 2011; Gabor et al., 2014; McGill et al., 1986). The characteristics of DOM are related to their source (e.g., throughfall, leaf leachate, soil pore water, hyporheic zones, and groundwater, see Inamdar et al. (2012) and Singh et al. (2014)). For each of these source reservoirs, molecular spectroscopic indices has shown systematic variation with location in the CZ. For example, the fluorescence index, a measure of OM provenance (McKnight et al., 2001), indicates microbially-derived DOM increases as a fraction of total DOM for waters exhibiting longer residence time in the CZ. Conversely, UV specific absorbance (i.e., absorbance at 254 nm normalized to dissolved organic C concentration), an indicator for DOM aromaticity (Weishaar et al., 2003), decreases systematically from soils to GW (Inamdar et al., 2011).

NOM composition changes during its journey through the CZ as different sources of DOM mix and they are acted on by microbial cycling and abiotic fractionation processes (processes that preferentially remove certain portions of DOM). The composition and character of NOM products in stream and groundwater are therefore source and process dependent (Boyer et al., 1997; Perdrial et al., 2014; Raymond and Saiers, 2010; van Verseveld et al., 2008). The fractionation of DOM by mineral surfaces, for example, strongly influences mobile OM composition. The relative retention of higher molar mass, aromatic fractions of DOM on soil particle surfaces has been investigated experimentally using metal (oxyhydr)oxides, smectite, and whole soils (Chorover and Amistadi, 2001; Guo and Chorover, 2003; Vázquez-Ortega et al., 2014). OM that is not attenuated is transported from soil as dissolved and colloidal products to influence surface- and groundwater composition. As mentioned C plays multiple roles in soils and is involved in many different soil biogeochemical reactions that are detailed in Section 6.3.

Biomolecules delivered to soil through autotrophic synthesis and microbial turnover exhibit varying degrees of lability to decomposition. When free from mineral and metal complexes, certain classes of biomolecules, such as lignin, exhibit greater kinetic stability that slows (but does not eliminate) microbial degradation relative to other classes of biomolecules, such as saccharides that are more readily mineralized to CO_2. Degradation rates of SOM, therefore, exhibit strong dependence on the relative concentrations of various biomolecular types (Berg and Meentemeyer, 2002; Talbot et al., 2012). However, as suggested earlier and discussed in more detail later, SOM can be protected from microbial processing by its interactions with other co-aggregated soil components and, therefore, even inherently labile biomolecular forms can be stabilized over long time scales (Houghton et al., 1998; Kaiser and Guggenberger, 2000; Kleber and Johnson, 2010; Lal, 2004; Schmidt et al., 2011). In both cases, molecular-scale processes are ultimately responsible for the formation of a global-scale C sink.

6.2.2 Nitrogen

Like carbon, nitrogen (N) is also greatly enriched in soils relative to parent rock (Sposito, 2008). This is because although it is abundant as N_2 gas in the Earth's atmosphere (78%), oxidizing it to soluble forms (e.g. N-fixation) requires enzymatic catalysis. Enzymatic catalysis of N-fixation can be accomplished by either free-living, N-fixing bacteria such as *Azotobacter* or by bacteria living in close symbiosis with roots of leguminous higher plants, such as *rhizobia*. N-fixation also can occur abiotically by lightning, but this pathway is small relative to those mediated microbially. Because the fixation process comes with a high metabolic cost, N-fixation is closely linked to the aerobic respiration of organic C (Schlesinger and Bernhardt, 2013).

The dominant soil N species and its mobility depend largely on environmental conditions. For example under conditions of low redox potential (E_h) that are prevalent in waterlogged soils, the reduced N-species, NH_4^+, dominates and, as a cation, can interact via *adsorption* (see Section 6.3) with negatively charged exchange sites on clay minerals or OM that predominate in temperate zone soils. Under oxic conditions, aerobic nitrification reactions mediated by nitrifying bacteria oxidize NH_4^+ to produce nitrate (NO_3^-) (Barsdate and Alexander, 1975; Kowalchuk and Stephen, 2001). As an anion, NO_3^- is subject to electrostatic repulsion from the same types of negatively charged surface sites that retain NH_4^+ and, as a result, it is readily transported through soil profiles that comprise dominantly negatively charged soil particles. Conversely, NO_3^- is adsorbed and NH_4^+ is mobilized in soils comprising predominantly positively charged surface sites, as is the case for highly weathered tropical soils (e.g., oxisols) that are dominated by Fe and Al (oxyhydr)oxides and kaolin group minerals. Organic N in soils is mostly present as peptide (amide) N and is, as an important component of SOM, closely linked to the fate of SOM and the reactive transport of DOM more generally. Although N is added to the soil profile from the atmosphere, it is recycled many times to meet CZ nutrient requirements (McNeill and Unkovich, 2007). Characteristic of the Anthropocene, global N cycling reflects a large human imprint; anthropogenic activities have altered the nitrogen cycle of many CZ ecosystems (Vitousek et al., 1997). To a greater extent than other elements, the global N cycle is affected by its delivery to soils as fertilizer or in the form of nitrogen oxides (NO_x) via wet and or dry deposition downwind from industrialized areas (Likens et al., 1972; Rice and Herman, 2012).

6.2.3 Lithogenic Elements

Elements derived from the weathering of rock, i.e., lithogenic elements, are transferred to the soil incipiently at the weathering front (i.e., bedrock–regolith interface), which resides at the bottom of the weathering profile (Brantley et al., 2007; Holbrook et al., 2014). As discussed in Chapter 4, weathering of primary minerals supplies the CZ with silicic acid and soluble metals (Group 1 and 2 cations, transition metal cations, nutrient and toxic metal(loid)s and nonmetals including P, S, Se and As and other trace metal(loid)s including the rare earth elements, etc.). Weathering consumes protons and/or CO_2 and O_2 in proton-promoted and oxidative weathering reactions, respectively. Rock-derived elements released to solution are either: (1) recycled in the CZ – as is the case for bio-essential lithospheric elements (e.g., Ca, K, and Si, Amundson et al. (2007)); (2) coprecipitated during the formation of secondary minerals (e.g., Al, Fe, and/or Si in clay minerals and/or metal (oxyhydr)oxides, (see Section 6.3); or (3) released to surface waters as dissipative products of CZ evolution (e.g., Na, Mg, bicarbonate).

Lithogenic element accretion in soils also results from aeolian deposition, which includes salts as well as mineral dust (Derry and Chadwick, 2007). Dust deposition can be the dominant soil source of lithogenic elements in arid regions where weathering rates are low, and it can also be an essential contributor to nutrient demands of ecosystems in highly weathered soils that have become depleted in elements such as P and Ca (Derry and Chadwick, 2007). Especially in nutrient poor soils, for example, intensely weathered soils in Hawaii (Carrillo et al., 2002; Derry and Chadwick, 2007) or soils on the Colorado plateau (Reynolds et al., 2006), deposition can supply the majority of nutrients. Atmospheric deposition from anthropogenic activities such as smelting operations and fossil fuel combustion contribute to the near surface accumulation of metals such as Mn and Pb (Herndon et al., 2010).

6.3 SOIL BIOGEOCHEMICAL REACTIONS AND THEIR IMPACT ON THE COMPOSITION OF ATMOSPHERE, SURFACE- AND GROUNDWATERS

6.3.1 Soil Fluids Determine the Reach of Biogeochemical Processes

Soils are variably saturated, aqueous heterogeneous systems where water and air serve as the fluid media for material exchange that fuels microbial activity, mineral weathering, nutrient and toxic element transformation, organic compound

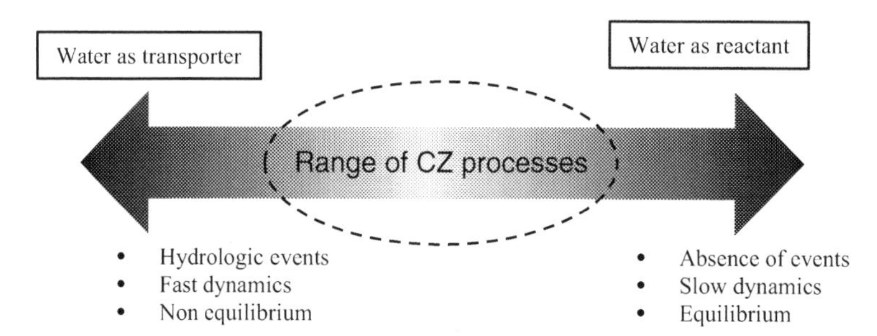

FIGURE 6.4 Conceptualization of two end member roles of water that impact soil geochemical processes, connects different Critical Zone (CZ) segments, and transfers external forcings to the CZ.

biodegradation, and their composite effects on soil formation. Both gaseous and aqueous phases of soil contain and transport reactants/products throughout different segments of the CZ and to external portions of the Earth System (e.g., atmosphere, hydrosphere; Perdrial et al. (2014)). Soils with high water throughput become depleted of mobile elements (Berner and Berner (2012) and references herein) (Fig. 6.4). In contrast, soils with low water throughput accumulate relatively soluble solids, including for example, carbonate, sulfate or even (under extremely arid conditions) nitrate and chloride minerals along with co-precipitated cations (Ca, Mg, Na) because precipitation is small relative to potential evapotranspiration and hence weathering products are not removed. Because of its functionality as a polar and weakly acidic solute, water is itself a reactant in many important CZ reactions and processes (Fig. 6.4). For example, water hydrates solids and solutes; self-dissociates/protonates as a weak acid; and serves as a reactant in several kinds of soil geochemical reactions, including water incorporation into secondary minerals, that are detailed in the following sections.

Like the above-ground atmosphere, the soil gas phase contains N_2, O_2, and CO_2 as well as other trace gases (Ar, NO, N_2O, NH_3, CH_4, H_2S), but their partial pressures can differ significantly over space and time, with much greater heterogeneity than that of the well-mixed, above-ground atmosphere. Globally relevant trace gases are produced and consumed by microbial activity in soils, with important feedbacks on atmospheric chemistry. Of particular importance are soil-derived "greenhouse" gases that contribute to the radiative forcing associated with global warming (N_2O, CH_4) that are produced when O_2 partial pressures limit its availability as a terminal electron acceptor in heterotrophic microbial respiration. The influences of soil (bio)geochemical processes on the chemical composition of atmospheric and surface water reservoirs are manifold. The impact of reactions within soils on reservoirs external to the CZ is termed here the "reach" of a soil biogeochemical process, and we include examples of these in the discussion of each biogeochemical process.

6.3.2 Element Speciation in Soil Pore Fluids

6.3.2.1 Overview

Water resides in soil pore spaces. In the unsaturated (vadose) zone, pore spaces are subject to fluctuations between wetting and drying conditions. After wetting events, as pore spaces become progressively drier because of evapotranspiration, solute concentrations increase, leading to precipitation of soluble salts including sulfates, carbonates and, under extremely arid conditions, even nitrates or chlorides. However, during wetting front propagation, as occurs following a rainfall or snowmelt event, these salts are rapidly re-solubilized as pores become progressively filled with aqueous solution. During such events, particularly in water-limited CZ systems, the increased availability of moisture often promotes a cascade of biological activity including root growth, root exudation and border cell production. This release of labile OM into soil pores, along with the downward percolation of DOM from the litter layer, drives heterotrophic microbial metabolism. As discussed later, the effects of such dynamics drive changes in the speciation of elements dissolved in soil solution and this, in turn, affects their transport and fate throughout and beyond the CZ.

6.3.2.2 Soil Geochemical Processes and Agents

The enhanced heterotrophic microbial activity that occurs in the presence of water during downward propagation of the wetting front results in the release of dissolved CO_2 plus carbonic acid (together denoted $H_2CO_3^*$) to soil solution because of the decomposition of SOM (i.e., CH_2O) (Huxman et al., 2004; Placella et al., 2012; Unger et al., 2010):

$$CH_2O_{(aq)} + O_{2(aq)} \rightarrow H_2CO_3{}^*{}_{(aq)} \tag{6.1}$$

$H_2CO_3^*$ is also produced as a result of root respiration of CO_2 followed by its dissolution (i.e., hydration) by water (Fig. 6.5). This dissolved carbon dioxide plus carbonic acid ($H_2CO_3^*$) is subject to two principal fates upon its formation as a result of enhanced biological activity. It can shed its hydration water and partition out of the soil solution and into the soil gas phase where it gives rise to partial pressures of $CO_{2(g)}$ that may be orders of magnitude higher than the above-ground atmosphere:

$$H_2CO_3{}^*{}_{(aq)} \rightarrow CO_{2(g)} + H_2O \tag{6.2}$$

OR carbonic acid may dissociate in the soil solution to form a reactive proton that contributes to acidifying soil solution, along with a bicarbonate ion that can be discharged to ground or surface waters.

$$H_2CO_3{}^*{}_{(aq)} \rightarrow HCO_3{}^-{}_{(aq)} + H^+{}_{(aq)} \tag{6.3}$$

These distinct fates highlight the "reach" of soil biogeochemistry to the atmosphere or surface waters external to the CZ, because they either result

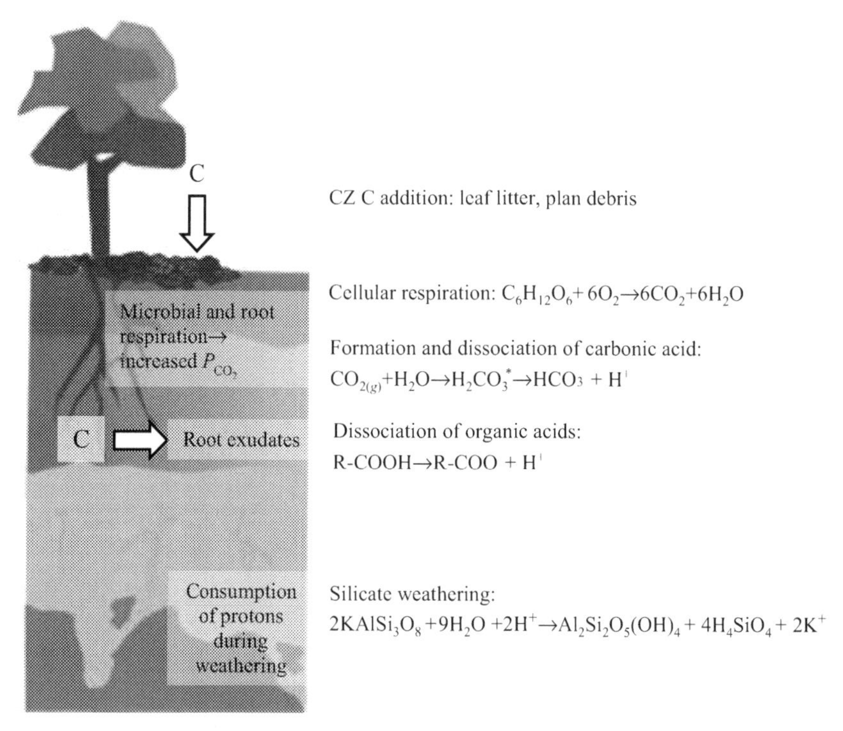

FIGURE 6.5 Schematic view of carbon (C) cycling in soils and its impact on the proton balance for biogeochemical reactions.

in $CO_{2(g)}$ return to the atmosphere Eq. (6.2) or its "consumption" via proton-promoted Eq. (6.3) weathering reactions (discussed later) that lead to the transport of bicarbonate ion out of the weathering zone and into rivers and oceans. The combinations of Eqs (6.1) and (6.2) describe the commonly observed "Birch effect" wherein soil wetting, particularly in water-limited systems, results in a pulsed release of $CO_{2(g)}$ from the soil to atmosphere (Fierer and Schimel, 2002).

Aqueous phase reactions that occur during pore filling are analogous to those pertaining to natural waters in general. These reactions involve water (solvent), and the dissolved solutes, which are inorganic and organic, and include cationic, anionic, and neutral molecules (Sposito, 2008). Initial wetting front propagation leads to simple hydration of cations and anions by water, and this is the dissolution reaction that brings "soluble salts" back into solution. As a dipolar solvent, water forms stable coordination complexes with cations through the electronegative oxygen and with anions through the electropositive hydrogen groups of the water molecule (Richens, 1997). These "hydration complexes" of, for example, Ca^{2+} or K^+ with water (i.e., $Ca(H_2O)_6{}^{2+}{}_{(aq)}$) are one example of metal-ligand complexation, where the metal (e.g., Ca^{2+}) forms a coordination complex with the electronegative oxygen of the ligand (H_2O).

Hydrated ions can undergo aqueous phase reaction with other ions in solution, including complexation and/or electron transfer (Richens, 1997). The former occurs when ligands form outer-sphere (hydration shell retained) or inner-sphere (hydration shell displaced) complexes with ligands other than water. For example, hydrated $Ca^{2+}_{(aq)}$ can form an outer-sphere complex (i.e., ion pair) with the sulfate anion to form the neutral $CaSO_4{}^0{}_{(aq)}$ species. Such reactions are highly relevant to the role of soils in CZ geochemistry, because, for example, the formation of a neutral dissolved species such as $CaSO_4{}^0{}_{(aq)}$ makes Ca more mobile than it would be as the free aquoion Ca^{2+}(aq), which can be adsorbed to negatively charged cation exchange sites. Hydrated polyvalent metal cations such as $Al^{3+}_{(aq)}$ or $Fe^{3+}_{(aq)}$ form strong (often inner-sphere) coordination complexes with dissolved organic ligands, including low molecular mass organic acids such as oxalic, phthalic or mallic acids. Again, such complexation with anionic ligands diminishes overall molecular charge on the metal–ligand complex, making the metal more mobile in the soil system, transporting it to depth, and/or to ground or surface waters. Indeed, the complexation of Al or Fe with organic ligands is a key reaction in podsolization: the process of Spodosol formation wherein metal-ligand complexes are "eluviated" from surface "E" horizons and precipitated in subsurface humic (Bh) or spodic (Bhs) horizons (Lundstrom et al., 2000).

6.3.2.3 *Reach*

Aqueous phase speciation has important implications for the fate and transport of lithogenic (e.g., Ca, Al, K, Fe) and biologically fixed atmogenic (e.g., C, N) elements in soils and throughout the CZ. Complexation reactions alter the charge, stereochemistry, and reactivity of ions in solution. The impacts are manifest in the genesis of soil profiles, such as the complexation, translocation, and aggregation of metal–ligand complexes. But they also impact more broadly the biogeochemistry of the CZ and beyond. For example, concentration-discharge relations of surface waters in many forested catchments show correlated catchment releases of polyvalent metals and DOM with characteristic stoichiometries of metal–ligand complexes (Chorover et al., 2011; Gangloff et al., 2014). Likewise, the partitioning of CO_2 between water- or gas-filled soil pores has important implications for the net exchange of carbon between land and atmosphere.

6.3.3 Adsorption–Desorption Reactions

6.3.3.1 *Overview*

Soils comprise a diversity of mineral and organic solids with reactive surface functional groups. Molecules that are dissolved in aqueous solution are termed "*adsorptive*" because they can potentially accumulate at these interfaces. The process of accumulation at interfaces, termed "*adsorption*," results in the formation of surface-associated species that are retained against leaching, and retarded in their transport by advected water. Such adsorbed species, "*adsorbate*", serve as a reservoir of various bioavailable nutrient, toxic or inert species subject to potential biological uptake or subsequent mobilization. The release of adsorbate back into solution (i.e., "*desorption*") occurs in response to pore water perturbations that are common in the open CZ system.

6.3.3.2 Soil Biogeochemical Process and Agents

Soils are very complex and heterogeneous mixture of fine particulate matter comprising minerals, microbial biomass, macrofauna and root tissue, as well as SOM in various stages of bio-decay, all admixed into multicomponent aggregates. Because of the small size of solid-phase weathering products such as secondary clays, a relatively small mass of soil can contain a massive amount of surface area. For example, one gram of soil rich in the secondary aluminosilicate, smectite, can contain up to hundreds of meters of highly reactive surface available for sorption reactions.

The majority of surface area in soil is composed of (1) layer silicate clays (e.g., vermiculite, montmorillonite, kaolinite), (2) metal (oxyhydr)oxides (e.g., gibbsite, goethite, ferrihydrite, birnessite), and (3) SOM. All of these solids comprise *surface functional groups*, that is, small groups of atoms combined into structural units that extend from the solid surface into contact with the solution phase. The types of surface functional groups on a given particle govern a particle's reactivity toward solution phase molecules, including nutrients and pollutants (Sposito, 2004).

Layer silicate clay surfaces are dominantly negatively charged as a result of isomorphic substitution of lower valence ions for higher valence ions in the crystal structure. For example, Al^{3+} for Si^{4+} substitution is common in the tetrahedral sheets of vermiculite and beidellite whereas Mg^{2+} for Al^{3+} is common in the octahedral sheets of montmorillonite. This isomorphic substitution gives rise to a negative *structural surface charge* that manifests at planar interlayer *siloxane* sites of these clay minerals, and that serves as important location for cation adsorption and retention against leaching. The permanent negative charge of siloxane sites must be balanced by an equal magnitude of oppositely charged adsorbate, which can undergo *cation exchange* upon a perturbation of solution chemistry, for example, as occurs during a rainfall or snowmelt event:

$$\boldsymbol{X}_2\mathrm{Ca}_{(s)} + 2\mathrm{Na}^+{}_{(aq)} \leftrightarrow 2\boldsymbol{X}\mathrm{Na}_{(s)} + \mathrm{Ca}^{2+}{}_{(aq)} \tag{6.4}$$

where, $\boldsymbol{X}$ represents one mole of negative structural surface charge. Equation (6.4) depicts a heterovalent ion exchange reaction where one mole of Ca^{2+} adsorbate is exchanged and released to solution upon adsorption of two moles of Na^+. Such cation exchange reactions in soils exert strong control over dynamic changes in pore water chemistry that occur during hydrologic events.

Whereas the metal (oxyhydr)oxides and SOM constituents do not contain significant permanent, structural charge, they have weakly acidic surface functional groups that undergo proton adsorption and desorption leading to *variable surface charge* as a function of soil pH. For example, the mineral goethite (α-FeOOH) comprises surface hydroxyl groups with a Brönsted acidity that confers a positively charged surface at acidic pH and a negatively charged surface at alkaline pH.

$$\equiv \text{Fe-OH}_{(s)} + \text{H}^+_{(aq)} \leftrightarrow \equiv \text{Fe-OH}_2{}^+_{(s)} \quad (6.5)$$

$$\equiv \text{Fe-OH}_{(s)} \leftrightarrow \equiv \text{Fe-O}^-_{(s)} + \text{H}^+_{(aq)} \quad (6.6)$$

where, the "≡" denotes the surface where the surface functional group (e.g., FeOH). It is noteworthy that the charge on such particles can be dynamic in the open CZ system because of the change in proton activity that may accompany respiration and carbonic acid production and dissociation Eq. (6.3). Particle surface charge then exerts strong control over the adsorption–desorption of charged ions in solution. For example, goethite can serve as an effective adsorbent for anions at low pH (<8):

$$\equiv \text{Fe-OH}_2{}^+\cdots\text{Cl}^-_{(s)} + \text{H}_2\text{PO}_4{}^-_{(aq)} \leftrightarrow \equiv \text{Fe-OH}_2{}^+\cdots\text{H}_2\text{PO}_4{}^-_{(s)} + \text{Cl}^-_{(aq)} \quad (6.7)$$

where, "···" denotes surface bond formation. In this case, the Cl^- adsorbate is subjected to desorption and displacement by $H_2PO_4{}^-$.

Such Fe (oxyhydr)oxides can serve as effective adsorbents for cations at high pH (>8):

$$\equiv \text{Fe-OH}_{(s)} + \text{Ca}^{2+}_{(aq)} \leftrightarrow \equiv \text{Fe-O}^- \cdots \text{Ca}^{2+}_{(s)} + \text{H}^+_{(aq)} \quad (6.8)$$

In addition to adsorption–desorption reactions of inorganic ions, both layer silicate clays and (oxyhydr)oxides are important adsorbents for SOM. These SOM adsorption reactions are central to the capacity of soils to retain the largest mass fraction of organic carbon at the Earth's surface. Adsorbed (exchangeable) cations, such as those depicted in Eq. (6.4), play a central role in SOM adsorption to layer silicate clays by bridging between the negatively charged siloxane sites of the mineral surface and the negatively charged sites (e.g., carboxyl, R-COO-) of SOM:

$$X\text{Ca}^+_{(s)} + \text{R-COO}^-_{(aq)} \leftrightarrow X\text{Ca}\cdots\text{OOC-R}_{(s)} \quad (6.9)$$

Adsorption to Fe and Al (oxyhydr)oxide surfaces is considered to be of particular importance for the retention of SOM. This is attributed to *ligand exchange* reactions at hydroxylated surfaces, where surface hydroxyl groups are exchanged with carboxyl groups of SOM enabling inner-sphere complexation of SOM with metals, for example,

$$\equiv \text{Fe-OH}_{(s)} + \text{R-COOH}_{(aq)} \leftrightarrow \equiv \text{Fe-OOC-R}_{(s)} + \text{H}_2\text{O} \quad (6.10)$$

Such adsorbate–adsorbent bonds formed between pedogenic clays and organic ligands are critically important to the stabilization of OM against degradation in soils.

Similar reactions to those shown in Eqs (6.6) and (6.8) occur on the surfaces of solid phase SOM since SOM comprises dominantly carboxylic (R-COOH) and phenolic (aromatic-OH) acid functional groups that likewise exhibit pH-dependent (variable) charge. In addition to its role in cation adsorption and exchange, OM is the principal sink for hydrophobic organic compounds (HOCs) introduced to soils, including many pollutants like polycyclic aromatic hydrocarbons deriving from fossil fuel combustion. These HOCs are adsorbed to OM via van der Waals interactions with uncharged moieties (aromatic rings and aliphatic chains) of the OM (Chorover and Brusseau, 2008).

6.3.3.3 Reach

Surface reactions in soils control the bioavailable pool of ions for plant and microbial uptake and represent a store of nutrients for vegetative growth. The composition of the exchange complex (e.g., nutrient cations such as K, Ca, and Mg versus toxic Al) feeds back to affect nutrient status and vegetation type, whereas adsorbate exchange reactions (e.g., Eq. 6.4) also control concentration-discharge relations observed in effluent surface waters. Adsorption of OM to clay particle surfaces leads to formation of heteroaggregates and the high spatial variation in geochemical conditions in soils (complex pore size distributions, steep biogeochemical gradients or proximal oxic and anoxic domains).

6.3.4 Reduction–Oxidation Reactions

6.3.4.1 Overview

Wetting front propagation into the unsaturated portion of the CZ, as occurs during rainfall or snowmelt events, leads to pore filling with water. Consumption of dissolved oxygen results from heterotrophic respiration Eq. (6.1) such that if bio-active soils remain saturated for prolonged periods, they become suboxic and a cascade of alternative (to O_2 reduction) terminal electron accepting processes (TEAPs) ensues (Borch et al., 2010). However, a diversity of TEAPs frequently occurs even in soils that have bulk oxic character because of their structural complexity and heteroaggregate structure (Blagodatsky and Smith, 2012; Renault and Stengel, 1994). Such soils comprise microsites with oxic, suboxic, and anoxic character often in close proximity (within millimeters) of each other. Steep biogeochemical gradients produce a thermodynamic driving force for reactions that would be mutually exclusive in well mixed systems. As a result, soils can host in close proximity a diverse array of microbes that use different TEAPs for respiration.

6.3.4.2 Soil Biogeochemical Process and Agents

Rainfall incident on the vegetated CZ infuses water and labile OM into microbially active soils where the oxidation of OM promotes the consumption of gaseous and dissolved O_2 in respiration Eq. (6.1). When molecular oxygen becomes limiting to respiration, facultative anaerobic microbes turn to anaerobic respiration, which is a form of respiration using electron acceptors other than oxygen. Whereas molecular oxygen is a strong oxidant, anaerobes use other weaker oxidants such as NO_3^-, Mn(IV), Fe(III), SO_4^{2-}, CH_2O, and CO_2. These terminal electron acceptors have progressively smaller reduction potentials than O_2, meaning that progressively less energy is released per mole of oxidized species. Anaerobic respiration is, therefore, in general energetically less efficient than aerobic respiration. However, the processes have important implications for soil geochemistry in the CZ.

Possible redox reactions are dictated by the state of the soil system and the corresponding Gibbs energies of reaction. Among the reactions that are possible thermodynamically, those that predominate at any given point in time are determined by redox kinetics, the latter being governed to a large degree by microbial catalysis. A sequence of TEAPs is observed typically along redox gradients both spatially and temporally in soil systems. This sequence corresponds closely to progressive decreases in the Gibbs energy of the full redox reaction. Since O_2 is sparingly soluble in water (0.25 mM at 25°C and atmospheric partial pressure) it can be depleted rapidly by microbial and root respiration in soils subjected to limited influx of air or oxygenated water. When that occurs, dissolved nitrate and available Mn(IV) solids are utilized as alternative electron acceptors during oxidation of OM (termed nitrate reduction or denitrification and Mn reduction, respectively). As these reactants become depleted, further reduction in redox potential (E_h) results in the successive use of Fe(III) solids (ferric reduction), SO_4^{2-} (sulfate reduction), and eventually OM itself (fermentation or methane fermentation) or CO_2 (methanogenesis). In each case, depletion of reactant oxidizing agents and accumulation of their reduced-form products (e.g., Mn^{2+}, Fe^{2+}, HS^-, CH_3COO^-, CH_4) diminishes the energy yield of a given TEAP as the Gibbs energy for the full reaction approaches zero (Chadwick and Chorover, 2001).

Importantly, redox reactions involving metals can bring about large shifts in pH as the metals undergo hydrolysis reactions. For instance, the oxidation of 1 mole of Fe^{2+} produces 2 moles of protons,

$$Fe^{2+}_{(aq)} + 0.25O_{2\,(g)} + 2.5H_2O_{(aq)} \leftrightarrow Fe^{III}(OH)_{3(s)} + 2H^+_{(aq)} \qquad (6.11)$$

thus, impacting adsorption-desorption reactions, microbial activity, and other pH dependent processes, including carbon availability (Hall and Silver, 2013). The reverse reaction leads to an increase in pH (Kogel-Knabner et al., 2010) and has been implicated in mobilizing OM (Buettner et al., 2014; Grybos et al., 2009; Pédrot et al., 2009), nutrients (Henderson et al., 2012), and metals (Davranche et al., 2011; Weber et al., 2009) as pH-driven changes in particle surface charge promote colloid dispersion (Ryan and Gschwend, 1994).

6.3.4.3 *Reach*

Production of CO_2 by a variety of TEAPs results in equilibrium partial pressures of CO_2 in soils that exceed that of the above-ground atmosphere by orders of magnitude. This drives mineral weathering reactions (see Section 6.3.5) and the associated long-term atmospheric CO_2 drawdown. While silicate weathering consumes CO_2, a potent greenhouse gas, N and C based TEAPs can also significantly promote radiative forcing. Methanogenesis, as occurs in soils subjected to prolonged anoxia, generates a greenhouse gas that is 30 times more effective than CO_2 at trapping heat in the atmosphere. Likewise, denitrification can produce gaseous intermediates that include N_2O, a greenhouse gas 300 times more effective than CO_2 at trapping heat in the atmosphere. Reduced forms of Fe^{2+} and Mn^{2+} are more soluble than their oxidized counterparts, and their removal from the soil by leaching results in a permanent loss in these lithogenic ions and their oxidizing capacity.

6.3.5 Mineral Weathering Reactions

6.3.5.1 *Overview*

The genesis of soil involves biogeochemical weathering (i.e., formation of secondary minerals from primary minerals) in addition to the physical disintegration of rock (e.g., fractures along grain boundaries, freeze-thaw, etc.). The focus of this section is on the former reactions, which are influenced by the full suite of reactions discussed earlier including those that affect (i) solution phase speciation (Section 6.3.2), (ii) adsorption–desorption (Section 6.3.3), and (iii) oxidation–reduction (Section 6.3.4). Mineral weathering in soil is driven by CZ open-system through-fluxes of freshwater and OM, and the resulting introduction of reducing equivalents, complexing ligands, and protons. As a result of mineral transformation reactions, primary minerals (large particles with low specific surface area) are transformed into secondary minerals (small particles with high specific surface area), creating key raw materials for soil fabric, while also releasing lithogenic macro- and micronutrients (e.g., Ca, Mg, K, Fe, Zn, etc.) to solution in support of plant and microbial growth.

6.3.5.2 *Soil Biogeochemical Process and Agents*

Geochemical weathering reactions in soil are termed *congruent* if they result in mineral dissolution with no secondary solid formed (i.e., only solution-phase products). Conversely, reactions are termed *incongruent* if they result in the formation of both solution-phase *and* solid-phase products. Incongruent weathering reactions are essential to the formation of soils because they result in the production of *secondary* (pedogenic) minerals from the weathering of *primary* minerals in rock. Because these secondary minerals are formed under the low temperature (T) and low pressure (P) conditions, most prevalent in the CZ, they tend to have smaller crystallite size and more disordered crystal structure than the primary silicates occurring in the igneous rock undergoing weathering. Their formation here also reflects the greater thermodynamic stability of secondary minerals under low P and T conditions. Secondary minerals do, of course, themselves undergo congruent or incongruent dissolution such that the mineralogical composition of soil evolves over pedogenic time scales (10^2–10^6 year) in response to climatic forcing and subject to thermodynamic and kinetic constraints.

CZ biota – including plants, microorganisms, and animals – can significantly affect the rates and trajectories of weathering reactions in soils through their impact on the solution-phase reactants – specifically the concentrations of protons, complexing ligands, and oxidizing or reducing equivalents. Atmospherically derived $CO_{2(g)}$ is supplemented by root and microbial respiration such that the high partial pressures of CO_2 enhance the thermodynamic favorability of primary silicate weathering reactions. For example, in the following reaction, plagioclase (a framework silicate common to igneous rock) undergoes *proton-promoted dissolution* in the presence of CO_2 and water to form the layer silicate clay kaolinite:

$$CaAl_2Si_2O_{8(s)} + CO_{2(g)} + 2H_2O + H^+ \rightarrow Al_2Si_2O_5(OH)_{4(s)} + HCO_3^- + Ca^{2+} \quad (6.12)$$

This reaction reflects several characteristic features of primary mineral weathering in addition to the transformation of primary (in this case framework) to secondary (layer-type) aluminosilicates. These additional characteristic features include water, CO_2, and/or proton consumption; bicarbonate production; and release of nonhydrolyzing (base) cations, such as Ca, Mg, and K, making them available for bio-uptake from solution or exchange sites. Hence, processes that enhance the chemical activities of water, CO_2, and protons will increase the thermodynamic driving force – and normally also the kinetics – of rock weathering (Brantley et al., 2004; Stumm, 1997; Ugolini and Sletten, 1991).

Biologic activity also influences weathering reactions through the production of metal-complexing organic ligands (Brantley et al., 2011; Chorover, 2012). These ligands include low molar mass aliphatic and aromatic organic acids, in addition to more specialized high-affinity species (e.g., siderophores). Complexing ligands promote mineral dissolution by sequestering lithogenic metals into metal–ligand complexes (thereby diminishing the solution-phase accumulation of free metals and transporting them from the site of dissolution). For example, the same plagioclase mineral may undergo dissolution in the presence of the common biogenic dicarboxylic acid, ethanedioic or "oxalic" acid ($C_2H_2O_4$) produced by plant roots and microbes:

$$CaAl_2Si_2O_{8(s)} + 2C_2H_2O_4 + 4H^+ \rightarrow 2(C_2O_4\text{-}Al)+ + 2Si(OH)_4^0 + Ca^{2+} \quad (6.13)$$

By forming a stable complex with Al released from plagioclase, oxalate (the conjugate base of oxalic acid) may preclude the reaction of Al with dissolved Si to form kaolinite, making the dissolution reaction congruent Eq. (6.13) rather than incongruent Eq. (6.12). Furthermore, prior work has shown that low molar mass ligands like oxalate can increase the kinetics of mineral dissolution, even under conditions that are far from equilibrium (Stumm and Morgan, 1996). The mechanism, termed *ligand-promoted dissolution*, has been shown to be surface-mediated and results from the adsorption of low molar mass ligands like oxalate to surface bound metals via a ligand exchange reaction (similar to Eq. 6.10), followed by dissociation to solution of the metal–ligand complex.

6.3.5.3 Reach

The larger the CZ, its geomorphic template, and the local climatic forcing have strong effects on mineral weathering reactions in soils. These CZ-external forcings dictate fluid flow and water–rock contact time, which in turn influence the degree of solution phase saturation with respect to mineral phases undergoing dissolution or precipitation. Such effects, however, can be characterized also in terms of geochemical reactions such as those for plagioclase dissolution or complexing ligands, reflecting the importance of water flux, proton, and/or CO_2 concentration. These weathering reactions are responsible for generating the secondary (as well as tertiary, quaternary, etc.) minerals in soil environments whose fine particle size and high specific surface area control much of the chemical reactivity of soil systems. They also exert ultimate control over the quality of water emanating from the CZ and entering into surface water bodies. Indeed, the chemistry of surface water is often used in an inverse modeling approach to assess predominating geochemical weathering reactions (Güler and Thyne, 2004; Yoo et al., 2009).

6.4 CLIMATE AS OVERARCHING CONTROL ON SOIL GEOCHEMISTRY AND ITS FEEDBACK TO ATMOSPHERE, SURFACE- AND GROUNDWATER COMPOSITION

Climate is one of the most important forcings impacting CZ evolution (Anderson et al., 2013), soil formation (Jenny, 1941; Volobuyev, 1964), and the types/amounts of dissipative products (Chorover et al., 2011; Rasmussen et al., 2011). But how is climatic forcing translated to the molecular scale where soil biogeochemical processes are impacted and how do they then feed back to the composition of atmosphere, surface- and groundwater? Because of current challenges of global climate change, impacts of climate on soils is an area of active research (Kirschbaum, 1995; Lal, 2004; Peng et al., 2014; Schuur et al., 2013; Zhang et al., 2013) and in many cases temperature and/or precipitation are investigated as climate proxies in isolation. In this section we will give examples illustrating the connection of global/regional climate to molecular scale soil biogeochemical processes that feed back to impact atmosphere, surface- and groundwaters (Fig. 6.6).

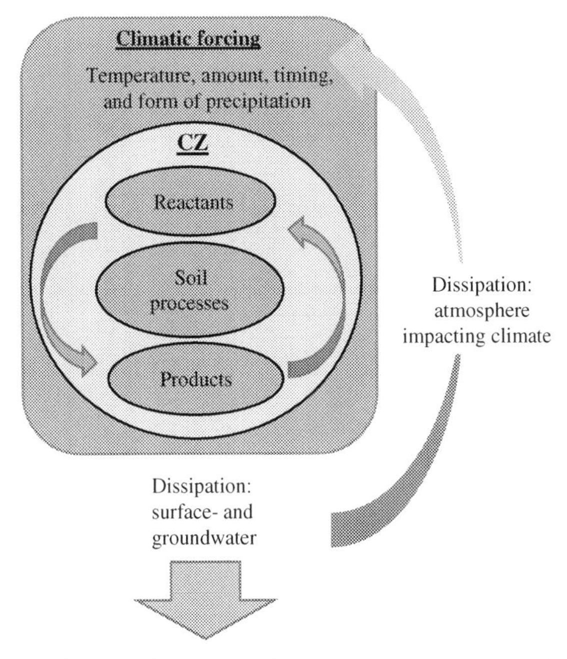

FIGURE 6.6 Nested factors impacting soil processes and their reach. Climatic forcing acts outside of the Critical Zone (CZ) to impact soil biogeochemical processes that in turn influence composition of atmosphere, surface- and groundwater. Changes in atmospheric composition in turn feeds back to climatic forcings.

6.4.1 Temperature

Temperature has an important bearing on the kinetics of all (bio)geochemical reactions (Stumm and Morgan, 1996). Soil biogeochemical reactions, such as weathering, in "cold" systems may proceed slower than in "warm" systems (Talibudeen, 1981) or may not receive the required activation energy to happen at all. Accordingly, less dissipative products are released to atmosphere and freshwaters in cold systems, than in warmer systems.

Increasing global temperatures have an important effect on the transition from snow to rain (Elsner et al., 2010; Kormos et al., 2013) and impact soil biogeochemistry and soil formation. A study by Rasmussen et al. (2007) showed the impact of rain–snow transition on soil development over andesitic bedrock along a climate gradient in the Sierra Nevada of California. Soils in the snowfall-dominated zone (cold) exhibited minimal soil development whereas soils dominated by rainfall (warm) showed the opposite. The authors attribute this pattern to the combined effect of temperature and water availability. In rain-dominated locations, soils are warm when water is available for weathering reactions, whereas in snow-dominated areas soils are cold when water is present.

6.4.2 Amount and Timing of Rainfall

Soil biogeochemical processes are directly impacted by the amount and timing of precipitation. A certain amount of water is necessary to allow for biogeochemical processes, such as microbial processing, weathering, sorption, exchange reactions, and redox reactions (Section 6.3). For example, soil respiration rates and CO_2 effluxes during the summer show a great sensitivity to moisture, therefore the predicted lengthening of growing seasons and increased summer rains due to global climate change may lead to increased dissipation of C to the atmosphere (Barron-Gafford et al., 2011; Curiel Yuste et al., 2003; Fierer and Schimel, 2002).

Water delivery also strongly impacts weathering and the dissipation of weathering derived solutes via surface waters (Gaillardet et al., 2011). Loss of solutes at the solid:aqueous interface are controlled by both thermodynamic and kinetic constraints. For solid phases with slow dissolution kinetics, changes in precipitation may have little influence on the loss of solutes. However, in many catchments, the concentration of solutes in surface- and groundwaters are similar across a wide range of water discharge rates. Catchments falling in this domain are described as chemostatic (Godsey et al., 2009; Maher, 2011) and the overall water flux through the CZ controls the mass of solute loss. Maher (2011) demonstrates this well for Si fluxes (Fig. 6.7a). The importance of water fluxes for other solute fluxes such as stream water dissolved organic carbon and dissolved inorganic carbon was shown by Perdrial et al. (2014) (Fig. 6.7b).

Rain can also induce threshold shifts in soil redox conditions (Chadwick and Chorover, 2001; Miller et al., 2001) with potential impacts on solute flux and composition of surface- and groundwaters (Thompson et al., 2006). This is most evident when rainfall exceeds the infiltration capacity of the soil and induces either seasonal or permanent water saturation within the soil profile. Although a greater propensity of anoxic conditions is proportional to rainfall, the response is

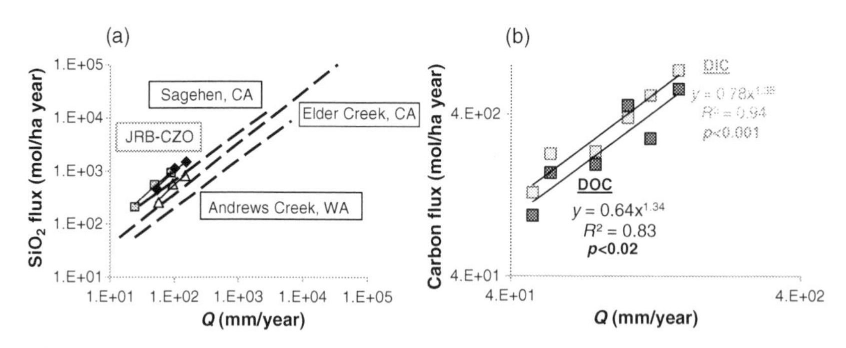

FIGURE 6.7 Importance of annual water fluxes on export of organic and inorganic dissipative products. (a) Correlation between specific discharge and SiO_2 effluxes from Jemez River Basin Critical Zone Observatory (JRB-CZO) catchments, overlain on correlations compiled by Maher (2011) for watersheds in CA and WA. (b) Relationship between dissolved organic and inorganic carbon (DOC, DIC) and annual specific discharge for JRB creeks, modified from Perdrial et al. (2014).

nonlinear (Chadwick and Chorover, 2001) due to feedbacks between aforementioned secondary mineral production and diminished soil hydraulic conductivity (Lohse et al., 2009; Vitousek and Chadwick, 2013). Reductive dissolution of Fe^{III}-solids to $Fe^{2+}_{(aq)}$ is evident in the common occurrence of rust colored iron oxide deposits throughout surface waters and springs in high rainfall areas produced as the $Fe^{2+}_{(aq)}$ is oxidized when exposed to oxygen (McKnight et al., 1992). Because metal redox reactions produce large shifts in pH (see Section 6.3.4), changes in rainfall patterns that alter soil redox conditions can have impacts on streams and ecosystems, potentially contributing to hot spots and hot moments across the landscape (McClain et al., 2003; Small et al., 2012).

6.4.3 Amount and Timing of Snowfall and Melt

Ecosystems that receive an important amount of precipitation in the form of snow (snow dominated) are particularly vulnerable towards global climate change and are increasingly studied (as of 2014, the Soiltrec homepage lists over 60 world Critical Zone Observatories (CZOs) (http://www.soiltrec.eu), of which at least ⅓ are dominated by a seasonal snowpack).

Seasonal snow cover exerts a varied and complex control on soil biogeochemistry because snow dominated systems are (at least) seasonally cold, which has aforementioned implications on geochemical reaction rates. Furthermore, these systems are impacted by the timing of snowfall and water delivery.

Soil biogeochemical processes depend on the availability of water in liquid form. In snow-dominated environments, the conversion of water from solid to liquid form represents an important threshold since soil frost is a major inhibitor of soil biogeochemical processes. Snow has a high insulation capacity, therefore soils typically do not freeze completely when protected by a thick snowpack that preserves liquid water for over-winter microbial processes (Brooks et al., 2011; Groffman et al., 2001, 2006). Such subnivial (under snowpack) processes play an important role in the soils N cycle (Brooks et al., 1999, 2011; Monson et al., 2006). For example, heterotrophic microbial activity under continuous snow cover reduces N loss allowing for better ecosystem N retention whereas the absence of microbial activity (due to late or missing snow) results in N loss (Brooks and Williams, 1999). Differences in snow cover can also be related to aspect (drier south-vs. wetter north-facing slopes) and can cause differences in solute export form montane environments (Hinckley et al., 2014).

Snow cover is equally important for C cycling since more than half of global organic carbon is stored in high latitude and/or high altitude regions that are at risk to turn from C sinks to sources (Brooks et al., 2011; Tarnocai et al., 2009). Estimates of winter CO_2 fluxes due to heterotrophic microbial activity are as high as 60% of the annual net ecosystem exchange (Brooks et al., 2011; Monson et al., 2002). Therefore timing and/or amount of winter precipitation is of prime importance for the extent of under snow processes and dissipation of

C and N species in gaseous form to impact atmospheric composition or liquid form to influence surface- and groundwaters.

It is easy to imagine that in snow-dominated systems, a pronounced disconnect exists between the biogeochemical generation of aqueous dissipative products and their actual dissipation to surface- and groundwaters. Given that liquid water is available at the weathering front, weathering can happen year round but weathering products are only removed during snowmelt. This process leads to the event-like introduction of solutes (weathering products, nutrients, C) (Boyer et al., 1997; Brooks and Williams, 1999; Perdrial et al., 2014) into surface- and groundwaters.

Introduction of meltwaters is not only important because of the transporting role of water but also for the reactant role of water. Meltwaters are typically not in equilibrium with mineral phases and therefore chemically aggressive. The introduction of meltwaters under saturated solutions promotes weathering and changes weathering rates in soils over the course of a year. Additionally, meltwaters contain accumulated dust and products of overwinter processes (e.g., subnivial microbial processing in organic rich horizons or water–rock interactions at the weathering front) that impact the composition of surface- and low residence time, shallow groundwater.

6.5 CONCLUSIONS

As a living porous medium that extends from weathering bedrock to the vegetated surface, soils comprise most of the interfacial area of the CZ. The aggregated structure of bioactive soil contains minerals, microbes, OM, gas, and liquid filled pore spaces all in close association. As a result, soils are characterized by steep geochemical gradients across relatively small time scales (e.g., following hydrologic events) and space scales (e.g., as reflected in distinct redox domains in soil aggregates). Hence, characterizing soil geochemistry tends to be even more challenging than the well-mixed atmospheric and surface water bodies that bound the CZ.

The gradients that drive soil geochemical reactions are sustained by continuous open-system throughputs of freshwater and NOM. The ensuing disequilibria result in mineral transformation reactions that favor the formation of high specific-surface-area minerals (layer silicate clays and metal (oxhydr)oxides) with reactive surface sites that contribute to the adsorption–desorption, oxidation–reduction and dissolution–precipitation reactions at highly localized scales. Nonetheless, the importance of these localized reactions is manifested across pedon to catchment to global scales through their regulation of such life-sustaining processes within the CZ, such as carbon stabilization, ecosystem nutrient retention, and contaminant attenuation.

As human activities now dominate the structure and function of many portions of the Earth's CZ, the Anthropocene poses tremendous challenges not only to the Earth's surface, but also to researchers engaged in understanding it (Crutzen, 2006; Goddéris and Brantley, 2013). Soils and their biogeochemical function are central to mediating changes that occur in this epoch, and it is important to understand how much capacity soil has to absorb changes while still providing life-sustaining services (Richter and Mobley, 2009; Richter and Yaalon, 2012).

REFERENCES

Amundson, R., Richter, D.D., Humphreys, G.S., Jobbágy, E.G., Gaillardet, J., 2007. Coupling between biota and Earth materials in the critical zone. Elements 3 (5), 327–332.

Anderson, R.S., Anderson, S.P., Tucker, G.E., 2013. Rock damage and regolith transport by frost: an example of climate modulation of the geomorphology of the critical zone. Earth Surf. Process. Land. 38 (3), 299–316.

Banwart, S., et al. 2011. Soil processes and functions in Critical Zone Observatories: hypotheses and experimental design. Vadose Zone J. 10 (3), 974–987.

Barron-Gafford, G.A., Scott, R.L., Jenerette, G.D., Huxman, T.E., 2011. The relative controls of temperature, soil moisture, and plant functional group on soil CO_2 efflux at diel, seasonal, and annual scales. J. Geophys. Res. Biogeosci. 116 (G1), G01023.

Barsdate, R., Alexander, V., 1975. The nitrogen balance of arctic tundra: pathways, rates, and environmental implications. J. Environ. Qual. 4 (1), 111–117.

Berg, B., Meentemeyer, V., 2002. Litter quality in a north European transect versus carbon storage potential. Plant Soil 242 (1), 83–92.

Berner, E.K., Berner, R.A., 2012. Global Environment: Water, Air and Geochemical Cycles. Princeton University Press, Princeton, NJ, pp. 444.

Blagodatsky, S., Smith, P., 2012. Soil physics meets soil biology: towards better mechanistic prediction of greenhouse gas emissions from soil. Soil Biol. Biochem. 47, 78–92.

Bockheim, J., Gennadiyev, A., Hammer, R., Tandarich, J., 2005. Historical development of key concepts in pedology. Geoderma 124 (1), 23–36.

Bolan, N.S., et al. 2011. Dissolved organic matter: biogeochemistry, dynamics, and environmental significance in soils. Adv. Agron. 110, 1–75.

Borch, T., et al. 2010. Biogeochemical redox processes and their impact on contaminant dynamics. Environ. Sci. Technol. 44 (1), 15–23.

Boyer, E.W., Hornberger, G.M., Bencala, K.E., McKnight, D.M., 1997. Response characteristics of DOC flushing in an alpine catchment. Hydrol. Process. 11 (12), 1635–1647.

Brantley, S.L., et al. 2004. Fe isotopic fractionation during mineral dissolution with and without bacteria. Geochim. Cosmochim. Acta 68 (15), 3189–3204.

Brantley, S.L., Goldhaber, M.B., Ragnarsdottir, K.V., 2007. Crossing disciplines and scales to understand the critical zone. Elements 3 (5), 307–314.

Brantley, S.L., et al. 2011. Twelve testable hypotheses on the geobiology of weathering. Geobiology 9 (2), 140–165.

Brooks, P.D., McKnight, D.M., Bencala, K.E., 1999. The relationship between soil heterotrophic activity, soil dissolved organic carbon (DOC) leachate, and catchment-scale DOC export in headwater catchments. Water Resour. Res. 35 (6), 1895–1902.

Brooks, P.D., Williams, M.W., 1999. Snowpack controls on nitrogen cycling and export in seasonally snow-covered catchments. Hydrol. Process. 13 (14–15), 2177–2190.

Brooks, P.D., et al. 2011. Carbon and nitrogen cycling in snow-covered environments. Geogr. Compass 5 (9), 682–699.

Buettner, S.W., Kramer, M.G., Chadwick, O.A., Thompson, A., 2014. Mobilization of colloidal carbon during iron reduction in basaltic soils. Geoderma 221, 139–145.

Carrillo, J.H., Hastings, M.G., Sigman, D.M., Huebert, B.J., 2002. Atmospheric deposition of inorganic and organic nitrogen and base cations in Hawaii. Global Biogeochem. Cycles 16 (4), 1076.

Chadwick, O.A., Brimhall, G.H., Hendricks, D.M., 1990. From a black to a gray box: a mass balance interpretation of pedogenesis. Geomorphology 3 (3–4), 369–390.

Chadwick, O.A., Chorover, J., 2001. The chemistry of pedogenic thresholds. Geoderma 100 (3–4), 321–353.

Chorover, J., 2011. Impact of soil physicochemical and biological reactions on transport of nutrients and pollutants in the critical zone. Handbook of Soil Sciences CRC Press, Boca Raton, FL, pp. 1–36.

Chorover, J., 2012. Impact of soil physicochemical and biological reactions on transport of nutrients and pollutants in the critical zone. In: Huang, P.M., Sumner, M. (Eds.), Handbook of Soil Science: Resource Management and Environmental Impacts. Academic Press, New York, NY, pp. 10.1–10.35.

Chorover, J., Amistadi, M.K., 2001. Reaction of forest floor organic matter at colloidal goethite, birnessite and smectite surfaces. Geochim. Cosmochim. Acta 65, 95–109.

Chorover, J., Brusseau, M.L., 2008. Kinetics of sorption–desorption. In: Brantley, S.L., Kubicki, J., White, A.F. (Eds.), Kinetics of Water–Rock Interaction. Springer, New York, pp. 109–149.

Chorover, J., et al. 2011. How water, carbon, and energy drive critical zone evolution: the Jemez-Santa Catalina Critical Zone Observatory. Vadose Zone J. 10 (3), 884–899.

Chorover, J., Kretzschmar, R., Garcia-Pichel, F., Sparks, D.L., 2007. Soil biogeochemical processes within the Critical Zone. Elements 3 (5), 321–326.

Churchman, G.J., 2010. The philosophical status of soil science. Geoderma 157 (3–4), 214–221.

Coffey, G.N., 1909. Physical principles of soil classification.

Crompton, E. 1960. The significance of the weathering/leaching ratio in the differentiation of major soil groups, with particular reference to some very strongly leached brown earths on the hills of Britain. Paper presented at the Trans. 7th int. Congr. Soil Sci., vol. 5, pp. 406–412.

Crutzen, P., 2006. The "Anthropocene". In: Ehlers, E., Krafft, T. (Eds.), Earth System Science in the Anthropocene. Springer, Berlin, Heidelberg, pp. 13–18.

Curiel Yuste, J., Janssens, I.A., Carrara, A., Meiresonne, L., Ceulemans, R., 2003. Interactive effects of temperature and precipitation on soil respiration in a temperate maritime pine forest. Tree Physiol. 23 (18), 1263–1270.

Daniels, R.B., Hammer, R.D., 1992. Soil Geomorphology. John Wiley & Sons, New York.

Davranche, M., et al. 2011. Rare earth element patterns: a tool for identifying trace metal sources during wetland soil reduction. Chem. Geol. 284 (1–2), 127–137.

Derry, L.A., Chadwick, O.A., 2007. Contributions from Earth's atmosphere to soil. Elements 3 (5), 333–338.

Dokuchajev, V., 1883. The Russian Chernozem Report to the Free Economic Society. Imperial Univ. of St. Petersburg, St. Petersburg, Russia.

Elsner, M.M., et al. 2010. Implications of 21st century climate change for the hydrology of Washington State. Climatic Change 102 (1/2), 225–260.

Essington, M.E., 2002. Soil and Water Chemistry: An Integrative Approach. CRC Press, Boca Raton, pp. 534.

Fierer, N., Schimel, J.P., 2002. Effects of drying-rewetting frequency on soil carbon and nitrogen transformations. Soil Biol. Biochem.V 34 (6), 777–787.

Gabor, R.S., Eilers, K., McKnight, D.M., Fierer, N., Anderson, S.P., 2014. From the litter layer to the saprolite: chemical changes in water-soluble soil organic matter and their correlation to microbial community composition. Soil Biol. Biochem. 68 (0), 166–176.

Gaillardet, J., et al. 2011. Orography-driven chemical denudation in the Lesser Antilles: evidence for a new feed-back mechanism stabilizing atmospheric CO_2. Am. J. Sci. 311 (10), 851–894.
Gangloff, S., Stille, P., Pierret, M.-C., Weber, T., Chabaux, F., 2014. Characterization and evolution of dissolved organic matter in acidic forest soil and its impact on the mobility of major and trace elements (case of the Strengbach watershed). Geochim. Cosmochim. Acta 130, 21–41.
Gislason, S.R., et al. 2009. Direct evidence of the feedback between climate and weathering. Earth Planet. Sci. Lett. 277 (1), 213–222.
Goddéris, Y., Brantley, S.L., 2013. Earthcasting the future critical zone. Elementa 1 (1), 000019.
Godsey, S.E., Kirchner, J.W., Clow, D.W., 2009. Concentration–discharge relationships reflect chemostatic characteristics of US catchments. Hydrol. Process. 23 (13), 1844–1864.
Groffman, P.M., et al. 2001. Colder soils in a warmer world: a snow manipulation study in a northern hardwood forest ecosystem. Biogeochemistry 56 (2), 135–150.
Groffman, P., Hardy, J., Driscoll, C., Fahey, T., 2006. Snow depth, soil freezing, and fluxes of carbon dioxide, nitrous oxide and methane in a northern hardwood forest. Global Change Biol. 12 (9), 1748–1760.
Grybos, M., Davranche, M., Gruau, G., Petitjean, P., Pedrot, M., 2009. Increasing pH drives organic matter solubilization from wetland soils under reducing conditions. Geoderma 154 (1–2), 13–19.
Güler, C., Thyne, G.D., 2004. Hydrologic and geologic factors controlling surface and groundwater chemistry in Indian Wells-Owens Valley area, southeastern California, USA. J. Hydrol. 285 (1–4), 177.
Guo, M.X., Chorover, J., 2003. Transport and fractionation of dissolved organic matter in soil columns. Soil Sci. 168 (2), 108–118.
Hall, S.J., Silver, W.L., 2013. Iron oxidation stimulates organic matter decomposition in humid tropical forest soils. Global Change Biol. 19 (9), 2804–2813.
Henderson, R., et al. 2012. Anoxia-induced release of colloid- and nanoparticle-bound phosphorus in grassland soils. Environ. Sci. Technol. 46 (21), 11727–11734.
Herndon, E.M., Jin, L., Brantley, S.L., 2010. Soils reveal widespread manganese enrichment from industrial inputs. Environ. Sci. Technol. 45 (1), 241–247.
Hilgard, E.W., 1882. Report on the relations of soil to climate. US Dep. Agric. Weather Bull. 3, 1–59.
Hinckley, E.-L.S., et al. 2014. Aspect control of water movement on hillslopes near the rain–snow transition of the Colorado Front Range. Hydrol. Process. 28 (1), 74–85.
Holbrook, W.S., et al. 2014. Geophysical constraints on deep weathering and water storage potential in the Southern Sierra Critical Zone Observatory. Earth Surf. Process. Land. 39 (3), 366–380.
Houghton, R.A., 2007. Balancing the global carbon budget. Annu. Rev. Earth Planet. Sci. 35 (1), 313–347.
Houghton, R.A., Davidson, E.A., Woodwell, G.M., 1998. Missing sinks, feedbacks, and understanding the role of terrestrial ecosystems in the global carbon balance. Global Biogeochem. Cycles 12 (1), 25–34.
Huggett, R., 1975. Soil landscape systems: a model of soil genesis. Geoderma 13 (1), 1–22.
Huxman, T.E., et al. 2004. Precipitation pulses and carbon fluxes in semiarid and arid ecosystems. Oecologia 141 (2), 254–268.
Inamdar, S., Singh, S., Dutta, S., Levia, D., Mitchell, M., Scott, D., Bais, H., McHale, P., 2011. Fluorescence characteristics and sources of dissolved organic matter for stream water during storm events in a forested Mid-Atlantic Watershed. J. Geophys. Res. G Biogeosci. 116 (3), G03043.

Inamdar, S., Finger, N., Singh, S., Mitchell, M., Levia, D., Bais, H., Scott, D., McHale, P., 2012. Dissolved organic matter (DOM) concentration and quality in a forested Mid-Atlantic Watershed, USA. Biogeochemistry 108 (1–3), 55–76.

Jenny, H., 1941. Factors of Soil Formation. McGraw-Hill, New York.

Kaiser, K., Guggenberger, G., 2000. The role of DOM sorption to mineral surfaces in the preservation of organic matter in soils. Org. Geochem. 31 (7–8), 711–725.

Kelleher, B.P., Simpson, A.J., 2006. Humic substances in soils: are they really chemically distinct? Environ. Sci. Technol. 40 (15), 4605–4611.

Kirschbaum, M.U.F., 1995. The temperature dependence of soil organic matter decomposition, and the effect of global warming on soil organic C storage. Soil Biol. Biochem. 27 (6), 753–760.

Kleber, M., Johnson, M.G., 2010. Advances in Understanding the Molecular Structure of Soil Organic Matter: Implications for Interactions in the Environment. In: Donald, L.S. (Ed.), Advances in Agronomy. Academic Press, pp. 77–142, Chapter 3.

Kogel-Knabner, I., et al. 2010. Biogeochemistry of paddy soils. Geoderma 157 (1–2), 1–14.

Kormos, P.R., et al. 2013. Soil, snow, weather, and sub-surface storage data from a mountain catchment in the rain-snow transition zone. Earth Syst. Sci. Data Discuss. 6 (2), 811–836.

Kowalchuk, G.A., Stephen, J.R., 2001. Ammonia-oxidizing bacteria: a model for molecular microbial ecology. Annu. Rev. Microbiol. 55 (1), 485–529.

Lal, R., 2004. Soil carbon sequestration impacts on global climate change and food security. Science 304 (5677), 1623–1627.

Lal, R., 2005. Forest soils and carbon sequestration. Forest Ecol. Manag. 220 (1–3), 242–258.

Leenheer, J.A., Rostad, C.E., Gates, P.M., Furlong, E.T., Ferrer, I., 2001. Molecular resolution and fragmentation of fulvic acid by electrospray ionization/multistage tandem mass spectrometry. Anal. Chem. 73 (7), 1461–1471.

Likens, G.E., Bormann, F.H., Johnson, N.M., 1972. Acid rain. Environ. Sci. Policy Sustain. Dev. 14 (2), 33–40.

Lohse, K., Brooks, P.D., McIntosh, J., Meixner, T., Huxman, T.E., 2009. Interactions between biogeochemistry and hydrologic systems. Annual Rev. Environ. Resour. 34, 65–96.

Lundstrom, U.S., van Breemen, N., Bain, D., 2000. The podzolization process. A review. Geoderma 94 (2–4), 91–107.

Maher, K., 2011. The role of fluid residence time and topographic scales in determining chemical fluxes from landscapes. Earth Planet. Sci. Lett. 312 (1–2), 48–58.

Marshall, C., Haseman, J., 1942. The quantitative evaluation of soil formation and development by heavy mineral studies: a Grundy silt loam profile. Soil Sci. Soc. Am. Proc. 7, 448–453.

McClain, M.E., et al. 2003. Biogeochemical hot spots and hot moments at the interface of terrestrial and aquatic ecosystems. Ecosystems 6 (4), 301–312.

McGill, W.B., Cannon, K.R., Robertson, J.A., Cook, F.D., 1986. Dynamics of soil microbial biomass and water-soluble organic C in Breton L after 50 years of cropping to two rotations. Can. J. Soil Sci. 66, 1–19.

McKnight, D.M., et al. 1992. Sorption of dissolved organic carbon by hydrous aluminum and iron oxides occurring at the confluence of Deer Creek with the Snake River, Summit County, Colorado. Environ. Sci. Technol. 26 (7), 1388–1396.

McKnight, D.M., et al. 2001. Spectrofluorometric characterization of dissolved organic matter for indication of precursor organic material and aromaticity. Limnol. Oceanogr. 46 (1), 38–48.

McNeill, A., Unkovich, M., 2007. The nitrogen cycle in terrestrial ecosystems. In: Marschner, P., Rengel, Z. (Eds.), Nutrient Cycling in Terrestrial Ecosystems (Soil Biology). Springer, Berlin, Heidelberg, pp. 37–64.

Miller, A.J., Schuur, E.A.G., Chadwick, O.A., 2001. Redox control of phosphorus pools in Hawaiian Montane Forest Soils. Geoderma 102 (3–4), 219–237.

Minasny, B., McBratney, A.B., Salvador-Blanes, S., 2008. Quantitative models for pedogenesis: a review. Geoderma 144 (1–2), 140–157.

Monson, R.K., et al. 2002. Carbon sequestration in a high-elevation, subalpine forest. Global Change Biol. 8 (5), 459–478.

Monson, R.K., et al. 2006. Winter forest soil respiration controlled by climate and microbial community composition. Nature 439 (7077), 711–714.

Pédrot, M., Dia, A., Davranche, M., 2009. Double pH control on humic substance-borne trace elements distribution in soil waters as inferred from ultrafiltration. J. Colloid Interf. Sci. 339 (2), 390–403.

Peng, J., Dan, L., Huang, M., 2014. Sensitivity of global and regional terrestrial carbon storage to the direct CO_2 effect and climate change based on the CMIP5 model intercomparison. PLoS ONE 9 (4), 1–17.

Perdrial, J.N., et al. 2014. Stream water carbon controls in seasonally snow-covered mountain catchments: impact of inter annual variability of water fluxes, catchment aspect and seasonal processes. Biogeochemistry 118 (1–3), 273–290.

Piccolo, A., 2002. The supramolecular structure of humic substances: a novel understanding of humus chemistry and implications in soil science. Adv. Agron. 75, 57–134.

Placella, S.A., Brodie, E.L., Firestone, M.K., 2012. Rainfall-induced carbon dioxide pulses result from sequential resuscitation of phylogenetically clustered microbial groups. Proc. Natl. Acad. Sci. USA 109 (27), 10931–10936.

Rasmussen, C., 2011. Thermodynamic constraints on effective energy and mass transfer and catchment function. Hydrol. Earth Syst. Sci. Discuss. 8 (4), 7319–7354.

Rasmussen, C., Tabor, N.J., 2007. Applying a quantitative pedogenic energy model across a range of environmental gradients. Soil Sci. Soc. Am. J. 71 (6), 1719–1729.

Rasmussen, C., Matsuyama, N., Dahlgren, R.A., Southard, R.J., Brauer, N., 2007. Soil genesis and mineral transformation across an environmental gradient on andesitic lahar in California. Soil Sci. Soc. Am. J. 71 (1), 225–237.

Rasmussen, C., Southard, R.J., Horwath, W.R., 2005. Modeling energy inputs to predict pedogenic environments using regional environmental databases. Soil Sci. Soc. Am. J. 69 (4), 1266–1274.

Rasmussen, C., et al. 2011. An open system framework for integrating critical zone structure and function. Biogeochemistry 102 (1–3), 15–29.

Raymond, P., Saiers, J., 2010. Event controlled DOC export from forested watersheds. Biogeochemistry 100 (1), 197–209.

Renault, P., Stengel, P., 1994. Modeling oxygen diffusion in aggregated soils: 1. Anaerobiosis inside the aggregates. Soil Sci. Soc. Am. J. 58 (4), 1017–1023.

Reynolds, R., Neff, J., Reheis, M., Lamothe, P., 2006. Atmospheric dust in modern soil on aeolian sandstone, Colorado Plateau (USA): variation with landscape position and contribution to potential plant nutrients. Geoderma 130, 108–123.

Rice, K.C., Herman, J.S., 2012. Acidification of Earth: an assessment across mechanisms and scales. Appl. Geochem. 27 (1), 1–14.

Richens, D.T., 1997. The Chemistry of Aqua Ions. John Wiley and Sons, West Sussex, England, pp. 592.

Richter, D.d., Mobley, M.L., 2009. Monitoring Earth's critical zone. Science 326 (5956), 1067–1068.

Richter, D.d., Yaalon, D.H., 2012. The changing model of soil revisited. Soil Sci. Soc. Am. J. 76, 766–778.

Runge, E.C.A., 1973. Soil development sequences and energy models. Soil Sci. 115, 183–193.

Ryan, J.N., Gschwend, P.M., 1994. Effect of solution chemistry on clay colloid release from an iron oxide-coated aquifer sand. Environ. Sci. Technol. 28 (9), 1717–1726.

Schaetzl, R.J., Schwenner, C., 2006. An application of the Runge energy model of soil development in Michigan's upper peninsula. Soil Sci. 17 (12), 152–166.

Schlesinger, W.H., Bernhardt, E.S., 2013. Biogeochemistry: An Analysis of Global Change. Academic Press, Waltham, MA; Kidlington, Oxford, 672 pp.

Schmidt, M.W., et al. 2011. Persistence of soil organic matter as an ecosystem property. Nature 478 (7367), 49–56.

Schuur, E., et al. 2013. Expert assessment of vulnerability of permafrost carbon to climate change. Clim. Change 119 (2), 359–374.

Shaler, N.S., 1892. The origin and nature of soils.

Simonson, R.W., 1968. Concept of soil. In: Norman, A.G. (Ed.), Advances in Agronomy. Academic Press, Waltham, Massachusetts, pp. 1–47.

Simpson, A.J., 2002. Determining the molecular weight, aggregation, structures and interactions of natural organic matter using diffusion ordered spectroscopy. Magn. Reson. Chem. 40, 72–82.

Singh, S., Inamdar, S., Mitchell, M., McHale, P., 2014. Seasonal pattern of dissolved organic matter (DOM) in watershed sources: influence of hydrologic flow paths and autumn leaf fall. Biogeochemistry 118 (1–3), 321–337.

Small, G.E., et al. 2012. Rainfall-driven amplification of seasonal acidification in poorly buffered tropical streams. Ecosystems 15 (6), 974–985.

Sparks, D.L., 2005. Toxic metals in the environment: the role of surfaces. Elements 1 (4), 193–197.

Sposito, G., 2004. The Surface Chemistry of Natural Particles. Oxford University Press, New York, NY, pp. 242.

Sposito, G., 2008. The Chemistry of Soils. Oxford University Press, New York.

Stumm, W., 1997. Reactivity at the mineral-water interface: dissolution and inhibition. Colloid Surface A 120 (1–3), 143–166.

Stumm, W., Morgan, J.J., 1996. Aquatic Chemistry. John Wiley & Sons, Inc, New York, pp. 1022.

Sutton, R., Sposito, G., 2005. Molecular Structure in Soil Humic Substances: The New View. Environ. Sci. Technol. 39 (23), 9009–9015.

Talbot, J., Yelle, D., Nowick, J., Treseder, K., 2012. Litter decay rates are determined by lignin chemistry. Biogeochemistry 108 (1–3), 279–295.

Talibudeen, O., 1981. Precipitation. In: Greenland, D.J., Hayes, M.H.B. (Eds.), The Chemistry of Soil Processes. John Wiley & Sons, New York, pp. 81–114.

Targulian, V.O., Sokolova, T.A., 1996. Soil as a biotic/abiotic natural system: a reactor, memory, and regulator of biospheric interactions. Eurasian Soil Sci. 29 (1), 30–41.

Tarnocai, C., et al. 2009. Soil organic carbon pools in the northern circumpolar permafrost region. Global Biogeochem. Cycles 23 (2), GB2023.

Thompson, A., Chadwick, O.A., Boman, S., Chorover, J., 2006. Colloid mobilization during soil iron redox oscillations. Environ. Sci. Technol. 40 (18), 5743–5749.

Trumbore, Susan, 2006. Carbon respired by terrestrial ecosystems – recent progress and challenges. Global Change Biol. 12 (2), 141–153.

Ugolini, F.C., Sletten, R.S., 1991. The role of proton donors in pedogenesis as revealed by soil solution studies. Soil Sci. 151 (1), 59–75.

Unger, S., Maguas, C., Pereira, J.S., David, T.S., Werner, C., 2010. The influence of precipitation pulses on soil respiration – assessing the Birch effect by stable carbon isotopes. Soil Biol. Biochem. 42 (10), 1800–1810.

USEPA, 2002. Onsite Wastewater Treatment Systems Manual. Office of Water, Washington, DC.

van Verseveld, W.J., McDonnell, J.J., Lajtha, K., 2008. A mechanistic assessment of nutrient flushing at the catchment scale. J. Hydrol. 358 (3–4), 268–287.

Vázquez-Ortega, A., Hernandez-Ruiz, S., Amistadi, M.K., Rasmussen, C., Chorover, J., 2014. Fractionation of dissolved organic matter by (oxy)hydroxide-coated sands: competitive sorbate displacement during reactive transport. Vadose Zone J. 13 (7), 1–13.

Vitousek, P.M., Chadwick, O.A., 2013. Pedogenic thresholds and soil process domains in basalt-derived soils. Ecosystems 16 (8), 1379–1395.

Vitousek, P.M., et al. 1997. Human alteration of the global nitrogen cycle: sources and consequences. Ecol. Appl. 7 (3), 737–750.

Volobuyev, V.R., 1964. Ecology of soils. Translated from the Russian edition (1963) by A. Gourevich. Israel Program for Scientific Translations, Jerusalem; Davey, New York, 1964. Institute of Soil Science and Agrochemistry: Academy of Sciences of the Azerbaidzan SSR.

Weber, F.-A., Voegelin, A., Kaegi, R., Kretzschmar, R., 2009. Contaminant mobilization by metallic copper and metal sulphide colloids in flooded soil. Nature Geosci. 2 (4), 267–271.

Weishaar, J.L., et al. 2003. Evaluation of specific ultraviolet absorbance as an indicator of the chemical composition and reactivity of dissolved organic carbon. Environ. Sci. Technol. 37 (20), 4702–4708.

Yasunari, T.J., et al. 2011. Cesium-137 deposition and contamination of Japanese soils due to the Fukushima nuclear accident. Proc. Natl. Acad. Sci. 108 (49), 19530–19534.

Yoo, K., Mudd, S.M., Sanderman, J., Amundson, R., Blum, A., 2009. Spatial patterns and controls of soil chemical weathering rates along a transient hillslope. Earth Planet. Sci. Lett. 288 (1/2), 184–193.

Young, I.M., Crawford, J.W., 2004. Interactions and self-organization in the soil–microbe complex. Science 304 (5677), 1634–1637.

Zachara, J.M., et al. 2007. Geochemical processes controlling migration of tank wastes in Hanford's vadose zone. Vadose Zone J. 6 (4), 985–1003.

Zhang, X., Zhang, G., Chen, Q., Han, X., 2013. Soil bacterial communities respond to climate changes in a temperate steppe. PLoS ONE 8 (11), 1–9.

Chapter 7

A Terrestrial Landscape Ecology Approach to the Critical Zone

Aniela Chamorro*, John R. Giardino*,**, Raquel Granados-Aguilar*, and Amy E. Price†

*High Alpine and Arctic Research Program, Department of Geology and Geophysics, Texas A&M University, College Station, Texas, USA; **Water Management and Hydrological Science Graduate Program, Texas A&M University, College Station, Texas, USA; †Department of Geology and Geophysics, Texas A&M University, College Station, Texas, USA

7.1 INTRODUCTION

Air travel today is no longer the enchantment it was in the past. Planes too crowded, too little legroom, pay-for entertainment, a coke or bottle of water, if you are lucky. What is one to do? Well, next time you fly make sure you have a seat next to a window so you have an uninterrupted view of Earth below. As your plane rolls down the runway and gains altitude, look out of the window. Your perspective quickly changes from an elevational view to a plan view or map view. You will notice as you gain altitude that the size of landscape features change. This is because the scale of your view is changing as you ascend. These two factors: scale and perspective (i.e., map view) are fundamental to landscape ecology, and in this respect are important to our understanding of the Critical Zone.

The idea of looking out of an airplane window to study the land below is not a new idea. Geologists and geographers have suggested the idea and written books to illustrate the features one sees below (Monkhouse, 1965; Dicum, 2004). However, they focused mostly on geology, landforms, and features constructed by humans. With our exercise of looking out of the window of the airplane, our goal is to obtain a visual appreciation of landscape ecology.

7.2 GOAL OF CHAPTER

The overall goal of this chapter is to provide a brief overview of landscape ecology and to show the contribution landscape ecology can make to the study of the Critical Zone. We think the three main contributions that can be made include the integration of 3D dimensionality in the analysis of the landscape system, contextualizing the Critical Zone processes both temporally and spatially, and investigating problems related with scaling. But, before we proceed

Developments in Earth Surface Processes, Vol. 19. http://dx.doi.org/10.1016/B978-0-444-63369-9.00007-0

to describe how the landscape can be described and studied, we need to answer the question: what is landscape ecology?

Landscape ecology is a rapidly developing interdisciplinary methodology (some now consider it an emerging discipline) that studies landscapes, its components, and the interactions that occur (Forman, 1995; Kirchhoff et al., 2013; Risser, 1984; Urban et al., 1987). The perspective of landscape ecology departs from that of traditional ecology because humans and their modification of the landscape and built environment are considered. Landscape ecology operates from a system perspective. And, this system point-of-view provides a way to analyze and study the landscape in terms of composition, structure, and function (Fortin and Agrawal, 2005; Moss, 2014). The analysis of the landscape can be carried out within a variety of landscape scales, spatial patterns, and levels of organization (Wu, 2006). We have not found any mention in the literature regarding the similarity of representation between landscape ecology and cartography, but we think it is interesting that the mapping of landscape-ecology components is accomplished via points, lines, and polygons. Traditional and automated cartography represent data on a map using points, lines, and polygons (Burrough, 1986).

Let us for a moment return to that airplane seat by the window. Below you stretch a mosaic of various landscapes in the arctic. The view out of the window is shown in Fig. 7.1. Study of the image shows a patchwork of different lines, clusters, and background. As you look at this landscape, the green color that makes up the background to the image is referred to as the matrix. This background is a combination of shrubs, grasses and trees that are somewhat homogeneous clusters of vegetation. The two small ponds represent patches; the lighter green clusters are also patches, as various types of patches exist. The green meadow has a stream flowing through it, which is a corridor; also, various other types of corridors exist.

FIGURE 7.1 An aerial view of the landscape showing different types of patches in a surrounding matrix. The green meadow has a stream flowing through it, which is a corridor. Matrix, patch, and corridors are the landscape structure used in landscape ecology.

Looking out of the window, it is easy to discriminate sharp boundaries between these apparently homogeneous patches. At a smaller scale, such boundaries and elements will disappear. Variation of scale will determine which processes are dominant between the elements of this landscape. The complexity and heterogeneity of this type of landscape at different scales of observation is the center of focus in landscape ecology.

Understanding the complexity of natural systems is a common goal of a variety of diverse disciplines in the Earth sciences. Nowadays, overlapping of numerous disciplines to address research questions in common for all spheres of science is the norm. In the study of landscapes at mesoscales, interdisciplinarity is a must. Landscape ecology and the Critical Zone approach in parallel are focused on understanding the landscape as a system with interdisciplinarity imbedded. The Critical Zone approach attempts to understand how climatic and anthropogenic perturbations will affect natural systems involving several disciplines with shared goals. Hence, landscape ecology and the Critical Zone approach have apparent similarities because they use a holistic vision that aspires to integrate the dynamic complexity of natural systems to solve current environmental problems (Lin, 2010; Vila Subirós et al., 2006).

Landscape ecology has developed scientific work during the last couple of decades using tools that can be valuable for studying natural systems. These studies (Liu et al., 2011; Urban, 2006) can enrich the study of the Critical Zone in addressing interactions and processes in natural systems from dynamic, temporal, and spatial perspectives. Some Critical Zone researchers, who would not normally refer to themselves as landscape ecologists, generally use the same framework and approaches in their research (Fortin and Agrawal, 2005). One important distinction arises between landscape ecology and Critical Zone research: landscape ecology focuses on distribution and changes in the horizontal plane, whereas Critical Zone research appears to focus more on vertical distributions and changes.

7.3 BOUNDARIES OF CRITICAL ZONE

The boundaries of the Critical Zone, as defined by the National Research Council (NRC, 2001) and National Science Foundation (Brantley et al., 2006) as extending, from the top of the vegetation to the bottom of the aquifer, are limited in perspective (Brantley et al., 2006). According to this original definition, the Critical Zone does not have strict horizontal boundaries, but those defined by the scale of the study.

Direct correlations between surface properties and subsurface processes are weak and difficult to link in the same spatial unit of the landscape (Moore et al., 1993; Rohdenburg, 1989; Verstappen, 2011). Still, heterogeneities occurring in the subsurface influence processes on the surface, beginning with hydrological implications of the spatial variation of these heterogeneities (Robinson et al.,

2008; Tague and Grant, 2009). At regional scales, species-habitat selection is controlled by environmental resources. Ecological- and hydrological-process interrelations are linked with the landscape pattern because catchment hydrology affects habitat through the direct dependence of animals to sources of water, or indirectly, by the dependence of being related with terrain, soil biota, and their interactions (Schröder, 2006). If one considers this zone from the opposite direction, the same is true. Vegetation patterns play an important role in determining, for example, the location of runoff, soil moisture, evapotranspiration, and infiltration (Scanlon et al., 2005).

7.4 CURRENT FOCUS OF RESEARCH IN LANDSCAPE ECOLOGY AND THE CRITICAL ZONE PROGRAM

In landscape ecology, key research questions are focused on understanding the interaction between spatial pattern and process. Research is centered on analysis of landscape pattern, land-use, and land-cover change; effects of landscape fragmentation; and connectivity on ecosystem processes (Wu, 2013a). Critical Zone questions are focused on process and feedbacks occurring at the interface of elements in the Critical Zone. Table 7.1 provides a summary showing the similarity and contrast between research questions addressed by landscape ecology and Critical Zone research.

Landscape ecology developed from the ideas of Carl Troll (1939) to understand biotic interaction in a spatial context. The main concern of landscape ecology is the reconnaissance of landscape patterns and their relation with ecological function and change. Function refers to the flux of energy, nutrients, and species among the elements of the landscape, whereas change refers to ecological dynamics of the mosaic as temporal functions.

Landscape ecology has two schools of thought: a geoecological approach and a bioecological approach. Most landscape ecologists have a biological sciences background. This preponderance of landscape ecologists with a bioecological perspective has resulted in criticism of the view that landscapes have a wider approach than biotic and abiotic, or a single biophysical focus (Kirchhoff et al., 2013; Moss, 2014; Wu, 2013a). The focus of study is the land as a system, including all its components, interactions, pathways, storages, and thresholds. Therefore, attention should focus on the properties of the landscape system proper rather than focusing on the factors acting upon the land (Moss, 2014).

One of the aspects that can be brought to Critical Zone research from landscape ecology is the recognition of the functioning relationships between individual components and the landscape system as a whole. Geomorphological processes may be used to understand these relationships, and the various tools geomorphologists use have the potential to integrate the knowledge of elementary units of the land with specific research questions (Anderson et al., 2012, 2013).

TABLE 7.1 Research Questions Addressed by Landscape Ecology and Critical Zone Research

Research topics	Landscape ecology questions (Fortin and Agrawal, 2005; Risser, 1984; Turner, 2005; Turner et al., 2013; Wagner and Fortin, 2005; Wu, 2013a; Wu and Hobbs, 2002):	CZ science questions (Anderson et al., 2010; Banwart et al., 2013; Lin, 2010; Raab et al., 2012)
1. Structure and processes	How are landscape metrics and spatial heterogeneity related with historical and present ecological processes, disturbances, fluxes of organisms, materials, and energy?	
2. Landscape system	Landscape complexity: nonlinear dynamics and identification of critical thresholds	The understanding of interaction and feedback occurring in the interface of elements of the Critical Zone.
3. Scale	Scaling: definition and identification of the right range of scales to examine pattern and process for a given question.	How is information transferred from a microscale to mesoscales, and macroscales?
	What are the algebraic rules for rescaling quantities?	
	How should we define and quantify spatial heterogeneity, given that, it is scale dependent?	
	How do ecological processes transfer between and across scales?	Identify how dominant processes and their control change throughout spatial and temporal scales.
	How can we account for uncertainty through space and time?	Search of quantitative studies in landform evolution.
4. Human inclusion	Causes, processes, and ecological function consequences of land use and land cover change.	What are the anthropogenic forces and the impacts of land use and management practices in the CZ?
	How can conventional natural resource management be enhanced through a landscape ecology approach?	

Both approaches share similarities that make them complementary.

A criticism of landscape ecology is the lack of a theoretical framework. This is an intellectual difficulty of interdisciplinary research and such criticisms may soon be leveled at Critical Zone research. Effective progress in studying the landscape cannot be made without a well-defined focus and a theoretical and methodological base to which interdisciplinarity adherently subscribes (Castri and Hadley, 1986; Moss, 2014). Like all disciplines of science, landscape ecology as well as Critical Zone research needs their own distinctive core and focus to strengthen the interdisciplinary-team approach (Raab et al., 2012; Moss, 2014). The ultimate result will be the creation of a base on which to develop a structured discipline (Moss, 2014). Focus on the landscape as a system could generate this necessary foundation.

One of the methodological shortcomings is the limited ability to replicate experiments at broad-scale. This limitation is explained by the vast number of variables affecting the natural system: multiple combinations of natural and anthropogenic forces produce landscape patterns. This has forced the landscape ecologist, as it will force Critical Zone researchers to use phenomenological and corroborative models over traditional experimentation (Turner, 2001).

Landscape ecology currently has tools to complement Critical Zone research. First, emphasis on the biotic component by landscape ecology enriches the focus of Critical Zone research in the abiotic components of the Critical Zone. Concepts of Critical Zone research come from the need to integrate the abiotic components and ecological processes by geologists and geochemists (Raab et al., 2012); ecological processes are the main focus of landscape–ecology research.

As we mentioned previously, horizontal emphasis in structure and pattern by landscape ecology can enrich Critical Zone research, which is mainly focused in the verticality of processes occurring in the Critical Zone. For example, the effect of patterns on lateral fluxes or transfer of matter between different homogeneous units of the landscape is only possible if the structure of the landscape is well defined. Also, advances in remote sensing (i.e., high-resolution digital-elevation models, satellite imagery and geophysical techniques to explore the subsurface, and geographic information systems technology (GIST)) also contribute to the study of landscape in three dimensions, incorporating vertical, horizontal, and temporal components in the structure of landscape ecology.

Table 7.1 shows common research questions that are being addressed by landscape ecology and Critical Zone research: identification of critical thresholds, dominant processes, and their relationship at different scales. These constituents have been addressed and continue to be addressed by landscape ecology, which has a five-decade head start in terms of discussion and development of methodologies. The potential of generating tridimensional conceptual models of the landscape system by combining the components of the horizontal vision of landscape ecology and the vertically focused Critical Zone could be a major step in defining more realistic conclusions.

7.5 DEVELOPMENT OF LANDSCAPE ECOLOGY AS A DISCIPLINE

Landscape ecology attempts to understand biotic interactions in a spatial and temporal context (Forman, 1995; Risser, 1984). The term "landscape ecology" was coined by Carl Troll, a European botanist and geographer, in 1939 (Troll, 1971). Landscape ecology is a branch of ecology that is concerned with the effects of disturbances (i.e., natural or anthropogenic) on ecological process in a spatial context. Landscape ecology is especially focused, but not exclusively, on mesoscales and macroscales (i.e., hundreds to thousands of square kilometers).

From the initial insights of Troll, the development of landscape ecology was influenced by two theoretical backgrounds: the general theory of island biogeography proposed by MacArthur and Wilson (1967) and later by metapopulation models (Hanski, 1998). Island biogeography theory articulated the idea of habitat fragmentation based on the analogy between fragments in the landscape and island dynamics. This idea has been criticized, because it considered that fragmentation occurring in insularity is the product of unique evolutionary processes (i.e., biogeographic processes), which is not applicable to habitat fragments in a continental context. For example, the observed critical influence of forest remnants over other surrounding remnant fragments in continental areas is impressive; however, the same influence is negligible in oceanic islands (Curtis and McIntosh, 1951; Haila, 2002; MacArthur and Wilson, 1967). Despite criticism, the framework of island biogeography theory paved the way for new approaches in conservation research, and it was the first step towards including spatial dimension in the study of habitat degradation (Haila, 2002).

By the 1980s, landscape ecology had gained acceptance as a new research focus where the assumptions of island theory of equilibrium, uniform space, and universal regularities were replaced by the investigation of heterogeneity driven by nonequilibrium, site-specific ecological processes (Haila, 2002). Landscape ecology reinforced the idea of fragments as patches embedded in a particular surrounding (i.e., matrix) rather than analogues to islands.

The terms patch and matrix come from the theory of metapopulations (Hanski and Gilpin, 1991). Landscape ecology uses metapopulation theory, and its basis to understand effects of pattern and structure in the distribution of populations and vice versa (Forman, 1995). Despite this influence in its development, landscape ecology differs from metapopulation theory because of the incorporation of two outlooks. One of these views began in Europe; it provided a focus on the integration of humans as part of the landscape system (Naveh and Lieberman, 1984; Turner, 2005). The second view of humans as integral parts of the landscape was enhanced with the incorporation of the human role in landscape development and management (Hanski and Gilpin, 1991).

As the theoretical basis of landscape ecology began to mature, the concepts of landscape ecology broadened; incorporating other disciplines from the geosciences and social sciences. This consideration of the entire landscape beyond biological and ecological sciences is critical in the differentiation of its influences and formation within the landscape-ecology framework.

Today, landscape ecology has a dual perspective, varying between a bio-ecological perspective and a geo-ecological perspective (Moss, 2014; Turner, 2005). The bio-ecological perspective, driven by ecological dogma (Forman, 1995), places emphasis on the concept of the spatial dimension of plant and animal populations (Moss, 2014; Turner, 2005). The geo-ecological perspective is based on geographic tenets, which focus on the concept of a geosystem and a systematic interpretation of the land (Haase and Richter, 1983; Rougerie and Beroutchachvili, 1991). In this context, land is understood in terms of landforms, soils, vegetation, and human activities and, more recently, incorporating energy and biogeochemical forces responsible for the formation, structure, and operation of distinct landscape units (Moss, 2014).

Both approaches share a common bond; in that understanding function is critical. To understand function, knowledge of processes is required, which incorporates a range of ecological, pedological, hydrological, geomorphic, lithospheric, and atmospheric processes (Moss, 2014). All of these processes occur in the vertical, horizontal, and temporal dimensions of the Critical Zone. They play fundamental, important roles in configuring the landscape system. Recently, the acknowledgment of the important role climate change can play in landscape ecology–ecosystem services and sustainability has been recognized (Wu, 2013a). Thus, it is easy to see the major role landscape ecology plays in the Critical Zone and is a fertile ground for Critical Zone research.

7.6 LANDSCAPE ECOLOGY DEFINITION AND FOCUS

In simple terms, landscape ecology studies the relationship between ecological processes and the environment. Spatial patterns are present in all environments; how these patterns are related with ecological processes is the main question of landscape ecology today. Forman and Godron (1986) defined landscape ecology as the study of physical-biological relationships that govern the units in a region, considering the vertical links in the system, inside of the spatial unit and horizontally, relating different spatial units. The definition and identification of spatial units are scale dependent.

The foci of landscape ecology have been highlighted by Forman and Godron (1986), Turner (2001), and Wu (2013a, 2013b). These include: (1) the flows of energy, mineral nutrients, and species among the elements in different spatial and temporal scales; and, (2) the ecological dynamics of the landscape mosaic through time, understanding the interaction between patterns and processes.

7.7 LANDSCAPE ECOLOGY CONCEPTS

Concepts in landscape ecology originated from theoretical ecology self-examinations, but their applications are pertinent to the analysis of an integrated natural system because landscapes consist of more than the biological component

(Fig. 7.2). Spatial pattern and heterogeneity in multiple scales are central in the theory and practice of landscape ecology (Wu, 2013a) (Fig. 7.3). This approach requires a researcher to focus attention on three components of a landscape as defined in the works of Forman (1995) and Forman and Godron (1986). The three components are:

- Structure: Spatial relationship between elements of landscape.
- Function: Energy, nutrients, and species fluxes among elements of landscape.
- Change: Ecological dynamic of mosaics with time.

From each of these components, several concepts commonly used in landscape ecology have emerged. Figure 7.3 illustrates how the concepts of spatial heterogeneity, pattern, and scale are interdependent and affect one another.

Because understanding and interpretation of results of studies in landscape ecology requires proper use and comprehension of the terminology used, we think

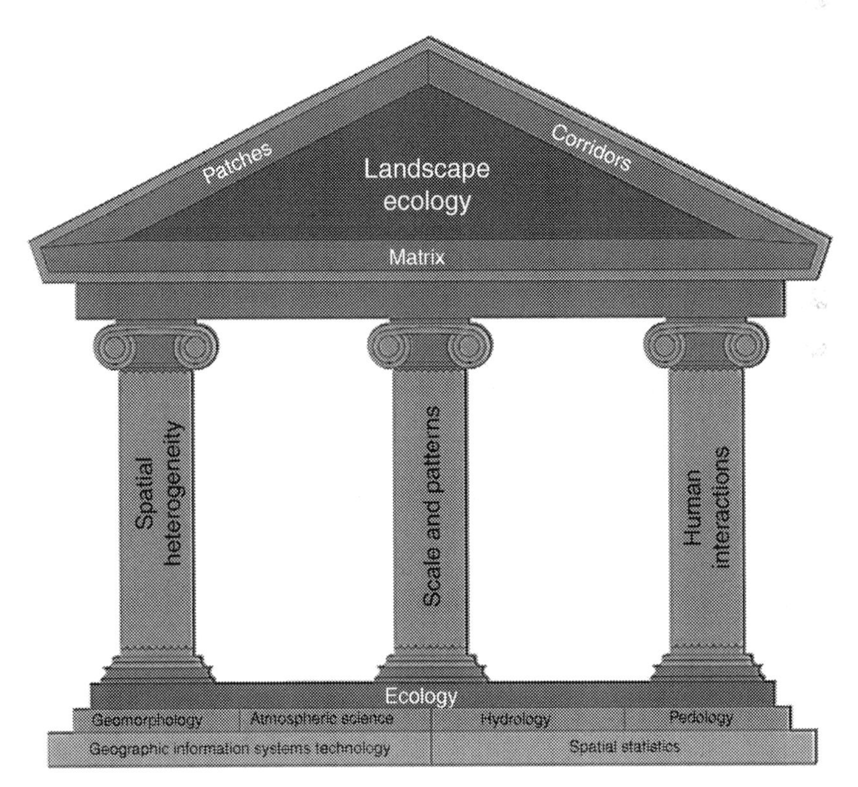

FIGURE 7.2 Landscape ecology is an emerging discipline with a focus on spatial heterogeneity, scale and patterns, and human–environment interactions. Landscape ecology uses a lens consisting of a matrix, patches, and corridors to study and explain the relationship between ecological processes and the environment. Environment is composed by elements that interact and that are studied by specific disciplines. The surface expressions of the landscapes are the product of all those underlying processes occurring in the abiotic realm.

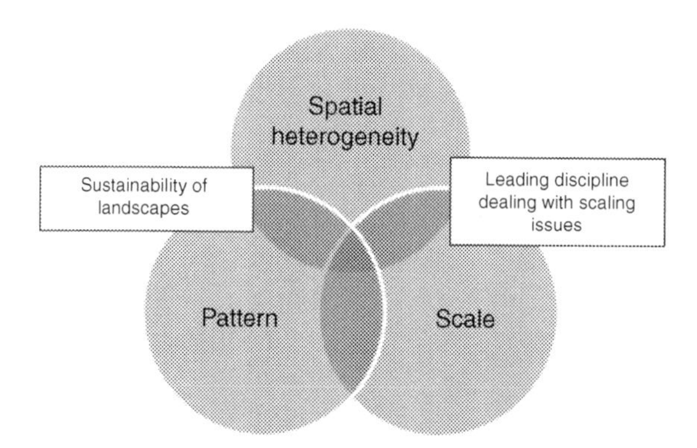

FIGURE 7.3 Landscape ecology emerges from ecology and geography. Essential terms in landscape ecology are spatial heterogeneity, pattern and scale (Wu, 2013a). Landscape ecology is distinguished by its advances in scaling issues and the focus on understanding heterogeneity. Pattern, as one of the pillars of landscape ecology, is location dependent. From the investigation of spatial heterogeneity, pattern, and scale, development of pertinent recommendations for sustainability of landscapes emerges.

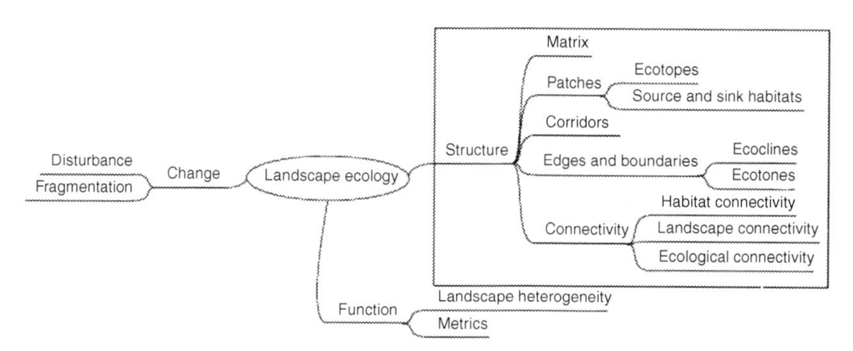

FIGURE 7.4 The linkages between change, structure, and function in landscape ecology can be seen with this Mind Map diagram.

it is necessary for the Critical Zone researcher to understand the main concepts of landscape ecology to ensure their proper use and application. Landscape ecology has three main components: structure, function, and change. Each of these main components is expressed in secondary functions or characteristics. To show the connection between the important concepts and terms used, we have created a mind map to illustrate these in Fig. 7.4. Study of Fig. 7.4 shows that scale, which is an important factor in landscape ecology, directly affects structure.

7.7.1 Structure

Structure is the spatial organization of the elements in a landscape. In the analysis of structure, emphasis is placed on the identification of patterns that involve the reconnaissance of relatively homogeneous areas (i.e., homogeneity is a first approximation of simplifying the real heterogeneity of landscapes into discrete units). The goal of this concept, also shared by geomorphologists, is to relate ecological processes through the investigation of the structure. Details regarding the systematization and definition of components of the landscape structure were described in Forman and Godron (1986) and in Forman (1995). Our goal here is only to introduce the core ideas of the study of structure.

7.7.2 Mosaic and Matrix

The basic element used in interpreting the landscape is the mosaic, which is a composite of the landscape elements. The matrix can be viewed as the background of the landscape, characterized by an extensive cover with a major control over processes (Forman, 1995). Hence, the correct identification of the matrix is based on its inherent connectivity, and it is based on the recognition of dominant processes or elements. Elements in the landscape can be differentiated by their substrate, internal dynamics, and human activities (Vila Subirós et al., 2006).

7.7.3 Patches

Patches are one of the elements that add complexity to the matrix. A patch is defined by Forman (1995, p. 39) as "...a relatively homogeneous nonlinear area that differs from its surrounding." Differentiation may be in nature or appearance (Turner, 2001) and depends on the research questions being posed. Different questions can imply different scales of work, characteristics relevant to the answer and, therefore, lead to different patch arrays, even for the same physical space (Fig. 7.5) (Cadenasso et al., 2003). Five different types of patches have been identified. These are spot disturbance patch, remnant patch, environmental resource patch, introduced patch, and ephemeral patch (Forman, 1995). Reconnaissance of patches is important because they are the basis for the study of landscape change. The patch seen in Fig. 7.5 is shown at three different scales.

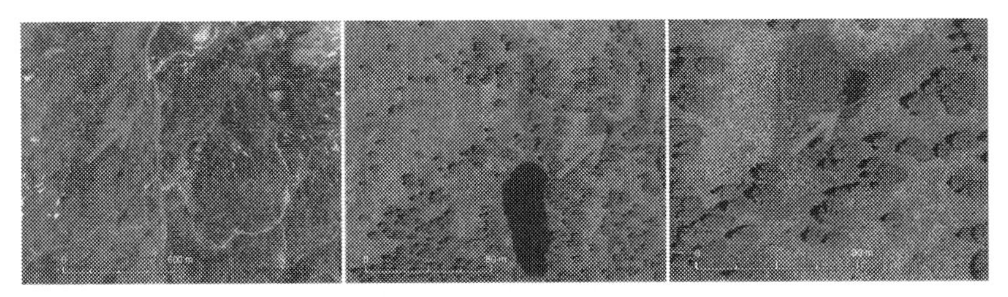

FIGURE 7.5 Scale plays an important role in landscape ecology. The three photographs of the same fen on the summit of Red Mountain Pass in Colorado show how the patch can be seen at different scales. In the left photograph the patch cannot be distinguished from the matrix as compared to the photograph on the right where the extent of the fen (patch) can be easily mapped. The arrows in each photograph point to the location of the patch (Google Earth® images).

This patch could also be studied from a Critical Zone perspective. Patches, matrices, and corridors all have Critical Zone dimensions.

The study of change inside patches is referred to as patch dynamics. Study of patch dynamics focuses on the effects of disturbances on the configuration of patches, which in turn, are essential for investigations about fragmentation.

Size, number, and shape (i.e., morphology) of patches, as well as their relative location with respect to other elements and other patches are the pillars for the analysis of structure. Identification and measurements of patches are made through computer algorithms, producing mapped patterns according to the size of the cells of the database. A continuous string of cells under the same category or classification is related to a string. Therefore, the concept of patch is affected by the grain size (scale) of the base data (Turner, 2001).

Some special classifications of patches exist, based on their function. Patches can be classified as habitat patches (i.e., source and sink habitats) or elementary units with relative homogeneity (i.e., ecotopes). Ecotopes and source/sink habitats classification are examined later in this chapter.

7.7.4 Corridors

Corridors are connectors that are present between, and among patches, and act as dispersal pathways, propagating energy and matter through the landscape (Fig. 7.6). In this way, connection occurs at the level of structure (i.e., patterns) or processes (Green et al., 2006). In plan shape, corridors are narrow strips of the land that differ from areas adjacent on both sides. (Turner, 2001). Some examples of corridors are rivers, creeks, trails, roads, highways, and funneling winds. Also, from a Critical Zone perspective, groundwater flow pathways can be classified as corridors.

Along with patches, corridors are habitats and their presence implies larger surfaces (i.e., edges), connectivity, or in some cases, they can act as barriers. This difference between serving as a pathway or barrier is because landscapes may be connected with respect to one process but not with respect to another (Green et al., 2006).

7.7.5 Ecotopes

The term ecotope comes from the term biotype in ecology. Ecotope is defined as the smallest spatial entity or element of the geographic landscape (Fig. 7.7). Definitions of ecotopes (classifications) are important because they reflect the distribution of changing successional forces, (i.e., to evaluate forest growth, agricultural assessment, vegetation mapping, landscape contamination, or landscape planning) (Troll, 2007). The ecotope is similar to the pedon in soil science (Birkeland, 1999), a third-order landform in geomorphology (Rohdenburg, 1989), or a pixel in remote sensing (Jensen, 2007). It is important to see the contribution the application of this concept can bring to the Critical Zone perspective.

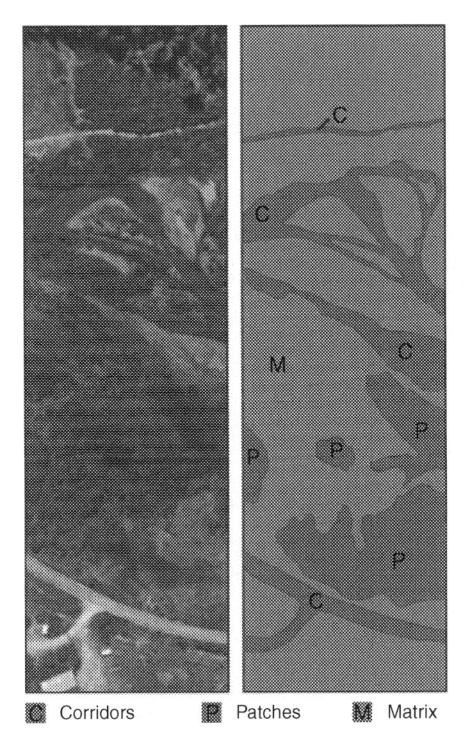

FIGURE 7.6 Example of the concepts of matrix, corridor, and patches. *(Google Earth® (2015). The idea of this diagram is based on Fischer and Lindenmayer (2007).)*

7.7.6 Ecoclines

Ecoclines are a type of ecological boundary defined by a gradual spatial and ecological change occurring between two systems. In their gradient zones, relatively heterogeneous communities are present, which are environmentally more stable than ecotones. An example of an ecocline is the gradient of vegetation communities responding to the environmental factor, temperature, which varies with elevation along a hillslope. Vegetation communities on the hillslope, although apparently stable, may change their position, or migrate up or down elevation. For example, with the onset of a glacial epoch, tree-line migrates to a lower elevation (Attrill and Rundle, 2002). Soil biota such as microorganisms can also change from the surface to depth along a gradient, signifying the presence of an ecocline.

7.7.7 Ecotones

Ecotones are abrupt changes in vegetation (Walker et al., 2003) or two adjacent, different and homogeneous community types, producing a narrow ecological zone between them (Attrill and Rundle, 2002). Ecotones are used to define ba-

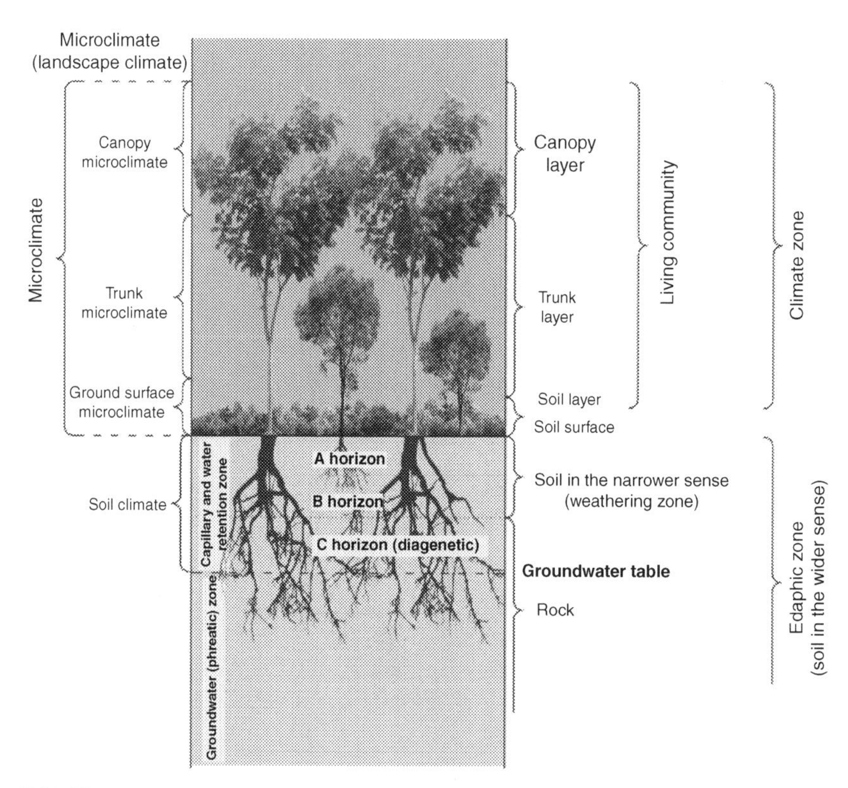

FIGURE 7.7 **Elements distinguished in an ecotope according to Troll (1955; published in English in 2007).** The concept of ecotope is very close to the concept of the Critical Zone: certain soil units that share common hydrological and pedological processes. *(Modified from Troll (2007).)*

sic units in landscape studies, and their identification relies on the sharpness of the vegetation transition, particularly ecological conditions and their causes (i.e., natural or anthropogenic environmental change, invasion or alteration of species present) (Walker et al., 2003). Ecotones are dynamic, sometimes with strong fluctuations. In the littoral zone of a lake, an ecotone is the transition zone of distinct aquatic communities that varies throughout the year according to seasonality (Attrill and Rundle, 2002). As Critical Zone research expands, scale will play a more and more important role. This will lead to the development of Critical Zone concepts similar to the ecotone concept in landscape ecology.

7.7.8 Source and Sink Habitats

Sources and sinks are terms related to rates of colonization and extinction within patches from the theory of metapopulations (Forman, 1995). Patches can be classified as sources or sinks; in a source-habitat patch, the dispersal of individuals is dominant compared with immigration (Forman, 1995). A source patch will produce offspring in excess of those necessary to maintain a stable population within the patch, generating emigration (Liu et al., 2011). Conversely, in a sink habitat, the patch immigration exceeds dispersal (Forman, 1995) because reproduction in these patches is inadequate to maintain population numbers. Populations in sink patches only persist if source patches replenish the necessary numbers to maintain population stability.

Sink and source systems are dynamic: sinks may turn into sources and vice versa with variations in environmental conditions or abundance. Like patches, sink and source systems are scale-dependent and site-specific. This specificity makes the data needed for sink–source classification difficult. Data about demography, population size, dispersal, and density in a patch are necessary. It is for this reason that most source–sink systems have relied on models. Only a few studies have gathered empirical information from real populations (Liu et al., 2011).

7.7.9 Edges or Boundaries

Edges and boundaries can play an important role in Critical Zone research. In landscape ecology, the outer portions of patches and corridors are the edges. The inner areas of the landscape elements are the interior or core (Forman, 1995). It is important to note that an edge can also be an outer layer that marks a gradual, rather than a discrete change between cover types. This concept can be applied in the vertical direction also. For example, as one transcends from the surface of Earth downward, the soil profile can be distinguished with abrupt changes from one horizon to another. In addition, the concept can be applied in Critical Zone research in both vertical and horizontal directions. For example, as one transects from one Critical Zone environment to another, the change may be subtle or abrupt,

A boundary is defined as the transition zone composed by the edges of two adjacent landscape elements (Forman, 1995). Boundaries are important components of spatially heterogeneous areas. They can be similar to the elements they separate, or they can be completely different. Boundaries differ in location or magnitude depending on the specific characteristic under observation, and their function is determined by the material, population, process, or energy affected by the boundary gradient (Cadenasso et al., 2003).

In Critical Zone research, boundaries are interfaces (e.g., hyporheic zones, soil–root interfaces, soil–horizon interfaces) where important interactions occur (e.g., soil–chemical reactions, soil adsorption–desorption reactions, biodegradation, weathering boundaries) (Buss and Moore, 2013; Lin, 2010). In this regard, we agree with the suggestions of Cadenasso et al. (2003) that boundaries are best construed in three dimensions, in agreement with the use of the term landscape in a system context. This definition also fits well into the 3D view of the Critical Zone.

Two types of ecological boundaries occur in the landscape ecology literature: ecoclines and ecotones. These two types are considered in further detail in subsequent sections.

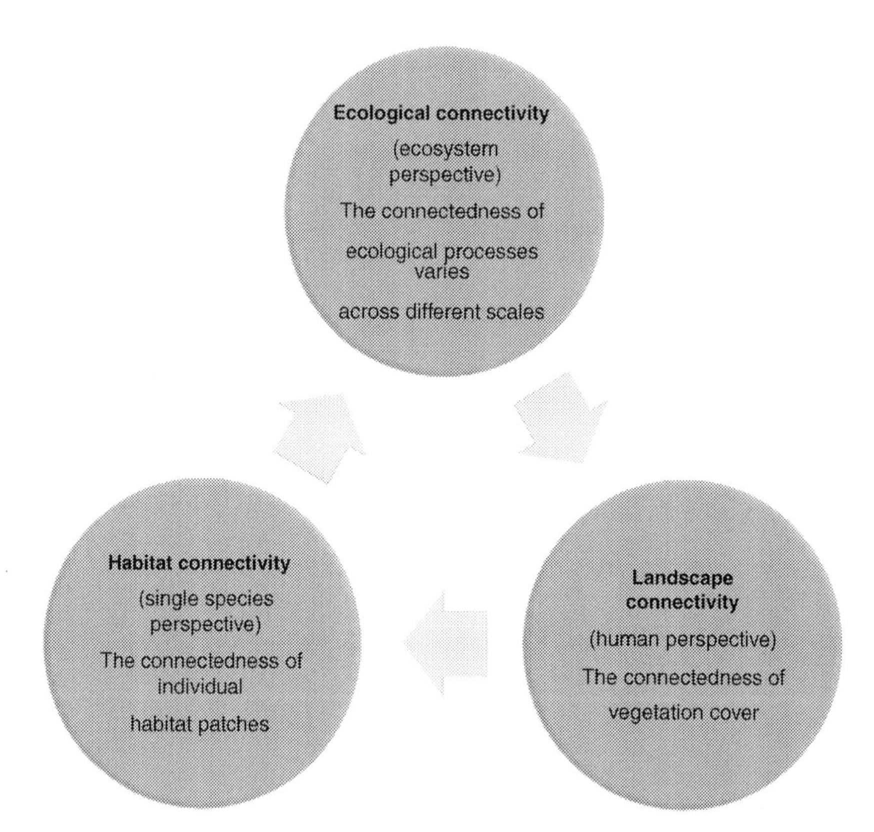

FIGURE 7.8 Relationship between the three connectivity concepts defined by Fischer and Lindenmayer (2007). Diagram is modified from Fischer and Lindenmayer (2007).

7.7.10 Connectivity

The three types of connectivity are: habitat, landscape, and ecological-based (Fig. 7.8). Habitat connectivity is the connectedness between patches of suitable habitat for a given individual species, as it is used by Turner (2001) and Moilanen and Hanski (2001). Landscape connectivity is the connectedness from a human perspective based on land-use cover (e.g., patches of native vegetation). Moreover, ecological connectivity is the connectedness of a specific ecological process at multiple scales (i.e., hydroecological flows, disturbance processes, trophic relationships, etc.). Forman (1995) refers to this third type (ecological) as behavioral or functional connectivity.

7.8 LANDSCAPE HETEROGENEITY

Wu (2006, p. 1) defined heterogeneity as "... a multiscaled structure composed of intertwining patchiness and gradients in space and time." In other words, heterogeneity can be recognized by distinct boundaries of a mosaic of patches and corridors distributed in the landscape, or as a gradient in which discrete boundaries cannot be recognized, and a fuzzy boundary is observed. Heterogeneity should not be misconceived as a random distribution of objects. Although uneven, a complex and unique structure exists (Forman, 1995). Indeed, the importance of heterogeneity is its situational location relative to its relationship with complexity and diversity. A heterogeneous landscape can be a natural occurrence where the landscape is formed by natural environmental gradients or products of anthropogenic activities, marked by different land-cover types. In the past, biogeography theory considered that disturbances would produce homogeneous landscapes; however, evidence exists that heterogeneous, modified landscapes support more species than similar but less heterogeneous modified landscapes, although exceptions exist (Fischer and Lindenmayer, 2007). Natural and anthropogenic sources of heterogeneity can be identified by their landscape signature. Natural activity favors irregular and curvilinear patterns, whereas intense, human activity leads to a simplification of forms and a linear character of the landscape (Vila Subirós et al., 2006).

Landscapes are three-dimensional entities. As such, landscape heterogeneity can be extended to a three-dimensional function. For example, in considering an individual landform or a soil pedon, the concept of heterogeneity can be used to define multiscaled structures composed of intertwining patchiness and gradients, which can be expressed as textural changes or lithologic changes in space and time.

7.8.1 Function

The concept of function is used to define the flows of energy and materials in food chains and cycles throughout the structure (landscape) (Forman, 1995). Function is also scale dependent based on the organism or abiotic element under study. It is possible to infer landscape function through an understanding of the relationship between structure and process (Turner, 2001; Wiens and Milne, 1989). Again, it is important to remember that landscape position affects ecosystem function. Horizontal and vertical spatial variations in abiotic variables (i.e., temperature, precipitation, horizontal distribution of soils) are constraints on ecosystem function (Turner, 1989).

7.8.2 Change

Landscapes are constantly subjected to change. Landscape change deals with alteration of the structure and function of the mosaic of a landscape. The object of focus is the dynamic aspect or transformation of the model through time (Turner, 1989). In the analysis of landscape change, emphasis is placed on the consequences of disturbances; the type of disturbance either natural or anthropogenic is irrelevant. This is one place where nature trumps humans, as natural disturbances are the major cause of landscape change (Coulson and Tchakerian, 2010).

The scale of time is an important component of the analysis. Whereas natural landscapes generally change in the geological dimension of time; change in the landscape caused by different economic experiences can lead to very rapid modifications in the landscape just between one generation and the next (Troll, 2007).

Thus, in dealing with changes in the Critical Zone it is imperative to keep in mind that change as well as function in landscape ecology emphasizes the implications of landscape heterogeneity. The spatial configuration of elements and the identification of lateral fluxes lead to the identification of processes and their effects on the system through time.

7.8.3 Habitat Disturbance

Changes in the landscape can be brought on by disturbance, which is an event that significantly alters the structure or function of an ecosystem (Forman, 1995), community, or population. A disturbance can alter resource availability or the physical environment (White and Pickett, 1985).

Disturbances are scale dependent (White and Pickett, 1985) and are linked to the study of patch dynamics in the questioning of how the disturbance influences the landscape and vice versa (Turner, 2001). Disturbances are widespread across the Critical Zone. For example, fire, severe air-pollution, volcanic eruptions, severe droughts, flood events, heavy snow, and windstorms, to name a few, may cause an increase or decrease in species, as well as directly or indirectly affecting pests and diseases. Disturbances are normal, periodic, but infrequent in a system. Some disturbance may decrease species numbers whereas others result in higher species numbers (Forman, 1995; Turner, 2001). Some species are affected by disturbance differently. Therefore, the temporal and spatial scale of study of effects of disturbance on a community of bryophyte along a stream bank consisting of boulders may be inappropriate for a forest community growing in the same site (White and Pickett, 1985).

In the majority of cases of disturbance, the effect on the structure of the landscape is an increase in mosaic heterogeneity, although exceptions do occur (Forman, 1995). It was originally assumed that large disturbances homogenize the landscape (Forman, 1995); however, today studies suggest that disturbances create spatial heterogeneity, often on multiple scales (Fischer and Lindenmayer, 2007; Turner, 2010).

Exceptions to the increase of heterogeneity by disturbances are caused by widespread events (e.g., fires) that create a more homogeneous landscape. At the same time, heterogeneity itself is a general inhibitor of the spread of disturbance.

The effects of disturbance depend on the state of the system before it was disturbed (Turner, 2001; White and Pickett, 1985). Also, how the system is structured (i.e., mosaic) can predispose the landscape to further disturbance. Added to that, climatic variability can amplify or mute anthropogenic effects and conversely, effects can compound climatic impacts.

In general, in scenarios of climate change or land-use change, heterogeneity plays an important role in the sustaining of ecosystems. Maintaining the inherent heterogeneity that characterizes a landscape, especially after a disturbance occurs, is important for protecting original habitats. Specifically, actions such as burning the remaining unburned patches for fire-suppression response should be avoided because this will destroy the seeds necessary for forest regeneration (influence among patches).

7.8.4 Habitat Fragmentation

Habitat fragmentation is a landscape modification that has generated concern as a result of its negative effects on biodiversity. A fragmented landscape is characterized by a strong contrast between vegetation patches and their surrounding matrix, commonly occurring in formerly forested areas (Fischer and Lindenmayer, 2007).

Fragmentation is the opposite of connectivity. Fragmentation from landscape modification can be interpreted from a pattern-oriented approach where the focus is the landscape pattern in terms of structure and its correlation with the occurrence of species. Its main limitation is the under-appreciation of the complexity of ecological processes and differences between species.

An additional interpretation of fragmentation occurs in the species-oriented approach, where a number of species are studied according to the needs of each individual species and interconnected processes affecting those species (Fischer and Lindenmayer, 2007).

7.8.5 Scale

Scale is defined as the spatial or temporal dimension of an object or process (Turner, 2001). Scale is a complex phenomenon. As scale is considered in both landscape ecology and Critical Zone research, numerous parameters must be considered. Some of the major parameters include: extent, grain, scaling factors, thresholds, and a hierarchical system. These parameters are exceptionally important in dealing with the overall application of scale. The various parameters are defined as: (1) extent refers to the horizontal as well as the vertical size of the overall area of the study. (2) Grain is defined as the finest resolution within a given data set (Turner, 2001). It refers to the coarseness in texture or granularity of spatial elements composing an area determined by the size of the patches that can be recognized. In this way, fine-grained landscapes are composed of small patches whereas a coarse-grained landscape is comprised mainly of large-sized patches (Forman, 1995). (3) Scaling is simply defined as the extrapolation of results or conclusions from a study completed at one scale to the scale of another study. Scaling can only occur if patterns and process are not fundamentally different at varying scales (Wiens and Milne, 1989). At the same time, the processes studied are important, depending on the scale studied (e.g., the importance of a process, when compared to another process, is scale dependent.) (4) Threshold is defined as a point, line, or zone where change occurs (Chorley and Kennedy, 1971). Several types of thresholds exist (i.e., chemical, physical, biological, etc.). Turner (2001) defined a critical threshold for landscape ecology as a point at which an abrupt change in quality, property, or phenomenon occurs.

Hierarchy in landscape ecology is a systematic approach of interconnections of discrete functional elements linked at two or more scales (Forman, 1995). For instance, from a catchment scale to the scale of an individual tree, each element has its own level of variability. The catchment scale contains the individual variability of each unit (i.e., tree), but at the same time it is a container of all those units. Landscape ecologists call this phenomenon a nested hierarchy, in which each level contains the level below. The linkages at each level are the processes. The concept of hierarchy has important implications for researchers dealing with any component of the Critical Zone. For example, Broxton et al. (2014) estimated snow-water inputs from a model of snow cover derived from distributions of tree canopies. A direct relationship exists between snowpack variability and tree-canopy variability (Broxton et al., 2014). Therefore, structure, and the specific spatial distribution of individual trees at a local or hillslope level, has important implications to hydrological processes at the watershed level in the Critical Zone.

Level of organization is another important concept that has direct implications for researchers dealing with Critical Zone processes. Level of organization is not synonymous with scale (Turner, 2001), but it is the description of place within a biotic hierarchy (e.g., from a single organism to an entire population). Processes occurring at each level of organization have their own spatial and temporal scales, which are referred to by Forman (1995) as domains of scale. In the Critical Zone, where biotic and abiotic factors are tied, the concept could be especially useful to select the correct level when investigating processes that will affect particular species, community, or resources. As an example, hydrological processes affecting colonies of bacteria in fens are completely different from the domain of scale of a study, where the focus of attention is mammals in the same area.

Scaling is a major concern in Critical Zone research. This factor becomes especially fundamental, as the focus of Critical Zone Observatories (CZOs) is to integrate the study of a specific area from a multiple-disciplinary approach. So, many of the initial lessons learned by the landscape ecologists have direct application to Critical Zone research.

Scale is a prominent topic in landscape ecology because scale is a condition for results of any analysis over a landscape, and determines if a certain conclusion can be extrapolated temporally and spatially. At fine-scales, landscape ecology investigates scale mainly by laboratory and plot experiments (Turner, 2001). At broad-scales, however, experimental results are difficult to replicate.

7.9 TOOLS FOR THE RECOGNITION OF STRUCTURE, PROCESS, AND CHANGE OF LANDSCAPES

Spatial patterns are the result of complex interplay between abiotic constraints, biotic interactions, and disturbances. This complexity is reflected in the narrow relationship between pattern and process (Schröder, 2006; Turner, 1989). Landscape ecology has developed several methods and models to investigate the landscape.

Quantifying spatial pattern is the first step towards the understanding of causes, processes, and consequences of spatial heterogeneity (Wu, 2013a). Numerous metrics can be measured to quantify landscape pattern, connectivity, and indicators of diversity: proportion of the cover type, number of cover types, and richness of landscapes. Common metrics of spatial configuration can be used to show probability of adjacency, connectivity, and contagion, which are used as indicators of spatial arrangement of cover types. Details about common calculations of metrics have been described in Turner (2001) and Forman (1995).

Multiple metrics can be used to obtain arrangement and configuration information using multivariable statistics, the objective being to relate variability in models. For instance, nutrient and sediment loading in streams are excellent examples of measurable quantities, which may be modeled, that fit the format of Critical Zone research (Jones et al., 2001). A third type of metric, fractals, provide descriptions of spatial distribution based on the idea of self-similarity throughout scale functions. Fractal analyses have been used to understand complexity and compare landscapes and their evolution through time (Green et al., 2006). The use of this type of fractal analysis is an open research area in Critical Zone research.

Fractals principles are employed in landscape ecology to study complexity. Although fractal dimensions can only be applied to certain types of data, the principle of self-similarity serves to characterize the relative importance of different processes at different scales. Special attention is placed on the study of landscape roughness. Roughness of landscapes is the product of natural processes operating at a finite range of scales (Green et al., 2006). Interesting examples of the application of fractals in the study of scale are the observed self-similarity in weathering patterns. Self-similarity was found among several orders of magnitude, from the natural texture of topography of a certain landscape to natural rock surfaces (Oleschko et al., 2004; Power and Tullis, 1991).

The availability of spatial data (i.e., remote sensing imagery) and geographic information system technology (GIST) has contributed to more accurate results in the study of patterns and have become primary resources in the metrics of landscape ecology. GIST resources range from map servers to sophisticated software packages, which are capable of data viewing, editing, raster and vector analysis, as well as modeling, and simulation. Common GISTs include: ESRI (ArcGIS®, ArcView®, and ArcMap® products), ERDAS®, and ENVI®. These are proprietary software; however, free and open-source desktop GIST exists: GRASS® (Neteler and Mitasova, 2004), ILWIS® (Rossiter, 1996), Quantum® or QGIS® (Jung, 2013;

Quantum, 2013) and SAGA® (Boyd and Foody, 2011). A complete list of gratis and open-source software is presented in Steiniger and Hay (2009).

All these GIST tools have been developed to quantitatively understand the spatial pattern of the landscape. Yet, quantitative methods in landscape ecology, unfortunately, do not provide quantification regarding the structure of the landscape. Instead, the methods deploy morphology information that helps to understand conditions and ecological processes by looking at the landscape in a holistic way. Nonetheless, the relationship between pattern and process remains a challenging and important area of research (Bell, 2012; Turner, 2001; Wu, 2013a).

Modeling is another fundamental tool used for pattern analysis. Models are used in landscape ecology to conduct experiments at the microlandscape scale. Models allow the manipulation of conditions that is not possible in field studies because of the large number of variables associated with complex and extensive landscapes. Regrettably, in most instances, complex modeling does not allow replication.

Models are important to landscape–ecology research, as they have robust application to studies of the Critical Zone. Three types of models have been used: (1) spatial models; (2) mechanistic and empirical models; and, (3) deterministic and stochastic models. Spatial models are used when explicit space (i.e., spatial pattern, spatial variation, spatial interaction among elements or processes) is an important determinant of the process under study (Turner, 2001). Mechanistic and empirical models are used when the spatial arrangement of the various components can be used to explain the whole. Empirical models are based on simple, correlative relationships (Turner, 2001). Both deterministic and stochastic models play essential roles in landscape ecology. A model is deterministic if the outcome is always the same for a group of input variables. If the model has uncertainty in the results, then, the model is stochastic (He and Mladenoff, 1999). Stochastic methods have the advantage of allowing statistical comparisons in multiple experiments or field observations.

Analytical and simulation models have a closed-form mathematical solution. If a mathematical solution is not possible, a model can be simulated using computer methods to obtain a representation of the system under study (Turner, 2001).

Percolation theory has the ability to study connectivity and fragmentation whereas self-organized criticality contributes to the study of scale-dependent phenomena. Self-organized criticality applies power-law statistics to observe self-similarity under the premise that estimation of large-scale process is possible through the understanding of phenomena occurring at small scales (Turner, 2001). Spatial statistics, ecological neighborhoods, and multiple regressions are other methods that have been used to investigate scale (Turner, 2001).

7.10 LANDSCAPE ECOLOGY IN THE CRITICAL ZONE

Up to this point, we have reviewed the basic concepts of landscape ecology. In this section, we examine the commonalities between landscape ecology and the Critical Zone. The lithospheric subsystem, the hydrological system, the anthropogenic, and the biological subsystems are interdependent. Unfortunately, the myopic description of the Critical Zone that has been created and adopted focuses attention on soil processes at the expense of reinforcing the strong linkages between all the systems of the Critical Zone. We here illustrate how the philosophical constraints of landscape ecology and the tools that have been created to study the various aspects of ecology of the landscape actually enrich the study of the Critical Zone.

If one compares scope, goals, and ambitions between landscape ecology and Critical Zone research, similarities between them can be noted. For example, (1) the object of study of both fields is the uppermost layer of the surface of Earth; (2) both landscape ecology and Critical Zone studies require multidisciplinarity and interdisciplinarity to confront research questions; (3) both fields have, and continue to develop, systemic, theoretical backgrounds, with emphasis on processes and thresholds; (4) the objects of study are positioned in a spatial and temporal context, making germane the consideration of scale; and (5) both fields include humans and their activities as part of the respective systems. These similarities are strong arguments that landscape ecology must be considered when one examines the Critical Zone.

Examination of research being produced at the various CZOs shows that soil is the dominant element of their approaches to systems. Soil is used as a unifying element, a natural chemical reactor that transports and transforms geological materials and biomass (Banwart et al., 2012). In landscape ecology, no consensus exists about a single element of the landscape system acting as the pivotal point or focus for the research.

Soil units have been used to define ecosystem units because soil creates strict boundaries rather than gradual, less distinct transitions. The biological structuring of habitats in scales smaller than the scale of ecotopes determined internal unit-heterogeneity: zones of distinct nitrogen concentrations or the impact of tree structure or the distribution and accumulation of soil moisture, fog, or snow cover. In contrast, Burrough et al. (1992) have recognized fuzzy, gradual boundaries in soils that can be related with these internal heterogeneities.

Similarities between the concept of ecotope and soil as a unifying element of the Critical Zone can enrich and complement research. Although in theory, landscape ecology and Critical Zone research focus on the vertical as well as the horizontal heterogeneity of their specific object of study (i.e., landscape or the Critical Zone), both approaches have commonality in their investigation. Whereas landscape ecology focuses on the horizontal component of the heterogeneity of landscapes under the concept of ecotone, the Critical Zone perspective focuses attention on the processes occurring vertically.

In a geological context, interfaces exist with a vertical connection, as with the aboveground biosphere and belowground with aquifers below. It is important to note that the focus of Critical Zone research is to examine the interfaces (e.g., surface–atmosphere interface, soil–vegetation interface or soil–bedrock interface) to understand the coupling between surface–subsurface dynamics. Surface dynamics, such as climate and land-use change, undoubtedly influence the subsurface processes, necessitating the inclusion of both surface and subsurface environments to obtain an unabridged vision of the unit affecting the dominant processes in the natural system (Lin, 2010).

Unlike landscape ecology, lateral heterogeneities and connections in Critical Zone research are mainly related to investigations in the field of hydrology (Banwart et al., 2012). Scalability and the patterns in a regional context are not seen as a priority by Critical Zone researchers. Rather, priority is placed on the understanding of processes occurring inside of the vertical component of the Critical Zone. Both landscape ecology and Critical Zone science demand a systemic view of the landscape, in which biotic–abiotic feedbacks are analyzed and linked through the soil structure (or the absence of it in bare units). One view complements the other, leading to a vision of the landscape system as a three-dimensional structure, allowing a more realistic understanding of ecological and physical processes (Fig. 7.9).

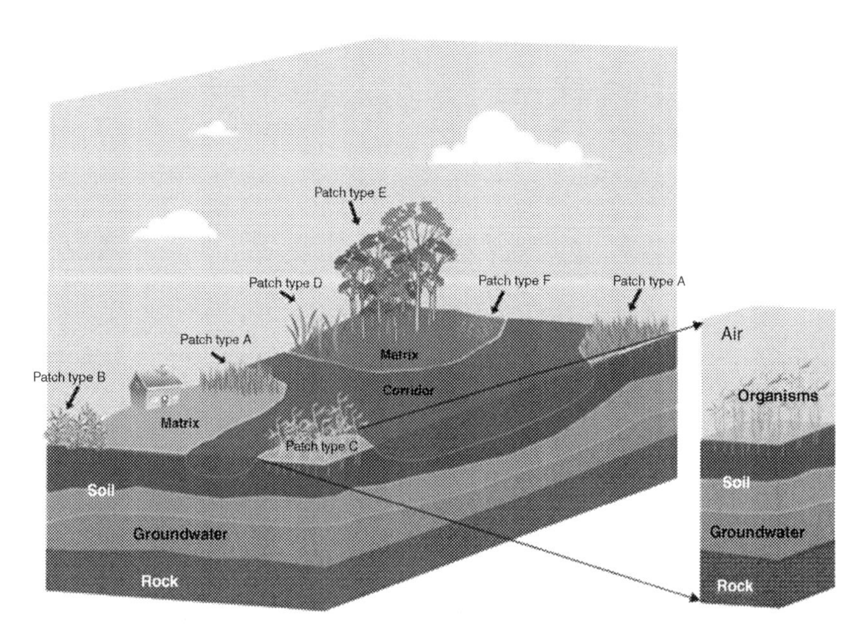

FIGURE 7.9 Complementary approaches: landscape ecology emphasizes the horizontality of landscapes at a regional scale, and the Critical Zone places more attention on the processes occurring in the vertical direction at a high-resolution scale. The combination of both approaches will lead to a better understanding of the landscape as a system. *(The idea of this diagram is based on Group (1998), Brantley et al. (2006), and Fischer and Lindenmayer (2007).)*

7.11 THE ROLE OF INTERDISCIPLINARITY IN CRITICAL ZONE RESEARCH

Interdisciplinarity is implicit in the CZO network. The multidisciplinary expertise is organized under an umbrella of a global network of Critical Zone observatories, mainly in the US and Europe, which consist of research field sites with goal-oriented research questions. At each site, the research team attempts to answer process-oriented questions (Brantley et al., 2007), understand various interdependencies and identify geochemical thresholds within the Critical Zone (Banwart et al., 2012; Brantley et al., 2006, 2007; Raab et al., 2012). In Critical Zone research, geochemical thresholds are mainly associated with processes in the water cycle and carbon cycle. Components of weathering, soil formation, erosion, and deposition; life processes and human impacts are also considered (Lin, 2010) but, to a lesser degree. Thresholds are searched for at the interfaces in the vertical direction. On the other hand, in landscape ecology, thresholds are searched in the horizontal direction, locating and defining edges in the analysis of patterns and function of landscape.

Even though Critical Zone research and landscape ecology have different approaches in their orientation of interdisciplinarity, we are pleased that they both bring unifying frameworks. One of the most important contributions of landscape ecology is the recognition of heterogeneity and complexity of landscapes and embracing all temporal and spatial ranges. This is an important fundamental view that needs to be incorporated in Critical Zone research.

Because landscape ecology is an extension of several disciplines, it has had an interdisciplinary perspective from its formation. This aspect of landscape ecology can serve as a guide for the integration sought by Critical Zone researchers to build conceptual or quantitative models needed to predict the outcome of the coupled processes within the Critical Zone in a variety of scales (Brantley et al., 2007). According to Critical Zone researchers, development of such models are an important step to the understanding of landscape as a system and to being able to observe in detail feedbacks of biotic and abiotic elements (Amundson et al., 2007; Anderson et al., 2008; Banwart et al., 2011; Brantley et al., 2007; Lin, 2010).

Another aspect that landscape ecology and Critical Zone research has in common is attention to process. Not only the present-day observable and measurable processes, but also past processes and the environmental conditions, which can be reconstructed from the investigation of subsurface features (Brantley et al., 2007; Rohdenburg, 1989). Geomorphology, through the reconstruction of the development of landscape before biotic elements are introduced, is one approach to focus investigation on processes (i.e., observing paleosols and sediments) (Anderson et al., 2008; Brantley et al., 2007; Moss, 2014; Raab et al., 2012; Rohdenburg, 1989).

Elucidating the abiotic aspects of processes can then lead to questions about how these processes in the Critical Zone nourish ecosystems and how they respond to change from external forcing. In the Critical Zone, besides the external forcing related to climate change, human impact is responsible for changing the composition of the hydrosphere and the lithosphere. But, this impact does not always bring negative outcomes. It has been shown in landscape–ecology investigations that disturbances may lead to a higher heterogeneity, resulting in a positive impact on ecosystem health (Turner et al., 2013). Interpretation of recurrent patterns of human intervention and their consequent impact on structure, function, and change under the perspective of landscape ecology may also contribute to modeling ambitions in the Critical Zone at various temporal and spatial scales.

Lastly, and perhaps most important, is the inclusion of human activities in the investigation of landscape systems. In landscape ecology, feedbacks between human activities and their physical environment are studied under the point-of-view of disturbances. These feedbacks between humans and the environment are occurring via the interface of elements in the Critical Zone, with processes related to: supply of sediment for soil formation, provision of nutrients for plants, removing atmospheric greenhouses gases, delivering clean recharge to aquifers, and producing soils providing habitat for fauna and microbial communities. These ecosystem services are the result of the important role of Critical Zone in providing nutrients and energy products by chemical reactions catalyzed by abiotic–biotic interactions, possibly trigged by disturbances (Brantley et al., 2007).

The holistic-approach used by landscape ecologists can contribute in two ways to viewing the Critical Zone as a system: observing patterns visually (e.g., from a plan-view perspective using aerial photographs and other remote sensing-type data) and from observations of particular ecological processes or landscape functions (e.g., movement of individuals or communities throughout the landscape). The concerns for the arrangements of landscape parts by landscape ecologist give spatial context to the particular process of interest.

In addition to pattern analysis, scaling is one of the main advances in landscape ecology (Wu, 2013a). Unfortunately, scale remains a scientific challenge in the geosciences disciplines (Brantley et al., 2007; Wu, 2013a). In the Critical Zone, temporal scales are challenging because the development of regolith profiles occurs over time-periods ranging from less than microseconds to more than millions of years. Meanwhile, spatial scales vary over 16 or more orders of magnitude, from the atomic scale to the global scale (Brantley et al., 2007). For example, the identification of mechanisms leading to subsurface heterogeneity and relating its importance in a regional context is one of these challenges. Application of this approach in Critical Zone research may yield ways to identify causes of variability (Lin, 2010).

Landscape ecologists can also benefit from Critical Zone research. Besides the already discussed analysis of landscapes from three dimensions, dissemination of fundamental concepts, research goals, and data available from the various CZOs around the world are valuable for landscape–ecology research.

The creation of the CZOs was based on the idea of encouraging a global alliance to study the consequences of global change and scientific research questions, as well as solution-driven investigations at specific locations. This type of approach requires coordination, involvement, and participation of a variety of scientific and nonscientific groups to carry out the research and to achieve multiple levels of required decision making. CZOs have built a cyber-structure, which provides access to data collected at the CZO locations. These types of data collections build strong foundations to stimulate debate among researchers from different disciplines interested in the overall Critical Zone concept. Landscape ecologists may also benefit from the Critical Zone approach, as it can amplify their research scope and their research approach.

A good example of the need of the interaction between Critical Zone research and landscape ecology is already starting to be seen as the debate about ecosystem services. Field et al. (2015) suggest that the assessment of ecosystem services can be expanded to the concept of Critical Zone services. According to the authors, ecosystem services, valuated from the perspective of surface processes need to be linked to the context of Critical Zone, in which development occurs in longer time scales and includes the interaction of more complex dynamics (Field et al., 2015). Patches and corridors, therefore, would be important structures to identify. First, human interactions occur mainly on the surface and, second, previous knowledge of the structure of the landscape is necessary to understand and elucidate the study of fluxes and inventories in the vertical direction. This is because an underground profile and its vertical components of a patch will look very different from one found in a corridor. In this case, landscape–ecology framework not only helps the assessment of services on a 3D view but will also give a landscape perspective to Critical Zone conclusions.

7.12 INTERDISCIPLINARITY IN LANDSCAPE ECOLOGY AND CRITICAL ZONE RESEARCH

One last, essential point that we want to introduce involves the important role that interdisciplinarity must play in the research focus of the Critical Zone. Landscape ecology and Critical Zone research are in search of an integrative way to combine different disciplines – basic and applied – to solve a set of specific scientific problems. A valid question is how to formulate an approach to address these concerns. Although the CZOs are based on the premise of teams of researchers focusing on one area to provide a body of research data that is derived and interpreted from an interdisciplinary perspective, examination of the literature being produced by the CZOs suggests that many of the studies are, in fact, strictly disciplinary in approach. Hopefully, in time, these data sets will serve as the catalyst for truly interdisciplinary-based studies.

Interdisciplinarity is a common concept mentioned as an integrative solution to construct and apply multiple perspectives to investigate complex issues

(Klein, 1984; Opdam et al., 2013; Wu, 2013b). But, interdisciplinarity is not the only type of integration of diverse disciplines. Castri and Hadley (1986) in discussing interdisciplinary research in ecology addressed the description of different types of integration, conceptualizing terms such as interdisciplinarity or multidisciplinarity. According to Castri and Hadley (1986) the term multidisciplinarity is used when several disciplines are involved but with no interaction; pluridisciplinarity, is the interaction but lack of coordination among disciplines; and unidirectional interdisciplinarity, is where interaction and coordination are imposed by a single discipline (see Fig. 7.10). Environmental disciplines tend towards unidirectional interdisciplinarity. The landscape system, which should be the primary focus, is neglected, leaning towards the expertise of the dominant discipline or dominance of a particular vision of a group of scientists (Moss, 2014).

A vision of goal-oriented interdisciplinarity and transdisciplinarity can be aligned to the potential of Critical Zone research. Goal-oriented interdisciplinarity is determined by the nature of an identified problem. Interaction and coordination occurs to attain a common objective. Transdisciplinarity, a superior

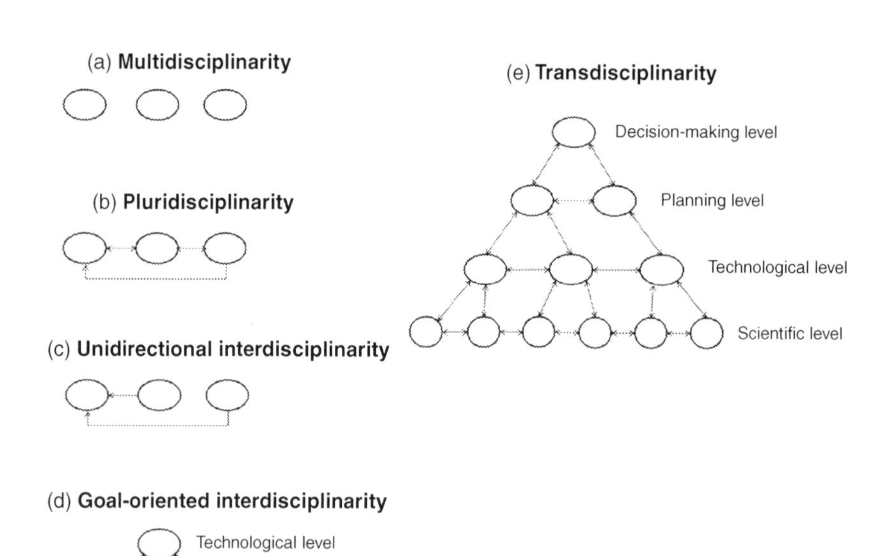

FIGURE 7.10 Types of discipline combination to address specific scientific or decision-making problems. (a) Several disciplines involved but with no interaction between them; (b) several interactions between disciplines but little or no coordination; (c) interactions and coordination are imposed by a single discipline; (d) interactions and coordination are determined by the nature of the complex problem to be tackle; and, (e) at a higher level, interactions involve not only scientific and technological disciplines, but also planners, administrators, and community. *(Modified from Castri and Hadley (1986).)*

level of interaction, exists when interaction involves not only the scientific and technological disciplines in stated goals, but also where investigators and users become involved in the processes (Moss, 2014; Wu, 2006).

Moving from multidisciplinarity towards transdisciplinarity requires an increment of the number of disciplines involved. As a consequence, Critical Zone research can be driven by research questions that display useful information for decision making. Wu (2006) in his discussion about cross-disciplinarily called this tendency "sustainability science," which focuses on the dynamic state between nature and society. In this respect, landscape ecology and Critical Zone programs funded by NSF are looking for transdisciplinarity, and consequently are significant contributors to sustainability science.

Kirchhoff et al. (2013) assertively criticized the vision of landscape ecology as a pure ecological subdiscipline. They asserted that landscapes cannot be the subject matter of ecology or any other natural research; rather, it is the ecosystem that is the object of study within landscape boundaries. According to Kirchhoff et al. (2013), landscapes are human aesthetic perceptions at certain scales. In this sense, reality is not independent of perception (Maturana and Varela, 1987) and although the term landscape can have a terminological discrepancy among disciplines, the phenomena of dominant processes (i.e., biotic or abiotic) at different scales is a reality, again, not independent of our perception (i.e., where we focus our attention).

Critical Zone research has a structure of transdisciplinarity, involving interactions not only of scientific and technological disciplines, but also with planners, and local populations (Brantley et al., 2007). Nonetheless, as a transdisciplinary exercise, Critical Zone research can also turn to unidirectional interdisciplinarity. Critical Zone research can benefit from common and clear objectives focused on the understanding of the landscape as an elementary unit and system. Thus, the focus should be on the transformation of the landscape system itself: problems, interventions, and disturbances will arise and should be addressed instead of the impacts from peripheral, yet still related, factors (Fig. 7.11) (Moss, 2014).

7.13 HOW TO DEAL WITH COMPLEXITY IN THE STUDY OF THE CRITICAL ZONE

Landscape systems are complex, which drives the fundamental question: how to deal with complexity? Raab et al. (2012) proposed that the tracing of early stages of landform evolution might be an answer. The assumption is that primary processes that occur in the development of a landscape impact later spatial structures. This approach could have important, direct results for both landscape ecology and Critical Zone research.

Dominant processes can be distinguished if they can be related to factors that direct the presence of certain biotic patterns. Interactions among newly emerging structures and related processes, including feedback processes among

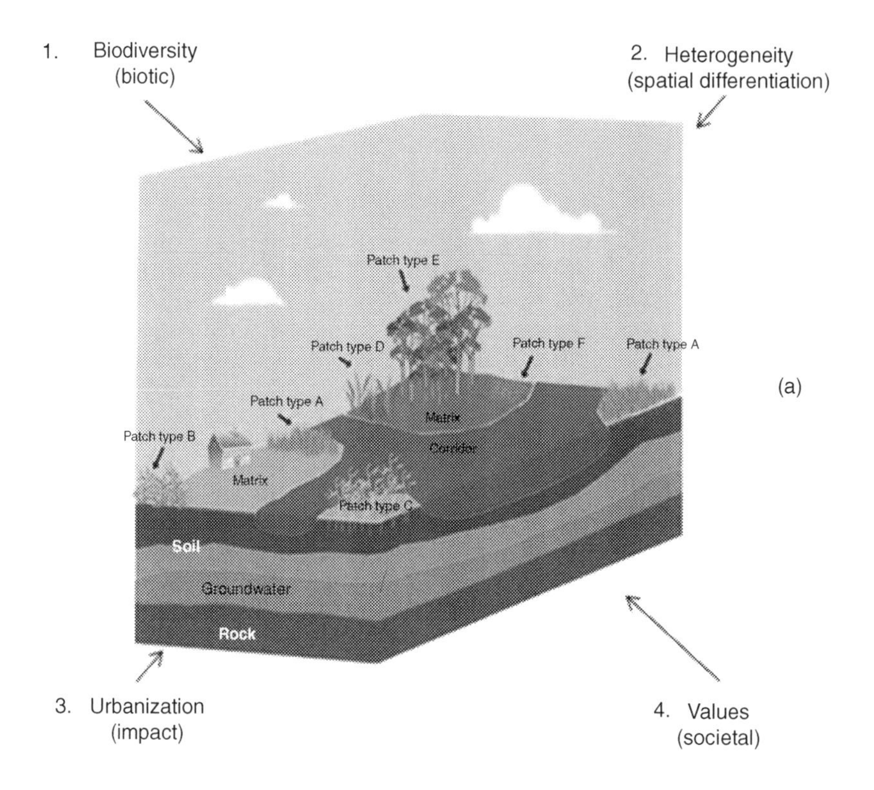

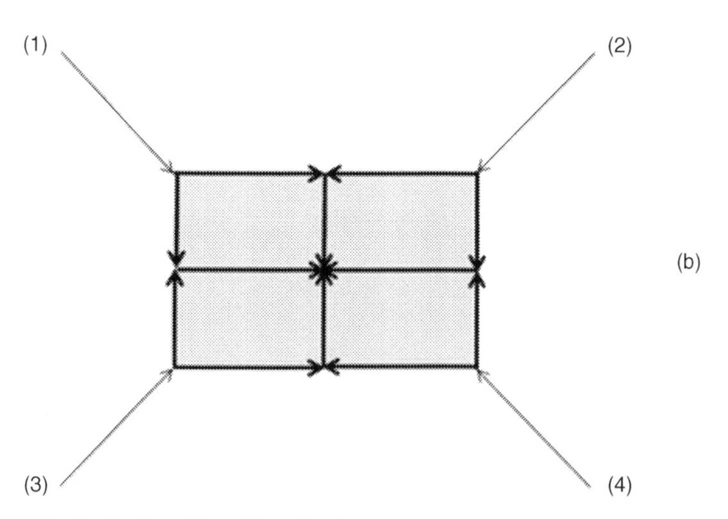

FIGURE 7.11 Goal-oriented interdisciplinarity versus the vision of the landscape as a system, and the consequent focus in the interactions and processes occurring within. (a) In the landscape system, each subtheme has its own emphasis (in brackets). The focus in this model is on the transformation subthemes. In (b), information concerning the transformation themes incorporated into the landscape system; the landscape system itself becomes the focus and the understanding of the landscape system itself is enhanced. *(Modified from: Moss (2014).)*

existing and new structures, would be much more visible if initial ecosystems are investigated (Raab et al., 2012). Thus, analysis of young ecosystems in their initial stages might be a way to disentangle the complex, process web and help to better understand ecosystem functioning in the Critical Zone (Schaaf et al., 2011).

Spatial structure has been suggested as an explanation of vegetation patterns in early stages of primary succession. Wiegleb and Felinks (2001) in their study regarding primary succession in post-mining landscapes at tailing sites 5–70 years old discovered that vegetation patterns do not bear strict relationships with explanatory environmental variables, such as pH, organic carbon, or available phosphate. Instead, spatial variables (i.e., structure) increase the variance explanation in multivariate analyses (Wiegleb and Felinks, 2001). In other words, spatial relationships in the structure were more important than specific heterogeneities in the composition of substrate. Roman and Gafta (2013) also studying vegetation patterns, examined 60-year-old tailings, and detected that spatial structure, and in particular the proximity of patches, plays an important role in the habitat of certain species. Spatial autocorrelation and, therefore, connectivity was critical for richness and diversity of species when compared with geomorphological variables (i.e., slope and aspect) and patch size (Roman and Gafta, 2013).

These ideas of describing primary stages of a succession chain of landscapes lead to the description of geomorphic processes acting in the early stages of development. The beginning of initial ecosystem development is abiotically dominated, particularly by the substrate (Raab et al., 2012). In the initial phase of ecosystem development, a wide range of geomorphic processes occur: mass movement, fluvial, wind and ice, erosion and sedimentation. These processes are decisive in the spatial heterogeneity itself and consequently, to the biota, hydrology, and other elements of the landscape (Raab et al., 2012). Thus, it appears that geomorphology is a crucial tool in deciphering complexity. The foremost question is: can complexity be addressed by studying early stages of landscape development? Thus, the ultimate point is to understand complexity of natural systems, which is the aim of landscape ecology and Critical Zone research.

Reconstruction of landscapes may serve not only in the understanding of transformational processes, but also future development. Geomorphological-landscape elements such as meadows, depressions, and high plains can be associated with units of very distinct ecological quality. This is because of the particular need of certain plant associations with a particular structure (i.e., rock composition, soil, aspect) and a particular process (i.e., weathering type, slope, aspect, groundwater flow).

Scale and linkages in landscape ecology are important components. It is important to note that complexity is linked to heterogeneity and heterogeneity is linked to scale (Wu, 2006). Identifying the correct scale at which to address a problem is a critical factor (Turner, 2001).

7.14 COMPLEXITY OF LANDSCAPES AND GEOMORPHOLOGY

Complexity is the result of a large number of interacting elements in a system. One way to deal with complex systems, therefore, is to reduce the large amount of factors to a subset that are actually relevant (Holling, 2001).

Landform evolution is one of the prime interests of the discipline of geomorphology. In the initial phases of ecosystem development, a wide range of geomorphic processes takes place (i.e., movement by gravity, water, wind and ice, deposition and sedimentation). And, with exception of extremely arid regions, the most important element of a new configuration of landscapes is water. Therefore, geomorphology and hydrology are crucial disciplines that have to be involved in the study of complexity. And in turn, the study of complexity through the deciphering of dominant processes is the aim of landscape ecology and Critical Zone research (see Table 7.1).

Another contribution of geomorphology in the study of Critical Zone is the value of mapping for monitoring and modeling in Critical Zone research. Lin (2010, p. 40) has pointed out that "An iterative loop of mapping, monitoring, and modeling (3 M) can provide a possible integrated and evolutionary approach to address the complexity and dynamics in the CZ." For example, modeling that considers the spatially distributed mapping of dominant processes has the potential to capture the internal flow dynamics (i.e., preferential vertical and lateral flow and thresholds of hydrological response) in a certain catchment and with biotic interactions and population dynamics.

7.15 CONCLUSIONS

The view of the Critical Zone has much promise as a research focus. By combining the horizontal perspective of landscape ecology with the vertical connectivity and perspective of the current operational procedures of the Critical Zone, a new perspective that can serve to enhance the impact on the understanding of the complete landscape and all of the processes can be achieved. The major focus of landscape ecology, which is based on the study of the landscape in terms of composition, structure, and function, adds the 3D dimensionality. We stress the point that transdisciplinarity can be the fundamental driver of future Critical Zone studies. The diagram we created displays the interconnectedness of the total landscape by combining the principles of geomorphology, atmospheric science, hydrology, pedology, ecology, geographic information system technology, and spatial statistics, along with the perspectives of spatial heterogeneity, scale and patterns, and human impacts and interactions for a complete understanding of the landscape both in horizontal and vertical directions.

REFERENCES

Amundson, R., Richter, D.D., Humphreys, G.S., Jobbágy, E.G., Gaillardet, J., 2007. Coupling between biota and earth materials in the critical zone. Elements 3 (5), 327–332.

Anderson, S.P., Bales, R.C., Duffy, C.J., 2008. Critical Zone Observatories: building a network to advance interdisciplinary study of Earth surface processes. Mineral. Mag. 72 (1), 7–10.

Anderson, S.P., Anderson, R.S., Tucker, G.E., 2012. Landscape scale linkages in critical zone evolution. C. R. Geosci. 344 (11), 586–596.

Anderson, R.S., Anderson, S., Aufdenkampe, A.K., Bales, R., Brantley, S., Chorover, J., Duffy, C.J., Scatena, F.N., Sparks, D.L., Troch, P.A., Yoo, K., 2010. Future Directions for Critical Zone Observatory (CZO) Science.

Anderson, R.S., Anderson, S.P., Tucker, G.E., 2013. Rock damage and regolith transport by frost: an example of climate modulation of the geomorphology of the critical zone. Earth Surf. Process. Land. 38 (3), 299–316.

Attrill, M., Rundle, S., 2002. Ecotone or ecocline: ecological boundaries in estuaries. Estuar. Coast. Shelf Sci. 55 (6), 929–936.

Banwart, S., Chorover, J., Gaillardet, J., Sparks, D., White, T., Anderson, S., Aufdenkampe, A., Bernasconi, S., Brantley, S., Chadwick, O., 2013. Sustaining Earth's Critical Zone Basic Science and Interdisciplinary Solutions for Global Challenges, vol. 47. University of Sheffield, Sheffield.

Banwart, S., Chorover, J., Sparks, D., White, T., 2011. Sustaining Earth's Critical Zone. Report of the International Critical Zone Observatory Workshop, U. Delaware, United States.

Banwart, S., Menon, M., Bernasconi, S.M., Bloem, J., Blum, W.E.H., Souza, D.M.d., Davidsdotir, B., Duffy, C., Lair, G.J., Kram, P., Lamacova, A., Lundin, L., Nikolaidis, N.P., Novak, M., Panagos, P., Ragnarsdottir, K.V., Reynolds, B., Robinson, D., Rousseva, S., de Ruiter, P., van Gaans, P., Weng, L., White, T., Zhang, B., 2012. Soil processes and functions across an international network of Critical Zone Observatories: introduction to experimental methods and initial results. C. R. Geosci. 344 (11–12), 758–772.

Bell, S., 2012. Landscape: Pattern, Perception and Process. Routledge.

Birkeland, P., 1999. Soils and Geomorphology. Oxford University Press, New York.

Boyd, D.S., Foody, G.M., 2011. An overview of recent remote sensing and GIS based research in ecological informatics. Ecol. Inform. 6 (1), 25–36.

Brantley, S., White, T., White, A., Sparks, D., Richter, D., Pregitzer, K., Derry, L., Chorover, J., Chadwick, O., April, R., 2006. Frontiers in Exploration of the Critical Zone: Report of a workshop sponsored by the National Science Foundation (NSF), October 24–26, 2005. Newark, DE, p. 30.

Brantley, S.L., Goldhaber, M.B., Ragnarsdottir, K.V., 2007. Crossing disciplines and scales to understand the critical zone. Elements 3 (5), 307–314.

Broxton, P.D., Harpold, A.A., Biederman, J.A., Troch, P.A., Molotch, N.P., Brooks, P.D., 2014. Quantifying the effects of vegetation structure on snow accumulation and ablation in mixed-conifer forests. Ecohydrology 8 (3).

Burrough, P., MacMillan, R., Deursen, W.v., 1992. Fuzzy classification methods for determining land suitability from soil profile observations and topography. J. Soil Sci. 43 (2), 193–210.

Burrough, P.A., 1986. Principles of Geographical Information Systems for Land Resources Assessment. Oxford University Press, New York.

Buss, H.L., Moore, O.W., 2013. Beyond the Regolith: A Deep Critical Zone Drilling Perspective on Weathering Profiles. School of Earth Sciences, University of Bristol, Denver, Colorado, UK.

Cadenasso, M.L., Pickett, S.T., Weathers, K.C., Jones, C.G., 2003. A framework for a theory of ecological boundaries. BioScience 53 (8), 750–758.

Castri, F.d., Hadley, M., 1986. Enhancing the credibility of ecology: is interdisciplinary research for land use planning useful? GeoJournal 13 (4), 299–325.

Chorley, R.J., Kennedy, B.A., 1971. Physical Geography: A Systems Approach. Prentice-Hall, London.

Coulson, R.N., Tchakerian, M.D., 2010. Basic Landscape Ecology. KEL Partners Incorporated, College Station, TX.

Curtis, J.T., McIntosh, R.P., 1951. An upland forest continuum in the prairie-forest border region of Wisconsin. Ecology 32 (3), 476–496.

Dicum, G., 2004. Window Seat: Reading the Landscape from the Air. Chronicle Books.

Field, J.P., Breshears, D.D., Law, D.J., Villegas, J.C., López-Hoffman, L., Brooks, P.D., Chorover, J., Barron-Gafford, G.A., Gallery, R.E., Litvak, M.E., Lybrand, R.A., McIntosh, J.C., Meixner, T., Niu, G.-Y., Papuga, S.A., Pelletier, J.D., Rasmussen, C.R., Troch, P.A., 2015. Critical zone services: expanding context, constraints, and currency beyond ecosystem services. Vadose Zone J. 14 (1).

Fischer, J., Lindenmayer, D.B., 2007. Landscape modification and habitat fragmentation: a synthesis. Global Ecol. Biogeogr. 16 (3), 265–280.

Forman, R., Godron, M., 1986. Landscape Ecology. J. Wiley & Sons, New York, 619 pp.

Forman, R.T., 1995. Land Mosaics: The Ecology of Landscapes and Regions. Cambridge University Press, Cambridge, UK.

Fortin, M.-J., Agrawal, A.A., 2005. Landscape ecology comes of age. Ecology 86 (8), 1965–1966.

Green, D.G., Klomp, N., Rimmington, G., Sadedin, S., 2006. Complexity in landscapes. Complexity in Landscape Ecology. Springer, New York, pp. 33–50.

Group, F.-F.I.S.R.W., 1998. Stream Corridor Restoration: Principles, Processes, and Practices. Published USA Government, October.

Haase, G., Richter, H., 1983. Current trends in landscape research. GeoJournal 7 (2), 107–119.

Haila, Y., 2002. A Conceptual genealogy of fragmentation research: from island biogeography to landscape ecology. Ecol. Appl. 12 (2), 321–334.

Hanski, I., 1998. Metapopulation dynamics. Nature 396 (6706), 41–49.

Hanski, I., Gilpin, M., 1991. Metapopulation dynamics: brief history and conceptual domain. Biol. J. Linn. Soc. 42 (1–2), 3–16.

He, H.S., Mladenoff, D.J., 1999. Spatially explicit and stochastic simulation of forest-landscape fire disturbance and succession. Ecology 80 (1), 81–99.

Holling, C.S., 2001. Understanding the complexity of economic, ecological, and social systems. Ecosystems 4 (5), 390–405.

Jensen, J.R., 2007. Remote Sensing of the Environment: An Earth Resource Perspective. Pearson Prentice Hall, Upper Saddle River, NJ, USA.

Jones, K.B., Neale, A., Nash, M., Van Remortel, R., Wickham, J., Riitters, K., O'Neill, R., 2001. Predicting nutrient and sediment loadings to streams from landscape metrics: a multiple watershed study from the United States Mid-Atlantic Region. Landscape Ecol 16 (4), 301–312.

Jung, M., 2013. LecoS-A QGIS plugin for automated landscape ecology analysis. 2167-9843, PeerJ PrePrints.

Kirchhoff, T., Trepl, L., Vicenzotti, V., 2013. What is landscape ecology? An analysis and evaluation of six different conceptions. Landscape Res. 38 (1), 33–51.

Klein, J.T., 1984. Interdisciplinarity and complexity: an evolving relationship. Structure 71, 72.

Lin, H., 2010. Earth's critical zone and hydropedology: concepts, characteristics, and advances. Hydrol. Earth Syst. Sci. 14, 25–45.

Liu, J., Hull, V., Morzillo, A.T., Wiens, J.A., 2011. Sources, Sinks and Sustainability. Cambridge University Press, Cambridge, UK.

MacArthur, R.H., Wilson, E.O., 1967. The Theory of Island iogeography. Princeton University Press, Princeton, NJ.

Maturana, H.R., Varela, F.J., 1987. The Tree of Knowledge: The Biological Roots of Human Understanding. New Science Library/Shambhala Publications, Boston, MA.

Moilanen, A., Hanski, I., 2001. On the use of connectivity measures in spatial ecology. Oikos 95 (1), 147–151.

Monkhouse, F.J., 1965. Landscape from the Air.

Moore, I.D., Gessler, P.E., Nielsen, G.A., Peterson, G.A., 1993. Soil attribute prediction using terrain analysis. Soil Sci. Soc. Am. J. 57 (2), 443–452.

Moss, M.R., 2014. Landscape ecology: the need for a discipline? Problemy Ekologii Krajobrazu 6 (6).

Naveh, Z., Lieberman, A., 1984. Landscape Ecology: Theory and Application. Springer, New York, p. 356.

Neteler, M., Mitasova, H., 2004. Open Source GIS: A GRASS GIS Approach. Springer, New York.

NRC, 2001. Basic Research Opportunities in Earth Science. National Academies Press, Washington, DC.

Oleschko, K., Parrot, J.-F., Ronquillo, G., Shoba, S., Stoops, G., Marcelino, V., 2004. Weathering: toward a fractal quantifying. Math. Geol. 36 (5), 607–627.

Opdam, P., Nassauer, J.I., Wang, Z., Albert, C., Bentrup, G., Castella, J.-C., McAlpine, C., Liu, J., Sheppard, S., Swaffield, S., 2013. Science for action at the local landscape scale. Landscape Ecol. 28 (8), 1439–1445.

Power, W.L., Tullis, T.E., 1991. Euclidean and fractal models for the description of rock surface roughness: J Geophys Res V96, NB1, Jan 1991, P415–P424. Int. J. Rock Mech. Mining Sci. Geomech. Abstracts 28 (6), A344.

Quantum, G., 2013. Quantum GIS Project.

Raab, T., Krümmelbein, J., Schneider, A., Gerwin, W., Maurer, T., Naeth, M.A., 2012. Initial ecosystem processes as key factors of landscape development — a review. Phys. Geogr. 33 (4), 305–343.

Risser, P.G., 1984. Landscape Ecology: Directions and Approaches, vol. 2. Illinois Natural History Survey.

Robinson, D.A., Campbell, C.S., Hopmans, J.W., Hornbuckle, B.K., Jones, S.B., Knight, R., Ogden, F., Selker, J., Wendroth, O., 2008. Soil moisture measurement for ecological and hydrological watershed-scale observatories: a review. Vadose Zone J. 7 (1), 358–389.

Rohdenburg, H., 1989. Landscape Ecology – Geomorphology, vol. 177. Catena, Germany.

Roman, A., Gafta, D., 2013. Proximity to successionally advanced vegetation patches can make all the difference to plant community assembly. Plant Ecol. Divers. 6 (2), 269–278.

Rossiter, D.G., 1996. A theoretical framework for land evaluation. Geoderma 72 (3–4), 165–190.

Rougerie, G., Beroutchachvili, N., 1991. Géosystèmes et paysages: bilan et méthodes.

Scanlon, B.R., Reedy, R.C., Stonestrom, D.A., Prudic, D.E., Dennehy, K.F., 2005. Impact of land use and land cover change on groundwater recharge and quality in the southwestern US. Global Change Biol. 11 (10), 1577–1593.

Schaaf, W., Bens, O., Fischer, A., Gerke, H.H., Gerwin, W., Grünewald, U., Holländer, H.M., Kögel-Knabner, I., Mutz, M., Schloter, M., 2011. Patterns and processes of initial terrestrial-ecosystem development. J. Plant Nutr. Soil Sci. 174 (2), 229–239.

Schröder, B., 2006. Pattern, process, and function in landscape ecology and catchment hydrology – how can quantitative landscape ecology support predictions in ungauged basins? Hydrol. Earth Syst. Sci. 10 (6), 967–979.

Steiniger, S., Hay, G.J., 2009. Free and open source geographic information tools for landscape ecology. Ecol. Infor. 4 (4), 183–195.

Tague, C., Grant, G.E., 2009. Groundwater dynamics mediate low-flow response to global warming in snow-dominated alpine regions. Water Resour. Res. 45 (7), W07421.

Troll, C., 1939. Bemerkungen zum Atlantischen Problem, geäußert im Anschluß an die drei ozeanographischen Beiträge. Geologische Rundschau 30 (3), 384–386.

Troll, C., 1971. Landscape ecology (geoecology) and biogeocenology — a terminological study. Geoforum 2 (4), 43–46.

Troll, C., 2007. The geographic landscape and its investigation. Found. Pap. Landscape Ecol., 71–101.

Turner, M.G., 1989. Landscape ecology: the effect of pattern on process. Ann. Rev. Ecol. Syst. 20, 171–197.

Turner, M.G., 2001. Landscape Ecology in Theory and Practice: Pattern and Process. Springer, New York.

Turner, M.G., 2005. Landscape ecology in North America: past, present, and future. Ecology 86 (8), 1967–1974.

Turner, M.G., 2010. Disturbance and landscape dynamics in a changing world. Ecology 91 (10), 2833–2849.

Turner, M.G., Donato, D.C., Romme, W.H., 2013. Consequences of spatial heterogeneity for ecosystem services in changing forest landscapes: priorities for future research. Landscape Ecol. 28 (6), 1081–1097.

Urban, D.L., 2006. Landscape ecology. In: Encyclopedia of Environmetrics, vol. 3.

Urban, D.L., O'Neill, R.V., Shugart, Jr., H.H., 1987. A hierarchical perspective can help scientists understand spatial patterns. BioScience 37 (2), 19–127.

Verstappen, H.T., 2011. Old and new trends in geomorphological and landform mapping. In: Smith, M.J., Paron, P., Griffiths, J.S. (Eds.), Geomorphological Mapping: Methods and Applications (Developments in Earth Surface Processes). Elsevier, Oxford.

Vila Subirós, J., Varga Linde, D., Llausàs i Pascual, A., Ribas Palom, A., 2006. Conceptos y métodos fundamentales en ecología del paisaje (landscape ecology). Una interpretación desde la geografía 48, 151–166.

Wagner, H.H., Fortin, M.-J., 2005. Spatial analysis of landscapes: concepts and statistics. Ecology 86 (8), 1975–1987.

Walker, S., Wilson, J.B., Steel, J.B., Rapson, G., Smith, B., King, W.M., Cottam, Y.H., 2003. Properties of ecotones: evidence from five ecotones objectively determined from a coastal vegetation gradient. J. Veg. Sci. 14 (4), 579–590.

White, P.S., Pickett, S.T., 1985. Natural disturbance and patch dynamics: an introduction. Ecol. Nat. Disturb. Patch Dyn., 3–13.

Wiegleb, G., Felinks, B., 2001. Predictability of early stages of primary succession in post-mining landscapes of Lower Lusatia, Germany. Appl. Veg. Sci. 4 (1), 5–18.

Wiens, J.A., Milne, B.T., 1989. Scaling of 'landscapes' in landscape ecology, or, landscape ecology from a beetle's perspective. Landscape Ecol. 3 (2), 87–96.

Wu, J., 2006. Landscape ecology, cross-disciplinarity, and sustainability science. Landscape Ecol. 21 (1), 1–4.

Wu, J., 2013a. Key concepts and research topics in landscape ecology revisited: 30 years after the Allerton Park workshop. Landscape Ecol. 28 (1), 1–11.

Wu, J., 2013b. Landscape ecology. Ecological Systems. Springer, New York, pp. 179–200.

Wu, J., Hobbs, R., 2002. Key issues and research priorities in landscape ecology: an idiosyncratic synthesis. Landsc. Ecol. 17 (4), 355–365.

Chapter 8

Ecohydrology and the Critical Zone: Processes and Patterns Across Scales

Georgianne Moore*, Kevin McGuire**, Peter Troch†, and Greg Barron-Gafford‡

*Ecosystem Science and Management, Texas A&M University, Texas, USA; **Forest Resources and Environmental Conservation, Virginia Water Resources Research Center, Virginia Tech, Blacksburg, Virginia, USA; †Hydrology and Water Resources, Biosphere 2, University of Arizona, Oracle, Arizona, USA; ‡School of Geography and Development, Biosphere 2, University of Arizona, Oracle, Arizona, USA

8.1 INTRODUCTION

Critical Zone development is the result of concurrent and interacting fluxes of radiant energy and mass transfer (Rasmussen et al., 2011). Ecosystems regulate these fluxes through their influence on biogeochemical cycling through and within the Critical Zone. Likewise, the Critical Zone exerts long-term feedbacks on ecosystems (Jenerette et al., 2012; Turnbull et al., 2012), making the interactions between vegetation and water central to understanding processes in the Critical Zone. Interactions between water and vegetation, or more broadly ecosystems, have recently fallen under the purview of ecohydrology.

Ecohydrology is concerned with the effects of hydrological processes on the distribution, structure, and function of ecosystems, and with the effects of biotic processes on elements of the water cycle (Nuttle, 2002). At the heart of ecohydrology is soil and soil moisture (Rodriguez-Iturbe, 2000). Thus, only by understanding how the distribution and patterns of plants within the soil affect soil moisture-driven Critical Zone processes will it be possible to decipher the Earth System. This effort begins with the root–soil–rock interface at the pore scale (Casper et al., 2003; Schwinning, 2010) and extends to the vegetation–atmosphere boundary at global scales (Zhang et al., 2010).

The study of ecosystems and hydrologic interactions certainly developed before the term "ecohydrology" was coined to describe a subdiscipline in hydrology (Bonell, 2002). An appreciation of the coupled nature between vegetation and the water cycle is illustrated by the historical texts of Horton (1933), Kittredge

Developments in Earth Surface Processes, Vol. 19. http://dx.doi.org/10.1016/B978-0-444-63369-9.00008-2

(1948), and Colman (1953) and traditional resource management disciplines such as forest hydrology, agricultural engineering, and rangeland management. This focus on vegetation and water interaction from a practical, management perspective alone, albeit useful, did not provide a theoretical framework to fully understand the dynamic coupling of climate–soil–vegetation systems within the water cycle (e.g., Eagleson, 1978). It was not until the turn of the twenty-first century that the term ecohydrology gained traction in the literature (Hannah et al., 2004; Asbjornsen et al., 2011). Increasing recognition of vegetation controls on water and vice versa has fused the disciplines of ecology and hydrology closer together.

Ecohydrological processes are essential for the development of the Critical Zone. The partitioning of water at the soil surface and the net input of water into soil, controlled by infiltration and evapotranspiration (ET), are fundamental controls on geomorphic and pedogenic processes that develop Critical Zone structure. Water fluxes moving through the Critical Zone are vital to the process of bedrock weathering and soil development, which produce feedbacks in ecohydrological processes (e.g., water storage and flow rates). Positive feedbacks between ecohydrology and Critical Zone development also lead to enhanced weathering (Graham et al., 2010; Chorover et al., 2011) and expansion of the rooting zone as soil develops, which increases soil volume and plant-available, water-holding capacity (Graham et al., 2010). Across all Critical Zone Observatories (CZOs), researchers are making advances in our understanding of the link between vegetation, weathering, and Critical Zone development (Brantley et al., 2011).

This chapter focuses on the ecohydrology of both energy- and water-limited environments and on the feedbacks between vegetation and water across scales. We describe water–vegetation interactions in the context of pattern, process, and scale, which interact in complex ways to shape the Critical Zone. Ecohydrology seeks to describe these interactions in ways that are both mechanistic and generalizable (McDonnell et al., 2007). Yet, both modelers and experimentalists alike are challenged by stochasticity, hysteresis, heterogeneity, nonlinearity, and time lags inherent in these systems and how dominant processes change across spatial and temporal scales. The need to conceptualize climate–soil–vegetation dynamics is paramount to solving problems related to desertification, water scarcity, climate change, and land degradation. We increasingly recognize the importance of ecosystem services that provide clean water. The study of "novel ecosystems" (e.g., Hobbs et al., 2009) is challenging our understanding of Critical Zone dynamics and much work is needed to understand how the Critical Zone will respond to altered climate regimes and land-cover change.

8.2 SCALES OF INTERACTION IN ECOHYDROLOGICAL PATTERNS AND PROCESSES

The issue of scale is critical throughout the biological and geophysical sciences and, thus, is relevant to how we understand, observe, and predict patterns and processes in ecohydrology (Wilby and Schimel, 1999). Ecosystems and the Critical Zone are heterogeneous in space and time, challenging how measurements and observations are collected on processes of interest. Furthermore, most measurement approaches have particular scale limitations, which therefore may require approaches to transfer information across different scales (Miller et al., 2004). Tools such as digital terrain models, remote sensing, and statistical or process models are often used to bridge information across scales or extrapolate processes to different scales, as discussed in Chapter 17 on geospatial science and technology. However, some processes that are important at one scale are not important, or are less important at other scales. Ecohydrological theories and models are characteristically scale-specific as well.

The dimensionality of fluxes generally considered by ecophysiologists (i.e., one dimensional) and hydrologists (i.e., two or three dimensional) has been mismatched (cf. Bond, 2003); however, the field of ecohydrology has embraced these two perspectives. For example, considerable effort has been made to understand controls on the spatial and temporal organization of soil moisture driven by vertical versus lateral fluxes in the landscape (Grayson et al., 1997). Plant-water uptake and soil properties (e.g., texture, water retention) provide local controls over one-dimensional fluxes (e.g., soil–plant–atmosphere) and surface- or subsurface-topographic gradients and physical properties provide nonlocal controls leading to higher-dimensional fluxes driven by lateral water flows on and within soil systems. This spatial organization of soil moisture and the development of water source and sink areas is coupled to patterns of vegetation cover, rooting patterns, vegetation structure, and ecophysiological processes (Emanuel et al., 2007, 2010; Guswa, 2012; Guswa and Spence, 2012; Naithani et al., 2013). Thus, as soil moisture is laterally redistributed due to soil-water potential gradients in catchments, downstream areas receive subsidies, which feedback to develop spatial patterns of vegetation and water uptake (Hwang et al., 2009, 2012; Thompson et al., 2011). For example, upslope areas with lower vegetation density compared to downslope areas are likely to increase availability of water and nutrients to downslope regions (Hwang et al., 2009). This suggests that catchment structure in terms of topography and geomorphology is an important factor controlling spatial patterns of vegetation water stress (Caylor et al., 2005; Emanuel et al., 2010). Likewise, runoff production in catchments is developed from portions of the catchment that are hydrologically connected to the stream network (Detty and McGuire, 2010; Jencso et al., 2010), which may be significantly influenced by upslope water subsidies caused by vegetation patterns (Thompson et al., 2011) and seasonal variations in transpiration (Hwang et al., 2012).

Critical limiting factors affecting ET at the leaf scale differ from those at the patch, hillslope, ecosystem, or regional scale. At the leaf scale, stomatal conductance is a primary driver of transpiration (Jarvis, 1981). Rates of stomatal conductance are strongly influenced by photosynthesis, which in turn is regulated by light, CO_2, and biochemicals such as RuBisCO (Ball et al., 1987). At the whole-plant scale, rooting depth may be constrained by plant type or by edaphic factors such as depth to bedrock (Canadell et al., 1996; Schulze et al., 1996). At the patch scale, leaf area and sapwood area are important scalars for transpiration, which is generally a function of vegetation type, age, and productivity (Moore et al., 2004). However, at the hillslope scale, topography affects the redistribution of soil moisture and defines the irradiance environment and soil drying potential causing differential rates of transpiration. At the watershed scale, estimates of ET are commonly derived from simple water balance (ET = precipitation − runoff). At the regional and global scale, estimates of ET tend to be derived from empirical observations at the point scale that are upscaled to regions using vegetation type and precipitation (Zhang et al., 2010). Using appropriate drivers at the appropriate scale remains an active area for research. Little is known about the thresholds that transition drivers from one scale to the next.

Scale and structure are important themes of ecohydrological systems. This chapter is organized such that these themes are featured as an organizing framework for describing the interactions of ecohydrology and the Critical Zone. First, patch scale is discussed where an emphasis is placed on vertical processes and patterns. Patch sizes can be as small as 1 m^2 up to 100 m^2 or more and are relatively homogeneous. Then, hillslope and catchment scale are discussed where the emphasis shifts more to lateral processes and patterns imposed by hillslope morphology. Hillslope subunits of catchments range in size from 100 m^2 to 1000 m^2 or more, whereas catchments typically encompass larger drainage networks of at least 1000 m^2.

8.3 ECOHYDROLOGICAL PROCESSES AND PATTERNS AT THE PATCH SCALE

8.3.1 Water, Energy, and Carbon Budgets at the Patch Scale

Transpiration and carbon assimilation is driven by available energy and plant-available water in the Critical Zone (Moore and Heilman, 2011). The energy used to transpire water, that is, latent heat, is partitioned from net radiation and sensible heat by factors such as albedo, leaf area, thickness of the boundary layer, and atmospheric vapor pressure deficit. Plant-available water, on the other hand, is constrained by precipitation and local properties of the Critical Zones; that is, soil water-holding capacity, depth to the saturated zone, and rooting depth. Transpiration and photosynthesis are further regulated by factors of plant physiology, including stomatal conductance and xylem hydraulic conductance.

However, critical-limiting factors commonly differ among ecosystems and climates (Calder, 1998). These are generally partitioned into two categories: water-limited and energy-limited. Ecosystems can be further classified into short-statured grasslands, croplands, shrublands, and tall-statured forests (evergreen or deciduous). In water-limited regions, short and tall vegetation differ in terms of rainfall interception, rooting depth, and total biomass. These factors also exist in energy-limited environments (Loescher et al., 2005), but ET in short and tall vegetation can also differ by advection and the way energy is absorbed or reflected. In energy-limited environments, plant- and canopy-centric

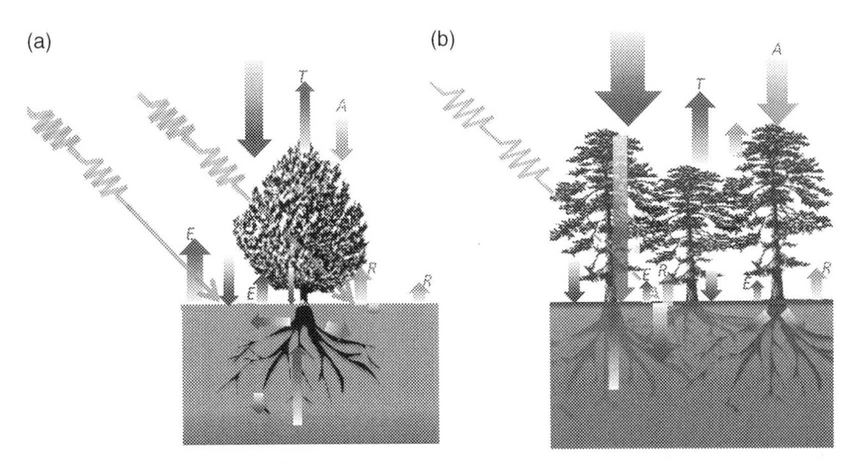

FIGURE 8.1 Water, energy, and carbon budgets at the patch scale in: (a) water-limited; and (b) energy-limited environments. Dark gray arrows (blue arrows in the web version) represent water fluxes, light gray arrows (red arrows in the web version) represent carbon fluxes, and thin gray arrows (orange arrows in the web version) represent incoming solar radiation. Arrow sizes illustrate the differences in magnitude between fluxes.

processes, such as canopy interception, stemflow, and foliar uptake, and the spatial heterogeneity in these processes, may have a profound influence on land–atmosphere fluxes (Loescher et al., 2002a). Through these mechanisms, vegetation alters the water balance in such systems in an entirely different manner than in dryland regions (Calder, 1998; Bonan, 2008) and likely differs with land cover (Pypker et al., 2005). In the following sections we review recent studies that highlight the contrasts in hydrological partitioning and ecological processes between water-limited and energy-limited systems (Fig. 8.1).

8.3.1.1 Water-Limited Regions

Arid and semiarid systems cover ~40% of the land surface of Earth and store ~15% of terrestrial organic carbon, representing the greatest areal extent and organic carbon pool of any terrestrial ecosystem, and are a significant driver of ecohydrological processes in the Critical Zone worldwide (Post et al., 1982; Asner et al., 2003; Lal, 2004; Reynolds et al., 2007). Further, the water-limited processes characteristic of these regions are globally distributed, with 97% of the terrestrial surface undergoing periods of at least a 1-month water deficit (Jenerette et al., 2012). The structure, integrity, and function of water-limited systems have been found to be highly sensitive to the amount and timing of precipitation, disturbance, and land-cover change (Schlesinger et al., 1990; Belnap and Gardner, 1993; Belnap, 1995; Weltzin et al., 2003; Huxman et al., 2004b; Ogle and Reynolds, 2004; Jenerette et al., 2006; Scott et al., 2009, Scott et al., 2010). The great expanse of drylands may impose substantial feedbacks to climate changes, making them globally important, yet they are comparatively understudied and underrepresented in Earth-System models. Dryland regions have large uncertainties in their functioning, role in biosphere processes, and future feedbacks to global climate-change scenarios (Schwalm et al., 2010; Jenerette et al., 2012; Schaefer et al., 2012).

Energy from solar radiation (Fig. 8.1) leads to higher rates of ET in water-limited systems (Rasmussen et al., 2011). In many dryland regions, a high fraction of annual precipitation, up to 100%, is evapotranspired. The fraction of water used for plant production varies greatly by land-use and land-cover type. Areas under vegetation in water-limited systems tend to remain cooler (e.g., reducing evaporative loss, augmenting subsurface root and microbial activity, and yielding higher soil respiratory fluxes) than intercanopy spaces. Increased energetics associated with subsurface biological activity (e.g., microbes, fungi, root exudates) can enhance biogeochemical weathering in areas with higher organic matter. Given the resulting mosaic of vegetative cover and seasonally bare soils, the largest source of stochasticity in Critical Zone function and development stems from horizontal variation in vegetation (Fig. 8.1).

A feedback exists between plant production in water-limited environments and dynamic soil characteristics. Soil water is not merely a result of the balance of inputs (precipitation, lateral flow) and outputs (drainage and ET). Processes such as mineralization, soil compaction, erosion, microbial-biomass production, root activity (including hydraulic redistribution), and others exert strong controls on plant cover and growth, and thus Critical Zone function and evolution. In fact, the lack of productivity and cover in intercanopy soils can counter Critical Zone development through soil loss via erosional processes and, eventually, a pathway towards desertification (Harman et al., 2014). A better understanding of Critical Zone dynamics will inevitably improve predictions of soil-water fluxes in these mixed mosaic systems.

8.3.1.2 Energy-Limited Regions

Many regions of the world are not considered water-limited and the dominant ecohydrological processes differ considerably (Fig. 8.1). Precipitation in wet environments tends to exceed ET rates, at least for much of the year, leading to high rates of percolation, runoff, and groundwater recharge. This poses a different set of dominant processes for such systems, which in ecohydrological terms are referred to as "energy-limited." In the context of the Critical Zone, "energy-limited" environments are typified by saturated or near-saturated soils, unrestricted plant-available water, or very low frequency of soil water deficit. Many of these regions harbor highly productive, yet highly biodiverse plant communities. Ultimately, much of the Sun's energy is captured and cycled within the biota. Nutrients are also tightly cycled within the biota. Conventional wisdom dictates that wet tropical forests retain most available nutrients within aboveground biomass (Proctor, 1987), although stochastic characteristics within these systems (e.g., aboveground and belowground heterogeneity) have hindered efforts to quantify nutrient budgets (Proctor, 2005). Critical Zone processes that exert strong controls on the biota in these environments include decomposition, leaching, and nutrient cycling. On a global scale, carbon cycling and land conversion are issues of fundamental concern in these regions.

In energy-limited environments, net radiation and atmospheric vapor pressure deficit explain much of the variation in plant ET (Loescher et al., 2005; Fisher et al., 2009). In regions with more than 2000 mm rainfall, annual ET approaches an energy-driven limit of approximately 1600 mm constrained by high humidity and cloud cover (Zhang et al., 2001; Bonell and Bruijnzeel, 2005). Rarely, if ever, do plants in these environments experience water deficits sufficient to suppress plant transpiration. However, additional factors occur such as waterlogging stress (such as in wetlands) and continuously wet surfaces (such as in tropical rainforests) that may stress plants to lower transpiration rates depending on species adaptations. Furthermore, plants can affect the depth to the water table, which in turn exerts controls on soil microbial activity and hydrologic fluxes. Thus, the presence of a water table presents challenges for characterizing the water balance in energy-limited systems compared with water-limited systems that generally do not root into the water table.

Evaporation of intercepted rainfall becomes a significant portion of the water balance in high rainfall regions (see Fig. 8.1, Holwerda et al., 2012), leading to irregular throughfall distribution and soil wetting (Loescher et al., 2002b). Intercepted rainfall or fog can also be absorbed directly by leaves via foliar uptake (Limm et al., 2009), which in some systems impacts the water balance. Where fog is prevalent and canopies are perpetually wet, cloud forests occur (Eugster et al., 2006). At the extreme, energy limitations lead to very slow-growing, dwarf vegetation (Bonell and Bruijnzeel, 2005).

Further, vegetation in high-rainfall regions contributes to "precipitation recycling." For example, one third of the precipitation that falls over the Amazon basin is supplied by recycled ET, which is further facilitated by the deep-rooted trees (e.g., Trenberth, 1999). Hence, ET has important influence on precipitation in tropical forests through land–atmosphere interactions. Anthropogenic effects of land cover change can also have profound impacts on the hydrologic cycle in these regions (Wohl et al., 2012). Additionally, climate change is predicted to modify the interaction between vegetation and the water cycle in humid regions by increasing atmospheric-moisture storage and altering the amount and timing of rainfall.

8.3.2 Patterns and Processes at the Patch Scale

Patch-scale ecohydrology is primarily concerned with vertical patterns and processes. Tall, dense canopies tend to capture all available energy primarily as latent heat, most of which is allocated to the upper, most exposed, portions of the canopy (Motzer, 2005). For this reason, after a rain event, dense forests dry out from the top down. It follows that upper canopies contribute a disproportionately large fraction of transpiration and carbon assimilation relative to lower canopy layers. However, the vertical distribution of latent heat, and how it partitions between wet-canopy evaporation and dry-canopy transpiration, has not been extensively studied (e.g., Klaassen, 2001).

At the patch scale, interrelationships between vegetation species and distribution and soil moisture are quite complex and involve not only understanding controls on transpiration and drainage in the soil, but also how the canopy structure itself affects interception and redistribution of precipitation. Spatial patterns of throughfall tend to persist through time (Keim et al., 2005; Gerrits et al., 2010), and this persistence can lead to wet and dry patterns of soil moisture that can affect recharge and runoff (Guswa and Spence, 2012). Keim et al. (2006) showed that the vegetation canopies can also reduce subsurface flows and, therefore, is an important process to include in models of catchment ecohydrology. At the Christiana CZO, Levia et al. (2011) demonstrated complex relationships between tree surface characteristics and geometry and the chemical constituents of stem flow.

Belowground, the distribution of roots influences patterns in root-zone moisture. The depth at which plants uptake water can extend quite deep (Bleby et al., 2010) and is known to differ between plant types (Jackson et al., 1996). Roots are generally concentrated in surface layers where water and nutrient uptake is driven by water potential gradients. However, when surface soil moisture is depleted, plants take up moisture from deeper in the soil profile or in rock (e.g., Wang et al., 1995; Jackson et al., 1999). Such water potential gradients can also lead to hydraulic redistribution of moisture by roots (Brooks et al., 2002) from wet to dry soil layers (i.e., either up or down and laterally). This process, which can even occur in senesced plants (Leffler et al., 2005), has important implications for extending water uptake and photosynthesis during dry periods (Domec et al., 2010), as well as influencing microbial activity.

8.4 ECOHYDROLOGICAL PROCESSES AND PATTERNS AT THE HILLSLOPE SCALE

A substantial body of ecohydrological research has examined plant–water interactions with respect to one-dimensional (vertical) water and nutrient fluxes at the patch scale, without incorporating lateral moisture redistribution of water and nutrients imposed by hillslope morphology. The hillslope scale forces integration between one- and two-dimensional conceptualizations of ecohydrological processes (Bond, 2003). Hillslopes provide the topological structures that connect patches on the landscape by gravitational fluxes organized by hillslope morphology. From the patch or pedon perspective, the hillslope provides nonlocal controls on water and nutrient fluxes, whereas local controls at the patch scale influence downslope patches in the ecohydrological system. For example, upslope patches with sparse vegetation may subsidize more water and nutrients to downslope patches than denser vegetated patches (Hwang et al., 2009; Emanuel et al., 2010; Thompson et al., 2011).

The basic elements that define hillslope morphology, such as shape, gradient, aspect, and slope complexity (i.e., nonuniformity) affect water and energy availability. Hillslope shape expresses the convergence or divergence of surface flowpaths in the planform (across slope) and profile (normal to slope) directions and therefore relates directly to soil moisture redistribution. Indices of hillslope shape have been used to predict tree (e.g., McNab, 1989) and grassland (e.g., Flores Cervantes et al., 2014) productivity or they are combined with other site variables in more complex models to predict site productivity (e.g., Iverson et al., 1997). Hillslope shape and gradient both affect runoff processes and erosion rates. Ludwig et al. (2005) found that vegetation patches on hillslopes in a semiarid landscape and runoff–erosion processes interact such that vegetation patches obstruct overland flow and create positive feedbacks for soil infiltration and vegetation productivity.

Hillslope aspect directly influences irradiance and hence energy availability for ET (Fig. 8.2). Aspect has a major influence on not only water availability, but also on other Critical Zone processes such as weathering rates and water-holding capacity. Geroy et al. (2011) found that soils that were derived from the same parent material, but occur on opposite north- and south-facing aspects, had different soil properties such as porosity, organic matter, and silt

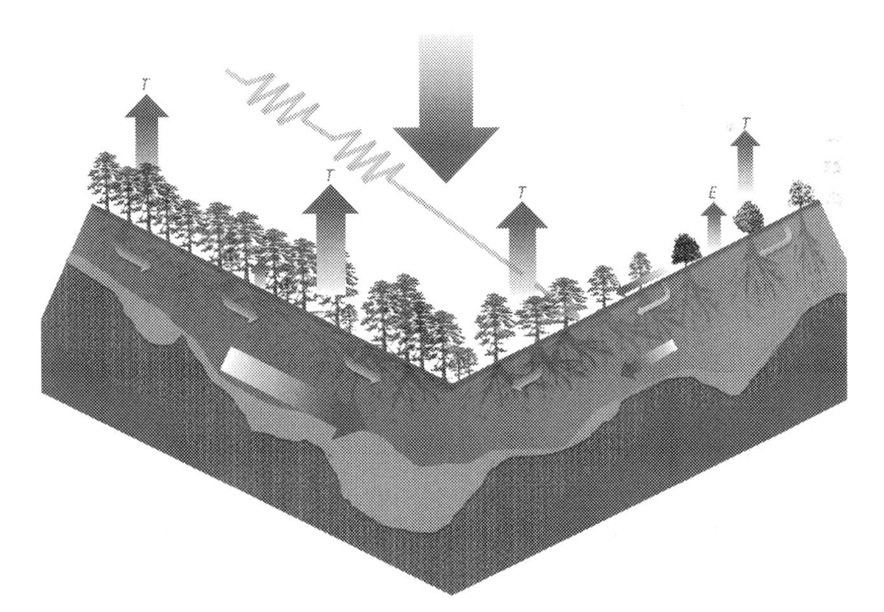

FIGURE 8.2 Water budgets at the hillslope scale in north-facing (left) and south-facing (right) aspects. Dark gray arrows (blue arrows in the web version) represent water fluxes and thin light gray arrows (orange arrows in the web version) represent incoming solar radiation. Arrow sizes illustrate the differences in magnitude between fluxes. Bedrock topography (dark gray [dark brown in the web version]) and depth to groundwater (light gray [blue in the web version]) interact with climate, vegetation, and soils to produce complex patterns in subsurface hydrology.

content, which led to higher water-holding capacity on the north-facing slope. These conditions co-evolved with different plant communities as well, where the north-facing slopes were occupied with fir species, which are less drought tolerant, and south-facing slopes were occupied by sagebrush. Complex interactions at the hillslope scale between topography, soil development, runoff processes, and vegetation create self-reinforcing, positive feedbacks in ecohydrological and Critical Zone processes that must be considered to develop a comprehensive understanding of ecohydrological patterns and processes (e.g., Gutierrez-Jurado et al., 2007).

8.4.1 Plant Controls on Water and Energy Budgets at the Hillslope Scale

The role of vegetation and how its associated ecophysiological properties affect hydrologic processes at the hillslope scale is complex (Asbjornsen et al., 2011). At hillslope scales, vegetation distribution and its effect on transpiration is commonly represented by vegetation indices (e.g., leaf area index or remotely sensed as a normalized difference vegetation index (NDVI)), vegetation height, topographic variables that control energy availability (e.g., slope and aspect), and/or soil properties (e.g., Gutierrez-Jurado et al., 2007; Hwang et al., 2009, 2012). Studies have shown, however, that age and species effects on sap flux, sapwood area, water potential, and hydraulic conductance are important, and these characteristics do not simply scale with vegetation indices, topographic variables, or soil properties (Ewers et al., 2002; Moore et al., 2004; Ford et al., 2007). Moreover, rooting and soil-depth distributions within hillslopes influence subsurface lateral flow and recharge patterns and are, thus, critical characteristics affecting ecohydrological processes. Tromp-van Meerveld and McDonnell (2006) found that spatial distribution of soil depth, which affected total water availability at the end of the dormant season and soil moisture content during the growing season, was related to spatial differences in basal area and species distribution on a hillslope. On this same hillslope, soil depth and moisture distribution controlled leakage to bedrock and partitioning of subsurface lateral flow (Tromp-van Meerveld et al., 2007). Hence, understanding interactions between vegetation and Critical Zone structure at these intermediate hillslope scales is important for bridging the gap between the patch scale, where ecophysiological characterization is most well-developed, to the catchment scale where water balance is best determined.

8.4.2 Patterns and Processes at the Hillslope Scale

The drivers behind ecohydrology at the hillslope scale are different between water-limited versus energy-limited environments. In water-limited systems, water availability will be the main driver, and thus consistent patterns in vegetation distribution across semiarid landscapes are directly related to spatial redistribution of soil water. In energy-limited systems, likely drivers are nutrient availability and/or photosynthetic active radiation distribution across the landscape. Interpretation of seemingly similar spatial patterns in vegetation characteristics across hillslopes, such as leaf area index or NDVI, need to be based on fundamental understanding about what drives such patterns.

Lateral redistribution via surface or subsurface flow is an important process that couples vegetation distribution and dynamics with hillslope-scale, hydrological partitioning (Fig. 8.2). Such redistribution happens in both semiarid and humid regions. In semiarid regions, overland flow and re-infiltration in local depressions and along rills and channels will favor denser vegetation in these lowland parts of the landscape. Niu et al. (2013) investigated the role of overland flow and re-infiltration (i.e., a process referred to as runoff–run on) on vegetation patterns and CO_2 fluxes in a semiarid catchment. They used a high-resolution, coupled, hydrologic, and land-surface model to simulate spatial variation of water, energy, and carbon fluxes across the catchment. They found that re-infiltration of overland flow along rills and channels resulted in wetter soils that sustained transpiration rates and net ecosystem exchange in these lowland parts of the catchment. The water subsidy related to this spatial redistribution of available water in this semiarid catchment provides plants with favorable conditions to produce more leaves, CO_2, and ET fluxes, and results in observable patterns of vegetation distribution. At the Southern Sierra CZO, soil-moisture distribution has been shown to differ greatly across elevational gradients because of altered snowpack and vegetation patterns (Bales et al., 2011; Graham and Hubbert, 2012; Hopmans et al., 2012).

Difference in energy availability due to aspect distribution in the catchment (e.g., north-facing vs. south-facing slopes) is another major driver of vegetation patterns on hillslopes (Fig. 8.2). Broxton et al. (2009) used hydrologic and isotopic data from eight catchments draining the resurgent dome in the Valles Caldera National Preserve in northern New Mexico, USA, to study the link between vegetation density and hydrologic response. These eight catchments have clear differences in aspect and thus differences in available energy throughout the year. These snow-dominated catchments also show clear patterns in vegetation density, with south-facing catchments having much less vegetation. Broxton et al. (2009) argued that such patterns result from differences in soil–water-balance climatology. North-facing catchments have less sublimation of the snow pack and thus more water available during spring-melt season. This results in wetter soils that sustain chemical weathering and deeper soils. Using stable water isotopes they found that the average transit time of water in north-facing catchments is significantly higher than in south-facing catchments, suggesting that water in north-facing catchments is available much longer for root-water uptake and plant transpiration. These processes can impact biogeochemical cycles, as has been shown at Boulder Creek CZO, where catchment aspect controlled annual discharge, which, in turn, is the dominant control on dissolved organic carbon effluxes from streams (Perdrial et al., 2014).

Irradiance differences between southerly or northerly exposures can drive differences in species composition, transpiration rate, and soil development, which all affect hydrologic processes, including the potential for downslope lateral drainage (Fig. 8.2). Shading of lower slope positions can also cause spatial variability in irradiance and reduce transpiration rates in those regions. Emanuel et al. (2010) showed that topographic variability (i.e., hillslope shape) could have significant implications for water stress. As water stress or limitation occurs during the growing season, ET can become decoupled from its atmospheric control (i.e., vapor pressure deficit) and transition to controls determined by hillslope topography and how water accumulated in hillslopes. Areas with tall vegetation, low drainage area, and steep slopes may experience the greatest water stress during the growing season. These upslope regions with higher water stress reduce soil moisture in downslope regions affecting lateral flow and transpiration.

In more humid systems, water redistribution is mainly driven by lateral subsurface flow in permeable soils developed on top of less permeable saprolite and bedrock. During the cold season when vegetation activity is suppressed due to temperature limitations and associated light limitations, rainfall infiltrates and percolates in the profile to form perched water tables that convey water downslope. This lateral redistribution of subsurface water is also associated with nutrient transport towards the riparian zones. Hwang et al. (2012) explored such hydrologic processes and their impact on emergent vegetation patterns in humid catchments. They computed the hydrologic vegetation index (HVI) (defined as the increase of NDVI per unit increase of the topographic wetness index; the topographic wetness index is defined by the ratio of upslope drainage area normalized by contour width and the local land surface slope (Beven and Kirkby, 1979), and is a measure of the potential wetness of location in the landscape for which it is computed) for several small forested catchments in Coweeta Hydrologic Laboratory and correlated HVI with several hydrologic signatures. They found strong correlations between HVI and average runoff coefficient and average Horton index. They also found strong correlations between HVI and the way these catchments release groundwater as baseflow. They argued that these results indicate that, without significant disturbance of the catchment ecosystems, the spatial organization of vegetation within catchments reflects the degree of dependency of ecosystems on flow pathways. Generally in heterogeneous terrain, more exposed areas such as slopes and ridge tops will limit water faster than low-lying areas, because of spatially varying energy inputs, aerodynamic effects caused by surface roughness, and topographically driven lateral drainage.

Thompson et al. (2011) hypothesized that spatial patterns in vegetation distribution along hillslopes are generated by flow convergence and result from a two-way coupling between vegetation driving ET and vegetation distribution reflecting water availability (Fig. 8.3). They implemented this hypothesis using a simple spatially explicit soil–water-balance model with nonlinear vegetation

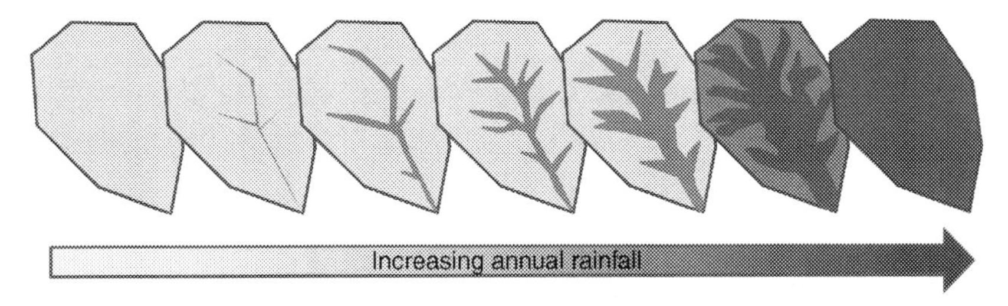

FIGURE 8.3 Catchment-scale trends in vegetation patterns are driven by spatial patterns in lateral transport of water from upslope contributing areas. Areas with little rainfall do not exhibit vegetation patterns because of a lack of lateral transport. With increasing annual rainfall, spatial patterns develop along ever-expanding drainage networks. Vegetation density in areas with high rainfall is no longer dependent on lateral transport, and canopy closure approaches 100% of the catchment area. *(Adapted from Thompson et al. (2011).)*

dynamics depending on water availability. They found that spatial organization was controlled by climate via the aridity index, flow convergence properties of the hillslopes, vegetation drought-resistance, and local controls on the competition between ET and lateral drainage. This self-organization of the vegetation across the landscape yielded spatial dependence in catchment-averaged hydrologic variables, water balance, and indices, such as the Horton index, describing hydrological partitioning.

8.5 ECOHYDROLOGICAL PROCESSES AND PATTERNS AT THE CATCHMENT SCALE

Catchment ecohydrology aims at understanding how ecosystems affect catchment water balance and how hydrological processes that operate across a catchment, in turn, affect ecosystems distribution and dynamics. Such understanding is needed to predict hydrological and ecological change during and after gradual or abrupt environmental disturbances, such as changing temperature and precipitation regimes related to climate change, land conversion, wildfires, and regional tree die-off.

At the catchment scale, ET has traditionally been determined as the difference between annual precipitation and annual water yield (e.g., Bosch and Hewlett, 1982). This annual water balance approach provides little information about internal watershed processes, but is useful for examining hydroclimate trends and responses to ecosystem disturbance (e.g., harvest, defoliation, species conversion, acid deposition) over long periods of time (e.g., see Ford et al., 2007; Green et al., 2013) across different catchments. However, at timescales less than a year and at spatial scales of small catchments, other approaches are necessary for examining how water uptake by plants affects soil moisture patterns and associated hydrological processes (e.g., recharge and streamflow generation). Techniques such as eddy-covariance and sap-flux measurements have been useful in understanding how plants affect hydrological processes within catchments (e.g., Oishi et al., 2010) and how water fluxes scale from leaves to catchments and relate to water balance (Asbjornsen et al., 2011).

The conservation equations that link ecological and hydrological processes at the catchment scale are similar to those formulated at the patch and hillslope scales, but closure relations are exceedingly difficult to quantify as a result of spatial heterogeneity and temporal variability of the processes involved. A bottom-up approach where patch-scale knowledge of the constitutive relations (i.e., linking, for example, soil moisture storage to plant transpiration) are scaled to whole-catchment stores and fluxes is very likely to become an impossible endeavor, and thus different methods need to be developed.

In the following sections, we introduce the conservation equations for water, energy, and carbon at the catchment scale and review different studies that focused on quantifying the effect of vegetation on catchment scale water balance. We then discuss emerging spatial patterns of vegetation that reflect different hydrological processes and review attempts to interpret such patterns to explain differences between catchment responses across climate gradients.

8.5.1 Plant Controls on Water and Energy Budgets at the Catchment Scale

Catchments integrate ecohydrological processes that occur at the patch and hillslope scales (Fig. 8.3). At the catchment scale the water balance can be written as:

$$\frac{\mathrm{d}S}{\mathrm{d}t} = P - Q - ET$$

where S is total water storage in the catchment, P is precipitation entering the catchment, Q is streamflow leaving the catchment, and ET is the combined land surface flux related to evaporation and transpiration. The water balance is coupled to the energy balance through this combined ET flux:

$$\frac{\mathrm{d}Q^*}{\mathrm{d}t} = R_\mathrm{n} - \lambda ET - H - G$$

where Q^* is the amount of heat energy stored in the catchment, R_n is net radiation flux, λET is latent heat flux, H is sensible heat flux, and G is ground heat flux. λ is the latent heat of vaporization of water. The water budget is also coupled to the carbon budget:

$$\frac{\mathrm{d}C}{\mathrm{d}t} = C_\mathrm{P} + \mathrm{NEE} + C_\mathrm{W} - C_\mathrm{Q}$$

where C is total carbon storage in the catchment, C_P is carbon import via precipitation, NEE is net ecosystem exchange and accounts for photosynthesis of vascular plants and cyanobacteria, as well as soil and plant respiration, C_W is carbon sequestration due to chemical weathering, and C_Q is carbon export via streamflow. NEE is directly related to catchment scale ET but this relationship is poorly understood (Troch et al., 2009).

These conservation equations form the basis of catchment ecohydrologic science. Hydrologists have traditionally used the water and energy balance to quantify land-surface hydrological partitioning, whereas ecologists typically start from the carbon and energy balance to study ecosystem dynamics. The relatively young science of catchment ecohydrology tries to synthesize knowledge gained in these separate disciplines to develop new fundamental knowledge about how plants affect catchment water balance and how hydrological partitioning at the catchment scale affects ecosystem dynamics.

Closing the conservation equations at the catchment scale is extremely difficult because all of the fluxes involved vary in space and time. Moreover, many of these fluxes are hard to quantify, even at the patch scale. The only fluxes that represent whole-watershed processes are related to streamflow and all other fluxes need to be estimated from local observations (e.g., a few rain gages across a catchment are averaged to estimate catchment scale precipitation) and, thus, involve assumptions about scaling. In the following section, we review recent studies that focus on the interactions between hydrological partitioning and ecological processes. Here we explore different alternative approaches to understand complex interactions between vegetation dynamics and catchment hydrological partitioning.

8.5.2 Patterns and Processes at the Catchment Scale

Investigating the effect of vegetation on catchment water balance is difficult because of the ubiquitous spatial heterogeneity present at larger scales (Fig. 8.3) and the lack of controlled experimentation at such scales. One alternative approach to the bottom-up aggregation of local knowledge is catchment intercomparison.

Using data from many catchments across different climates, Budyko (1950) found that the average annual water balance was first-order controlled by the aridity index. The aridity index is defined as the ratio of average potential ET (i.e., a measure of available energy in the catchment) to average precipitation (i.e., a measure of water availability). When the aridity index is high, most rainfall leaves the catchment as ET; whereas, when the aridity index is low most rainfall/precipitation leaves the catchment as runoff or streamflow. This remarkably simple spatial pattern in hydrological partitioning has motivated several researchers to explore different ecohydrological processes that potentially can explain Budyko's hypothesis.

Horton (1933) investigated the inter-annual variability of the water balance of a catchment in the NE USA and found that the ratio between catchment annual ET, estimated from the water balance, and catchment wetting, estimated by subtracting quick runoff from annual rainfall, was remarkably constant between years. He hypothesized that "the natural vegetation of a region tends to develop to such an extent that it can utilize the largest possible proportion of the available soil moisture supplied by infiltration" (Horton, 1933, p. 456).

Motivated by Horton's observation, Troch et al. (2009) examined the annual water balance of 89 catchments in the conterminous USA. They confirmed Horton's observation that the amount of annual ET versus the amount of water availability in the soil remains relatively constant between years across a range of ecosystem types and spatial scales, and termed the average ratio as the Horton index. They also found that vegetation becomes more efficient in its water-use as water availability decreases during the driest years. The latter is consistent with conclusions reached by Huxman et al. (2004a) using rain-use efficiency data from different biomes across climates. They showed that rain-use efficiency at individual sites converges to a common value during the driest years, independent of biome type.

Voepel et al. (2011) using data from 312 catchments in the USA, showed that the average Horton index of catchments varies predictably with aridity index, but also depends strongly on catchment average slope, suggesting that water redistribution plays an important role in plant-water availability. They also showed that the variability in annual Horton index is a better predictor of catchment maximum NDVI (a surrogate for vegetation productivity) than annual precipitation, potential ET or their ratio, the aridity index. Brooks et al. (2011) using data from a subset used in Voepel et al. (2011), observed negative correlations between annual Horton Index and maximum annual NDVI values, indicating water limitation during dry years in most catchment ecosystems. In nine of the wettest catchment ecosystems; however, NDVI values increased as the Horton index increased, suggesting greater vegetation productivity under drier conditions. These results demonstrate that catchment-scale, hydrologic partitioning provides information on both the fractions of precipitation available to and used by vegetation. Consequently, catchment-scale partitioning provides useful information for scaling point observations and quantifying regional ecohydrological response to climate or vegetation change.

Using data from more than three hundred catchments in the conterminous United States and a simple water balance model, Gentine et al. (2012) explored the ecohydrological controls on the average annual water balance. Their model showed that aboveground transpiration efficiency and belowground rooting structure of the vegetation (two parameters in the model) depend on the aridity index and the phase lag between peak seasonal potential ET and precipitation. The vertical and/or lateral extent of the rooting zone (a measure of storage capacity in the catchment) exhibited a maximum when peak radiation and precipitation are out of phase (e.g., in Mediterranean climate). This suggests that plant strategies in Mediterranean climates have adapted, compared to other catchments, to deal with intra-annual variability of water availability. These conclusions are consistent with findings from Milly (1994) who argued that rooting depth and climate forcing are connected and result in predictable long-term hydrological partitioning (i.e., expression of dominant form of water loss).

8.6 IMPACTS AND FEEDBACKS ACROSS SCALES

8.6.1 Impacts of Land Use Change Across Scales

The most fundamental way by which vegetation affects the water, energy, and carbon budgets is through establishment, growth, and community assembly. Thus, understanding ecohydrological processes within the Critical Zone is crucial to predict impacts of land-use changes from both natural and anthropogenic disturbance, successional processes, and climate variation. Only by understanding the critical drivers of ET can we understand the impacts of land use change on the water cycle. According to Moore and Heilman (2011), in a given localized area, transpiration will vary between contrasting vegetation patches if: (1) more energy is partitioned into latent heat in one patch than the other; (2) one patch has access to more available water than the other; or (3) available water is depleted faster in one patch than the other, provided such differences in water use are maintained over long timescales.

As vegetation establishes within any landscape, a fundamental change in the patterns of water, energy, and carbon cycling begins (Jenny, 1941; Berner, 1992), but key uncertainties remain in the magnitude, timing, and impacts in those shifts (Schwalm et al., 2010; Keenan et al., 2012). Understanding the degree to which vegetation alters patterns of inputs, whether they are energy, precipitation, or assimilated carbon, as well as losses such as interception and transpiration, is fundamental in closing the water, energy, and carbon budgets. Catchments integrate these changes across scales as an accumulation of patch-level responses (e.g., Wilcox et al., 2006). Thus, differences are manifested at the catchment scale only if a sufficient proportion of a catchment is affected.

Many paired catchment experiments have elucidated potential impacts of land-use change on water balance. Brown et al. (2005) summarized results from many such studies to reach some general conclusions about impacts of afforestation, deforestation, forest regrowth, and forest conversion across different climate zones. Afforestation is the conversion from short-statured vegetation to forest. Deforestation is the result of forest harvest that may be permanently or temporarily converted to short-statured vegetation. In many cases, the forest is then allowed to regrow into the same (forest regrowth) or a contrasting type of forest (forest conversion). Hydrologic impacts of these practices are driven by the magnitude and duration of vegetation changes.

Whereas the general consensus remains that reductions in vegetation cover can increase water yield and increases in vegetation cover can reduce water yield, results are highly unpredictable. Additional constraints are related to soils and climate. Water-yield changes are the greatest in high-rainfall areas (Brown et al., 2005). A 100% conversion from forest to grass is expected to have diminishing impacts on site ET where mean annual rainfall drops below 1000 mm (Zhang et al., 2001). Furthermore, streamflow changes caused by alterations of vegetation cover of less than 20% are likely to be undetectable (Bosch and Hewlett, 1982; Stednick, 1996).

Impacts of forest clearing are rapid, followed by a gradual recovery to pretreatment conditions over a period of years if regrowth is permitted (Brown et al., 2005). Suppression of regrowth can prolong the period where water yield changes are apparent, illustrating the coupled human-natural system dimensions more closely associated with Critical Zone evolution in the Anthropocene, as discussed in Chapter 19. It is also possible for the ET of regrowth vegetation to exceed that of the original forest, particularly old-growth forests (Moore et al., 2004), leading to a period where water yield of regrowth is lower than pretreatment conditions (e.g., Hornbeck et al., 1997).

In high-elevation, snow-dominated catchments, forests play a central role in regulating the partitioning of snow and snowmelt through complex ecohydrological interactions, which can alter the amount, timing, and duration of snowpack melting during spring and summer (Molotch et al., 2009). Recently, forests across the world have been subject to increased stresses as a result of warmer temperatures, drought, and fire suppression (Allen et al., 2010). Over the last decade, these stresses have facilitated epidemic-level populations of bark beetles throughout the forests of the western United States and Canada (Breshears et al., 2005; Fettig et al., 2007; Raffa et al., 2008). The simultaneous and expansive nature of these outbreaks has gained widespread attention, and has been identified as an indicator of future high-severity forest disturbance under global climate change (Allen et al., 2010). Investigators at Catalina–Jemez and Boulder CZOs have reported that insect-related tree mortality altered canopy interception and sublimation of snow in that region (Biederman et al., 2014).

The influence of subalpine forest vegetation on ET and snow-sublimation processes portends changes in catchment water budgets following severe bark beetle outbreak and tree mortality. However, conflicting potential changes are evident. For example, reductions in both interception and sublimation from the canopy, and transpiration by beetle-killed trees, may act to increase streamflow, whereas increased snowpack energy input under a beetle-killed canopy may increase snowpack sublimation and reduce streamflow. Empirical evidence for the net response of streamflow to bark-beetle outbreak is provided by studies of streamflow change following a spruce-beetle (*Dendroctonus rufipennis*) outbreak at two sites in central Colorado in the 1940s (Love, 1955; Bethlahmy, 1974, 1975), and following an outbreak of mountain pine beetle at one site in southern Montana in the 1970s (Potts, 1984). In all catchments, annual streamflow values were higher than expected by an average of 10–30%, and corresponded to increased snowmelt season streamflow. Postoutbreak, increases in seasonal low-flow (fall and winter) and peak streamflow, and earlier snowmelt timing, are also reported. These streamflow changes varied among catchments and in time, and were assumed to be dependent on catchment aspect and postoutbreak climatological conditions, notably precipitation, though limited climate data were available. Results provided a framework for expected water-budget changes following beetle outbreak, with decreased ET/canopy sublimation losses and increased streamflow dominating, and reduced transpiration and snow interception by beetle-killed trees acting as the underlying streamflow change mechanisms. This conclusion agrees with those of higher snowpack accumulation under beetle-killed plots relative to intact forest plots (Boon, 2007, 2009), and higher annual streamflow following forest harvest treatments at the Fraser Experimental Forest in the Colorado Rockies (Troendle and King, 1985, 1987).

Despite the general agreement between previous studies, questions related to streamflow response to recent outbreaks remain, notably those involving streamflow change under varied climate. Observed and predicted trends in temperature and precipitation may amplify or dampen the expected streamflow increase documented in historical studies following recent and future bark beetle outbreak episodes.

Guardiola-Claramonte et al. (2012) investigated the effect of regional piñon pine die-off due to prolonged drought in the Four-Corners region of southwest USA on catchment water yield, expressed as the ratio between average annual runoff coefficient after and before the timing of die-off. Counter to general expectation of increasing water yields after the die-off of a dominant woody species that dominated selected catchments in the area, they found that water yield was reduced by as much as 50%, depending on severity of impact in specific catchments. Catchments that were not impacted exhibited the same average annual runoff coefficient as before but catchments with significant impact exhibited reduced water yield that could not be explained by changes in temperature and precipitation regimes after tree mortality. They speculated that enhanced understory growth of grasses and forbes made possible by reduced competition for water and light in these semiarid catchments resulted in enhanced interception, reduced overland flow, and increased soil-water availability due to runoff–run on processes.

8.6.2 Feedbacks of Vegetation on the Water Cycle at Continental Scales

Changes in land–atmosphere feedbacks can alter the role of precipitation recycling in the water cycle. Improvements in ET parameterizations simulated by coupled atmosphere–land models are needed to better capture those dynamics. It remains unknown whether altered rainfall patterns would generate feedbacks that would diminish our ability to accurately predict ET from energy-partitioning alone. For example, the same amount of rainfall distributed into more frequent events would likely increase the proportion of inputs lost via evaporation of intercepted rainfall and increased time intervals of wet-canopy conditions. On the other hand, the same amount of rainfall extended over longer, less frequent but more intense periods would also likely increase evaporation per unit rainfall and possibly increase time intervals of wet-canopy conditions as well. Thus, even though ET predictions may seem simplified when soil-water availability is unlimited, the frequency and intensity of rainfall is another less understood determinant of ET.

8.6.3 Conclusions and Future Directions

This chapter highlights the many ecohydrological factors that dominate Critical Zone processes at patch, hillslope, and catchment scales. However, it should be emphasized that our ability to integrate such processes across scales is limited. Addressing current problems related to anthropogenic disturbance and altered climate regimes will ultimately require additional innovative approaches for scaling observations from patch to landscape. These challenges are confounded by the inherent dynamic nature of hydrology, which is dominated by stochastic events. Experiments and targeted observations that test process behavior at different scales, like those at CZOs, are needed to continue to elucidate the mechanisms underlying patterns, improve projections, and inform Critical Zone management. As illustrated throughout this chapter, vegetation has the ability to change the Critical Zone in profound ways that we have yet to fully understand or interpret. This is a vital area for future research to combat food and water scarcity.

REFERENCES

Allen, C.D., Macalady, A.K., Chenchouni, H., Bachelet, D., McDowell, N., Vennetier, M., Kitzberger, T., Rigling, A., Breshears, D.D., Hogg, E.H., Gonzalez, P., Fensham, R., Zhang, Z., Castro, J., Demidova, N., Lim, J.H., Allard, G., Running, S.W., Semerci, A., Cobb, N., 2010. A global overview of drought and heat-induced tree mortality reveals emerging climate change risks for forests. Forest Ecol. Manag. 259, 660–684.

Asbjornsen, H., Goldsmith, G.R., Alvarado-Barrientos, M.S., Rebel, K., Van Osch, F.P., Rietkerk, M., Chen, J., Gotsch, S., Tobon, C., Geissert, D.R., Gomez-Tagle, A., Vache, K., Dawson, T.E., 2011. Ecohydrological advances and applications in plant-water relations research: a review. J. Plant Ecol. 4, 192.

Asner, G.P., Archer, S., Hughes, R.F., Ansley, R.J., Wessman, C.A., 2003. Net changes in regional woody vegetation cover and carbon storage in Texas Drylands, 1937–1999. Global Change Biol. 9, 316–335.

Bales, R.C., Hopmans, J.W., O'Geen, A.T., Meadows, M., Hartsough, P.C., Kirchner, P., Hunsaker, C.T., Beaudette, D., 2011. Soil moisture response to snowmelt and rainfall in a Sierra Nevada Mixed-Conifer Forest. Vadose Zone J. 10, 786–799.

Ball, J.T., Woodrow, I.E., Berry J.A., 1987. A model predicting stomatal conductance and its contribution to the control of photosynthesis under different environmental conditions. In: International Photosynthesis Congress. Photosynthesis Research.

Belnap, J., 1995. Surface disturbances: their role in accelerating desertification. Environ. Monit. Assess. 37, 39–57.

Belnap, J., Gardner, J.S., 1993. Soil microstructure in soils of the Colorado plateau – the role of the cyanobaterium microcoleus vaginatus. Great Basin Nat. 53, 40–47.

Berner, R.A., 1992. Weathering, plants, and the long-term carbon cycle. Geochim. Cosmochim. Acta 56, 3225–3231.

Bethlahmy, N., 1974. More streamflow after a bark beetle epidemic. J. Hydrol. 23, 185–189.

Bethlahmy, N., 1975. A Colorado episode: beetle epidemic, ghost forests, more streamflow. Northwest Sci. 49, 95–105.

Beven, K.J., Kirkby, M.J., 1979. A physically based, variable contributing area model of basin hydrology. Hydrol. Sci. 24, 45.

Biederman, J.A., Brooks, P.D., Harpold, A.A., Gutmann, E., Gochis, D.J., Reed, D.E., Pendall, E., 2014. Multi-scale observations of snow accumulation and peak snowpack following widespread, insect-induced lodgepole pine mortality. Ecohydrology 7, 150–162.

Bleby, T.M., Mcelrone, A.J., Jackson, R.B., 2010. Water uptake and hydraulic redistribution across large woody root systems to 20 m depth. Plant Cell Environ. 33, 2132–2148.

Bonan, G.B., 2008. Forests and climate change: forcings, feedbacks, and the climate benefits of forests. Science 320, 1444–1449.

Bond, B., 2003. Hydrology and ecology meet – and the meeting is good. Hydrol. Process. 17, 2087–2089.

Bonell, M., 2002. Ecohydrology – a completely new idea? Hydrol. Sci. 47, 809–810.

Bonell, M., Bruijnzeel, L.A., 2005. Forests, Water and People in the Humid Tropics: Past, Present and Future Hydrological Research for Integrated Land and Water Management. Cambridge University Press, New York.

Boon, S., 2007. Snow accumulation and ablation in a beetle-killed pine stand in Northern Interior British Columbia, BC. J. Ecosyst. Manag. 8, 1–13.

Boon, S., 2009. Snow ablation energy balance in a dead forest stand. Hydrol. Process. 23, 2600–2610.

Bosch, J.M., Hewlett, J.D., 1982. A review of catchment experiments to determine the effect of vegetation changes on water yield and evapo-transpiration. J. Hydrol. 55, 3–23.

Brantley, S.L., Megonigal, J.P., Scatena, F.N., Balogh-Brunstad, Z., Barnes, R.T., Bruns, M.A., Van Cappellen, P., Dontsova, K., Hartnett, H.E., Hartshorn, A.S., Heimsath, A., Herndon, E., Jin, L., Keller, C.K., Leake, J.R., McDowell, W.H., Meinzer, F.C., Mozdzer, T.J., Petsch, S., Pett-Ridge, J., Pregitzer, K.S., Raymond, P.A., Riebe, C.S., Shumaker, K., Sutton-Grier, A., Walter, R., Yoo, K., 2011. Twelve testable hypotheses on the geobiology of weathering. Geobiology 9, 140–165.

Breshears, D.D., Cobb, N.S., Rich, P.M., Price, K.P., Allen, C.D., Balice, R.G., Romme, W.H., Kastens, J.H., Floyd, M.L., Belnap, J., Anderson, J.J., Myers, O.B., Meyer, C.W., 2005. Regional vegetation die-off in response to global-change-type drought. Proc. Natl. Acad. Sci. USA 102, 15144–15148.

Brooks, J.R., Meinzer, F.C., Coulombe, R., Gregg, J., 2002. Hydraulic redistribution of soil water during summer drought in two contrasting Pacific Northwest coniferous forests. Tree Physiol. 22, 1107–1117.

Brooks, P.D., Troch, P.A., Durcik, M., Gallo, E., Schlegel, M., 2011. Quantifying regional scale ecosystem response to changes in precipitation: not all rain is created equal. Water Resour. Res, 47 (10), W00J08.

Brown, A.E., Zhang, L., McMahon, T.A., Western, A.W., Vertessy, R.A., 2005. A review of paired catchment studies for determining changes in water yield resulting from alterations in vegetation. J. Hydrol. 310, 28–61.

Broxton, P.D., Troch, P.A., Lyon, S.W., 2009. On the role of aspect to quantify water transit times in small mountainous catchments. Water Resour. Res. 45 (8), W08427.

Budyko, M.I., 1950. Climatic factors of the external physical-geographical processes (in Russian). Gl Geofiz. Observ. 19, 25–40.
Calder, I.R., 1998. Water use by forests, limits and controls. Tree Physiol. 18, 625–631.
Canadell, J., Jackson, R.B., Ehleringer, J.R., Mooney, H.A., Sala, O.E., Schulze, E.D., 1996. Maximum rooting depth of vegetation types at the global scale. Oecologia 108, 583–595.
Casper, B.B., Schenk, H.J., Jackson, R.B., 2003. Defining a plant's belowground zone of influence. Ecology 84, 2313–2321.
Caylor, K.K., Manfreda, S., Rodriguez-Iturbe, I., 2005. On the coupled geomorphological and ecohydrological organization of river basins. Adv. Water Resour. 28, 69–86.
Chorover, J., Troch, P.A., Rasmussen, C., Brooks, P.D., Pelletier, J.D., Breshears, D.D., Huxman, T.E., Kurc, S.A., Lohse, K.A., McIntosh, J.C., Meixner, T., Schaap, M.G., Litvak, M.E., Perdrial, J., Harpold, A., Durcik, M., 2011. How water, carbon, and energy drive critical zone evolution: the Jemez–Santa Catalina Critical Zone Observatory. Vadose Zone J. 10, 884–899.
Colman, E., 1953. Vegetation and Watershed Management. Ronald Press Co., New York.
Detty, J.M., McGuire, K.J., 2010. Topographic controls on shallow groundwater dynamics: implications of hydrologic connectivity between hillslopes and riparian zones in a till mantled catchment. Hydrol. Process. 24, 2222–2236.
Domec, J.C., King, J.S., Noormets, A., Treasure, E., Gavazzi, M.J., Sun, G., McNulty, S.G., 2010. Hydraulic redistribution of soil water by roots affects whole-stand evapotranspiration and net ecosystem carbon exchange. New Phytol. 187, 171–183.
Eagleson, P.S., 1978. Climate, soil, and vegetation .1. Introduction to water-balance dynamics. Water Resour. Res. 14, 705–712.
Emanuel, R.E., D'Odorico, P., Epstein, H.E., 2007. A dynamic soil water threshold for vegetation water stress derived from stomatal conductance models. Water Resour. Res, 43 (3), W03431.
Emanuel, R.E., Epstein, H.E., McGlynn, B.L., Welsch, D.L., Muth, D.J., D'Odorico, P., 2010. Spatial and temporal controls on watershed ecohydrology in the northern Rocky Mountains. Water Resour. Res. 46 (11), W11553.
Eugster, W., Burkard, R., Holwerda, F., Scatena, F.N., Bruijnzeel, L.A.S., 2006. Characteristics of fog and fogwater fluxes in a Puerto Rican elfin cloud forest. Agric. Forest Meteorol. 139, 288–306.
Ewers, B.E., Mackay, D.S., Gower, S.T., Ahl, D.E., Burrows, S.N., Samanta, S.S., 2002. Tree species effects on stand transpiration in northern Wisconsin. Water Resour. Res, 38 (7), 1103.
Fettig, C.J., Klepzig, K.D., Billings, R.F., Munson, A.S., Nebeker, T.E., Negron, J.F., Nowak, J.T., 2007. The effectiveness of vegetation management practices for prevention and control of bark beetle infestations in coniferous forests of the western and southern United States. Forest Ecol. Manag. 238, 24–53.
Fisher, J.B., Malhi, Y., Bonal, D., Da Rocha, H.R., De Araujo, A.C., Gamo, M., Goulden, M.L., Hirano, T., Huete, A.R., Kondo, H., Kumagai, T., Loescher, H.W., Miller, S., Nobre, A.D., Nouvellon, Y., Oberbauer, S.F., Panuthai, S., Roupsard, O., Saleska, S., Tanaka, K., Tanaka, N., Tu, K.P., Von Randow, C., 2009. The land-atmosphere water flux in the tropics. Global Change Biol. 15, 2694–2714.
Ford, C.R., Hubbard, R.M., Kloeppel, B.D., Vose, J.M., 2007. A comparison of sap flux-based evapotranspiration estimates with catchment-scale water balance. Agric. Forest Meteorol. 145, 176–185.
Flores Cervantes, J.H., Istanbulluoglu, E., Vivoni, E.R., Holifield Collins, C.D., Bras, R.L., 2014. A geomorphic perspective on terrain-modulated organization of vegetation productivity: analysis in two semiarid grassland ecosystems in Southwestern United States. Ecohydrology 7, 242–257.

Gentine, P., D'Odorico, P., Lintner, B.R., Sivandran, G., Salvucci, G., 2012. Interdependence of climate, soil, and vegetation as constrained by the Budyko curve. Geophys. Res. Lett. 39 (19), L19404.

Geroy, I.J., Gribb, M.M., Marshall, H.P., Chandler, D.G., Benner, S.G., McNamara, J.P., 2011. Aspect influences on soil water retention and storage. Hydrol. Process. 25, 3836–3842.

Gerrits, A.M.J., Pfister, L., Savenije, H.H.G., 2010. Spatial and temporal variability of canopy and forest floor interception in a beech forest. Hydrol. Process. 24, 3011–3025.

Graham, R.C., Hubbert, K.R., 2012. Comment on "Soil Moisture Response to Snowmelt and Rainfall in a Sierra Nevada Mixed-Conifer Forest". Vadose Zone J, 11 (4), doi: 10.2136/vzj2012.0004.

Graham, R.C., Rossi, A.M., Hubbert, K.R., 2010. Rock to regolith conversion: producing hospitable substrates for terrestrial ecosystems. GSA Today 20, 4–9.

Grayson, R.B., Western, A.W., Chiew, F.H.S., Bloschl, G., 1997. Preferred states in spatial soil moisture patterns: local and nonlocal controls. Water Resour. Res. 33, 2897–2908.

Green, M.B., Bailey, A.S., Bailey, S.W., Battles, J.J., Campbell, J.L., Driscoll, C.T., Fahey, T.J., Lepine, L.C., Likens, G.E., Ollinger, S.V., Schaberg, P.G., 2013. Decreased water flowing from a forest amended with calcium silicate. Proc. Natl. Acad. Sci. USA 110, 5999–6003.

Guardiola-Claramonte, M., Troch, P.A., Breshears, D.D., Huxman, T.E., Switanek, M.B., Durcik, M., Cobb, N.S., 2012. Decreased streamflow in semi-arid basins following drought-induced tree die-off: a counter-intuitive and indirect climate impact on hydrology. J. Hydrol. 414, 560.

Guswa, A.J., 2012. Canopy vs. roots: production and destruction of variability in soil moisture and hydrologic fluxes. Vadose Zone J. 11 (3), doi: 10.2136/vzj2011.0159.

Guswa, A.J., Spence, C.M., 2012. Effect of throughfall variability on recharge: application to hemlock and deciduous forests in western Massachusetts. Ecohydrology 5, 563–574.

Gutierrez-Jurado, H.A., Vivoni, E.R., Istanbulluoglu, E., Bras, R.L., 2007. Ecohydrological response to a geomorphically significant flood event in a semiarid catchment with contrasting ecosystems. Geophys. Res. Lett. 34.

Hannah, D.M., Wood, P.J., Sadler, J.P., 2004. Ecohydrology and hydroecology: a 'new paradigm'? Hydrol. Process. 18, 3439–3445.

Harman, C.J., Lohse, K.A., Troch, P.A., Sivapalan, M., 2014. Spatial patterns of vegetation, soils, and microtopography from terrestrial laser scanning on two semiarid hillslopes of contrasting lithology. J. Geophys. Res. Biogeosci. 119, 163–180.

Hobbs, R.J., Higgs, E., Harris, J.A., 2009. Novel ecosystems: implications for conservation and restoration. Trends Ecol. Evol. 24, 599–605.

Holwerda, F., Bruijnzeel, L.A., Scatena, F.N., Vugts, H.F., Meesters, A.G.C.A., 2012. Wet canopy evaporation from a Puerto Rican lower montane rain forest: the importance of realistically estimated aerodynamic conductance. J. Hydrol. 414, 1–15.

Hopmans, J.W., Bales, R.C., O'Geen, A.T., Hunsaker, C.T., Beaudette, D., Hartsough, P.C., Malazian, A., Kirchner, P., Meadows, M., 2012. Response to "Comment on 'Soil Moisture Response to Snowmelt and Rainfall in a Sierra Nevada Mixed-Conifer Forest'". Vadose Zone J. 11 (4), doi: 10.2136/vzj2012.0004r.

Hornbeck, J.W., Martin, C.W., Eagar, C., 1997. Summary of water yield experiments at Hubbard Brook Experimental Forest, New Hampshire. Can. J. Forest Res. 27, 2043–2052.

Horton, R.E., 1933. The role of infiltration in the hydrologic cycle. In: Transactions of the American Geophysical Union. National Research Council of the National Academy of Sciences, Washington, DC, pp. 446–460.

Huxman, T.E., Smith, M.D., Fay, P.A., Knapp, A.K., Shaw, M.R., Loik, M.E., Smith, S.D., Tissue, D.T., Zak, J.C., Weltzin, J.F., Pockman, W.T., Sala, O.E., Haddad, B.M., Harte, J., Koch, G.W., Schwinning, S., Small, E.E., Williams, D.G., 2004a. Convergence across biomes to a common rain-use efficiency. Nature 429, 651–654.

Huxman, T.E., Snyder, K.A., Tissue, D., Leffler, A.J., Ogle, K., Pockman, W.T., Sandquist, D.R., Potts, D.L., Schwinning, S., 2004b. Precipitation pulses and carbon fluxes in semiarid and arid ecosystems. Oecologia 141, 254–268.

Hwang, T., Band, L., Hales, T.C., 2009. Ecosystem processes at the watershed scale: extending optimality theory from plot to catchment. Water Resour. Res, 45 (11), W11425.

Hwang, T., Band, L.E., Vose, J.M., Tague, C., 2012. Ecosystem processes at the watershed scale: hydrologic vegetation gradient as an indicator for lateral hydrologic connectivity of headwater catchments. Water Resour. Res, 48 (6), W06514.

Iverson, L.R., Dale, M.E., Scott, C.T., Prasad, A., 1997. A Gis-derived integrated moisture index to predict forest composition and productivity of Ohio forests (U.S.A.). Landsc. Ecol. 12, 331–348.

Jackson, R.B., Canadell, J., Ehleringer, J.R., Mooney, H.A., Sala, O.E., Schulze, E.D., 1996. A global analysis of root distributions for terrestrial biomes. Oecologia 108, 389–411.

Jackson, R.B., Moore, L.A., Hoffmann, W.A., Pockman, W.T., Linder, C.R., 1999. Ecosystem rooting depth determined with caves and DNA. Proc. Natl. Acad. Sci. 96, 11387–11392.

Jarvis, P.G., 1981. Stomatal conductance, gaseous exchange and transpiration. In: Grace, J., Ford, E. D., Jarvis, P.G. (Eds), Plants and Their Atmospheric Environment: The 21st Symposium of the British Ecological Society. Blackwell Scientific Publications, Edinburgh, UK, pp. 175–203.

Jencso, K.G., McGlynn, B.L., Gooseff, M.N., Bencala, K.E., Wondzell, S.M., 2010. Hillslope hydrologic connectivity controls riparian groundwater turnover: implications of catchment structure for riparian buffering and stream water sources. Water Resour. Res, 46 (10), W10524.

Jenerette, G.D., Barron-Gafford, G.A., Guswa, A.J., McDonnell, J.J., Villegas, J.C., 2012. Organization of complexity in water limited ecohydrology. Ecohydrology 5, 184–199.

Jenerette, G.D., Wu, J.G., Grimm, N.B., Hope, D., 2006. Points, patches, and regions: scaling soil biogeochemical patterns in an urbanized arid ecosystem. Global Change Biol. 12, 1532–1544.

Jenny, H., 1941. Factors of Soil Formation: A System of Quantitative Pedology. McGraw-Hill, New York.

Keenan, T.F., Baker, I., Barr, A., Ciais, P., Davis, K., Dietze, M., Dragon, D., Gough, C.M., Grant, R., Hollinger, D., Hufkens, K., Poulter, B., Mccaughey, H., Raczka, B., Ryu, Y., Schaefer, K., Tian, H.Q., Verbeeck, H., Zhao, M.S., Richardson, A.D., 2012. Terrestrial biosphere model performance for inter-annual variability of land-atmosphere CO_2 exchange. Global Change Biol. 18, 1971–1987.

Keim, R.F., Meerveld, H.J.T.V., McDonnell, J.J., 2006. A virtual experiment on the effects of evaporation and intensity smoothing by canopy interception on subsurface stormflow generation. J. Hydrol. 327, 352–364.

Keim, R.F., Skaugset, A.E., Weiler, M., 2005. Temporal persistence of spatial patterns in throughfall. J. Hydrol. 314, 263–274.

Kittredge, J., 1948. Forest Influences: The Effects of Woody Vegetation on Climate, Water, and Soil, with Applications to the Conservation of Water and the Control of Floods and Erosion. McGraw-Hill, New York.

Klaassen, W., 2001. Evaporation from rain-wetted forest in relation to canopy wetness, canopy cover, and net radiation. Water Resour. Res. 37, 3227–3236.

Lal, R., 2004. Carbon sequestration in dryland ecosystems. Environ. Manag. 33, 528–544.

Leffler, A.J., Peek, M.S., Ryel, R.J., Ivans, C.Y., Caldwell, M.M., 2005. Hydraulic redistribution through the root systems of senesced plants. Ecology 86, 633–642.

Levia, D.F., Van Stan, J.T., Siegert, C.M., Inamdar, S.P., Mitchell, M.J., Mage, S.M., McHale, P.J., 2011. Atmospheric deposition and corresponding variability of stemflow chemistry across temporal scales in a mid-Atlantic broadleaved deciduous forest. Atmos. Environ. 45, 3046–3054.

Limm, E.B., Simonin, K.A., Bothman, A.G., Dawson, T.E., 2009. Foliar water uptake: a common water acquisition strategy for plants of the redwood forest. Oecologia 161, 449–459.

Loescher, H.W., Powers, J.S., Oberbauer, S.F., 2002a. Spatial variation of throughfall volume in an old-growth tropical wet forest, Costa Rica. J. Trop. Ecol. 18, 397–407.

Loescher, H.W., Powers, J.S., Oberbauer, S.F., 2002b. Spatial variation of throughfall volume in an old-growth tropical wet forest, Costa Rica. J. Trop. Ecol. 18, 949.

Loescher, H.W., Gholz, H.L., Jacobs, J.M., Oberbauer, S.F., 2005. Energy dynamics and modeled evapotranspiration from a wet tropical forest in Costa Rica. J. Hydrol. 315, 274–294.

Love, L.D., 1955. The effect on stream flow of the killing of spruce and pine by the Engelmann spruce beetle. Trans. Am. Geophys. Union 36, 113–118.

Ludwig, J.A., Wilcox, B.P., Breshears, D.D., Tongway, D.J., Imeson, A.C., 2005. Vegetation patches and runoff–erosion as interacting ecohydrological processes in semiarid landscapes. Ecology 86, 288–297.

McDonnell, J.J., Sivapalan, M., Vache, K., Dunn, S., Grant, G., Haggerty, R., Hinz, C., Hooper, R., Kirchner, J., Roderick, M.L., Selker, J., Weiler, M., 2007. Moving beyond heterogeneity and process complexity: a new vision for watershed hydrology. Water Resour. Res. 43, W07301.

McNab, W.H., 1989. Terrain shape index: quantifying effect of minor landforms on tree height. Forest Sci. 35 (1), 91–104.

Miller, J.R., Turner, M.G., Smithwick, E.A.H., Dent, C.L., Stanley, E.H., 2004. Spatial extrapolation: the science of predicting ecological patterns and processes. Bioscience 54, 310–320.

Milly, P.C.D., 1994. Climate, soil-water storage, and the average annual water-balance. Water Res. Res. 30, 2143–2156.

Molotch, N.P., Brooks, P.D., Burns, S.P., Litvak, M., Monson, R.K., McConnell, J.R., Musselman, K., 2009. Ecohydrological controls on snowmelt partitioning in mixed-conifer sub-alpine forests. Ecohydrology 2, 129–142.

Moore, G.W., Heilman, J.L., 2011. Ecohydrology bearings - invited commentary: proposed principles governing how vegetation changes affect transpiration. Ecohydrology 4, 351–358.

Moore, G.W., Bond, B.J., Jones, J.A., Phillips, N., Meinzer, F.C., 2004. Structural and compositional controls on transpiration in 40-and 450-year-old riparian forests in western Oregon, USA. Tree Physiol. 24, 481–491.

Motzer, T., 2005. Micrometeorological aspects of a tropical mountain forest. Agric. Forest Meteorol. 135, 230–240.

Naithani, K.J., Baldwin, D.C., Gaines, K.P., Lin, H., Eissenstat, D.M., 2013. Spatial distribution of tree species governs the spatio-temporal interaction of leaf area index and soil moisture across a forested landscape. PLoS ONE 8 (3), e58704.

Niu, G.-Y., Troch, P.A., Paniconi, C., Scott, R.L., Durcik, M., Zeng, X., Huxman, T., Goodrich, D., Pelletier, J., 2013. An integrated modelling framework of catchment-scale ecohydrological processes: 2. The role of water subsidy by overland flow on vegetation dynamics in a semi-arid catchment. Ecohydrology. 7 (2), 815–827.

Nuttle, W.K., 2002. Eco-hydrology's past and future in focus. EOS 83 (19), 205–212.

Ogle, K., Reynolds, J.F., 2004. Plant responses to precipitation in desert ecosystems: integrating functional types, pulses, thresholds, and delays. Oecologia 141, 282–294.

Oishi, A.C., Oren, R., Novick, K.A., Palmroth, S., Katul, G.G., 2010. Interannual invariability of forest evapotranspiration and its consequence to water flow downstream. Ecosystems 13, 421–436.

Perdrial, J.N., McIntosh, J., Harpold, A., Brooks, P.D., Zapata-Rios, X., Ray, J., Meixner, T., Kanduc, T., Litvak, M., Troch, P.A., Chorover, J., 2014. Stream water carbon controls in seasonally snow-covered mountain catchments: impact of inter-annual variability of water fluxes, catchment aspect and seasonal processes. Biogeochemistry 118, 273–290.

Post, W.M., Emanuel, W.R., Zinke, P.J., Stangenberger, A.G., 1982. Soil carbon pools and world life zones. Nature 298, 156–159.

Potts, D.F., 1984. Hydrologic impacts of a large-scale mountain pine-beetle (*Dendroctonus ponderosae* Hopkins) epidemic. Water Resour. Bull. 20, 373–377.

Proctor, J., 1987. Nutrient cycling in primary and old secondary rain-forests. Appl. Geogr. 7, 135–152.

Proctor, J., 2005. Rainforest mineral nutrition: the 'black box' and a glimpse inside it. In: Bonell, M., Bruijnzeel, L.A. (Eds.), Forests, Water and People in the Humid Tropics. Cambridge University Press, Cambridge, pp. 422–446.

Pypker, T.G., Bond, B.J., Link, T.E., Marks, D., Unsworth, M.H., 2005. The importance of canopy structure in controlling the interception loss of rainfall: examples from a young and an old-growth Douglas-fir forest. Agric. Forest Meteorol. 130, 113–129.

Raffa, K.F., Aukema, B.H., Bentz, B.J., Carroll, A.L., Hicke, J.A., Turner, M.G., Romme, W.H., 2008. Cross-scale drivers of natural disturbances prone to anthropogenic amplification: the dynamics of bark beetle eruptions. Bioscience 58, 501–517.

Rasmussen, C., Troch, P.A., Chorover, J., Brooks, P., Pelletier, J., Huxman, T.E., 2011. An open system framework for integrating critical zone structure and function. Biogeochemistry 102, 15–29.

Reynolds, J.F., Stafford Smith, D.M., Lambin, E.F., Turner, B.L., Mortimore, M., Batterbury, S.P.J., Downing, T.E., Dowlatabadi, H., Fernandez, R.J., Herrick, J.E., Huber-Sannwald, E., Jiang, H., Leemans, R., Lynam, T., Maestre, F.T., Ayarza, M., Walker, B., 2007. Global desertification: building a science for dryland development. Science 316, 847–851.

Rodriguez-Iturbe, I., 2000. Ecohydrology: a hydrologic perspective of climate-soil-vegetation dynamics. Water Resour. Res. 36, 3–9.

Schaefer, K., Schwalm, C.R., Williams, C., Arain, M.A., Barr, A., Chen, J.M., Davis, K.J., Dimitrov, D., Hilton, T.W., Hollinger, D.Y., Humphreys, E., Poulter, B., Raczka, B.M., Richardson, A.D., Sahoo, A., Thornton, P., Vargas, R., Verbeeck, H., Anderson, R., Baker, I., Black, T.A., Bolstad, P., Chen, J.Q., Curtis, P.S., Desai, A.R., Dietze, M., Dragoni, D., Gough, C., Grant, R.F., Gu, L.H., Jain, A., Kucharik, C., Law, B., Liu, S.G., Lokipitiya, E., Margolis, H.A., Matamala, R., McCaughey, J.H., Monson, R., Munger, J.W., Oechel, W., Peng, C.H., Price, D.T., Ricciuto, D., Riley, W.J., Roulet, N., Tian, H.Q., Tonitto, C., Torn, M., Weng, E.S., Zhou, X.L., 2012. A model-data comparison of gross primary productivity: results from the North American Carbon Program site synthesis. J. Geophys. Res. Biogeosci., 117 (G3), G03010.

Schlesinger, W.H., Reynolds, J.F., Cunningham, G.L., Huenneke, L.F., Jarrell, W.M., Virginia, R.A., Whitford, W.G., 1990. Biological feedbacks in global desertification. Science 247, 1043–1048.

Schulze, E.D., Mooney, H.A., Sala, O.E., Jobbagy, E., Buchmann, N., Bauer, G., Canadell, J., Jackson, R.B., Loreti, J., Oesterheld, M., Ehleringer, J.R., 1996. Rooting depth, water availability, and vegetation cover along an aridity gradient in Patagonia. Oecologia 108, 503–511.

Schwalm, C.R., Williams, C.A., Schaefer, K., Arneth, A., Bonal, D., Buchmann, N., Chen, J.Q., Law, B.E., Lindroth, A., Luyssaert, S., Reichstein, M., Richardson, A.D., 2010. Assimilation exceeds respiration sensitivity to drought: a FLUXNET synthesis. Global Change Biol. 16, 657–670.

Schwinning, S., 2010. Ecohydrology bearings - invited commentary – the ecohydrology of roots in rocks. Ecohydrology 3, 238–245.

Scott, R.L., Jenerette, G.D., Potts, D.L., Huxman, T.E., 2009. Effects of seasonal drought on net carbon dioxide exchange from a woody-plant-encroached semiarid grassland. J. Geophys. Res. Biogeosci. 114 (G4), G04004.

Scott, R.L., Hamerlynck, E.P., Jenerette, G.D., Moran, M.S., Barron-Gafford, G.A., 2010. Carbon dioxide exchange in a semidesert grassland through drought-induced vegetation change. J. Geophys. Res. Biogeosci. 115 (G3), G03026.

Stednick, J.D., 1996. Monitoring the effects of timber harvest on annual water yield. J. Hydrol. 176, 79–95.

Thompson, S.E., Harman, C.J., Troch, P.A., Brooks, P.D., Sivapalan, M., 2011. Spatial scale dependence of ecohydrologically mediated water balance partitioning: a synthesis framework for catchment ecohydrology. Water Resou. Res. 47 (10), W00J03.

Trenberth, K.E., 1999. Atmospheric moisture recycling: role of advection and local evaporation. J. Climate 12, 1368–1381.

Troch, P.A., Martinez, G.F., Pauwels, V.R.N., Durcik, M., Sivapalan, M., Harman, C., Brooks, P.D., Gupta, H., Huxman, T., 2009. Climate and vegetation water use efficiency at catchment scales. Hydrol. Process. 23, 2409–2414.

Troendle, C.A., King, R.M., 1985. The effect of timber harvest on the Fool Creek watershed, 30 years later. Water Resour. Res. 21, 1915–1922.

Troendle, C.A., King, R.M., 1987. The effect of partial and clearcutting on streamflow at Deadhorse Creek, Colorado. J. Hydrol. 90, 145–157.

Tromp-van Meerveld, H.J., McDonnell, J.J., 2006. On the interrelations between topography, soil depth, soil moisture, transpiration rates and species distribution at the hillslope scale. Adv. Water Resour. 29, 293–310.

Tromp-van Meerveld, H.J., Peters, N.E., McDonnell, J.J., 2007. Effect of bedrock permeability on subsurface stormflow and the water balance of a trenched hillslope at the Panola Mountain Research Watershed, Georgia, USA. Hydrol. Process. 21, 750–769.

Turnbull, L., Wilcox, B.P., Belnap, J., Ravi, S., D'Odorico, P., Childers, D., Gwenzi, W., Okin, G., Wainwright, J., Caylor, K.K., Sankey, T., 2012. Understanding the role of ecohydrological feedbacks in ecosystem state change in drylands. Ecohydrology 5, 174–183.

Voepel, H., Ruddell, B., Schumer, R., Troch, P.A., Brooks, P.D., Neal, A., Durcik, M., Sivapalan, M., 2011. Quantifying the role of climate and landscape characteristics on hydrologic partitioning and vegetation response. Water Resour. Res. 47 (10), W00J09.

Wang, Z.Q., Newton, M., Tappeiner, J.C., 1995. Competitive relations between Douglas-Fir and Pacific Madrone on shallow soils in a Mediterranean Climate. Forest Sci. 41, 744–757.

Weltzin, J.F., Loik, M.E., Schwinning, S., Williams, D.G., Fay, P.A., Haddad, B.M., Harte, J., Huxman, T.E., Knapp, A.K., Lin, G.H., Pockman, W.T., Shaw, M.R., Small, E.E., Smith, M.D., Smith, S.D., Tissue, D.T., Zak, J.C., 2003. Assessing the response of terrestrial ecosystems to potential changes in precipitation. Bioscience 53, 941–952.

Wilby, R.L., Schimel, D.S., 1999. Scales of interaction in eco-hydrological relations. In: Baird, A.J., Wilby, R.L. (Eds.), Eco-Hydrology: Plants and Water in Terrestrial and Aquatic Environments. Routledge, New York, pp. 39–77.

Wilcox, B.P., Owens, M.K., Dugas, W.A., Ueckert, D.N., Hart, C.R., 2006. Shrubs, streamflow, and the paradox of scale. Hydrol. Process. 20, 3245–3259.

Wohl, E., Barros, A., Brunsell, N., Chappell, N.A., Coe, M., Giambelluca, T., Goldsmith, S., Harmon, R., Hendrickx, J.M.H., Juvik, J., McDonnell, J., Ogden, F., 2012. The hydrology of the humid tropics. Nat. Climate Change 2, 655–662.

Zhang, L., Dawes, W.R., Walker, G.R., 2001. Response of mean annual evapotranspiration to vegetation changes at catchment scale. Water Resour. Res. 37, 701–708.

Zhang, K., Kimball, J.S., Nemani, R.R., Running, S.W., 2010. A continuous satellite-derived global record of land surface evapotranspiration from 1983 to 2006. Water Resour. Res. 46 (9), W09522.

Chapter 9

Rivers in the Critical Zone

Ellen Wohl
Department of Geosciences, Colorado State University, Fort Collins, Colorado, USA

9.1 INTRODUCTION

At the simplest level, a river is a "gutter down which flow the ruins of continents," to quote Leopold et al. (1964). This metaphor focuses on river channels as conveyances for water, solutes, and sediments entering a channel network from adjacent uplands. A more comprehensive approach is to treat a river channel as part of a *river system* that includes the adjacent floodplains and hyporheic zone and provides habitat for diverse aquatic and riparian plants and animals, as well as a wide variety of ecosystem services. In this context, the *floodplain* is defined as a low-relief sedimentary surface adjacent to the active channel that results from river erosion and deposition and is flooded relatively frequently (typically at intervals <10 years). The *hyporheic zone* is the portion of unconfined, near-stream aquifers with flow paths that originate and terminate in the active channel (Gooseff, 2010).

River systems adjust through time and space in response to changes in the fundamental drivers of water, sediment, base level, and substrate. Water and sediment enter a discrete portion of a channel from upland and upstream sources and from atmospheric inputs. *Base level* refers to the downstream point that represents the lowest level to which a river flows. If base level drops, the channel incises. If base level rises, sediment accumulates along the streambed, causing a rise in bed elevation. *Substrate* is the material composing the channel boundaries, which varies from clay through sand and boulders to bedrock, and the riverside vegetation that can strongly influence erosional resistance of the channel boundaries.

River systems integrate diverse fluxes occurring within the drainage basin and beyond the basin boundaries, including fluxes of energy from the sun and Earth's interior, and water, sediment, solutes, organic matter, nutrients, contaminants, and organisms moving across the surface, between the atmosphere and the surface, and between the surface and the subsurface (Fig. 9.1). Because river systems are so closely connected to other components of the Critical Zone and react so readily to changes in fluxes, river systems exemplify the basic definition of the Critical Zone as the "heterogeneous, near-surface environment in

Developments in Earth Surface Processes, Vol. 19. http://dx.doi.org/10.1016/B978-0-444-63369-9.00009-4

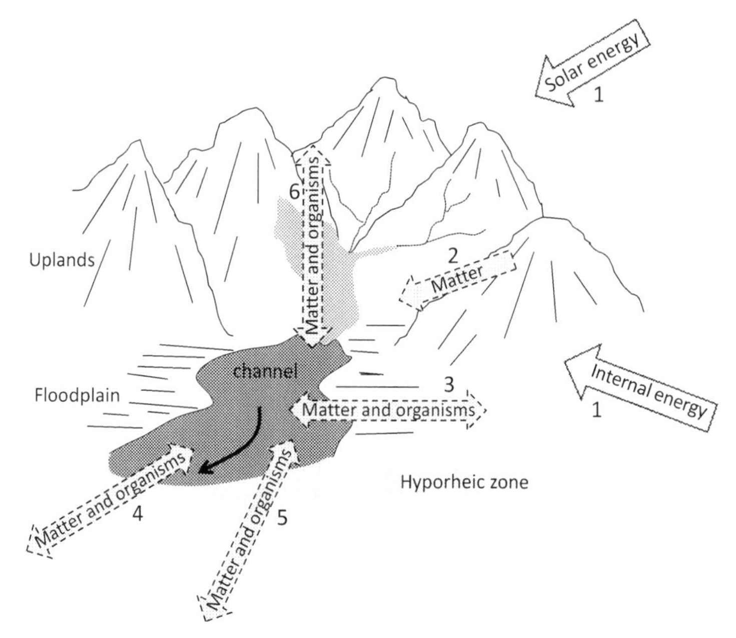

FIGURE 9.1 Schematic illustration of how rivers integrate fluxes within and beyond a drainage basin. Numbers refer to: (1) inputs of solar energy and internal planetary energy from radioactive decay; (2) fluxes of matter (water, mineral sediment, solutes, organic matter) from uplands; (3) fluxes of matter and organisms between the channel and floodplains, or lateral connectivity; (4) downstream fluxes of matter, and upstream and downstream movements of organisms within the channel network, or longitudinal connectivity; (5) fluxes of matter and organisms between water in the channel and subsurface (hyporheic, groundwater) zones, or vertical connectivity; and (6) fluxes of matter (precipitation, aeolian dust, volatiles) and organisms (insects) between the channel and the atmosphere.

which complex interactions involving rock, soil, water, air, and living organisms regulate the natural habitat and determine the availability of life-sustaining resources" (NRC, 2001).

Interactions require some form of connectivity. *Connectivity* can be defined as the transfer of matter or energy between two different landscape compartments (Jain and Tandon, 2010; Wainwright et al., 2011; Fryirs, 2013). Connectivity occurs via physical contact between two compartments, as when sediment eroded from a hillslope directly enters a river channel. Connectivity also occurs via transfer of matter or energy between two physically disconnected compartments, as when aeolian dust eroded from a desert surface is transported long distances and deposited directly on a river floodplain (Wester et al., 2014). *Disconnectivity* is any process or structure limiting physical contact or transfer between compartments (Wester et al., 2014). An alluvial fan can create sediment disconnectivity between a tributary and mainstem river, for example, by storing sediment transported by the tributary for periods more than 10^3 years (Harvey, 2002).

Numerous forms of connectivity that are relevant to rivers have been described. Hydrologic, sediment, and geochemical connectivity refer to transfers of water (Bracken et al., 2013), particulates eroded from rocks (Harvey, 1997; Fryirs et al., 2007), and solutes, respectively (Michel et al., 2000; Anderson et al., 2012). River connectivity describes water-mediated fluxes within a channel network (Ward, 1997; Jaeger and Olden, 2012), such as the seasonal movements of fish between a channel and the nursery habitat of the adjacent floodplain (Junk et al., 1989; Bayley, 1991; Sparks, 1995). Biological connectivity refers to the dispersal of organisms or plant propagules between suitable habitats or between isolated populations for breeding (Erős et al., 2012). Hydrochory (water dispersal of plant propagules), for example, reflects river flow that transports propagules and river-margin complexity that results in propagules being trapped in potential germination sites (Nilsson et al., 1991; Merritt and Wohl, 2002). Landscape connectivity describes fluxes between individual landforms (Brierley et al., 2006). Landscape connectivity of water, for example, reflects processes such as runoff and infiltration that govern downslope movement of water into channels (Tetzlaff et al., 2007). Structural connectivity describes the extent to which landscape units are physically linked to one another, including landscape units <1 m in scale (With and Crist, 1995). As the spatial extent of rills expands after a wildfire, for example, structural connectivity increases between the uplands and the river network (Wester et al., 2014). Functional connectivity describes process-specific interactions between multiple landscape units, such as runoff moving downslope between patches of grass and patches of exposed soil (Kimberly et al., 1997; Wainwright et al., 2011). Any form of connectivity can be characterized with respect to magnitude (rate or volume flux per unit time), duration (time span of connectivity), frequency (number of times connectivity is present within some longer time span), and spatial extent (Wohl, 2014b).

The characteristics of (dis) connectivity govern interactions within the Critical Zone. Rivers are an integral part of Critical-Zone connectivity because rivers both reflect and influence diverse forms of connectivity. The spatial zonation of processes and landforms within drainage basins, for example, *reflects* connectivity of transferring precipitation and sediment inputs from hillslopes to channels and downstream to oceans. Cascade channels – small, steep headwater channel segments that lack bedforms – occur in portions of the river network where channels have limited ability to transport the coarse sediment introduced from adjacent uplands; in other words, portions of the river network with limited longitudinal sediment connectivity. In contrast, lower gradient headwater channels with finer bed sediment develop pool-riffle sequences that reflect greater ability of stream flows to sort sediment into bedforms. River system geometry *influences* transfers of water, sediment and nutrients between channels and floodplains and between surface flow and groundwater. Flood peaks are transmitted efficiently along narrow valley segments with minimal floodplain development, for example, whereas wider valley segments with extensive floodplains attenuate flood peaks.

9.2 HOW RIVERS REFLECT CONNECTIVITY

9.2.1 Spatial Zonation of Drainage Basins and Formation of Channels

River systems exhibit spatial zonation of process and form that has been conceptualized as: three basic zones of a drainage basin (upper basin production, mid-basin transfer, lower basin deposition; Schumm, 1977); geomorphic process domains (spatially identifiable areas of a drainage basin characterized by distinct suites of geomorphic processes; Montgomery, 1999); and River Styles (reaches with consistent planform, channel and floodplain geomorphic units, and bed-material texture; Brierley and Fryirs, 2005). Each conceptualization recognizes that inputs of matter and energy from the atmosphere and from Earth's interior interact with a geologic framework to create physical forms – hillslopes, river channels, valley bottoms – that regulate connectivity and interactions (Fig. 9.1).

At the scale of an entire drainage basin, the formation and evolution of a river channel network reflects connectivity. Uplands can be longitudinally differentiated into: hillslopes governed by diffusive processes of water and sediment movement; unchanneled valley heads in which topographic concavities on slopes begin to concentrate water and sediment movement without crossing the threshold for channel head formation; and channels (Fig. 9.2).

A critical contributing area must be exceeded before surface or subsurface runoff can concentrate sufficiently to erode longitudinally continuous channels (Montgomery and Dietrich, 1988, 1989). The critical area for formation of

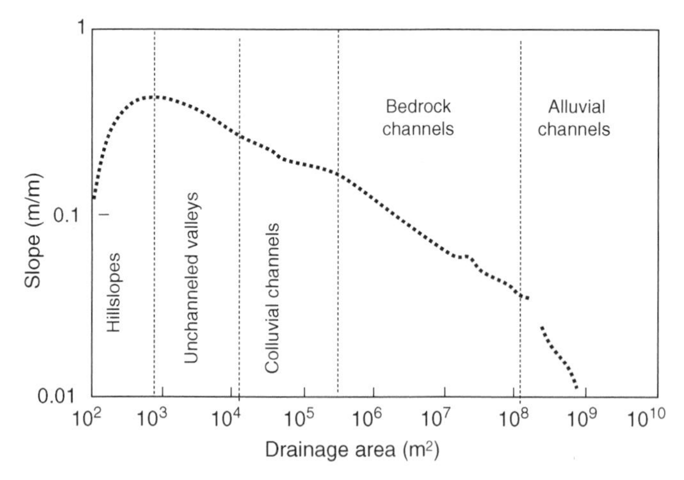

FIGURE 9.2 Plot of log-bin averaged drainage area versus slope for the Olympic Mountains of Washington, USA. The plot shows the mean slope of individual 10 m grid cells for each 0.1 log interval in drainage area. Dashed vertical lines divide the plot into areas that reflect different geomorphic zones within the landscape, or process domains. The sequence of colluvial, bedrock, and alluvial channel segments is idealized and can alternate downstream. For example, the Danube River alternately flows through large, lowland basins with alluvial channel substrate and mountain ranges with bedrock channel substrate. *(After Montgomery (2001, Figure 9.5A).)*

channel heads differs among drainage basins as a function of at least four factors. First, precipitation inputs govern how much water is available. Second, infiltration capacity of the surface influences whether surface runoff or subsurface throughflow occurs in response to precipitation. Here, the hydrologic connectivity of water moving across the surface, between the surface and subsurface, and within the subsurface determines the minimum distance downslope from the drainage divide that is necessary to sufficiently concentrate water, erode sediment, and initiate a channel. Microroughness in the form of large clasts or clumps of vegetation can disrupt surface hydrologic connectivity (Bergkamp, 1998), for example, limit runoff concentration, and prevent formation of channel heads. Third, erosional resistance of the near-surface substrate influences critical area for channel heads. Erosional resistance results from substrate grain size, cohesion, and vegetation. Fourth, the slope of the surface, together with water volume, controls the flow energy available to erode the surface (Moody and Kinner, 2006).

Water and sediment move downslope within channels and the upper portion of a channel network can be predominantly shaped by nonwater flows, such as sediment-laden debris flows. This portion of a channel network is typically designated as colluvial. Channels can be differentiated based on the dominant flow process (colluvial versus alluvial) and the dominant channel substrate (bedrock versus alluvial). Each of these types of channels reflects different forms of connectivity (Fig. 9.3). Colluvial channels are so closely coupled with adjacent

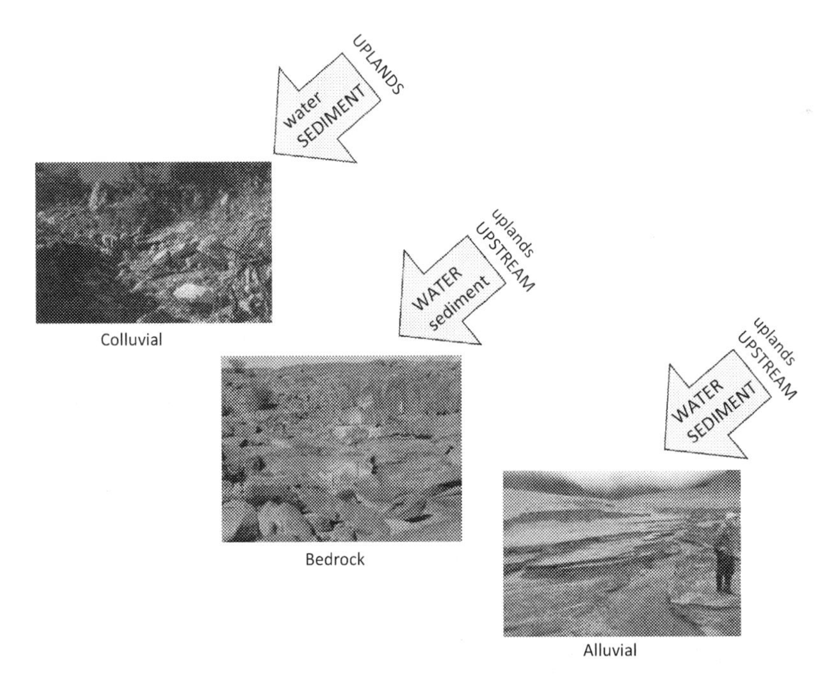

FIGURE 9.3 Schematic illustration of differing levels of connectivity among colluvial, bedrock, and alluvial channels, respectively, and adjacent uplands and upstream portions of the river network. The relative font size reflects the strength of the connectivity.

uplands that they regularly receive large inputs of sediment, as well as water. In alluvial channel segments, hydrologic connectivity to adjacent uplands and upstream portions of the channel network is stronger than upland-sediment connectivity, so that water flows predominantly shape channel geometry. Bedrock channel segments have a transport capacity that exceeds sediment supply, which can also be conceptualized as hydrologic connectivity that exceeds sediment connectivity to uplands and to upstream portions of the channel network. Alluvial channel segments have a sediment supply that exceeds transport capacity, which equates to more balance between hydrologic and sediment connectivity with uplands and upstream channel segments. Bedrock channels typically have less vertical connectivity between surface and hyporheic flow than do alluvial channels.

9.2.2 Reach-Scale Classifications of River Systems

A reach is typically defined as a length of river with consistent valley and channel geometry (downstream gradient, valley-bottom width, channel planform, bedforms, and substrate). On a large river, a reach can be 10^2–10^4 m long; on a smaller river, individual reaches of consistent geometry might be only 10^1–10^2 m long. Because river systems commonly exhibit substantial downstream variability in geometry, and associated water and sediment fluxes, reach-scale classifications are widely used to understand river process and form.

Several, complementary classification systems are used to differentiate aspects of reach-scale valley and channel geometry. Valley segments can be classified based on lateral confinement, defined as the ratio of active channel width to valley-bottom width (Wohl et al., 2012). Lateral confinement typically correlates with downstream gradient: narrow valley bottoms are commonly steep, whereas wider valley bottoms generally have a lower gradient. Channel geometry can be classified based on planform into straight, meandering, braided, and anastomosing (Fig. 9.4a). Channels with downstream gradients exceeding 1% can be classified based on dominant bedform type into cascade, step-pool, plane-bed, pool-riffle, and dune-ripple (Montgomery and Buffington, 1997; Fig. 9.4b). Channels can also be classified based on predominant bed substrate into bedrock, gravel-bed, sand-bed, and fine cohesive, bed channels.

These classifications can be combined to describe a single river reach in a manner that implies not only differences in geometry between reaches, but also differences in process and connectivity. Consider a laterally unconfined valley segment with a meandering, sand-bed, pool-riffle channel, for example, and a laterally confined valley segment with a straight, bedrock, step-pool channel. The bedforms within the sinuous sand-bed channel will respond quickly (at periods of minutes to hours) to changes in water and sediment inputs. Individual floods are likely to change cross-sectional channel geometry and planform geometry. The laterally unconfined valley segment is probably a depositional zone that limits longitudinal connectivity by storing water, sediment, and other

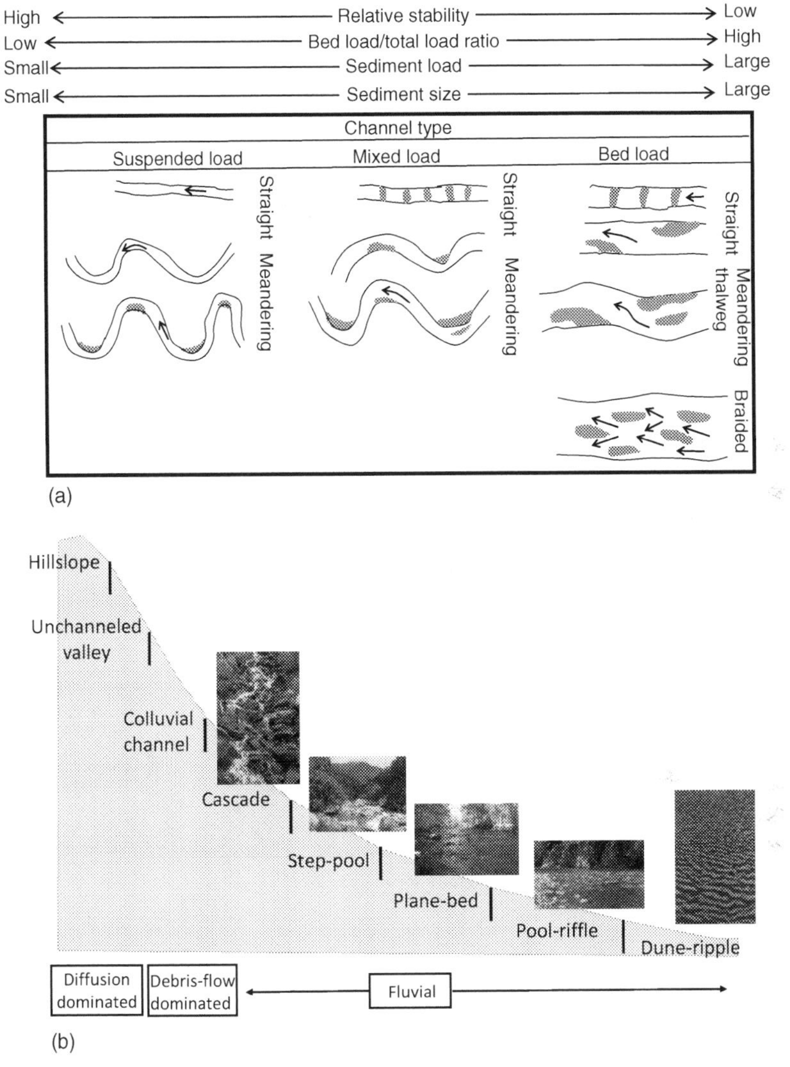

FIGURE 9.4 (a) Simple classification of channel planform into straight (single flow path, sinuosity < 1.5), meandering (single flow path, sinuosity ≥ 1.5) or braided (multiple flow paths). Relative levels of channel stability, the ratio of bed load to total sediment load, the amount of sediment transported, and the grain size of the sediment in transport are shown along the upper horizontal axis. Gray shading within plan view drawing of each channel segment indicates riffles and bars; arrows indicate flow paths. After Schumm (1981). (b) Classification of steep channel segments based on bedforms. The downstream progression shown here reflects the usual pattern of decreasing reach-scale gradient downstream. This progression can be altered if a steep segment occurs downstream from a lower gradient segment. After Montgomery and Buffington (1997).

material in the floodplain. However, this valley segment is likely to be highly connected laterally and vertically because of relatively frequent overbank flows onto the floodplain and hyporheic exchange between the channel and the underlying alluvium. In contrast, the bedrock channel has a much higher erosional threshold, so that bedforms and cross-sectional and planform geometry are much less responsive to changes in water and sediment discharge at the timescale of individual floods. The straight, narrow channel will have high longitudinal connectivity, but minimal lateral connectivity because of the lack of a well-developed floodplain and minimal vertical connectivity because of the lack of porous and permeable substrate to support a hyporheic zone.

9.2.3 Temporal Scales of Connectivity

Biogeochemists coined the phrase *hot moments* to describe a short period of time with disproportionately high reaction rates relative to longer intervening time periods (McClain et al., 2003). Diverse studies indicate that most fluxes and forms of connectivity are not evenly distributed through time and space. Greater than 75% of the long-term sediment flux from mountain rivers in Taiwan occurs during typhoon floods that occupy less than 1% of the time (Kao and Milliman, 2008). About half of the suspended sediment discharged by rivers of the Western Transverse Ranges of California, USA comes from the 10% of the drainage underlain by weakly consolidated bedrock (Warrick and Mertes, 2009). An estimated 17–35% of the total particulate organic carbon flux to the world's oceans comes from high-standing islands in the southwest Pacific, even though these islands constitute only about 3% of Earth's landmass (Lyons et al., 2002).

Disturbance regime is a term sometimes used to describe the temporal nonuniformity of fluxes within a river system. Ecologists describe a *disturbance* as "any relatively discrete event in time that disrupts ecosystem, community, or population structure and changes resources, substrate availability, or the physical environment" (White and Pickett, 1985, p. 7). A flood or debris flow is an obvious example of a disturbance that affects river systems. *Disturbance regime* "describes the spatial pattern and statistical distribution of events in terms of frequency, magnitude, and duration of associated changes in the physical environment" (Montgomery, 1999, p. 402). At the scale of a river reach, the disturbance regime results from both the sources of water, sediment, and other materials that originate outside the river network – for example, precipitation inputs and sediment inputs from adjacent uplands – and the position of the river reach within the drainage basin. A semiarid mountainous river network of the Colorado Front Range, USA, provides an example of how disturbance regime varies downstream in response to elevational changes in hydroclimatology, wildfires, slope stability, and topography (Fig. 9.5). The disturbance regime characterizes both the type of disturbance affecting a river reach and the magnitude and frequency at which disturbances occur.

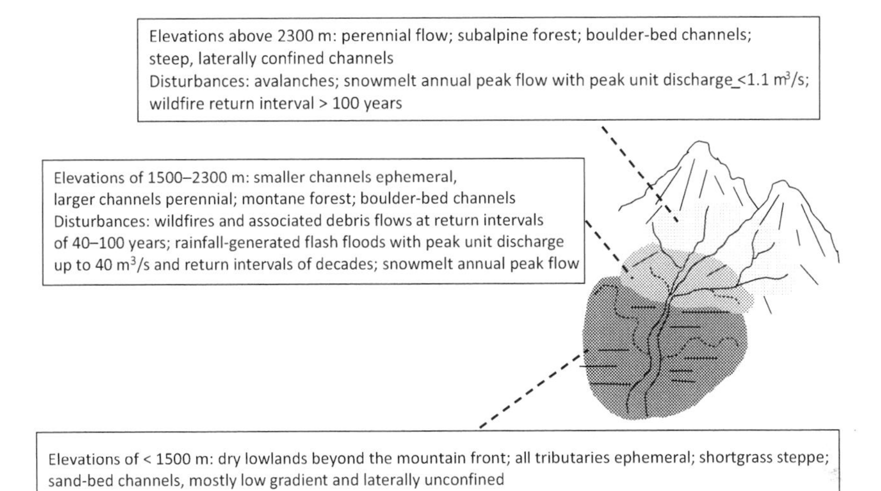

FIGURE 9.5 Schematic illustration of downstream variations in disturbance regime within a river network in the Colorado Front Range, USA. *(Peak unit discharge values from Jarrett (1990) and fire return intervals from Veblen and Donnegan (2005).)*

9.2.4 River Adjustments to Changes in Connectivity

A river system can exhibit various forms of stability or equilibrium. *Equilibrium* typically refers to a condition of no net change and is thus highly dependent on the spatial and temporal scale being considered (Schumm and Lichty, 1965; Mayer, 1992). A large flood can cause extensive bank erosion and channel widening, for example, but the river can again narrow as riparian vegetation encroaches during subsequent, smaller flows, so that channel width does not change when averaged over longer time intervals (Friedman and Lee, 2002). Geomorphologists distinguish static equilibrium, which is typically present only over short time and small spatial scales, as a situation in which no change occurs in the parameter of interest. Fluctuations are more likely to occur over longer time periods, although mean conditions may not change, creating steady-state equilibrium. Dynamic equilibrium describes fluctuations about a progressively changing mean condition and dynamic metastable equilibrium describes a scenario in which progressive change occurs unevenly through time (Fig. 9.6).

Uneven rates of change can reflect the existence of a *threshold*, or abrupt change in river process or form. External thresholds are crossed as a result of changes in variables external to the river system, such as water or sediment inputs or base level. Internal thresholds can be crossed in the absence of changes in external variables (Schumm, 1979), as in the case of ephemeral channels in arid regions, which typically alternate between incision and aggradation even in the absence of changes in water and sediment inputs (Patton and Schumm, 1975).

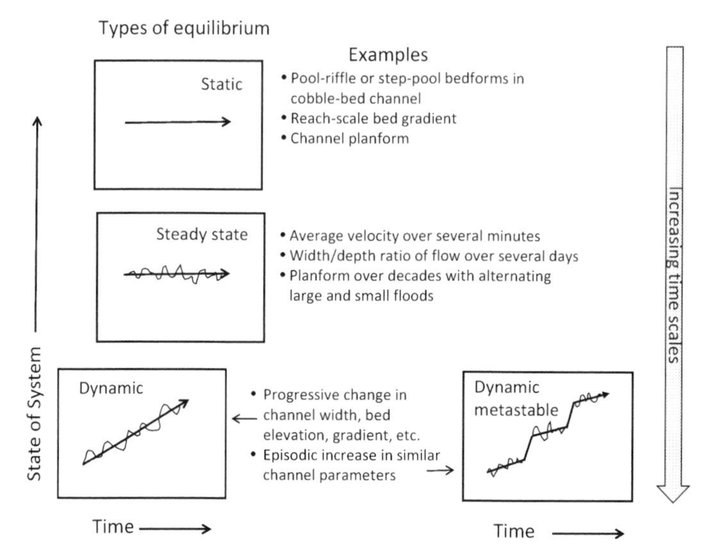

FIGURE 9.6 Different forms of equilibrium illustrated via plots of channel condition through time. Over the shortest periods and smallest areas, variables such as channel planform or reach-scale gradient are static and unchanging. Over progressively longer periods and larger areas, these variables can fluctuate about a steady-state mean or a progressively changing mean. *(After Wohl (2014b, Figure 1.3).)*

Equilibrium implies that process and/or form in a river system will change in response to changes in water or sediment inputs or base level. The interactions among variables that create these changes can take the form of *self-enhancing feedbacks* that promote continuing change, as when a stationary piece of wood in a stream traps other wood pieces in transport, creating a logjam that then traps more wood (Wohl and Beckman, 2014). Interactions can also occur as *self-arresting feedbacks* that limit change, as when a tributary fan laterally constricts the main channel, causing a flow acceleration that enhances transport capacity and erodes the lateral constriction (Kieffer, 1989).

Sometimes the interactions among variables can cause a *complex response* in which a river system undergoes multiple, asynchronous changes after a single external perturbation (Schumm, 1973; Schumm and Parker, 1973). Channelization involves straightening and sometimes widening and deepening a channel in order to increase conveyance and limit overbank flooding. By concentrating flow within the channel and increasing the gradient, channelization typically causes a channel to erode, starting at the downstream end of the channelized reach. Erosion occurs in multiple stages that exemplify complex response: for a cross section at the downstream end of the channelized reach, the channel bed first cuts down, then the channel banks erode and the channel widens, followed by aggradation as sediment enters the cross section from actively eroding reaches upstream, another round of erosion as the upstream reaches stabilize, and, eventually, stability of the cross section. This sequence of adjustments,

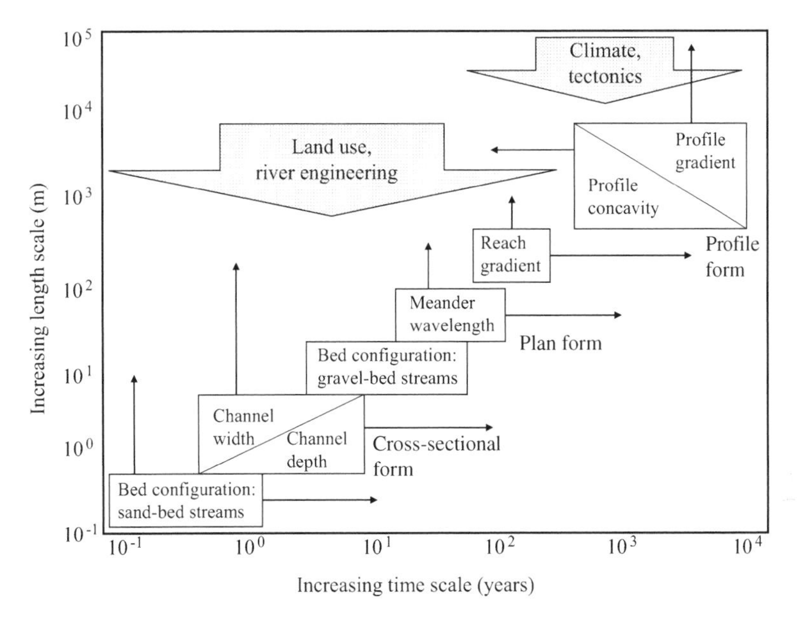

FIGURE 9.7 Illustration of the different forms and temporal and spatial scales of river adjustment. *(After Knighton (1998, Figure 5.3) and Wohl (2014b, Figure 5.18).)*

described in *channel evolution models*, can require more than a hundred years to complete (Simon and Castro, 2003; Simon and Rinaldi, 2006). *Lag times* between the change in an external variable and the response of the river can be very short in small channels with readily deformable boundaries, such as sand beds, or can extend over millions of years for large rivers formed in bedrock (Fig. 9.7).

For the most part, adjustments in process and form within a river system occur in response to changes in the primary external variables of water and sediment inputs and base level. The degree to which changes in external variables are transmitted to river systems reflects various forms of landscape connectivity. Increased sediment yields from hillslopes as a consequence of land cover change may be directly and quickly transmitted to headwater channels that are closely coupled to adjacent uplands, for example, whereas the lowermost portion of the river network could be buffered from increased sediment yields because sediment is stored in floodplains (Starkel, 1988).

9.2.5 Rivers as Paleoenvironmental Indicators

Depositional landforms associated with river systems reflect connectivity and changes in connectivity through time. Floodplains accumulate and retain sediment in a manner that reflects upstream supplies of sediment relative to the ability of the river to transport sediment. If sediment inputs increase or river transport capacity declines, the volume or average grain size of floodplain sediment can increase. This change will be reflected in floodplain stratigraphy. Large accumulations of floodplain sediment can also cause the floodplain to rise above an elevation regularly inundated by peak flows. Alternatively, if sediment inputs decrease or river transport capacity increases, the river can cut down into the floodplain until peak flows no longer spread across the floodplain. In either scenario, the floodplain becomes a relict feature known as a terrace and the stratigraphy of the terrace can be used to infer past sediment inputs and levels of connectivity (Pazzaglia, 2013).

Floodplains and terraces record both sustained changes in connectivity associated with climatic changes, tectonic uplift, or land use, and episodic changes caused by discrete floods. Hereford (2002) documented widespread floodplain aggradation in the southwestern United States associated with decreased occurrence of large floods during the Little Ice Age (circa 1400–1880 AD). Nanson (1986) described sustained accumulation of floodplain sediment during moderate floods along rivers in southeastern Australia, followed by wide scale floodplain erosion during infrequent, large floods.

Alluvial fans and deltas develop at locations where river transport capacity declines, either because of a change in lateral valley confinement (alluvial fan) or mixing of river flow with a body of standing water (delta). The location of active deposition continually shifts across alluvial fans and deltas, leaving a complex stratigraphic record of sediment volume and grain-size distribution within the river system through time. Pierce and Meyer (2008), for example, used alluvial fan deposits to infer changes in fire-related debris flows over the past 2000 years in the western United States.

9.3 HOW RIVERS INFLUENCE CONNECTIVITY

Rivers influence connectivity within drainage basins and at larger spatial scales by influencing fluxes of material. A nice example comes from work on the Indus River in Pakistan, where Burbank et al. (1996) demonstrated a feedback between river incision and adjacent hillslopes over tens of thousands of years. Tectonic uplift creates relative base-level fall and this triggers incision of the Indus River into bedrock. River incision locally over-steepens adjacent hillslopes, which then become unstable and introduce large volumes of sediment to the river via landslides and debris flows. Increased sediment inputs accumulate on the valley floor, limiting channel incision until the river has transported the accumulated sediment downstream and again exposed bedrock in the channel bed. The rate of river incision thus influences sediment inputs to the channel network and the rate of river sediment transport influences sediment flux to the ocean.

Another example comes from the influence of rivers on the global carbon cycle. Current estimates are that nearly half of the terrestrially derived organic carbon entering inland freshwaters is released to the atmosphere (Aufdenkampe et al., 2011) (Fig. 9.8). Headwater streams tend to be net sources of CO_2 to the atmosphere, but large river floodplains that are seasonally inundated are also

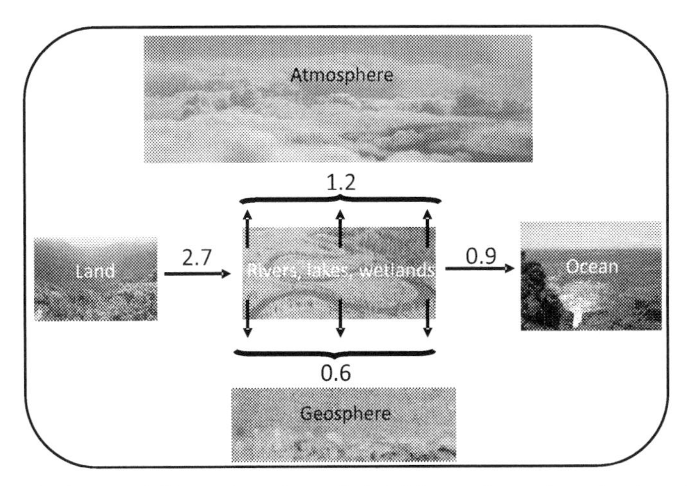

FIGURE 9.8 Schematic illustration of annual global fluxes of organic carbon in petagrams. *(After Aufdenkampe et al. (2011, Figure 9.3).)*

substantial sources of CO_2 emissions (Cole et al., 2007). CO_2 evades to the atmosphere across the stream surface, so the extent of surface water, water temperature and chemistry, and biological activity within the channel, hyporheic zone, and floodplain – all of which are influenced by river process and form – interact with terrestrial carbon inputs to govern CO_2 evasion from rivers. Factors that promote retention of dissolved and particulate organic carbon along rivers can promote biological uptake of carbon by plants, microbes, and aquatic macroinvertebrates at timescales of minutes to days (Battin et al., 2008, 2009). Such factors include physically complex channel boundaries associated with logjams (Beckman and Wohl, 2014), beaver dams (Naiman et al., 1986), obstructions or bedforms such as pools and riffles that drive hyporheic exchange (Kasahara and Wondzell, 2003; Fanelli and Lautz, 2008), and channel-floodplain connectivity that promotes overbank flow and floodplain deposition (Wohl et al., 2012). Retention of carbon in river systems over longer time spans can result in carbon storage in the geosphere in the form of channel, floodplain, and delta sediments (Walter and Merritts, 2008; Hoffmann et al., 2009; Cierjacks et al., 2010). In addition to physical complexity, carbon retention correlates with river discharge (Hilton et al., 2008, 2011; West et al., 2011). In other words, river process and form strongly influence the rate at which organic carbon moves through a channel network – carbon connectivity – and thus partitioning of carbon among the atmosphere, geosphere, and oceans.

9.4 HUMAN ALTERATIONS OF RIVER CONNECTIVITY

River process and form and the characteristics of river (dis)connectivity strongly influence Critical-Zone interactions across a much broader spatial context than individual drainage basins. Rivers also strongly influence habitat and resources for diverse organisms that may never actually inhabit a river network. Wood exported from rivers to the oceans, for example, substantially impacts the biogeochemistry of marine environments (Holmes et al., 2002; Bourgeois et al., 2011) and is essential to marine ecosystems on the shore, in the nearshore, in the open sea, and on the deep ocean floor (Maser and Sedell, 1994). Driftwood along coastlines contributes moisture and nutrients that enhance growth of coastal plants, traps sediment, and limits coastal erosion. In the nearshore environment, wood can remain intact for many years (Gonor et al., 1988), despite the ability of wood-boring crustaceans and bivalve mollusks to rapidly attack and disperse wood before the microbial decomposition of its constituents. Floating trees in the ocean disperse attached organisms and attract pelagic fish, and communities of organisms form food webs in association with drift logs (Gonor et al., 1988). Wood sinking to the ocean floor provides an energy base and increases the productivity of benthic communities. A single sunken log serves as a focus of abundant deep-sea life for many decades (Turner, 1977). Wood is more common on the deep-sea floor off the mouths of rivers and wooded coastlines, where it is associated with a large number and high diversity of deep-sea animals (Köhler et al., 2009).

The example of wood export from rivers to oceans is also very useful for illustrating how human alterations of river process and form have altered connectivity and interactions across the Critical Zone. For the past few hundred years, human activities have directly and indirectly removed wood from river networks (Montgomery et al., 2003; Wohl, 2014a). Naturally recruited wood continues to be directly removed from channels for flood control, navigation, reducing potential damage to infrastructure along rivers, and esthetic reasons. Removal or alteration of riparian and upland forests has indirectly reduced wood recruitment to rivers. Channel engineering such as dredging, channelization, and bank stabilization creates more uniform channels that limit wood retention within the channel. Flow regulation and levee construction reduce channel-floodplain connectivity and limit deposition of river-transported wood on floodplains. The net effect of human alterations of drainage basins has been to reduce wood recruitment to rivers, wood storage within channels and floodplains, and wood export to oceans.

Some forms of environmental alteration increase river connectivity. Examples include flow diversions from one drainage basin to another and removal of naturally occurring obstructions within channels, such as logjams and beaver dams. Reduced river connectivity, however, is by far the most substantial effect of human alterations (Table 9.1) (Kondolf et al., 2006). This occurs primarily via dams and diversions.

Dams typically physically block any upstream or downstream movement along a river system. Although water is eventually released downstream from dams, the downstream hydrograph is commonly severely altered relative to natural flow patterns on the river. Alteration of the hydrograph in turn alters downstream patterns of solute, sediment, and organism movement. The longitudinal

TABLE 9.1 Examples of Human Manipulations of Landscapes that Alter River Connectivity

Manipulation	Effects on river connectivity	Sample references
Within the river corridor (channels and floodplain)		
Flow regulation (dams, diversions)	Changes the magnitude, duration, frequency, seasonality, and rate of rise and fall of peak flows, as well as changing other components of hydrograph. These changes alter movement of solutes, sediment, and organisms downstream, upstream (organisms), between the channel and floodplain, and between the channel and the subsurface.	Graf (2001); Vörösmarty et al. (2004); Nilsson et al. (2005); Poff et al. (2007); Walter and Merritts (2008)
Levee construction	Reduces lateral connectivity by limiting channel–floodplain fluxes of water, solutes, sediment, and organisms. Reduces vertical connectivity by limiting infiltration of overbank flow.	Sparks (1995); Blanton and Marcus (2009, 2013)
Floodplain drainage	Reduces vertical connectivity by limiting infiltration from floodplain. Reduces lateral connectivity by limiting fluxes of water, solutes, sediment, and organisms from floodplain to channel.	Krause et al. (2011); Andrews et al. (2014)
Groundwater withdrawal	Reduces vertical connectivity by limiting fluxes of water and solutes between the channel and subsurface. Commonly causes channel incision that can increase longitudinal connectivity and reduce lateral connectivity.	Falke et al. (2010)
Channelization	Commonly involves straightening channels, enlarging channel cross-sectional area, and making cross-sectional geometry more uniform. Increases longitudinal connectivity, but can reduce lateral connectivity by reducing overbank flows and can reduce vertical connectivity by removing bedforms and variation in bed substrate that can promote hyporheic exchange.	Rhoads et al. (2003); Hohensinner et al. (2004)
Bank stabilization	Typically creates more regular, uniform banks, which can increase longitudinal connectivity.	Habersack and Piégay (2008)

(Continued)

TABLE 9.1 Examples of Human Manipulations of Landscapes that Alter River Connectivity *(cont.)*

Manipulation	Effects on river connectivity	Sample references
Wood removal	Removal of individual pieces of wood and logjams within channels increases longitudinal connectivity, but can reduce lateral connectivity by increasing channel conveyance and limiting overbank flow and bank erosion. Wood removal can also reduce vertical connectivity by decreasing hyporheic exchange.	Montgomery et al. (2003); Wohl (2014a)
Beaver trapping	Removal of beaver dams can increase longitudinal connectivity, but reduces lateral and vertical connectivity.	Naiman et al. (1988); Pollock et al. (2014)
Outside the river corridor		
Changes in land cover (deforestation, agriculture, urbanization)	By changing hydrologic processes such as interception, evaporation, transpiration, infiltration, and runoff, changes in land cover alter the movement of water, solutes, and sediment from uplands into channel networks. Altered vegetation cover typically equates to greater water and sediment fluxes into channels, whereas urbanization typically causes increased water flux, reduced sediment flux, and changes in the type of solutes entering rivers (e.g., synthetic chemicals rather than nutrients).	Syvitski et al. (2005); Scanlon et al. (2007)
Topographic engineering	Alteration of topography for mining, construction of transportation corridors, or urbanization, can substantially alter upland stability and surface cover, and hence fluxes of water, solutes, and sediment into river networks.	Hooke (2000); Palmer et al. (2010); Hooke et al. (2012)
Climate change	Effects vary by region, but fluxes of water, solute, and sediments from uplands into river networks, as well as longitudinal, lateral, and vertical connectivity within river networks, are likely to be altered.	Hauer et al. (1997); Goode et al. (2012); Knight and Harrison (2013)

movement of sediment and organisms is completely blocked by most dams. Dams indirectly reduce lateral and vertical connectivity by changing the abundance of flow: severely reduced flood peaks downstream from a dam, for example, may eliminate channel-floodplain connectivity because flow is not high enough to overtop the channel banks and inundate the floodplain.

Diversions may involve a physical structure built across the channel that directly limits longitudinal connectivity, or diversions may draw water into a ditch or pipe without an obstruction across the main channel. In either case, alteration of flow downstream from the diversion affects longitudinal, lateral, and vertical connectivity of not only water, but also solutes, sediment, and organisms within the river system (Caskey et al., in press).

Other human alterations that reduce river connectivity include artificial levees, which limit or eliminate channel-floodplain connectivity, and removal of naturally occurring obstructions such as logjams or beaver dams that create local backwaters and facilitate overbank flows and channel-floodplain connectivity. Although the effect of any single logjam or beaver dam typically extends only a few channel widths upstream, the cumulative effect of removing millions of these structures from river systems is undoubtedly substantial (Naiman et al., 1986; Wohl, 2014a).

Numerous case studies over the past several decades demonstrate that human alterations of connectivity can substantially alter river process and form within a limited spatial area or single drainage basin (e.g., Gregory, 2006; James and Marcus, 2006). Since circa 2000, an increasing number of papers also emphasize the global-scale, cumulative effects of human alterations of connectivity. People have transformed the global water system by constructing millions of dams, causing a doubling to tripling of the residence time of continental runoff in rivers and a 600–700% increase in water stored in channels (Vörösmarty et al., 2004). Anthropogenic nitrogen sources are greater than natural sources in Asia, North America, and Europe (Boyer et al., 2006) and human activities increasingly dominate the global nitrogen budget (Galloway et al., 2004). Human-induced soil erosion has increased sediment transport by rivers, but retention within reservoirs has reduced the flux of sediment reaching the world's coasts by 20% (Syvitski et al., 2005). People have modified about half of Earth's land surface (Hooke et al., 2012) and an estimated 83% of Earth's ice-free land area is directly influenced by humans (Sanderson et al., 2002). We now move more sediment than any natural process (Hooke, 2000). As the effects of human-induced climate change become more evident and well documented, it can be argued that no river system is beyond the influence of at least indirect human alteration (Goudie, 2006; Wohl, 2013b).

9.5 RIVER RESTORATION

People have been directly and indirectly altering rivers for millennia (Wohl, 2000). Crops and grazing typically alter water and sediment yields to rivers and these alterations date to the early Neolithic (circa 7000 years ago) in regions as diverse as Poland (Starkel, 1988) and China (Mei-e and Xianmo, 1994). Floodplain wetlands have been drained since at least the sixteenth century in Europe (Girel, 1994). Direct river engineering has an equally long history: the earliest known dam was built circa 2800 BC in Egypt (Smith, 1971). Rivers in Europe have been channelized since about 1750 (Petts, 1989) and levee construction dates back 3500 years in China (Clark, 1982).

River restoration is another form of indirect and direct river engineering, undertaken to improve hydrologic, geomorphic, and/or ecological processes within a river system (Wohl et al., 2005). River restoration covers a wide variety of activities, including direct modifications of channels and floodplains, and indirect modifications of water, solute, and sediment inputs to river systems (Bennett et al., 2011). Bernhardt and Palmer (2011) distinguish restoration undertaken to restore connectivity by removing or retrofitting infrastructure such as dams or levees and restoration designed to reconfigure river systems through channel engineering or planting vegetation. River restoration also occurs at a variety of spatial scales, from a limited length of channel (Bernhardt et al., 2005) to an entire watershed (Warne et al., 2000; Bloesch and Sieber, 2003).

Although some examples of river restoration are designed to limit connectivity in order to prevent dispersal of pollutants or nonnative organisms (Wohl, 2013a), the majority of restoration activities are undertaken to restore connectivity (Kondolf et al., 2006), typically by reducing the effects of levees, diversions, and dams. Experimental flow releases such as those on the Colorado River below Glen Canyon Dam in Arizona, USA (Melis, 2011) can be used to enhance river connectivity altered by dams. The most complete restoration of connectivity altered by a dam involves completely removing the dam. As of 2014, an estimated 1150 dams have been removed in the United States, mostly during the preceding 20 years (Lovett, 2014). Many of these are smaller dams <5-m tall, but in the past few years Glines Canyon Dam (64-m tall), Condit Dam (38-m tall), Marmot Dam (14-m tall), and other tall dams in the western United States have also been removed. Among the primary objectives in removing dams is to restore longitudinal connectivity for various species of fish (Raabe and Hightower, 2014).

Removing a dam can immediately restore lost hydrologic connectivity, although restoration of sediment connectivity can require a longer adjustment period as sediment stored upstream from the dam is mobilized and dispersed downstream. The large amounts of sediment released by dam removal can endanger downstream aquatic organisms or result in aggradation and river planform change, although progressive notching of a dam, rather than instantaneous removal, can be used to control the rate of sediment release (MacBroom and Schiff, 2013). More problematic is a situation in which the sediment contains contaminants (Evans and Gottgens, 2007; Woelfle-Erskine et al., 2012), such as the 1970s removal of a dam on the Hudson River in New York that resulted in downstream dispersal of PCBs (Lovett, 2014).

The cumulative effects of dam removal in the United States have largely been positive, with measurable increases in the populations of fish and other river organisms (Hogg et al., 2013). At the same time that dams are being removed or operated to allow environmental flow releases in the United States and Europe, substantial numbers of very large dams are being built in Asia, Africa, and Latin America (Wohl, 2011).

Even where a dam is not removed, other options exist to restore connectivity along regulated rivers. Chief among these are *environmental flows* (Tharme, 2003), or flow releases designed to mimic ecologically important aspects of the natural flow regime (Poff et al., 1997), including the magnitude, frequency, duration, timing, and rate of change of discharge. Environmental flows are commonly tied explicitly to restoring longitudinal, lateral, and vertical connectivity within river systems (Bunn and Arthington, 2002) and typically involve experimental flood releases (Shafroth et al., 2009; Konrad et al., 2011). Connectivity can also be restored by structural manipulation of the river corridor, such as setting back artificial levees in order to connect at least some of the historical floodplain to the channel (e.g., Europe's Room for the River program http://www.ruimtevoorderivier.nl/room-for-the-river-programme).

Manipulating connectivity may also be a component of river restoration aimed primarily at maintaining or improving water quality. In the United States, water quality standards mandated by the US Environmental Protection Agency and various states drive such restoration. In the European Union, the Water Framework Directive adopted by the European Parliament in 2000 drives water quality restoration. Many aspects of water quality are closely tied to longitudinal connectivity (e.g., minimum flows to maintain water temperature or dissolved oxygen; Bernhardt et al., 2005), lateral connectivity (e.g., suspended sediment deposition on floodplains; Gumiero et al., 2013), and vertical connectivity (e.g., hyporheic exchange that promotes denitrification; Lautz and Siegel, 2007). Consequently, river restoration commonly includes an explicit objective of influencing diverse aspects of river connectivity.

9.6 RIVERS IN THE CRITICAL ZONE

River flow constitutes much less than 1% of freshwater on Earth (Berner and Berner, 1987). River channels and floodplains also cover a relatively small portion of Earth's surface. Rivers, however, are disproportionately important, relative to their water fluxes and surface area, for landscape evolution, for fluxes of water, sediment, solutes, organic matter, nutrients, and contaminants, and for biota, including humans. In the context of landscape evolution, the rate at which rivers cut down as base level falls or accumulate sediment as base level rises, and the spatial extent over which these river adjustments occur, govern sediment production and movement from adjacent uplands (Whipple, 2004). With respect to fluxes, humans move more sediment than natural processes such as hillslope failures, glaciers, and rivers (Hooke, 2000) and our activities dominate the introduction of nitrogen into the environment over much of the planet (Boyer et al., 2006). For diverse plant and animal species and human communities, river systems provide vital habitat, dispersal corridors, water supplies, navigation, power generation, and waste disposal. River systems are the preeminent conduits of connectivity across diverse spatial and temporal scales of the Critical Zone.

REFERENCES

Anderson, S.P., Anderson, R.S., Tucker, G.E., 2012. Landscape scale linkages in critical zone evolution. C. R. Geosci. 344, 586–596.

Andrews, C.S., Miranda, L.E., Goetz, D.B., Kroger, R., 2014. Spatial patterns of lacustrine fish assemblages in a catchment of the Mississippi alluvial valley. Aquat. Conserv. 24, 634–644.

Aufdenkampe, A.K., Mayorga, E., Raymond, P.A., Melack, J.M., Doney, S.C., Alin, S.R., Aalto, R.E., Yoo, K., 2011. Riverine coupling of biogeochemical cycles between land, oceans, and atmosphere. Front. Ecol. Environ. 9, 53–60.

Battin, T.J., Kaplan, L.A., Findlay, S., Hopkinson, C.S., Marti, E., Packman, A.I., Newbold, J.D., Sabater, F., 2008. Biophysical controls on organic carbon fluxes in fluvial networks. Nat. Geosci. 1, 95–100.

Battin, T.J., Luyssaert, S., Kaplan, L.A., Aufdenkampe, A.K., Richter, A., Tranvik, L.J., 2009. The boundless carbon cycle. Nat. Geosci. 2, 598–600.

Bayley, P.B., 1991. The flood pulse advantage and the restoration of river-floodplain systems. Regul. River. 6, 75–86.

Beckman, N.D., Wohl, E., 2014. Carbon storage in mountainous headwater streams: the role of old-growth forest and logjams. Water Resour. Res. 50, 2376–2393.

Bennett, S.J., Simon, A., Castro, J.M., Atkinson, J.F., Bronner, C.E., Blersch, S.S., Rabideau, A.J., 2011. The evolving science of stream restoration. In: Simon, A., et al. (Eds.), Stream Restoration in Dynamic Fluvial Systems: Scientific Approaches, Analyses, and Tools. AGU, Washington, DC, pp. 1–8. Geophys. Monogr. Ser. 194.

Bergkamp, G., 1998. A hierarchical view of the interactions of runoff and infiltration with vegetation and microtopography in semiarid shrublands. Catena 33, 201–220.

Berner, E.K., Berner, R.A., 1987. The Global Water Cycle: Geochemistry and Environment. Prentice Hall, Englewood Cliffs, New Jersey.

Bernhardt, E.S., Palmer, M.A., 2011. River restoration – the fuzzy logic of repairing reaches to reverse watershed scale degradation. Ecol. Appl. 21, 1926–1931.

Bernhardt, E.S., Palmer, M.A., Allan, J.D., Alexander, G., Barnas, K., Brooks, S., Carr, J., Clayton, S., Dahm, C., Follstad-Shah, J., Galat, D., Gloss, S., Goodwin, P., Hart, D., Hassett, B., Jenkinson, R., Katz, S., Kondolf, G.M., Lake, P.S., Lave, R., Meyer, J.L., O'Donnell, T.K., Pagano, L., Powell, B., Sudduth, E., 2005. Synthesizing U.S. river restoration efforts. Science 308, 636–637.

Blanton, P., Marcus, W.A., 2009. Railroads, roads and lateral disconnection in the river landscapes of the continental United States. Geomorphology 112, 212–227.

Blanton, P., Marcus, W.A., 2013. Transportation infrastructure, river confinement, and impacts on floodplain and channel habitat, Yakima and Chehalis Rivers, Washington, USA. Geomorphology 189, 55–65.

Bloesch, J., Sieber, U., 2003. The morphological destruction and subsequent restoration programmes of large rivers in Europe. Arch. Hydrobiol. 14 (Suppl. 147/3), 363–385.

Bourgeois, S., Pruski, A.M., Sun, M.-Y., Buscail, R., Lantoine, F., et al. 2011. Distribution and lability of land-derived organic matter in the surface sediments of the Rhone prodelta and the adjacent shelf (Mediterranean Sea, France): a multi proxy study. Biogeosciences 8, 3107–3125.

Boyer, E.W., Howarth, R.W., Galloway, J.N., Dentener, F.J., Green, P.A., Vörösmarty, C.J., 2006. Riverine nitrogen export from the continents to the coasts. Global Biogeochem. Cycles 20, GB1S91.

Bracken, L.J., Wainwright, J., Ali, G.A., Tetzlaff, D., Smith, M.W., Reaney, S.M., Roy, A.G., 2013. Concepts of hydrological connectivity: research approaches, pathways and future agendas. Earth Sci. Rev. 119, 17–34.

Brierley, G.J., Fryirs, K.A., 2005. Geomorphology and River Management: Applications of the River Styles Framework. Blackwell Publishing, Oxford, UK.

Brierley, G.J., Fryirs, K., Jain, V., 2006. Landscape connectivity: the geographic basis of geomorphic applications. Area 38, 165–174.

Bunn, S.E., Arthington, A.H., 2002. Basic principles and ecological consequences of altered flow regimes for aquatic biodiversity. Environ. Manag. 30, 492–507.

Burbank, D.W., Leland, J., Fielding, E., Anderson, R.S., Brozovic, N., Reid, M.R., Duncan, C., 1996. Bedrock incision, rock uplift and threshold hillslopes in the northwestern Himalayas. Nature 379, 505–510.

Caskey, S.T., Blaschak, T.S., Wohl, E., Schnackenberg, E., Merritt, D.M., Dwire, K.A., 2015. Downstream effects of stream flow diversion on channel characteristics and riparian vegetation in the Colorado Rocky Mountains, USA. Earth Surf. Process. Land. 40, 586–598.

Cierjacks, A., Kleinschmit, B., Babinsky, M., Kleinschroth, F., Markert, A., Menzel, M., Ziechmann, U., Schiller, T., Graf, M., Lang, F., 2010. Carbon stocks of soil and vegetation on Danubian floodplains. J. Plant Nutr. Soil Sci. 173, 644–653.

Clark, C., 1982. Flood. Time-Life Books, Alexandria, VA.

Cole, J.J., Prairie, Y.T., Caraco, N.F., McDowell, W.H., Tranvik, L.J., Striegl, R.G., Duarte, C.M., Kortelainen, P., Downing, J.A., Middelburg, J.J., Melack, J., 2007. Plumbing the global carbon cycle: integrating inland waters into the terrestrial carbon budget. Ecosystems 10, 171–184.

Erős, T., Olden, J.D., Schick, R.S., Schmera, D., Fortin, M.J., 2012. Characterizing connectivity relationships in freshwaters using patch-based graphs. Landscape Ecol. 27, 303–317.

Evans, J.E., Gottgens, J.F., 2007. Contaminant stratigraphy of the Ballville Reservoir, Sandusky River, NW Ohio: implications for dam removal. J. Great Lakes Res. 33, 182–193.

Falke, J.A., Bestgen, K.R., Fausch, K.D., 2010. Streamflow reductions and habitat drying affect growth, survival, and recruitment of brassy minnow across a Great Plains riverscape. Trans. Am. Fish. Soc. 139, 1566–1583.

Fanelli, R.M., Lautz, L.K., 2008. Patterns of water, heat, and solute flux through streambeds around small dams. Ground Water 46, 671–687.

Friedman, J.M., Lee, V.J., 2002. Extreme floods, channel change, and riparian forests along ephemeral streams. Ecol. Monogr. 72, 409–425.

Fryirs, K.A., 2013. (Dis)connectivity in catchment sediment cascades: a fresh look at the sediment delivery problem. Earth Surf. Process. Land. 38, 30–46.

Fryirs, K.A., Brierley, G.J., Preston, N.J., Spencer, J., 2007. Catchment scale (dis)connectivity in sediment flux in the upper Hunter catchment, New South Wales, Australia. Geomorphology 84, 297–316.

Galloway, J.N., Dentener, F.J., Capone, D.G., Boyer, E.W., Howarth, R.W., Seitzinger, S.P., Asner, G.P., Cleveland, C.C., Green, P.A., Holland, E.A., Karl, D.M., Michaels, A.F., Porter, J.H.,

Townsend, A.R., Vörösmarty, C.J., 2004. Nitrogen cycles: past, present, and future. Biogeochemistry 70, 153–226.
Girel, J., 1994. Old distribution procedure of both water and matter fluxes in floodplain of western Europe: impact on present vegetation. Environ. Manag. 18, 203–221.
Gonor, J.J., Sedell, J.R., Benner, P.A., 1988. What we know about large trees in estuaries, in the sea, and on coastal beaches. In: Maser, C., Tarrant, R.F., Trappe, J.M., Franklin, J.F. (Eds.), From the Forest to the Sea: A Story of Fallen Trees. USDA Forest Service General Technical Report PNW-GTR-229, Portland, Oregon, pp. 83–112.
Goode, J.R., Luce, C.H., Buffington, J.M., 2012. Enhanced sediment delivery in a changing climate in semi-arid mountain basins: implications for water resource management and aquatic habitat in the northern Rocky Mountains. Geomorphology 139-140, 1–15.
Gooseff, M.N., 2010. Defining hyporheic zones – advancing our conceptual understanding and operational definitions of where stream water and groundwater meet. Geogr. Compass 4, 945–955.
Goudie, A.S., 2006. Global warming and fluvial geomorphology. Geomorphology 79, 384–394.
Graf, W.L., 2001. Damage control: restoring the physical integrity of America's rivers. Ann. Assoc. Am. Geogr. 91, 1–27.
Gregory, K.J., 2006. The human role in changing river systems. Geomorphology 79, 172–191.
Gumiero, B., Mant, J., Hein, T., Elso, J., Boz, B., 2013. Linking the restoration of rivers and riparian zones/wetlands in Europe: sharing knowledge through case studies. Ecol. Eng. 56, 36–50.
Habersack, H.M., Piégay, H., 2008. River restoration in the Alps and their surroundings: past experience and future challenges. In: Habersack, H., Piégay, H., Rinaldi, M. (Eds.), Gravel-bed Rivers VI: From Process Understanding to River Restoration. Elsevier, Amsterdam, pp. 703–737.
Harvey, A.M., 1997. Coupling between hillslope gully systems and stream channels in the Howgill Fells, northwest England: temporal implications. Geomorphologie 3, 3–19.
Harvey, A.M., 2002. Coupling between hillslopes and channels in upland fluvial systems: implications for landscape sensitivity, illustrated from the Howgill Fells, northwest England. Catena 42, 225–250.
Hauer, F.R., Baron, J.S., Campbell, D.H., Fausch, K.D., Hostetler, S.W., Leavesley, G.H., Leavitt, P.R., McKnight, D.M., Stanford, J.A., 1997. Assessment of climate change and freshwater ecosystems of the Rocky Mountains, USA and Canada. Hydrol. Process. 11, 903–924.
Hereford, R., 2002. Valley-fill alluviation during the Little Ice Age (ca. A.D. 1400–1880), Paria River basin and southern Colorado Plateau, United States. Geol. Soc. Am. Bull. 114, 1550–1563.
Hilton, R.G., Galy, A., Hovius, N., Chen, M.-C., Horng, M.-J., Chen, H., 2008. Tropical-cyclone-driven erosion of the terrestrial biosphere from mountains. Nat. Geosci. 1, 759–762.
Hilton, R.G., Galy, A., Hovius, N., Horng, M.-J., Chen, H., 2011. Efficient transport of fossil organic carbon to the ocean by steep mountain rivers: an orogenic carbon sequestration mechanism. Geology 39, 71–74.
Hoffmann, T., Glatzel, S., Dikau, R., 2009. A carbon storage perspective on alluvial sediment storage in the Rhine catchment. Geomorphology 108, 127–137.
Hogg, R., Coghlan, S.M., Zydlewski, J., 2013. Anadromous sea lampreys recolonize a Maine coastal river tributary after dam removal. Trans. Am. Fish. Soc. 142, 1381–1394.
Hohensinner, H., Habersack, H., Jungwirth, M., Zauner, G., 2004. Reconstruction of the characteristics of a natural alluvial river-floodplain system and hydromorphological changes following human modifications: the Danube River (1812–1991). River Res. Appl. 20, 25–41.
Holmes, R.M., McClelland, J.W., Peterson, B.J., Shiklomanov, I.A., Shiklomanov, A.I., et al. 2002. A circumpolar perspective on fluvial sediment flux to the Arctic Ocean. Global Biogeochem. Cycles 16, 1098.

Hooke, R.L., 2000. On the history of humans as geomorphic agents. Geology 28, 843–846.
Hooke, R.L., Martin-Duque, J.F., Pedraza, J., 2012. Land transformation by humans: a review. GSA Today 22, 4–10.
Jaeger, K.L., Olden, J.D., 2012. Electrical resistance sensor arrays as a means to quantify longitudinal connectivity of rivers. River Res. Appl. 28, 1843–1852.
Jain, V., Tandon, S.K., 2010. Assessment of (dis)connectivity and its application to the Ganga River dispersal system. Geomorphology 118, 349–358.
James, L.A., Marcus, W.A., 2006. The human role in changing fluvial systems: retrospect, inventory and prospect. Geomorphology 79, 152–171.
Jarrett, R.D., 1990. Hydrologic and hydraulic research in mountain rivers. Water Resour. Bull. 26, 419–429.
Junk, W.J., Bayley, P.B., Sparks, R.E., 1989. The flood pulse concept in river-floodplain systems. Special Publ. Can. J. Fish. Aquat. Sci. 106, 110–127.
Kao, S.L., Milliman, J.D., 2008. Water and sediment discharge from small mountainous rivers, Taiwan: the roles of lithology, episodic events, and human activity. J. Geol. 116, 431–448.
Kasahara, T., Wondzell, S.M., 2003. Geomorphic controls on hyporheic exchange flow in mountain streams. Water Resour. Res. 39, 1005.
Kieffer, S.W., 1989. Geologic nozzles. Rev. Geophys. 27, 3–38.
Kimberly, A.W., Gardner, R.H., Turner, M.G., 1997. Landscape connectivity and population distributions in heterogeneous environments. Oikos 78, 151–169.
Knight, J., Harrison, S., 2013. The impacts of climate change on terrestrial Earth surface systems. Nat. Climate Change 3, 24–29.
Knighton, D., 1998. Fluvial Forms and Processes: A New Perspective. Arnold, London.
Köhler, S.J., Buffam, I., Seibert, J., Bishop, K.H., Laudon, H., 2009. Dynamics of stream water TOC concentrations in a boreal headwater catchment: controlling factors and implications for climate scenarios. J. Hydrol. 373, 44–56.
Kondolf, G.M., Boulton, A.J., O'Daniel, S., Poole, G.C., Rahel, F.J., Stanley, E.H., Wohl, E., Bång, A., Carlstrom, J., Cristoni, C., Huber, H., Koljonen, S., Louhi, P., Nakamura, K., 2006. Process-based ecological river restoration: visualizing three-dimensional connectivity and dynamic vectors to recover lost linkages. Ecol. Soc. 11 (2), http://www.ecologyandsociety.org/vol11/iss2/art5/.
Konrad, C.P., Olden, J.D., Lytle, D.A., Melis, T.S., Schmidt, J.C., Bray, E.N., Freeman, M.C., Gido, K.B., Hemphill, N.P., Kennard, M.J., McMullen, L.E., Mims, M.C., Pyron, M., Robinson, C.T., Williams, J.G., 2011. Large-scale flow experiments for managing river systems. BioScience 61, 948–959.
Krause, B., Culmsee, H., Wesche, K., Bergmeier, E., Leuschner, C., 2011. Habitat loss of floodplain meadows in north Germany since the 1950s. Biodivers. Conserv. 20, 2347–2364.
Lautz, L.K., Siegel, D.I., 2007. The effect of transient storage on nitrate uptake lengths in streams: an inter-site comparison. Hydrol. Process. 21, 3533–3548.
Leopold, L.B., Wolman, M.G., Miller, J.P., 1964. Fluvial Processes in Geomorphology. W.H. Freeman, San Francisco.
Lovett, R.A., 2014. Rivers on the run. Nature 511, 521–523.
Lyons, W.B., Nezat, C.A., Carey, A.E., Hicks, D.M., 2002. Organic carbon fluxes to the ocean from high-standing islands. Geology 30, 443–446.
MacBroom, J.G., Schiff, R., 2013. Sediment management at small dam removal sites. Rev. Eng. Geol. 21, 67–79.
Maser, C., Sedell, J.E. (Eds.), 1994. From the Forest to the Sea: The Ecology of Wood in Streams, Rivers, Estuaries, and Oceans. St. Lucie Press, Delray Beach, Florida, USA.

Mayer, L., 1992. Some comments on equilibrium concepts and geomorphic systems. Geomorphology 5, 277–295.
McClain, M.E., Boyer, E.W., Dent, C.L., Gergel, S.E., Grimm, N.B., Groffman, P.M., Hart, S.C., Harvey, J.W., Johnston, C.A., Mayorga, E., McDowell, W.H., Pinay, G., 2003. Biogeochemical hot spots and hot moments at the interface of terrestrial and aquatic ecosystems. Ecosystems 6, 301–312.
Mei-e, R., Xianmo, Z., 1994. Anthropogenic influences on changes in the sediment load of the Yellow River, China, during the Holocene. Holocene 4, 314–320.
Melis, T.S. (Ed.), 2011. Effects of three high-flow experiments on the Colorado River ecosystem downstream from Glen Canyon Dam. US Geological Survey Circular 1366, Reston, VA.
Merritt, D.M., Wohl, E.E., 2002. Processes governing hydrochory along rivers: hydraulics, hydrology, and dispersal phenology. Ecol. Appl. 12, 1071–1087.
Michel, R.L., Campbell, D., Clow, D., Turk, J.T., 2000. Timescales for migration of atmospherically derived sulphate through an alpine/subalpine watershed, Loch Vale, Colorado. Water Resour. Res. 36, 27–36.
Montgomery, D.R., 1999. Process domains and the river continuum. J. Am. Water Resour. As. 35, 397–410.
Montgomery, D.R., 2001. Slope distributions, threshold hillslopes, and steady-state topography. Am. J. Sci. 301, 432–454.
Montgomery, D.R., Buffington, J.M., 1997. Channel-reach morphology in mountain drainage basins. Geol. Soc. Am. Bull. 109, 596–611.
Montgomery, D.R., Dietrich, W.E., 1988. Where do channels begin? Nature 336, 232–234.
Montgomery, D.R., Dietrich, W.E., 1989. Source areas, drainage density, and channel initiation. Water Resour. Res. 25, 1907–1918.
Montgomery, D.R., Collins, B.D., Buffington, J.M., et al., 2003. Geomorphic effects of wood in rivers. In: Gregory, S.V., Boyer, K.L., Gurnell, A.M. (Eds.), The Ecology and Management of Wood in World Rivers. American Fisheries Society Symposium 37, Bethesda, MD, pp. 21–47.
Moody, J.A., Kinner, D.A., 2006. Spatial structure of stream and hillslope drainage networks following gully erosion after wildfire. Earth Surf. Process. Land. 31, 319–337.
Naiman, R.J., Melillo, J.M., Hobbie, J.E., 1986. Ecosystem alteration of boreal forest streams by beaver (*Castor canadensis*). Ecology 67, 1254–1269.
Naiman, R.J., Johnston, C.A., Kelley, J.C., 1988. Alteration of North American streams by beaver. BioScience 38, 753–762.
Nanson, G.C., 1986. Episodes of vertical accretion and catastrophic stripping: a model of disequilibrium flood-plain development. Geol. Soc. Am. Bull. 97, 1467–1475.
Nilsson, C., Gardfjell, M., Grelsson, G., 1991. Importance of hydrochory in structuring plant communities along rivers. Can. J. Bot. 69, 2631–2633.
Nilsson, C., Reidy, C.A., Dynesius, M., Revenga, C., 2005. Fragmentation and flow regulation of the world's large river systems. Science 308, 405–408.
National Research Council, 2001. Basic Research Opportunities in the Earth Sciences. National Academies Press, Washington, DC.
Palmer, M.A., Bernhardt, E.S., Schlesinger, W.H., Eshleman, K.N., Foufoula-Georgiou, E., Hendryx, M.S., Lemly, A.D., Likens, G.E., Loucks, O.L., Power, M.E., White, P.S., Wilcock, P.R., 2010. Mountaintop mining consequences. Science 327, 148–149.
Patton, P.C., Schumm, S.A., 1975. Gully erosion, northwestern Colorado: a threshold phenomenon. Geology 3, 88–90.

Pazzaglia, F.J., 2013. Fluvial terraces. In: Wohl, E. (Ed.), Treatise on Fluvial Geomorphology. Academic Press, San Diego, pp. 379–412.

Petts, G.E., 1989. Historical analysis of fluvial hydrosystems. In: Petts, G.E. (Ed.), Historical Change of Large Alluvial Rivers: Western Europe. John Wiley and Sons, Chichester, pp. 1–18.

Pierce, J., Meyer, G., 2008. Long-term fire history from alluvial fan sediments: the role of drought and climate variability, and implications for management of Rocky Mountain forests. Int. J. Wildland Fire 17, 84–95.

Poff, N.L., Allan, J.D., Bain, M.B., Karr, J.R., Prestegaard, K.L., Richter, B.D., Sparks, R.E., Stromberg, J.C., 1997. The natural flow regime. BioScience 47, 769–784.

Poff, N.L., Olden, J.D., Merritt, D.M., Pepin, D.M., 2007. Homogenization of regional river dynamics by dams and global biodiversity implications. Proc. Natl. Acad. Sci. 104, 5732–5737.

Pollock, M.M., Beechie, T.J., Wheaton, J.M., Jordan, C.E., Bouwes, N., Weber, N., Volk, C., 2014. Using beaver dams to restore incised stream ecosystems. BioScience 64, 279–290.

Raabe, J.K., Hightower, J.E., 2014. Assessing distribution of migratory fishes and connectivity following complete and partial dam removals in a North Carolina river. N. Am. J. Fish. Manag. 34, 955–969.

Rhoads, B.L., Schwartz, J.S., Porter, S., 2003. Stream geomorphology, bank vegetation, and three-dimensional habitat hydraulics for fish in Midwestern agricultural streams. Water Resour. Res. 39, 1218.

Sanderson, E.W., Jaiteh, M., Levy, M.A., Redford, K.H., Wannebo, A.V., Woolmer, G., 2002. The human footprint and the last of the wild. BioScience 52, 891–904.

Scanlon, B.R., Jolly, I., Sophocleous, M., Zhang, L., 2007. Global impacts of conversions from natural to agricultural ecosystems on water resources: quantity versus quality. Water Resour. Res. 43, W03437.

Schumm, S.A., 1973. Geomorphic thresholds and complex response of drainage systems. In: Morisawa, M. (Ed.), Fluvial Geomorphology. Publications in Geomorphology, SUNY Binghamton , pp. 299–310.

Schumm, S.A., 1977. The Fluvial System. John Wiley and Sons, New York.

Schumm, S.A., 1979. Geomorphic thresholds: the concept and its applications. Trans. Inst. Br. Geogr. 4, 485–515.

Schumm, S.A., 1981. Evolution and response of the fluvial system, sedimentologic implications. Soc. Econ. Palentologists Mineral. Special Pub. 31, 19–29.

Schumm, S.A., Lichty, R.W., 1965. Time, space, and causality in geomorphology. Am. J. Sci. 263, 110–119.

Schumm, S.A., Parker, R.S., 1973. Implications of complex response of drainage systems for Quaternary alluvial stratigraphy. Science 243, 99–100.

Shafroth, P.B., Wilcox, A.C., Lytle, D.A., Hickey, J.T., Andersen, D.C., Beauchamp, V.B., Hautzinger, A., McMullen, L.E., Warner, A., 2009. Ecosystem effects of environmental flows: modelling and experimental floods in a dryland river. Freshwater Biol. 55, 68–85.

Simon, A., Castro, J., 2003. Measurement and analysis of alluvial channel form. In: Kondolf, G.M., Piegay, H. (Eds.), Tools in Fluvial Geomorphology. John Wiley and Sons, Chichester, pp. 291–322.

Simon, A., Rinaldi, M., 2006. Disturbance, stream incision, and channel evolution: the roles of excess transport capacity and boundary materials in controlling channel response. Geomorphology 79, 361–383.

Smith, N., 1971. A History of Dams. Peter Davies, London.

Sparks, R.E., 1995. Need for ecosystem management of large rivers and their floodplains. BioScience 45, 168–182.

Starkel, L., 1988. Tectonic, anthropogenic and climatic factors in the history of the Vistula River valley downstream of Cracow. In: Lang, G., Schluchter, C. (Eds.), Lake, Mire, and River Environments During the Last 15,000 Years. A.A. Balkema, Rotterdam, pp. 161–170.

Syvitski, J.P.M., Vörösmarty, C.J., Kettner, A.J., Green, P., 2005. Impact of humans on the flux of terrestrial sediment to the global coastal ocean. Science 308, 376–380.

Tetzlaff, D., Soulsby, C., Bacon, P.J., Youngson, A.F., Gibbins, C., Malcolm, I.A., 2007. Connectivity between landscapes and riverscapes – a unifying theme in integrating hydrology and ecology in catchment science? Hydrol. Process. 21, 1385–1389.

Tharme, R.E., 2003. A global perspective on environmental flow assessment: emerging trends in the development and application of environmental flow methodologies for rivers. River Res. Appl. 19, 397–441.

Turner, R.D., 1977. Wood, mollusks, and deep-sea food chains. Bull. Am. Malacol. Union 1976, 13–19.

Veblen, T.T., Donnegan, J.A., 2005. Historical range of variability for forest vegetation of the national forests of the Colorado Front Range. Final report, USDA Forest Service Agreement 1102-0001-99-033, Rocky Mountain Region, Golden, CO.

Vörösmarty, C., Lettenmaier, D., Leveque, C., Meybeck, M., Pahl-Wostl, C., Alcamo, J., Cosgrove, W., Grassl, H., Hoff, H., Kabat, P., Flansigan, F., Lawford, R., Naiman, R., 2004. Humans transforming the global water system. Eos Trans. AGU 85, 509–520.

Wainwright, J., Turnbull, L., Ibrahim, T.G., Lexarta-Artza, I., Thornton, S.F., Brazier, R.E., 2011. Linking environmental regimes, space and time: interpretations of structural and functional connectivity. Geomorphology 126, 387–404.

Walter, R.C., Merritts, D.J., 2008. Natural streams and the legacy of water-powered mills. Science 319, 299–304.

Ward, J.V., 1997. An expansive perspective of riverine landscapes: pattern and process across scales. River Ecosyst. 6, 52–60.

Warne, A.G., Toth, L.A., White, W.A., 2000. Drainage-basin-scale geomorphic analysis to determine reference conditions for ecologic restoration – the Kissimmee River, Florida. Geol. Soc. Am. Bull. 112, 884–899.

Warrick, J.A., Mertes, L.A.K., 2009. Sediment yield from tectonically active semiarid Western Transverse Ranges of California. Geol. Soc. Am. Bull. 121, 1054–1070.

West, A.J., Lin, C.-W., Lin, T.-C., Hilton, R.G., Liu, S.-H., Chang, C.-T., Lin, K.-C., Galy, A., Sparkes, R.B., Hovius, N., 2011. Mobilization and transport of coarse woody debris to the oceans triggered by an extreme tropical storm. Limnol. Oceanogr. 56, 77–85.

Wester, T., Wasklewicz, T., Staley, D., 2014. Functional and structural connectivity within a recently burned drainage basin. Geomorphology 206, 362–373.

Whipple, K.X., 2004. Bedrock rivers and the geomorphology of active orogens. Annu. Rev. Earth Planet. Sci. 32, 151–185.

White, P.S., Pickett, S.T.A., 1985. Natural disturbance and patch dynamics: an introduction. In: Pickett, S.T.A., White, P.S. (Eds.), The Ecology of Natural Disturbance and Patch Dynamics. Academic Press, New York, pp. 3–9.

With, K.A., Crist, T.O., 1995. Critical thresholds in species' response to landscape structure. Ecology 76, 2446–2459.

Woelfle-Erskine, C., Wilcox, A.C., Moore, J.N., 2012. Combining historical and process perspectives to infer ranges of geomorphic variability and inform river restoration in a wandering gravel-bed river. Earth Surf. Process. Land. 37, 1302–1312.

Wohl, E.E., 2000. Anthropogenic impacts on flood hazards. In: Wohl, E.E. (Ed.), Inland Flood Hazards: Human, Riparian, and Aquatic Communities. Cambridge University Press, Cambridge, UK, pp. 104–141.

Wohl, E., 2011. A World of Rivers: Environmental Change on Ten of the World's Great Rivers. University of Chicago Press, Chicago, IL.

Wohl, E., 2013a. Wide Rivers Crossed: The South Platte and the Illinois of the American Prairie. University Press of Colorado, Boulder, CO.

Wohl, E., 2013b. Wilderness is dead: whither critical zone studies and geomorphology in the Anthropocene? Anthropocene 2, 4–15.

Wohl, E., 2014a. A legacy of absence: wood removal in US rivers. Prog. Phys. Geogr. 38, 637–663.

Wohl, E., 2014b. Rivers in the Landscape: Science and Management. Wiley-Blackwell, Chichester, UK.

Wohl, E., Beckman, N., 2014. Leaky rivers: implications of the loss of longitudinal fluvial disconnectivity in headwater streams. Geomorphology 205, 27–35.

Wohl, E., Angermeier, P.L., Bledsoe, B., Kondolf, G.M., MacDonnell, L., Merritt, D.M., Palmer, M.A., Poff, N.L., Tarboton, D., 2005. River restoration. Water Resour. Res. 41, W10301.

Wohl, E., Dwire, K., Sutfin, N., Polvi, L., Bazan, R., 2012. Mechanisms of carbon storage in mountainous headwater rivers. Nat. Commun. 3, 1263.

Chapter 10

Characteristic and Role of Groundwater in the Critical Zone

Quanrong Wang* and Hongbin Zhan**
*School of Environmental Studies, China University of Geosciences, Wuhan, Hubei, China;
**Department of Geology and Geophysics, Texas A&M University, College Station, Texas, USA

10.1 INTRODUCTION

Earth's Critical Zone (ECZ) refers to a fragile skin of the planet from the outer extent of vegetation to the lower limits of groundwater (NRC, 2001), including the land surface, vegetation, water bodies through the pedosphere, unsaturated vadose zone, and saturated zone, as shown in Fig. 10.1. Within ECZ supporting all life on Earth, a series of complex physical, chemical, biological, and geological processes occurred to control the energy, water, and carbon cycles, the fate and transport of nutrients and pollutants, and atmospheric composition (Anderson et al., 2007; Chorover et al., 2007). Understanding such processes in ECZ could provide more accurate scheme for the presently critical issues like climate change, environmental quality, sustainability, ecosystem services, water resources, and so on (Anderson et al., 2007, 2008).

Comparing with the atmospheric and surface water bodies during the water cycle, groundwater might be the largest reservoir and an active component of the hydrologic system (Leung et al., 2011). However, due to the inherent slow response time, many previous studies simplified or ignored the effect of groundwater when studying the climate change or the land surface processes, including the exchanges of heat, water, CO_2, and other trace constituents in the land surface (Yang, 2004). In order to substantially advance the science of ECZ, the systems-based Critical Zone Observatories (CZOs) have been established globally. For instance, US National Science Foundation firstly established three systems-based CZOs in 2007, and subsequently funded three additional CZOs in 2009 (Anderson et al., 2008). The German Helmholtz Association funded four CZOs to investigate the effect of global change on terrestrial ecosystems and associated socioeconomics (Anderson et al., 2008). Although the research in the Earth-surface processes in CZOs were started only a few years ago, a lots of interesting findings have been reported, and they demonstrated that groundwater

Developments in Earth Surface Processes, Vol. 19. http://dx.doi.org/10.1016/B978-0-444-63369-9.00010-0

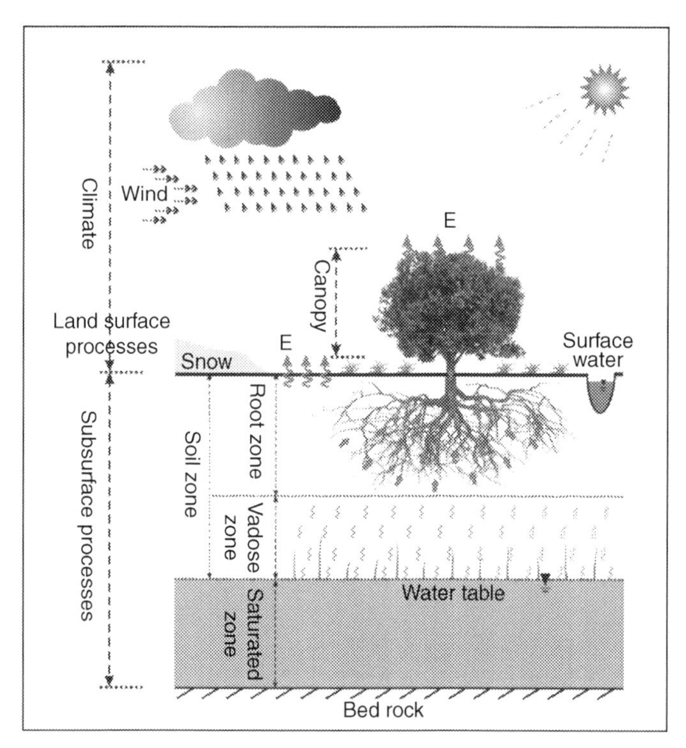

FIGURE 10.1 Schematic diagram of Earth-surface processes in ECZ. *(Modified from Maxwell and Miller (2005).)*

might have very strong interactions with land surface processes and climate change, especially when the water table was shallow (Lin et al., 2011).

10.2 ROLE OF GROUNDWATER IN ECZ

Theoretically, almost all kinds of the Earth-surface processes are somewhat related to water, and water cycle, as verified by numerous previous studies (Brantley et al., 2007; Leung et al., 2011; Rasmussen et al., 2011). For example, a rising groundwater table could increase soil moisture, evapotranspiration, and stream flow, and then influence the regional climate change; it may be related to the heat exchange as well, since evapotranspiration needs to take up energy from its surroundings and cools the environment. Subsequently, such water movement and energy exchange affect the other Earth-surface processes. In contrary, a declining groundwater table could lead to an opposite effect.

A special issue of Vadose Zone Journal, entitled "Interdisciplinary Sciences in Critical Zone Observatories," partially reported some new findings observed in CZOs, which were established to understand the links between the Earth-surface processes in helping improve forecasts of environmental change (Lin et al., 2011).

Lin et al. (2011) summarized the scientific advancements of these publications, in which five papers showed that the soil moisture played an important role in controlling the partitioning of hydrologic fluxes. For instance, Bales et al. (2011) found that about one-third of annual evapotranspiration may be from water storage below the 1-m soil depth. Swarowsky et al. (2011) concluded that vegetation, clay pan occurrence, topographic index, and solar radiation were significantly correlated to the properties of the soil moisture and stream flow. Graham and Lin (2011), Takagi and Lin (2011), and Kuntz et al. (2011) demonstrated the impacts of shallow water table on the spatial variability of soil moisture. In addition to the special issue publication in Vadose Zone Journal, there are also many other papers reporting the new findings of the Earth-surface processes in ECZ. Jones (1985) pointed out that the chemical weathering processes, which produced saprolite aquifer horizon in the high African Plateau, were largely controlled by circulating groundwater. Lo and Famiglietti (2010) demonstrated that many studies, which found that soil moisture increased after adding a groundwater component, was not exactly correct. Calmels et al. (2011) found that the deep groundwater contributed about 16% of the river discharge in Liwu River, Taiwan, and it provided about 40% of the cation weathering flux to the river. Shi et al. (2013) pointed out that latent heat fluxes were strongly correlated with water table depth, especially near the stream. Condon et al. (2013) found that heterogeneity did not fundamentally alter the connection between groundwater and land surface processes, but impacted the extent and location of the Critical Zone.

Climate change could impact the hydrologic cycle, land cover, and land–atmosphere feedbacks. Different kinds of water movements between atmospheric, surface, and subsurface water occuring in land surface also potentially alters the climate, as they are critical in the generation of cloud and moisture within the cloud. Leung et al. (2011) employed a regional climate model with and without surface water–groundwater interactions to assess the interplay between the climate–soil–vegetation and the water table dynamics. The results showed that vegetation properties, like the minimum stomatal resistance and root depth, could affect the groundwater table dynamics (Leung et al., 2011). In contrary, the groundwater table dynamics could also alter the land surface response and the climate change (Leung et al., 2011). Maxwell and Miller (2005) pointed out that both the traditional land surface models (LSMs) and the groundwater models (GWMs) could not exactly simulate the Earth-surface processes, since the treatment of LSMs–GWMs connection was too simple and excluded some vital processes. Therefore, Maxwell and Miller (2005) coupled a state-of-the-art LSM and a variably saturated GWM (ParFlow) as a single-column model, which has beem applied to interpret the data from Valdai, Russia, and the results revealed that the cold processes such as frozen soil and freeze/thaw processes could impact the groundwater table dynamics. Subsequently, Maxwell and Kollet (2008) employed this coupled model to simulate the water and energy flows in a changing climate for the southern Great Plains, USA, and concluded that the groundwater control was critical to understand processes of recharge and drought in a changing climate. Ferguson and Maxwell (2010) analyzed integrated watershed response and groundwater–land surface feedbacks in the Little Washita River watershed in North America, and found that the climate change and groundwater had strong correlationships. Rihani et al. (2010) concluded that the terrain shape, subsurface heterogeneity, vegetation type, and climatological region were strongly correlated with the water table depth, and the vegetative land cover had a large effect on the energy balance but a small effect on streamflow and water table dynamics.

Understanding the climate and water cycle is also critical to the long-term biogeochemical cycles and vegetation dynamics, since an accurate modeling of these two processes depends on credible knowlege of the temperature and soil moisture (Lin et al., 2011; Newman et al., 2006). Actually, biogeochemical processes may influence the water cycle. To study such complex processes, several new hybrid disciplines have been established, such as "Ecohydrology" for the feedbacks between the biological communities on the water cycle (Nuttle, 2002; Rodriguez-Iturbe, 2000), "Hydropedology" for the pedologic and hydrologic processes (Lin, 2003; Minasny et al., 2013; Yang et al., 2014). These new interdisciplinary researches would provide a systematic approach to the study of the water cycle, climate, earth surface, and subsurface environments, and so on. Up to date, many new findings of the relationship between the biogeochemistry and hydrology have been observed in CZOs, which have been partially summarized by Lin et al. (2011), including Chorover et al. (2011), Williams et al. (2011), Jin et al. (2011), Befus et al. (2011), Andrews et al. (2011), Zacharias et al. (2011), and Banwart et al. (2011). The results showed that the interplay between hydrology and biogeochemistry was complex. Additionally, Kim et al. (2014) pointed out that the hydrochemical dynamics of groundwater in the weathered bedrock zone was highly related to the structure of ECZ and groundwater flow processes.

In summary, ECZ is an earth surface system coupling vegetation, regolith, and groundwater, which are essential to sustaining life on the planet, within which the complex interactions among physical, chemical, and biological processes occur, and groundwater is an important factor.

10.3 MODELS OF EARTH-SURFACE PROCESSES

Understanding interactions among different Earth-surface processes remains a challenge to earth system sciences, since it may require interdisciplinary approaches, and the coupling of climate change models, hydrologic models, water quality models, ecosystem models, population/land use models, and so on (Anderson et al., 2007; Brantley et al., 2007; Maxwell and Kollet, 2008). Climate models may represent atmosphere, ocean, land surface, sea ice, and so on. Hydrologic models are rainfall-runoff models used to predict surface water, like stream flow, resulting from precipitation. Groundwater flow models describe the flow in the subsurface media, and they may be incorporated into the hydrologic models, since the interactions between groundwater and surface water are strong when the water table is shallow. Water quality models are used to answer the question that how water quality varies temporally and spatially due to the environment loadings. Ecosystem models could be applied to determine where individual species of plants and animals could flourish. Biogeochemistry models could simulate changes in basic ecosystem processes, and biogeography models are used to describe the shifts in the geographic distribution of major plant species and communities. Population/land use models refer to the projected temporal and spatial changes in land use and population. In addition to these models, infrastructure models may be needed sometimes, which are used to predict the performance of components of the urban water infrastructure.

Actually, many previous models have considered the interactions among the Earth-surface processes, but most of them only concentrated on mass or energy exchange in a localized part of the water cycle, by simplifying or ignoring the effect of the other parts during the water cycle. For instance, the traditional LSM used for modeling climate projection has considered the groundwater effect through the lower boundary condition, in which the flux was zero or the soil moisture content was assumed constant. GWMs employed the evaporation and precipitation coefficients, occurring in the upper boundary conditions, to describe the effect of climate change on groundwater flow. They ignored soil heating, runoff, snow, and root-zone uptake. Some physically based groundwater–river models incorporate various aquifer properties, but interaction with the atmosphere is usually not considered (Vionnet et al., 1997). From a mass/energy cycle point of view, the physical, chemical, biological, and geological processes in ECZ may affect each other (Maxwell and Miller, 2005). For example, the shallow groundwater table maintains elevated soil moisture in the root zone (Chen and Hu, 2004), which influences land surface processes through capillary rise, such as evapotranspiration, runoff, and infiltration. These factors may further affect the processes of rock weathering and the production of soil, sustainability of agricultural and ecological systems, flux of carbon, climate change, the long-term evolution of topography, or some geologic hazards including flood, debris flow hazards, and so on (Maxwell and Kollet, 2008; Niu et al., 2007; Yeh and Eltahir, 2005; York et al., 2002).

When modeling climate, land surface processes always supply the model's lower boundary condition (Eagleson, 1982; Steiner et al., 2005), so many LSMs have been designed for weather and climate models. LSMs are always the upper boundary of the subsurface processes model, through which the climate and subsurface processes are connected, as shown in Fig. 10.1. As for the mechanics of the physical, chemical, and biological subsurface processes, many investigations have been undertaken and the robustness of the associated mathematic models has been widely tested. However, the research on the coupling model of LSMs and GWM has many gaps, and is still ongoing. In the following section, we will briefly introduce the mathematic model of LSMs and GWM, and then summarize the recent studies of the coupling models.

10.3.1 The Model of LSMs

The early climate models usually employed a leaky-bucket parameterization to represent land surface hydrology (Manabe et al., 1965), which did not consider the processes of water movement (Yang et al., 1995). Subsequently, many sophisticated soil–vegetation–atmosphere transfer schemes have been developed to describe the land surface processes, such as biosphere–atmosphere transfer scheme (BATS) (Dickinson et al., 1986), simple biosphere model (Sellers et al., 1986), common land model (CLM) (Dai et al., 2003). Yang (2004) reviewed the historical development of LSMs. Recently, the ecological and biogeochemical processes have been included in LSMs (Bonan, 1996; Dai and Zeng, 1997; Oleson et al., 2004).

CLM was developed as a multi-institutional code (Dai et al., 2003), by combining the best features of three existing successful LSMs: the BATS developed by Dickinson et al. (1986), the LSM by Bonan (1996), and the snow model by Dai and Zeng (1997). The overall structure includes three elements: the core single-column soil–snow–vegetation biophysical code, the land boundary data, and the scaling procedures within a climate model required to interface atmospheric model grid-square inputs to land single-column processes (Dai et al., 2003). CLM includes many state variables, such as the temperature of snow and soil, ice lens mass and liquid mass, leaf temperature, canopy water storage, nondimensional snow age, snow-layer number, and snow-layer thickness (Dai et al., 2003; Zeng et al., 2002). It has been coupled with the National Center for Atmospheric Research (NCAR) community climate model (CCM3) for the climate modeling (Zeng et al., 2002). Dai et al. (2003) pointed out that the CLM was better than the LSM of Bonan (1996) in modeling the surface air temperature, the annual cycle of runoff, and snow mass. In addition, Dai et al. (2003) have established ties to groups performing carbon cycle and ecological modeling.

The physical descriptions of CLM may be divided into several aspects, such as surface albedo, turbulent fluxes, photosynthesis and stomatal resistance, temperatures, water balance, and snow compaction. The water balance includes canopy water storage, snow water, soil moisture, and runoff. Canopy water is based on a simple mass balance determined by the interception of precipitation and the dew condensation and loss from evaporation. Snow water is calculated by a simple explicit scheme, which permits the liquid water over the holding capacity of snow to percolate into the underlying layer. The soil moisture movement is governed by infiltration, runoff, gradient diffusion, gravity, and soil water extraction through roots for canopy transpiration. The time rate of change in soil water content is defined as (Maxwell and Miller, 2005):

$$\frac{\partial w_{\mathrm{liq},j}}{\partial t} = \left[q_{j-1} - q_j\right] - f_{\mathrm{root},j} E_{\mathrm{tr}} + \left[M_{\mathrm{il}} \Delta z\right]_j, \tag{10.1}$$

$$\frac{\partial w_{\mathrm{ice},j}}{\partial t} = \left[q_{j-1,\mathrm{ice}} - q_j\right] - f_{\mathrm{root},j} E_{\mathrm{tr}} + \left[M_{\mathrm{il}} \Delta z\right]_j, \tag{10.2}$$

where $w_{\text{liq},j} = (\rho_{\text{liq}}\theta_{\text{l}}\Delta z)_j$ and $w_{\text{ice},j} = (\rho_{\text{i}}\theta_{\text{i}}\Delta z)_j$ are the liquid and ice mass in each of the j soil layers; ρ is the density; θ is the volumetric soil moisture content; the subscript l and i represent liquid and ice, respectively; M_{il} is the ice to liquid phase change; q_j is the water mass flux at each layer interface; E_{tr} is the transpiration; $f_{\text{root},j}$ is the root fraction for the j layer; and Δz is the vertical discretization of the soil column.

Soil water flux is calculated by Darcy's law (Dai et al., 2003; Maxwell and Miller, 2005),

$$q = -K\left(\frac{\partial \psi}{\partial z} - 1\right), \tag{10.3}$$

where the soil negative potential ψ changes with soil water content and soil texture based on relationships proposed by several previous studies such as Clapp and Hornberger (1978) and Cosby et al. (1984). The unsaturated hydraulic conductivity is given as $K = K_{\text{sat}}{}^{2B+3}$, where B is a fitting constant which depends on the soil texture, and the saturated hydraulic conductivity at depth (K_{sat}) is based on an exponential assumption,

$$K_{\text{sat}} = K_{\text{sat},0}\exp\left(-\frac{z}{z_{\text{L}}}\right), \tag{10.4}$$

where $K_{\text{sat},0}$ is the surface saturation hydraulic conductivity and z_{L} is the length scale for the decrease in K_{sat}.

Runoff is calculated from surface and base flow for saturated and unsaturated regions, using the principles of TOPMODEL (TOPography-based hydrological MODEL) (Beven, 1997; Wigmosta and Lettenmaier, 1999), which simulates the groundwater table at a quasi-equilibrium state (Stieglitz et al., 1997; Walko et al., 2000). Total runoff is the sum of the surface runoff (R_{s}) and base flow (R_{b}), which are computed for saturated (f_{sat}) and unsaturated ($1-f_{\text{sat}}$) regions separately (Dai et al., 2003; Maxwell and Miller, 2005):

$$R_{\text{s}} = (1 - f_{\text{sat}})\overline{w}_{\text{s}}^{4}G_{w} + f_{\text{sat}}G_{w}, \tag{10.5}$$

$$R_{\text{b}} = (1 - f_{\text{sat}})K_{\text{d}}\overline{w}_{\text{b}}^{2B+3} + f_{\text{sat}}10^{-5}\exp(-z_{\text{w}}), \tag{10.6}$$

in which G_w is the effective net liquid input to the upper soil layer; $\overline{w}_{\text{s}}$ and $\overline{w}_{\text{b}}$ are surface and bottom soil thickness weighted soil wetness, respectively; f_{sat} is the fraction of the watershed that is saturated, given by (Dai et al., 2003; Maxwell and Miller, 2005):

$$f_{\text{sat}} = w_{\text{fact}}\exp(-z_{\text{w}}), \tag{10.7}$$

where w_{fact} is the fraction of the watershed having an exposed water table. The mean, dimensionless water table depth, z_w, is given by (Dai et al., 2003; Maxwell and Miller, 2005):

$$z_w = f_z\left(z_{bot} - \sum_{j=1,n\,soil} s_j \Delta z_j\right), \tag{10.8}$$

where f_z is a water table depth scaling parameter, z_{bot} is the depth of the bottom of the domain, K_d is the saturated hydraulic conductivity at the bottom layer.

Generally, the LSM lower boundary is assumed to be zero flux or the soil moisture content is set to a constant value. From Eqs 10.1–10.2 of this model, one can see that the groundwater model was highly simplified, and it cannot be used to properly describe groundwater movement and solute transport under actual field conditions.

Another notable point of LSM is the runoff model, in which CLM used the principles of TOPMODEL. Actually, there are many other hydrologic models which could simulate the runoff in the watershed scales equally well or probably even better. These include VIC (variable infiltration capacity model, Liang et al., 1994, 2003; Liang and Xie, 2001), SWAT (soil and water assessment tool, Arnold et al., 1998; Neitsch et al., 2004; Srinivasan et al., 1998), HMS (hydrologic model system, Yu et al., 1999), SVAT (soil–vegetation–atmosphere transfer model, Franks et al., 1997), DHSVM (distributed hydrology–soil–vegetation model, Wigmosta et al., 1994), AVSWAT (ArcView GIS-SWAT, Di Luzio et al., 2004), WetSpa (Liu and De Smedt, 2004), MIKE SHE (Graham and Butts, 2005), OpenMI (open modelling interface, Moore and Tindall, 2005), OMS (object modeling system, Ahuja et al., 2005; David et al., 2013), and so on. The models of the runoff processes in such hydrologic models may be closer to reality than the one used in TOPMODEL.

10.3.2 The Model of Groundwater Flow

Considering the three-dimensional Darcian flow of water in a variably saturated rigid porous medium, the governing equation can be described as:

$$\frac{\rho}{\rho_0} F \frac{\partial H}{\partial t} + \frac{\theta}{\rho_0} \frac{\partial \rho}{\partial C} \frac{\partial C}{\partial t} = \nabla\left[K\left(\nabla H + \frac{\rho}{\rho_0} \nabla z\right)\right] + \frac{\rho^*}{\rho_0} w, \tag{10.9}$$

where H is the pressure head; t is the time; K is the hydraulic conductivity tensor; z is the potential head; w is the source and/or sink; ρ is the water density at chemical concentration C; ρ_0 is the referenced water density at zero chemical concentration; ρ^* is the density of either the injection fluid or the withdrawn water; ∇ is the differential operator; F is the storage coefficient expressed as:

$$F = \alpha' \frac{\theta}{n} + \beta' \theta + n \frac{dS}{dh}; \qquad (10.10)$$

where S is the saturation; θ is the moisture content; n is the porosity of the medium; α' is the modified compressibility of the medium, expressed as:

$$\alpha' = \alpha \rho_0 g;$$

where α is the coefficient of consolidation of the media; β' is the modified compressibility of water:

$$\beta' = \beta \rho_0 g;$$

where β is the compressibility of water $\beta = \frac{1/\rho}{\partial \rho / \partial p}$, and p is the fluid pressure.

The hydraulic conductivity K is given by

$$K = \frac{\rho / \rho_0}{\mu / \mu_0} K_{s0} k_r,$$

where μ is the dynamic viscosity of water at chemical concentration C; μ_0 is the referenced dynamic viscosity at zero chemical concentration; k_r is the relative permeability; K_{s0} is the referenced saturated hydraulic conductivity tensor.

The governing Eq. 10.9 is the modified Richards equation, which is derived by Lin et al. (1997), and the detailed derivation can be seen in Appendix A of Lin et al. (1997). Obviously, one can see that Eq. 10.9 contains concentration C, which is used to describe the variable density flow (like seawater intrusion). Following Frind (1982) by neglecting the second term on the right side of Eq. 10.9, the FEMWATER software package employs the Galerkin finite element method to solve Eq. 10.9 along with the initial and boundary conditions (Lin et al., 1997). Yeh et al. (1994) employed a similar assumption and developed a software package named 3DFEMFAT, a three-dimensional (3D) finite element model of flow and transport through saturated–unsaturated media, in which the governing equation was based on the Richards equation. Voss (1984) developed SUTRA (saturated–unsaturated transport) for 3D saturated–unsaturated, variable-density groundwater flow with solute or energy transport, based on a hybridization of finite-element and integrated-finite-difference methods employed in a weighted residual framework. Another widely used software for 3D groundwater flow in the variably saturated aquifer is HYDUS-3D, in which the governing equation is the modified Richards equation by assuming the density constant. In HYDUS-3D, the governing flow equation is solved numerically using the Galerkin-type linear finite element schemes. ParFlow is a groundwater flow code developed at Lawrence Livermore National Laboratory (Ashby and Falgout, 1996). It solves two different sets of groundwater equations in two modes: (1) steady-state, fully saturated flow using a parallel, multigrid-preconditioned conjugate gradient solver, or (2) transient, variably saturated flow using a parallel, and globalized Newton method coupled to the multigrid-preconditioned linear solver. Parflow solves the variably saturated Richards equation in three dimensions.

In the saturated aquifer, the governing equation of the variable-density flow is given by:

$$\rho S_s \frac{\partial H}{\partial t} + \theta \frac{\partial \rho}{\partial C}\frac{\partial C}{\partial t} = \nabla\left[\rho \frac{\mu_0}{\mu} K_{s0}\left(\nabla H + \frac{\rho - \rho_0}{\rho_0}\nabla z\right)\right] + \rho^* w, \tag{10.11}$$

where μ_0 and μ are dynamic viscosities of water and fluid (the water with solute, such as the seawater). The SEAWAT software used a finite-difference method to solve the governing equation of the variable-density flow Eq. 10.1 by coupling the MODFLOW and MT3DMS codes. The finite-difference representation of the viscosity terms in Eq. 10.11 is given in Thorne et al. (2006).

Assuming the constant density of groundwater, the governing equation of groundwater flowing in a saturated aquifer can be described as:

$$\begin{aligned}
&\frac{\partial}{\partial x}\left(K_{xx}\frac{\partial H}{\partial x}\right) + \frac{\partial}{\partial x}\left(K_{xy}\frac{\partial H}{\partial y}\right) + \frac{\partial}{\partial x}\left(K_{xz}\frac{\partial H}{\partial z}\right)\\
&+\frac{\partial}{\partial y}\left(K_{yx}\frac{\partial H}{\partial x}\right) + \frac{\partial}{\partial y}\left(K_{yy}\frac{\partial H}{\partial y}\right) + \frac{\partial}{\partial z}\left(K_{yz}\frac{\partial H}{\partial z}\right)\\
&+\frac{\partial}{\partial z}\left(K_{zx}\frac{\partial H}{\partial x}\right) + \frac{\partial}{\partial z}\left(K_{zy}\frac{\partial H}{\partial y}\right) + \frac{\partial}{\partial z}\left(K_{zz}\frac{\partial H}{\partial z}\right) + w = S_s\frac{\partial H}{\partial t},
\end{aligned} \tag{10.12}$$

where $H(x, y, z, t)$ is hydraulic head; S_s is specific storage; K_{xx}, K_{xy}, K_{xz}, K_{yx}, K_{yy}, K_{yz}, K_{zx}, K_{zy}, and K_{zz} are the components of a second order symmetrical tensor of hydraulic conductivity

$$K_{s0} = \begin{vmatrix} K_{xx} & K_{xy} & K_{xz} \\ K_{yx} & K_{yy} & K_{yz} \\ K_{zx} & K_{zy} & K_{zz} \end{vmatrix}.$$

When the axes of the principal hydraulic conductivity tensor are aligned with the coordinate axes, Eq. 10.12 becomes:

$$\frac{\partial}{\partial x}\left(K_x\frac{\partial H}{\partial x}\right) + \frac{\partial}{\partial y}\left(K_y\frac{\partial H}{\partial y}\right) + \frac{\partial}{\partial z}\left(K_z\frac{\partial H}{\partial z}\right) + w = S_s\frac{\partial H}{\partial t}, \tag{10.13}$$

where K_x, K_y, and K_z are values of principal saturated hydraulic conductivity along the x, y, and, z coordinate axes, respectively.

Equations 10.12 and 10.13 are partial differential equations, which can yield solutions when subject to the boundary conditions and the following initial condition.

$$H(x,y,z,t=0)=H_0(x,y,z), \tag{10.14}$$

where H_0 represents a known function of x, y, and z.

Generally, there are three types of boundary conditions. The first-type (Dirichlet) boundary condition is applied when the value of the dependent variable at the boundary is known

$$H\big|_{B_1}=H_1(x,y,z,t),\,(x,y,z)\in B_1, \tag{10.15}$$

where H_1 presents a known function of x, y, z, and t; B_1 is the domain of the first-type boundary condition. The second-type boundary condition is

$$K\frac{\partial H}{\partial n}\bigg|_{B_2}=q(x,y,z,t),\,(x,y,z)\in B_2, \tag{10.16}$$

where $q(x, y, z, t)$ is a known flux at the boundary; n is the outward normal direction to the boundary surface, B_2. The third-type boundary condition is a mixture of the first-type boundary condition and the second-type boundary condition. It applies when potential and normal flux components at the boundary are related to each other. Equations 10.12–10.16 consist of the mathematic model of three-dimensional groundwater flow in a saturated aquifer. By comparing Eqs 10.9, 10.11, and 10.13, one can see that the model of saturated groundwater flow with a constant density is the simplest, and many software packages have been developed for such a model, including MODFLOW using a finite-difference method (McDonald and Harbaugh, 1988), FEFLOW using a finite-element method, and so on.

10.3.3 Water Quality Model

Water quality model is a kind of solute transport model, in which the solute represents the contaminant or some chemicals. With the processes of advection, dispersion/diffusion, adsorption, decay, biodegradation, and source/sink, the governing equations for solute transport are derived based on the continuity of mass and flux laws:

$$\begin{aligned}\theta\frac{\partial C}{\partial t}+\rho_b\frac{\partial S_C}{\partial t}+V\cdot\nabla C-\nabla\cdot(\theta D\nabla C)&=-\left(\alpha'\frac{\partial H}{\partial t}+\lambda\right)(\theta C+\rho_b S_C)\\&-(\theta K_w C+\rho_b K_s S_C)+wC_{in}-\frac{\rho^*}{\rho}wC\\&+\left(F\frac{\partial H}{\partial t}+\frac{\rho_0}{\rho}V\cdot\nabla\left(\frac{\rho}{\rho_0}\right)-\frac{\partial\theta}{\partial t}\right)C,\end{aligned} \tag{10.17}$$

$S_C = K_d C$ for linear isotherm,

$$S_C = \frac{S_{max} K_L C}{1 + K_L C} \text{ for Langmuir isotherm,}$$

$S_C = K_L C^n$ for Freundlich isotherm,

where ρ_b is bulk density of the medium; C is the material concentration in aqueous phase; S is the material concentration in adsorbed phase; C_{in} is the material concentration in the source; V is the specific discharge (or Darcy velocity); D is the dispersion coefficient tensor; λ is the decay constant; K_w is the first order biodegradation rate constant through dissolved phase; K_s is the first order biodegradation rate through adsorbed phase; F is the storage coefficient; K_d is the distribution coefficient; S_{max} is the maximum concentration of medium in the Langmuir nonlinear isotherm; n is the power index in the Freundlich nonlinear isotherm; K_L is the coefficient in the Langmuir or Freundlich nonlinear isotherm. The FEMWATER software employed the hybrid Lagrangian–Eulerian finite element method to solve Eq. 10.17 along with the initial and boundary conditions.

In a saturated aquifer, the governing equation of solute transport becomes:

$$\theta \frac{\partial C}{\partial t} = \nabla \cdot (\theta D \nabla C) - V \cdot \nabla C + wC_{in} + \sum R_n, \tag{10.18}$$

where ΣR_n represents the chemical reaction term, including the aqueous-solid surface reaction (sorption) and first-order rate reaction, the other types of chemical reaction, and so on.

One of the typical software package for solving this problem is MT3DMS developed by Zheng and Wang (1999), in which the Darcy velocity V is computed by MODFLOW. MT3DMS has a multicomponent program structure that can accommodate add-on reaction packages for modeling general biological and geochemical reactions. Konikow et al. (1996) developed a 3D method-of-characteristics solute-transport model (MOC3D) simulating 3D solute transport in flowing groundwater. Similar to MT3DMS, MOC3D is integrated with MODFLOW. MOC3D employed the method-of-characteristics to solve the transport equation on the basis of the hydraulic gradients computed with MODFLOW for a given time step. One point to note is that the governing equation of solute transport is a type of advection–dispersion equation (ADE), which can also be used to describe the heat movement. Therefore, the simulation of heat transport may be conducted by modifying the computer program of solute transport, owing to the mathematical similarities between heat and solute transport. Hecht-Méndez et al. (2010) evaluated MT3DMS for simulating heat transport against the analytical solutions, FEFLOW codes and SEAWAT codes, and results showed that the solutions of MT3DMS and other approaches agreed with each other very well. Both FEFLOW and SEAWAT are designed to simulate heat transport in aquifers. In addition, the FEMWATER, SUTRA, HYDUS-3D, 3DFEMFAT, HST3D, and FEFLOW software packages are capable of solving ADEs with different initial and boundary conditions.

Table 10.1 presents the summary for the software package mentioned in this study from three aspects, mathematic model, the capability of the

TABLE 10.1 Summary of the Software Simulating Groundwater Flow and Solute or Energy Transport

Software	Model	Capability	Method
FEMWATER	3D modified Richards equation for groundwater flow and 3D ADE for transport	Density-dependent flow and solute transport in variably saturated media	Galerkin finite element method for flow, the hybrid Lagrangian–Eulerian finite element methods for transport
SUTRA	3D Richards equation for groundwater flow and 3D ADE for transport	Density-dependent flow, solute, or heat transport in variably saturated media	The hybrid method of finite-element and integrated-finite-difference method
HYDUS-3D	3D Richards equation for flow and 3D ADE for heat and solute transport	Variably saturated flow, solute, or heat transport	Galerkin-type linear finite element schemes
3DFEMFAT	3D modified Richards equation for groundwater flow and 3D ADE for transport	Density-dependent flow, solute, or heat transport in variably saturated media	Finite-element method for flow, and the hybrid Lagrangian–Eulerian finite-element approach for transport
HST3D	3D Darcian equation for flow and 3D ADE for heat and solute transport	Density-dependent flow, solute, or heat transport in saturated media	Finite-difference method
MODFLOW	3D Darcian equation for flow	Saturated flow	Finite-difference method
FEFLOW	3D Richards equation for flow and 3D ADE for heat and solute transport	Density-dependent and temperature-dependent flow, solute, or heat transport in variably saturated media	Finite-element method
MT3DMS	3D ADE for heat and solute transport	Solute or heat transport in saturated media	Include the standard finite-difference method, several mixed Eulerian–Lagrangian methods, and a third-order TVD method

(Continued)

TABLE 10.1 Summary of the Software Simulating Groundwater Flow and Solute or Energy Transport *(cont.)*

Software	Model	Capability	Method
MOC3D	3D ADE for solute transport	Solute transport in saturated media	Eulerian–Lagrangian localized adjoint method algorithm incorporating an implicit-in-time difference approximation for the dispersive and sink terms
AQUA3D	3D Darcian equation for flow and 3D ADE for solute transport	Flow and solute transport in saturated media	Galerkin finite-element method

software, and the numerical method solving the mathematic model. FEMWATER, SUTRA, HYDUS-3D, 3DFEMFAT, and FEFLOW can be used to solve the problems of flow and transport in variable-saturated aquifers. AQUA3D and HST3D are capable of handling flow and transport in saturated aquifers. MODFLOW is a very powerful tool to simulate groundwater flow in saturated aquifers, while MT3DMS and MOC3D are for transport, which are integrated into MODFLOW for flow velocity information. Each software package has its own advantages, for instance HST3D considered the effect of the fluid density, temperature, and mass fraction of solute in the fluid phase on pressure change (Kipp, 1987). In addition to these software packages, there are still many other software packages for simulating flow and transport, such as PHAST1.2 (Parkhurst et al., 2004), RT3D, TOUGH2, TOUGHREACT, AQUA3D, and so on.

10.4 COUPLED MODELS

Up to date, the mechanisms of mass flux and energy exchange in a subregion of ECZ seems clear in the previous studies, and many mathematic models have been developed to describe the processes occurring in the subregion. However, integrating the models simulating the processes in a sublocal region over ECZ scale is challenging for the actual problems, due to the temporal and spatial scale discrepancy of processes in each subregion. The velocity of water flowing laterally and vertically through aquifers is substantially slower than that in rivers/streams. By reviewing the literature, we find that a number of approaches have been proposed to simulate the process in ECZ, and we introduce some representative ones in this study.

10.4.1 Coupling Model of CLM and ParFlow

Maxwell and Miller (2005) presented a comprehensive modeling approach by integrating subsurface flow and overland flow processes in a coherent, numerical model framework. They coupled CLM by Dai et al. (2003) and a variably saturated ParFlow by Ashby and Falgout (1996) to interpret the effect of a dynamic water table on simulated watershed flow, in which Parflow is an isothermal, variably saturated groundwater flow solver driven by external boundary conditions (Ashby and Falgout, 1996). CLM can be used to simulate the energy and water balance at the land surface, and it is initially designed as a truly community-developed land component of the community climate system model. Two models were coupled at the land surface by replacing the soil column/root-zone soil moisture formulation in CLM with the ParFlow formulation (Eq. 10.1 was replaced by Eq. 10.9). All processes within CLM, except for equations predicting soil moisture, are preserved with the original CLM. The CLM simulation of soil moisture, replaced by ParFlow's formulation, results in a continuous hydrologic scheme, especially at the bottom layer of CLM. The results of soil moisture simulations by ParFlow are passed into CLM to calculate the infiltration, evaporation, and root uptake fluxes. These fluxes are feedback to ParFlow again, in which they are treated as water fluxes into or out of the model. The detailed information can be seen in Maxwell and Miller (2005).

In a series of papers, Maxwell and co-workers evaluted the coupling model of CLM and ParFlow through the field data and scenario models, to demonstrate that incorporating groundwater leads to more realistic patterns of soil moisture and runoff on the landscape. For example, Maxwell and Miller (2005) applied the coupling model of LSM and ParFlow to the synthetic data and data from the Project for Intercomparison of Landsurface Parameterization Schemes (PILPS), version 2(d), 18-year dataset from Valdai, Russia. Ferguson and Maxwell (2010) used this model to analyze integrated watershed response and groundwater–land surface feedbacks in the Little Washita River watershed, located within the southern Great Plains region of North America, under observed and perturbed climate conditions. Maxwell and Kollet (2008) compared three scenario simulations with modified atmospheric forcing in terms of temperature and precipitation with a simulation of present-day climate. Rihani et al. (2010) used idealized simulations to investigate the role of terrain and subsurface heterogeneity on the interactions between groundwater dynamics and land surface energy fluxes by the coupling model of LSM and ParFlow. Condon et al. (2013) analyzed the impact of aquifer characterization on land energy fluxes, using coupling model of LSM and ParFlow, and conclued that heterogeneity did not fundamentally change the connection between groundwater and land surface processes.

In summary, most studies on the coupled model of CLM and ParFlow demostrate that the sensitivity of hydrologic response to climate change is related to the feedbacks between groundwater and land surface processes, especially in regions with a shallow water table.

10.4.2 Coupled Land–Atmosphere Simulation Program (CLASP)

To study potential feedbacks between different water reservoirs, Gutowski et al. (2002) developed a Coupled Land–Atmosphere Simulation Program (CLASP) that represented hydrologic processes extending from the atmosphere through the land surface to aquifers and, ultimately, river systems. CLASP consists of three modules: the atmospheric column model (ATMOS); a soil–vegetation–atmosphere transfer (SVAT) model that computes the exchange of water, energy, and momentum between the land surface and the atmosphere; and a groundwater/surface water (GW/SW) model that simulates the flow of surplus water from SVAT through a subsurface and river drainage network.

Gutowski et al. (2002) used observations from the First International Satellite Land Surface Climatology Project (ISLSCP) Field Experiment (FIFE) (Sellers et al., 1992) to evaluate further CLASP behavior. Subsequently, York et al. (2002) pointed out that CLASP I by Gutowski et al. (2002) utilized an idealized, quasi two-dimensional representation of aquifers, with restricted linkages between groundwater and surface water. York et al. (2002) developed CLASP II using a fully three-dimensional groundwater model (MODFLOW) to describe the groundwater flow, and used a 9-year historical record in the Mill Creek catchment in northeastern Kansas to evaluate CLASP II. They found that the CLASP II model captured monthly and yearly hydrologic trends in streamflow, aquifer levels, evapotranspiration, and soil moisture.

10.4.3 Flux-PIHM Model

Flux-PIHM was developed by Shi et al. (2013) by coupling an LSM and the Penn State Integrated Hydrologic Model (PIHM) (Qu, 2004; Qu and Duffy, 2007). The LSM was adapted from the Noah LSM (Chen and Dudhia, 2001; Ek et al., 2003). PIHM was a multiprocess and multiscale hydrologic model, and it could simulate evapotranspiration, infiltration, recharge, overland flow, groundwater flow, and channel routing in a fully coupled scheme (Qu, 2004; Qu and Duffy, 2007). Channel flow and overland flow were respectively governed by the 1D and 2D Saint-Venant equations (Saint-Venant, 1871). Groundwater flow was described using the Richards equation, in which the unsaturated hydraulic conductivities were calculated using the van Genuchten equations. Above two models were coupled by exchanging water table depth, soil moisture, infiltration rate, recharge rate, net precipitation rate, and evapotranspiration rate. Adding the capability of simulating surface energy balance to PIHM, the Flux-PIHM improved the prediction of hourly evapotranspiration, discharge, and peak discharge events, especially after extended dry periods.

To evaluate the coupled model of Flux-PIHM, Shi et al. (2013) employed this model to interpret the data from the Shale Hills watershed in central Pennsylvania, and the results showed that the predictions of the hydrologic data and surface energy balance date agreed with observations very well. Shi et al. (2013) concluded that annual average sensible and latent heat fluxes were strongly correlated with water table depth, especially near the stream.

10.4.4 MM5-VIC Model

MM5 is a regional climate model, the fifth generation Pennsylvania State University-NCAR Mesoscale Model, and has been widely used to simulate the hydroclimate conditions in widely different climate regimes (Leung et al., 2003; Qian and Leung, 2007). VIC is a macroscale hydrologic model based on the three-layer VIC model (Cherkauer and Lettenmaier, 2003; Liang et al., 1994, 1999, 2003; Liang and Xie, 2001). When simulating the dynamic movement of groundwater table, VIC employs the two-way coupling approach by Liang et al. (2003). This approach solves the soil moisture profile using Richards equation to the unsaturated zone only, treating the groundwater table as a moving boundary. Leung et al. (2011) pointed out that this metheod was better than model of TOPMODEL when simulating the dynamic interactions between surface water and groundwater. To assess the effect of climate, soil, and vegetation on groundwater table dynamics, and its potential feedbacks to the climate, MM5 and VIC were coupled by Leung et al. (2011) through a surface layer parameterization. In the surface layer parameterization, the mass and energy fluxes between the land surface and the atmosphere were determined by the exchange coefficients for momentum, heat, and moisture (Troen and Mahrt, 1986).

Leung et al. (2011) employed the MM5-VIC model to check the interactions bwteen the climate and the groundwater table variation, and they concluded that precipitation and vegetation properties (including minimum stomatal resistance, and root depth and fraction) could influence the spatial and temporal variations of groundwater table. By comparing two climate simulations with and without groundwater table dynamics, results showed that the dry bias by the MM5-VIC model is slightly reduced in the summer precipitation over the central and eastern United States, and the groundwater table dynamics could provide important feedbacks to atmospheric processes. Such feedbacks were stronger in regions with deeper groundwater table.

10.5 SUMMARY AND CONCLUSIONS

By reviewing the literature on the research of Earth-surface processes in ECZ, we find that there are many new findings demonstrating that the correlations between groundwater table depth and Earth-surface processes are very strong, especially when the water table is very shallow, such as less than 5 m below the surface. Many coupled mathematic models have been developed for modeling the Earth-surface processes. However, the current knowledge of interactions among climate, land surface processes, and subsurface processes (especially the groundwater table) is still limited. Some important research problems of ECZ related to groundwater may contain several aspects as follows.

1. *The mechanism of energy, water, and carbon cycles in ECZ.* Many pervious studies of the mechanism of the complex physical, chemical, biological, and geological processes were primarily conducted at local temporal and spatial scales. Whether or not these theories work at the large temporal and spatial scales like ECZ is still debatable.
2. *Coupled models of integrated processes in ECZ.* Multiscale coupled models are needed when describing the integrated processes in ECZ, especially for the biological processes. The mathematic models of the pore, core, field, and watershed scales might be vastly different. Therefore, unified multiscale coupled models are much needed in the future.
3. *Solution of the coupled models in ECZ.* Theoretically, different processes in ECZ affect each other directly or indirectly. Integrating the models of climate, land surface processes, and subsurface processes remains a significant challenge, due to the much difference of temporal and spatial scales among these processes, and the demand for fast and accurate computational methods when solving large scale coupled models. For instance, climate modeling is from the regional to global spatial scales and from seasonal to decadal timescales, while subsurface hydrology is always from the hillslope to catchment spatial scales and from day to century timescales.

In summary, when groundwater is a concern, improving the understanding of integrated processes over multiple spatial and temporal scales in ECZ and finding efficient and accurate computational methodology of dealing with the resulting coupled models remain the central focus of future studies.

ACKNOWLEDGMENTS

This research was partially supported by Program of the China Postdoctoral Science Foundation funded Project (No. 2014M560635), National Basic Research Program of China (973) (Nos. 2011CB710600, 2011CB710602), National Basic Research Program of China (863) (No. 2012AA062602), National Natural Science Foundation of China (Nos. 41120124003, 41172281, 41372253), and 1:5 million hydrogeological investigations in Jianghan Plain (No. 12120114069301).

REFERENCES

Ahuja, L.R., Ascough Ii, J.C., David, O., 2005. Developing natural resource models using the object modeling system: feasibility and challenges. Adv. Geosci. 4, 29–36.

Anderson, S.P., von Blanckenburg, F., White, A.F., 2007. Physical and chemical controls on the critical zone. Elements 3 (5), 315–319.

Anderson, S.P., Bales, R.C., Duffy, C.J., 2008. Critical Zone Observatories: building a network to advance interdisciplinary study of Earth surface processes. Mineral. Mag. 72 (1), 7–10.

Andrews, D.M., Lin, H., Zhu, Q., Jin, L., Brantley, S.L., 2011. Hot spots and hot moments of dissolved organic carbon export and soil organic carbon storage in the Shale Hills Catchment. Vadose Zone J. 10 (3), 943–954.

Arnold, J.G., Srinivasan, R., Muttiah, R.S., Williams, J.R., 1998. Large area hydrologic modeling and assessment – part 1: model development. J. Am. Water Resour. Assoc. 34 (1), 73–89.

Ashby, S.F., Falgout, R.D., 1996. A parallel multigrid preconditioned conjugate gradient algorithm for groundwater flow simulations. Nucl. Sci. Eng. 124, 145–159.

Bales, R.C., Hopmans, J.W., O'Geen, A.T., Meadows, M., Hartsough, P.C., Kirchner, P., Hunsaker, C.T., Beaudette, D., 2011. Soil moisture response to snowmelt and rainfall in a Sierra Nevada mixed-conifer forest. Vadose Zone J. 10 (3), 786–799.

Banwart, S., Bernasconi, S.M., Bloem, J., Brandao, M., Brantley, S., Chabaux, F., Duffy, C., Kram, P., Lair, G., Lundin, L., Nikolaidis, N., Novak, M., Panagos, P., Ragnarsdottir, K.V., Reynolds, B., Rousseva, S., de Ruiter, P., van Gaans, P., van Riemsdijk, W., White, T., Zhang, B., 2011. Soil processes and functions in Critical Zone Observatories: hypotheses and experimental design. Vadose Zone J. 10 (3), 974–987.

Befus, K.M., Sheehan, A.F., Leopold, M., Anderson, S.P., Anderson, R.S., 2011. Seismic constraints on critical zone architecture, Boulder Creek Watershed, Front Range, Colorado. Vadose Zone J. 10 (3), 915–927.

Beven, K., 1997. Process, heterogeneity and scale in modelling soil moisture fluxes. In: Sorooshian, S., Gupta, H., Rodda, J. (Eds.), Land Surface Processes in Hydrology. NATO ASI Series. Springer, Berlin, Heidelberg, pp. 191–213.

Bonan, G.B., 1996. A Land Surface Model (LSM version 1.0) for ecological, hydrological, and atmospheric studies: Technical description and user's guide, NCAR Technical Note NCAR/TN-417+STR. National Center for Atmospheric Research, Boulder, Colorado. 150 pp.

Brantley, S.L., Goldhaber, M.B., Ragnarsdottir, K.V., 2007. Crossing disciplines and scales to understand the critical zone. Elements 3 (5), 307–314.

Calmels, D., Galy, A., Hovius, N., Bickle, M., West, A.J., Chen, M.C., Chapman, H., 2011. Contribution of deep groundwater to the weathering budget in a rapidly eroding mountain belt, Taiwan. Earth Planet. Sci. Lett. 303 (1–2), 48–58.

Chen, F., Dudhia, J., 2001. Coupling an advanced land surface–hydrology model with the Penn State-NCAR MM5 modeling system. Part I: model implementation and sensitivity. Mon. Weather Rev. 129 (4), 569–585.

Chen, X., Hu, Q., 2004. Groundwater influences on soil moisture and surface evaporation. J. Hydrol. 297 (1–4), 285–300.

Cherkauer, K.A., Lettenmaier, D.P., 2003. Simulation of spatial variability in snow and frozen soil. J. Geophys. Res. Atmos. 108 (D22), 8858.

Chorover, J., Kretzschmar, R., Garcia-Pichel, F., Sparks, D.L., 2007. Soil biogeochemical processes within the critical zone. Elements 3 (5), 321–326.

Chorover, J., et al. 2011. How water, carbon, and energy drive critical zone evolution: the Jemez–Santa Catalina Critical Zone Observatory. Vadose Zone J. 10 (3), 884–899.

Clapp, R.B., Hornberger, G.M., 1978. Empirical equations for some soil hydraulic properties. Water Resour. Res. 14 (4), 601–604.

Condon, L.E., Maxwell, R.M., Gangopadhyay, S., 2013. The impact of subsurface conceptualization on land energy fluxes. Adv. Water Resour. 60, 188–203.

Cosby, B.J., Hornberger, G.M., Clapp, R.B., Ginn, T.R., 1984. A statistical exploration of the relationships of soil moisture characteristics to the physical properties of soils. Water Resour. Res. 20, 682–690.

Dai, Y.J., Zeng, Q.C., 1997. A land surface model (IA94) for climate studies. Part I: formulation and validation in of-line experiments. Adv. Atmos. Sci. 14, 433–460.

Dai, Y., Zeng, X., Dickinson, R.E., Baker, I., Bonan, G.B., Bosilovich, M.G., Denning, A.S., Dirmeyer, P.A., Houser, P.R., Niu, G., Oleson, K.W., Schlosser, C.A., Yang, Z.L., 2003. The common land model. Bull. Am. Meteorol. Soc. 84 (8), 1013–1023.

David, O., Ascough, II, J.C., Lloyd, W., Green, T.R., Rojas, K.W., Leavesley, G.H., Ahuja, L.R., 2013. A software engineering perspective on environmental modeling framework design: the object modeling system. Environ. Modell. Softw. 39, 201–213.

Di Luzio, M., Srinivasan, R., Arnold, J.G., 2004. A GIS-coupled hydrological model system for the watershed assessment of agricultural nonpoint and point sources of pollution. Trans. GIS 8 (1), 113–136.

Dickinson, R.E., Henderson-Sellers, A., Kennedy, P.J., Wilson, M.F., 1986. Biosphere atmosphere transfer scheme (BATS) for the NCAR community climate model. NCAR Technical Note, NCAR/TN-275 + STR.

Eagleson, P.S., 1982. Land Surface Processes in Atmospheric General Circulation Models. Cambridge University Press, Cambridge, UK.

Ek, M.B., Mitchell, K.E., Lin, Y., Rogers, E., Grunmann, P., Koren, V., Gayno, G., Tarpley, J.D., 2003. Implementation of Noah land surface model advances in the National Centers for Environmental Prediction operational mesoscale Eta model. J. Geophys. Res. Atmos. 108 (D22), 8851.

Ferguson, I.M., Maxwell, R.M., 2010. Role of groundwater in watershed response and land surface feedbacks under climate change. Water Resour. Res. 46 (10), W00F02.

Franks, S.W., Beven, K.J., Quinn, P.F., Wright, I.R., 1997. On the sensitivity of soil-vegetation-atmosphere transfer (SVAT) schemes: equifinality and the problem of robust calibration. Agric. Forest Meteorol. 86 (1–2), 63–75.

Frind, E.O., 1982. Simulation of long-term transient density-dependent transport in groundwater. Adv. Water Resour. 5 (2), 73–88.

Graham, D.N., Butts, M.B., 2005. Flexible integrated watershed modelling with MIKE SHE. In: Singh, V.P., Frevert, D.K. (Eds.), Watershed Models. CRC Press, Boca Raton, pp. 245–272.

Graham, C.B., Lin, H.S., 2011. Controls and frequency of preferential flow occurrence: a 175-event analysis. Vadose Zone J. 10 (3), 816–831.

Gutowski, Jr., W.J., Vorosmarty, C.J., Person, M., Otles, Z., Fekete, B., York, J., 2002. A coupled land-atmosphere simulation program (CLASP): calibration and validation. J. Geophys. Res. Atmos. 107 (D16), 4283.

Hecht-Méndez, J., Molina-Giraldo, N., Blum, P., Bayer, P., 2010. Evaluating MT3DMS for heat transport simulation of closed geothermal systems. Ground Water 48 (5), 741–756.

Jin, L., Andrews, D.M., Holmes, G.H., Lin, H., Brantley, S.L., 2011. Opening the "back box": water chemistry reveals hydrological controls on weathering in the Susquehanna Shale Hills Critical Zone Observatory. Vadose Zone J. 10 (3), 928–942.

Jones, M.J., 1985. The weathered zone aquifers of the basement complex areas of Africa. Q. J. Eng. Geol. Hydrogeol. 18 (1), 35–46.

Kim, H., Bishop, J.K.B., Dietrich, W.E., Fung, I.Y., 2014. Process dominance shift in solute chemistry as revealed by long-term high-frequency water chemistry observations of groundwater flowing through weathered argillite underlying a steep forested hillslope. Geochim. Cosmochim. Acta 140, 1–19.

Kipp, K.L., 1987. HST3D: A computer code for simulation of heat and solute transport in three-dimensional ground-water flow systems. US Geological Survey Water-Resources Investigations Report 86-4095, Reston, Virginia, USA.

Konikow, L.F., Goode, D.J., Hornberger, G.Z., 1996. A three-dimensional method-of-characteristics solute-transport model (MOC3D). US Geological Survey Water-Resources Investigation Report 96-4267, Reston, Virginia, USA.

Kuntz, B.W., Rubin, S., Berkowitz, B., Singha, K., 2011. Quantifying solute transport at the Shale Hills Critical Zone Observatory. Vadose Zone J. 10 (3), 843–857.

Leung, L.R., Huang, M., Qian, Y., Liang, X., 2011. Climate–soil–vegetation control on groundwater table dynamics and its feedbacks in a climate model. Climate Dynam. 36 (1-2), 57–81.

Leung, L.R., Qian, Y., Bian, X., Hunt, A., 2003. Hydroclimate of the western United States based on observations and regional climate simulation of 1981–2000. Part I: seasonal statistics. J. Clim. 16 (12), 1892–1912.

Liang, X., Xie, Z.H., 2001. A new surface runoff parameterization with subgrid-scale soil heterogeneity for land surface models. Adv. Water Resour. 24 (9–10), 1173–1193.

Liang, X., Lettenmaier, D.P., Wood, E.F., Burges, S.J., 1994. A simple hydrologically based model of land surface water and energy fluxes for general circulation models. J. Geophys. Res. Atmos. 99 (D7), 14415–14428.

Liang, X., Wood, E.F., Lettenmaier, D.P., 1999. Modeling ground heat flux in land surface parameterization schemes. J. Geophys. Res. Atmos. 104 (D8), 9581–9600.

Liang, X., Xie, Z.H., Huang, M.Y., 2003. A new parameterization for surface and groundwater interactions and its impact on water budgets with the variable infiltration capacity (VIC) land surface model. J. Geophys. Res. Atmos. 108 (D16), 8613.

Lin, H., 2003. Hydropedology: bridging disciplines, scales, and data. Vadose Zone J. 2 (1), 1–11.

Lin, H.C.J., Richards, D.R., Talbot, C.A., Yeh, G.T., Cheng, J.R., Cheng, H.P., Jones, N.L., 1997. FEMWATER: A three-dimensional finite element computer model for simulating density-dependent flow and transport in variably saturated media. Technical Report CHL-97-12. U.S. Army Corps of Engineers, Waterways Experiment Station, Vicksburg, Mississippi, USA.

Lin, H., Hopmans, J.W., Richter, D.d., 2011. Interdisciplinary sciences in a global network of Critical Zone Observatories. Vadose Zone J. 10 (3), 781–785.

Liu, Y.B., De Smedt, F., 2004. WetSpa extension, a GIS-based hydrologic model for flood prediction and watershed management. Documentation and User Manual, Department of Hydrology and Hydraulic Engineering, Vrije Universiteit Brussel, Belgium. http://www.vub.ac.be/WetSpa/downloads/WetSpa_manual.pdf (accessed 22.03.15).

Lo, M.H., Famiglietti, J.S., 2010. Effect of water table dynamics on land surface hydrologic memory. J. Geophys. Res. Atmos. 115 (D22), D22118.

Manabe, S., Smagorinsky, J., Strickler, R.F., 1965. Simulated climatology of a general circulation model with a hydrologic cycle. Mon. Weather Rev. 93, 769–798.

Maxwell, R.M., Kollet, S.J., 2008. Interdependence of groundwater dynamics and land-energy feedbacks under climate change. Nature Geosci. 1 (10), 665–669.

Maxwell, R.M., Miller, N.L., 2005. Development of a coupled land surface and groundwater model. J. Hydrometeorol. 6 (3), 233–247.

McDonald, M.G., Harbaugh, A.W., 1988. A modular three-dimensional finite-difference ground-water flow model. US Geological Survey Techniques of Water-Resources Investigations, Book 6, Chapter A1, Reston, Virginia, USA.

Minasny, B., Whelan, B.M., Triantafilis, J., McBratney, A.B., 2013. Pedometrics research in the vadose zone – review and perspectives. Vadose Zone J. 12 (4), doi:10.2136/vzj2012.0141.

Moore, R.V., Tindall, C.I., 2005. An overview of the open modelling interface and environment (the OpenMI). Environ. Sci. Policy 8 (3), 279–286.

Neitsch, S.L., Arnold, J.G., Kiniry, J.R., Srinivasan, R., Williams, J.R., 2004. Soil and water assessment tool input/output file documentation, Version 2005. Blackland Research Center, Texas Agricultural Experiment Station, Texas, USA. http://swat.tamu.edu/media/1291/swat2005io.pdf (accessed 22.03.15.).

Newman, B.D., Wilcox, B.P., Archer, S.R., Breshears, D.D., Dahm, C.N., Duffy, C.J., McDowell, N.G., Phillips, F.M., Scanlon, B.R., Vivoni, E.R., 2006. Ecohydrology of water-limited environments: a scientific vision. Water Resour. Res. 42 (6), W06302.

Niu, G.Y., Yang, Z.L., Dickinson, R.E., Gulden, L.E., Su, H., 2007. Development of a simple groundwater model for use in climate models and evaluation with gravity recovery and climate experiment data. J. Geophys. Res. Atmos. 112 (D7), D07103.

National Research Council, 2001. Basic Research Opportunities in Earth Sciences. The National Academies Press, Washington, DC, USA.

Nuttle, W.K., 2002. Eco-hydrology's past and future in focus. Eos Trans. AGU 83, 205–212.

Oleson, K.W., Dai, Y., Bonan, G., Bosilovich, M., Dickinson, R., Dirmeyer, P., Hoffman, F., Houser, P., Levis, S., Niu, G., Thornton, P., Vertenstein, M., Yang, Z., Zeng, X., 2004. Technical description of the Community Land Model (CLM), NCAR Technical Note. NCAR/TN-461+STR, National Center for Atmospheric Research, Boulder, Colorado.

Parkhurst, D.L., Kipp, K.L., Engesgaard, P., Charlton, S.R., 2004. PHAST – a program for simulating groundwater flow, solute transport, and multicomponent geochemical reactions. US Geological Survey Techniques and Methods 6-AB, Reston, Virginia, USA.

Qian, Y., Leung, L.R., 2007. A long-term regional simulation and observations of the hydroclimate in China. J. Geophys. Res. Atmos. 112 (D14), D14104.

Qu, Y., 2004. An integrated hydrologic model for multi-process simulation using semi-discrete finite volume approach. PhD Dissertation. Department of Civil and Environmental Engineering, The Pennsylvania State University. PA, USA.

Qu, Y., Duffy, C.J., 2007. A semidiscrete finite volume formulation for multiprocess watershed simulation. Water Resour. Res. 43 (8), W08419.

Rasmussen, C., Troch, P.A., Chorover, J., Brooks, P., Pelletier, J., Huxman, T.E., 2011. An open system framework for integrating critical zone structure and function. Biogeochemistry 102 (1–3), 15–29.

Rihani, J.F., Maxwell, R.M., Chow, F.K., 2010. Coupling groundwater and land surface processes: idealized simulations to identify effects of terrain and subsurface heterogeneity on land surface energy fluxes. Water Resour. Res. 46 (12), W12523.

Rodriguez-Iturbe, I., 2000. Ecohydrology: a hydrologic perspective of climate–soil–vegetation dynamics. Water Resour. Res. 36 (1), 3–9.

Saint-Venant, B., 1871. Theory of unsteady water flow with application to floods and to propagation of tides in river channels. Proc. French Acad. Sci 73, 148–154.

Sellers, P.J., Mintz, Y., Sud, Y.C., Dalcher, A., 1986. A simple biosphere model (SiB) for use within general circulation models. J. Atmos. Sci. 43 (6), 505–531.

Sellers, P.J., Hall, F.G., Asrar, G., Strebel, D.E., Murphy, R.E., 1992. An overview of the First International Satellite Land Surface Climatology Project (ISLSCP) Field Experiment (FIFE). J. Geophys. Res. Atmos. 97 (D17), 18345–18371.

Shi, Y., Davis, K.J., Duffy, C.J., Yu, X., 2013. Development of a coupled land surface hydrologic model and evaluation at a Critical Zone Observatory. J. Hydrometeorol. 14 (5), 1401–1420.

Srinivasan, R., Ramanarayanan, T.S., Arnold, J.G., Bednarz, S.T., 1998. Large area hydrologic modeling and assessment – Part II: model application. J. Am. Water Resour. Assoc. 34 (1), 91–101.

Steiner, A.L., Pal, J.S., Giorgi, F., Dickinson, R.E., Chameides, W.L., 2005. The coupling of the common land model (CLM0) to a regional climate model (RegCM). Theor. Appl. Climatol. 82 (3–4), 225–243.

Stieglitz, M., Rind, D., Famiglietti, J., Rosenzweig, C., 1997. An efficient approach to modeling the topographic control of surface hydrology for regional and global climate modeling. J. Climate 10 (1), 118–137.

Swarowsky, A., Dahlgren, R.A., Tate, K.W., Hopmans, J.W., O'Geen, A.T., 2011. Catchment-scale soil water dynamics in a Mediterranean-type oak woodland. Vadose Zone J. 10 (3), 800–815.

Takagi, K., Lin, H.S., 2011. Temporal dynamics of soil moisture spatial variability in the Shale Hills Critical Zone Observatory. Vadose Zone J. 10 (3), 832–842.

Thorne, D., Langevin, C.D., Sukop, M.C., 2006. Addition of simultaneous heat and solute transport and variable fluid viscosity to SEAWAT. Comput. Geosci. 32 (10), 1758–1768.

Troen, I.B., Mahrt, L., 1986. A simple model of the atmospheric boundary layer; sensitivity to surface evaporation. Boundary-Layer Meteorol. 37 (1–2), 129–148.

Vionnet, L.B., Maddock, T., Goodrich, D.C., 1997. Investigations of stream-aquifer interactions using a coupled surface-water and ground-water flow model. HWR No. 1997-020. Department of Hydrology and Water Resources, The University of Arizona.

Voss, C.I., 1984. SUTRA: A finite-element simulation model for saturated-unsaturated, fluid-density-dependent ground-water flow with energy transport or chemically-reactive single-species solute transport. US Geological Survey Water-Resources Investigations Report 84-4369, Reston, Virginia, USA.

Walko, R.L., Band, L.E., Baron, J., Kittel, T.G.F., Lammers, R., Lee, T.J., Ojima, D., Pielke, Sr., R.A., Taylor, C., Tague, C., Tremback, C.J., Vidale, P.L., 2000. Coupled atmosphere–biophysics–hydrology models for environmental modeling. J. Appl. Meteorol. 39 (6), 931–944.

Wigmosta, M.S., Lettenmaier, D.P., 1999. A comparison of simplified methods for routing topographically driven subsurface flow. Water Resour. Res. 35 (1), 255–264.

Wigmosta, M.S., Vail, L.W., Lettenmaier, D.P., 1994. A distributed hydrology–vegetation model for complex terrain. Water Resour. Res. 30 (6), 1665–1679.

Williams, M.W., Barnes, R.T., Parman, J.N., Freppaz, M., Hood, E., 2011. Stream water chemistry along an elevational gradient from the continental divide to the foothills of the Rocky Mountains. Vadose Zone J. 10 (3), 900–914.

Yang, Z.L., 2004. Modeling land surface processes in short term weather and climate studies. In: Zhu, X., Li, X., Cai, M., Zhou, S., Zhu, Y., Jin, F.F., Zou, X., Zhang, M. (Eds.), Observation, Theory and Modeling of Atmospheric Variability. World Scientific Series on Meteorology of East Asia. World Scientific, New Jersey, USA, pp. 288–313.

Yang, Z.L., Dickinson, R.E., Henderson-Sellers, A., Pitman, A.J., 1995. Preliminary study of spin-up processes in land surface models with the first stage data of Project for Intercomparison of Land Surface Parameterization Schemes Phase 1(a). J. Geophys. Res. 100 (D8), 16553–16578.

Yang, F., Zhang, G.L., Yang, J.L., Li, D.C., Zhao, Y.G., Liu, F., Yang, R.M., Yang, F., 2014. Organic matter controls of soil water retention in an alpine grassland and its significance for hydrological processes. J. Hydrol. 519, 3086–3093.

Yeh, P.J.F., Eltahir, E.A.B., 2005. Representation of water table dynamics in a land surface scheme. Part II: subgrid variability. J. Climate 18 (12), 1881–1901.

Yeh, G.T., Cheng, J.R., Cheng, H.P., 1994. 3DFEMFAT: A 3-dimensional finite element model of density-dependent flow and transport through saturated-unsaturated media, version 2.0. Department of Civil and Environmental Engineering, Pennsylvania State University, University Park, PA, USA.

York, J.P., Person, M., Gutowski, W.J., Winter, T.C., 2002. Putting aquifers into atmospheric simulation models: an example from the Mill Creek Watershed, northeastern Kansas. Adv. Water Resour. 25 (2), 221–238.

Yu, Z., Lakhtakia, M.N., Yarnal, B., White, R.A., Miller, D.A., Frakes, B., Barron, E.J., Duffy, C., Schwartz, F.W., 1999. Simulating the river-basin response to atmospheric forcing by linking a mesoscale meteorological model and hydrologic model system. J. Hydrol. 218 (1–2), 72–91.

Zacharias, S., Bogena, H., Samaniego, L., Mauder, M., Fu, R., Putz, T., Frenzel, M., Schwank, M., Baessler, C., Butterbach-Bahl, K., Bens, O., Borg, E., Brauer, A., Dietrich, P., Hajnsek, I., Helle, G., Kiese, R., Kunstmann, H., Klotz, S., Munch, J.C., Papen, H., Priesack, E., Schmid, H.P., Steinbrecher, R., Rosenbaum, U., Teutsch, G., Vereecken, H., 2011. A network of terrestrial environmental observatories in Germany. Vadose Zone J. 10 (3), 955–973.

Zeng, X., Shaikh, M., Dai, Y., Dickinson, R.E., Myneni, R., 2002. Coupling of the common land model to the NCAR community climate model. J. Climate 15, 1832–1854.

Zheng, C.M., Wang, P.P., 1999. MT3DMS: A modular three-dimensional multispecies transport model for simulation of advection, dispersion, and chemical reactions of contaminants in groundwater systems, Contract Report SERDP-99-1, US Army Corps of Engineers, Engineering Research and Development Center, Vicksburg, Mississippi, USA.

Chapter 11

A Review of Mass Movement Processes and Risk in the Critical Zone of Earth

Netra R. Regmi*, John R. Giardino**, Eric V. McDonald***, and John D. Vitek**

**Department of Soil, Water and Environmental Science, University of Arizona, Tucson, Arizona, USA; **High Alpine and Arctic Research Program, Department of Geology and Geophysics, Texas A&M University, College Station, Texas, USA; ***Desert Research Institute, Division of Earth and Ecosystem Sciences, Reno, Nevada, USA*

11.1 INTRODUCTION

The Critical Zone is defined as "the heterogeneous, near-surface environment, where complex interactions involving rock, soil, water, air, and living organisms regulate natural habitats and determine the availability of life-sustaining resources" (National Research Council, 2001, p. 89). The zone extends from the top of the canopy to the bottom of the groundwater aquifer. Mass movement, one of the main processes of the Critical Zone in areas that have varying elevations connected by slopes, represents complex hydrologic, atmospheric, cryogenic, soil-geomorphic, and biogeochemical feedbacks in response to climatic, tectonic, and anthropogenic forcing at a range of spatial and temporal scales. Mass movement occurs in all climatic regions and plays a major role in the sustainability of terrestrial life and environment.

Processes of mass movement are the dominant modes of mobilization (i.e., erosion), transport, and deposition of materials in the Critical Zone. Mass movement is the result of a variety of geomorphic, geologic, and hydrologic factors, which predispose hillslopes towards instability. Events such as earthquakes, intense rainfalls, and snowmelt trigger various types of mass movement, which yield a large amount of mass (i.e., sediment) from the hillslopes and play a significant role in the modification of the hillslope by transporting mass from a slope to the base below. The velocity of movement of slope material downslope ranges from extremely slow or imperceptible to extremely rapid movement (5 m/s). The sediment transported by these processes moves downslope and is deposited along valley floors, in lakes, along floodplains, or into oceans by

Developments in Earth Surface Processes, Vol. 19. http://dx.doi.org/10.1016/B978-0-444-63369-9.00011-4

fluvial, glacial, and/or anthropogenic processes. The volume of debris mobilized by mass movement depends on a combination of the spatial distribution and frequency of triggering events, the number of failures triggered in a given event, the probability distribution of mass-movement volume for such a triggering event, and the flux of the debris from mass movement into the channel network.

The study of mass movement is one of the major interests among geomorphologists and engineers. Literature on mass movement exists from the beginning of the nineteenth century (Hubbard, 1908; Van Horn, 1909; Sharpe, 1938) to recent (Brunsden, 1993; Shroder et al., 2011; Roering, 2012). This chapter briefly discusses the types, mechanics, and causes of mass movement with detailed discussions of causes and mechanics of slides, creeps, topples, and falls. The chapter also discusses spatial and temporal distribution of landslides in terms of frequency, magnitude, and sediment yield. We specifically focus on the contribution to the evolution of the Critical Zone in mountain environments.

11.2 MASS MOVEMENT IN THE CRITICAL ZONE OF EARTH

The Critical Zone of Earth is a highly dynamic, somewhat heterogeneous and complex region consisting of a number of distinct layers vertically below and above the surface of the ground. Each of the layers below the ground (i.e., soil/regolith, rock, root zone, vadose zone, permafrost, and aquifer zone), in general, has distinctive physical, chemical, and biological characteristics and exhibits characteristic responses and feedbacks to perturbations in the layers above the ground (i.e., atmosphere, climate, landscapes, and vegetation) as well as hydrosphere, biosphere, cryosphere, pedosphere, and lithosphere (Arnold et al., 1990; Anderson et al., 2012). Major interfaces in the Critical Zone include: land surface–atmosphere interface, soil–vegetation interface, vadose zone and groundwater interface, surface water–groundwater interface, soil–stream interface, and soil–bedrock interface. These interfaces are unique and important controls of the processes in the Critical Zone and relationships among landscape, soil, water, ecosystem, and climate. The interactions of the components in these interfaces determine the balance of energy and mass and result in the evolution of the environment in the Critical Zone. For example, fluctuation in groundwater has a significant influence on the energy balance in the soil surface and subsurface, and, thereby, on the nature and properties of soil (Narasimhan, 2005; Bales et al., 2011), which in turn control mass movement, erosion, and ecosystem processes (Regmi et al., 2010a; Lauber et al., 2008; Richter et al., 2014). Important interfaces also exist within the soil including: the soil–horizons interface, the soil–root interface, the macropore–matrix interface, the microbe–aggregate interface, the water–air interface, and the soil–water table interface. Slight perturbations in these interfaces also trigger change in soil properties, such as moisture and shear strength and can impact stability of the hillslope, transport of sediment and water, and soil–water reaction processes (Lin et al., 2008; Lin and Zhou, 2008).

The Critical Zone is an open system where transfer of mass and energy across scales drive its evolution and functioning (Rasmussen, 2012). The changes in the zone are generally irreversible and cumulative (Lin, 2010), which may be associated with either slow, more gradual, or extreme abrupt changes in energy (i.e., solar radiation) and/or mass (i.e., water and sediment). Major processes that induce an abrupt change that is pervasive and long lasting in this environment include geologic processes such as, uplift and weathering of rocks, erosion, transport, and deposition of weathered products that eventually deposit in rivers, lakes and oceans, water and wind that serve as key conduits as well as agents for mass and energy transfer (National Research Council, 2001; Richter, 2007; Richter and Mobley, 2009; Anderson et al., 2012, 2013) and anthropogenic forces capable of reshaping the surface of Earth (Clark et al., 2004). In addition, biologic processes that are vastly different in space and timescales (e.g., occur in shorter timescales and in small areas) compared to geologic processes are also critical to understanding the evolution and dynamic functions of the Critical Zone. Biologic processes control the production and consumption of food and energy in the ecosystem, and the accumulation and decomposition of organic materials in soils, which in turn regulate the energy, nutrients, and trophic relations, and determine the overall nature of an ecosystem. The Critical Zone, therefore, is a coupled system of geologic, hydrologic, and biologic processes over vastly different spatial and temporal scales.

Mass movement, one of the major geologic processes in the Critical Zone, occurs as a result of a coupled interaction of geologic, hydrologic, atmospheric, biologic, and anthropogenic forces and plays a critical role in the development and evolution of the Critical Zone by transferring mass and energy across a range of temporal and spatial scales vertically as well as horizontally within the Critical Zone (Larsen et al., 2010; De Graff et al., 2011). For example, mass movement generally occurs on steep topography where focused water flow on the surface and in the subsurface; the slope material is fully saturated, and the forces that resist the slope movement are minimal. The following sections discuss the relationships of mass movement with the evolution and function of the Critical Zone by providing a detailed discussion on the forces in the Critical Zone that are responsible for the movement of materials from the slope; the causes and characteristics of different types of mass movement; and the roles of mass movement in the evolution of the Critical Zone.

11.3 FORCES IN MASS MOVEMENT

The various types of mass movement (i.e., movement of snow, soil, debris, vegetation, and rock) are similar in that, all initiate when the shear stress (i.e., driving force) tends to displace slope material by exceeding the resisting strength (i.e., resisting force). No slope failure will occur as long as slope material maintains its internal resistance at a level greater than the driving force. The principal driving force in slope failure is gravity and the principal resisting force is the shear strength of the slope material. Other factors that assist movement of mass by reducing the shear strength of slope materials are called triggers, and include forces associated with the atmosphere, biosphere, hydrology, slope morphology, earthquake, and human causes.

11.3.1 Physical and Mineralogical Properties of Soil

11.3.1.1 Shear Strength

Shear strength of slope material is a fundamental property that governs the stability of a hillslope. It resists the stresses generated by gravitational force. Soil strength (S) is described as the function of three major parameters including effective normal stress or net force per unit area acting perpendicular to a slip surface (σ'), cohesion (c), and internal-friction angle (φ) of the slope material, and can be described as:

$$S = c + \sigma' \tan\phi.$$

The importance of normal stress is its capacity to hold material together, thereby increasing the internal resistance to shear. The forces contributing to the total normal stress (σ) perpendicular to the slip surface include weight of the soil/regolith (hereafter soil and regolith terms are used equivalent) column above the failure plane, infiltrated water, moisture content, and any slope surcharge, such as vegetation. The effective normal stress (σ') takes into account the total normal stress (σ), and the pressure exerted by water and air in the pores and interstices (μ). The pore-water pressure can add or detract from the value of total stress depending on the amount of moisture in the soil (Ritter, 1986). In completely dry soil, the pore pressure is atmospheric ($\sigma' = \sigma$, assuming the role of atmospheric pressure is negligible). Below the water table, where soil is saturated and the pore pressure is greater than atmospheric pressure, the pore-water pressure detracts from the total normal stress, so that the effective normal stress is lower ($\sigma' = \sigma - \mu$). Whereas, above the groundwater table, where the soils are commonly unsaturated, the pore-water pressure is negative (i.e., suction) and the effective normal stress is higher ($\sigma' = \sigma + \mu$).

Cohesion depends on bonding of fine-grained particles (i.e., clays and fine silts), physical properties of soils such as soil-moisture content, grain-size distribution and relative density, and root strength. Increasing soil moisture reduces soil cohesion and thus, soil strength. Rogers and Selby (1980) showed that the cohesion component of two landslide soils rich in clay and silty clay decreased by 18% and 70%, respectively, after saturation. But, in unsaturated or a less moisture-content condition, cohesion typically increases proportionately to clay content because of the greater surface-contact area associated with clay particles. Absorption of water and ions by clay minerals creates a binding structure among the particles. In vegetated landscapes, cohesion is supplemented by the contribution of rooting strength.

The internal angle of friction represents the degree of interlocking of individual grains or aggregates and is influenced by shape, size, and packing. Angular particles have greater internal angles of friction compared to rounded particles because of greater inter-locking capabilities. Similarly, compacted soils have a greater internal angle of friction because of rearrangement of the particles in a tighter packing configuration. The internal angle of friction is most important in characterizing the shear strength of cohesionless soils. For example, the shear strength of sand, a truly cohesionless soil, mainly depends on inter-granular friction and grain interlocking.

11.3.2 Soil Mineralogy

The mineralogy and chemistry of soil, particularly soils rich in clay, affect many physical and engineering properties that determine the stability of cohesive soils (Torrance, 1999; Duzgoren-Aydin et al., 2002). High-clay content is particularly important in generating slump-earthflow movement, deep-seated soil creep, and shallow slides and flows in nonsensitive clay soils. Major types of clay minerals that can influence the stability of slopes include kaolinite (i.e., halloysite), smectite (i.e., montmorillonite and beidellite), hydrous mica (i.e., illite and vermiculite), chlorite, and amorphous clays. Clays impact stability because the mineralogic structure, cation concentration, water content, and response to vibration can decrease soil strength (Sidle and Ochiai, 2006). Many studies have reported the role of clay minerals in the occurrence of landslides generated from soil-mantled as well as bedrock slopes (Ciolkosz et al., 1979; Parry et al., 2000; Ambers, 2001; Chigira and Yokoyama, 2005). For example, halloysite allows water to penetrate between clay layers and smectite can readily expand or contract, depending on the amount of water and cations present. Shrinkage in such soil effectively reduces the stability of the slope by reducing the length of the failure surface over which shearing resistance is mobilized (Rogers and Selby, 1980). Illite and chlorite are nonswelling clays but associate with quick clays, which have a nature of changing suddenly from a fairly stable, brittle solid to a liquid of negligible strength under vibration because of lower cation concentrations within the soil matrix, coupled with naturally high-water contents (Eden and Mitchell, 1970). Amorphous clays are susceptible to landslide because of low bulk density, high plastic and liquid limits, low plasticity index, and high water-holding capacity.

11.3.3 Atmosphere–Earth Surface Interaction

Various climatic factors contribute to the instability of a slope. The intensity and duration of rainfall and the fluctuation in atmospheric pressure have been strongly linked to the initiation and movement of the landslides (Caine, 1980; Buma, 2000; Schulz et al., 2009). Rainfall leads to the buildup of water pressure in the ground, as well as producing additional load surcharge, and fluctuation in atmospheric pressure influences the flow of water in sediment pores.

The role of rainfall in mass movement has been studied in terms of total rainfall over a specific interval of time (i.e., annual); short-term rainfall intensity (i.e., daily or monthly); antecedent-storm precipitation; and storm duration (Caine, 1980; Fuchu et al., 1999; Dhakal and Sidle, 2003; Dahal and Hasegawa, 2008). The rate and intensity of precipitation influence the generation of pore-water pressure that reduces the frictional strength of slope material. Studies have demonstrated relationships between shallow landslides and long- (Keefer et al., 1987; Glade, 1998; Pasuto and Silvano, 1998; Fuchu et al., 1999) and short-term rainfall (Larsen and Simon, 1993; Fuchu et al., 1999). Other studies suggest that combined effect of mean- and maximum-hourly intensity of a storm, and duration and total amount of rainfall influence slope failures (Dhakal and Sidle, 2003). In all rainstorms, most failures occur after some threshold of cumulative rainfall and maximum hourly rainfall intensity (Caine, 1980; Dahal and Hasegawa, 2008). In addition, spatial patterns of rainfall and snow accumulation are found closely associated with initiation of landslides (Minder et al., 2009). Typically, higher mountain elevations experience larger volumes of rain and snow, and, thus, a larger frequency of landslides.

Fluctuations in atmospheric pressure have also been linked to the variations in the pore-water pressure that trigger mass movement (Schulz et al., 2009). A study of the Slumgullion landslide, located in the San Juan Mountains of Colorado by Schulz et al. (2009), suggests that tidal change in air pressure causes air and water in the sediment pores to flow vertically, altering the frictional stress of the shear surface; upward fluid flow during periods of low, atmospheric pressure is most conducive to sliding. Schulz et al. (2009) also suggested that tidally modulated changes in shear strength may also affect the stability of other landslides worldwide, and that the rapid variations in pressure associated with some fast-moving storm systems could trigger a similar response.

11.4 TYPES AND CHARACTERISTICS OF MASS MOVEMENT

Classifications of mass movement proposed by Varnes (1978), Hutchinson (1988), and Cruden and Varnes (1996), are the most widely accepted classifications in use today. Although schemes of classifications vary to some degree, all identify principal types of mass movement as falls, topples, slides, flows, lateral spreads, and assorted combinations based on the type of movement, the type of materials involved, and amount of fluid present.

Mass movement exhibits a wide range of characteristics, including the effect of gravity, the rate of movement, amount of water present (i.e., moisture content), and depth to the slip surface. The variability in these characteristics among different types of mass movement is the process–response feedback of the type and thickness of slope material, slope morphology, water content, and frequency and magnitude of triggering events. In general, falls and topples are primarily the result of gravity, and only occur on steep slopes whereas, lateral spread and creep are mostly driven by environmental conditions, moisture, and soil mineralogy, and primarily occur on more gentle slopes. The movement of creep is slow and sometimes imperceptible over a short-temporal framework whereas a rock avalanche or snow avalanche is extremely rapid. Similarly, the amount of moisture present significantly varies in different types of mass movement. In general, flows require a greater interaction with water than falls, topples, and slides. The amount of moisture also varies significantly in different types of flows (i.e., earthflows, mudflows, and debris flows). Slides exhibit distinct surfaces of slip with a wide range of depths to the slip surface whereas, flows may not have distinct surfaces of shear and the depth to the deformed/undeformed boundary can vary widely. In these viewpoints, types of mass movement can also be identified as gravity-driven, slow to rapid, dry to wet, and shallow to deep seated. The following sections discuss the characteristics of major mass-movement types including falls, topples, slides, and flows.

11.4.1 Falls (Gravity-Driven)

Falls are abrupt movement of geologic materials, such as rocks, boulders, debris, and soil (Fig. 11.1). A fall begins with the detachment of a mass of rock, debris, or soil from a steep slope along a surface along which little or no shear displacement takes place. The separation of mass occurs along discontinuities,

FIGURE 11.1 An example of rock fall from south China's Guangxi Zhuang Autonomous Region. The fall occurred in November 24, 2008 and killed five people. *(Photograph adapted from China daily (http://www.chinadaily.com.cn/china/2008-11/24/content_7233838.htm).)*

such as fractures, joints, and bedding planes or along weak surfaces in soil, such as cracks. The material then descends mainly through the air by free fall, bouncing, or rolling once it lands on the adjoining surface below. Falls are strongly influenced by gravity, and the movement ranges from very rapid to extremely rapid. Falls are very common in steep slopes of rock, soil, and debris in upland areas. Major causes of failure include mechanical weathering (i.e., fracturing and jointing) of rock, soil development along the surface of discontinuity, the presence of interstitial water, slope morphology, and stratigraphy of erodible and erosion-resistant lithology. Falls may be activated by undercutting of the toe of a slope, loss of internal strength of slope material because of weathering, mechanical breakup of slope material by freeze–thaw processes and diurnal-temperature fluctuations in the rock surface, and surface and subsurface hydrology (Schumm and Chorley, 1964; Day, 1997). Major triggers include frost shattering (Gardner, 1983; Matsuoka and Sakai, 1999; Arosio et al., 2013), seismic shaking (Bull et al., 1994; Bull and Brandon, 1998; Wieczorek et al., 2000; Vidrih et al., 2001), and human activity (Barnard et al., 2001; Brunsden, 2001).

Rock falls are among the most common type of mass movement in mountain areas worldwide. Steepened-alpine hillslopes, modified by glacial and periglacial processes and exposed to frequent alternating freezing and thawing, generate low-magnitude, high-frequency rock falls (Douglas, 1980; Hungr et al., 1999; Matsuoka and Sakai, 1999; Jomelli and Francou, 2000). Rock material released during detachment from steep slopes, descends downslope in three different modes of motion: freefall through the air; bouncing on the slope surface; and rolling on the slope surface. The type of motion depends mostly on the gradient of the hillslope. A falling rock can go through one, two, or all three types of movement until it comes to rest as the slope changes rapidly or continuously from steep to gentle. Freefalls largely occur on very steep slopes (i.e., >70°) (Dorren, 2003; Ritchie, 1963); bounce occurs primarily on moderate slopes (i.e., 45–70° (Dorren, 2003); and roll occurs on gentle slopes (i.e., <30°).

Rock falls can be extremely hazardous because of the velocity. Various factors that control the velocity of rock falls include size of the rocks; material covering the slope, such as soil, scree, and vegetation; surface roughness; and the gradient of the hillslope. Velocity of falling rocks also depends on frictional forces that act on the rock during transport over slope surfaces. The frictional force is dependent on the shape of the rock as well as the characteristics of the slope surface. Surface characteristics of the slope can be characterized in terms of surface roughness, which is controlled by the type, frequency, and magnitude of surface materials, and erosional and depositional features (Regmi et al., 2014), or in terms of dynamic angle of friction (Kirkby and Statham, 1975). Such characteristics of a slope can significantly vary within short distances and play a major role in the dynamics of falling rocks.

11.4.2 Topples (Gravity-Driven)

A topple is the forward rotation out of the slope of a mass of soil, rock, or debris about a point or axis below the center of gravity of the displaced mass (Fig. 11.2). Toppling movements are the results of forces that cause an overturning moment about a pivot point. In most of the cases, these forces are directly related to the weight of the toppling block, thrust of adjacent blocks and water pressure in the planes of discontinuity (Goodman and Bray, 1976). Topples range from extremely slow to extremely rapid. Three principal types of toppling mechanisms have been identified: block toppling, flexural toppling, and block-flexural toppling (Goodman and Bray, 1976; Evans, 1981).

Flexural toppling is a mode of failure involving the bending of interacting rock columns formed by a single set of steeply dipping discontinuities, such as bedding planes, foliation, or joints (Fig. 11.2a). The rock columns bend forward

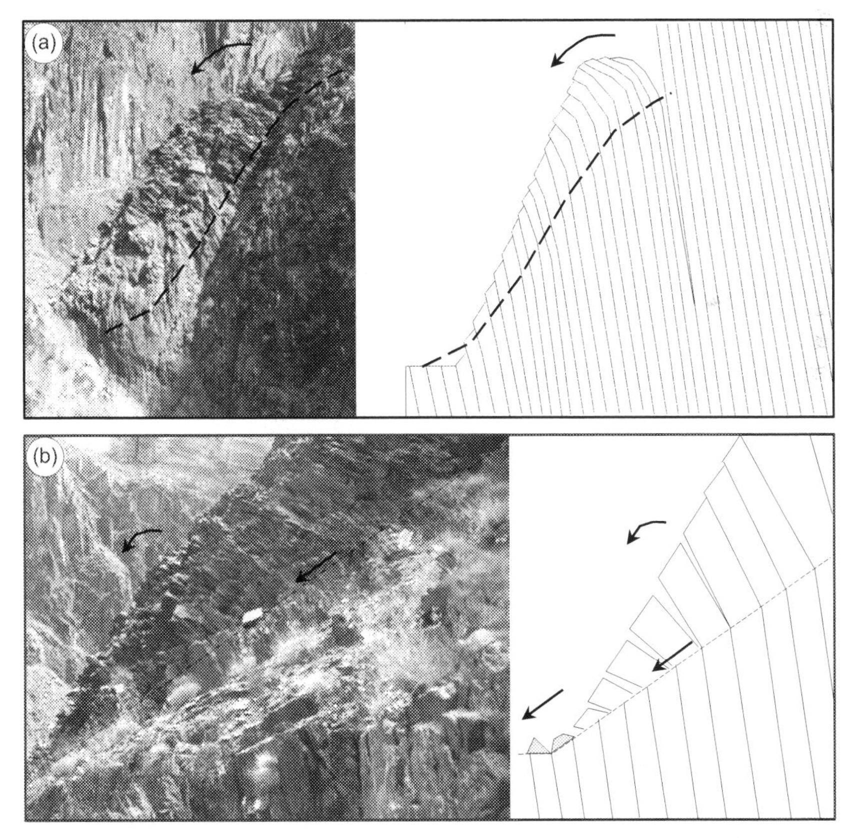

FIGURE 11.2 Photographs showing topples in Grand Canyon, Colorado. (a) A flexural toppling. (b) Block toppling with a distinct basal detachment surface. *(Photographs adapted from webpage of David J Rogers (http://web.mst.edu/~rogersda/grand_canyon_research/granite_gorge_toppling.htm).)*

like cantilever beams under their own weight and transfer the load to the underlying columns, thus, giving rise to tensile and compressive bending stresses. Failure initiates when the tensile stress in the toe column exceeds the tensile strength of the rock. The process is facilitated by the bending of columns only, without discrete hinge surfaces (McAffee and Cruden, 1996). The bending is accommodated by the low strength of the rock and by slippage between the steep faces of the columns. Slopes undergoing flexural toppling often self-stabilize once the discontinuities rotate to a sufficiently shallow dip angle (Goodman and Bray, 1976). Slow, flexural topples are quite common in schist, phyllite, slate, gneiss, and sedimentary rocks.

Block toppling, a brittle mechanism of toppling, includes sliding and toppling of rock columns along a pre-existing basal failure plane that is formed by widely spaced and steep, vertical discontinuities combined with a nearly horizontal and roughly orthogonal cross-joints. The basal failure divides a stronger rock mass into blocks of finite height and provides release surfaces for rotation of the blocks forward out of and away from the slope (Fig. 11.2b). The movement of blocks is primarily driven by mass. The stability depends on the location of the center of gravity of the blocks relative to their bases. Once the blocks begin to tip, the stabilizing forces acting on the bases decrease, and after exceeding a threshold, blocks tend to fail catastrophically.

Block-flexure topples exhibit transitional failure mechanisms between block and flexural toppling (Goodman and Bray, 1976). These topples involve several, less prominent hinge surfaces, combined with bending of long columns (McAffee and Cruden, 1996), as a result of the accumulated motion along numerous cross-joints. Sliding is distributed along several joint-surfaces in the toe of the displaced mass whereas sliding and overturning occur in close association through the rest of the mass (Goodman and Bray, 1976).

11.4.3 Slides (Shallow to Deep-Seated)

A slide is a downslope movement of mass of soil, rock, or debris along a well-defined surface-of-rupture known as a slip or shear surface (Fig. 11.3). The material moves as a coherent mass along a nearly straight-line path downslope, but it is likely to disintegrate with distance from the initial point of failure. Slides have two categories based on the styles of movement: (1) translational slides and (2) rotational slides. The rates of movement of translational and rotational slides vary from long-term creep, catastrophic movement that is preceded by long-term creep, to catastrophic movement with no-creep phase (Petley and Allison, 1997). Both types of slides show one of two styles of movement during the accelerating phases: (1) the style in which 1/velocity relates with time by a linear relationship; and (2) the style in which the relationship of 1/velocity and time shows an asymptotic form, trending toward steady-state movement rates. The linear form occurs in landslides in which crack propagation (i.e., heat-surface generation) is the dominant process, whereas the second style

FIGURE 11.3 Examples of translational and rotational slides. (a) A translational debris slide in Somerset of west-central Colorado. (b) A translational rock slide in Syangja of western Nepal. (c) A rotational landslide (soil slide) in Laconchita, California (adapted from USGS).

occurs where movement is taking place across existing planes of weakness or as a result of ductile-deformation processes (Petley et al., 2002).

Slides in which mass-of-slope material moves downslope on a largely planar surface are known as translational slides or planar slides. Translational slides form mostly on steep slopes where the shearing resistance along the slip surface remains low, and the slides attain a long runout distance. The surface develops mostly along faults, joints, bedding planes and is the result of variations in shear strength between beds or the contact between firm bedrock and overlying regolith (Varnes, 1978). Translational slides in steep slopes are generally the result of intense or long-duration rainfall or rapid snowmelt (Selby, 1982; Regmi et al., 2010a). Rock slides and debris (or sediment) slides are the most common types of translational slides. Rock slides usually occur along discontinuities such as bedding planes, fractures, and joints. Slopes where the discontinuities lie parallel to the ground surface are the most susceptible to generating rock slides. Debris slides occur mostly on shallow shear-surfaces at or near the regolith–bedrock interface where the strength and permeability of the slope materials change significantly (Petley et al., 2002; Regmi et al., 2010a). Translational slides also can produce movement of a nearly intact mass of very large blocks of rock or soil over a planar failure surface where the mass of sliding material is only partially deformed (i.e., during initial transport). This type, known as block gliding, occurs mostly on shallow buried bedrock (Rogers and Selby, 1980). A limited occurrence of intact transitional slides on slopes composed of soil is considered to be the result of a significant increase in soil-water (i.e., pore) pressure that causes a decrease in soil strength and disintegration of the soil block along a planar surface. Soil-block-slides that form at the upper part of the slope can transition into a broken slide in mid-slope and into a debris flow in the lower slope because of the increase in soil deformation and the surface and subsurface water that frequently increases with downslope transport.

Rotational slides involve movement of the displaced soil, debris, and/or rock mass on a shear surface that is concave upwards in the direction of movement. Rotational movement of soil, sediment, or fill is referred to as slump. The displaced mass rotates about an axis, which is parallel to the slope. Backward rotation of the sliding mass often causes fracturing, also called tension cracks, on the headscarp, which can lead to the sliding of a new block. Seepage along the initial-failure plane and fractures can cause water ponding in the depressed part of the backward-tilted displaced block (Selby, 1982), and thereby multiple events of sliding of the displaced block (Petley et al., 2002).

11.4.3.1 Shallow Versus Deep-Seated

Mass movements with discrete surfaces of slip are commonly known as shallow landslides and deep-seated or large landslides (Guzzetti et al., 2008; Larsen et al., 2010; Regmi et al., 2013b). Translational slides are commonly shallow features (i.e., the depth of slip surface located within the tree-root zone) and rotational slides are generally deep-seated features (i.e., the depth of the slip surface is located below tree roots). In practice, other features of mass movement, which may not have distinct surfaces of slip, such as flows and creeps, also have been included in shallow and deep-seated landslide categories based on the depth of the boundary between deformed and undeformed layers.

Shallow landslides commonly occur at valley heads, bedrock and colluvial hollows, and inner gorges. These landforms are concave and promote high rates of soil accumulation and surface- and subsurface-water convergence. These types of slides are, in general, rapid events that coincide with either rain or snowmelt. Deep, rotational failures are typically triggered by long- and short-term build-up of pore-water pressure and seismic activities in high-relief topography of weak bedrock (i.e., shale, phyllite, weak volcanic rocks), thick soil, mechanically weak deposits (i.e., glacial deposits), and often clay-rich rocks (Schmidt and Montgomery, 1995; Regmi et al., 2013b). The rate of deep-seated landslide movement may be slow or incremental; however, the movements in the majority of the cases exhibit subtle indicators, such as tension cracks and tipped and deformed trees. The movement may involve rapid displacement of large blocks or groups of blocks (Petley et al., 2002) and the formation of debris flows. Deep-seated landslides can remain active for long periods of time, particularly during wet periods, which may span years to decades, or can attain a state of dormancy or inactivity after the initial failure (Sidle and Ochiai, 2006).

FIGURE 11.4 A photograph of an arctic hillslope showing solifluction lobes formed by frost creeping.

11.4.4 Flows (Slow to Rapid and Dry to Wet)

A flow is a viscous movement of slope materials where surfaces of shear are short-lived, closely spaced, and usually not preserved (Figs 11.4 and 11.5). Materials in a flow disintegrate immediately after the failure and move downslope. The lower boundary of the displaced mass may be a surface along which appreciable differential movement has taken place or a thick zone of distributed shear. The distribution of velocities in the displacing mass resembles that in a flow of viscous liquid (Cruden and Martin, 2013) and suggests that the progression of slides to flows depends on water content, mobility, and evolution of the movement. The moisture content can range from dry to totally saturated and the rate of movement can range from extremely slow (<16 mm/year) to extremely rapid (5 m/s) (Cruden and Varnes, 1996). Flows can occur in bedrock, but they are extremely slow in relation to flows in unconsolidated materials. Depending on the rate of movement and the type of materials involved various types of flows can be identified (Table 11.1). The following section focuses on creep, earthflow/mudflow, and debris flow.

11.4.4.1 Creep

Creep is slow and imperceptible but continuous movement of slope materials (Fig. 11.4). The hillslope material can be rock and soil/regolith. Rock creep involves gradual deformation of the rock mass along joints and fractures without developing a distinct failure plane. Soil creep is the most studied process of hillslope creep. Soil creep generally occurs on very gentle slopes consisting of fine-grained sediments rich in expansive clay, such as lacustrine silts

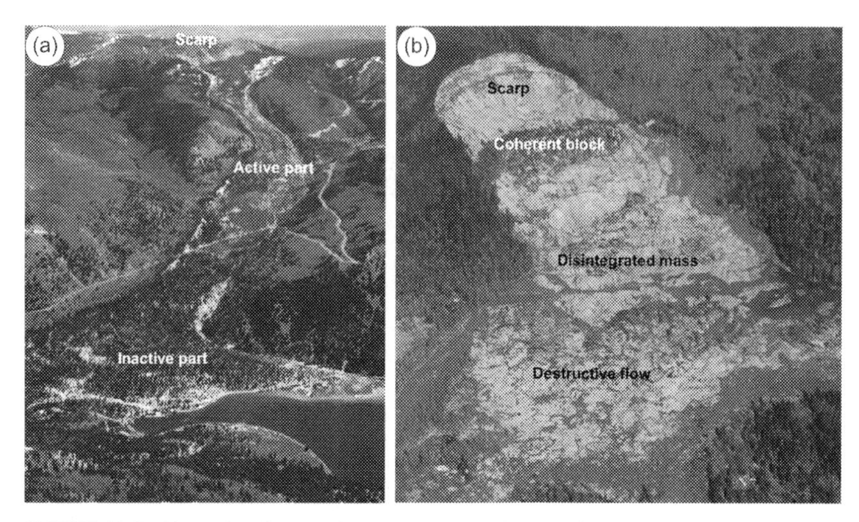

FIGURE 11.5 Examples showing flows. (a) Slumgullion earthflow in Colorado. The flow has 3.9-km length, 1.46 km^2 area, ~20×10^6 m^3 volume, 6–7 m/year velocity in the narrowest region, and 1–2 m/year velocity in the head and toe region (Parise and Guzzi, 1992). (b) A deep-seated mudslide occurred in Washington in 2014. *(Photograph adapted from Business Insider, http://www.businessinsider.com.)*

TABLE 11.1 Classification of Flows (Varnes, 1978; Hungr et al., 2001)

Rate of movement	Bedrock	Debris (<80% sand and finer)	Earth (>80% sand and finer)
Rapid and higher (>1.5 m/day)	Rock flows (creep, slope sagging)	Debris flows, debris avalanche	Wet sand and silt flow, rapid earth flows, loess flows, dry sand flows
Less than rapid (<1.5 m/day)		Solifluction, soil creep, block stream	Earth flow

(Owen, 1991) and is considered the result of repeated vertical expansion and contraction of soil particles during wet and dry periods. Soil particles, when wet, expand because of increase in size and weight, and when dry, they contract. As a result, the soil slowly moves downslope, and the movement may result in folding, bending, bulging, or other manifestations of plastic response. The rate of movement is extremely slow and continuous, typically less than ~0.3 m per decade (Cruden and Varnes, 1996). The rate varies, however, and primarily depends on slope gradient. Rates of movement range from a few millimeter per year on 3° slope to ~70 mm/year on a 40° slope (Schumm, 1967), and may remain constant over time. Nonlinear relationships have been suggested between the rate of creep and the hillslope gradient (Andrews and Bucknam, 1987;

Vanasch et al., 1989). Other factors responsible for the variations in the rate of movement include physical and mineralogical characteristics of the soil, moisture content, microclimate, and disturbances (Schumm, 1967). Soil deformation is the result of the change in physical properties of sensitive clay (i.e., consistency and permeability) because of progressive heating and cooling, wetting and drying, freezing and thawing.

Creep occurs in various cold, mountain environments. Creep, including frost creep and gelifluction, are the dominant types of mass movement processes in cold regions (i.e., tundra region). These processes produce distinctive lobate and terrace-like landforms, which are easy to recognize whereas, active, but difficult to distinguish from earthflow and mudflow lobes after surface modification by the other processes (Benedict, 1976). Large frost creep and gelifluction features are currently active in many tundra environments that experience only deep-seasonal freezing (Black, 1976). Creep also occurs in high alpine and arctic environments associated with permafrost (Fig. 11.4).

11.4.4.2 Earthflow/Mudflow

Earthflows are viscous-fluid flows that generally occur in fine-grained materials or clay-bearing rocks on moderate slopes (Fig. 11.5). These features represent a transition between a soil-slide and a mudflow. The flow mass can be dry or wet and the slip surfaces within the moving mass are commonly invisible and constitute that which the engineering community may consider to be "distributed shear". Velocity of movement can vary from a slow movement (i.e., a few meters or less per day) that can persist for several days, months, or years to a rapid flow that occurs within minutes to hours. The surface topography of earthflows is commonly wrinkled and hummocky. Slow earthflows are mostly dry, and move at rates ranging from 10^{-10} m/s to 10^{-3} m/s (Cruden and Varnes, 1996) primarily by sliding on discrete basal- and lateral-slip surfaces. Earthflows may continue to move for many years after initiation until they attain a slope gradient where the shear strength of the slope material is sufficient to stabilize the slope. Slow earthflows form in fine-grained soils or rocky soils that are supported by a silt-clay matrix that enhances plastic deformation. The Slumgullion landslide in Colorado (Fig. 11.5a) is an example of well-known slow type of earthflow (Coe et al., 2003). Wet earthflows (Fig. 11.5b) are mostly rapid and generally occur in highly sensitive clay (i.e., quick clay) deposits (Varnes, 1978; Cruden and Varnes, 1996). Highly sensitive slope materials liquefy in the presence of triggering factors such as excess precipitation, elevated groundwater pressure, and seismic shaking; and rapidly flow over long distances. The long run-out distance results from low post-failure shear strength of the sensitive material (Locat and Demers, 1988; Torrance, 1999). The overall movement could be translational, rotational, progressive, and or retrogressive (Baum et al., 2003). The velocity can reach as high as several meters per second (Cruden and Varnes, 1996). Mudflows or mudslides are quite similar to earthflows (Skempton and Hutchinson, 1969; Keefer and Johnson, 1983) in mechanics. Mudflows usually flow rapidly;

however, they contain more water and plastic debris formed from clay or silt relative to earthflows (Hungr et al., 2001).

11.4.4.3 Debris Flows

A debris flow is a rapid downslope movement of slope materials that may include a combination of loose soil, rock, organic matter, air, and water, all of which are mobilized and transported as slurry. Debris flows are a mass of poorly sorted sediment and commonly contain a large amount of coarse fragments (<50% fines). Debris flows often result from large amounts of precipitation or rapid snowmelt that saturates, erodes, and mobilizes loose soil or rock on steep slopes along zones of topographic convergence (Regmi et al., 2010a). Deposition of debris primarily occurs at the base of steep slopes, commonly associated with channel confluence with river valleys, and lake and ocean shorelines. Debris flows commonly exhibit characteristics of rock avalanches and sediment-laden water floods (Iverson, 1997). Solid-grain forces dominate the physics of avalanches; fluid forces dominate the physics of floods; whereas, solid-grain forces and fluid forces in concert influence the motion of debris flows. Debris flows can be one of the most hazardous types of mass movement because of the combination of fluid- and solid-grain forces, pose destructive power, and can exert great impulsive loads on objects they encounter. These flows are fluid enough to travel long distances in channels with modest slopes and to inundate vast areas quickly. They can denude vegetation, clog drainage ways, damage structures, and endanger humans, structure, and animals. Large debris flows can exceed 10^9 m^3 in volume and release more than 10^{16} J of potential energy (Iverson, 1997).

11.5 CAUSES OF MASS MOVEMENT

11.5.1 Biosphere–Earth Surface Interaction

Vegetation plays positive and negative roles in the occurrence of mass movement and erosion. Vegetation can increase stability by reducing soil moisture through evapotranspiration, and through root cohesion of the soil mantle. Evapotranspiration leads to drier soil conditions, and, thus, lower pore-water pressure. During storms or snowmelt, root cohesion increases the shear strength of the soil (O'Loughlin and Ziemer, 1982; Riestenberg and Sovonickdunford, 1983). Vegetation plays a negative role by increasing the load on a slope, infiltration capacity of soil, and wind shear (Greenway, 1987; Phillips and Watson, 1994).

Vegetation cover strongly influences seasonal soil–water balance, particularly when water is limited by low precipitation or evapotranspiration demands are high. The soil–water budget in a vegetated landscape can be influenced by the canopy interception of rainfall or snow and subsequent loss to the atmosphere by evaporation from the vegetation canopy, transpiration of infiltrated water, and evaporation from the soil/regolith surface. These factors are very important for the stability of a slope in a vegetated landscape.

Tree roots produce apparent slope cohesion, also known as root strength, by root-fiber reinforcement of soil and rock mass. Root systems stabilize hillslopes by bonding or anchoring unstable soil mantles to more stable subsoil or bedrock. On the other hand, tree roots can destabilize hillslopes by creating preferential flow paths for groundwater and increasing the conductivity of subsurface soil. The role of root strength in the stability of hillslopes has widely been studied, particularly to understand the response of a landscape after timber harvesting (Sidle, 1992). Timber harvesting reduces the shear strength of slope material because of the reduction in root strength, which in turn decreases the threshold of movement for a landslide-triggering event. Several investigators have observed an increased frequency of shallow, translational landslides occurring about 3–15 years after clear-cutting (Bishop and Stevens, 1964; Swanson and Dyrness, 1975). Other studies that compared landslides in forested and nonforested landscapes (De Graff, 1979; Froehlich and Starkel, 1993) also found lower frequency of landslides in forested terrain compared to other types of land cover.

The weight (surcharge) of trees increases the normal and slope-parallel stress components on a potential sliding surface. The role of vegetation load in destabilizing the slope has been considered minimal or nonexistent if the gradient of the slope is gentle (i.e., slope is less than angle of internal friction) and if the total soil weight above a potential failure plane greatly exceeds the weight of vegetation (O'Loughlin and Ziemer, 1982). The surcharge can destabilize the slope, however, if the soil depth is considerably shallow and or if the slope gradient exceeds the internal friction angle of soil.

Wind generates shear stress on the surface of forested landscape because of tree-induced wind drag. The shear stress has the potential to decrease the stability of hillslopes that are steep and have a thin soil mantle. Although the role of wind shear on landsliding has not been studied well, based on theoretical and experimental perspectives, some studies presumed that steep slopes with shallow soils are potential candidates for initiation of shallow landslides because of tree-wind throw (Greenway, 1987; Millard, 2003). Generally, the wind-drag coefficient or wind-driven shear stress remains highest near forest edges and openings (Greenway, 1987), and, thus, the areas are susceptible to movement. Other mechanism of wind-induced destabilization of a slope may include the breakage of the tree trunk and disturbance of the surface and subsurface soil.

11.5.2 Surface and Subsurface Hydrology

Groundwater fluctuation can initiate shallow and deep-seated landslides. Effective infiltration, patterns of deep-water circulation, and increases in hydrostatic levels or pore-water pressures over time all play major roles in mass movement (Hutchinson, 1970; Cappa et al., 2004; Montety et al., 2007). The most significant hydrologic processes include spatial and temporal distribution of precipitation (i.e., rain and snow), evapotranspiration, overland flow, interception and water recharge into soils, and lateral and vertical movement of water within the soil. The relative rates of these spatially and temporally distributed processes depend on various factors such as, intensity, duration, and frequency of precipitation; micro- and macroscale slope morphology; vegetation cover and density; and soil physical properties (Anderson and Burt, 1977; Montgomery et al., 1997; Sulebak et al., 2000). These processes can collectively determine the transient level of groundwater in different portions of the hillslope and, thus, the potential for slope failure during rainstorms, snowmelt, or prolonged periods of water accretion. Among these factors, soil physical properties such as infiltration and water-holding capacity primarily determine the rate of water movement into and through the hillslope material. These properties mainly depend on the microscale soil characteristics, such as porosity, hydraulic conductivity, pore-size distribution, and preferential-flow network. Vegetation cover, the density, and surface roughness contributed by micro- and macrotopography control the rate of overland flow and interception.

11.5.3 Slope Morphology

11.5.3.1 Slope Gradient

The majority of the research worldwide on landslides has found that frequency of landslides relates to slope gradient alone or in concert with other environmental factors. Landslides, particularly shallow, frequently occur on steep slopes. Large frequencies and magnitudes of landslides have been reported from areas that are tectonically active or shaped by glaciation. Tectonic uplift promotes streams to incise and dissect the landscape by lowering the base-level whereas, glaciers alone or in conjunction with tectonic uplift carve and widen the valley. Both processes, therefore, contribute significantly to oversteepening of hillslopes, and, thus, mass movement. Various studies also indicate that different types of mass movement have different slope thresholds for occurrence. For example, soil creep has been documented on slopes as gentle as 1.3° (Finlayson, 1981) to as steep as 25° (Burroughs et al., 1976). Sidle and Ochiai (2006) compiled slope gradient data of mass movement from many locations worldwide and found that the lower limit of slope gradient ranges from ~48° to ~65° for debris avalanches, debris slides, and debris flows; ~20° to ~68° for shallow, rapid slides, and debris flows; ~4° to ~25° for earth flows; and ~4° to ~18° for slumps. The variability in the threshold is related to the type of mechanism, presence of moisture content, surface and subsurface hydrology, and the physical properties of soil and rock.

11.5.3.2 Slope Curvature

Forms or curvature of a slope influence the dynamics of surface and subsurface hydrology as well as formation and accumulation of soil. Hillslopes can be categorized as divergent (convex), planar, and convergent (concave) based on the curvature. Concave slopes better support formation and accumulation of soil in comparison to divergent slopes. Soil thickness in concave hillslopes may be significantly greater than on convex hillslopes (Dietrich and Dunne, 1978; Dengler and Montgomery, 1989). Surface and subsurface water tend to converge and accumulate in a concave hillslope, and thereby promote a rapid increase in pore-water pressure during storms or periods of snowmelt (Montgomery et al., 1997; Regmi et al., 2010a). Because of this effect, a perched-water table is common in these landforms whereas, surface and subsurface water tend to diverge and rapidly flow downslope in convex slopes and results in slow and insignificant increase in pore-water pressure and absence of perched water table. These characteristics lead to the condition that convex hillslopes are generally most stable in steep terrain, followed by planar hillslopes, and then concave hillslopes, which are the least stable. For example, geomorphic hollows, valley heads, and zones of topographic convergence, mostly studied concave hillslope landforms, commonly exhibit evidence of repeated episodic mass movement, particularly shallow and rapid landslides (Marron, 1985; Reneau et al., 1986; Regmi et al., 2010a). Convex landforms, such as intervening ridges and side slopes, exhibit the least frequency of mass movement. Planar landforms, such as stream inner-gorges exhibit intermediate characteristics between concave and convex landforms.

11.5.3.3 Slope Aspect

Slopes having different aspects receive different annual-solar radiation, which strongly affects the process of rock weathering and soil development, dynamics of surface and subsurface hydrology, and growth of vegetation. In the Northern Hemisphere, mostly in areas below 45°N latitude, north-facing slopes receive lower annual-solar radiation and maintain higher and less variable moisture compared to south-facing slopes, which receive comparatively more annual-solar radiation and generally remain drier and experience more frequent periods of wetting and drying (Parsons, 1988; Burnett et al., 2008; Istanbulluoglu et al., 2008). The higher moisture levels in north-facing slopes support greater weathering and, thus, deeper soil development, greater vegetation density, greater organic-soil depth, and, thus, greater infiltration capacity. Therefore, in the Northern Hemisphere higher frequency of mass movement can be expected on north-facing slopes compared to south-facing slopes, and in the Southern Hemisphere the opposite relationship can be expected. Our ongoing study of erosion and mass movement on Santa Catalina Island of California also found significantly larger frequency of rills, gullies, and small-sized landslides on north-facing slopes compared to south-facing slopes. Some studies, however, have reported the opposite or no trends of landslides with slope aspect (Marston et al., 1998; Dai and Lee, 2002; Ayalew and Yamagishi, 2005). One or more factors, such as prior mass movement, pattern and direction of frontal precipitation, topography, lithology and structure, direction of seismic propagation, faults, and soil-drainage properties could be associated with such exceptions.

11.5.4 Human Impact

Humans modify the equilibrium of the natural slope mostly by reducing vegetation cover, increasing slope gradient, and overloading slopes. Reduction in root strength by improper management of land, timber harvesting, forest conversion, grazing, recreation and human-induced fire modifies slope hydrology and soil properties. These changes in slope can trigger slope erosion and mass movement in the short- and long-term (Smith, 1996). Similarly, oversteepening and overloading slopes during road and trail construction, urban development, mining, and residence development on hillsides alter slope morphology, hydrology, and physical properties of slope materials and influence the stability of a slope (Sidle and Ochiai, 2006; Regmi et al., 2010c).

11.5.5 Earthquake

Seismic vibration resulting from earthquakes can trigger various types of mass movement, such as lateral spreads, liquefaction, falls, slides and flows, in a range of spatial scales in all types of terrain (i.e., gentle to highly dissected topography) and cause widespread destruction. Seismic events commonly influence the stability of a large region (hundreds of square kilometers) and generate multiple landslides of variable magnitude in a very short period of time (i.e., few minutes or hours) (Bommer and Rodriguez, 2002; Keefer, 2002). Landslides triggered by earthquakes cause the largest loss of life because of their sudden occurrence. For example, earthquake-induced, destructive landslides that occurred in central China in December 1920, are associated with the loss of ~180,000 people (Derbyshire et al., 2000; Sidle and Ochiai, 2006).

11.6 SPATIAL AND TEMPORAL SCALE OF MASS MOVEMENT

Landslides occur across a variety of spatial and temporal scales (Fig. 11.6). The size of landslides can range from tens of square meters (i.e., rock falls, soil slips) to tens of square kilometers (i.e., catastrophic rock avalanches and debris flows). Frequency of landslides for a given event in a given area can also vary from few slides to thousands of individual slides. Moreover, a single event can generate different types of landslides with different transport characteristics (i.e., slow to rapid). Initiation of a landslide can also vary greatly ranging from instantaneously following a specific trigger such as an earthquake, an intense rainfall-event, an explosion, or undercutting event or initiation can be a delayed response to critical-triggering conditions.

The spatial and temporal responses of landslides and the occurrence of specific types of landslides can be associated with one or more factors described above and the spatial and temporal characteristics of triggering events. The major causes for different types and frequencies of mass movement, however, can be considerably different. For example, rock avalanches and instantaneous rock and debris falls, slides, and flows with long-transport distances (i.e., run-out)

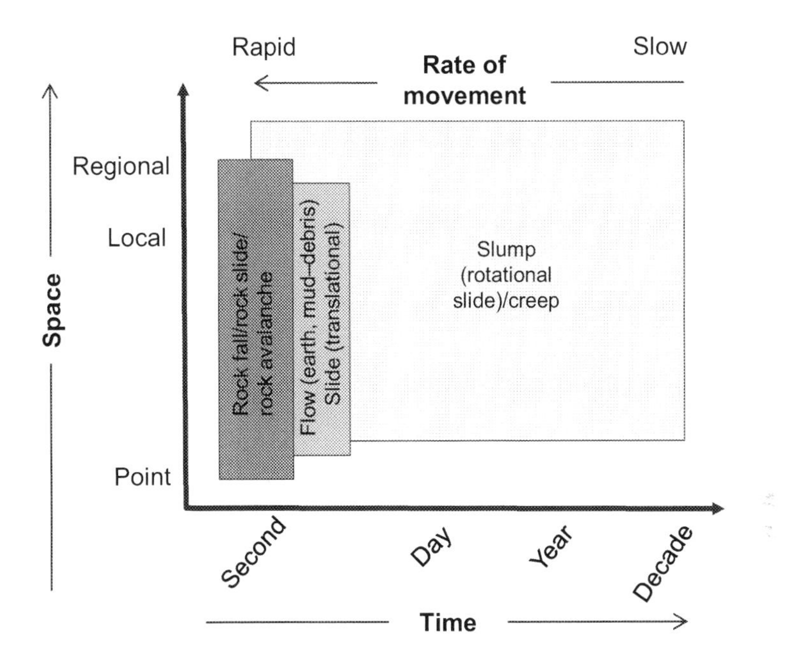

FIGURE 11.6 **A generalized schematic diagram showing spatial, temporal, and velocity scales of landslide occurrence.** *(Diagram modified after Glade and Crozier (2005, Fig. 3.1, p.75).)*

are generally restricted to high-relief steep mountainous areas. High-relief steep slopes provide potential and kinetic energy for slope failure and long run-out length. On the other hand, creep and rotational failures (i.e., slumps) can occur on gentle topography. A number of research efforts have been performed to understand such spatial-temporal characteristics of landslides in various parts of the world.

11.6.1 Frequency and Magnitude

For at least four decades, researchers have been investigating the magnitudes (i.e., area or volume) and temporal as well as spatial frequencies of mass movement worldwide to understand the role in sediment yield, landscape evolution, and as hazards (Noever, 1993; Stark and Hovius, 2001; Stark and Guzzetti, 2009; Larsen et al., 2010; Regmi et al., 2013b). Almost all investigators obtained a power relationship between the frequencies and magnitudes of landslides related to a single triggering event or multiple triggering events. Small landslides occur frequently, but the frequency of occurrence of large landslides is very low. The relationship characterizes the physical condition of the hillslope (Hovius et al., 1997; Pelletier, 1997), the strength of the material that limits the number of small slides and the overall geometry of the slope that limits the number of very large slides (Pelletier, 1997; Guzzetti et al., 2002), and self-organized criticality (Noever, 1993).

11.6.2 Sediment Yield

Assessment of the total volume of sediments yielded by mass movement is important to understand the mechanics and hazards of mass movement as well as contribution to the evolution of mountain environments. Relationships exist between landslide area and the volume of the sediments produced (Table 11.2). The areas (A) and volumes (V) of landslides relate by a power equation:

$$V = \alpha A^{\gamma}$$

where α is the intercept of the relationship curve in volume-axis and γ is the slope of the curve, which generally ranges from 1.09 to 1.95 (Table 11.2). These values depend on the type of landslide, depth of the regolith, bedrock lithology and surface-soil characteristics, and the frequency and magnitude of the triggers. Larsen et al. (2010) suggested that, in general, the γ values for the soil slides range from 1.1 to 1.3 and for the bedrock slides the value ranges from 1.3 to 1.6.

11.7 CONTRIBUTION TO THE EVOLUTION OF MOUNTAIN ENVIRONMENTS

Integrated erosional processes of mass movement and sediment transport from water, ice, and wind transfer considerable volumes of slope material and influence mountain environments. Large landslides can displace large volumes of materials, greatly damage or modify the environment of the Critical Zone in and around the landslides, and leave geomorphic signatures for many years (Korup, 2005; Guthrie and Evans, 2007). The frequency of occurrence of large landslides, however, is very low. In contrast, small landslides are considerably more frequent but displace only very small amounts of slope material at a time. Therefore, the corresponding damage or modification occurs across a smaller area of the Critical Zone environment and the geomorphic signature exists for a short period of time (Guthrie and Evans, 2007). An important question then is in determining the minimum size and magnitude of landslides that will significantly impact the Critical Zone environment of a mountainous landscape over time. This question can be answered by examining the frequency and geomorphic work of landslides. Landslides that occur frequently and do much of the geomorphic work significantly impact the Critical Zone environment. Geomorphic work of landslides has been quantified by taking the product of magnitude (i.e., size) and frequency or probability of landslide occurrence (Guthrie and Evans, 2007; Regmi et al., 2013b). Guthrie and Evans (2007) suggests that landslides having a size of 1,000–3,000 m^2, and 10,000–20,000 m^2 for coastal British Columbia, and Northridge, California, respectively, do much of the geomorphic work and occur frequently, thereby contributing the most to the evolution of mountain environments. Although much research has been undertaken

TABLE 11.2 Area Volume Relationships of Various Types of Mass Movement

Equation	Area range (m^2)	Sources	Location	Mean volume (m^3)	Mass movement type
$V = 0.02 \times A^{1.45}$	$85–1.6 \times 10^5$	Regmi et al. (2013b)	Western Colorado	2×10^4	Mixed (debris, soil, and rock slides)
$V = 0.37 \times A^{1.40}$	NA	Larsen et al. (2010)	Oregon and Washington	NA	Soil slides
$V = 0.08 \times A^{1.43}$	$170–5.5 \times 10^6$	Guzzetti et al. (2008)	Central Italy	3.9×10^5	Shallow and large
$V = 0.02 \times A^{1.36}$	NA	Simonett (1967)	Central Italy	3.2×10^5	Shallow and large
$V = 0.05 \times A^{1.5}$	$100–1 \times 10^6$	Hovius et al. (1997)	Central Italy	5.4×10^5	Shallow and large
$V = 0.02 \times A^{1.95}$	$5 \times 10^4–5 \times 10^6$	Korup (2005)	Central Italy	1.43×10^5	Large
$V = 0.03 \times A^{1.39}$	30–900	Innes (1983)	Scottish Highlands	NA	Shallow
$V = 0.15 \times A^{1.09}$	$1124–4.09 \times 10^6$	Guthrie and Evans (2004a, 2004b)	British Columbia	NA	Shallow and large

to document morphology and mechanics of landsides, susceptibility and slope stability of landslides, research into the role of magnitudes and frequencies of landslides on the evolution of mountain environments is still needed.

11.8 SUSCEPTIBILITY/SLOPESTABILITY MAPPING

Landslides on steep slopes are always a major concern because they can affect lives and inflict economic losses. Identification of the hazardous areas associated with landslides is an important geomorphologic component of disaster management and an important basis for promoting safe, human occupation, infrastructure development, and environmental protection in mountainous environments. Many studies have been undertaken to assess susceptibility to landslides through qualitative or knowledge-based and quantitative approaches. The qualitative approaches are based entirely on the judgment based on *a priori* knowledge of the expert carrying out the susceptibility or hazard assessment. The input data are usually derived from field-geomorphologic mapping, topographic maps, and interpretation of aerial photographs and satellite imagery. The expert then makes *a priori* assumptions about those sites where landslides have occurred, assigns weights to the factor causing landslides, and develops decision rules to map areas where landslides are likely to occur again (Rupke et al., 1988; Regmi et al., 2010c).

Quantitative approaches are based partly on field observations, geomorphologic mapping, an expert's *a priori* knowledge, and partly on statistical computation of the weight or probabilities of occurrence of a landslide, or a factor of safety of a slope. The quantitative-based methods include statistical approaches, such as neural network, fuzzy logic, weight of evidence, logistic regression, and decision-tree-based approaches (Nefeslioglu et al., 2008; Regmi et al., 2010a, 2010b, 2013a; Pradhan, 2013). Statistical approaches focus on relationships of existing landslides with factors causing landslides and use the established relationship to develop models of susceptibility/hazards to landslides.

Deterministic methods mainly focus on slope geometry, shear-strength data (i.e., cohesion and angle of internal friction), and pore-water pressure related data (Montgomery and Dietrich, 1994; Gokceoglu and Aksoy, 1996; Xie et al., 2004). A deterministic approach considers angle of slope, strength of the slope material, structure (i.e., rock discontinuities, rock and soil stratification), moisture content of the slope material, and depth of the groundwater table in physics-based equations to determine an index of the stability of the slope, such as the factor of safety. This method usually ignores factors such as local and regional climate, human disturbances, stream erosion, and geologic structure, which also can contribute to landslides. Furthermore, this method only provides the stability of a slope at the time of data collection. It does not account for changes or modifications of the factors that cause landslides, the spatial and temporal frequencies of the landslides, and the magnitudes of the landslides.

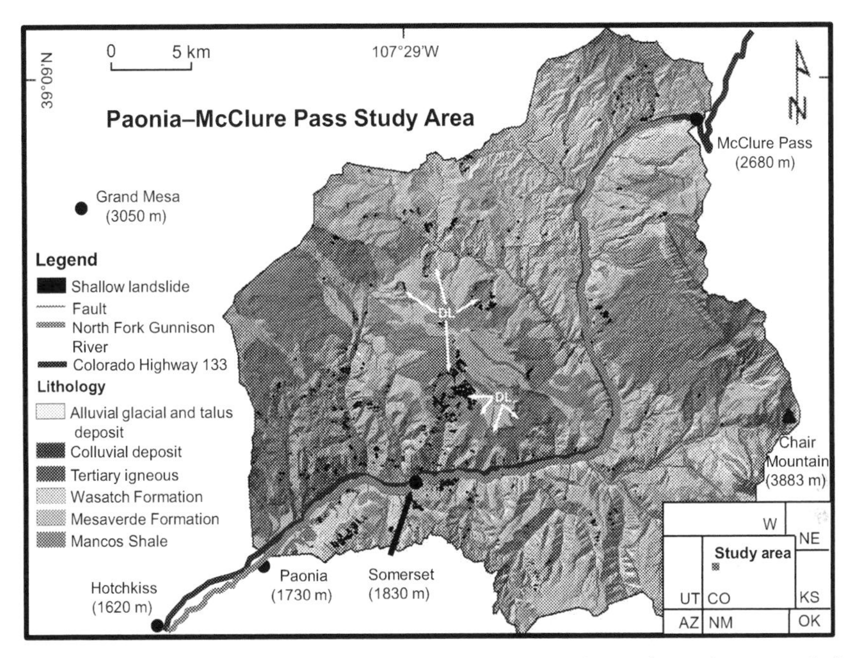

FIGURE 11.7 Location of Paonia–McClure Pass study area. The area is mostly composed of alluvial, glacial, talus, and colluvial deposits of Quaternary period. Major rock types in the area include: tertiary igneous rocks; claystone, mudstone, and sandstone lenses of Tertiary Wasatch Formation; sandstone, mudstone, and shale of Upper Cretaceous Mesaverde Formation; and shale and mudstone of Upper Cretaceous Mancos Shale. A detailed lithologic description can be found in Regmi (2010) and Dunrud (1989). W, Wyoming; NE, Nebraska; UT, Utah; CO, Colorado; KS, Kansas; AZ, Arizona; NM, New Mexico; OK, Oklahoma.

11.9 A CASE STUDY: CHARACTERISTICS OF LANDSLIDES IN WESTERN COLORADO

The following is a summary of our research on landslides in an area located between Paonia and McClure Pass (815 km^2) in western Colorado (Fig. 11.7). The climate of the area is predominantly semiarid with 1.8°C (minimum) to 18°C (maximum) average annual temperatures, 400 mm average annual precipitation, and 1220 mm average annual snowfall based on 1905–2005 data of Paonia 1SW climatic station (Western Regional Climate Center, 2014). The area has rugged topography and a dendritic drainage pattern. The main lithology in the area consists of igneous intrusive rocks, including dikes of basalt and gabbros; sedimentary rocks including sandstone, mudstone, and shale; and Quaternary deposits of glacial, colluvial, alluvial, and mixed origin (Dunrud, 1989). Sandstone and plutonic rocks have steep slopes and mudstone, shale, alluvial, and colluvial deposits have medium and gentle slopes. Most of the hilly mountains have steep slopes and flat mesa like tops, whereas the high-mountain highlands have sharp ridges and steep slopes sculpted into horns, arêtes, and cirques from alpine glaciation (Fig. 11.7). Major processes involved in the evolution of these hillslopes are the incision of the North Fork of Gunnison River and its streams and tributaries, Pleistocene glaciation, and mass movement from the coupling effects of snowmelt, rainfall, and river erosion.

The basin of the North Fork Gunnison River is one of the areas affected by shallow and deep-seated landslides in Colorado, and listed as the sixth most critical site on the 2002 priority-list plan for Colorado landslide hazard mitigation (Rogers, 2003). Landslides in the area have damaged Colorado Highway 133 numerous times in the past. The total cost of direct landslide losses and excess maintenance along this river corridor is estimated to be at least $1 million a year (Rogers, 2003). We studied shallow and deep-seated landslides using field surveys, aerial-photographic interpretation, and analysis of digital topographic data. Shallow landslides are modern (<100 years) and small in size (<160,000 m^2), whereas deep-seated landslides are old or paleo-landslides (hundreds to thousands of years) and larger in size (>160,000 m^2).

Shallow landslides (735), identified as debris flows (155), debris slides (86), rock slides (139), and soil slides (324), were mapped and the characteristics were studied in detail. The smallest landslide mapped is ~85 m^2 and the largest landslide mapped is ~160,000 m^2. These landslides are composed of sandstone, mudstone, and colluvial deposits, and are located in close proximity to rivers and roads. Debris flows mostly occur in areas of topographic convergence covered by unconsolidated deposits. These areas include inner gorges of first order streams, valley heads, and bedrock and colluvial hollows. Debris slides mostly occur in planar and concave slopes covered by unconsolidated alluvium and colluviums. These areas mostly include the inner gorges of higher-order streams. Rock slides mostly occur in steep and convex to planar-bedrock slopes. Soil slides, the most dominant type of landslide, are distributed everywhere but are smaller in size in comparison to the other types.

In most of the shallow landslides, the boundary between the soil and the underlying bedrock is abrupt. The regolith is cohesionless, has a low bulk density, and contains abundant fragments of rocks. The underlying rock is highly fractured, gently dipping and has considerable cohesion, as well as frictional strength. We consider that most of the landslides in the area are probably the result of the rock fractures and pore-water pressure generated by subsurface flow because loose regolith and fractured bedrock conduct large volumes of translocated water (Montgomery et al., 1997). The main reason for the rock fractures is sufficient moisture coupled with considerable depth of frost penetration in winter. Furthermore, many landslides occurred between the boundary of hard, competent rocks (i.e., sandstone and plutonic rock) and soft, friable rocks (i.e., mudstone and shale). An interface of differential shear strength also contributed to these landslides (Fig. 11.8). Interpretation of aerial photographs, acquired in 1993 and 2005, identified only 30 landslides that occurred between 1993 and 2005. Information that describes the occurrence of shallow landslides prior to 1993 is limited. Debris flows occurred throughout the area during intense

FIGURE 11.8 A photograph showing distribution of different types of shallow landslides in nearby Somerset village in North Fork Gunnison watershed.

rainfalls in 1975, 1983, 1984, 1985, 1986, and 1987, respectively (Rogers, 2003). Based on these events and the lack of vegetation, we assumed that many of the medium and small, shallow landslides were initiated during these intense rainfalls. One large, shallow landslide in the area has been dated to the 1940s (Rogers, 2003). The maximum ages of the shallow landslides were estimated as ~100 years (Regmi et al., 2013b) based on the ages of some landslides recorded by the Colorado Geological Survey (Rogers, 2003) and empirical equations developed by Guthrie and Evans (2007).

Deep-seated landslides occur on the gentle slope of higher elevations and along the edges of the upland plateaus (Fig. 11.7). The surfaces of these landslides are densely vegetated and incised by the channels and indicate the landslides are very old (probably hundreds to thousands of years). The headscarps of some of the large landslides are still active and produce shallow landslides. Both types of landslides contribute to the upslope propagation of steep edges of upland plateaus and steep heads of rivers and tributaries. Many surface and subsurface faults, colluvium and glacial deposits exist nearby the locations of the deep-seated landslides. The activities along these faults, in concert with groundwater level associated with intense precipitation and snowmelt during Late Pleistocene to Late Holocene, could have related to the occurrence of most of these landslides. The rest of the chapter discusses the detail characteristics of only shallow landslides.

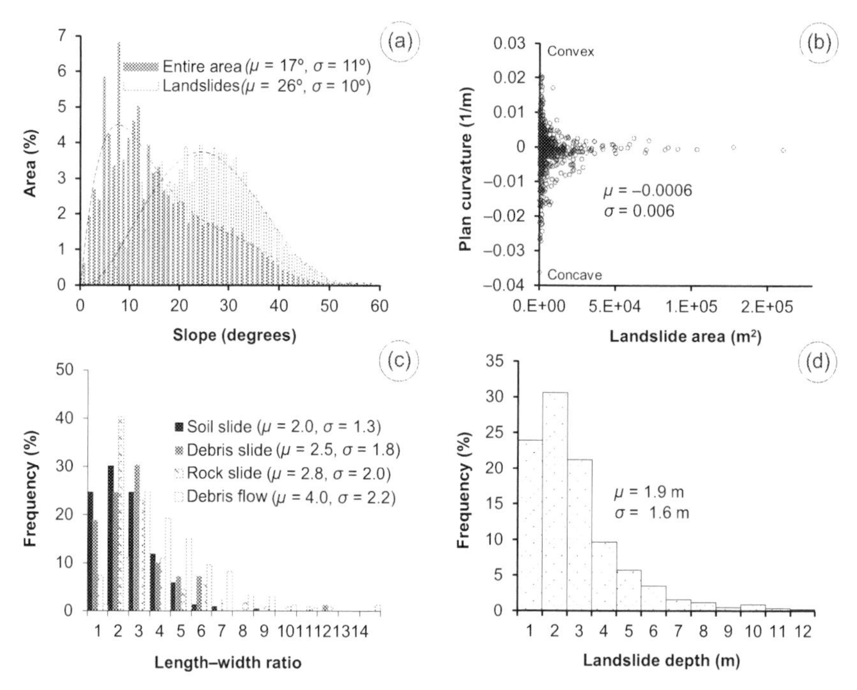

FIGURE 11.9 Geomorphic characteristics of shallow landslides determined based on the analysis of 10 m USGS digital elevation data and 1 m NAIP aerial photographs. (a) Histograms showing distribution of the slope on the landslide surface and entire area. Dashed lines represent sixth-order polynomial fit. The average slope of the landslide surface is ~26° and entire area is ~17°. (b) A scatter plot showing areas and plan curvatures of the landslides. (c) Histograms showing runout length to width ratios of different types of landslides. (d) A histogram showing the depths of landslides. The average depth of the landslide is ~1.9 m. μ, mean; σ, standard deviation. *(Figures modified after Regmi et al. (2013b, Fig. 12.5, p. 593).)*

11.9.1 Characteristics of Shallow Landslides

11.9.1.1 Geomorphic Characteristics

The slope, curvature, runout length, width, area, and volume of shallow landslides were studied in the field, as well as from the interpretation of aerial photographs and digital topographic data to understand the characteristics of different types of landslides, and the contribution in sediment yield and the evolution of the hillslope environment. The average slopes of landslides and areas devoid of landslides are 26° and 17°, respectively (Fig. 11.9a). Among these landslides, small- and medium-sized landslides occur in convergent and divergent parts of the slope whereas larger landslides occur in planar slopes (Fig. 11.9b). Most of the smaller landslides are soil slides and larger landslides are rock slides. Average runout length–width ratios were found different for different types of landslides. Debris flows have the largest runout length-to-width ratio and the soil slides have the smallest runout, length-to-width ratio (Fig. 11.9c). The depth to the slip surface of each landslide is determined based on the analysis of elevation data, which results in the average thickness of all landslides as ~1.9 m (Fig. 11.9d).

11.9.1.2 Areas and Volumes

The volumes of the landslides were estimated based on the field surveys, aerial photographic interpretation, and analysis of 10-m horizontal resolution digital elevation data (see Regmi et al., 2013b for detail). Similar to the other studies, this study shows that the areas and the volumes of landslides are related by a power relationship (Fig. 11.10a). An empirical equation was obtained to estimate the volumes of landslides based on the least square regression ($R^2 = 0.87$) of areas (A) and the estimated volumes (V) of 735 landslides.

$$V = 0.0254 \times A^{1.45}$$

The area–volume relationships of landslides described in some literature (Table 11.2) are compared with our results. The result is not significantly

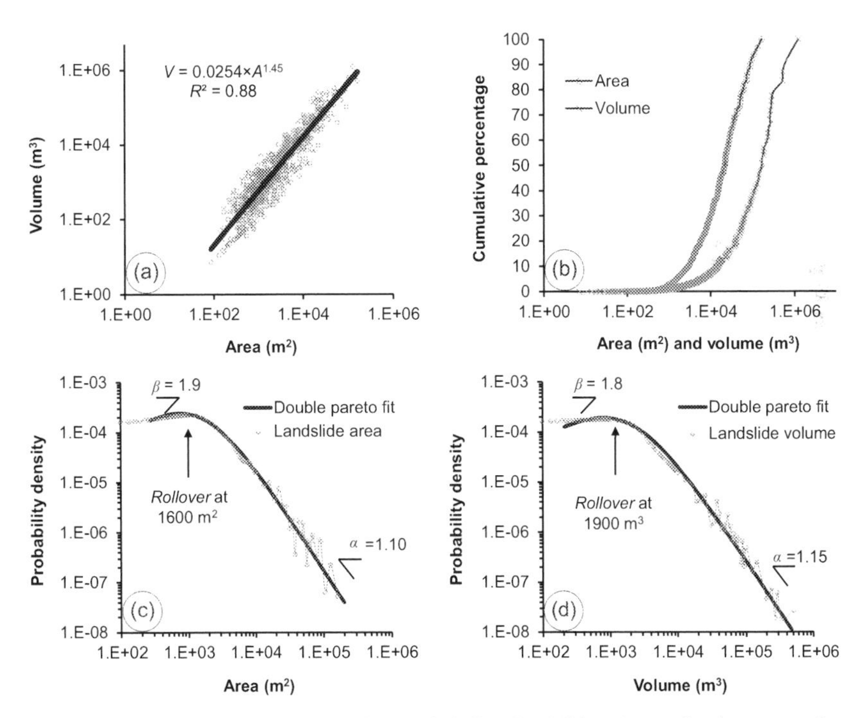

FIGURE 11.10 Frequency and magnitude of shallow landslides. Area and volume were determined based on the analysis of 10 m USGS digital elevation data and 1 m NAIP aerial photographs. (a) Area–volume relationship. (b) Cumulative percentage of landslide areas and volumes. (c) Probability density of landslide areas. (d) Probability density of landslide volumes. The slopes of the relationship curves (c, d) were determined based on the comparison with fitted double-pareto curves (e.g., Stark and Hovius, 2001). *(Figures modified after Regmi et al. (2013b, Figs 12.8 and 12.9, pp. 597–598).)*

different from Hovius et al. (1997), Guzzetti et al. (2008), Innes (1983), and Simonett (1967), whereas the result is significantly different than those obtained by Korup (2005) and Guthrie and Evans (2004a). The equation obtained by Korup (2005) is for very large landslides ($A > 1\ km^2$) and may be the reason for the results being different than this study.

The areas of recorded landslides ranged from 85 to $1.6 \times 10^5\ m^2$. The total area eroded by all landslides is $4.8 \times 10^6\ m^2$ with the average area of 6600 m^2 and a standard deviation of $1.36 \times 10^4\ m^2$. The total volume of the soil displaced by all landslides is $1.4 \times 10^7\ m^3$ with the average volume of 20,000 m^3 and standard deviation of $7 \times 10^4\ m^3$. These measurements of landslide areas and volumes are the lower estimate because only distinguishable landslides were considered in the analysis. Large landslides move most of the sediments on the landscape. Only 36 landslides have a volume larger than $1 \times 10^5\ m^3$, and these landslides contribute 62% of the total volume of all landslides. The three largest landslides (landslide area, 1×10^5–$1.6 \times 10^5\ m^2$) mapped in the area account for ~17% of the total landslide volume; 58 landslides (landslide area, 20,000–$1 \times 10^5\ m^2$) mapped in the area account for ~54% of the total landslide volume; 346 landslides (landslide area, 1,600–20,000 m^2) mapped in the area account for 28% of the total landslide volume and 328 landslides (landslide area, 85–1,600 m^2) mapped in the area account for~1.4% of the total landslide volume. These data suggest that the large landslides contribute much of the sediments in the study area. Based on the estimation of the ages of landslides as 100 years, the volume of the soil displaced by landslides each year is ~$1.2 \times 10^5\ m^3$ and the contribution of shallow landslides to the rate of denudation for the entire area (815 km^2) is at least ~0.15 mm/year.

11.9.1.3 Frequency, Magnitude, and Geomorphic Work

We developed frequency–magnitude curves showing the cumulative percentage distribution of sizes of landslides (area and volume) (Fig. 11.10b), and probability density function versus the sizes of landslides (Fig. 11.10c and d). All curves show power-law relationships between frequencies and magnitudes of landslides. A rollover effect, however, exists in the relationship curves (Fig. 11.10c and d) and suggests that landslides show negative power scaling for medium to large landslides and positive power scaling for small landslides separated by a rollover point. Comparison of the landslide area pdf curve with a double-pareto curve (Stark and Hovius, 2001) shows that the rollover occurs at 1600 m^2 with power scaling for large and medium landslides as $\alpha = 1.1$ and for small landslides as $\beta = 1.9$. Similarly, the probability plot for landslide volume rolls at 1900 m^3 with the power scaling value for large and medium landslides as $\alpha = 1.15$ and the scaling for smaller landslides as $\beta = 1.9$. Both curves show small values of α and large values of β indicating that the distribution has a long tail, which means the mass movement process in the area is debris dominated. Although 328 landslides have an area less than the area at rollover point (1600 m^2), the total area of these landslides is only 6% of the total area of all landslides

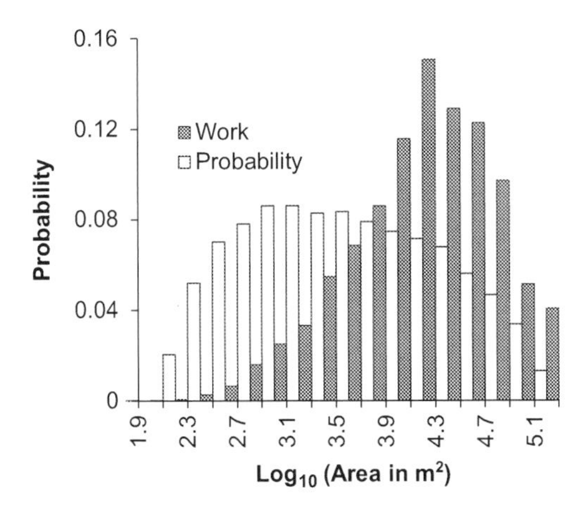

FIGURE 11.11 Histograms showing the probability distribution of the landslide areas in logarithmic intervals and the work (probability × area) performed by the landslides of each logarithmic interval. Work is scaled to fit the graph. *(Figure modified after Regmi et al. (2013b, Fig. 12.10, p. 598).)*

and the total volume of the material displaced by these landslides is only 1.5% of the total volume of debris mobilized by all landslides (Fig. 11.10b). This indicates that the slope material mobilized by small landslides is very insignificant in comparison to the amount of slope material mobilized by medium and large landslides.

These characteristics of landslides raise a question about what magnitude of landslides influence the environment of Critical Zone by modifying landscape cover over time. This question can be answered in one way by quantifying the geomorphic work of landslides as the product of landslide probability and landslide area (Guthrie and Evans, 2007). The landslides, which do much of the geomorphic work and occur frequently contribute mostly to the evolution of the Critical Zone environments. The probability of occurrence and geomorphic works computed by landslides were plotted against the size of landslides in $\log_{10}$ scale (Fig. 11.11). The plot suggests that the landslides of size ranging from 1,600 m^2 to 20,000 m^2 have a high probability of occurrence and do high geomorphic work, and, thus, are the important landslides for the evolution of Critical Zone environment in the mountainous landscape of Paonia–McClure Pass area.

11.9.2 Map of Susceptibility to Shallow Landslides

Three statistical methods (weight of evidence, fuzzy logic, and logistic regression) were employed to develop the most appropriate approach for mapping areas susceptible to landslides. The detail application of these methods for the Paonia–McClure Pass area can be found in Regmi et al. (2010a, 2010b, 2013a).

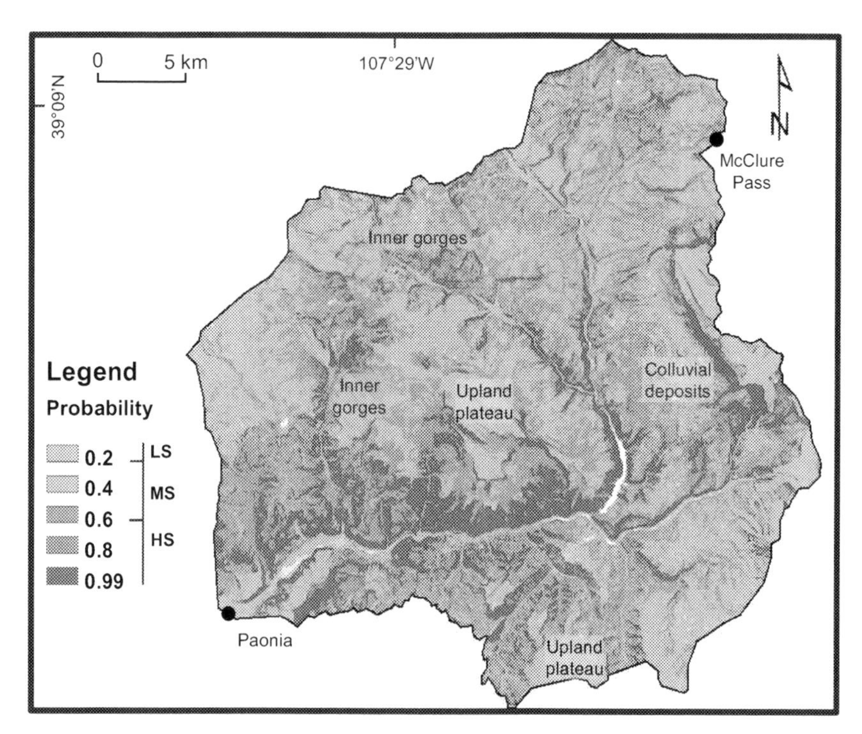

FIGURE 11.12 Susceptibility to landslides based on logistic regression approach. High susceptibility (HS) is classified as landslide probability >0.6, medium susceptibility (MS) is classified as landslide probability 0.2–0.6, and low susceptibility (LS) is classified as landslide probability <0.2. Numbers indicate percent of the total area. *(Figure modified after Regmi et al. (2013a, Fig. 12.10, p. 255).)*

The 17 factors causing landslides, including hydrologic, geologic, soil, geomorphic, and anthropogenic, were mapped from field surveys, aerial photographs, existing data archives, and 10-m horizontal resolution USGS digital elevation data. The relationships of all the factors with landslides were evaluated, and the factors determined significant for the susceptibility to landslides were used to develop models of landslide susceptibility. Results suggest that each of the three approaches is applicable for mapping landslide susceptibility in western Colorado; however, the logistic regression approach predicted the most of the observed landslides (Fig. 11.12). The weight of evidence predicted ~78% of the observed landslides (Regmi et al., 2010a), the fuzzy logic predicted ~80% of the observed landslides (Regmi et al., 2010b), and the logistic regression predicted 86% of the observed landslides (Regmi et al., 2013a). The logistic regression approach was then used to develop models of mapping susceptibility to individual types of landslides, and the prediction accuracy is 88% for debris flows, 92% for debris slides, 91% for rock slides, and 83% for soil slides, respectively. We believe prediction of debris slides is the highest because these slides mostly occur in concave to planar steep slopes covered by unconsolidated deposits. Prediction of rock slides and debris flows are also excellent because rock slides mostly occur on steep slopes, region of convex to planar curvature, and they are larger in size. Debris flows predominantly occur in zones of topographic convergence and areas covered by unconsolidated soil. The prediction accuracy for soil slides is the lowest because they are smaller in size and occur in diverse environments. This implies that different types of landslides have well-defined environments of occurrence, and consideration of these characteristics in susceptibility mapping significantly improves the prediction of landslides.

Field survey and the overlay analysis of landslides and landslide causing factors suggest that the major factors predisposing the landscape to shallow landslides include mudstone and shale, shrubland and woodland, and fine- to medium-grained nonplastic to low plastic soils, 20–40° slopes, south-facing slopes, topographic convergence, within 350 m distance of fault, and within 100 m distance of rivers and roads. Observed and predicted landslides occur on the slopes of the inner gorges of the North Fork Gunnison River and its associated streams that are incising into upland plateaus, valley heads, and bedrock and colluvial hollows. These characteristics of landslides represent potential for the first order prediction of the landslides in this landscape. The geomorphic and spatial characteristics of landslides, determined from this study, will help the decision makers to develop future plans for managing streams, transportation, and land use given the potential destruction from landslides that may result.

11.10 IMPACT OF MASS MOVEMENT IN THE CRITICAL ZONE

The Critical Zone is an open system, where transfer of mass and energy takes place in a range of spatial and temporal scales. Mass movement serves as a mechanism that yields inputs and transports sediment and debris from higher elevations to lower elevations within the Critical Zone. Mass movement features also act as short- and long-term sinks for sediment, ice, debris, and water resulting in unstable landscapes. The displacement of slope-mass, and their transport and deposition in concert with fluvial and glacial processes results in distinct surface topography, such as rills, gullies, topographic hollows, and undulating or hummocky landscape (Glade, 2003; Regmi et al., 2013b); slump-block topography (Baum and Odum, 1996); and colluvial fans, debris fans, talus cones, rock glaciers, and alluvial fans (Giardino, 1979; Giardino and Vitek, 1988a, 1988b; Harvey, 1984; Burger et al., 1999; McDonald et al., 2003; Giardino et al., 2011; Janke et al., 2013). The spatial scale of these landforms ranges from tens of square meter (i.e., rills and gullies) to hundreds of square kilometers (i.e., alluvial fans, bajadas), and depends on the frequency, magnitude, and the type of the mass-movement events.

The identification of mass movement-associated topography is, therefore, a key element for mapping mass-movement hazards in the Critical Zone. The topography created by mass movement can be identified and mapped, based on the analysis of aerial photographs, remotely sensed images, and high-resolution elevation data (Van Westen and Getahun, 2003; Roering et al., 2005). Tonal, textural, and vegetation characteristics are the parameters commonly used to identify mass-movement features on aerial photographs and remotely sensed images. Whereas, the variation in surface geometries, such as slope, curvature, relief, and surface roughness are the parameters commonly used to identify and map mass-movement features from elevation data. In addition, high-resolution LiDAR topographic data have been used to map various mass-movement signatures, head or crown scar, detached grass or vegetation, and deformed soil layers (Glenn et al., 2006; Ardizzone et al., 2007). Techniques of dendrogeomorphology have been used to map slow to moderate rates of movement of slopes in the Critical Zone (Shroder, 1980; Giardino et al., 1984; Bollati et al., 2012; Stoffel et al., 2013).

Mass movement occurs in all environments. Thus, mass movement and humans play a major role of interacting in the Critical Zone. To help control and minimize the impact of mass movement on humans and their infrastructures major approaches in use today to control mass movement and associated damages in the Critical Zone include: forestation and slope bioengineering, slope modification by grading and excavation, use of geotextile mats and grids, and installation of engineering structures (Cruden and Varnes, 1996). Vegetation increases the stability of a slope by intercepting precipitation, and by reducing runoff and excessive infiltration, as well as the binding strength added by root growth. Slope bioengineering or planting of carefully selected species of vegetation increases the strength of the slope materials. Grading and excavation increase the stability of a slope by reducing its overall gradient and eliminating unconsolidated surface materials. However, the approach completely destroys the natural Critical Zone environment, and may only be effective where adequate space is available and no habitat and infrastructure will be affected. An appropriate alternative to the grading approach could be the development of reinforced earth embankments with grids or mats of geotextiles placed in horizontal layers and covered with soil. The technique maintains steep slopes by allowing regrowth of the vegetation.

A number of engineered structures have been used to stabilize a slope (Wyllie and Mah, 2004; Duncan and Wright, 2005). The major purpose of the engineered structures is to stabilize a slope primarily by increasing the strength of slope-mass, reducing the amount of mass transfer from a hillslope, and modifying the surface and subsurface drainages. Approaches that can increase the strength of the slope mass include covering or spraying steep slopes with concrete covered with a wire mesh, bolting fractured rocks, and nailing soil by drilling holes horizontally in the slope and anchoring long steel rods into the slope with concrete. These techniques mostly use hexagonal wire mesh, cable nets, or high-tensile strength steel mesh draped or anchored over a slope face to slow erosion, apply active retention force (if anchored) to retain rocks and soil on a slope, and control the descent of falling rocks/soils and restrict them to the catchment area.

Different types of barriers, such as earthen barrier, concrete barrier, structural walls and retaining walls, flexible barriers, attenuators, and ditches, have been used to prevent the transfer of mass from a hillslope, to control the trajectory of failing rocks/soils, reduce their energy, and/or provide a catchment (Volkwein et al., 2011). Earthen barriers, constructed of natural soil, rocks, or mechanically stabilized earth, are easy to repair and can withstand large kinetic energies and repeated impacts. Concrete barriers are rigid, relatively cheap, and fast to install and are applicable for the protection of a slope from low-energy impacts. Structural walls and retaining walls are also rigid and can be used to intercept falling rocks/soils and restrict them to a prescribed catchment area. These barrriers can withstand significant kinetic energies and repeated impacts. Flexible barriers are fences made up of wire ring or high strength wire mesh with high energy-absorption capacity, supported by steel posts and anchor ropes with a braking system. They are expensive to construct and prone to damage by high-energy rock fall events, but very effective on high- to low-energy events. Attenuators are similar to flexible barriers, but the fence is not attached at the bottom. These flexible barriers allow rocks to move beneath the fence and direct them into a catchment area. Ditches and ditch–barrier combinations are the structures useful to contain failed slope materials (i.e., rock fall). They can be placed along the roadside or slope base to reduce the amount of slope material moving away from the slope.

Unstable slopes also have been stabilized by controlling surface and subsurface drainage (Cedergren, 1997). Installation of drainage pipes into the slope is the most-used technique for drainage control. Drainage pipes allow water to flow more readily and to avoid increases in fluid pressure, and thereby reduce the possibility of liquefaction and the weight of slope-mass resulting from the weight of water trapped as soil moisture or free water in voids. Thus, the interaction between humans and mass movement in the Critical Zone is unavoidable, but humans have developed various techniques to control or minimize the impact of mass movement on human lives and human-built infrastructure.

11.11 LIVING WITH MASS MOVEMENT IN THE CRITICAL ZONE

A March 2014 mudslide in Oso, Washington (Fig. 11.5b), along the banks of the Stillaguamish River claimed the lives of 43 people (Keaton et al., 2014). In May 2014, Afghanistan experienced a massive loess landslide that claimed the lives of over 500 people (Petley, 2014). Over the last 4 years, the number of recorded mass movements was 303 in 2010; 214 in 2011; 220 in 2012; and 237 in 2013. The total number of lives that were lost during these 4 years is 2318 (Petley, 2012, 2014). The significant loss of life contributes to the hazardous nature of mass movement. When people are not impacted, only geomorphic work is accomplished because rocks and sediments are moved to lower elevations. In total, mass movement causes one to two billion dollars worth of damage in the USA annually in addition to lives lost. Mass movement is the most devastating of all natural events on Earth.

"Watch for Falling Rocks" is a common sign in areas with significant relief where mass movement can occur. Other signs warn of avalanches and debris flows. Because such events are infrequent, the threat is minimal but it is not zero, hence people are killed and/or injured when an event suddenly occurs and they are in the wrong place at the right time. In many places, warning signs do not exist, nor are local residents aware of the type of mass movement that may impact their lives. When buying property or building a home or business, assessing the safety of the site may prevent future problems. An old adage, "let the buyer beware," is true relative to any mass movement or any other natural phenomena that becomes a risk. Consult your local city engineer, county, state, or federal agencies to acquire knowledge that you need about the dynamic nature of your environment. Numerous publications are available to assist with your decision process (Keaton et al., 2014).

11.12 SUMMARY

Mass movement, one of the major processes in the Critical Zone of Earth, represents complex hydrologic, atmospheric, cryogenic, soil-geomorphic, and biogeochemical feedbacks in response to climatic, tectonic, and anthropogenic forcing at a range of spatial and temporal scales. Mass movement plays significant roles in the evolution of Critical Zone environments by direct and indirect influence on surface and subsurface hydrology, slope morphology, physical characteristics of slope materials, ecology, and biologic communities. Identification of types, mechanics and causes of mass movement, therefore, is very important for protection of Critical Zone environments, safe occupation of human and other biologic communities, and infrastructure development. In addition, frequency, magnitude, sediment yield, geomorphic work, and susceptibility of mass movement in an area over time are very important attributes of mass movement that need to be considered in all management practices in the Critical Zone.

REFERENCES

Ambers, R.K.R., 2001. Relationships between clay mineralogy, hydrothermal metamorphism, and topography in a Western Cascades watershed, Oregon, USA. Geomorphology 38, 47–61.

Anderson, M.G., Burt, T.P., 1977. Automatic monitoring of soil-moisture conditions in a hillslope spur and hollow. J. Hydrol. 33, 27–36.

Anderson, S.P., Anderson, R.S., Tucker, G.E., 2012. Landscape scale linkages in Critical Zone evolution. C. R. Géosci. 344, 586–596.

Anderson, R.S., Anderson, S.P., Tucker, G.E., 2013. Rock damage and regolith transport by frost: an example of climate modulation of the geomorphology of the Critical Zone. Earth Surf. Process. Land. 38, 299–316.

Andrews, D.J., Bucknam, R.C., 1987. Fitting degradation of shoreline scarps by a nonlinear diffusion model. J. Geophys. Res. Solid Earth Planets 92, 12857–12867.

Ardizzone, F., Cardinali, M., Galli, M., Guzzetti, F., Reichenbach, P., 2007. Identification and mapping of recent rainfall-induced landslides using elevation data collected by airborne Lidar. Nat. Hazards Earth Syst. Sci. 7, 637–650.

Arnold, R.W., Szabolcs, I., Targulian, V.O., 1990. Global soil change. Report of an IIASA-ISSS-UNEP task force on the role of soil in global change. International Institute for Applied Systems Analysis, Laxenburg, Austria.

Arosio, D., Longoni, L., Mazza, F., Papini, M., Zanzi, L., 2013. Freeze-thaw cycle and rockfall monitoring. In: Margottini, C., Canuti, P., Sassa, K. (Eds.), Landslide Science and Practice, 2. Springer, Berlin, Heidelberg, pp. 385–390.

Ayalew, L., Yamagishi, H., 2005. The application of GIS-based logistic regression for landslide susceptibility mapping in the Kakuda-Yahiko Mountains, Central Japan. Geomorphology 65, 15–31.

Bales, R.C., Hopmans, J., O'Geen, A.T., Meadows, M., Hartsough, P.C., Kirchner, P., Hunsaker, C.T., Beaudette, D., 2011. Soil moisture response to snowmelt and rainfall in a Sierra Nevada Mixed-Conifer Forest. Vadose Zone J. 10, 786–799.

Barnard, P.L., Owen, L.A., Sharma, M.C., Finkel, R.C., 2001. Natural and human-induced landsliding in the Garhwal Himalaya of northern India. Geomorphology 40, 21–35.

Baum, R.L., Odum, J.K., 1996. Geologic map of slump-block deposits in part of the Grand Mesa Area, Delta and Mesa Counties, Colorado. US Geological Survey Open-File Report 96-017, 12 p., 2 oversize plates, 1:24,000.

Baum, R.L., Savage, W.Z., Wasowski, J., 2003. Mechanics of earthflows. In: Picarelli, L. (Ed.), Proceedings of the International Workshop on Occurrence and Mechanisms of Flow-Like Landslides in Natural Slopes and Earthfills. Associazione Geotechnica Italiana, Bologna, Sorrento, Italy, pp. 185–190.

Benedict, J.B., 1976. Frost creep and gelifluction features: a review. Quaternary Res. 6, 55–76.

Bishop, D.M., Stevens, M.E., 1964. Landslides on logged areas southeast Alaska. US forest Service Department of Agriculture Research Paper NOR-1, Juneau, Alaska, p. 18.

Black, R.F., 1976. Features indicative of permafrost. Annu. Rev. Earth Planet. Sci. 4, 75–94.

Bollati, I., Della Seta, M., Pelfini, M., Del Monte, M., Fredi, P., Lupia Palmieri, E., 2012. Dendrochronological and geomorphological investigations to assess water erosion and mass wasting processes in the Apennines of Southern Tuscany (Italy). Catena 90, 1–17.

Bommer, J.J., Rodriguez, C.E., 2002. Earthquake-induced landslides in Central America. Eng. Geol. 63, 189–220.

Brunsden, D., 1993. Mass movement – the research frontier and beyond – a geomorphological approach. Geomorphology 7, 85–128.

Brunsden, D., 2001. A critical assessment of the sensitivity concept in geomorphology. Catena 42, 99–123.

Bull, W.B., Brandon, M.T., 1998. Lichen dating of earthquake generated regional rock fall events, southern Alps, New Zealand. GSA Bull. 110, 60–83.

Bull, W.B., King, J., Kong, F.C., Moutoux, T., Phillips, W.M., 1994. Lichen dating of coseismic landslide hazards in alpine mountains. Geomorphology 10, 253–264.

Buma, J., 2000. Finding the most suitable slope stability model for the assessment of the impact of climate change on a landslide in southeast France. Earth Surf. Process. Land. 25, 565–582.

Burger, K.C., Degenhardt, J.J., Giardino, J.R., 1999. Engineering geomorphology of rock glaciers. Geomorphology 31, 93–132.

Burnett, B.N., Meyer, G.A., McFadden, L.D., 2008. Aspect-related microclimatic influences on slope forms and processes, northeastern Arizona. J. Geophys. Res. 113, F03002, doi:10.1029/2007JF000789.

Burroughs, E.R., Chalfant, G.R., Townsend, M.A., 1976. Slope Stability in Road Construction: A Guide to the Construction of Stable Roads in Western Oregon and Northern California. US Department of the Interior, Bureau of Land Management, Portland, OR, p. 102.

Caine, N., 1980. The rainfall intensity-duration control of shallow landslides and debris flows. Geogr. Ann. 62A, 23–27.

Cappa, F., Guglielmi, Y., Soukatchoff, V.M., Mudry, J., Bertrand, C., Charmoille, A., 2004. Hydromechanical modeling of a large moving rock slope inferred from slope levelling coupled to spring long-term hydrochemical monitoring: example of the La Clapie're landslide (Southern Alps, France). J. Hydrol. 291, 67–90.

Cedergren, H.R., 1997. Seepage, Drainage, and Flow Nets, vol. 16. John Wiley & Sons, New York, p. 496.

Chigira, M., Yokoyama, O., 2005. Weathering profile of non-welded ignimbrite and the water infiltration behavior within it in relation to the generation of shallow landslides. Eng. Geol. 78, 187–207.

Ciolkosz, E.J., Petersen, G.W., Cunningham, R.L., 1979. Landslide-prone soils of southwestern Pennsylvania. Soil Sci. 128, 348–352.

Clark, W.C., Crutzen, P.J., Schellnhuber, H.J., 2004. Science for global sustainability: toward a new paradigm. In: Schellnhuber, H.J., Crutzen, P.J., Clark, W.C., Claussen, M., Held, H. (Eds.), Earth System Analysis for Sustainability. The MIT Press, Cambridge, MA, pp. 1–28.

Coe, J.A., William, L.E., Jonathan, W.G., William, Z.S., Jill, E.S., Michael, J.A., John, D.K., Philip, S.P., David, J.L., Sylvain, D., 2003. Seasonal movement of the Slumgullion landslide determined from global positioning system surveys and field instrumentation, July 1998–March 2002. Eng. Geol. 68, 67–101.

Cruden, D.M., Martin, C.D., 2013. Assessing the stability of a natural slope. In: Wu, F., Qi, S. (Eds.), Global View of Engineering Geology and the Environment. Taylor & Francis Group, London, pp. 17–26.

Cruden, D.M., Varnes, D.J., 1996. Landslide types and processes. In: Turner, A.K., Schuster, R.L. (Eds.), Landslides Investigation and Mitigation. Transportation Research Board, Washington, DC, pp. 36–75.

Dahal, R.K., Hasegawa, S., 2008. Representative rainfall thresholds for landslides in the Nepal Himalaya. Geomorphology 100, 429–443.

Dai, F.C., Lee, C.F., 2002. Landslide characteristics and, slope instability modeling using GIS, Lantau Island, Hong Kong. Geomorphology 42, 213–228.

Day, R.W., 1997. Case studies of rockfall in soft versus hard rock. Environ. Eng. Geosci. 3, 133–140.

De Graff, J.V., 1979. Initiation of shallow mass movement by vegetative-type conversion. Geology 7, 427–429.

De Graff, J.V., Sidle, R.C., Ahmad, R., Scatena, F.N., 2011. Recognizing the importance of tropical forests in limiting rainfall-induced debris flows. Environ. Earth Sci. 67, 1225–1235.

Dengler, L., Montgomery, D.R., 1989. Estimating the thickness of colluvial fill in unchanneled valleys from surface topography. Bull. Assoc. Eng. Geol. 26, 333–342.

Derbyshire, E., Wang, J.T., Meng, X.M., 2000. A treacherous terrain: background to natural hazards in northern China with special reference to the history of landslide in Gansu Province. In: Derbyshire, E. et al. (Eds.), Landslides in the Thick Loess Terrain of North-West China. John Wiley & Sons, New York, NY, pp. 1–19.

Dhakal, A.S., Sidle, R.C., 2003. Long-term modelling of landslides for different forest management practices. Earth Surf. Process. Land. 28, 853–868.

Dietrich, W.E., Dunne, T., 1978. Sediment budget for a small catchment in mountainous terrain. Z. Geomorph. N. F., Suppl. Bd. 29, 191–206.

Dorren, L.K.A., 2003. A review of rockfall mechanics and modelling approaches. Prog. Phys. Geogr. 27, 69–87.

Douglas, G.R., 1980. Magnitude frequency study of rockfall in Co. Antrim, North Ireland. Earth Surf. Process. Land. 5, 123–129.

Duncan, J.M., Wright, S.G., 2005. Soil Strength and Slope Stability. John Wiley & Sons, p. 297.

Dunrud, R.C., 1989. Geologic map and coal stratigraphic framework of the Paonia area, Delta and Gunnison Counties, Colorado. US Geological Survey, Coal Investigations Map C-115, Scale 1:50,000.

Duzgoren-Aydin, N.S., Aydin, A., Malpas, J., 2002. Distribution of clay minerals along a weathered pyroclastic profile, Hong Kong. Catena 50, 17–41.

Eden, W.J., Mitchell, R.J., 1970. The mechanics of landslides in Leda clay. Can. Geotech. J. 7, 285–296.

Evans, R.S., 1981. An analysis of secondary toppling rock failures – the stress redistribution method. J. Eng. Geol. Hydrogeol. 14 (2), 77–86.

Finlayson, B., 1981. Field-measurements of soil-creep. Earth Surf. Process. Land. 6, 35–48.

Froehlich, W., Starkel, L., 1993. The effects of deforestation on slope and channel evolution in the tectonically active Darjeeling Himalaya. Earth Surf. Process. Land. 18, 285–290.

Fuchu, D., Lee, C.F., Wang, S.J., 1999. Analysis of rainstorm-induced slide-debris flows on natural terrain of Lantau Island, Hong Kong. Eng. Geol. 51, 279–290.

Gardner, J.S., 1983. Rockfall frequency and distribution in the Highwood Pass area, Canadian Rocky Mountains. Z. Geomorph. 27, 311–324.

Giardino, J.R., 1979. Rock glacier mechanics and chronologies: Mt.Mestas, Colorado. PhD dissertation, Department of Geography, University of Nebraska, p. 244.

Giardino, J.R., Vitek, J.D., 1988a. The significance of rock glaciers in the glacial-periglacial landscape continuum. J. Quaternary Sci. 3, 97–103.

Giardino, J.R., Vitek, J.D., 1988b. Interpreting the internal fabric of a rock glacier. Geogr. Ann. A Phys. Geogr. 70, 15–25.

Giardino, J.R., Shroder, J.F., Lawson, M.P., 1984. Tree-ring analysis of movement of a rock-glacier complex on Mount Mestas, Colorado, USA. Arctic Alpine Res. 16, 299–309.

Giardino, J.R., Regmi, N.R., Vitek, J.D., 2014. Rock Glaciers. In: Singh, V.P., Singh, P., Haritashya, U.K. (Eds.), Encyclopedia of Snow, Ice and Glaciers. Encyclopedia of Earth Sciences Series. Springer, Netherlands, pp. 943–948.

Glade, T., 1998. Establishing the frequency and magnitude of landslide-triggering rainstorm events in New Zealand. Environ. Geol. 35, 160–174.

Glade, T., 2003. Landslide occurrence as a response to land use change: a review of evidence from New Zealand. Catena 51, 297–314.

Glade, T., Crozier, M.J., 2005. A review of scale dependency in landslide hazard and risk analysis. In: Glade, T. et al. (Eds.), Landslide Hazard and Risk. Wiley, Chichester, UK, pp. 75–138.

Glenn, N.F., Streutker, D.R., Chadwick, D.J., Thackray, G.D., Dorsch, S.J., 2006. Analysis of LiDAR-derived topographic information for characterizing and differentiating landslide morphology and activity. Geomorphology 73, 131–148.

Gokceoglu, C., Aksoy, H., 1996. Landslide susceptibility mapping of the slopes in the residual soils of the Mengen region (Turkey) by deterministic stability analyses and image processing techniques. Eng. Geol. 44, 147–161.

Goodman, R.E., Bray, J.W., 1976. Toppling of rock slopes. In: Rock Engineering for Foundations and Slopes. American Society of Civil Engineers, Geotechnical Engineering Division Conference, Boulder, CO, pp. 201–234

Greenway, D.R., 1987. Vegetation and slope stability. In: Anderson, M.G., Richards, K.S. (Eds.), Slope Stability, Geotechnical Engineering and Geomorphology. John Wiley & Sons, Chichester, UK, pp. 187–230.

Guthrie, R.H., Evans, S.G., 2004a. Magnitude and frequency of landslides triggered by a storm event, Loughborough Inlet, British Columbia. Nat. Hazards Earth Syst. Sci. 4, 475–483.

Guthrie, R.H., Evans, S.G., 2004b. Analysis of landslide frequencies and characteristics in a natural system, Coastal British Columbia. Earth Surf. Process. Land. 29, 1321–1339.

Guthrie, R.H., Evans, S.G., 2007. Work, persistence, and formative events: the geomorphic impact of landslides. Geomorphology 88, 266–275.

Guzzetti, F., Malamud, B.D., Turcotte, D.L., Reichenbach, P., 2002. Power-law correlations of landslide areas in central Italy. Earth Planet. Sci. Lett. 195, 169–183.

Guzzetti, F., Ardizzone, F., Cardinali, M., Galli, M., Reichenbach, P., Rossi, M., 2008. Distribution of landslides in the Upper Tiber River basin, central Italy. Geomorphology 96, 105–122.

Harvey, A.M., 1984. Debris flows and fluvial deposits in Spanish Quaternary alluvial fans: implications for fan morphology. Can. Soc. Petrol. Geol. Mem. 10, 123–132.

Hovius, N., Stark, C.P., Allen, P.A., 1997. Sediment flux from a mountain belt derived by landslide mapping. Geology 25, 231–234.

Hubbard, G.D., 1908. Two notable landslides. Ohio Naturalist 8, 287–289.

Hungr, O., Evans, S.G., Hazzard, J., 1999. Magnitude and frequency of rock falls and rock slides along the main transportation corridors of southwestern British Columbia. Can. Geotech. J. 36, 224–238.

Hungr, O., Evans, S.G., Bovis, M.J., Hutchinson, J.N., 2001. A review of the classification of landslides of the flow type. Environ. Eng. Geosci. 7, 221–238.

Hutchinson, J.N., 1970. A coastal mudflow on the London clay cliffs at Beltinge, North Kent. Geotechnique 24, 412–438.

Hutchinson, J.N., 1988. General report: morphological and geotechnical parameters of landslides related to geology and hydrology. In: Bonnard, C. (Ed.), Landslides. Fifth International Symposium on Landslides, Lausanne, Switzerland, pp. 3–35.

Innes, J.L., 1983. Lichenometric dating of debris-flow deposits in the Scottish Highlands. Earth Surf. Process. Land. 8, 579–588.

Istanbulluoglu, E., Yetemen, O., Vivoni, E.R., Gutierrez-Jurado, H.A., Bras, R.L., 2008. Eco-geomorphic implications of hillslope aspect: inferences from analysis of landscape morphology in central New Mexico. Geophys. Res. Lett. 35, L14403, doi:10.10292008GL034477.

Iverson, R.M., 1997. The physics of debris flows. Rev. Geophys. 35, 245–296.

Janke, J.R., Regmi, N.R., Giardino, J.R., Vitek, J.D., 2013. Rock glaciers. In: Shroder, J. (Editor in Chief), Giardino, R., Harbor, J. (Eds.), Treatise on Geomorphology. Academic Press, San Diego, CA, vol. 8, Glacial and Periglacial Geomorphology, pp. 238–273.

Jomelli, V., Francou, B., 2000. Comparing the characteristics of rockfall talus and snow avalanche landforms in an Alpine environment using a new methodological approach: Massif des Ecrins, French Alps. Geomorphology 35, 181–192.

Keaton, J.R., Wartman, J., Anderson, S., Benoît, J., deLaChapelle, J., Gilbert, R., Montgomery, D.R., 2014. The 22 March 2014 Oso Landslide, Snohomish County, Washington. Geotechnical Extreme Events Reconnaissance: Turning Disaster into Knowledge, 172 p., plus appendix.

Keefer, D.K., 2002. Investigating landslides caused by earthquakes – a historical review. Surv. Geophys. 23, 473–510.

Keefer, D.K., Johnson, A.M., 1983. Earth flows: morphology, mobilization and movement. US Geological Survey Professional Paper 1264, p. 56.

Keefer, D.K., Wilson, R.C., Mark, R.K., Brabb, E.E., Brown, W.M., Ellen, D.S., Harp, E.L., Wieczorek, G.F., Alger, C.S., Zatkin, R.S., 1987. Real-time landslide warning during heavy rainfall. Science 238, 921–925.

Kirkby, M.J., Statham, I., 1975. Surface stone movement and scree formation. J. Geol. 83, 349–362.

Korup, O., 2005. Distribution of landslides in southwest New Zealand. Landslides 2, 43–51.

Larsen, M.C., Simon, A., 1993. A rainfall intensity-duration threshold for landslides in a humid-tropical environment, Puerto-Rico. Geogr. Ann. A Phys. Geogr. 75, 13–23.

Larsen, I.J., Montgomery, D.R., Korup, O., 2010. Landslide erosion controlled by hillslope material. Nat. Geosci. 3, 247–251.

Lauber, C.L., Strickland, M.S., Bradford, M.A., Fierer, N., 2008. The influence of soil properties on the structure of bacterial and fungal communities across land-use types. Soil Biol. Biochem. 40, 2407–2415.

Lin, H., 2010. Earth's Critical Zone and hydropedology: concepts, characteristics, and advances. Hydrol. Earth Syst. Sci. 14, 25–45.

Lin, H.S., Zhou, X., 2008. Evidence of subsurface preferential flow using soil hydrologic monitoring in the Shale Hills Catchment. Euro. J. Soil Sci. 59, 34–49.

Lin, H.S., Singha, K., Chittleborough, D., Vogel, H.-J., Mooney, S., 2008. Advancing the emerging field of hydropedology. EOS Trans. 89, 490.

Locat, J., Demers, D., 1988. Viscosity, yield stress, remolded strength, and liquidity relationships for sensitive clay. Can. Geotech. J. 25, 799–806.

Marron, D.C., 1985. Colluvium in bedrock hollows on steep slopes, Redwood Creek Drainage-Basin, Northwestern California. In: Jugerius, P.D. (Ed.), Soil and Geomorphology (Catena Supplement 6). Catena-Verlag, Braunschweig, pp. 59–68.

Marston, R.A., Miller, M.M., Devkota, L.P., 1998. Geoecology and mass movement in the Manaslu-Ganesh and Langtang-Jugal Himals, Nepal. Geomorphology 26, 139–150.

Matsuoka, N., Sakai, H., 1999. Rockfall activity from an alpine cliff during thawing periods. Geomorphology 28, 309–328.

McAffee, R.P., Cruden, D.M., 1996. Landslides at rock glacier site, Highwood Pass, Alberta. Can. Geotech. J. 33, 685–695.

McDonald, E.V., McFadden, L.D., Wells, S.G., 2003. Regional response of alluvial fans to the Pleistocene-Holocene climatic transition, Mojave Desert, California. Geol. Soc. Am. Special Pap. 368, 189–206.

Millard, T., 2003. Schmidt Creek sediment sources and the Johnstone Strait Killer Whale Rubbing Beach. British Columbia Forest Service, Forest Research Technical Report TR-025, Nanaimo, Canada.

Minder, J.R., Roe, G.H., Montgomery, D.R., 2009. Spatial patterns of rainfall and shallow landslide susceptibility. Water Resour. Res. 45, W04419.

Montety, V.D., Marc, V., Emblanch, C., Malet, J.P., Bertrand, C., Maquaire, O., Bogaard, T.A., 2007. Identifying the origin of groundwater and flow processes in complex landslides affecting black marls: insights from a hydrochemical survey. Earth Surf. Process. Land. 32, 32–48.

Montgomery, D.R., Dietrich, W.E., 1994. A physically-based model for the topographic control on shallow landsliding. Water Resour. Res. 30, 1153–1171.

Montgomery, D.R., Dietrich, W.E., Torres, R., Anderson, S.P., Heffner, J.T., Loague, K., 1997. Hydrologic response of a steep, unchanneled valley to natural and applied rainfall. Water Resour. Res. 33, 91–109.

Narasimhan, T.N., 2005. Pedology: a hydrogeological perspective. Vadose Zone J. 4, 891–898.

National Research Council, 2001. Basic Research Opportunities in Earth Science. National Academy Press, Washington, DC, p. 154.

Nefeslioglu, H.A., Gokceoglu, C., Sonmez, H., 2008. An assessment on the use of logistic regression and artificial neural networks with different sampling strategies for the preparation of landslide susceptibility maps. Eng. Geol. 97, 171–191.

Noever, D.A., 1993. Himalayan Sandpiles. Phys. Rev. E 47, 724–725.

O'Loughlin, C.L., Ziemer, R.R., 1982. The importance of root strength and deterioration rates upon edaphic stability in steepland forests. Proceedings of I.U.F.R.O. Workshop P.1.07-00 Ecology of Subalpine Ecosystems as a Key to Management. August 2–3, 1982, Corvallis, Oregon. Oregon State University, Corvallis, OR. pp. 70–78.

Owen, L.A., 1991. Mass movement characteristics in the Karakoram Mountains: their sedimentary characteristics, recognization and role in Karakoram landform evolution. Z. Geomorph. N.F. 35, 401–424.

Parise, M., Guzzi, R., 1992. Volume and shape of the active and inactive parts of the Slumgullion landslide, Hinsdale County, Colorado. US Geological Survey Open-File Report 92-216, p. 29.

Parry, S., Campbell, S.D.C., Churchman, G.J., 2000. Kaolin-rich zones in Hong Kong saprolites – their interpretation and engineering significance. In: International Society for Rock Mechanics International Symposium. November 19–24, 2000, Melbourne, Australia.

Parsons, A.J., 1988. Hillslope Form. Routledge, London.

Pasuto, A., Silvano, S., 1998. Rainfall as a trigger of shallow mass movements. A case study in the Dolomites, Italy. Environ. Geol. 35, 184–189.

Pelletier, J.D., 1997. Kardar-Parisi-Zhang scaling of the height of the convective boundary layer and fractal structure of Cumulus cloud fields. Phys. Rev. Lett. 78, 2672–2675.

Petley, D.N., 2012. Global patterns of loss of life from landslides. Geology 40, 927–930.

Petley, D.N, 2014. Fatal landslides in the first seven months of 2014. AGU Blogosphere, http://blogs.agu.org/landslideblog/2014/10/06/.

Petley, D.N., Allison, R.J., 1997. The mechanics of deep-seated landslides. Earth Surf. Process. Land. 22, 747–758.

Petley, D.N., Bulmer, M.H., Murphy, W., 2002. Patterns of movement in rotational and translational landslides. Geology 30, 719–722.

Phillips, C.J., Watson, A.J., 1994. Structural tree root research in New Zealand: a review. Manaaki Whenua Press, Lincoln, New Zealand, p. 71.

Pradhan, B., 2013. A comparative study on the predictive ability of the decision tree, support vector machine and neuro-fuzzy models in landslide susceptibility mapping using GIS. Comput. Geosci. 51, 350–365.

Rasmussen, C., 2012. Thermodynamic constraints on effective energy and mass transfer and catchment function. Hydrol. Earth Syst. Sci. 16, 725–739.

Regmi, N.R., 2010. Hillslope dynamics in the Paonia-McClure Pass area, Colorado, USA. PhD thesis, Texas A&M University, College Station, TX, USA.

Regmi, N.R., Giardino, J.R., Vitek, J.D., 2010a. Modeling susceptibility to landslides using the weight of evidence approach: Western Colorado, USA. Geomorphology 115, 172–187.

Regmi, N.R., Giardino, J.R., Vitek, J.D., 2010b. Assessing susceptibility to landslides: using models to understand observed changes in slopes. Geomorphology 122, 25–38.

Regmi, N.R., Giardino, J.R., Vitek, J.D., Dangol, V., 2010c. Mapping landslide hazards in Western Nepal: comparing qualitative and quantitative approaches. Environ. Eng. Geosci. 16, 127–142.

Regmi, N., Giardino, J., McDonald, E., Vitek, J., 2013a. A comparison of logistic regression-based models of susceptibility to landslides in western Colorado, USA. Landslides 11, 247–262.

Regmi, N.R., Giardino, J.R., Vitek, J.D., 2013b. Characteristics of landslides in western Colorado, USA. Landslides 11, 589–603.

Regmi, N.R., McDonald, E.V., Bacon, S.N., 2014. Mapping Quaternary alluvial fans in the southwestern United States based on multiparameter surface roughness of LiDAR topographic data. J. Geophys. Res. Earth Surf. 119, 12–27.

Reneau, S.L., Dietrich, W.E., Dorn, R.I., Berger, C.R., Rubin, M., 1986. Geomorphic and paleoclimatic implications of latest Pleistocene radiocarbon-dates from colluvium-mantled hollows, California. Geology 14, 655–658.

Richter, D.D., 2007. Humanity's transformation of Earth's soil: pedology's new frontier. Soil Sci. 172, 957–967.

Richter, D.D., Mobley, M.L., 2009. Monitoring Earth's Critical Zone. Science 326, 1067–1068.

Richter, D., Bacon, A.R., Billings, S.A., Binkley, D., Buford, M., Callaham, M.A., Curry, A.E., Fimmen, R.L., Grandy, A.S., Heine, P.R., Hofmockel, M., Jackson, J.A., LeMaster, E., Li, J., Markewitz, D., Mobley, M.L., Morrison, M.W., Strickland, M.S., Waldrop, T., Wells, C.G., 2014. Evolution of soil, ecosystem, and Critical Zone research at the USDA FS Calhoun Experimental Forest. In: Hayes, D.C., Stout, S.L., Crawford, R.H., Hoover, A.P. (Eds.), USDA Forest Service Experimental Forests and Ranges. Springer, New York, pp. 405–433.

Riestenberg, M.M., Sovonickdunford, S., 1983. The role of woody vegetation in stabilizing slopes in the Cincinnati area, Ohio. Geol. Soc. Am. Bull. 94, 506–518.

Ritchie, A.M., 1963. Evaluation of Rockfall and its Control. Highway Research Board, National Research Council, Washington, DC.

Ritter, D.F., 1986. Process Geomorphology. Wm. C. Brown Publishers, Dubuque, IO, p. 603.

Roering, J., 2012. Tectonic geomorphology: landslides limit mountain relief. Nat. Geosci. 5, 446–447.

Roering, J.J., Kirchner, J.W., Dietrich, W.E., 2005. Characterizing structural and lithologic controls on deep-seated landsliding: implications for topographic relief and landscape evolution in the Oregon Coast Range, USA. Geol. Soc. Am. Bull. 117, 654–668.

Rogers, W.P., 2003. Critical landslides of Colorado – a year 2002 review and priority list. Colorado Geological Survey, Open-File Report OF-02-16, 1 map.

Rogers, N.W., Selby, M.J., 1980. Mechanisms of shallow translational landsliding during dry summer storms, North Island, New Zealand. Geogr. Ann. A V 62, 11–20.

Rupke, J., Cammeraat, E., Seijmonsbergen, A.C., Vanwesten, C.J., 1988. Engineering geomorphology of the Widentobel Catchment, Appenzell and Sankt-Gallen, Gallen, Switzerland – a geomorphological inventory system applied to geotechnical appraisal of slope stability. Eng. Geol. 26, 33–68.

Schmidt, K.M., Montgomery, D.R., 1995. Limits to relief. Science 270, 617–620.

Schulz, W.H., Kean, J.W., Wang, G., 2009. Landslide movement in southwest Colorado triggered by atmospheric tides. Nat. Geosci. 2, 863–866.

Schumm, S.A., 1967. Rates of surficial rock creep on hillslopes in western Colorado. Science 155, 560–562.

Schumm, S.A., Chorley, R.J., 1964. The fall of threatening rock. Am. J. Sci. 262, 1041–1064.

Selby, M.J., 1982. Hillslope Materials and Processes. Oxford University Press, Oxford, p. 264.

Sharpe, C.F., 1938. Landslides and Related Phenomena. Columbia University Press, New York, NY, p. 137.

Shroder, J.F., 1980. Dendrogeomorphology review and new techniques of tree-ring dating. Prog. Phys. Geogr. 4, 161–188.

Shroder, J.F., Owen, L.A., Seong, Y.B., Bishop, M.P., Bush, A., Caffee, M.W., Copland, L., Finkel, R.C., Kamp, U., 2011. The role of mass movements on landscape evolution in the Central Karakoram: discussion and speculation. Quaternary Int. 236, 34–47.

Sidle, R.C., 1992. A theoretical model of the effects of timber harvesting on slope stability. Water Resour. Res. 28, 1897–1910.

Sidle, R.C., Ochiai, H., 2006. Landslides: Processes, Prediction, and Land Use. American Geophysical Union, Washington, DC, p. 317.

Simonett, D.S., 1967. Landslide distribution and earthquakes in the Bewani and Torricelli Mountains, New Guinea. In: Jennings, J.N., Mabutt, J.A. (Eds.), Landform Studies from Australia and New Guinea. Cambridge University Press, Cambridge, pp. 64–84.

Skempton, A.W., Hutchinson, J.N., 1969. Stability of natural slopes and embankment foundations. 7th International Conference on Soil Mechanics and Foundation Engineering, Mexico City.

Smith, K., 1996. Environmental Hazards: Assessing Risk and Reducing Disaster. Routledge, London, p. 478.

Stark, C.P., Guzzetti, F., 2009. Landslide rupture and the probability distribution of mobilized debris volumes. J. Geophys. Res. Earth Surf. 114, F00A02. doi:10.1029/2008JF001008.

Stark, C.P., Hovius, N., 2001. The characterization of landslide size distributions. Geophys. Res. Lett. 28, 1091–1094.

Stoffel, M., Butler, D.R., Corona, C., 2013. Mass movements and tree rings: a guide to dendrogeomorphic field sampling and dating. Geomorphology 200, 106–120.

Sulebak, J.R., Tallaksen, L.M., Erichsen, B., 2000. Estimation of areal soil moisture by use of terrain data. Geogr. Ann. A Phys. Geogr. 82A, 89–105.

Swanson, F.J., Dyrness, C.T., 1975. Impact of clear-cutting and road construction on soil erosion by landslides in the western Cascade Range, Oregon. Geology 3, 393–396.

Torrance, J.K., 1999. Physical, chemical and mineralogical influences on the rheology of remoulded low-activity sensitive marine clay. Appl. Clay Sci. 14, 199–223.

Van Horn, F.R., 1909. A recent landslide in a shale bank near Cleveland accompanied by buckling. Science 29, 626.

Van Westen, C.J., Lulie Getahun, F., 2003. Analyzing the evolution of the Tessina landslide using aerial photographs and digital elevation models. Geomorphology 54, 77–89.

Vanasch, T.W.J., Deimel, M.S., Haak, W.J.C., Simon, J., 1989. The viscous creep component in shallow clayey soil and the influence of tree load on creep rates. Earth Surf. Process. Land. 14, 557–564.

Varnes, D.J., 1978. Slope movement types and processes. In: Schuster, R.L., Krizek, R.J. (Eds.), Landslides: Analysis and Control. Special Report 176, Transportation Research Board, Washington, DC, pp. 11–33.

Vidrih, R., Ribičič, M., Suhadolc, P., 2001. Seismogeological effects on rocks during the 12 April 1998 upper Soča Territory earthquake (NW Slovenia). Tectonophysics 330, 153–175.

Volkwein, A., Schellenberg, K., Labiouse, V., Agliardi, F., Berger, F., Bourrier, F., Dorren, L.K.A., Gerber, W., Jaboyedoff, M., 2011. Rockfall characterisation and structural protection – a review. Nat. Hazards Earth Syst. Sci. 11, 2617–2651.

Western Regional Climate Center, 2014. http://www.wrcc.dri.edu/cgi-bin/cliMAIN.pl?copaon.

Wieczorek, G.F., Snyder, J.B., Waitt, R.B., Morrissey, M.M., Uhrhammer, R.A., Harp, E.L., Norris, R.D., Bursik, M.I., Finewood, L.G., 2000. Unusual July 10, 1996, rock fall at Happy Isles, Yosemite National Park, California. Geol. Soc. Am. Bull. 112, 75–85.

Wyllie, D.C., Mah, C., 2004. Rock Slope Engineering: Civil and Mining. Spon Press, New York, p. 431.

Xie, M.W., Esaki, T., Zhou, G.Y., 2004. GIS-based probabilistic mapping of landslide hazard using a three-dimensional deterministic model. Nat. Hazards 33, 265–282.

Chapter 12

The Impact of Glacial Geomorphology on Critical Zone Processes

Kevin R. Gamache*, John R. Giardino*, Netra R. Regmi**, and John D. Vitek*
*High Alpine and Arctic Research Program, Department of Geology and Geophysics, Texas A&M University, College Station, Texas, USA; **Department of Soil, Water and Environmental Sciences, University of Arizona, Tucson, Arizona, USA

12.1 INTRODUCTION

The powerful erosional and depositional processes of glaciers have created much of the alpine, Arctic and Antarctic landscapes of Earth. Glaciers are one part of the five subsystems of Earth, namely the cryogenic regime or subsystem. Throughout geological time, glaciers have advanced and waned, but beginning in the Pleistocene, glaciers have had a massive effect on the present landscapes of many areas of North America, South America, Europe, Asia, and both Polar Regions. In addition to their sculpturing and depositing powers, glaciers have served as sinks for much of the freshwater on the planet, and in the past, covered as much as 30% of the land surface.

In dealing with glacial environments, it is difficult to visualize this environment as part of the Critical Zone if we use the general definition of the Critical Zone, as extending from the top of the canopy to the bottom of the aquifer (NRC, 2001). Thus, it might be easy to dismiss or exclude the glacial environment as being part of the Critical Zone. In Polar Regions, no trees, shrubs or plants exist on glaciers, so where is the top of the canopy? We place the top of the canopy in the lower extent of the stratosphere. In alpine areas, numerous landforms make up the glacial environment. For example, various glacial deposits surround glaciers, often underlie glaciers, and even form atop glaciers. They have soil profiles and support plant life. Thus, we define the top of the canopy in this environment as extending from the top of trees, bushes, shrubs or grasses growing on various glacial deposits.

What about the bottom of the Critical Zone? Although much of the terrain beneath glaciers is frozen, some areas below glaciers remain in an unfrozen state as the result of latitudinal location or pressure melting. Thus, water in the

Developments in Earth Surface Processes, Vol. 19. http://dx.doi.org/10.1016/B978-0-444-63369-9.00012-6

liquid state can and does exist. Lakes beneath Antarctic glaciers, for instance, also contain life. Complex interactions involving rock, soil, water, air, and living organisms exist. Glaciers themselves serve as stores of freshwater. The water supply for life sustaining resources in the City of Boulder, Colorado, for example, is the Arapaho glacier and its surrounding watershed. Nevertheless, it is water either in a frozen or liquid state. We have not found a definition that requires water to be in a liquid state to qualify as an aquifer.

If we use the original definition of the Critical Zone as "... the heterogeneous, near surface environment in which complex interactions involving rock, soil, water, air and living organisms regulate the natural habitat and determine availability of life sustaining resources" (NRC, 2001, p. 2) then glacial environments are very much a part of the Critical Zone.

12.2 GOAL OF THIS CHAPTER

Earth has a history of frequent local and global scale glaciations, which range from many hundreds of million years to recent. At the present time, most of the glaciers on Earth are located in Antarctica, ~98% of which is covered by ice with an average thickness of 1.6 km; in the Arctic, the principal ice sheet that overlies Greenland has an area of 1,755,637 km^2 and a thickness of up to 3,200 m; and individual valley glaciers, such as those of Vatnajökull, which is the largest glacier (~1000-m thick) in Iceland (Gregory, 2010) (Fig. 12.1).

Today, mountain glaciers, ice sheets and ice caps generate glacial processes that directly affect about 10% of the land surface of Earth, primarily in high-latitude regions (Gregory, 2010). This has not always been the case; however, in the past three million years more than 30% of the land surface was covered by glacial ice (Gregory, 2010). Earth was encapsulated in ice, around 750 M BP, a period referred to as Snowball Earth (Walker, 2003). Glaciers have ebbed and flowed since Snowball Earth ended. The glacial landscapes of today represent the most recently created landscapes on Earth. Although the Antarctic ice sheet has existed for at least 34 million years, extensive ice and permafrost did not exist prior to the Pliocene. At present 91.4% of the volume of glacier ice is in the Antarctic ice sheet and 8.3% is in Greenland, leaving just 0.3% for all the remaining glaciers in North America, South America, continental Europe, Asia and New Zealand (Gregory, 2010) (Fig. 12.2).

Chris Rapley, former Director of the British Antarctic Survey, recognized these regions as being most sensitive to global change and providing a super-sensitive early warning system for Earth (Rapley, 1999, 2006; Gregory, 2010). For that reason, understanding the basal and near-bed response of glaciers and ice sheets, and of glacier-bed interactions, is extremely important to glacial geomorphologists (Theakstone, 1982). Because glaciers do play a role in the Critical Zone, the major goal of this chapter is to provide detailed descriptions of processes and relationships in glacial areas, so that one can understand the important role these landscapes play in the Critical Zone. Although our focus

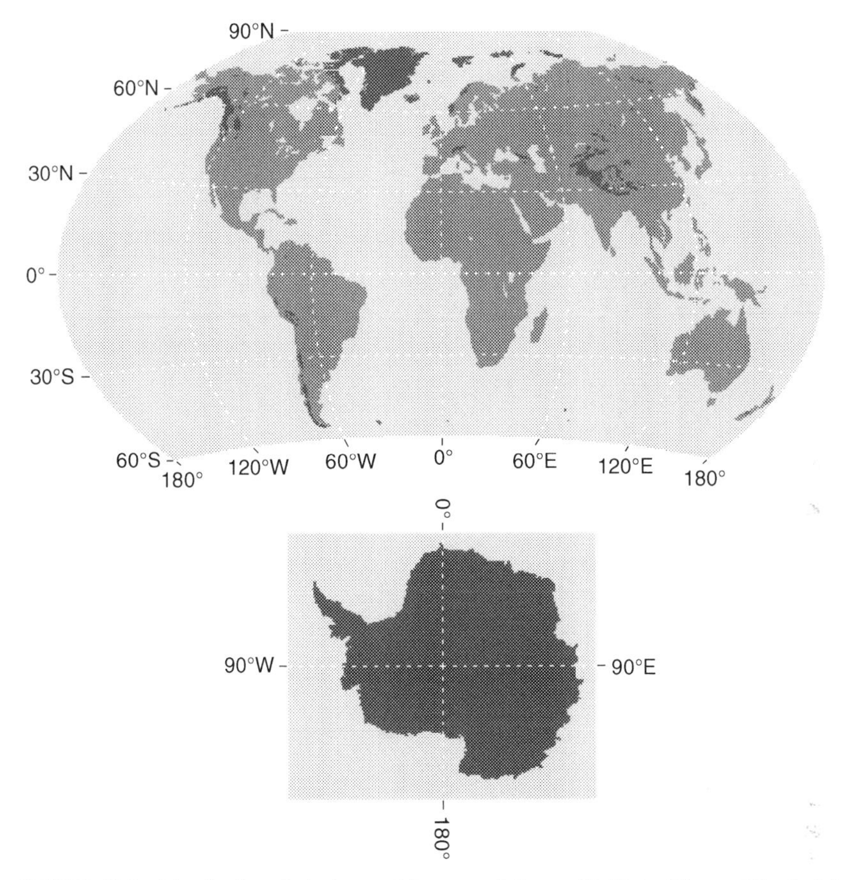

FIGURE 12.1 Distribution of glaciers and ice around the world. Most of the world's glacial ice is found in Antarctica and Greenland, but glaciers are found on nearly every continent. *(Image/photo courtesy of the National Snow and Ice Data Center, University of Colorado, Boulder.)*

is on erosional and depositional processes of glaciers, glacier ecosystems, and hydrology, we link the processes with risks associated with glaciated regions and the response of glaciers to changes in climate.

12.3 GLACIER MASS BALANCE

Today Critical Zone research is being driven by six overarching science questions. Of these six big-science questions, one that is directly related to glaciers, especially mass-balance studies, is focused on establishing in what manner theory and data can be linked from molecular to global scale to explain past transformations of the surface of Earth and predict the rate of development and planetary control of the Critical Zone (Banwart et al., 2012). Study and understanding of the processes and rates of growth and decline of glaciers is fundamental to answering the big science question. Glacier mass balance is expressed as the

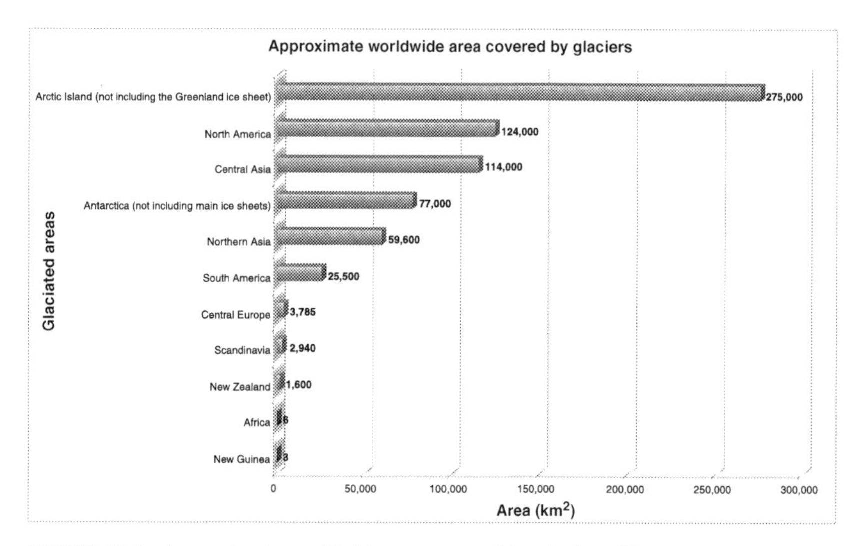

FIGURE 12.2 Approximate worldwide area covered by glaciers. These measurements were compiled by the World Glacier Monitoring Service, and published in the 2008 Publication, Global Glacier Changes: facts and figures (Roer et al., 2008). *(Figure was adapted from National Snow and Ice Data Center, University of Colorado, Boulder.)*

change in the volume of a glacier system by the gain or loss of ice on a temporal basis. The volume of a glacier is the function of the volume of ice it receives from snow accumulation and the amount of ice it loses by various processes, including sublimation and melting. Mass balance of a glacier can be described in terms of two processes as: accumulation and ablation of snow and ice.

Snow and ice can be accumulated directly from precipitation, avalanches, windblown snow, and in minor amounts, from hoar frost, a layer of ice crystals formed by vapor transfer (i.e., sublimation followed by deposition) within dry snow beneath the snow surface. Accumulation can take place in the interior of the glacier when precipitation falls as rain and both meltwater and rainwater percolate through the snowpack and then refreeze. Accumulation can also take place at the base of a glacier or ice sheet by liquid-water freezing. The accumulated snow and ice can be transported downslope, as the glacier flows, until they reach a point where they are lost to the system, either by surface melting, surface meltwater runoff, sublimation, avalanching, wind erosion, and calving (the breakaway of iceblocks and icebergs).

These processes are collectively known as ablation. Other processes of ablation include subaqueous melting, and melting within the ice and at the ice bed, which moves towards the terminus. Accumulation usually occurs over the entire glacier, but the rate of accumulation may change with elevation. Colder air temperatures at higher elevations of a glacier may result in more precipitation falling as snow, whereas, warmer air temperatures at lower elevations of a glacier may result in more precipitation falling as rain.

The area where there is more accumulation than ablation (more mass gained than lost) is referred to as the accumulation zone. Similarly, ablation usually occurs over the entire glacier, but the part of the glacier that has more ablation than accumulation is the zone of ablation. Zones of accumulation and ablation are separated by a line where accumulation is equal to ablation and is referred to as the equilibrium line altitude (i.e., elevation). The snowline is commonly considered as the equilibrium line on many mountain glaciers. Glaciers losing more mass than they receive will be in a negative mass balance and so will be reduced in volume. The reduction of volume can result in the reduction of the thickness as well as the length of the glacier. Many mistakenly refer to this process as glacial retreat. The simple fact is glaciers do not retreat; they melt. Glaciers gaining more mass than they lose will be in positive mass balance and will grow and increase in volume. The glacier snout appears to moves backward in melting glaciers and the snout appears to move forward in expanding glaciers. Glaciers in equilibrium lose and gain approximately the same volume of snow and ice. They will neither increase nor decrease in depth and length.

Accumulation and ablation of glacier ice vary over time and space, and depend on various factors including: elevation, aspect, relief, season, local and regional climate, latitude, and long-term change in climate (Kaser and Georges, 1999; Oerlemans, 2005; Oerlemans and Fortuin, 1992). Glaciers tend to be colder at the top than at the bottom. As a result, of this temperature profile, accumulation is greater in the upper extent of the glacier, and ablation is greater in the lower extent of the glacier. In the short-term, mass balance of a glacier varies throughout the year; glaciers typically receive more accumulation in the winter and experience more ablation in the summer (Dyurgerov and Meier, 1999). In the long-term, the dynamics of a glacier is primarily the function of change in climate (Haeberli and Beniston, 1998; Maisch, 2000). Study of glaciers around the world suggests that mass balances of glaciers have generally decreased during the past few decades, possibly as a result of current global warming (Oerlemans, 2005).

Differences in the mass balance of a glacier due to changing climate causes adjustments in morphology to a new steady-state condition. The time it takes for the morphology of a glacier to adjust is known as its *response time*. The response time of a glacier depends on its mass-balance gradient, which is a critical factor controlled by the climatic regime in which the glacier is located. Mass-balance gradients measure the degree of change in net mass balance as a function of elevation. Net mass balance is generally at a maximum at the head of the glacier, zero at the equilibrium line, and minimum at the front or toe of the glacier. Glaciers with steeper mass-balance gradients have greater sensitivity to climate change than glaciers with less steep mass-balance gradients. Mass-balance gradients increase with increasing humidity; therefore, glaciers in wetter climates are more sensitive to changes in mass balance than glaciers in dry climates (Kuhn, 1981; Oerlemans and Fortuin, 1992). Glaciers in temperate regions, such as in the New Zealand Alps, receive very high amounts of precipitation, thus have a steeper mass-balance gradient (Chinn, 1996). These wet glaciers may have a shorter response time and higher climate sensitivity than cold and dry-polar glaciers that receive little accumulation and have relatively low rates of ablation (Rignot and Thomas, 2002). These glaciers may respond more slowly to climate change.

The mass balance of a glacier over time can be measured by various techniques including: (1) installation of stakes, GPS and automatic weather stations, across a glacier (Andreassen et al., 2005; Van de Wal et al., 2008); (2) probing snowpits and crevasse stratigraphy (Pelto, 1990); and (3) analysis of time series remotely sensed images and aerial photographs (Bishop et al., 2004; Berthier et al., 2007). Stakes provide measurements of rates of accumulation and ablation at the stake locations on the glacier surface. This approach is time-consuming, often arduous, and thus challenging. Use of weather stations is key to understanding the fluxes of energy on the glacier, and the use of GPS is key to determining the movement dynamics of the glacier. A remote-sensing approach is generally the best approach for determining changes in the mass balance of a glacier because it is the most cost-effective approach and precludes commonly arduous fieldwork. However, the remote-sensing approach always requires ground validation of mass-balance measurements.

12.4 GLACIAL CHRONOLOGY AND QUATERNARY GLACIATION

Earth has a history of frequent local and global scale glaciations at different time intervals and for various durations. Five major known ice ages have occurred in history (Hambrey and Harland, 1985). Ice ages are the periods of long-term reduction in the temperature of the surface of Earth and the atmosphere, which promotes expansion of continental and polar ice sheets and alpine glaciers. Within each ice age, a series of severe glacial and interglacial periods (more temperate) existed. The oldest unambiguous ice age in the history of Earth appears to be during the Proterozoic Eon in early Earth history (Table 12.1).

Early Proterozoic (2400–2100 Ma). The Neoproterozoic glacial epoch (1000–544 Ma) represents the most severe ice age in the history of Earth (Hoffman et al., 1998; Rieu et al., 2007). During this time, Earth was fully or almost completely covered with ice, a period referred to as Snowball Earth (Walker, 2003). Earth is currently in an interglacial period of the Quaternary.

The Quaternary Ice Age represents the last ~2.5 Ma of geological time. Quaternary glaciation, more refined at the epoch stage as Pleistocene glaciation, is a series of glacial events separated by interglacial events. During this time, ice sheets expanded in Antarctica and Greenland and fluctuating ice sheets occurred in other parts of the world, also; a notable example being the Laurentide ice sheet (Huybrechts, 2002). Almost all of the present glacial landscapes worldwide are the result of Quaternary glaciation. Glacial activity during this

TABLE 12.1 Global Quaternary Glacial Chronology

Epoch	Chronose-	Global region					Glacial/		
Anthro-pocene	quences/ glacial sequence	Alps	Midwest US	British Isles	Northern European	South American	interglacial period	Period	Marine Isotope Stage (MIS)
Holocene	Subatlantic							2.5 ka–1.50 a	
	Subboreal							5 ka–2.5 ka	
	Atlantic						Interglacial	8 ka–5 ka	MIS 1–14 ka, end of the Younger Dryas marks the start of the Holocene, continuing to the present
	Preboreal							12ka–9 ka	
Pleisto-cene	1st	Würm	Wisconsin	Devensian	Weichselian	Llanquihue	Glacial	12–71	MIS 2–29 ka, near last glacial maximum MIS 3–57 ka MIS 4–71 ka MIS 5a–82 ka MIS 5b–87 ka MIS 5c–96 ka MIS 5d–109 ka
		Riss–Würm	Sangamonian	Ipswichian	Eemian	Valdivia	Interglacial	115–130	MIS 5e–123 ka
	2nd	Riss	Illinoian	Wolstonian	Saalian	Santa Maria	Glacial period	130–200	MIS 6–191 ka
		Mindel–Riss	Yarmouthian	Hoxnian	Holsteinian		Interglacial(s)	374–424	MIS 10–374 ka MIS 11–424 ka
	3rd–6th	Mindel	Kansan	Anglian	Elsterian	Rio Lico	Glacial period(s)	424–478	MIS 12–478 ka
		Günz–Mindel	Altonian	Cromerian	Waalian		Interglacial(s)	478–533–563	MIS 13–533 ka MIS 14–563 ka
	7th–8th	Günz	Nebraskan	Beestonian	Elburonian	Caracol	Glacial period	621–676	MIS 15–621 ka MIS 16–676 ka

A summary of the global Quaternary glacial chronology and correlates each glacial/interglacial period with their respective Marine Isotope Stages. Derived from Ehlers and Gibbard (2008); Ehlers et al. (2011); and Frenzel (1992).

period resulted in intense erosion and deposition of glacial sediments over large parts of the continents, modification of river systems, changes in sea level, creation of millions of glacial and pluvial lakes, climatic variability, and isostatic adjustment of the crust (Berger, 1978; Currey, 1990; Lambeck and Chappell, 2001). The last glacial period of the Quaternary ended approximately 11,000 years BP and currently Earth is in an interglacial period. Climatic warming caused the ice sheets from the last glacial period to begin to disappear. Remnants of the last glacial period, however, still exist in Greenland, Antarctica, and some alpine mountain regions.

The current glaciation (i.e., Pleistocene glaciation) began at least ~1.5 Ma in the Northern Hemisphere and continues to the Holocene (Jones, 2005). Evidence suggests climates in the Northern Hemisphere were glacial ~70,000 BP, with two major short-term glacial periods ~25,000 and ~70,000 BP separated by a short period (<10,000 years) of climatic conditions associated with minimum glacial extent (Jones, 2005). Evidence further suggests that approximately one-third of the total land area of Earth was covered by ice at the Pleistocene glacial maximum. The largest ice sheet, known as the Laurentide of North America, covered the vast majority of Canada and the northern United States. The period from ~130,000 BP to ~70,000 BP is considered a period of interglaciation, and the period before that is considered glacial, which has slightly more ice coverage than the last glaciation (Jones, 2005).

Minor glacial fluctuations have been reported for the past few hundred years. The period from 700BP to 100 BP has been termed the "Little Ice Age," and was marked by observed glacier advances, especially in Europe (Hagen et al., 1993). Since then, however, most glaciers have shown a steady decline in extent.

12.5 GLACIAL FEATURES IN THE CRITICAL ZONE

A glacier is a persistent mass of snow and ice that covers an area >0.1 km^2, deforms under its own weight, shows evidence of downslope movement, and exhibits a fluid flow (Gregory, 2010). The mass of ice in a glacier forms by the recrystallization of snow or other forms of solid precipitation. Snow in the accumulation zone is transformed to firn (snow which has survived one summer melt season) and later to glacial ice where regrowth of ice crystals and elimination of air passages results in densities of 0.83–0.91 kg/m^3. This may occur in only a year in valley glaciers, but can take several thousand years in ice-sheet environments (Gregory, 2010).

Water and sediment are major components of glacial systems. The routes taken by the ice, commonly determine many of the resultant landscape features and landforms associated with the glacier environment. Sediment is carried on, in, and under the ice, and water can flow on, in, below, and around glaciers (Gregory, 2010). Sediment can be moved long distances from its source region.

12.6 TYPES OF GLACIERS

Glaciers exhibit a range of shapes, sizes, and morphology as a function of climate, topography, and location (Benn and Evans, 2014). For example glaciers range in size from a niche glacier to large ice sheets. Additionally, the movement mechanics and associated processes of mountain glaciers are mostly controlled by the topography. The movement mechanics and processes of ice sheets and ice caps are mostly independent of the topography. Therefore, for the first order classification, glaciers have been categorized based on the relation to the topography (Sugden and John, 1976). We here adopt this approach and classify glaciers as: (1) unconstrained by the topography and (2) constrained by the topography.

12.6.1 Glaciers Unconstrained by Topography

12.6.1.1 Ice Sheet and Ice Cap

Ice sheets or ice caps, in the form of broad domes, are glaciers that bury the underlying topography with ice radiating outwards as a sheet. They have major patterns of ice flow that are largely independent to the bed topography (Gregory, 2010). The difference between an ice sheet and an ice cap is the area of coverage. Ice sheets generally cover $>50{,}000\,km^2$ and ice caps generally cover $<50{,}000\,km^2$ (Benn and Evans, 2014). Ice masses presently covering most of Antarctica and Greenland are larger than the threshold and are referred to as ice sheets, whereas ice masses in Nordaustlandet (Svalbard), Ellesmere Island, Baffin Island, and Iceland are mostly smaller than the threshold and are designated as ice caps (Benn and Evans, 2014). The Antarctic ice sheets cover around 98% of the continent (Laybourn-Parry et al., 2012). Ice sheets and ice caps form from the interplay of intensive accumulation of frozen water, mainly snow, and the insufficient melt during summer (Müller and Koch, 2012). Ice sheets of the northern hemisphere were characterized by the greatest dynamics known from about 120,000 BP down to about 10,000 BP. Several phases of ice-sheet growth and decay occurred during this time (Laybourn-Parry et al., 2012; Müller and Koch, 2012).

These large masses of ice exert a tremendous weight on the underlying bedrock and cause it to be depressed. Upon melting, the weight gets removed and the land slowly rebounds, a process known as isostatic adjustment. The Hudson Bay area of Canada and the Baltic Sea between Finland and Norway are regions in which the rate of rebound has been measured. Although slow, the change signifies the impact that the presence of an ice mass had on the bedrock.

Ice sheets and ice caps can be further categorized as ice domes, outlet glaciers, and ice streams. Ice domes are upstanding areas of ice sheets or ice caps where the movement of ice is relatively slow. Outlet glaciers and ice streams move rapidly out from the interiors of ice sheets and ice caps as channeled ice. These types of glaciers are constrained partially by the surrounding topography and exposed bedrock.

12.6.1.2 Ice Shelves

We do not discuss ice shelves in detail, as they are associated more with ocean areas adjacent to the Earth terrestrial environments that is, the Critical Zone.

12.6.2 Glaciers Constrained by Topography

12.6.2.1 Ice Field

Ice fields develop in areas that are gentle in slope, but have locally fretted topography and sufficient elevation for the accumulation of ice. They appear similar to ice caps in terms of size, but differ in shape, and relation to the topography. They do not exhibit a dome-like structure, and the underlying surface topography influences the ice flow. Examples of well-known ice fields worldwide include the Columbia Icefields in the Canadian Rocky (Luckman et al., 1997), the ice fields of the St. Elias Mountains in the Canadian Yukon Territory/Alaska (Calkin, 1988), the Tien Shan/Kunlun Shan ice fields in China (Yi et al., 2002), and the Patagonian Ice Fields (Llibourty, 1998). Large valley glaciers drain all of these ice fields.

12.6.2.2 Valley Glaciers

Valley glaciers form when ice is discharged from an ice field or a cirque into a deep bedrock valley. They mostly occur in mountain valleys in Polar Regions (Gregory, 2010) and alpine mountains, and may develop a simple, single-branched planform, or a dendritic network (Penck, 1905). The form of the network and the glacial flow is commonly strongly influenced by the topography itself, the amounts of debris cover, and bedrock lithology and structure (Gregory, 2010). Bedrock slopes beneath valley glaciers are relatively steeper than the slopes beneath ice sheets and ice caps. In addition, the surrounding topography in valley glaciers is mostly ice-free over steepened slopes that generally feed snow and ice to the glacier surface (Benn and Evans, 2014) (Fig. 12.3).

12.6.2.3 Cirque Glaciers

Cirque glaciers are partly enclosed by steep headwalls and may remain separated from the main valley glaciers or ice caps (Gregory, 2010). They form in bowl-shaped depressions, also known as bedrock hollows or cirques, located on the side of, or near mountains. They characteristically form by the accumulation of snow and ice avalanching from upslope areas. The size of cirque glaciers ranges from glaciers that are completely limited within hosting bedrock hollows, to glaciers that form the heads of large valley glaciers. The stability of a cirque glacier depends on various factors including, the size of the depression and the morphology of the surrounding topography, wind frequency and magnitude, and the availability of snow and ice in the cirque and surrounding slopes (Giardino et al., 1987; Sugden et al., 1999). The morphology of cirques and

FIGURE 12.3 Blanca Peak (right center of photo) in the Sangre de Cristo Mountains, south-central Colorado is 4372 m in elevation. Glaciated valleys and arêtes are prominently displayed in this scene. *(Photo by John D. Vitek (October, 2005).)*

bedrock hollows can shelter snow from wind, and if the snow persists through summer months, it becomes glacier ice. Snow may also be sheltered from wind on the leeward slope of a mountain and can be a major source of snow for the cirque glacier. Rock falls from upslope areas also play an important role in sheltering the snow and ice from sunlight. If enough rock falls onto the glacier, it may become a rock glacier (Giardino et al., 1987, 2014; Giardino and Vitek, 1988; Janke et al., 2013). If a cirque glacier advances far enough, it may become a valley glacier, and if the climate warms sufficiently to cause ablation, the valley glacier may form a cirque glacier.

12.6.2.4 Piedmont Glacier

Piedmont glaciers develop by the process of valley glaciers debouching onto lowland areas after travelling through bedrock troughs and spreading out at the foot of mountain ranges. Some examples of piedmont glaciers include Malaspina Glacier in Alaska (Sharp, 1958) and Skeioararjokull glacier in Iceland (Sigurdsson, 1998). Many piedmont glaciers occur in the Canadian High Arctic, where subpolar glaciers debouch from plateau ice fields onto U-shaped valleys (Evans, 1990) (Fig. 12.4).

FIGURE 12.4 Malaspina Glacier in southeastern Alaska is considered a classic example of a piedmont glacier. Piedmont glaciers occur where valley glaciers exit a mountain range onto broad lowlands, are no longer laterally confined, and spread to become wide lobes. *(NASA Earth Observatory.)*

12.6.2.5 Small-Sized Glaciers

Masses of glacier ice also occur as ice fields or isolated glaciers. The smallest glacier ice-masses are known as ice aprons. They form by thin accumulations of snow and ice on mountainsides that is sometimes referred to as hanging glaciers. Similar patches of ice and snow occupying depressions on less precipitous terrain are commonly referred to as glacierets. They form as a result of snow drifting and avalanching from upslope areas. Glacierets that form because of ice avalanching from icefalls at steep plateau edges have been called fall glaciers. If a niche or rock bench in a mountain or valley side controls the location of an ice body, it is termed a niche glacier. Similar types of snow and ice accumulations in small depressions along coasts are referred to as ice fringes (Fig. 12.5).

12.7 EROSIONAL PROCESSES AND FORMS

Although the body of literature concerning glacial-erosional processes and landforms is expanding, it is small compared to that concerning the processes and landforms of glacial deposition (Glasser and Bennett, 2004). The relationship of glacial erosional processes and landforms to former ice sheets determines the continuity of subglacial deformed layers, associated landforms, and sediments. Glacial erosion generates landscape evolution and relief production and moreover, can be used to assess paleoenvironmental factors (Glasser and Bennett, 2004).

FIGURE 12.5 **Arapaho Glacier, located along the Front Range, Colorado, in 2003.** *(Photo Courtesy of NASA Earth Observatory Program.)*

12.8 EROSIONAL PROCESS

Major processes of glacial erosion include: quarrying (also known as plucking), crushing abrasion, and mechanical and chemical erosion by glacial meltwater (Glasser and Bennett, 2004; Gregory, 2010). Glacial erosion creates a suite of landforms that are frequently observed in areas formerly occupied by ice sheets and glaciers (Glasser and Bennett, 2004). Quarrying involves the fracturing or crushing of bedrock beneath the glacier; and the entrainment of this fractured or crushed rock (Glasser and Bennett, 2004). Fracturing of bedrock occurs where a glacier flowing over bedrock creates pressure differences in the underlying rock, causing stress fields that are commonly sufficient to induce rock fracture (Glasser and Bennett, 2004; Morland and Morris, 1977). Plucking is particularly effective where a glacier flows over rock. Pressure exerted at the base of the ice melts the ice, which can refreeze in cracks in the underlying bedrock. Removing the pressure permits water to refreeze to the glacier and plucking occurs. Crushing occurs when the pressure exerted by basal rock fragments crushes the bedrock surface. Crushing develops crescentic fractures

FIGURE 12.6 Exposed bedrock sculptured by glacial ice, Svartisen Glacier, Norway. *(Photo by John D. Vitek (July 1988).)*

called chattermarks, which indicate the direction of motion of the glacier. Abrasion involves the wearing down of rock surfaces by the grinding effect of rock fragments frozen into the base of glaciers. It occurs when bodies of subglacial sediment slide over bedrock (Glasser and Bennett, 2004), and produces smoothed bedrock surfaces that often exhibit parallel sets of scratches, called striations (1–10 mm diameter) and fine silt-sized particles (0.1 mm) known as rock flour (Fig. 12.6).

The rate of glacial abrasion depends on ice velocity, presence of basal rock debris, hardness of the abrading rock fragment, removal of the rock flour, and thermal and pressure regime of the glacier (Boulton, 1976, 1982; Hallet, 1979, 1981; Schweizer and Iken, 1992). The rate of abrasion initially increases as glacial pressure and ice velocity increase, but decreases as the pressure becomes too great and ice at the base of the glacier begins to melt and release rock fragments from ice-rock fragment mixture. Abrasion increases with the increase in basal debris concentration. Furthermore, abrasion occurs only if the abrading rock fragments are harder than the bedrock. In addition, rock flour on the ice–rock bed interface needs to be flushed away by a constant supply of meltwater to sustain the process of abrasion. The thermal regime of a glacier also exerts a strong influence on the nature of erosion. The rate of abrasion beneath very cold glaciers (i.e., polar glaciers) tends to be negligible because of the lack of basal sliding; and in this condition, the adhesion of cold ice to bedrock, facilitates plucking (Benn and Evans, 2014). Once plucked, the blocks become the grinding tools that can cause abrasion.

Meltwater erosion beneath ice sheets and glaciers may be caused by either mechanical or chemical processes (Glasser and Bennett, 2004). The effectiveness of meltwater as an erosional agent is dependent upon the susceptibility of the bedrock involved in erosion, the velocity and turbulent flow of the discharge regime, and the quantity of sediment in transport (Glasser and Bennett, 2004). The streams of meltwater that flow along the base of a glacier erode rock through the combined action of abrasion, hydraulic action, attrition, and solution (Hallet, 1981; Herman et al., 2011). Water at the base of a thick glacier generally remains under high hydrostatic pressure, which causes meltwater streams to have greater flow rates and erosive potential than the surface streams.

12.9 EROSIONAL FORMS

Landforms developed by glacial erosion have a range of shapes, sizes, and morphology. We have categorized landforms developed by glacial erosion into three categories based on the sizes: (1) microscale forms; (2) intermediate scale forms; and (3) macroscale forms.

12.9.1 Micro-Scale Erosional Forms

Major micro-scale forms of glacial erosion include striations, micro-crag and tails, bedrock gouges, and cracks. Striations are lines or scratches on rock surfaces, usually no more than a few millimeters in depth, produced by the process of glacial abrasion (Glasser and Bennett, 2004). These small grooves or scratches on bedrock surfaces, up to several meters long, are commonly associated with polished-bedrock surfaces (Glasser and Bennett, 2004). Striations demonstrate that ice sheets contained a significant amount of basal debris, experienced basal melting and a flow through basal sliding, and transported rock debris and sediments (Glasser and Bennett, 2004). Small tails of rock protected from glacial abrasion in the lee of resistant grains or mineral crystals on the surface of a rock are called micro-crag and tail. They are important for the reconstruction of former ice sheets because they provide clear evidence of the orientation and the direction of ice flow (Glasser and Bennett, 2004).

Principal types of micro-scale cracks, gouges, and indentations created by glacial erosion on bedrock surfaces include: chattermarks, crescentic gouges, and crescentic cracks (Glasser and Bennett, 2004). Chattermarks and crescentic gouges generally consist of a shallow bedrock furrow with a crescentic outline (Glasser and Bennett, 2004). The convexity of the crescent is turned towards the direction of ice flow in crescentic gouges, whereas in chattermarks it is turned backwards (Glasser and Bennett, 2004). Chattermarks are commonly associated with larger bedrock grooves engraved on bedrock surfaces (Glasser and Bennett, 2004). Crescentic cracks, the vertical fracture of the rock without the removal of bedrock fragments, are normally curved in plane with the concavity turned towards the direction of flow (Glasser and Bennett, 2004).

12.9.2 Intermediate Scale Landforms

Intermediate scale landforms of glacial erosion, typically between 1 m and 1 km in size, have high relief on a wide range of scales, relatively smooth undulating surfaces, and upstanding streamline protrusions (Glasser and Bennett, 2004; Hindmarsh, 1999). These landforms may form parallel to the flow-direction of ice, and may be symmetric or asymmetric in down-glacier shape. These features have been referred to as roche moutonnées, rock drumlins, tadpole rocks and streamlined hills, ice-molded forms, whaleback and stoss and lee forms, and bedrock knolls depending on the size, shape, orientation with respect to the direction of flow, and morphology (i.e., asymmetry characteristics) (Fairchild, 1907; Flint, 1971, Dionne, 1987). Roche moutonnées are glaciated bedrock surfaces, usually in the form of rounded knobs, where the upstream and downstream slopes lie perpendicular to the general flow-direction of the former ice mass, and have a down-glacier asymmetry. The upstream side of a roche moutonnée forms a gentle, polished, and striated slope because of glacial scouring, and the downstream side resulted in a steep, irregular, and jagged slope because of glacial plucking.

Drumlins are features similar to roche moutonnées, where an upstream side is steep and the downstream side is gentle, which is likely the result of the reduction in ice mass velocity (Benn and Evans, 2014). A crag and tail is an elongated, tapered ridge of till extending downstream formed commonly by the selective erosion of softer strata.

12.9.3 Large-Scale Erosional Forms

Large-scale landforms developed by the glacial erosion include: glacial troughs, fjords, rock basins, knock and lochain, glacial lakes, and cirques. The glaciated landscape, particularly in mountains, exhibits distinguishing U-shaped (cross-section) valleys up to several thousand meters thick and tens of kilometers long; typically known as glacial troughs (Benn and Evans, 2014). They are the result of confining a glacier within valley walls, so that the glacier lacks uniform erosion and tends to deepen and widen the valley floor. This process transforms a commonly V-shaped stream valley into a more or less straight U-shaped valley because the U-shape provides relatively less frictional resistance to the relatively more viscous moving glacier (Boulton, 1982). The walls of U-shaped glacial valley may be almost vertical and striated by boulders dragged by the glacier. The valley floor may be covered with till or moraines. Because thickness of the ice is the dominant factor in the deepening process, smaller tributary glaciers erode the troughs less rapidly than the trough erosion of the main glacier. As a result the tributary troughs appear as valleys hanging on the wall of main glacial valley. When the glaciers melt, postglacial streams may form waterfalls from the mouths of the hanging valleys (Fig. 12.7).

FIGURE 12.7 The U-shaped Huerfano River valley north of Blanca Peak, Sangre de Cristo Mountains, south central Colorado. *(Photo by John D. Vitek (August 1978).)*

Fjords, or fiords, are long narrow arms of the sea that commonly extend far inland. They result from marine inundation of glaciated valleys associated with rising sea levels. Fjords commonly are deeper in the middle and upper reaches than at the seaward end because of the greater erosional power of the glaciers closer to the source (Rafferty, 2011). Fjords are also U-shaped. The visible walls of fjords may rise vertically for hundreds of meters from the edge of the water, and close to the shore the water may be many hundreds of meter deep. The depth of a fjord may range up to more than a kilometer. For example, Sognefjord in Norway is ~1308 m deep (Manzetti and Stenersen, 2010). The great depths of these submerged valleys are probably the result of the erosion from the thick glaciers that formed in these valleys.

12.10 GLACIAL TRANSPORT AND DEPOSITION

Glaciers transport debris that has been plucked away or bits of rock that have been broken off, or fallen onto the glacier. Glacial debris can be transported in three ways: (1) on top of the glacier; (2) within the glacier; and (3) underneath the glacier. Glaciers deposit materials on the surfaces beneath, in surrounding areas, and at the margins, which become depositional features with a range of shapes, sizes and morphologies. Debris in the glacial environment may be deposited directly from the glacier ice (till) or from several associated processes involving glacial meltwater (outwash), and the resulting deposits are commonly termed as glacial drift.

As the ice in a glacier moves from the zone of accumulation into the zone of ablation, it transports debris located beneath, within, and above the glacier toward its terminus or toward the outer margins where the ice velocity decreases. As the ice melts, the debris (till) that was originally frozen into the ice commonly forms a rocky and/or muddy blanket over the glacier margin. Typically, it is a nonstratified mixture of rock fragments and boulders in a fine-grained sandy or muddy matrix. Till occurs in various locations of a glacier, and depending on the mechanism of formation and location, can be categorized into different types. The deposit of glacial debris laid down more or less in place, as the ice melts, is called melt-out till or ablation till. In many cases, the glacial debris located between a moving glacier and its bedrock bed is severely sheared, compressed, and over-compacted; this type of deposit is called lodgement till. Tills often contain rock fragments and boulders that glaciers use to abrade the bedrock surface. If these rocks and boulders are different than the bedrock on which they are deposited, they are called erratics. Erratics are useful to determine the direction of ice movement and source region for the material (Fig. 12.8).

Meltwater deposits, also called glacial outwash, can form in channels directly beneath or in front of the melting glacier or in lakes and streams in front of its margin. Outwash deposits generally consist of bedded, laminated, or stratified drift, with the individual layers composed of relatively well-sorted sediments (Benn and Evans, 2014). The grain size of individual deposits depends on the availability of different sizes of debris and also on the velocity of the current and the distance from the head of the stream. Larger boulders are deposited closer to the glacier margin, and the grain sizes of deposited material decreases with increasing distance from the glacier. The finest fractions, such as clay and silt, may be deposited in glacial lakes or ponds or transported all the way to the ocean.

FIGURE 12.8 The massive sediment in the background is a poorly sorted till, deposited at the base of the now-thinning and retreating casement glacier. The boulders in the foreground are a lag deposit resulting from the removal of the finer grain-sizes in the till by the ice-marginal stream. Glacier Bay National Park, Alaska. USGS Photo.

12.10.1 Major Depositional Landforms of Valley Glaciers

As a glacier moves downslope in a valley, it picks up rock debris from the valley walls and floor, and transports the sediment in, on, or under the ice. But, when the sediment–ice mixture reaches the lower parts of the glacier where ablation is the dominant process, the glacier deposits sediments along its margins during melting. The deposit is called a moraine. The size of the moraine depends on the size of the glacier, and the amount of material deposited by the glacier over time. Large valley glaciers are capable of forming moraines a few hundred meters high and many hundreds of meters wide (Bennett, 2001). If the position of the glacier margin is constant for an extended period of time, large moraines at the margin of the glaciers may develop because of the accumulations of larger amount of glacial debris (till). Depending on the location of formation with respect to the glacier, moraines can be categorized as end moraine, lateral moraine, and recessional moraine. Linear accumulations of till immediately in front or terminus of the glacier are called end moraines. The end moraine of the largest extent formed by the glacier during a period of given glaciation is referred to as the terminal moraine. Moraines formed along the valley slopes next to the side margins of the glacier are termed lateral moraines. The successive melting of a glacier can produce a series of recessional moraines; the joining of lateral moraines of two glaciers creates medial moraines (Figs 12.9 and 12.10).

FIGURE 12.9 Complex moraine system adjacent to Fortuna Bay, South Georgia. *(Photo by John D. Vitek (November 2013).)*

FIGURE 12.10 Glacial tarn formation at the edge of a recessional moraine in a cirque near Grinnell Glacier, Glacier National 2013. *(Photo by John D. Vitek.)*

Another major depositional landform of a valley glacier is a flute. Flutes are narrow, elongated, straight, parallel ridges of till which generally develop close to the lower margin of glaciers. They form when glaciers accumulate large volumes of debris beneath, as a result of the glacier gliding on a bed of pressurized muddy till. When basal ice flows around a bedrock knob or a boulder lodged in the substrate, a cavity can form in the ice on the lee side of the obstacle, and any pressurized mud present under the glacier may then be injected into this cavity and deposited as an elongate tail of till, or flute. The size of flutes range from a few centimeters to tens of meters in height and tens of centimeters to kilometers in length (Benn and Evans, 2014). The size of the flute mainly depends on the size of the obstacle and availability of subglacial debris. Flutes occur in valley and continental glaciers; however, very large flutes generally are also associated with continental ice sheets.

12.10.2 Depositional Landforms of Continental Glaciers

Many depositional landforms associated with continental ice sheets are similar to the landforms of valley glaciers in terms of the formation. For example, the similar processes in both types of glaciers form terminal, end, and recessional moraines; however, the sizes associated with ice sheets are much larger. Morainic ridges of an ice sheet may extend for hundreds of kilometers, with hundreds of meters height and several kilometers width (Benn and Evans, 2014). In addition to moraine ridges, continental glaciers also deposit more or less continuous, thin (less than 10 m) sheets of till over large areas (Benn and Evans, 2014). Known as ground moraines, these deposits form undulating topography (hummocky) of low relief, with alternating small till mounds and depressions where swamps or lakes typically occur. Flutes are common features in areas covered by ground moraine.

Continental glaciers develop streamlined, elongated mounds of sediment usually close to the edge of an ice sheet, called drumlins. Drumlins occur in groups of tens to hundreds, forming large drumlin fields (see Fig. 12.10, a map of the drumlins in central Wisconsin mapped by Atwood (1940). The long axis of a drumlin usually aligns parallel to the direction of regional ice flow. These features are generally asymmetric, where the stoss side is steeper than the lee side. Some drumlins consist entirely of till and display a fabric where long axes of the individual rocks and sand grains are aligned parallel to the direction of ice flow, whereas others have bedrock cores draped with till (Fig. 12.11).

12.10.3 Meltwater Deposits

Much of the debris in the glacial environment of valley and continental glaciers is transported, reworked, and laid down by water. Whereas meltwater streams form glaciofluvial deposits, glaciolacustrine sediments accumulate at the margins and bottoms of glacial lakes and ponds.

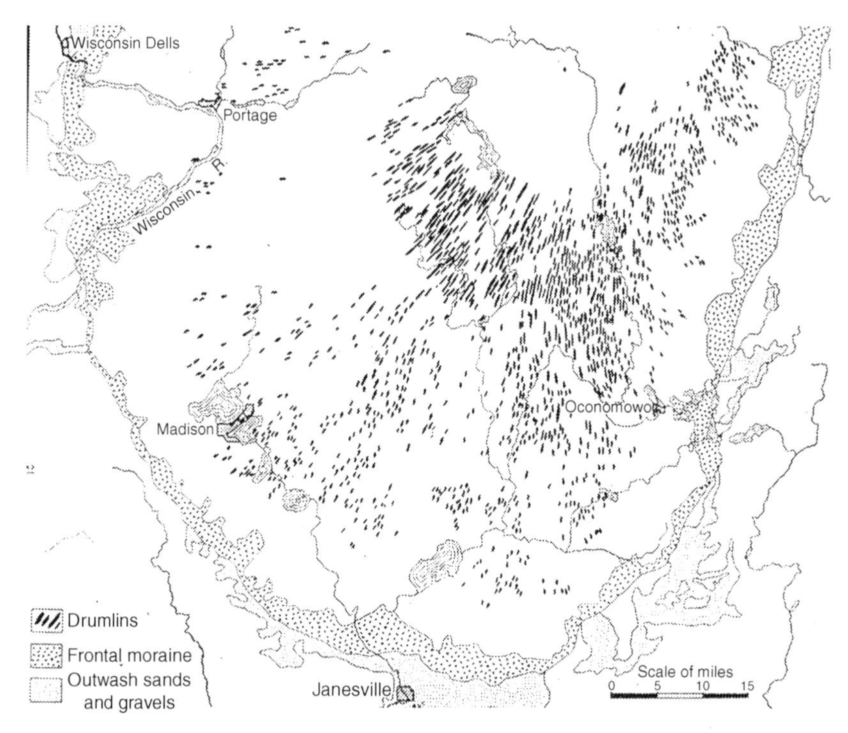

FIGURE 12.11 Map of drumlins showing the direction of ice motion in this portion of Wisconsin during the last advance of ice into this area (Atwood, 1940).

12.10.4 Glaciofluvial Deposits

Meltwater streams at the snout of a valley glacier or along the margin of an ice sheet are generally laden with debris and have relatively high velocities. Beyond the glacier margin, once the walls of an ice tunnel no longer confine the water, it spreads out, and loses velocity. As a result, some of the load is deposited, which may cause the stream to separate into multiple channels separated by sand and gravel bar deposits. The process develops braided stream networks where the deposits usually consist of lenses of fine-grained, cross-bedded sands interbedded laterally and vertically with stringers of coarse, bouldery gravel referred to as outwash. In addition, the amount of sediment laid down by the stream generally (Fig. 12.12) decreases and becomes thinner with distance from the ice margin. The morphology of the outwash deposits depend on the surrounding topography. Where valleys are deep enough not to be buried by the glaciofluvial sediments, as is the case in most mountainous regions, the resulting elongate, planar outwash deposits are termed valley trains. In low-relief areas, the deposits of several ice-marginal streams may merge to form outwash plains, or sandurs, which are wide plains. If the ice margin stabilizes at

FIGURE 12.12 Braided channel flowing away from glacial snout creating a valley train deposit. Alaska, 2008. *(Photo by John R. Giardino.)*

a recessional position during glacial melting, additional valley trains or sandurs may form. The new sandur may lie at a lower elevation than the older sandur because of the downstream thinning of the outwash, and as a result, flat-topped remnants of the older plain may be left along the valley sides and develop terraces. Again, these structures can be used to reconstruct the positions of ice margins through time.

Glaciofluvial streams that flow over the terminus of a glacier often deposit stratified drift in the channels and in depressions on the ice surface. These deposits, known as kames or kame moraines, commonly develop isolated mounds of bedded sands and gravels. Kames sometime form between the lateral margin of a glacier and the valley wall and develop terrace-like structure known as kame terraces. Streams also deposit stratified drift in subglacial or englacial tunnels, and as the ice melts away, these sinuous channel deposits may be left as long sinuous gravel ridges called eskers. The length can range from few hundred meters to hundreds of kilometers, and the height can range from few to tens of meters. Eskers have been used extensively as sources of sand and gravel for the construction industry.

Steep-sided depressions, such as kettles, potholes, or ice pits are typical of many glacial and glaciofluvial deposits. Kettles form where till or outwash is deposited around ice blocks that are separated from active glacier because of ablation. When the ice melts, depressions are bordered by masses of glacial deposits. Lakes formed in kettles are called kettle lakes, and if a sandur or valley train contains kettles, it is referred to as a pitted outwash plain.

12.10.5 Glaciolacustrine Deposits

Glacial and proglacial lakes commonly occur on a glaciated landscape. Lakes on glaciated landscapes are formed by two major mechanisms: first, by the process of erosion, and second, by damming streams by ice or by glacial deposits, or by a combination of these processes. When a stream from a glacier enters a standing body of water, it deposits its bedload. The stream deposits coarser gravel and sand directly at the mouth of the stream as steeply inclined foreset beds. The finer, suspended silt and clay are transported farther into the lake, where they deposit as relatively flat-lying bottomset beds. As the sediment builds farther into the lake, the river deposits a thin veneer of subhorizontal gravelly topset-beds over the foreset units. The entire body of this deposit is comprised of foreset–topset complex.

During warmer summer months, meltwater streams carry silt and clay into the lakes, and the silt settles out of suspension more rapidly than the clay. A thicker, silty summer layer is, thus, deposited. During winter, as the surface of the lake freezes and the meltwater discharge ceases, the clays contained in the lake water slowly settle out of suspension to form a thin winter clay layer. Such lacustrine deposits with annual silt and clay "couplets" are known as varves. These varves serve as good tools to date the age of deposit.

12.10.6 Glacier Hydrology

Glaciers are dynamic systems. They store and mobilize snow, ice, water, and sediments, which characterize the overall system of glacial hydrology. Water on the surface of the glacier disappears through crevasses and holes in the glaciers, and flows emerge from the glacier snout. The hydrological system of a glacier can be categorized as: (1) supraglacial hydrology; (2) englacial hydrology; (3) subglacial hydrology; and (4) proglacial drainage systems.

Supraglacial, or surface water on a glacier, is formed by the melting of ice (ablation) during the summer. The water flows off the glacier into crevasses and develops a network of channels, which are commonly sinuous where water can flow at rates of up to several meters per second (Cuffey and Paterson, 2010). Surface melt occurs in firn, the transitional state between snow and ice, and can pond above the impermeable ice. Saturation of the firn all the way to the surface creates a swamp zone where pools of standing water may form. The zone moves up glacier as the melt season progresses and the surface drains more rapidly as more ice is exposed, and the firn zone is filled with water (Cuffey and Paterson, 2010). Such processes of subglacial hydrology are considered responsible for large lakes on the surface of an ice sheet during the summer season in Greenland, and much of the coastal meltwater runoff in Antarctic coastal areas and ice shelves (Scambos et al., 2009).

Englacial or within-glacier hydrological processes are controlled by crevasses, large tensional structures, and moulins, circular vertical shafts, which allow water to penetrate into the ice. Surface water cascades down into the ice sheets via crevasses and moulins and forms numerous water pockets and channels where water can remain trapped for some time (Fountain and Walder, 1998).

Subglacial hydrological processes (hydrology underneath a glacier or ice sheet) are critically important in understanding the flow of Antarctic glaciers, where basal meltwater flows through the networks of large subglacial drainage, some of which extend up to hundreds of kilometers (Smith et al., 2007), and facilitate glacial erosion and movement of ice. Water in this system comes from two major sources: first by the basal melting of ice, and second by the downward percolation and flow of supraglacial and englacial water. Basal melting is primarily the result of geothermal heating and pressure of the ice mass. Water reaching the base of the ice sheets can be ponded in subglacial lakes, or flow through subglacial channels. Beneath the Antarctic ice sheet, these subglacial drainage channels are commonly connected to numerous subglacial lakes (Siegert et al., 2005; Smith et al., 2009).

The proglacial area of temperate glaciers is characterized by abundant meltwater runoff from the glacier. The impoundment of proglacial meltwater in the deepened glacier basin may develop proglacial lakes. In addition, rapid exit of abundant meltwater from a glacier can form large braided-river plains, or sandurs, where glacial streams deposit, redeposit glacial sediments and rework glacial landforms. Examples of proglacial lakes occur in front of San Jose Glacier in James Ross Island, and braided streams are common characteristics in the northern Antarctic Peninsula (Carrivick et al., 2012) (Fig. 12.12).

12.11 GLACIER ECOSYSTEM

Scientists have determined life exists in glaciers (Anesio and Laybourn-Parry, 2012; Kohshima, 1984). For example, a new species of cold-tolerant midge has been found in Himalayan glacier (Kohshima, 1984), microbial life has been known to occur in deep Greenland basal ice and Antarctica ice sheets (Price, 2000; Tung et al., 2006), bacteria and archaea have been found in the waters and sediments of Lake Whillans in Antarctic Ice sheet (Christner et al., 2014), and methanogenesis have been found in subglacial sediments from Robertson Glacier, Canadian Rockies (Boyd et al., 2010). Recent work suggests that moss plants can survive for centuries underneath glaciers (La Farge et al., 2013). Glacial surfaces receive sufficient sunlight, liquid water and direct contact with the atmosphere during summer. It has been suggested that temperatures far below zero do not present an absolute obstacle to microbial activity (Bakermans and Skidmore, 2011; Junge et al., 2006). These conditions create an environment favorable for the growth of life in three habitats including englacial habitat where rock-eating and other microbes live in thin films of water on the surface of debris entrained in the ice and the network of veins forming between individual ice crystals (Price, 2000; Tung et al., 2006), interface between ice and the atmosphere, and the surface of the ice. Keeping this in mind, snow that disappears at the peak of summer in the Northern Hemisphere (Nghiem et al., 2012), can contribute to the environment potential for a massive bloom of bacteria and microbes. In addition, summer conditions can be sufficiently gentle for some plants and animals to thrive on the ice surface. For example, mosses or ice worms (Porter et al., 2008) can increase the complexity of the icy food chain. Therefore, the, glacial ecosystem is one of the important topics for the study of the biological function in this Critical Zone. In addition, the topic is important because organic carbon trapped under the ice can be metabolized by microbes to form methane (Boyd et al., 2010), a potent greenhouse gas. Studies suggest that up to 21,000 petagrams (Pg) of organic carbon (10 times the permafrost carbon stock) might be trapped under the ice of Antarctica and that microbial conversion of this carbon to methane could be a major feedback in climate change (Wadham et al., 2012). In addition, subglacial microbes seem to greatly accelerate the weathering of minerals (Montross et al., 2013), and play a major role in the subglacial geochemical processes, and the glacier mass balance (Irvine-Fynn et al., 2012). In addition, ice algae (Yallop et al., 2012) and aggregates of microbes bound to minerals called cryoconite (Edwards et al., 2011) accelerate the rates of melting surface ice.

12.12 LIVING IN THE CRITICAL ZONE OF GLACIATED LANDSCAPE

Glacial environments have significant relevance in global climate warming. Major components of these environments including snow, river and lake ice, sea ice, and frozen ground invoke positive feedback mechanisms that amplify global climate change and variability. For example, a decrease in snow and ice extent reduces the values of albedo and increases heat adsorption. Likewise, the thaw degradation of debris-covered glaciers may release large quantities of greenhouse gases (carbon and methane) to the atmosphere, which have an important feedback to climatic warming (e.g., Aleina et al., 2013; Schneider von Deimling et al., 2012; Vonk et al., 2013).

Processes of erosion and deposition in glacial environments result in a unique landscape. Erosion by wind, nivation, frost weathering, meltwater (e.g., Thorn, 2009) will modify much of the Critical Zone environment. Seasonal patterns of stream discharge and sediment deposition in these environments are distinctive. Most of the high stream flows tend to occur during summer periods when snowmelt occurs (e.g., Mol et al., 2000; Vandenberghe, 2002). The short-lived peak discharges, in general, leave peculiar depositional features particularly in proglacial environments, such as poorly developed, shallow braided channels and large quantities of gravel and boulder deposits. Strong winds in glacial areas commonly move large quantities of loose sediment and soil (löess). The process dominantly occurs during (Fig. 12.13) dry summer months because

FIGURE 12.13 Loess deposits form steep cliffs at the Troy L. Pewe Climatic Change Permafrost Reserve, Fairbanks, Alaska. *(Photograph by J.R.Giardino (2008).)*

sediments will dry out, meltwater from snow and ice decreases, which results in an increase in stream deposits available for aeolian transport. The process creates various landforms including sand dunes, löess accumulations, and sand sheets. Ventifacts, small rock fragments sculpted by sand blowing across a surface, are indicative of the role of wind in this environment (Knight, 2008). The Sand Hills of western Nebraska are the result of wind activity associated with Pleistocene glaciers.

Natural processes in these environments can create high levels of risk for anthropogenic activities, including oil, gas, and mineral exploration, defense, tourism, transportation, and social and economic change. Most high-altitude and high-latitude cold environments on Earth remain covered by ice and snow in the winter, which then melts during the summer. Some of these events can be life-threatening and some affect the quality of life, infrastructure, natural resources, habitat, and agriculture. Most of the glacial landscapes in the world contain vast natural resources, notably hydrocarbons, gold, diamonds, iron, copper, and zinc as well as large reservoirs of water and construction materials. The exploitation of these resources in climatically rigorous and remotely located areas requires the knowledge of the processes, landforms, and potential hazards in the area to develop sound geotechnical, and engineering designs for constructions, including drilling, blasting, excavation of roads, buildings, bridges, pipelines, and other infrastructure.

12.13 IMPLICATIONS FOR THE 21ST CENTURY

In 2005, a mass of Antarctic ice, comparable to the size of California, briefly melted and separated from the main mass of ice, prompting the National Aeronautics and Space Administration (NASA) to report it as the most significant Antarctic melting in the past 30 years (Gregory, 2010). Such a phenomenon suggests that glaciers are one of the most dynamic systems on Earth. Glaciers are large, very dynamic, storage sites of water and sediment. They constantly exchange mass and energy with the atmosphere, hydrosphere, and other parts of the system of Earth. Glaciers gain mass by the accumulation of snow and ice on the surface, and lose mass by melting (ablation), iceberg calving, and many other processes. Therefore, the evolution of glacier mass depends on the balance between accumulation and ablation of snow and ice, which in turn depends on climate and local topographic factors (Kaser and Georges, 1999; Oerlemans, 2005; Oerlemans and Fortuin, 1992). Similarly glacier temperature evolves by the balance between energy input and output at the surface and the bed, and the heat generated within the glacier by ice flow. Both of these glacier characteristics strongly couple with the climate, and provide feedback responses to change in climate.

Glacier mass balance and thermal structure impact many other parts of glacial and extra glacial environments at the local and global scale. For example, changes in snow and ice storage at the catchment-scale influence stream discharge and, thus, water resources and the risks from floods, whereas at the global scale, fluctuations in glacier volume directly impact the change in sea level (Bamber and Payne, 2004). Similarly, the fluctuation in glacier temperature, as a result of change in energy balance, influences many processes associated with glaciers, including surface ablation, englacial, and subglacial water flow, rates of glacier motion, and patterns of glacial erosion and deposition. For example, an increase in temperature increases the rates of snow and ice melt, subglacial flow, and glacial movements, which in turn increases the rates of erosion, deposition, and flooding. The processes of glaciers and ice sheets can also have significant influence on the atmosphere, in particular weather and climate modification at a local or global scale. Therefore, research on the links between glacier mass balance, thermal regime, and climate should be one of the prime focuses of research in the twenty-first century. Such research allows scientists to reconstruct long-term environmental changes from evidence of past glacier fluctuations, as well as future prediction of the environment in the glacial regions as a feedback response of current trends of global warming and change in climate. Fields of higher priority for research may include: (1) the better understanding of the influence of climate change in glacial and associated environments, to predict variety of potential natural events which may become hazardous, such as that associated with short- and long-term stream flow and water resources, seasonal and interannual climatic variability, local and global scale sea-level change; and (2) to make decisions on how to adapt to climate change.

12.14 SUMMARY

This chapter provides a quick overview of the processes that operate in glacial environments. These processes create a variety of landforms that modify local environments. Sediments are generated through erosion, moved to lower elevations and impact other geomorphic systems, especially the fluvial, aeolian and coastal systems. Whereas the amount of land impacted by glaciers continues to shrink (except in Antarctica and Greenland), the loss of ice means less water for human uses, including irrigation and industry. More water can also cause sea level to rise, which will impact every coastal city throughout the world. The economic impact of this rise will be staggering. To exclude and/or ignore the glacial environment as a component of the Critical Zone is a mistake that leads to humans being complacent in terms of learning how to respond to changes in the glacial system. As we continuously seek more resources to satisfy human needs, more exploration and development of glaciated regions will be undertaken. Without full knowledge of the processes and forms present in these areas, mistakes can be made that will ultimately increase the costs of goods and services. Throughout this chapter, we have linked the various processes with the potential inputs and how they can impact research in the Critical Zone such that human use of these regions can proceed efficiently.

REFERENCES

Aleina, F.C., Brovkin, V., Muster, S., Boike, J., Kutzbach, L., Sachs, T., Zuyev, S., 2013. A stochastic model for the polygonal tundra based on Poisson-Voronoi diagrams. Earth Syst. Dynam. 4 (2), 187–198.

Andreassen, L.M., Elvehøy, H., Kjøllmoen, B., Engeset, R.V., Haakensen, N., 2005. Glacier mass-balance and length variation in Norway. Ann. Glaciol. 42 (1), 317–325.

Anesio, A.M., Laybourn-Parry, J., 2012. Glaciers and ice-sheets as a biome. Trends Ecol. Evol. 27 (4), 219–225.

Atwood, W.W., 1940. The Physiographic Provinces of North America. Ginn and Company, Boston.

Bakermans, C., Skidmore, M., 2011. Microbial respiration in ice at subzero temperatures (−4 degrees C to −33 degrees C). Environ. Microbiol. Rep. 3 (6), 774–782.

Bamber, J., Payne, T., 2004. Mass Balance of the Cryosphere: Observations and Modelling of Contemporary and Future Changes. Cambridge University Press, Cambridge; New York.

Banwart, S., Menon, M., Bernasconi, S.M., Bloem, J., Blum, W.E., de Souza, D.M., Davidsdotir, B., Duffy, C., Lair, G.J., Kram, P., 2012. Soil processes and functions across an international network of Critical Zone Observatories: introduction to experimental methods and initial results. C.R. Geosci. 344 (11), 758–772.

Benn, D., Evans, D.J., 2014. Glaciers and Glaciation. Routledge, London.

Bennett, M.R., 2001. The morphology, structural evolution and significance of push moraines. Earth Sci. Rev. 53 (3), 197–236.

Berthier, E., Arnaud, Y., Kumar, R., Ahmad, S., Wagnon, P., Chevallier, P., 2007. Remote sensing estimates of glacier mass balances in the Himachal Pradesh (Western Himalaya, India). Remote Sens. Environ. 108 (3), 327–338.

Bishop, M.P., Olsenholler, J.A., Shroder, J.F., Barry, R.G., Raup, B.H., Bush, A.B.G., Copland, L., et al. 2004. Global Land Ice Measurements from Space (GLIMS): remote sensing and GIS investigations of the Earth's cryosphere. Geocarto Int. 19 (2), 57–84.

Berger, A., 1978. Long-term variations of daily insolation and Quaternary climatic changes. J. Atmos. Sci. 35 (12), 2362–2367.

Boulton, G., 1976. The origin of glacially fluted surfaces – observations and theory. J. Glaciol. 17, 287–309.

Boulton, G.S., 1982. Processes and Patterns of Glacial Erosion. Springer, Netherlands.

Boyd, E.S., Skidmore, M., Mitchell, A.C., Bakermans, C., Peters, J.W., 2010. Methanogenesis in subglacial sediments. Environ. Microbiol. Rep. 2 (5), 685–692.

Calkin, P.E., 1988. Holocene glaciation of Alaska (and adjoining Yukon Territory, Canada). Quaternary Sci. Rev. 7 (2), 159–184.

Carrivick, J.L., Davies, B.J., Glasser, N.F., Nyvlt, D., Hambrey, M.J., 2012. Late-Holocene changes in character and behaviour of land-terminating glaciers on James Ross Island. Antarctica J. Glaciol. 58 (212), 1176–1190.

Chinn, T.J., 1996. New Zealand glacier responses to climate change of the past century. N. Z. J. Geol. Geophys. 39 (3), 415–428.

Christner, B.C., Priscu, J.C., Achberger, A.M., Barbante, C.F., Carter, S.P., Christianson, K., Michaud, A.B., Mikucki, J.A., Mitchell, A.C., Skidmore, M.L., Vick-Majors, T.J., Team, W.S., 2014. A microbial ecosystem beneath the West Antarctic ice-sheet (vol. 512, p. 310, 2014). Nature 514 (7522), 394.

Cuffey, K.M., Paterson, W.S.B., 2010. The Physics of Glaciers, fourth ed. Academic Press, Amsterdam, p. 704.

Currey, D.R., 1990. Quaternary palaeolakes in the evolution of semidesert basins, with special emphasis on Lake Bonneville and the Great Basin, USA. Palaeogeogr. Palaeoclimatol. Palaeoecol. 76 (3), 189–214.

Dionne, J.C., 1987. Tadpole rock (rocdrumlin): a glacial streamline moulded form. In: Rose, J., Menzies, J. (Eds.), Drumlin Symposium, Balkema, Rotterdam, pp. 149–159.

Dyurgerov, M.B., Meier, M.F., 1999. Analysis of winter and summer glacier mass balances. Geogr. Ann. A Phys. Geogr. 81 (4), 541–554.

Edwards, A., Anesio, A.M., Rassner, S.M., Sattler, B., Hubbard, B., Perkins, W.T., Young, M., Griffith, G.W., 2011. Possible interactions between bacterial diversity, microbial activity and supraglacial hydrology of cryoconite holes in Svalbard. ISME J. 5 (1), 150–160.

Ehlers, J., Gibbard, P.L., Hughes, P.D., 2011. Quaternary Glaciations-Extent and Chronology: A Closer Look. Elsevier, Amsterdam.

Ehlers, J., Gibbard, P.L., 2008. Quaternary glaciations extent and chronology.

Evans, D.J., 1990. The last glaciation and relative sea level history of northwest Ellesmere Island, Canadian High Arctic. J. Quaternary Sci. 5 (1), 67–82.

Fairchild, H.L., 1907. Drumlins of central New York. NY State Museum Bull. 111, 391–443.

Flint, R.F., 1971. Glacial and Quaternary Geology. Wiley, New York, NY, p. 892.

Fountain, A.G., Walder, J.S., 1998. Water flow through temperate glaciers. Rev. Geophys. 36 (3), 299–328.

Frenzel, B., 1992. Atlas of Paleoclimates and Paleoenvironments of the Northern Hemisphere. Geographical Research Institute; Hungarian Academy of Sciences; Gustav Fischer Verlag, Budapest; Stuttgart; Jena; New York.

Giardino, J., Regmi, N., Vitek, J., 2014. Rock Glaciers. In: Singh, V., Singh, P., Haritashya, U. (Eds.), Encyclopedia of Snow, Ice and Glaciers. Encyclopedia of Earth Sciences Series. Springer, Netherlands, pp. 943–948.

Giardino, J.R., Shroder, J.F., Vitek, J.D. (Eds.), 1987. Rock Glaciers. Springer, London, p. 355.

Giardino, J.R., Vitek, J.D., 1988. The significance of rock glaciers in the glacial-periglacial landscape continuum. J. Quaternary Sci. 3 (1), 97–103.

Glasser, N.F., Bennett, M.R., 2004. Glacial erosional landforms: origins and significance for palaeoglaciology. Prog. Phys. Geogr. 28 (1), 43–75.

Gregory, K.J., 2010. Earth's Land Surface: Landforms and Processes in Geomorphology. SAGE Publications, London.

Haeberli, W., Beniston, M., 1998. Climate change and its impacts on glaciers and permafrost in the Alps. Ambio 27, 258–265.

Hagen, J.O., Liestøl, O., Roland, E., Jørgensen, T., 1993. Glacier atlas of Svalbard and Jan Mayen. Meddelelser NR. p. 129, Oslo.

Hallet, B., 1979. A theoretical model of glacial abrasion. J. Glaciol. 23, 39–50.

Hallet, B., 1981. Glacial abrasion and sliding: their dependence on the debris concentration in basal ice. Ann. Glaciol. 2 (1), 23–28.

Hambrey, M.J., Harland, W.B., 1985. The Late Proterozoic Glacial Era. Palaeogeogr. Palaeocl. Palaeoecol. 51 (1–4), 255–272.

Herman, F., Beaud, F., Champagnac, J.-D., Lemieux, J.-M., Sternai, P., 2011. Glacial hydrology and erosion patterns: a mechanism for carving glacial valleys. Earth Planet. Sci. Lett. 310 (3), 498–508.

Hindmarsh, R.C.A., 1999. On the numerical computation of temperature in an ice-sheet. J. Glaciol. 45 (151), 568–574.

Hoffman, P.F., Kaufman, A.J., Halverson, G.P., Schrag, D.P., 1998. A Neoproterozoic snowball earth. Science 281 (5381), 1342–1346.

Huybrechts, P., 2002. Sea-level changes at the LGM from ice-dynamic reconstructions of the Greenland and Antarctic ice-sheets during the glacial cycles. Quaternary Sci. Rev. 21 (1), 203–231.

Irvine-Fynn, T.D.L., Edwards, A., Newton, S., Langford, H., Rassner, S.M., Telling, J., Anesio, A.M., Hodson, A.J., 2012. Microbial cell budgets of an Arctic glacier surface quantified using flow cytometry. Environ. Microbiol. 14 (11), 2998–3012.

Janke, J.R., Regmi, N.R., Giardino, J.R., Vitek, J.D. 2013, Rock Glaciers. In: Treatise on Geomorphology, Academic Press, San Diego, 238-273.

Jones, P.N., 2005. Respect for the Ancestors: American Indian Cultural Affiliation in the American West. Bauu Institute, Colorado.

Junge, K., Eicken, H., Swanson, B.D., Deming, J.W., 2006. Bacterial incorporation of leucine into protein down to-20 degrees C with evidence for potential activity in sub-eutectic saline ice formations. Cryobiology 52 (3), 417–429.

Kaser, G., Georges, C., 1999. On the mass balance of low latitude glaciers with particular consideration of the Peruvian Cordillera Blanca. Geogr. Ann. A Phys. Geogr. 81 (4), 643–651.

Knight, J., 2008. The environmental significance of ventifacts: a critical review. Earth Sci. Rev. 86 (1–4), 89–105.

Kohshima, S., 1984. A novel cold-tolerant insect found in a Himalayan glacier. Nature 310 (5974), 225–227.

Kuhn, M.C., 1981. Process and fundamental considerations of selected hydrometallurgical systems. Society of Mining Engineers of American Institute of Mining, Metallurgical, and Petroleum Engineers, New York, NY.

La Farge, C., Williams, K.H., England, J.H., 2013. Regeneration of little ice age bryophytes emerging from a polar glacier with implications of totipotency in extreme environments. Proc. Natl. Acad. Sci. USA 110 (24), 9839–9844.

Lambeck, K., Chappell, J., 2001. Sea level change through the last glacial cycle. Science 292 (5517), 679–686.

Laybourn-Parry, J., Tranter, M., Hodson, A.J., 2012. Ecology of Snow and Ice Environments. Oxford University Press, Oxford.

Llibourty, L., 1998. Glaciers of Chile and Argentina. Geol. Surv. Prof. Pap. 1386, 1103.

Luckman, B.H., Briffa, K.R., Jones, P., Schweingruber, F., 1997. Tree-ring based reconstruction of summer temperatures at the Columbia Icefield, Alberta, Canada, AD 1073–1983. Holocene 7 (4), 375–389.

Maisch, M., 2000. The long-term signal of climate change in the Swiss Alps: Glacier retreat since the end of the Little Ice Age and future ice decay scenarios. Geogr. Fis. Dinam. Quat. 23, 139–151.

Manzetti, S., Stenersen, J.H.V., 2010. A critical view of the environmental condition of the Sognefjord. Mar. Pollut. Bull. 60 (12), 2167–2174.

Mol, J., Vandenberghe, J., Kasse, C., 2000. River response to variations of periglacial climate in mid-latitude Europe. Geomorphology 33 (3–4), 131–148.

Montross, S.N., Skidmore, M., Tranter, M., Kivimaki, A.L., Parkes, R.J., 2013. A microbial driver of chemical weathering in glaciated systems. Geology 41 (2), 215–218.

Morland, L., Morris, E., 1977. Stress in an elastic bedrock hump due to glacier flow. J. Glaciol. 18, 67–75.

Müller, J.A., Koch, L., 2012. Ice-sheets: Dynamics, Formation and Environmental Concerns. Earth Sciences in the 21st Century. Nova Science Publisher's, Inc., Hauppauge, NY, 212 p.

National Research Council, 2001. Committee on Basic Research Opportunities in the Earth. Basic Research Opportunities in Earth Science. 0-309-07133-X, National Research Council, Washington, DC.

Nghiem, S.V., Hall, D.K., Mote, T.L., Tedesco, M., Albert, M.R., Keegan, K., Shuman, C.A., DiGirolamo, N.E., Neumann, G., 2012. The extreme melt across the Greenland ice-sheet in 2012. Geophys. Res. Lett. 39, 39.

Oerlemans, J., 2005. Extracting a climate signal from 169 glacier records. Science 308 (5722), 675–677.

Oerlemans, J., Fortuin, J.P., 1992. Sensitivity of glaciers and small ice caps to greenhouse warming. Science 258 (5079), 115–117.

Pelto, M. (1990). Annual balance of North Cascade, Washington glaciers predicted from climatic records. Eastern c, 201.

Penck, A., 1905. Glacial features in the surface of the Alps. J. Geol. 13 (1), 1–19.

Porter, P.R., Evans, A.J., Hodson, A.J., Lowe, A.T., Crabtree, M.D., 2008. Sediment-moss interactions on a temperate glacier: Falljokull. Iceland. Ann. Glaciol. 48, 25–31.

Price, P.B., 2000. A habitat for psychrophiles in deep Antarctic ice. Proc. Natl Acad. Sci. USA 97 (3), 1247–1251.

Rafferty, J.P., 2011. Landforms. Britannica Educational Publishing, Chicago.

Rapley, C., 1999. Invited keynote address: global change and the polar regions. Polar Res. 18 (2), 117–118.

Rapley, C., 2006. The Antarctic Ice Sheet and Sea Level Rise. Avoiding Dangerous Climate Change. Cambridge University Press, Cambridge, pp. 25–27.

Rieu, R., Allen, P.A., Plotze, M., Pettke, T., 2007. Climatic cycles during a Neoproterozoic "snowball" glacial epoch. Geology 35 (4), 299–302.

Rignot, E., Thomas, R.H., 2002. Mass balance of polar ice sheets. Science 297 (5586), 1502–1506.

Roer, I., Zemp, M., van Woerden, J., 2008. Global glacier changes: facts and figures. UNEP/Earthprint.

Scambos, T., Fricker, H.A., Liu, C.C., Bohlander, J., Fastook, J., Sargent, A., Massom, R., Wu, A.M., 2009. Ice shelf disintegration by plate bending and hydro-fracture: satellite observations and model results of the 2008 Wilkins ice shelf break-ups. Earth Planet. Sci. Lett. 280 (1–4), 51–60.

Schneider von Deimling, T., Meinshausen, M., Levermann, A., Huber, V., Frieler, K., Lawrence, D.M., Brovkin, V., 2012. Estimating the near-surface permafrost-carbon feedback on global warming. Biogeosciences 9 (2), 649–665.

Schweizer, J., Iken, A., 1992. The role of bed separation and friction in sliding over an undeformable bed. J. Glaciol. 38, 77–92.

Sharp, R.P., 1958. Malaspina Glacier, Alaska. Geol. Soc. Am. Bull. 69 (6), 617–646.

Siegert, M.J., Carter, S., Tabacco, I., Popov, S., Blankenship, D.D., 2005. A revised inventory of Antarctic subglacial lakes. Antarctic Sci. 17 (3), 453–460.

Sigurdsson, O., 1998. Glacier variations in Iceland 1930–1995. Jokull 45, 3–26.

Smith, A.M., Murray, T., Nicholls, K.W., Makinson, K., Aoalgeirsdottir, G., Behar, A.E., Vaughan, D.G., 2007. Rapid erosion, drumlin formation, and changing hydrology beneath an Antarctic ice stream. Geology 35 (2), 127–130.

Smith, B.E., Fricker, H.A., Joughin, I.R., Tulaczyk, S., 2009. An inventory of active subglacial lakes in Antarctica detected by ICESat (2003–2008). J. Glaciol. 55 (192), 573–595.

Sugden, D.E., John, B.S., 1976. Glaciers and Landscape: A Geomorphological Approach. Edward Arnold, London.

Sugden, D.E., Summerfield, M.A., Denton, G.H., Wilch, T.I., McIntosh, W.C., Marchant, D.R., Rutford, R.H., 1999. Landscape development in the Royal Society Range, southern Victoria Land, Antarctica: stability since the mid-Miocene. Geomorphology 28 (3), 181–200.

Theakstone, W.H., 1982. Glacial Geomorphology. Prog. Phys. Geogr. 6 (2), 261–274.

Thorn, E.C., 2009. Holocene microweathering rates and processes on ice-eroded bedrock, Roldal area, Hardangervidda, southern Norway. Geological Society, London, Special Publications January 1, 2009, vol. 320, pp. 29–49.

Tung, H.C., Price, P.B., Bramall, N.E., Vrdoljak, G., 2006. Microorganisms metabolizing on clay grains in 3-km-deep Greenland basal ice. Astrobiology 6 (1), 69–86.

Van de Wal, R.S.W., Boot, W., Van den Broeke, M.R., Smeets, C.J.P.P., Reijmer, C.H., Donker, J.J.A., Oerlemans, J., 2008. Large and rapid melt-induced velocity changes in the ablation zone of the Greenland ice sheet. Science 321 (5885), 111–113.

Vandenberghe, J., 2002. Periglacial sediments: do they exist? Geological Society, London, Special Publications January 1, 2011, vol. 354, pp. 205–212.

Vonk, J.E., Mann, P.J., Dowdy, K.L., Davydova, A., Davydov, S.P., Zimov, N., Spencer, R.G.M., Bulygina, E.B., Eglinton, T.I., Holmes, R.M., 2013. Dissolved organic carbon loss from Yedoma permafrost amplified by ice wedge thaw. Environ. Res. Lett. 8 (3), 9, 035023.

Wadham, J.L., Arndt, S., Tulaczyk, S., Stibal, M., Tranter, M., Telling, J., Lis, G.P., Lawson, E., Ridgwell, A., Dubnick, A., Sharp, M.J., Anesio, A.M., Butler, C.E.H., 2012. Potential methane reservoirs beneath Antarctica. Nature 488 (7413), 633–637.

Walker, G., 2003. Snowball Earth. Three Rivers Press, New York, p. 269.

Yallop, M.L., Anesio, A.M., Perkins, R.G., Cook, J., Telling, J., Fagan, D., MacFarlane, J., Stibal, M., Barker, G., Bellas, C., Hodson, A., Tranter, M., Wadham, J., Roberts, N.W., 2012. Photophysiology and albedo-changing potential of the ice algal community on the surface of the Greenland ice-sheet. ISME J. 6 (12), 2302–2313.

Yi, C., Li, X., Qu, J., 2002. Quaternary glaciation of Puruogangri – the largest modern ice field in Tibet. Quaternary Int. 97, 111–121.

Chapter 13

Periglacial Processes and Landforms in the Critical Zone

Taylor Rowley*, John R. Giardino*,**, Raquel Granados-Aguilar**, and John D. Vitek*,**

*High Alpine and Arctic Research Program, Water Management and Hydrological Science Graduate Program, Texas A&M University, College Station, Texas, USA; **High Alpine and Arctic Research Program, Department of Geology and Geophysics, Texas A&M University, College Station, Texas, USA

13.1 INTRODUCTION

The surface of Earth consists of abundantly different geomorphic environments. These environments have been created by various geologic and geomorphic processes, which form distinctive suites of landforms. The geomorphic processes affect mainly the terrestrial surface of Earth, including the interface between the solid and fluid parts, but also extend to depths below the surface. These processes are best studied using an interdisciplinary approach (Fig. 13.1).

In 2001, the US National Research Council (NRC) and the US National Science Foundation (NSF) created a new focus on the upper part of Earth from what they defined as the top of the canopy to the bottom of the aquifer. The term Critical Zone was coined to describe this zone (NRC, 2001; NSF, 2005). The Critical Zone is highly variable in thickness, as one moves spatially around the planet. The geomorphic surface of Earth was assumed to be somewhat homogenous. We are not sure why this assumption was made as soil is defined as the link between all interfaces in the Critical Zone. Although differences in geomorphic processes have been acknowledged, regrettably, important and critical differences in the various geomorphic environments have been minimal with regard to the periglacial environment. The exception has been some focus on alpine periglacial processes coming out of the Boulder Creek Critical Zone Observatory (CZO). Global change is occurring. We think the influence of global change will have a significant impact.

This chapter addresses this issue discussing the distinct periglacial geomorphic processes that occur in portions of the Critical Zone, as well as identifying current trends in periglacial geomorphology. CZOs are established in various environments, but unfortunately, no CZO is currently located in a periglacial

Developments in Earth Surface Processes, Vol. 19. http://dx.doi.org/10.1016/B978-0-444-63369-9.00013-6

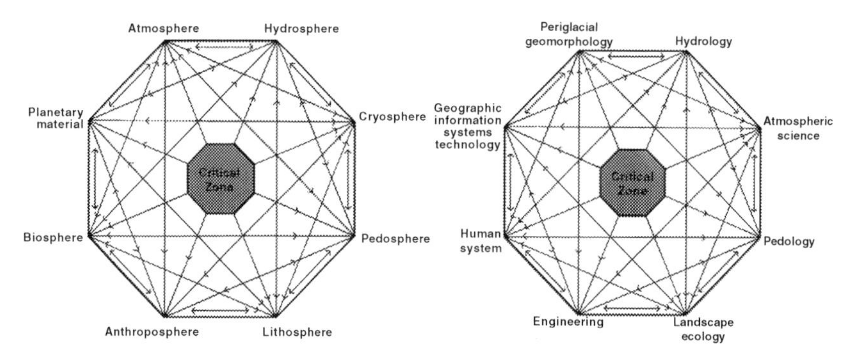

FIGURE 13.1 Interactions between the systems of Earth in the Critical Zone on the left. Interdisciplinarity in the study of the Critical Zone on the right.

environment. We aim to eliminate this exclusion, as we think it is one of the most important geomorphic landscapes because of its susceptibility to the impact of global climatic change. Thus, it must be considered in any discussion of the Critical Zone. We also think that the nature of the periglacial environment is changing, so it is essential that permafrost, as a thermal concept, cannot be the defining factor for periglacial geomorphology. Periglacial geomorphology is process driven and must be recognized as a major contributor in the evolution of the cryosphere.

Today, with global climate change at the forefront of much scientific debate, the periglacial environment plays an even more crucial role in the Critical Zone. This environment is extremely sensitive to changing climate, as many processes operating in this realm are temperature- and precipitation-driven. The inclusion of this landscape and its geomorphic processes are essential for future development of research in the Critical Zone.

13.2 GOAL OF THIS CHAPTER

This chapter reviews basic concepts of periglacial environments and integrates these ideas with current trends in research on the Critical Zone. Understanding the relationships between the fields, and applying interdisciplinary techniques to tackle research questions, will ultimately strengthen the research agenda in the Critical Zone. In this chapter we examine the interactions of periglacial processes with other phenomena and the impact on the Critical Zone. The term periglacial has complex meanings and applications. Before discussing the periglacial environment in the Critical Zone, broad connotations of the periglacial environment are discussed.

13.3 WHAT DOES PERIGLACIAL MEAN?

The term periglacial was initially introduced to portray the climatic conditions and geomorphic forms bordering late Pleistocene ice sheets (Lozinski, 1909, 1912). Since that time, the expression has gone through extensive alterations.

Unfortunately, today no universally acknowledged definition exists. Zeuner (1959) suggested that climate is the major factor in maintaining periglacial environments and that periglacial areas should be determined by mean annual temperature. In other words, climate is the dominant driver. French (1976) then suggested that the term should be broader and include environments where climatic processes result in rigorous frost action as the dominant driving process. Where permafrost is not present, the active layer is simply seasonal, frozen ground dominated by short-term frost action processes. Although no consensus exists on the definition of periglacial, we follow French (2013) and define periglacial as referring to a range of cold, nonglacial processes. Periglacial geomorphology is therefore a subdiscipline of geomorphology encompassing cold, nonglacial and azonal processes and landforms.

French (1976) specifically identified four categories of periglacial environments: (1) high arctic climates with large seasonal but small diurnal temperature fluctuations. Such conditions are present in the Canadian arctic. (2) Continental subarctic climates with large seasonal but small diurnal temperature fluctuations. The interior regions of Alaska are representative of this. (3) Alpine climates in the middle latitudes with large seasonal and diurnal temperature fluctuations. The high elevations of the Rocky Mountains and the Alps are typical. (4) Further severe climates, which are widely distributed across Earth with small seasonal and diurnal temperature fluctuations. Examples include isolated mountains, such as Mount Kilimanjaro, subarctic islands, and various mountains in the Andes of South America.

In general terms, the periglacial environment extends from high-latitude polar regions; through high-elevational, mid-latitude alpine regions to low-latitude, high-elevational alpine environments around Earth. Frozen ground dominates the periglacial, at short- and long-term timescales. Periglacial regions can be divided into two provinces: areas that are underlain with permafrost; and areas with no permafrost, but dominated by frost-action processes. From his work in Alaska, Péwé (1975) suggested that permafrost is not a necessary prerequisite, but it is virtually ubiquitous in periglacial environments. Thus, frost action processes do occur daily in both provinces within the top centimeters of frozen ground, whereas seasonal and annual processes occur in permafrost-covered areas. Expanding on our general locational factors, frozen ground and frost action contribute to the geomorphic features seen across the periglacial landscape.

The periglacial landscape is dominated by unique geomorphic processes, including frost wedging, solifluction, frost creep, and frost heaving. Wind also plays an important role in periglacial environments (Seppälä, 2004). These processes result in distinct periglacial landforms that include patterned ground, pingos, palsas, thermokarst, and rock glaciers. Frost action, physical and chemical weathering, mass wasting, and fluvial processes can dominate the types of landforms present. Fig. 13.2 shows the extent of permafrost on Earth.

(a)

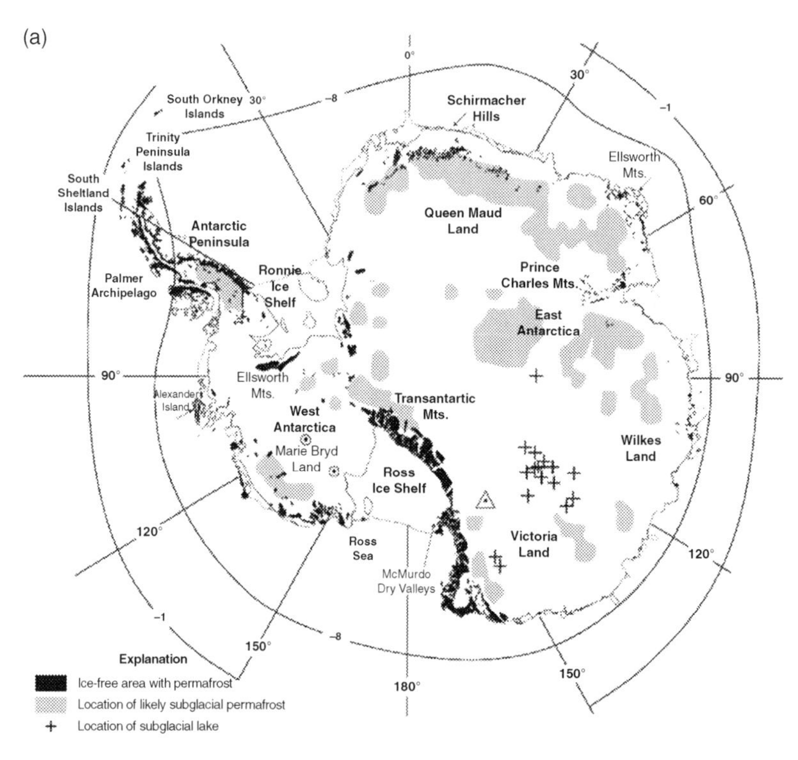

(b)

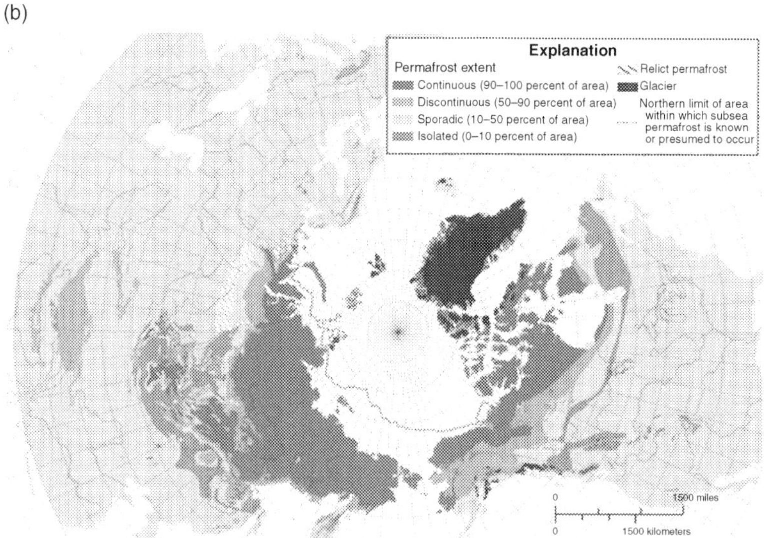

FIGURE 13.2 The two polar projections (a and b) show the generalized distribution of permafrost in the Southern and Northern Hemispheres. *(The maps are from Heginbottom et al. (2012).)*

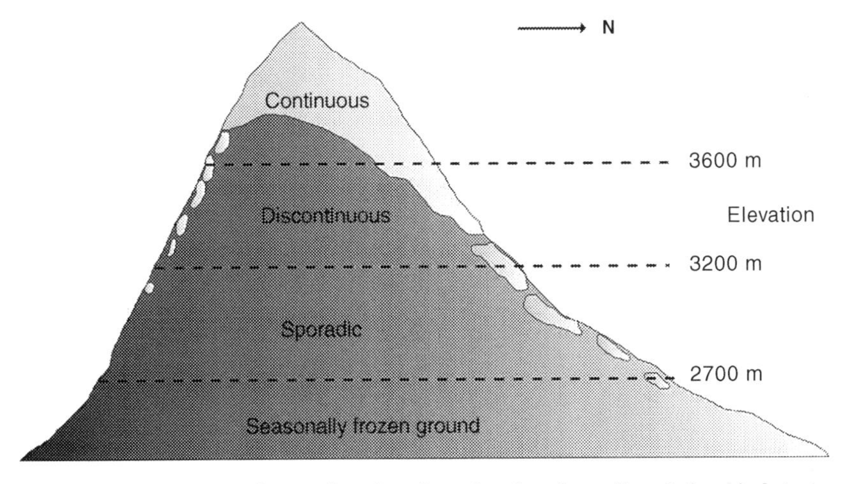

FIGURE 13.3 The type of permafrost based on elevation shows the relationship between elevation and the occurrence of continuous, discontinuous, sporadic, and seasonally frozen ground. *(Diagram modified from Heginbottom et al. (2012).)*

13.4 DESCRIPTION OF PERMAFROST

Today, approximately 25% of the terrestrial surface of Earth is underlain with permafrost. Permafrost occurs globally in polar environments and in some alpine environments. As pointed out, permafrost can occur in mid-latitude and tropical regions at high elevation (Fig. 13.3); the important factor is minimal thermal flux (Price, 1972). Soil moisture, snow cover, air temperature, aspect, and elevation are all factors that determine the development of local permafrost. The most favorable conditions, however, are characterized by areas with low heat fluxes. Large landmasses, high latitude areas as well as some high-elevation environments also facilitate the formation of permafrost (French and Harbor, 2013). In all of these environments, if the ground remains perennially frozen with temperatures at 0°C or less for two consecutive years, permafrost is considered to be in a stable state (Muller, 1943; Permafrost Subcommittee, 1988).

Where permafrost is present, typically an active layer, or a near-surface layer is susceptible to seasonal thaw. Consequently, frozen permafrost has a transient active layer; a boundary between frozen and unfrozen ground. Depth of the active layer can differ annually, as ambient air temperatures fluctuate and freeze–thaw cycles vary. As temperatures change, the volume of ice content also changes. In summer months, thawing allows moisture (i.e., liquid water) to seep to greater depths and refreeze at deeper depths whereas in winter months, any unfrozen water will migrate toward the freezing plane at the surface where it accumulates, causing permafrost to grow or expand in depth (Fig. 13.4). Permafrost is limited in space and depth, however, by the geothermal gradient. Heat

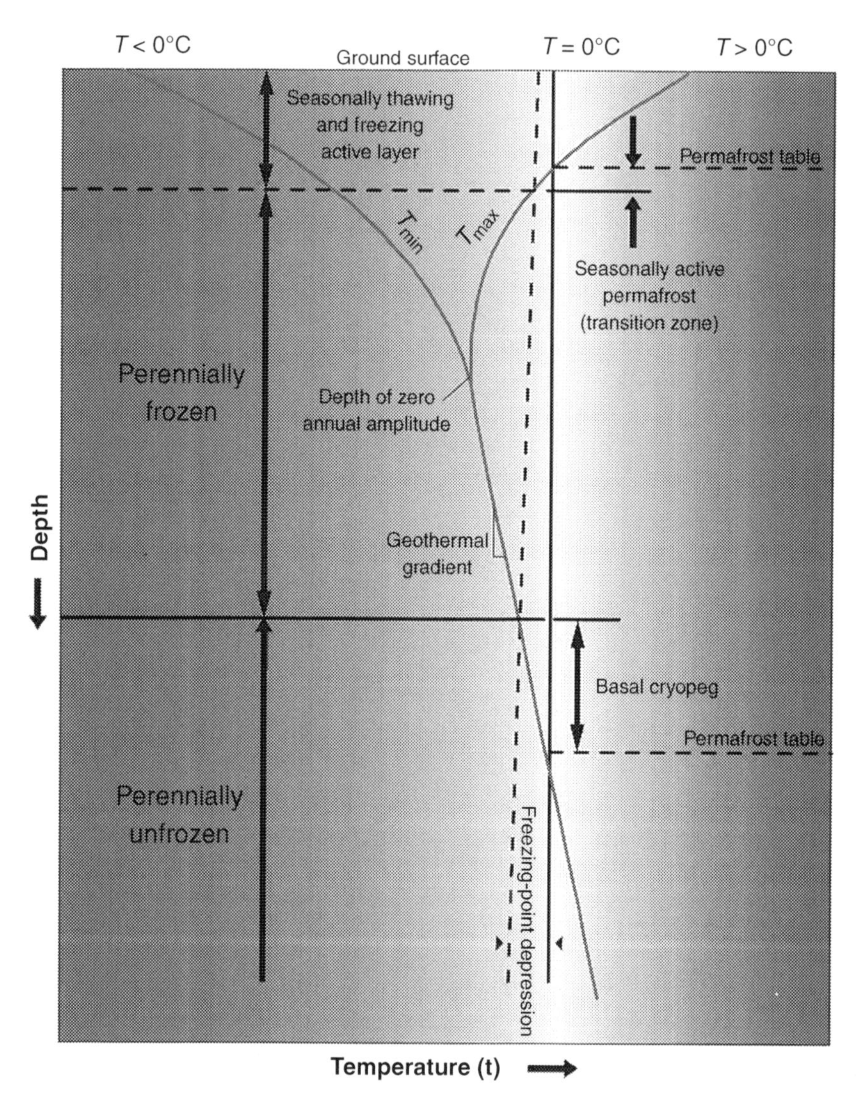

FIGURE 13.4 The effect of the geothermal gradient limits the maximum depth of permafrost. *(Diagram is modified from Heginbottom et al. (2012).)*

from the interior of Earth counteracts the growth of permafrost and limits the depth it attains. Geothermal variations and ground material with higher conductivities are responsible for variable depths across regions with permafrost (Walker 1986; Seppälä, 1998; Ritter et al., 2011; French and Harbor, 2013).

Three classes of permafrost occur globally: continuous, discontinuous, and sporadic (Fig. 13.5). The depth of permafrost also varies across Earth, from less than a meter in depth to more than 1500 m in depth in parts of Siberia (Costard and Gautier, 2007; Marchenko and Etzelmüller, 2013). Much of the

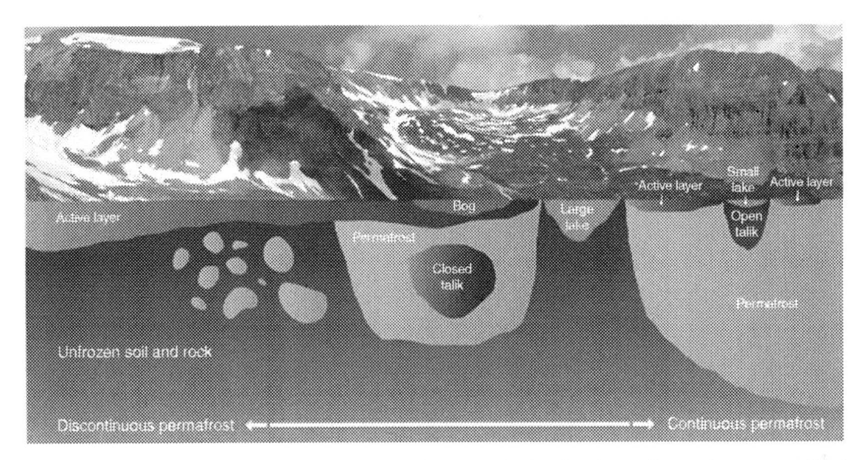

FIGURE 13.5 The different types of permafrost and distributions. *(Part of image modified from www.PhysicalGeography.net; background image from the San Juan Mountains by J. Giardino (2013).)*

permafrost present today formed during cold, glacial periods has been able to persist through periods of warming. Unfortunately, today many areas of permafrost are undergoing rapid melting as the result of shifting of seasons responsible for accelerated thawing and shrinking of areas current underlain with permafrost (IPCC, 2014). Much evidence is being collected that suggests warming of the climate (Clow, 2010; Janke et al., 2012; Briggs et al., 2014). Giardino and Vitek (unpublished data) have been monitoring 32 sites in the San Juan Mountains since the 1980s, and these sites have shown a continual warming trend. This warming trend is impacting permafrost in the high latitudes, also (Fig. 13.6). A review on permafrost by Dobinski (2011) summarized the majority of issues necessary to understand this phenomenon. Recent advances in mountain permafrost research using new technology have expanded knowledge of the distribution of permafrost in mountainous regions (Etzelmüller, 2013).

The categories of permafrost are dependent on ground temperatures and depth. In areas where permafrost is present, areas of continuous and discontinuous permafrost occur. Continuous permafrost covers 90–100% of the specific land area underlain by permafrost, having mean annual surface temperatures around -8°C. These temperatures are conducive to maintaining permafrost coverage. Discontinuous permafrost covers 50–90% of land areas in permafrost environments, and sporadic permafrost covers less than 50% of the land area in permafrost environments. Mean annual surface temperatures are approximately -5°C for discontinuous permafrost areas, whereas mean annual surface temperatures approaching 0°C frequently give rise to sporadic permafrost (IPA, 2014). The dependency on stable mean air temperature is essential for maintaining permafrost in a continuous state. As warming temperatures approach 0°C,

FIGURE 13.6 Permafrost has been exposed as a result of erosion in this area of Northwest Territory, Canada. *(Photograph by J. Vitek (1983).)*

thawing occurs, resulting in discontinuous or sporadic areas of permafrost. The distribution of permafrost is contingent on climate and pre-existing land-surface characteristics. In high-latitude polar regions, permafrost is more conditional on the mean-annual surface temperatures, whereas in lower-latitude high elevation alpine regions, the extent of permafrost depends on climate, but it is influenced by other factors including relief, aspect, snow-cover, and vegetation (Walker, 1986; Janke et al., 2012).

Mountain permafrost is highly sensitive to changing air temperatures because they affect the thawing depth of the annual active layer as well as the time and speed of the refreezing process mainly in the winter. The long-term ecological research site of Niwot Ridge and the Critical Zone Observatory Green Lakes in Colorado, with the high alpine tundra climate and vegetation offer ideal conditions to study changes of mountain permafrost. The sites provide high quality climate data, together with studies on permafrost since the 1979s, which makes these places rather unique in the United States (Leopold et al., 2014).

13.5 PERIGLACIAL LANDFORMS AND ASSOCIATED PROCESSES

Landforms that are characteristic of periglacial environments vary from small- to largescale. These various landforms are driven primarily by permafrost and freeze–thaw processes. Many of these landforms are restricted solely to periglacial environments, whereas, some of the other landforms are rather ubiquitous. Many of the periglacial landforms occur in areas where permafrost is present, which allows these geomorphic landforms to be unique and be considered as a separate group in periglacial geomorphology. Although frost action and mass movement are not restricted to periglacial regions, many of the resulting geomorphic features are unique to this environment as a result of the presence of ground ice and permafrost.

13.6 GROUND ICE

Ground ice is fundamental in the periglacial environment and is present in differing shapes and sizes. Ground ice varies between massive buried glacial ice to smaller-scale *in situ* ice-segregation processes. Because of the variability in the types of ground ice, various types of cryostructures have been identified and used to provide insight into ground-freezing processes that produce various types of landforms. The distribution and proportions of sediments and ice within the frozen ground determine the cryostructure. Shur and Jorgenson (1998) identified distinct cryostructures that can be used for classification through field observations. These distinct cryostructures are important for researchers studying the vertical changes in specific regions of the Critical Zone in periglacial environments. Each structure helps to indicate, to the observer, water migration toward freezing planes and water content during the freezing process (French, 2013).

13.7 SEGREGATED ICE

Segregated ice is common across periglacial landscapes, as the process is driven by daily freeze–thaw cycles (Fig. 13.7). As soil-water begins to freeze, the water will solidify in place and create a type of cementing agent, or it will migrate toward a freezing plane, as the result of suction-potential in the soil pores (Washburn, 1979). The type of segregated ice that forms is dependent on the texture of the soil. Coarser soils generally have soil-water that freezes in the pores where the ice serves as a cementing agent whereas fine-grained soils are susceptible to the action of pore-water suction, which draws water toward the freezing plane. If the soil is impermeable, like many clays, the ice will accumulate as lenses at the threshold where suction-potential and impermeability meet (Ritter et al., 2011). Many times, complex landforms are created, as the result of ice segregation resulting in expansion of the ice and increases in associated pressures (Murton et al., 2006; Rempel, 2007, 2011).

13.8 ICE WEDGES

Ice wedges (Fig. 13.8), first described by Leffingwell (1919), are a category of ground ice that result from frost cracking. The resulting fractures penetrate into the active layer and permafrost table. A new fracture is typically only millimeters wide, but can extend to meters deep. During winter months, this crack is exposed and open; however, during summer months, the fracture is filled with snow and water as the result of the onset of thawing. In continuous permafrost

FIGURE 13.7 (a) The segregated ice is pushing soil particles and stones to the surface. On slopes, this results in the downslope movement of materials as the ice pushes material up, then on thawing, gravity draws the particle in a downslope direction. (b) Shows an example of ice formed slightly below the surface. Note the slight mounding that is occurring. *(Photographs courtesy of Bill Shields and Lon and Susan Rollison.)*

areas, ground temperatures at the base of the crack are less than 0°C allowing the accumulated water and snow to freeze, forming ice veins. The original fracture becomes a zone of weakness, which is susceptible to additional fracturing during subsequent winters. This process facilitates the fracture to grow in width and depth and ice to accumulate from the original ice vein. Ice wedges can

FIGURE 13.8 Ice wedge in permafrost exposed by placer mining near Livengood approximately 50 miles northwest of Fairbanks. *(USGS Photograph by T.L. Pewe (1949).)*

grow to diameters in excess of 100 m and are often connected in a polygonal network. Ice-wedge polygons are indicators of a sustaining permafrost environment (Walker, 1986; Watanabe et al., 2013).

Larger ice wedges commonly form in a polygonal pattern across the periglacial landscape. These polygons can range in size, but are commonly between 8–18 m wide. Ice wedges border the polygons reaching depths of 10 m and widths varying between 2 m and 3 m. These wedges form as ground freezes and contracts creating high tensile stress that eventually exceeds the tensile strength of the soil, thus cracking occurs (Murton, 2013; Kokelj et al., 2014). The cracks vary in size at first, but weathering and the continuation of freeze–thaw cycle allows the wedges to become larger. Some erosion can occur as the area is locally not in geomorphic equilibrium, which will allow the crack to grow. A larger crack means that more snow and ice can fill the space often resulting in abundant vein ice as any meltwater follows the fracture line. The summer thaw may not affect the vein ice in the

FIGURE 13.9 A typical palsa, which stands 5.5-m high. *(USGS Photograph by Harald Svensson.)*

bottom of the fracture as it is closer to the permafrost table. The vein ice can then continue to accumulate, creating more ice wedges (Lachenbruch, 1962).

13.9 FROST MOUNDS

Seasonal frost mounds are ice-cored mounds or blisters in the active layer formed by hydraulic pressure in the subpermafrost layer during winter months. Seasonal frost mounds are irregularly shaped whereas blisters have an oval shape. These mounds generally last less than a decade because of the instability of the ice cores. Seasonal frost mounds differ from perennial frost mounds because of the shorter development time and ice instability (Yoshikawa, 2013; Morse and Burn, 2014).

Perennial frost mounds are also known as palsas, lithalsas, or pingos. These mounds have ice cores as well, but form differently than the seasonal mounds. Palsas and lithalsas (Fig. 13.9) do not have intrusive ice, or ice that forms as a result of local groundwater. Palsas are the result of segregated ice-lens accumulation by cryosuction, and have an overlying layer of peat. These frost mounds are typically found on discontinuous permafrost (Michel and Van Everdingen, 1994). Lithalsas are palsas without peat cover (Pissart et al., 2011). These exist in a smaller range than palsas, commonly occurring in oceanic climate regimes. The two mounds are relatively small compared to the similarly structured pingo; typically less than 3 m (Yoshikawa, 2013). Palsas generally leave no trace after thawing, whereas lithalsas leave circular depressions with embankments bordering; typically consisting of sediments as a result of solifluction (Wolfe et al., 2014).

13.10 PINGOS

Pingos (Fig. 13.10) are typically larger than palsas, reaching heights greater than 50 m (Walker, 1986). The defining characteristic of these mounds is the presence of intrusive ice throughout most of the core. An accompanying ice lens may be present toward the top of the mound, above the ice core. Two types

FIGURE 13.10 **A typical open-system pingo.** *(Photograph by J. Vitek (1983). Northwest Territory, Canada.)*

of Pingos exist; open and closed, and they are present in both continuous and discontinuous permafrost areas. Development of pingos in these two categories are dominated by hydraulic pressure for an open pingo, and hydrostatic pressure for a closed pingo (Mackay, 1979).

In an open system, a pingo forms as a result of artesian pressure induced from locally higher topographies. Open system pingos commonly occur at the base of a hillslope where hydraulic pressures are high, allowing the groundwater to inject into the pingo. As the groundwater approaches the top of the pingo, it freezes and accumulates, causing the pingo to grow. Because groundwater is essential for the creation and growth of an open system pingo, these often occur in areas of discontinuous permafrost where groundwater can flow freely.

In a closed system, a pingo forms as a result of hydrostatic pressure. Closed system pingos commonly form in drained shallow lake basins. Where surface water is present, a thermal gradient lowers the permafrost table, leaving an unfrozen layer termed a talik. As a surface water basin drains, the residual water in the saturated soil is exposed to the atmosphere where it freezes. As residual pore water freezes, cryostatic pressure pushes remaining unfrozen water in the talik toward the top of the pingo where it also freezes; eventually creating the inner ice core dominant in pingos.

Mackay (1998) noted that pingos are dynamic and can pulse vertically. He also noted that the distinctive feature of these mounds is an ice core, and the creation is dependent on the expansion of pore-water in the soil of the mound (Mackay, 1979). Jones et al. (2012) observed the distribution of pingos in

FIGURE 13.11 **Typical landscape north of Nome, Alaska.** *(Photograph by J. Giardino (2005).)*

northern Alaska, noting the morphological characteristics as well. The analysis of distribution and characteristics suggest that pingos are extremely dynamic and individualistic and a wide range of heights, diameters, slopes, and spacing exist.

13.11 PATTERNED GROUND

Patterned ground is the distinct morphological feature of periglacial landscapes (Fig. 13.11). It consists of mostly symmetrical geometries displayed across the ground surface in relation to local frost action and cryogenic processes. Although this phenomena is distinct across the periglacial landscape, much disagreement exists among researchers, past and present, as to how and why the patterns form (Washburn, 1956; Nicholson, 1976; Peterson and Krantz, 2003; 2008; Feuillet, 2011; Frost et al., 2013; Warburton, 2013).

The patterns occur in the top layer, or active layer in permafrost areas. Areas of no permafrost have seasonal frost layers where freeze–thaw cycling occurs. Patterns emerge as a result of surface disturbances caused by thermal anomalies and freeze processes such as frost heave (Marr, 1969). Frost heave will disturb the frost layer as ice lenses accumulate and protrude, causing unstable soil conditions. Once these ice lenses begin to thaw, pore water pressures increase, destabilizing the soil, thus, increasing the potential for mass movements. Other types of ground ice and cryogenic processes can also lead to the formation of patterned ground. Characteristically the geometry of the

FIGURE 13.12 Large stone polygons at 3822 m in the Sangre de Cristo Mountains, CO. *(Photograph by J. Vitek (1975).)*

patterned ground can provide insight into the underlying cryogenic processes (Warburton, 2013).

Several geometries emerge as a result of varying cryogenic processes including polygons, circles, stripes, nets, and steps. Washburn (1956) classified patterned ground on the basis of two criteria; (1) being the geometric shape, and (2) being whether the composing material was sorted or unsorted; a result of frost sorting.

Sorted patterned ground (Fig. 13.12) is bounded by stones surrounding an inner core of finer sediment. This sorted patterned ground occurs in groups, and reaches diameters of only about 10 m, whereas unsorted patterned ground can have diameters of 100 m. Unsorted geometries are not bound by coarse sediments, and typically accompany ice wedge polygons (Ritter et al., 2011).

13.12 THERMOKARST

Thermokarst features are topographic depressions created in a variety of shapes and sizes as a result of thawing ground ice. Thermokarst is present in areas where the thermal equilibrium has shifted, allowing for the thaw of ground ice. This process can be a result of lateral erosion where ground ice becomes exposed, or lateral degradation where warm surface water penetrates into the adjacent shore, causing taliks to form and ground ice to thaw (Jorgenson, 2013; Bouchard et al., 2014).

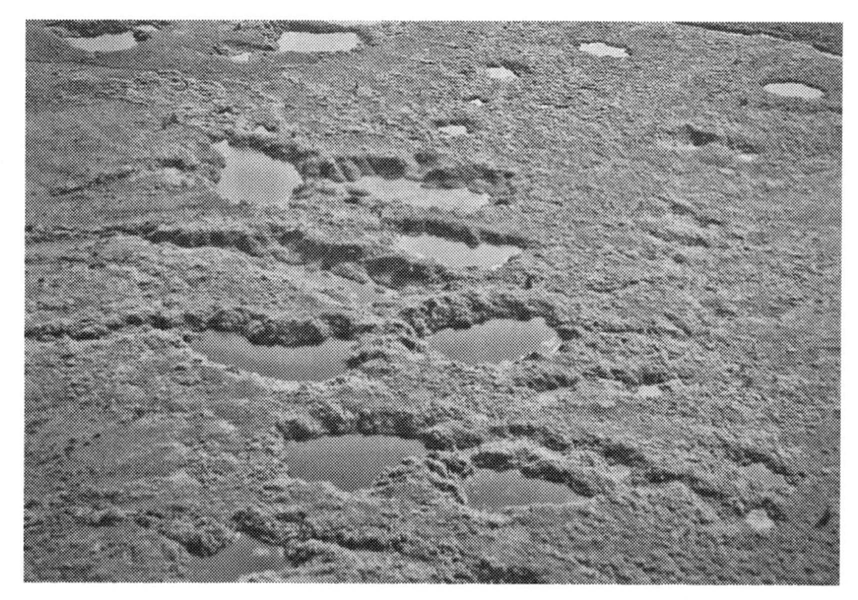

FIGURE 13.13 Thermokarst lakes near Cape Krusenstern, Alaska. *(Photograph by J. Giardino (2005).)*

A thermokarst lake (Fig. 13.13), or an alas, is a result of subsidence as ground ice below decays. The lake will continue to grow as the warmer surface water is in contact with the frozen ground of the shoreline. This will result in further ground ice decay, and the expansion of the lake. Thermal gradients also accelerate bank erosion in rivers where ground ice is present. Although the thermal gradients are a major factor in the development of thermokarst features, atmospheric thermal gradients do not necessarily increase thermokarst development. A study performed by Burn and Smith (1990) concludes that local land disturbances are more of a factor in determining where a thermokarst feature will exist rather than climate alone. Twenty-two distinct thermokarst landforms have been identified based on their topographical characteristics, including beaded streams (Fig. 13.14) formed by the melting of ice wedges, collapsed pingos, and thermokarst fens forming as a result of rapid thaw of lowland deposits from groundwater springs (Jorgenson et al., 2008; Jorgenson, 2013).

13.13 DESCRIPTION OF SURFACE TO NEAR-SURFACE, FROST-ACTION PROCESSES

Whether permafrost is present or not, freeze–thaw cycles can occur daily in periglacial areas. Short-term (i.e., diurnal) frost action processes affect the upper few centimeters nearest to the surface as diurnal freezing and thawing occur. Processes include frost wedging, frost heaving, and frost sorting. All of these processes result in the formation of various landforms.

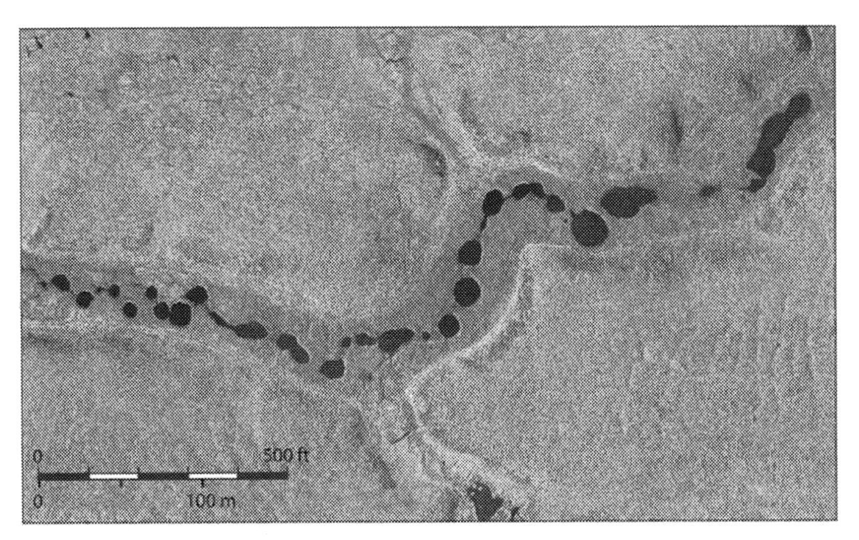

FIGURE 13.14 Thermokast features can be connected via stream activity forming beaded streams. *(USGS Photograph by M.T. Jorgenson (2004).)*

13.13.1 Frost Wedging

Frost wedging, or frost shattering, is the result of rapid freezing of water in cracks and pores of the parent material, causing fracture. This rapid freezing, near the surface, will close the atmospheric interface from the soil, which in turn increases pressures that will lead to fracturing. The size and shape of these wedges and fractures are dependent on rock type, porosity, water availability, and number of freeze–thaw cycles (Walker, 1986). Mellor (1970) noted that high-water content increased the likelihood of rock shatter because of increased strain. Angular debris is commonly the result of frost shatter and accounts for much of the talus occurring throughout the periglacial landscape (Ritter et al., 2011). Murton et al. (2006) suggest, however, that fracturing may be the result of ice segregation instead of expansion resulting from freezing pore water. Although fracturing can result from ice segregation, the topic remains poorly understood and understudied.

13.13.2 Frost Heave

Frost heave occurs as a result of capillary gradients that are created as water passes through porous media toward a freezing front. Lenses and layers of ice accumulate as more water is drawn toward the freeze plane. The freezing plane is irregular because of dissimilar ice accumulations and variable sizes of material in the soil resulting in a varying surface topography. As ice accumulates and lenses grow, surrounding soil and rock are forced upward, and often sorted with coarser material transported toward the surface. Varying rates of heaving depend on the dynamics of the system. Price (1972) noted that particles can move upward ~5 cm/year. Vitek et al. (2008) demonstrated that horizontal motion of stones in

FIGURE 13.15 Small sorted polygon in the Sangre de Cristo Mountains of Southern Colorado, elevation 3870 m. Scale is 6 in. (15.24 cm). *(Photography by J. Vitek (2004).)*

the centers of stone polygons (Fig. 13.15) can fluctuate up to 29 cm in 30 years, but all samples averaged 7.4 cm over the 30-year period. A closed system is less susceptible to volume increases compared to an open system. In an open system, if a freezing front is present, it will draw water toward it. This additional force in the open system accounts for increased heaving rates (Nixon, 1991).

13.13.3 Frost Sorting

Heterogeneous sediments will commonly sort vertically and laterally. Two main theories exist to explain this process. Washburn (1973) described frost-pull as the ground freezes, all sediment is lifted as the result of expansion, and as thaw occurs, the cohesive fine soils sink more readily than the larger clasts. This repetitive motion allows the larger clasts to remain near the surface. Frost-push, however, suggests that the larger clasts are more conductive, and will change temperatures more readily than the neighboring smaller clasts and fine-grain sediments. Large clasts will, therefore, cool rapidly allowing freezing planes to form adjacent to the clast, thus pushing the stones vertically toward the surface (Fig. 13.16). Mackay (1984) suggested that freezing planes move upward with a rising permafrost table, thus supporting Washburn's frost-push theory.

Sorting may also be a result of convective circulation when the ground is thawed. Hallet and Waddington (1991) suggested that convection is driven by

FIGURE 13.16 **Sorted stone polygons in Alaska.** *(Photograph by J. Vitek (1983).)*

differing buoyancies in the soil. As repetitive freeze–thaw actions occur, soil that is deeper become less dense than the soil near the surface. This process occurs because the thawing moves from the surface and travels downward. The top layer can become very compact, whereas at depth, it is less dense because less time is available for compaction. This is the result of it being thawed for less time than the layers above. Density differences force the less dense, finer-grained sediment toward the surface, whereas coarser material migrates toward the edges of the circulation. This process is common in sorted patterned ground.

13.14 MASS MOVEMENT

The evolution of the surface of Earth is a result of dynamic interactions in the Critical Zone. Some of these mechanisms result in mass movement, characteristically creating risks to the inhabitants of diverse regions or environments, being more frequent in mountainous areas. Mass movement occurs when gravity and slope instabilities initiate the regolith and soils to move downslope (Millar, 2013) (Fig. 13.17).

The periglacial environment is not exempt from these movements. Landslides, debris flows, and avalanches involving surface and ground ice, or rock can occur abundantly (Fig. 13.18). This can be attributed to active freeze–thaw processes and permafrost melt contributing to soil pore-water pressure changes.

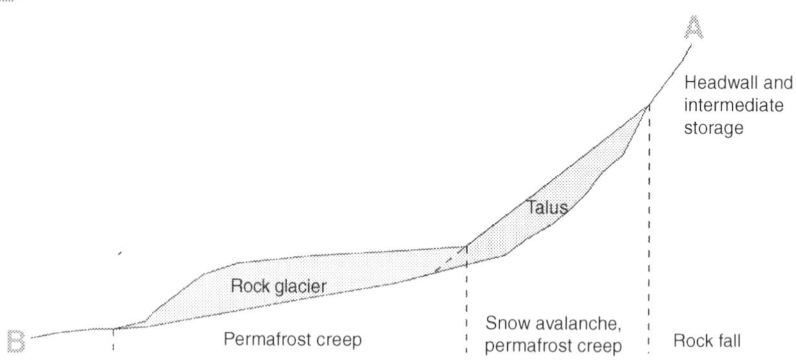

FIGURE 13.17 A conceptual profile of a mountain slope and associated mass movement in the periglacial environment (based on Müller et al., 2014). *(Background photograph of the Talkeetna Mountains, Alaska courtesy of USGS (2004).)*

These changes result in slope instabilities across the landscape. In this particular environment, slope failure can occur at much lower angles than those usually predicted using a slope stability analysis in a different environment. This occurs because of the uncertainties associated with the soils under consideration.

FIGURE 13.18 An earthquake-triggered debris flow shunted the Savage River to the opposite bank River in Alaska. *(Photograph by M. F. Giardino (2008).)*

Important properties of these soils present a heterogeneity in the distribution (such as soil-water pressure and seepage) that, in addition to the complexity of the freeze–thaw cycling effects, make the mechanical response of the soils difficult to predict (Millar, 2013). As a result, numerous types of mass movement, both slow and fast, occur throughout the periglacial realm.

13.15 SOLIFLUCTION

Solifluction (sometimes termed gelifluction in periglacial environments) is the slow flow of saturated soil downslope indicating no frozen ground is present in the moving layer (Washburn, 1979). Movement of these saturated soils can be initiated by thawing, creating excess pore-pressure in the soil, resulting in slope movements (Mackay, 1981) (Fig. 13.19). The relief of the slope, depth of thaw, and water content are significant factors in solifluction rates (Walker, 1986; Hjort et al., 2014).

Solifluction reaches its maximum potential in the late spring and summer months when thaw saturates soils. Saturated soils have increased pore pressures, resulting in unstable conditions because of a lack of friction and cohesion. Solifluction can occur on slope less than 1°, but it is more common on slope gradients between 5° and 20°. With more relief, soil saturation typically decreases as the water drains down the slope as runoff (Walker, 1986). Stratified

FIGURE 13.19 **Solifluction lobes show saturated conditions on this slope.** *(Photograph by J. Giardino (2008).)*

and multilayered slope sediments in the Colorado Front Range are attributed to solifluction (Voelkel et al., 2011). Such deposits will impact the movement of water and development of soils.

13.16 DETACHMENT LAYERS

Where slope failure occurs, it is common to observe a slide scar (Fig. 13.20) in the upper part of the slope, from which the material detached uncovering the shear plane, as well as a colluvial silty-clay deposit at the bottom that usually expresses the compression as a result of the accumulation process (Harris and Lewkowicz, 1993; Millar, 2013).

The local condition of permafrost and trigger factors, such as climate, vegetation presence, thawing, and mechanical properties of the soil, as well as the favoring action of solifluction determine the frequency of incidence of these active layer failures. If the motion mechanism is flowing instead of sliding, authors refer to these events with terms such as skin or earth flows (Carter and Galloway, 1981; Lewkowicz, 2007; Cogley and McCann, 1976; Lewkowicz and Harris, 2005).

13.17 RETROGRESSIVE-FALL SLUMPING

In periglacial regions, the soil is frequently covered by snow and ice. Such coverage provides additional water, which can result in increasing rates of erosion in periglacial regions as noted by French (2013). He pointed out that failure

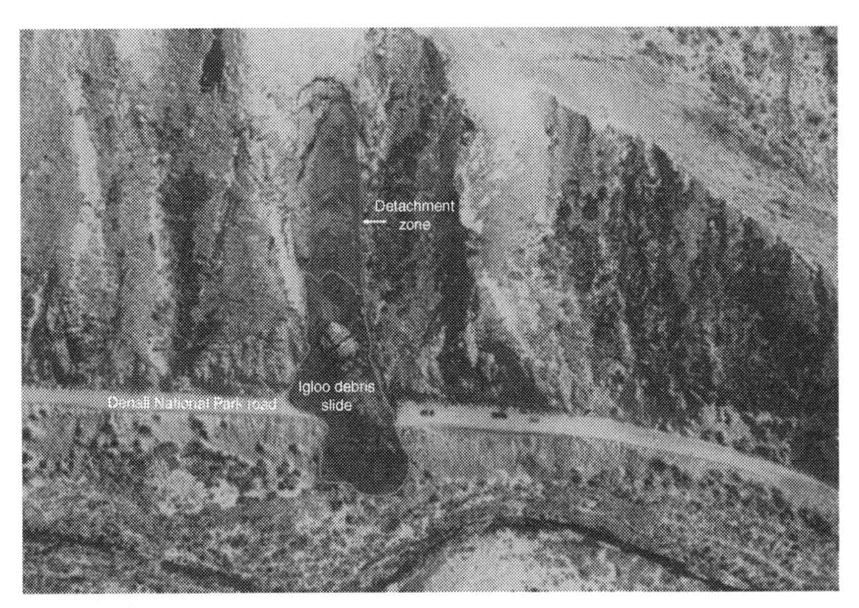

FIGURE 13.20 **Igloo debris slide, which occurred in October 2013 on the Denali Park road.** *(NPS photograph.)*

caused by the melting of ground ice is a major contributor to the occurrence of this process. Diverse geomorphic processes involving the movement of sediments and thawed material, in some cases triggered by anthropogenic activity, occur in the formation of retrogressive thaw slumps. Thaw slumps (Fig. 13.21) are commonly attributed to the melting of permafrost under soil in an accelerated failure that exhibits curved headwalls and slumped bottoms (Fraser and Burn, 1997; Harry and MacInnes, 1988; Mackay and Terasmae, 1963; Burn, 2000; Millar, 2013).

13.18 SNOW AVALANCHES AND SLUSH FLOWS

Two processes that occur in highly mountainous areas are snow avalanches and slush flows (Figs 13.22 and 13.23). The first refers to a rapid flow of snow down steep slopes, but can be referred to as a dirty avalanche where it includes a combination of soil, boulders, and vegetation (Bartsch et al., 2009; Rapp, 1960; White, 1981). Factors determining the occurrence of avalanches include temperature changes, internal structure of snowpack, and the underlying lithology (McClung and Schweizer, 1999).

Slush flows occur along first-order headwater channels where the snowpack is saturated by snow and mud; generally following rainfall or thawing (Larocque et al., 2001; Rapp, 1960; Washburn and Goldthwait, 1958).

FIGURE 13.21 **Retrogressive slumping along the coast of Herschel Island, Yukon.** *(Photograph courtesy of Hugues Lantuit.)*

13.19 ROCK FALLS

Freeze–thaw cycling, water content, rock mechanics, and weathering play important roles in the formation of microcracks in bedrock faces. These processes result in penetration of water in the pores of the rock where temperature changes cause the water to freeze and thaw repeatedly; generating stress through continuous volumetric changes. The process eventually leads to the separation of fragments of rock that fall and tend to accumulate below, forming a debris slope that is then subject to a variety of geomorphic processes (Matsuoka, 2001, 2008; Rapp, 1957, 1960; Sass, 2005; Wilson, 2009; ; Walder and Hallet, 1985) (Matsuoka and Murton, 2008 Fig. 13.24).

Rock falls are the fastest type of mass movement and can present risks because of the high energy involved in the movement. Gardner (1983) noted that rock falls is an accretionary process and calculates rates of accretion in six areas across the front range of the Canadian Rocky Mountains. Other authors have also observed accretion rates and noted the conditions to determine major triggers for their specific study areas (Matsuoka and Sakai, 1999; Jomelli and Francou, 2000; Perret et al., 2006). Field observations and remote technologies are being used to try to quantify and predict rock falls because they can be extremely hazardous if human activity is nearby (Abellán et al., 2010).

FIGURE 13.22 Snow avalanches, Colorado. *(Photograph by J. Giardino (2014).)*

FIGURE 13.23 Slush flow (also referred to as fluvial talus) in the San Juan Mountains, Colorado. *(Photograph by J. Giardino (2009).)*

FIGURE 13.24 Rock fall off the headwall of a glacial cirque in Governor Basin, Colorado. *(Photograph by J. Giardino (2010).)*

13.20 ROCK GLACIERS

High alpine systems are continuously modified by natural periglacial processes and increasingly affected by anthropogenic interactions (civil infrastructure). Rock glaciers are common units of these systems. Located at the foot of rock free faces, they generally mark the mountain permafrost and are identifiable by its lobe or tongue-shaped forms (20–100-m thick when active) (Humlum, 2000) (Fig. 13.25). Characterization of rock glaciers includes: debris input, ice content, rates of flow (flowing rock glaciers resemble viscous substances like the pahoehoe lava), size, position on the hillslope, microrelief (characterized by a surface consisting of poorly sorted, angular, blocky debris; where transverse and longitudinal ridges and furrows form perpendicular and parallel, respectively, to the direction of movement), and distribution which is determined by local environmental variables (lithology, geographic location, microclimates, topography), generally considered as characteristic of continental environments (Troll, 1973; Barsch and Caine, 1984; Barsch, 1977, 1993, 1996; Walker, 1993; Beniston, 2000; Konrad et al., 1999; Giardino and Vick, 1987; Barsch et al., 1979; Haeberli et al., 1999; Haeberli, 1985, 2000; Burger et al., 1999; Capps, 1910; Wahrhaftig and Cox, 1959; Benedict, 1973; Washburn, 1979; Martin and Whalley, 1987; Giardino et al., 1987, 2011; Ives, 1940; Potter, 1972; Barsch, 1987; White, 1987; Kääb and Weber, 2004; Janke, 2005; Calkin et al., 1987; Ackert, 1998; Humlum, 2000).

FIGURE 13.25 A tongue-shaped rock glacier overlies the Upper Camp Bird mine shaft #3 in the San Juan Mountains. *(Photograph by J. Giardino (2012).)*

Giardino and Vitek (1988) created a unique way of considering the temporal conditions of a rock glacier (Fig. 13.26). Their figure suggested that a rock glacier can form from either a glacial or periglacial origin. The diagram traces the route of rock-glacier formation through time. In their discussion, Giardino and Vitek (1988) made a strong distinction between process and form. This distinction has been blurred by many researchers who refer to the process via the form.

Thermal conditions cause spoon-shaped depressions in rock glaciers. The common profile of a rock glacier is concave toward the head or rooting zone, with a gradual transition turning convex with a steep front slope at the toe. The traditional classification separates rock glaciers by process with their basal stress as: active (1.0–2.0 bars) and inactive (<1.0 bars). Processes acting in rock glaciers serve to identify their origin. It is said to be of periglacial origin when cemented ice (pore-ice or ice lenses) occurs with a hydrostatic pressure-driven movement, and glacial when it has an iced-core internal structure (Washburn, 1979; Vitek and Giardino, 1987; Barsch, 1977, 1987, 1996; Wahrhaftig and Cox, 1959; Barsch and King, 1975; Ikeda and Matsuoka, 2002; Haeberli, 1985; Giardino, 1979).

Indicators of rock-glacier degradation are: slumping surface morphology, frontal activity, internal ice content, downslope movement variation, or temperature of the materials (Francou et al., 1999).

Haeberli (2005) and Berthling et al. (2013) stated that it is not possible to separate the origin of rock glaciers into purely glacial or periglacial without taking into account the permafrost–glacier interactions that explain these landforms.

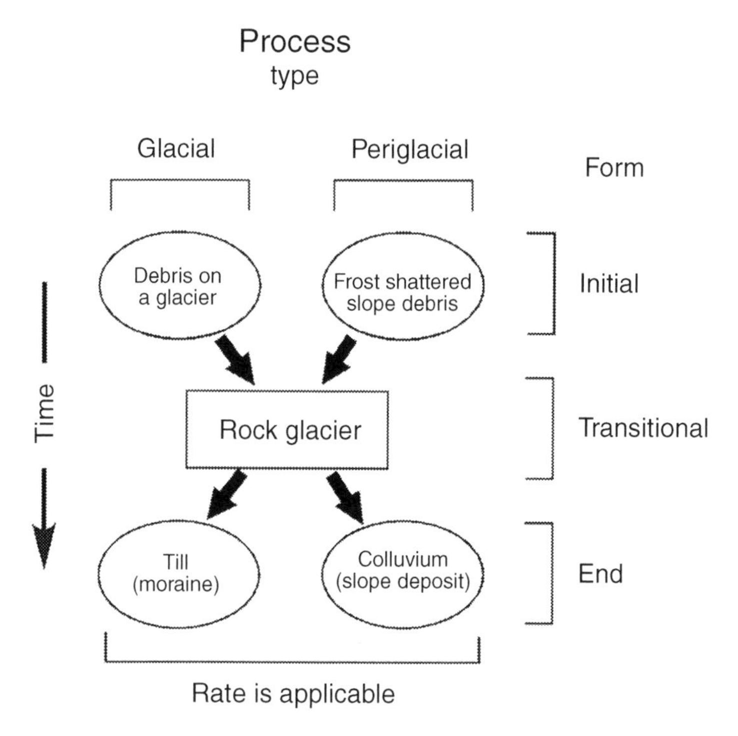

FIGURE 13.26 The concept of a periglacial landscape continuum involves space, time, and processes. A rock glacier is a transitional form that can develop from two distinct processes and progress to distinct end-members. Rate of movement is related to the process, not the form (Giardino and Vitek, 1988).

The permafrost environment receives ice, water, and debris in various ways by glacier influence, with the change of the ground-surface energy balance being the most important one. On the other hand, permafrost provides boundary conditions as well as conditions dictating the mass balance of the glaciers.

Rock glaciers play an important role in the periglacial Critical Zone. They serve as a transport mechanism and a sink for material. Numerous engineering problems have occurred as humans have attempted to use rock glaciers as borrow sources for road construction, built ski towers on a landform that is dynamic, and have attempted to mine underneath a rock glacier, which resulted in collapse of the mine tunnel and death of the miners. The Critical Zone–glacier interaction is covered in depth in another chapter of this volume.

13.21 THE IMPACT OF THE PERIGLACIAL CRITICAL ZONE ON HUMAN ACTIVITY

The role of polar and mountainous regions in the climate of the planet is of tremendous importance; however, local and global anthropogenic practices are rapidly altering these regions. Globally, temperature increases are changing climatic norms, whereas locally, humans are altering the environment to fit their own needs (IPA, 2014). As a result of global change, the periglacial environment is one of the most susceptible regions to rapid change. This change is occurring daily in these environments. Thus, understanding the process that shape the periglacial environment is vitally important in studies of the Critical Zone.

For millennia, people have lived in periglacial regions. Anthropogenic activities accompany climatic and geologic forces acting on the Critical Zone as humans alter and adapt to the harsh environment. Periglacial regions are littered with geologic engineering successes and failures, as humans attempt to manage the resources of this respective landscape. Humans are trying to survive as a part of the environment, and are securing the natural resources including oil, gas, timber, game and fish, and minerals. The extraction of these resources leads to complications of engineering problems that are commonly accomplished with unique solutions. Problems requiring engineering solutions are ubiquitous across all environments and landscapes; however, problems specific to the periglacial environment require specific approaches because of the complex, geotechnical challenges associated with human interactions within periglacial environments.

The Critical Zone initiative is fundamental to producing an interdisciplinary view of site-specific locations. From these site-specific locations, a mass of data will be captured. These data will provide a system's approach in continued research in the Critical Zone of these harsh environments. In the next sections, we will examine the interactions of humans and the periglacial Critical Zone. For convenience of the reader we have grouped these activities as: housing, infrastructure, water resources, oil and gas, mining, recreation, and the impact of global change. The important point regarding research in the Critical Zone of a periglacial region is the focus on the system. The inputs, outputs, storages (short- and long-term), pathways, and thresholds must be considered.

13.21.1 Buildings

Continuous permafrost is the dominant cause of unique engineering solutions. Permafrost is dependent on the ground–atmosphere interface remaining in thermal equilibrium, so that permafrost can remain static. When disturbed, the entire permafrost system must adjust to the change; further compounding engineering-related problems (Nash, 2011) (Fig. 13.27). An excellent visual example of the impact of global change is the Village of Shishmaref, which is located on a mile-wide barrier island about 1000 km northwest of Anchorage, Alaska. Global change is resulting in late freezing of the Bering Sea, which results in higher rates of erosion during the winter. In the past, the sea would be frozen for longer periods of time and, thus, winter storm waves did not occur. In addition, warming temperatures and exposure resulting from erosion by wave action have resulted in thawing of permafrost in the area. The Village of Shishmaref is being moved inland to avoid further destruction (Fig. 13.28).

FIGURE 13.27 **Building on permafrost requires specific engineering solutions.** This damaged structure in Dawson City, Canada, shows what can happen when the warm interior of a building causes the permafrost underneath to thaw. *(Photography courtesy of Andrew Walker.)*

FIGURE 13.28 **The structure, which was built on permafrost has been undermined by increased erosion by wave action, as the permafrost is being exposed and melts.** *(Photograph by J. Giardino (2007).)*

Typical approaches to areas with continuous permafrost include passive and active approaches. In some cases, the presence of permafrost is ignored; typically only in areas where permafrost is discontinuous. An active approach is the excavation of a top portion of the permafrost and subsequent replacement with permafrost resistant material. This process typically includes obtaining local construction materials with more suitable geotechnical properties. Materials are commonly obtained from nearby streams, lakes, or outcrops. Shiklomanov and Nelson (2013) have suggested that these practices have proven to be successful, but have consequences. When a foreign material is introduced, thermal equilibrium is again offset. If this aspect is not recognized and addressed, issues may include deeper penetration of thawing through the foreign material, a decrease of surface albedo increasing the absorption of radiation, and the increase of the flow of heat. These changes can result in thermokarst and the alteration of drainage patterns (Lunardini, 1981; Walker and Everett, 1987; Ferrians et al., 1969; Nelson et al., 2002; Nelson, 2003).

Research regarding engineering problems in periglacial areas has been carried out for some time (Brown, 1967). Recently, new methods have been developed to allow structures to be constructed on permafrost with minimum disruption of the permafrost (McFadden, 2000, 2001).

A more passive approach is to preserve the permafrost with ventilation, insulation, or vegetation removal (Fig. 13.29). The removal of vegetation decreases

FIGURE 13.29 Ventilation tubes used on a building in Kotzebue, Alaska, to keep underlying permafrost in dynamic equilibrium. *(Photograph courtesy of Rawan Maki (2011).)*

local insulation and allows the ground and atmosphere to interact directly to accelerate achievement of a thermal equilibrium.

13.21.2 Infrastructure

Roads and railways connect humans and resources to each other. The first road connecting Alaska and Canada (i.e., ALCAN Highway) was constructed during World War II; a nearly 3000-km highway. Although the completion of the road was a success, the effects of the periglacial landscape it sits on were not considered (Twichell, 1992). Placing a surface that limits interactions within the Critical Zone is detrimental to the foreign object. The Critical Zone must reacclimate and achieve a new thermal equilibrium. This adjustment causes local frost heaving and subsidence along the road as the Critical Zone adjusts.

Adjusting occurs across many spatial scales as a result of local and global changes. The access road to Denali National Park has been adjusting from the time it was constructed. Local degradation of the road has occurred as well as degradation surrounding the road. Park officials think climate change is the main driver of this change. As the road degrades, insulation techniques are applied where road sections are rebuilt (Vinson and Lofgren, 2003; Morris et al., 2014). In 2013, the Park Road was closed by a major landslide. The frequency of landslides appears to be increasing (Dr. Denny Capps, personal communication, 2015). Vincent et al. (2013) are building a model aptly named, ADAPT, to better model the effect engineering projects have on a permafrost landscape. The goal of the project is to better predict how the landscape will react to a foreign structure, and to better plan for how to build structures that have minimal effect on the total landscape.

Pipelines are scattered across periglacial environments as oil extraction is commonly a major export from these regions. In Alaska, oil is abundant and transportation of the resource via pipeline is essential to allow the resource to be accessed. Extreme planning when building a pipeline must be undertaken so as to minimally disturb underlying permafrost. The landscape in which the pipeline overlays was examined in detail to determine the best way to lay the pipe. The temperature of the transported oil exceeds 60°C and affects underlying permafrost even where not in direct contact. In areas of well-drained unfrozen ground, the pipeline is buried (Fig. 13.30), and in some areas it gets refrigerated, depending on local conditions. Where not buried, the pipeline is commonly suspended above the ground on numerous pilings. These pilings are sometimes refrigerated to ensure that the ground below experiences minimal effects of the increased temperatures of the pipeline (Fig. 13.31). A spill would have drastic impacts across the landscape and to the companies responsible, so extreme measures are taken to minimize the potential for any accidents to occur because of melting permafrost. Because the climate is changing, a need exists for closer monitoring and planning of additional and future pipelines (Oswell, 2011; Ritter et al., 2011; Dimov and Dimov, 2014).

FIGURE 13.30 The zigzag pattern of the Alaska pipeline is a design feature allowing the pipeline to expand and contract as ambient temperatures change, as well as absorbing energy from earthquakes. The pipeline not only zigzags, but it is mounted on sliders as it crosses the Denali fault. The structure allows the ground to slide during an earthquake without rupturing the pipeline. *(Photograph by J. Giardino (2008).)*

FIGURE 13.31 Alaska pipeline north of Fairbanks, Alaska. The vertical risers supply support as well as provide cooling for permafrost below the pipeline. *(Photograph by J. Giardino (2008).)*

13.21.3 Utilities

In regions underlain by permafrost, access to drinking water and removal of waste is generally problematic. These can cause health and welfare problems. Obtaining a source of water can be difficult because fresh water is characteristically frozen. Unfrozen taliks can be tapped, but generally cannot provide a continuous supply of water. The installation of a well for any groundwater source is also unlikely because it is extremely difficult to drill and maintain the well in a permafrost area. Once a source is found, the delivery of the resource can be problematic. Insulated and heated pipelines transport water to larger communities (O'Brien and Whyman, 1977; Ritter et al., 2011).

Waste removal is another issue in this region. Utilidors are insulated above-ground storage containers holding water, waste water, and electrical cables; serving as a protector from the harsh environment (Fig. 13.32). These approaches provide a way to store the waste, but the more pressing issue is the final disposal. Sewage lagoons are commonly used, as well as basic treatment, before disposal into nearby waterways (Alter, 1969). Tsigonis (2002) patented a sewage treatment system for cold climates above or within frozen ground that keeps permafrost intact. This energy-efficient, aerobic system involves constant air circulation to prevent freezing while keeping the sewage at the correct temperature for maximum aerobic activity.

13.21.4 Water Resources

Periglacial hydrology must consider the presence of surface and groundwater, and consider that although they interact, they also affect the landscape in

FIGURE 13.32 Utilidor system in Inuvik, Northwest Territory, Canada. Also notice the building is not on the ground but elevated on piers. *(Photograph by J. Vitek (1983).)*

different ways. Although the periglacial environment consists of mainly frozen liquids on the surface, unfrozen water does flow for a few months of the year. Hydraulic conductivities are low in these regions because frozen or thawed saturated soils prevail and limit infiltration. As a result, overland flow dominates after precipitation. Nival fluvial systems typically dominate, but climate models are predicting increases in unfrozen precipitation leading to more pluvial systems (Dugan et al., 2009, 2012). A shift toward this precipitation-dominated system would lengthen seasonal flows, increase flooding potential, and increase sediment yield (Ritter et al., 2011).

Although much groundwater is frozen as permafrost in the periglacial environment, a warming climate is leading to permafrost degradation and an increase in groundwater. This has implications throughout the landscape, including increased thermokarst and other landforms dependent on hydraulic pressures, increased stream discharges and base flows, and increased solute leaching (Okkonen and Kløve, 2011). Across all disciplines, groundwater and surface water interactions are minimally documented. A special need exists in the periglacial environment to link the two because both acting together have major implications on the degradation of permafrost (Woo et al., 2008; Lyon and Destouni, 2010; Lindborg and Bosson, 2013).

13.21.5 Mining

Mining operations in periglacial environments can have huge impacts on the surface as well as underlying permafrost. Surface mining results in alterations of mountain slopes whose specific shapes and aspects facilitate the accumulation of snow and ice, and the maintenance of existing thermal conditions to preserve permafrost. We are not against mining, but think that Critical Zone researchers need to be aware of some of the negative aspects that are associated with large, mining operations. These aspects of concern include: (1) disturbance of the delicate steady-state creep of the rock-ice mixture, which may lead to the collapse of the structure and ultimately the destruction of landforms in periglacial environments; (2) explosions can impact and cause permafrost and other ice structures to collapse; and (3) transporting the materials needs roads across the permafrost and results in increased sediment transport and meltwater discharge. All this can lead to destruction of water storage capacity in the area (Fig. 13.33).

Open-pit mining can produce residues, dust, and overburden rock-waste, which can be deposited on the surface of permafrost and lead to a quickening of the melting of the permafrost. The dust and residues can alter the albedo of the surface, which can change the rates of absorption of energy. This can lead to accelerated melting of permafrost. Contaminates from mining can also result in deposition of acidic chemicals and heavy metals on the surface and in drainage systems. The contaminated waters can migrate to depths in the permafrost and cause contamination and enhanced melting as a result of these geochemical processes (Taillant, 2012; Brenning, 2008).

FIGURE 13.33 **Red Dog Mine, located in the DeLong Mountains in the remote western Brooks Range ~140 km north of Kotzebue, is the largest source for zinc and a significant source of lead in the world.** *(Photograph by J. Giardino (2007).)*

Mining also plays an important role in watersheds. Many watersheds underlain with permafrost operate as water-basin regulators. Thus, these watersheds play a hydrologic and strategic role in the sustainability of ecosystems in these areas.

Rock glaciers have also served as borrow sources for road construction. For example, attempts to use rock glaciers on Mount Mestas, Colorado, for borrow sources for the construction of La Veta Pass failed when excavation into the rock glaciers encountered an ice-cemented interior. The initial excavations resulted in melting of the ice matrix to a considerable depth (Johnson, 1967; Giardino, 1979).

13.21.6 Recreation

A changing climate will impact recreation in permafrost regions. Historically permanent glaciers on which skiing occurs year-round are shrinking; reducing recreation time (Scott and McBoyle, 2007; Scott et al., 2008; Moen and Fredman, 2007). Ice climbing, trekking, and general cold climate tourism are also on the decline. Hikers and climbers are experiencing route closures because ice is lacking or conditions are too hazardous (Chiarle and Mortara, 2008). As periglacial areas change, tourism may be reduced at some tourist attractions. Another example from March, 2015, involves the annual Iditarod dog sled race. For only the second time in the history of the race, the starting line was moved

400 km north to Fairbanks from Anchorage, Alaska because of a lack of snow; the first occurrence being in 2003 (Imam, 2015).

Apart from, direct lack of snow and earlier melting are the dangers that premature melting may cause. Premature melting offsets balance throughout the Critical Zone, triggering changes across all spatial scales (Anderson et al., 2012). Kääb et al. (2005) identified many potential hazards in permafrost-covered regions resulting from a warming climate. The authors described failures of ice dams or overtopping, breaching of moraine dams, and increased surface runoff as major causes of flooding in these regions. The authors also noted that changes in permafrost will likely increase rock falls, avalanches, debris flows, and landslides. An increase in local hazards has the potential to initiate a chain reaction that could result in more immediate risks (Kääb, 2008). Furthermore, Noetzli et al. (2003) worked to identify causes of rock falls in high alpine regions. Their research identified the influence permafrost has on the amount and timing of rock falls to better understand and possibly predict future occurrences. Rock falls and avalanches are commonly a major concern regarding safety, especially for ski resorts, as they try to ensure safety for patrons. Predictive models could significantly reduce risk for resorts because levels of risk are increasing as climate continues to change.

13.21.7 A Changing Climate

One of the largest contributions that studying change in periglacial regions in the Critical Zone perspective can make is a system's view of the impacts human interventions are having on local parameters across periglacial regions. These interventions, as well as global, anthropogenic influences are drastically affecting periglacial regions. These regions are extremely sensitive to changes because of dependence on maintaining a thermal equilibrium. Periglacial regions span elevations from sea level to over ~8 km. A range this large requires a system to adapt, regardless of changing climate.

Researchers from the Boulder Creek CZO are applying the Critical Zone perspective to investigate changing permafrost in the Front Range of Colorado (Leopold et al., 2010, 2014). High-elevation environments, specifically, are susceptible to greater amounts of erosion and weathering as rates of river erosion increase to eventually reach new base levels (Dethier and Lazarus, 2006). High-elevation periglacial environments are, thus, predetermined to undergo changes to reach equilibrium. Anderson et al. (2012) argued that erosion and weathering are not controls of shaping the environment alone, but factors that play a part in evolving it as a rapid step function. They argue that the Critical Zone is the connecting factor that allows the landscape to evolve, simultaneously not gradually, and not only locally but also globally. This notion is directly applied to the periglacial landscape, as adjustments constantly occur at various spatial scales, and often have chain-reaction effects (Fig. 13.34). Because of these aspects, periglacial processes and environments need to be included within the definition of the Critical Zone.

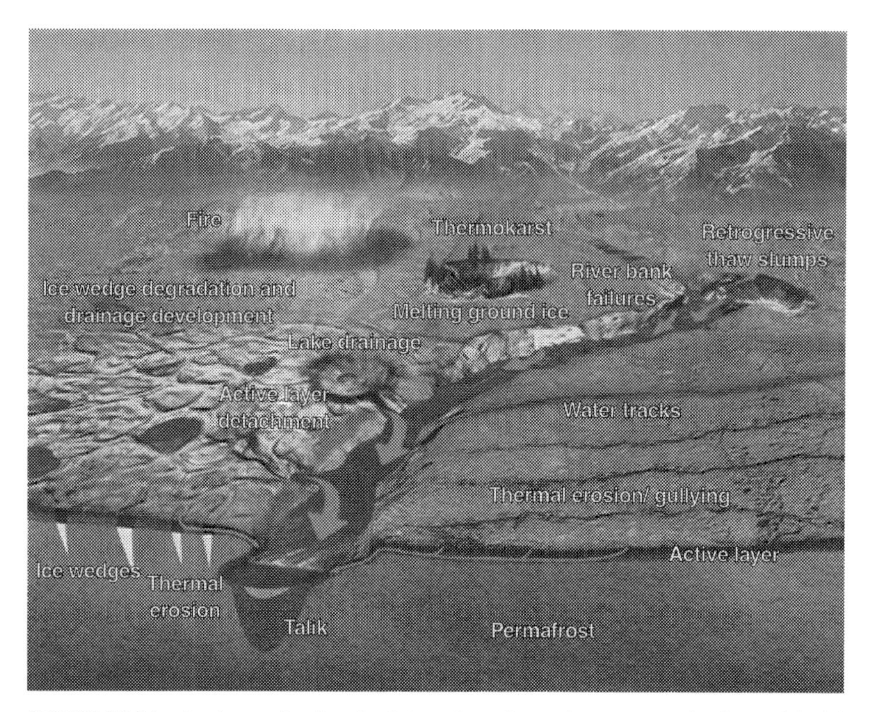

FIGURE 13.34 A scheme showing the interaction of complex processes in the periglacial environment. Thick arrows represent the flux of sediment and thin arrows denote water fluxes. *(Modified from Rowland et al. (2010).)*

Part of the Critical Zone is the atmospheric interface, which is the driver of the changing climate as global temperatures rise. The periglacial region is arguably most susceptible to these changes because the landscape is moving toward equilibrium within the system and the surrounding environments, while also changing to maintain its thermal equilibrium. Jorgenson et al. (2010) explained that the resilience of permafrost to climate change depends on various interactions within the Critical Zone across periglacial regions.

Mean annual air temperatures are a significant driver of permafrost resilience, and minor increases will severely alter the periglacial landscape as permafrost continues to degrade. As permafrost is the most prominent feature in the periglacial environment, any change will begin a chain reaction of feedback loops that will alter a vast portion of the landscape.

A known result of thawing permafrost is the release of a significant amount of carbon into the atmosphere. The major unknown variable is the timing and how much will be released. Schuur and Abbott (2011) gathered data from numerous scientists researching various aspects of permafrost to determine the most probable rates of thawing and how much carbon would be released. Results are staggering; predicting a possible release of 232–380 billion tons of

FIGURE 13.35 A thermokarst slump as a result of local permafrost thaw, Denali National Park. *(Courtesy of NPS, Denali (2013).)*

carbon by the year 2100 (Schuur and Abbott (2011)). Although fossil-fuel emission is the immediate driver of climate change, if thawing rates of permafrost increase, emissions from thawing soils may surpass any anthropogenic emissions.

Deforestation and forest fires are another source of carbon emissions. Although significantly less activity occurs in periglacial regions compared to other parts of the globe, it should be noted that deforestation in periglacial regions is more devastating because it disrupts thermal equilibrium within the Critical Zone. Vegetation acts as an insulator for permafrost-covered areas. If this "blanket" is removed without any insulation replacement, this Critical Zone interface must seek a new balance. As global temperatures increase, this balance is often skewed toward ground temperatures warming from higher atmospheric temperatures.

Permafrost thaw also results in changes in thermokarst. As permafrost melts, differential settling and subsidence occur across the landscape (Fig. 13.35). In some areas, thermokarst lakes are becoming abundant as a result of thawing (Yoshikawa and Hinzman, 2003). In other areas, these lakes are disappearing as the water in the lakes is draining into the recently thawed subsurface and contributing to groundwater (Smith et al., 2005; Kirpotin et al., 2008). Lake distributions vary depending on hydrological, topographical, and historical landscape conditions (Grosse et al., 2008).

Shifts in temperatures have also affected the oceans adjacent to periglacial regions. This change has led to severe increases in coastal erosion. Jones et al. (2009) looked at a segment of Alaskan coastline along the Beaufort Sea. The

FIGURE 13.36 Coastal erosion on Alaska's Arctic Coast. *(Courtesy of USGS Alaska Division (2014).)*

authors found by analyzing historical aerial photographs that erosion rates have doubled between 1955 and 2007. They explained the cause as a combined result of increased sea temperatures, sea level rise, and increased waves linked to storm activity (Fig. 13.36).

Climate change may, in turn, provide enhanced potential for agriculture in northern lands. This climatic potential is offset somewhat, however, by limitations imposed by the presence of ice-rich permafrost. Agriculture in subpolar lands typically involves forest clearing as an initial step. In ice-rich sediments, the removal of the insulating vegetation can increase permafrost thaw; resulting in severe subsidence and thermokarst features, making agriculture difficult to manage and maintain. In Fairbanks, Alaska, for example, a field was cleared in 1908 at the University of Alaska Experimental Farm. Within 20 years, differential subsidence created mounds 0.9–2.5-m high and (20–50 ft.) 6–15 m in diameter (Péwé, 1983).

Every change in equilibrium has an effect on the entire system, as the Critical Zone is interconnected with itself across the landscape. All Critical Zone processes result in positive and negative feedback loops. For example, permafrost thawing can lead to a shift in surface water discharge patterns along with increased thermal erosion. This change can in turn result in increased erosion along the banks and possibly adjacent hillslopes. A river's path may meander away from a once-vegetated area, causing vegetation to decay and desiccate;

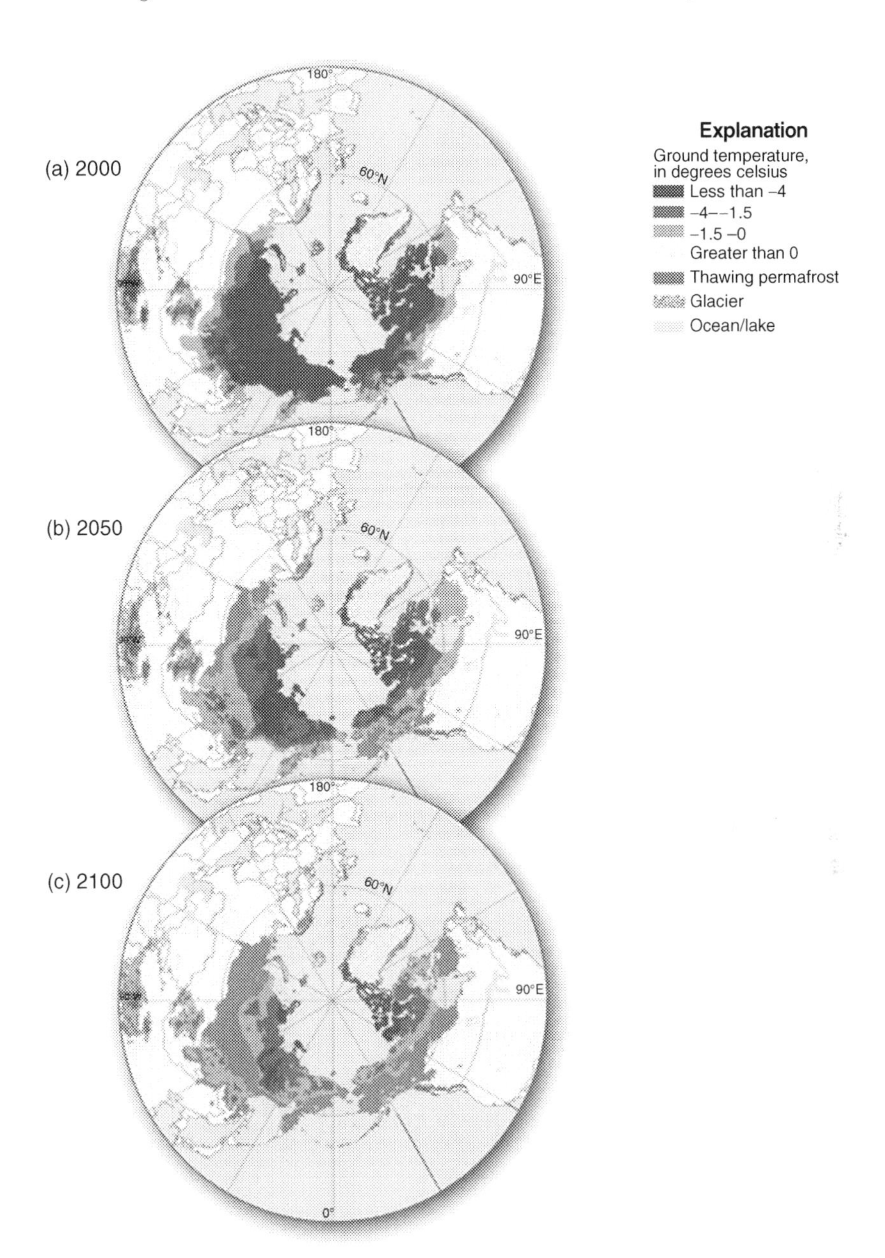

FIGURE 13.37 The circumpolar permafrost temperatures are modeled for mean annual temperature at the permafrost surface for: (a) 2000; (b) 2050; and (c) 2100. *(The map is from Heginbottom et al., 2012.)*

possibly leading to a wildfire. This wildfire has the potential to eliminate more vegetation; removing insulation for the remaining permafrost, ultimately repeating the cycle (Rowland et al., 2010).

Climatic, geologic, and anthropogenic forces all influence the equilibrium. The natural balance between climatic and geologic forces is being disrupted by anthropogenic interventions. The International Panel on Climate Change (IPCC, 2014) stated that global warming is caused by anthropogenic activity, and indicated that the Arctic Region is the most vulnerable area of Earth to the effects of global heating. Heginbottom et al. (2012) indicated that at present, there is no model representing all parameters affecting the distribution and degradation of permafrost in a changing climate. Fig. 13.37 displays a model used from the Permafrost Laboratory, Geophysical Institute, University of Alaska – Fairbanks assuming certain specific climatic and landscape conditions. The model attempts to quantify permafrost degradation in response to climate change to the year 2100 (Romanovsky et al., 2007). This model indicates rapid permafrost degradation as a result of the three forces acting together, accelerating change to the point that the periglacial environment cannot keep pace. It is, thus, essential to understand Critical Zone processes within the periglacial environment, and protect this delicate environment while it is still in existence.

13.22 SUMMARY AND CONCLUSIONS

This chapter provides the background discussion on the various processes that are operational in periglacial environments and the resulting landforms within the Critical Zone. The description of each process and landform provides a quick overview of the topic. Further investigation should be undertaken using the associated references to obtain an in-depth understanding of periglacial processes and landforms. Although brief, this chapter provides the framework to build upon the various interactions within the Critical Zone in the periglacial environment, as well as across all geomorphic landscapes. This framework provides a foundation for further interdisciplinary research undertaken throughout the Critical Zone, highlighting this delicate environment.

Although the human population of the periglacial realm of the Critical Zone is small in numbers, change in the processes and forms throughout the mountainous and polar regions with periglacial processes have a major impact on Earth. The delicate balance within this Critical Zone for a thermal equilibrium at the surface, if stability is not maintained, will activate processes and change landforms that can and will impact human use and their presence of these regions. Resource extraction, be it petroleum, gold, zinc, or other resources, must strive to maintain the natural thermal environment. Failure to do so will simply add to the costs of the resources obtained and result in severe degradation of the fragile periglacial environment. More research is needed on a variety of topics within the periglacial realm of the Critical Zone to limit the mistakes of human activity that will have broader system implications.

REFERENCES

Abellán, A., Calvet, J., Vilaplana, J.M., Blanchard, J., 2010. Detection and spatial prediction of rockfalls by means of terrestrial laser scanner monitoring. Geomorphology 119 (3), 162–171.

Ackert, Jr., R.P., 1998. A rock glacier/debris-covered glacier system at Galena Creek, Absaroka Mountains, Wyoming. Geogr. Ann. A Phys. Geogr. 80 (3–4), 267–276.

Alter, A.J., 1969. Sewerage and Sewage Disposal in Cold Regions. Corps of Engineers, U.S. Army Cold Regions Research and Engineering Laboratory, Hanover, NH, p. 107.

Anderson, S.P., Anderson, R.S., Tucker, G.E., 2012. Landscape scale linkages in Critical Zone evolution. C. R. Geosci. 344 (11), 586–596.

Barsch, D., 1977. Nature and importance of mass-wasting by rock glaciers in alpine permafrost environments. Earth Surf. Process. 2 (2–3), 231–245.

Barsch, D., 1987. The problem of the ice-cored rock glacier. Rock GlaciersAllen and Unwin, London, pp. 45–53.

Barsch, D., 1993. Periglacial geomorphology in the 21st century. Geomorphology 7 (1), 141–163.

Barsch, D., 1996. Rock Glaciers: Indicators for the Present and Former Geocryology in High Mountain Environments. Springer-Verlag, Berlin, p. 331.

Barsch, D., Caine, N., 1984. The nature of mountain geomorphology. Mountain Res. Dev. 4(4), 287–298.

Barsch, D., King, L., 1975. An attempt to date fossil rock glaciers in Grison, Swiss Alps. Quaestiones Geographicae 2, 5–13.

Barsch, D., Fierz, H., Haeberli, W., 1979. Shallow core drilling and bore-hole measurements in the permafrost of an active rock glacier near the Grubengletscher, Wallis, Swiss Alps. Arctic Alpine Res. 11(2), 215–228.

Bartsch, A., Gude, M., Gurney, S.D., 2009. Quantifying sediment transport processes in periglacial mountain environments at a catchment scale using geomorphic process units. Geogr. Ann. A Phys. Geogr. 91 (1), 1–9.

Benedict, J.B., 1973. Origin of rock glaciers. J. Glaciol. 12, 520–522.

Beniston, M., 2000. Environmental change in mountains and uplands. Arnold/Hodder and Stoughton/Chapman and Hall Publishers; Oxford University Press, London, UK; New York, USA, p. 172.

Berthling, I., Schomacker, A., Benediktsson, Í.Ö., 2013. The glacial and periglacial research frontier: where from here? In: Giardino, J.R., Harbor, J. (Eds.), Treatise on Geomorphology, vol. 8; Glacial and Periglacial Geomorphology. Academic Press, San Diego, CA, pp. 479–499.

Bouchard, F., Francus, P., Pienitz, R., Laurion, I., Feyte, S., 2014. Subarctic thermokarst ponds: investigating recent landscape evolution and sediment dynamics in thawed permafrost of northern Québec (Canada). Arctic Antarctic Alpine Res. 46 (1), 251–271.

Brenning, A., 2008. The impact of mining on rock glaciers and glaciers: examples from Central Chile. In: Orlove, B.S., Wiegandt, E., Luckman, B. (Eds.), Darkening Peaks: Glacial Retreat, Science and Society. University of California Press, Berkeley, pp. 196–205, Chapter 14.

Briggs, M.A., Walvoord, M.A., McKenzie, J.M., Voss, C.I., Day-Lewis, F.D., Lane, J.W., 2014. New permafrost is forming around shrinking Arctic lakes, but will it last? Geophys. Res. Lett. 41 (5), 1585–1592.

Brown, R.J.E., 1967. Permafrost investigation in British Columbia and Yukon Territory. National Research Council of Canada, Division of Building Research, Ottawa. Technical Paper No. 253. p. 55.

Burger, K.C., Degenhardt, J.J., Giardino, J.R., 1999. Engineering geomorphology of rock glaciers. Geomorphology 31 (1), 93–132.

Burn, C.R., 2000. The thermal regime of a retrogressive thaw slump near Mayo, Yukon Territory. Canadian J. Earth Sci. 37 (7), 967–981.

Burn, C.R., Smith, M.W., 1990. Development of thermokarst lakes during the Holocene at sites near Mayo, Yukon Territory. Permafrost Periglacial Process. 1 (2), 161–175.

Calkin, P.E., Haworth, L.A., Ellis, J.M., 1987. Rock Glaciers of Central Brooks Range, Alaska, USA. Rock Glaciers. Allen and Unwin, London, pp. 65–82.

Capps, S.R., 1910. Rock glaciers in Alaska. J. Geol., 18(4), 359–375.

Carter, L.D., Galloway, J.P., 1981. Earth flows along Henry Creek, northern Alaska. Arctic 34 (4), 325–328.

Chiarle, M., Mortara, G., 2008. Geomorphological impact of climate change on alpine glacial and periglacial areas. Examples of processes and description of research needs. Interpraevent. 2, pp. 111–122.

Clow, D.W., 2010. Changes in the timing of snowmelt and streamflow in Colorado: a response to recent warming. J. Climate 23, 2293–2306.

Cogley, J.G., McCann, S.B., 1976. An exceptional storm and its effects in the Canadian High Arctic. Arctic Alpine Res., 8(1), 105–110.

Costard, F., Gautier, E., 2007. The Lena River: hydromorphodynamic feature in a deep permafrost zone. In: Gupta, A. (Ed.), Large Rivers: Geomorphology and Management. John Wiley & Sons Ltd, England, p. 227.

Dethier, D.P., Lazarus, E.D., 2006. Geomorphic inferences from regolith thickness, chemical denudation and CRN erosion rates near the glacial limit, Boulder Creek catchment and vicinity, Colorado. Geomorphology 75 (3), 384–399.

Dimov, L.A., Dimov, I.L., 2014. Selection of placement method for oil pipelines constructed on permafrost. Soil Mech. Found. Eng. 51 (5), 258–262.

Dobinski, W., 2011. Permafrost. Earth Sci. Rev. 108 (3–4), 158–169.

Dugan, H.A., Lamoureux, S.F., Lafrenière, M.J., Lewis, T., 2009. Hydrological and sediment yield response to summer rainfall in a small high Arctic watershed. Hydrol. Process. 23 (10), 1514–1526.

Dugan, H.A., Lamoureux, S.F., Lewis, T., Lafrenière, M.J., 2012. The impact of permafrost disturbances and sediment loading on the limnological characteristics of two high Arctic lakes. Permafrost Periglacial Process. 23 (2), 119–126.

Etzelmüller, B., 2013. Recent advances in mountain permafrost research. Permafrost Periglacial Process. 24 (2), 99–107.

Ferrians, Jr., O.J., Kachadoorian, R., Greene, G.W., 1969. Permafrost and related engineering problems in Alaska. US Geological Survey Professional Paper, 678. p. 37.

Feuillet, T., 2011. Statistical analyses of active patterned ground occurrence in the Taillon Massif (Pyrénées, France/Spain). Permafrost Periglacial Process. 22 (3), 228–238.

Francou, B., Fabre, D., Pouyaud, B., Jomelli, V., Arnaud, Y., 1999. Symptoms of degradation in a tropical rock glacier, Bolivian Andes. Permafrost Periglacial Process. 10 (1), 91–100.

Fraser, T.A., Burn, C.R., 1997. On the nature and origin of "muck" deposits in the Klondike area, Yukon Territory. Can. J. Earth Sci. 34 (10), 1333–1344.

French, H.M., 1976. The Periglacial Environment. Longman, London, p. 341.

French, H.M., 2013. The Periglacial Environment. John Wiley & Sons, p. 478.

French, H., Harbor, J., 2013. The development and history of glacial and periglacial geomorphology. In: Giardino, J.R., Harbor, J. (Eds.), Treatise on Geomorphology, vol. 8; Glacial and Periglacial Geomorphology. Academic Press, San Diego, CA, pp. 1–18.

Frost, G.V., Epstein, H.E., Walker, D.A., Matyshak, G., Ermokhina, K., 2013. Patterned-ground facilitates shrub expansion in Low Arctic tundra. Environ. Res. Lett. 8 (1), 1–2.

Gardner, J.S., 1983. Accretion rates on some debris slopes in the Mt. Rae area, Canadian Rocky Mountains. Earth Surf. Process. Land. 8 (4), 347–355.

Giardino, J.R., 1979. Rock glacier mechanics and chronologies: Mount Mestas, Colorado. University of Nebraska, Nebraska, p. 220.

Giardino, J.R., Vick, S.G., 1987. Geologic engineering aspects of rock glaciers. In: Giardino, J.R., Shroder, J.F., Vitek, J.D. (Eds.), Rock Glaciers. Allen and Unwin, Boston, pp. 265–288.

Giardino, J.R., Vitek, J.D., 1988. The significance of rock glaciers in the glacial-periglacial landscape continuum. J. Quaternary Sci. 3 (1), 97–103.

Giardino, J.R., Shroder, J.F., Vitek, J.D. (Eds.), 1987. Rock Glaciers. Allen & Unwin, London, p. 355.

Giardino, J.R., Regmi, N.R., Vitek, J.D., 2011. Rock glaciers. In: Singh, V.P., Singh, P., Haritashya, U.K. (Eds.), Encyclopedia of Ice Snow and Glaciers. Springer, pp. 435–447.

Grosse, G., Romanovsky, V., Walter, K., Morgenstern, A., Lantuit, H., Zimov, S., 2008. Distribution of thermokarst lakes and ponds at three yedoma sites in Siberia. In: Kane, D.L., Hinkel, K.M. (Eds.), Proceedings of the Ninth International Conference on Permafrost, vol. 1, University of Alaska Fairbanks, Institute of Northern Engineering, Fairbanks, AK, June 29–July 3, 2008. pp. 551–556.

Haeberli, W., 1985. Creep of mountain permafrost: internal structure and flow of alpine rock glaciers. Mitteilungen der Versuchsanstalt fur Wasserbau, Hydrologie und Glaziologie an der ETH Zurich 77, 5–142.

Haeberli, W., 2000. Modern research perspectives relating to permafrost creep and rock glaciers: a discussion. Permafrost Periglacial Process. 11 (4), 290–293.

Haeberli, W., 2005. Investigating glacier-permafrost relationships in high-mountain areas: historical background, selected examples and research needs. In: Harris, C., Murton, J.B. (Eds.), Cryospheric Systems: Glaciers and Permafrost. Geological Society, London, Special Publications, 242, pp. 29–37.

Haeberli, W., Frauenfelder, R., Hoelzle, M., Maisch, M., 1999. On rates and acceleration trends of global glacier mass changes. Geogr. Ann. A Phys. Geogr. 81 (4), 585–591.

Hallet, B., Waddington, E.D., 1991. Buoyancy forces induced by freeze-thaw in the active layer: implications for diapirism and soil circulation. Periglacial Geomorphology. John Wiley and Sons, New York, NY, 251–279.

Harris, C., Lewkowicz, A.G., 1993. Form and internal structure of active-layer detachment slides, Fosheim Peninsula, Ellesmere Island, Northwest Territories, Canada. Canadian J. Earth Sci. 30 (8), 1708–1714.

Harry, D.G., MacInnes, K.L., 1988. The effect of forest fires on permafrost terrain stability, Little Chicago-Travaillant Lake area, Mackenzie Valley, NWT. Current Research, Part D, 91–94.

Heginbottom, J.L., Brown, J., Humlum, O., Svensson, H., Williams., 2012. Permafrost and periglacial environments. In: Williams, Jr., R.S.,Ferrigno, J.G., (Ed.). State of the Earth's Cryosphere at the Beginning of the 21st Century – Glaciers, Global Snow Cover, Floating Ice, and Permafrost and Periglacial Environments: US Geological Survey Professional Paper 1386-A, pp. 429–489.

Hjort, J., Ujanen, J., Parviainen, M., Tolgensbakk, J., Etzelmüller, B., 2014. Transferability of geomorphological distribution models: evaluation using solifluction features in subarctic and Arctic regions. Geomorphology 204, 165–176.

Humlum, O., 2000. The geomorphic significance of rock glaciers: estimates of rock glacier debris volumes and headwall recession rates in West Greenland. Geomorphology 35 (1), 41–67.

Ikeda, A., Matsuoka, N., 2002. Degradation of talus-derived rock glaciers in the Upper Engadin, Swiss Alps. Permafrost Periglacial Process. 13 (2), 145–161.

Imam, J. 2015: Race to Move North. https://www.cnn.com (March 9th, 2015).

International Permafrost Association. 2014. What is permafrost? Available from: http://ipa.arcticportal.org/ (accessed 20.02.2015).

IPCC, 2014. Climate Change 2014: Impacts, Adaptation, and Vulnerability. Part A: Global and Sectoral Aspects. Contribution of Working Group II to the Fifth Assessment Report of the Intergovernmental Panel on Climate Change. Cambridge University Press, UK, p. 1132.

Ives, R.L., 1940. Rock glaciers in the Colorado front range. Geol. Soc. Am. Bull. 51 (9), 1271–1294.

Janke, J.R., 2005. Long-term flow measurements (1961–2002) of the Arapaho, Taylor, and Fair rock glaciers, Front Range, Colorado. Phys. Geog. 26 (4), 313–336.

Janke, J.R., Williams, M.W., Evans, A., 2012. A comparison of permafrost prediction models along a section of Trail Ridge Road, Rocky Mountain National Park, Colorado, USA. Geomorphology 138 (1), 111–120.

Johnson, R.B., 1967. Rock Streams on Mount mestas, Sangre de Cristo Mountains, Southern Colorado. USGS Professional Paper 575. US Government Printing Office. 217–220.

Jomelli, V., Francou, B., 2000. Comparing the characteristics of rockfall talus and snow avalanche landforms in an Alpine environment using a new methodological approach: Massif des Ecrins, French Alps. Geomorphology 35 (3), 181–192.

Jones, B.M., Arp, C.D., Jorgenson, M.T., Hinkel, K.M., Schmutz, J.A., Flint, P.L., 2009. Increase in the rate and uniformity of coastline erosion in Arctic Alaska. Geophys. Res. Lett. 36 (3), 5.

Jones, B.M., Grosse, G., Hinkel, K.M., Arp, C.D., Walker, S., Beck, R.A., Galloway, J.P., 2012. Assessment of pingo distribution and morphometry using an IfSAR derived digital surface model, western Arctic Coastal Plain, Northern Alaska. Geomorphology 138 (1), 1–14.

Jorgenson, M.T., 2013. Themokarst Terrains. In: Giardino, R., Harbor, J. (Eds.), Treatise on Geomorphology, vol. 8; Glacial and Periglacial Geomorphology. Academic Press, San Diego, CA, p. 314.

Jorgenson, M.T., Shur, Y.L., Osterkamp, T.E., 2008. Thermokarst in Alaska. In: Kane, D.L., Hinkel, K.M. (Eds.), Proceedings of the Ninth International Conference on Permafrost, 29 June–3 July 2008, Fairbanks, Alaska. Institute of Northern Engineering, University of Alaska Fairbanks, Fairbanks, Alaska, pp. 869–876.

Jorgenson, M.T., Romanovsky, V., Harden, J., Shur, Y., O'Donnell, J., Schuur, E.A., Kanevskiy, M., Marchenko, S., 2010. Resilience and vulnerability of permafrost to climate change: resilience and vulnerability in response to climate warming. Can. J. Forest Res. 40 (7), 1219–1236.

Kääb, A., 2008. Remote sensing of permafrost-related problems and hazards. Permafrost Periglacial Process. 19 (2), 107–136.

Kääb, A., Weber, M., 2004. Development of transverse ridges on rock glaciers: field measurements and laboratory experiments. Permafrost Periglacial Process. 15 (4), 379–391.

Kääb, A., Huggel, C., Fischer, L., Guex, S., Paul, F., Roer, I., Weidmann, Y., 2005. Remote sensing of glacier- and permafrost-related hazards in high mountains: an overview. Nat. Hazards Earth Syst. Sci. 5 (4), 527–554.

Kirpotin, S., Polishchuk, Y., Zakharova, E., Shirokova, L., Pokrovsky, O., Kolmakova, M., Dupre, B., 2008. One of the possible mechanisms of thermokarst lakes drainage in West-Siberian North. Int. J. Environ. Stud. 65 (5), 631–635.

Kokelj, S.V., Lantz, T.C., Wolfe, S.A., Kanigan, J.C., Morse, P.D., Coutts, R., Molin-Giraldo, N., Burn, C.R., 2014. Distribution and activity of ice wedges across the forest-tundra transition, western Arctic Canada. J. Geophys. Res. Earth Surf. 119 (9), 2032–2047.

Konrad, S.K., Humphrey, N.F., Steig, E.J., Clark, D.H., Potter, N., Pfeffer, W.T., 1999. Rock glacier dynamics and paleoclimatic implications. Geology 27 (12), 1131–1134.

Lachenbruch, A.H., 1962. Mechanics of thermal contraction cracks and ice-wedge polygons in permafrost. Geol. Soc. Am. Special Pap. 70, 1–66.

Larocque, S.J., Hétu, B., Filion, L., 2001. Geomorphic and dendroecological impacts of slush-flows in central Gaspé Peninsula (Québec, Canada). Geogr. Ann. A Phys. Geogr. 83 (4), 191–201.

Leffingwell, E.de K., 1919. The Canning River region, northern Alaska. US Government Printing Office, vol. 109, p. 251.

Leopold, M., Voelkel, J., Dethier, D., Williams, M., Caine, N., 2010. Mountain permafrost – a valid archive to study climate change? Examples from the Rocky Mountains Front Range of Colorado, USA. Nova Acta Leopoldina 112 (384), 281–289.

Leopold, M., Völkel, J., Dethier, D.P., Williams, M.W., 2014. Changing mountain permafrost from the 1970s to today – comparing two examples from Niwot Ridge, Colorado Front Range, USA. Z. Geomorph. Suppl. 58 (1), 137–157.

Lewkowicz, A.G., 2007. Dynamics of active-layer detachment failures, Fosheim Peninsula, Ellesmere Island, Nunavut, Canada. Permafrost Periglacial Process. 18 (1), 89–103.

Lewkowicz, A.G., Harris, C., 2005. Frequency and magnitude of active-layer detachment failures in discontinuous and continuous permafrost, northern Canada. Permafrost Periglacial Process. 16 (1), 115–130.

Lindborg, T., Bosson, E., 2013. Hydrology and ecosystems in a periglacial catchment – mass balance calculations on landscape level. In: EGU General Assembly Conference Abstracts, vol. 15, p. 4172.

Lozinski, W.V., 1909. Über die mechanische Verwitterung der Sandsteine im gemässigten Klima. Bulletin International de L'Academie des Sciences de Cracovie class des Sciences Mathematique et Naturalles 1, 1–25.

Lozinski, W.V., 1912. Die periglaziale fazies der mechanischen verwitterung. Comptes Rendus, XI Congres Internationale Geologie, Stockholm 1910, 1039–1053.

Lunardini, V.J., 1981. Heat Transfer in Cold Climates. Van Nostrand Reinhold Company, New York, NY, p. 731.

Lyon, S.W., Destouni, G., 2010. Changes in catchment-scale recession flow properties in response to permafrost thawing in the Yukon river basin. Int. J. Climatol. 30 (14), 2138–2145.

Mackay, J.R., 1979. Pingos of the Tuktoyaktuk Peninsula area, Northwest Territories. Géographie Phys. Quaternaire 33 (1), 3–61.

Mackay, J.R., 1981. Active layer slope movement in a continuous permafrost environment, Garry Island, Northwest Territories, Canada. Can. J. Earth Sci. 10, 26–41.

Mackay, J.R., 1984. The frost heave of stones in the active layer above permafrost with downward and upward freezing. Arctic Alpine Res. 16, 439–446.

Mackay, J.R., 1998. Pingo growth and collapse, Tuktoyaktuk Peninsula area, western arctic coast, Canada: a long-term field study. Géographie Phys. Quaternaire 52 (3), 271–323.

Mackay, J.R., Terasmae, J., 1963. Pollen diagrams in the Mackenzie delta area, NWT. Arctic 16 (4), 228–238.

Marchenko, S., Etzelmüller, B., 2013. Permafrost: formation and distribution, thermal and mechanical properties. In: Giardino, J.R., Harbor, J. (Eds.), Treatise on Geomorphology, vol. 8; Glacial and Periglacial Geomorphology. Academic Press, San Diego, CA, p. 211.

Marr, J.W., 1969. Cyclical change in a patterned-ground ecosystem, Thule, Greenland. In: Péwé, T.L. (Ed.), The Periglacial Environment, Past and Present. McGill- Queen's University Press, Montreal, pp. 177–201.

Martin, H.E., Whalley, W.B., 1987. Rock glaciers. Part 1: rock glacier morphology: classification and distribution. Prog. Phys. Geogr. 11 (2), 260–282.

Matsuoka, N., 2001. Direct observation of frost wedging in alpine bedrock. Earth Surf. Process. Land. 26 (6), 601–614.

Matsuoka, N., Murton, J., 2008. Frost weathering: recent advances and future directions. Permafrost Periglacial Process. 19 (2), 195–210.

Matsuoka, N., Sakai, H., 1999. Rockfall activity from an alpine cliff during thawing periods. Geomorphology 28 (3), 309–328.

McClung, D., Schweizer, J., 1999. Skier triggering, snow temperatures and the stability index for dry-slab avalanche initiation. J. Glaciol. 45 (150), 190–200.

McFadden, T., 2000. Design manual for new foundations on permafrost. Permafrost Technology Foundation, p. 94.

McFadden, T. 2001. A design manual for stabilizing foundations on permafrost. Permafrost Technology Foundation, p. 181.

Mellor, M., 1970. Phase composition of pore water in cold rocks. US Army Corps of Engineers, CRREL, Res. Rep. 292, 59.

Michel, F.A., Van Everdingen, R.O., 1994. Changes in hydrogeologic regimes in permafrost regions due to climatic change. Permafrost Periglacial Process. 5 (3), 191–195.

Millar, S., 2013. Mass movement processes in the periglacial environment. In: Giardino, J.R., Harbor, J. (Eds.), Treatise on Geomorphology, vol. 8; Glacial and Periglacial Geomorphology. Academic Press, San Diego, CA, pp. 375–387.

Moen, J., Fredman, P., 2007. Effects of climate change on alpine skiing in Sweden. J. Sustain. Tourism 15 (4), 418–437.

Morris, J., Fresco, N., Krutikov, L., Timm, K., Winfee, R., Rice, B., Giddens, J., 2014. State of change: climate change in AK's national parks. National Park Service, University of Alaska, Fairbanks, SNAP Program, 57.

Morse, P.D., Burn, C.R., 2014. Perennial frost blisters of the outer Mackenzie Delta, western Arctic coast, Canada. Earth Surface Process. Land. 39 (2), 200–213.

Muller, S.W., 1943. Permafrost or permanently frozen ground and related engineering problems. Special Report, Strategic Engineering Study, Intelligence Branch Chief of Engineers, 62, p. 136. Second printing, 1945, p. 230. Reprinted in 1947, J.W. Edwards, Ann Arbor, Michigan, p. 231.

Müller, J., Gärtner-Roer, I., Kenner, R., Thee, P., Morche, D., 2014. Sediment storage and transfer on a periglacial mountain slope (Corvatsch, Switzerland). Geomorphology 218, 35–44.

Murton, J.B., Peterson, R., Ozouf, J.C., 2006. Bedrock fracture by ice segregation in cold regions. Science 314 (5802), 1127–1129.

Murton, J.B., 2013. Ice wedges and ice wedge casts. In: Elias, S.A. (Ed.), Permafrost and Periglacial Features. Elsevier, Amsterdam, pp. 436–451.

Nash, A., 2011. Permafrost: A building problem in Alaska. Cooperative Extension Service, University of Alaska, Fairbanks, p. 4.

Nelson, F.E., 2003. (Un)frozen in time. Science 299, 1673–1675.

Nelson, F.E., Anisimov, O.A., Shiklomanov, N.I., 2002. Climate change and hazard zonation in the circum-Arctic permafrost regions. Nat. Hazards 26 (3), 203–225.

Nicholson, F.H., 1976. Patterned ground formation and description as suggested by low arctic and subarctic examples. Arctic Alpine Res. 8, 329–342.

Nixon, J.F., 1991. Discrete ice lens theory for frost heave in soils. Can. Geotech. J. 28 (6), 843–859.

Noetzli, J., Hoelzle, M., Haeberli, W., 2003. Mountain permafrost and recent Alpine rock-fall events: a GIS-based approach to determine critical factors. In: Proceedings of the Eighth International Conference on Permafrost. vol. 2, pp. 827–832.

National Research Council, 2001. Basic Research Opportunities in Earth Science. National Academy Press, Washington DC, USA, p. 168.

NSF Advisory Committee for Environmental Research and Education (NSF AC-ERE), 2005. Complex Environmental Systems: Pathways to the Future. Arlington, VA, USA, p. 12.

O'Brien, E.T., Whyman, A., 1977. Insulated and heat traced polyethylene piping systems: a unique approach for remote cold regions. In: Utilities Delivery in Arctic Regions, Environmental Protection Service, Environment Canada. Ottawa, Ontario. Report No. EPS 3-WP-77-1. pp. 309–339.

Okkonen, J., Kløve, B., 2011. A sequential modelling approach to assess groundwater–surface water resources in a snow dominated region of Finland. J. Hydrol. 411 (1), 91–107.

Oswell, J.M., 2011. Pipelines in permafrost: geotechnical issues and lessons. Can. Geotech. J. 48 (9), 1412–1431.

Permafrost Subcommittee, Associate Committee on Geotechnical Research, 1988. Glossary of Permafrost and Related Ground-ice Terms, Ottawa, p. 111.

Perret, S., Stoffel, M., Kienholz, H., 2006. Spatial and temporal rockfall activity in a forest stand in the Swiss Prealps – a dendrogeomorphological case study. Geomorphology 74 (1), 219–231.

Peterson, R.A., Krantz, W.B., 2003. A mechanism for differential frost heave and its implications for patterned-ground formation. J. Glaciol. 49 (164), 69–80.

Peterson, R.A., Krantz, W.B., 2008. Differential frost heave model for patterned ground formation: corroboration with observations along a North American arctic transect. J. Geophys. Res. Biogeosci. 113 (G3), 1–3.

Péwé, T.L., 1975. Quaternary geology of Alaska. US Geological Survey professional paper number 835, p. 145.

Péwé, T.L., 1983. The periglacial environment in North America during Wisconsin time. Late-Quater. Environ. US 1, 157–189.

Pissart, A., Calmels, F., Wastiaux, C., 2011. The potential lateral growth of lithalsas. Quaternary Res. 75 (2), 371–377.

Potter, N., 1972. Ice-cored rock glacier, Galena Creek, northern Absaroka Mountains, Wyoming. Geol. Soc. Am. Bull. 83 (10), 3025–3058.

Price, L.W., 1972. The periglacial environment, permafrost, and man. Association of American Geographers. Commission on College Geography. Washington D.C., 88.

Rapp, A., 1957. Studies of debris cones in Lapland and Spitbergen. Z. Geomorph. 1, 179–200.

Rapp, A., 1960. Recent development of mountain slopes in Kärkevagge and surroundings, northern Scandinavia. Geogr. Ann. 42, 65–200.

Rempel, A.W., 2007. Formation of ice lenses and frost heave. J. Geophys. Res. Earth Surf. 112, F02S21.

Rempel, A.W., 2011. Microscopic and environmental controls on the spacing and thickness of segregated ice lenses. Quaternary Res. 75 (2), 316–324.

Ritter, D.F., Kochel, R.C., Miller, J.R., 2011. Periglacial processes and landforms, fifth ed. Process Geomorphology; Chapter 11Waveland Press, Illinois, 399–470.

Romanovsky, V.E., Gruber, S., Jin, H., Marchenko, S.S., Smith, S.L., Trombotto, D., Walter, K.M., 2007. Frozen ground, in Global outlook for ice and snow: Arendal, Norway, United Nations Environment Programme, pp. 181–200.

Rowland, J.C., Jones, C.E., Altmann, G., Bryan, R., Crosby, B.T., Geernaert, G.L., Hinzman, L.D., Kane, D.L., Lawrence, D.M., Mancino, A., Marsh, P., McNamara, J.P., Romanovsky, V.E., Toniolo, H., Travis, B.J., Trochim, E., Wilson, C.J., 2010. Arctic landscapes in transition: responses to thawing permafrost. Eos Trans. AGU 91 (26), 229–230.

Sass, O., 2005. Spatial patterns of rockfall intensity in the northern Alps. Z. Geomorph. 138, 51–65.

Schuur, E.A., Abbott, B., 2011. Climate change: high risk of permafrost thaw. Nature 480 (7375), 32–33.

Scott, D., McBoyle, G., 2007. Climate change adaptation in the ski industry. Mitig. Adapt. Strat. Global Change 12 (8), 1411–1431.

Scott, D., Dawson, J., Jones, B., 2008. Climate change vulnerability of the US Northeast winter recreation–tourism sector. Mitig. Adapt. Strat. Global Change 13 (5–6), 577–596.

Seppälä, M., 1998. New permafrost formed in peat hummocks (pounus), Finnish Lapland. Permafrost Periglacial Process. 9 (4), 367–373.

Seppälä, M., 2004. Wind as a geomorphic agent in cold climates. Cambridge University Press, Cambridge, UK, p. 358.

Shiklomanov, N.I., Nelson, F.E., 2013. Mass movement processes in the periglacial environment. In: Schroder, J., Giardino, R., Harbor, J. (Eds.), Treatise on Geomorphology, vol. 8; Glacial and Periglacial Geomorphology. Academic Press, San Diego, CA, pp. 375–387.

Shur, Y., Jorgenson, T., 1998. Cryostructure Development on the Floodplain of the Colville River Delta, Northern Alaska. In: Permafrost. Seventh International Conference, June 23–27, 1998, Proceedings, Yellowknife, Canada, Centre d'études Nordiques, Université Laval, Collection Nordicana, 57, pp. 993–999.

Smith, L.C., Sheng, Y., MacDonald, G.M., Hinzman, L.D., 2005. Disappearing arctic lakes. Science 308 (5727), 1429.

Taillant, J.D., 2012. The Periglacial Environment and Mining Sector in Argentina: The National Glacier Law and Frozen Grounds (English Translation). Of CEDHA's Glaciers and Mining Series, 86.

Troll, C., 1973. High mountain belts between the polar caps and the equator: their definition and lower limit. Arctic Alpine Res. 5, A19–A27.

Tsigonis, R.C., 2002. Apparatus and method for treating sewage in cold climates. Google Patents, 489.

Twichell, Heath, 1992. Northwest Epic: The building of the Alaska Highway. St. Martin's Press, New York, NY, p. 368.

Vincent, W.F., Lemay, M., Allard, M., Wolfe, B.B., 2013. Adapting to permafrost change: a science framework. Eos Trans. AGU 94 (42), 373–375.

Vinson, T.S., Lofgren, D., 2003. Denali park access road icing problems and mitigation options. In: Proceedings of the 8th International Conference on Permafrost. AA Balkema. pp. 331–336.

Vitek, J.D., Giardino, J., 1987. Rock glaciers: a review of the knowledge base. In: Giardino, J.R., Shroder, J.F., Vitek, J.D. (Eds.), Rock Glaciers. Allen and Unwin, Boston, pp. 1–26.

Vitek, J.D., Regmi, N.R., Humbolt, D., Giardino, J.R., 2008. Stone polygons in southern Colorado, USA: Observations of surficial activity – 1975–2004. In: Kane, D.L., Hinkel, K.M. (Eds.), Ninth International Permafrost Conference, vol. 2, pp. 1857–1862.

Voelkel, J., Huber, J., Leopold, M., 2011. Significance of slope sediments layering on physical characteristics and interflow within the Critical Zone: examples from the Colorado Front Range, USA. Appl. Geochem. 26 (Suppl.), S143–S145.

Wahrhaftig, C., Cox, A., 1959. Rock glaciers in the Alaska Range. Geol. Soc. Am. Bull. 70 (4), 383–436.

Walder, J., Hallet, B., 1985. A theoretical model of the fracture of rock during freezing. Geol. Soc. Am. Bull. 96 (3), 336–346.

Walker, H.J., 1986. Periglacial environments. In: Fookes, P.G., Vaughan, P.R. (Eds.), A Handbook of Engineering Geomorphology. Chapman and Hall, New York, NY, pp. 82–96.

Walker, H.J., 1993. Geomorphology: the research frontier and beyond – introduction. Geomorphology 7 (1), 1–7.

Walker, D.A., Everett, K.R., 1987. Road dust and its environmental impact on Alaskan taiga and tundra. Arctic Alpine Res. 19, 479–489.

Warburton, J., 2013. Patterned ground and polygons. In: Giardino, J.R., Harbor, J. (Eds.), Treatise on Geomorphology, vol. 8; Glacial and Periglacial Geomorphology. Academic Press, San Diego, CA, pp. 298–303.

Washburn, A.L., 1956. Classification of patterned ground and review of suggested origins. Geol. Soc. Am. Bull. 67 (7), 823–866.

Washburn, A.L., 1973. Periglacial processes and environments. Arnold. London. pp. 4-7.

Washburn, A.L., 1979. Geocryology. Edward Arnold, London, 117(5), pp. 503–504.

Washburn, A.L., Goldthwait, R.P., 1958. Slushflows. Bull. Geol. Soc. Am. 69, 1657–1658.

Watanabe, T., Matsuoka, N., Christiansen, H.H., 2013. Ice- and soil-wedge dynamics in the Kapp Linné Area, Svalbard, investigated by two-and three-dimensional GPR and ground thermal and acceleration regimes. Permafrost Periglacial Process. 24 (1), 39–55.

White, S.E., 1981. Alpine mass movement forms (noncatastrophic): classification, description, and significance. Arctic Alpine Res. 13, 127–137.

White, S.E., 1987. Differential movement across transverse ridges on Arapaho rock glacier, Colorado Front Range, USA. Rock GlaciersAllen and Unwin, London, pp. 145–149.

Wilson, P., 2009. Rockfall talus slopes and associated talus-foot features in the glaciated uplands of Great Britain and Ireland: periglacial, paraglacial or composite landforms? Geological Society, London, Special Publications, 320(1), pp. 133–144.

Wolfe, S.A., Stevens, C.W., Gaanderse, A.J., Oldenborger, G.A., 2014. Lithalsa distribution, morphology and landscape associations in the Great Slave Lowland, Northwest Territories, Canada. Geomorphology 204, 302–313.

Woo, M.K., Kane, D.L., Carey, S.K., Yang, D., 2008. Progress in permafrost hydrology in the new millennium. Permafrost Periglacial Process. 19 (2), 237–254.

Yoshikawa, K., Hinzman, L.D., 2003. Shrinking thermokarst ponds and groundwater dynamics in discontinuous permafrost near Council, Alaska. Permafrost Periglacial Process. 14 (2), 151–160.

Yoshikawa, K., 2013. Pingos. In: Giardino, J.R., Harbor, J. (Eds.), Treatise on Geomorphology, vol. 8; Glacial and Periglacial Geomorphology. Academic Press, San Diego, CA, pp. 291–297.

Zeuner, F.E., 1959. The Pleistocene Period: Its Climate, Chronology and Faunal Successions. Hutchinson, London, p. 447.

Chapter 14

The Critical Zone in Desert Environments

Vatche Tchakerian* and Patrick Pease**
*Department of Geography and Geology & Geophysics, Texas A&M University, College Station, Texas, USA; **Department of Geography, University of Northern Iowa, Cedar Falls, Iowa, USA

14.1 INTRODUCTION

Desert environments (arid lands/drylands) constitute the most widespread terrestrial biome on Earth covering about 35% of the land areas of the world (Fig. 14.1) and are home to over 20% of the world's population. Rainfall scarcity, higher temperatures and evapotranspiration, lower humidity, and a general paucity of vegetation cover characterize deserts. The global arid zone is found owing to a unique combination of atmospheric, geologic, and geomorphic conditions, and, as a result of these factors, a very distinct Critical Zone (CZ) is encountered. The CZ ranges from the top of the vegetation biome down to the bottom of the aquifer and includes regolith and both the weathering and soil profile (Fig. 14.2). For geomorphic studies, quantifying the origin, transport, and deposition of regolith/sediment is crucial for understanding and predicting the origin of the CZ, particularly in desert environments. The types, frequency, and magnitude of geomorphic processes that produce and transport sediment in arid lands are highly variable in space and time. Understanding the CZ in desert environments will ultimately lead to a better management of the unique ecosystems, soils, and agriculture potential in the global arid zone. In this chapter, we review (1) the main features of the CZ in arid lands; (2) the geomorphic and geologic factors influencing the formation of the CZ in deserts, such as weathering and hydrology; (3) a numerical model of regolith formation; and (4) case studies that illustrate the connection between certain desert geomorphic landforms and the transport and deposition of regolith. For the CZ in deserts, we will only focus on the regolith (see Fig. 14.2). We use the term regolith following Lin (2010) to include all materials above fresh, unweathered bedrock (R horizon), which is equivalent to a broad definition of the soil.

Developments in Earth Surface Processes, Vol. 19. http://dx.doi.org/10.1016/B978-0-444-63369-9.00014-8

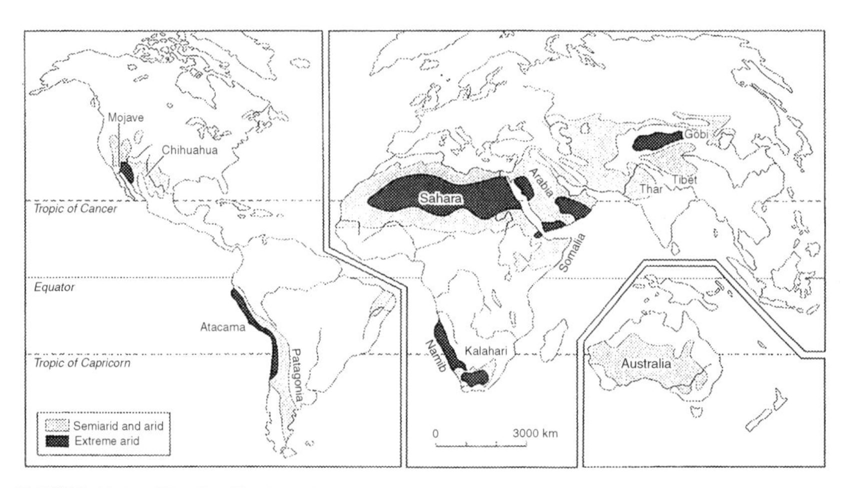

FIGURE 14.1 **The distribution of the world's arid and semiarid areas.** *(From Goudie, (2002).)*

14.2 MAIN FEATURES OF THE CRITICAL ZONE IN ARID LANDS

The CZ is defined by the US National Research Council (NRC, 2001) as "a heterogeneous, near surface environment in which complex interactions involving rock, soil, water, air, and living organisms regulate the natural habitat and determine availability of life sustaining resources." This holistic, integrative, interdisciplinary, and multiscale approach has opened up new horizons for research (Fig. 14.3) and led to a number of reports such as Committee on Challenges and Opportunities in Earth Surface Processes, by the National Research Council (NRC, 2001).

14.2.1 The Nature and Dynamics of Desert Soils and Surfaces

Desert soils and surfaces in arid lands commonly encompass large areas of bare rock, varnished stone pavements, and coarse-weathered mantles. They also exhibit less weathering and leaching (especially chemical alteration) and hence desert soils tend to have coarser textures, shallow soil profiles, high concentrations of salts/evaporites and/or carbonates, and aeolian material (especially dust). Lower or diminished rates of erosion in arid lands favors the preservation of paleosols (or abrupt soil boundaries), especially in areas dominated by quiet tectonic activity and/or ancient shields, such as in the Sahara and Australian deserts. Desert surfaces and soils are also highly susceptible to anthropogenic-induced stresses and soil and land degradation and desertification are a constant threat in these environments.

Laity (2008) summarizes the key factors influencing soil development in drylands. These include vegetation (organic content, biotic environment of the soil); aeolian input (dust, salts); faunal activity (burrowing and soil structure);

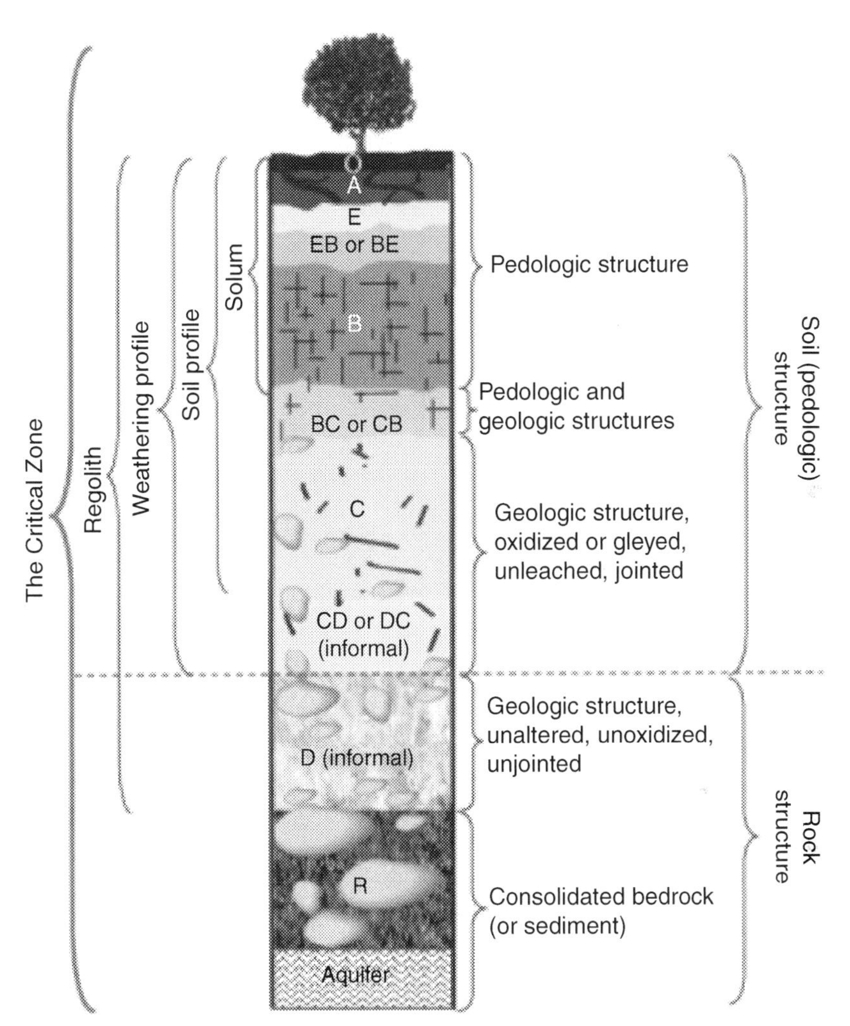

FIGURE 14.2 Concepts of the CZ, regolith, weathering, and soil profiles. This study emphasizes the geomorphology of the regolith. *(Modified from Lin (2010).)*

geologic time and climate – the latter based on precipitation (which controls weathering rates and the translocation of minerals and nutrients); and temperature (which controls the rate of decay of organic matter). Climate in turn is the primary determinant of key mechanical and chemical weathering processes and the evolution of hillslopes and thus regolith and soil formation. Understanding (using both field and mathematical modeling) weathering and hillslope processes and sediment mobilization and transport, thus becomes one of the key components in the transformation of the CZ in desert environments.

Desert soils occupy approximately 46 million km^2, or 31.5% of the Earth's surface, with Aridisols, Entisols, and Inceptisols constituting over 85% of the

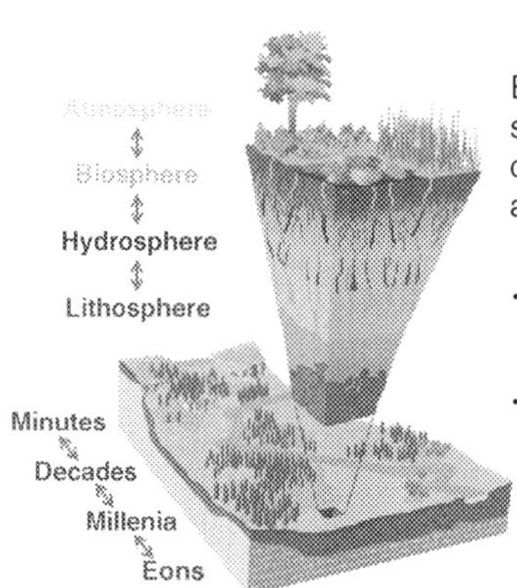

FIGURE 14.3 The main components of the holistic approach to CZ science. *(From http://semanticommunity.info/@api/deki/files/29763/09.)*

global desert soils (Dregne, 1976). The majority of desert soils with relatively well-developed soil horizons fall into the Aridisol classification. Entisols and Inceptisols either exhibit no distinct horizons (Entisols) or weakly developed soil profiles (Inceptisols). Aridisols tend to have shallow soil horizons, typically tan or gray in color and contain low organic matter (<3%), because of low biomass use. The low translocation rates of water, chemicals, and nutrients, combined with low rates of weathering, lead to high alkalinity values and concentration of salts (such as gypsum) and/or carbonates (such as calcium carbonate). They also tend to exhibit low clay content; although some desert soils tend to have an enriched vesicular (Av) horizon formed as a result of aeolian addition of fine-grained materials (e.g., dust) and associated soluble salts, carbonates, and iron oxides, such as those found under certain stone pavements developed on basaltic rocks (Yaalon and Ganor, 1973; McFadden, 2013). The presence of clay rich argillic horizons in desert soils is most likely a relic indicator from a wetter geologic period (such as the Pleistocene) and can be used for palaeoenvironmental reconstruction.

Desert soils tend to be poorly developed with low levels of organic matter owing to low plant productivity and reduced microorganism activity. Soil fertility, organic matter, and vegetation cover are closely linked. Organic matter content affects water retention in soils; low organic matter leads to dryer soils and less vegetation. Desert soil fertility exhibits strong spatial heterogeneity and is strongly influenced by existing plants. Schlesinger and Pilmanis (1998) referred to "islands of fertility" noting that nutrient levels tended to be higher in soils with shrubs as compared to nonvegetated patches of soil. The existence of vegetation creates a positive feedback whereby the shrubs provide organic input while shielding the underlying soil from erosional losses from rain splash and wind. Vegetation patches also reduce soil evaporation rates. High evaporation is problematic because it not only dries the soil but intense evaporation also concentrates soluble salts near the surface leading to the development of saltpans. Existing vegetation also promotes deposition of airborne dust and nutrients, serving as deposition traps. Although desert precipitation is limited, small spatial, or temporal variations in precipitation can have significant impacts on soil development. Even modest increases in precipitation result in the growth of soil carbon levels (Post et al., 1982) as vegetation expands. The inverse relationship holds true for temperature. For any given precipitation level, soil carbon decreases with increasing temperature (Post et al., 1982). This is the result of increased evapotranspiration and soil drying as well as increased organic matter mineralization (Alvarez and Lavado, 1998).

14.3 GEOMORPHIC AND GEOLOGIC FACTORS INFLUENCING THE CZ IN DESERTS

The lack of vegetation and often high-relief topography found in desert environments leads to dynamic geomorphic environments. Local regolith weathering can often supply sediment that is reworked by wind, water, or hillslope processes. The production, transport, and deposition of sediments within desert environments are an important component of understanding the nature of the CZ in arid lands.

14.3.1 Desert Weathering Processes

Since the production of regolith and thus the formation of CZ (and soils) are highly dependent on the breakdown of rocks, understanding the balance between mechanical, chemical, and biological contributions (i.e., processes) to weathering determines the various components (i.e., features) in the CZ (Fig. 14.4). In the following section, we review some of the fundamental processes associated with weathering in arid lands.

Mechanical (or physical) weathering processes in arid lands include insolation weathering or thermoclastis, salt weathering, slaking (wetting and drying), and to a lesser degree, frost weathering (Tchakerian, 2015). Insolation weathering involves the mechanical breakdown of rocks owing primarily to diurnal or annual heating and cooling cycles. Greater than 30°C daily temperature fluctuations have been described in the interior of continental deserts, such as in the Dasht-E Lut Desert (mostly a stone pavement desert) of Iran, as measured by satellites using Land Skin Temperatures (LST; http://earthobservatory.nasa.gov/Features/HottestSpot/page2.php).

LST is a measure of heating of the land surface, where solar energy is absorbed and the resulting thermal wavelength emissions are measured. The accumulated surface temperature is often significantly hotter than air temperature and a better measure of insolation weathering on exposed surfaces, such as hillslopes and rocks. The hottest LST's are likely to occur where the skies are clear, the

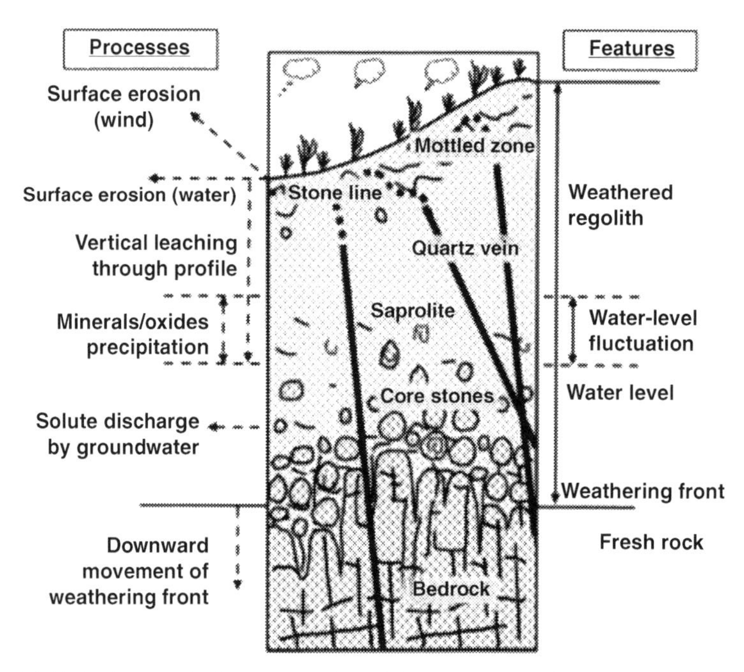

FIGURE 14.4 **A conceptual representation of the processes and features associated with weathering profiles from humid climatic conditions.** *(From Tijani et al. (2006).)*

soil is dry, and winds are light, and the land surface (rocks and hillslopes) has a low albedo (it should absorb most light and reflect little). Rocky, continental interior deserts offer the most ideal geographical location for thermoclastis to operate because of the surface characteristics and low humidity. The volumetric change from contraction and expansion of minerals associated with diurnal (and annual) temperature fluctuations, most likely leads to microfractures and cracks at grain boundaries and breakdown of rocks into regolith (Gómez-Heras et al., 2006; Luque et al., 2011; Gunzburger and Merrien-Soukatchoff, 2011). The previously described process is especially important among silicate rocks (such as granites), where temperature fluctuations will lead to granular disintegration of the biotite minerals, whereby the remaining quartz and feldspars become part of the weathered granites (gruss). Thermoclastis is also believed to make rocks more susceptible to other mechanical, chemical, and biological weathering processes (Tchakerian, 2015).

Salt weathering refers to the breakdown of rock through three interrelated physical and chemical changes produced by the thermal expansion of salt crystals, hydration, and the growth of salt crystals. Many salts exhibit a high coefficient of volumetric expansion from heating and cooling (such as sodium chloride – table salt). Hydration largely involves the addition (and subtraction) of water in the crystal lattice such as the gypsum to anhydrite cycle. As water is absorbed, a volumetric change takes place, exerting pressure against the rock minerals, ultimately leading to their disintegration. The most effective form of salt weathering takes place when salt crystals grow in the pores, cracks, and grain boundaries of the rock (Goudie, 1986; Rodriguez-Navarro and Doehne, 1999; Viles and Goudie, 2007). When saline solutions in rock pores become saturated as a result of changes in temperature and/or evaporation, salt crystals begin to form and considerable pressures are generated. Hence salt weathering will depend upon the stress produced by crystal growth and the tensile strength of the rock. Certain salts when subjected to repeat cycles of wetting and drying (slaking) can also expand volumetrically and exert additional stresses and pressures on the surrounding minerals and grains. In some interior continental deserts or in mountainous environments, frost weathering as a result of water freezing in cracks and pores (up to 9% volumetric expansion) can also contribute to the mechanical disintegration of mineral grains.

Although moisture is limited in time and space in arid lands, precipitation does occur and there is water available (rain, fog, dew, groundwater, etc.) for chemical weathering and decomposition. Chemical weathering processes such as hydrolysis and solution are common, although limited in their effectiveness owing primarily to the shallow percolation of water in the soil or the weathered mantle, very high rates of evaporation, reduced chemical reaction rates because of the high alkalinity of most desert soils, and the paucity of organic acids in the weathering zone to assist in chemical decomposition. Similarly, biologic weathering in the global arid zone is also limited to certain environments.

Weathering by algae and lichens, as well as by burrowing animals does take place but its efficacy is restricted (Viles, 1995; Chen et al., 2000; Büdel et al., 2004). The excretion of acids by growing lichens can etch quartz, feldspars and ferromagnesian minerals, dissolve limestone, and contribute towards rock varnish formation. Certain cryptobiotic crusts (biologic/organic soil crusts) on desert surfaces can also contribute to the weathering of the underlying rocks and sediments. In many parts of the global arid zone and particularly in the deserts of North Africa, Asia, and Australia, ancient weathered regolith and soil profiles have been preserved as duricrusts from past geologic periods and currently make up much of the CZ (Goudie, 2002). These surface and subsurface accumulations of iron, aluminum, silica, calcium carbonate, and other chemically precipitated substances are typically referred to as duricrusts (such as calcretes and laterites). They are formed pedogenically (within the soil) and nonpedogenic (deposition by groundwater and in lacustrine or fluvial environments).

14.3.2 Desert Hydrology and the CZ

In addition to the weathering processes that influence the formation of the CZ, desert hydrologic processes also play an instrumental role in the dynamics of the CZ in arid lands, particularly the role of moisture both below and above the surface. The spatial and temporal variability of precipitation is one of the key metrics influencing the availability of moisture in the weathering zone. Deserts are characterized by low frequency, high magnitude, short but intense rainfalls, and most storms occur as discrete convective cells, enhanced only by orographic factors (presence of mountains). Individual storms are unlikely to affect the entire drainage network and thus the infiltration of water in the CZ is highly varied in space and time. Owing to the sparse vegetation cover in deserts, interception of rainfall is relatively low, but this is negated by excessive evapotranspiration, rapid surface runoff, and thus the infiltration of moisture in the weathering zone is severely limited. Infiltration length tends to be shallow and slow, since most of the moisture only reaches low depths, and with the exception of alluvial channels, subsurface water movement is low and further complicated by regolith chemistry. Interflow (flow between layers of soil) is usually insignificant because of the thin nature of most desert soils. Any remaining moisture in the regolith will very slowly make its way to the groundwater, a process that might take decades for the moisture to reach the water table.

Another major constraint for infiltration or translocation of water further into the desert weathering zone or regolith is the nature of overland flow or runoff. Overland flow dominates since most open desert surfaces (excluding bedrock) tend to have a cemented or crusted surface owing to armoring or sealing as a result of infrequent rainfall; thus, surface runoff tends to be high (the armored surface only broken during major storms) and less water infiltrates down to the regolith (higher infiltration is possible under alluvial or sandy ephemeral channels in deserts). The development of soil itself can have an impact on water infiltration. Soil development is strongly influenced by the depth of water infiltration, which in turn affects vegetation growth (Hamerlynck et al., 2002). Studying plant growth on alluvial fan surfaces of different ages in the Mojave Desert, Hamerlynck et al. (2002) revealed that younger surfaces with coarser texture and more weakly developed soil lead to deeper infiltration, which affected plant canopy volume. Similarly Young et al. (2004) studied desert pavements in the Mojave to compare how pedologic development impacts soil hydraulic characteristics. They found that as desert soil developed Av horizons and increased silt and clay content through dust deposition and local weathering, infiltration was drastically reduced. In summary, infiltration of water into the CZ is highly variable in time and space in deserts and largely controlled by the surface characteristics, which range from bedrock and stone pavements to sands and organic crusts. In the case of desert groundwater (bottom of the CZ), recharge takes a very long time and is controlled largely by surface and subsurface lithology, fracture patterns, porosity, and permeability, and the magnitude and frequency of precipitation. Most of the fresh groundwater found today in deserts antedates the formation of deserts and thus most major aquifers are fossil with very slow rates of recharge.

14.4 NUMERICAL MODELING OF SOIL AND REGOLITH MOVEMENT ON DESERT HILLSLOPES

Development of soil has been examined and modeled for well over a century (cf. McFadden, 2013). Many conceptual models have been developed to explain the mass balance of inputs, outputs, and processes involved. One of the key components of soil development is the relationship between supply of new material, including regolith weathering, and erosional losses. A general conceptual model for soil accumulation shows the mass balance between inputs and losses (Egli et al., 2014; Phillips et al., 2005) where soil thickness (T) over time is a function:

$$T = (W + B) + (A + O + V) - (E + L + C_{\mathrm{surf+sub}})$$

where T=soil thickness; W=bedrock weathering rate; B=bioturbation; A=sediment accretion; O=organic matter accumulation; V=volumetric changes; E=erosional losses; L=subsurface leaching; C=other losses including bio-uptake, fire, etc. from both the surface and subsurface.

Although conceptual, the model provides a framework by which to understand the variables of soil development. Note that the first five factors involve inputs of mass or density changes whereas the remaining variables represent losses of mass and volume. In a desert environment, rates of regolith development can vary from other locations. Chemical weathering rates can be slowed by the dry climate and mechanical weathering often dominates leading to slow W. However, the high degree of aeolian activity in desert environments can result in significant inputs of windblown sand and dust (A). Although fauna bioturbation can be significant, the accumulation of plant organic matter is usually very slow. Likewise, erosional losses can be high due to the lack of vegetation and ease of fluvial erosion, hillslope processes, and deflation.

One of the key variables in the formation of desert soils is the availability of regolith (weathered rock), as well as the geologic time required for its formation. Thus, understanding the processes that replenish the soil with disaggregated mineral grains or regolith becomes crucial. Since in deserts the formation of regolith and soil from bedrock requires very long timescales and tectonic and climatic stability, a proper field and theoretical understanding of weathering becomes very important. The rock to regolith to soil transformation (or weathering processes) can be modeled using a simplified set of mathematical expressions for mass conservation as proposed for general landscape evolution by NRC's Committee on Challenges and Opportunities in Earth Surface Processes summary report (2001). Using a profile through a hillslope (Fig. 14.5), a coupled regolith/soil (h) and bedrock (b) two-layer model can be used to explain all the variables involved in the production and routing of soils and sediment (Heimsath et al., 1997, 2001; NRC, 2001; Roering, 2008; Egli et al., 2014).

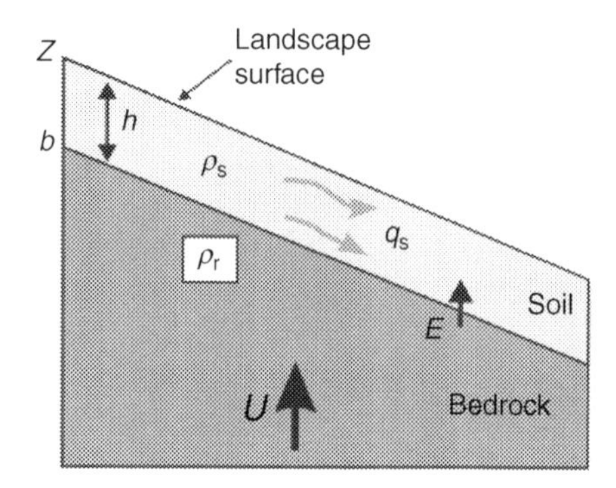

FIGURE 14.5 A theoretical hillslope profile with uplift (U), it's conversion to soil or regolith (E), soil thickness (h), its transport (q_s), above the datum (z). *(From NRC (2001).)*

In this model, (h) represents the vertical thickness of a layer of transportable soil/sediment/regolith, underlain by a bedrock layer with a thickness (b). The elevation of the land surface above the datum is $z=b+h$. The coupled mass conservation equations are:

$$\frac{\partial z}{\partial t}=\frac{\partial b}{\partial t}+\frac{\partial h}{\partial t} \tag{14.1}$$

$$\frac{\partial b}{\partial t}=U-\varepsilon \tag{14.2}$$

$$\frac{\partial h}{\partial t}=\frac{\rho r}{\rho s}\varepsilon-\nabla.qs \tag{14.3}$$

The nature and dynamics of the landscape surface (z) will depend on changes to the thickness of (b) and (h) with time (t). The bedrock thickness (b) is a balance between the rate of uplift (U) and the rate of conversion of bedrock to soil or regolith (ε). In case of subsidence, (U) will be negative. The transportable regolith or sediment (h) is a balance between the weathered bedrock debris and the erosional volume flux denoted as $-\nabla q_s$ (a vector quantity of sediment flux), with ρ_r and ρ_s being the bulk densities of rock and regolith/soil/sediment, respectively. Chemical weathering processes are left out of the model because they can affect ρ_r and ρ_s and thus the conversion of bedrock to regolith is assumed to proceed mainly by mechanical weathering processes (Roering, 2008), the dominant weathering process in most desert environments.

One criticism of the model is that it does not account for external inputs including aeolian deposits or organics. In particular, dust deposition has been suggested as a major input source for mineral material in desert soils (Wells et al., 1987; McFadden, 2013) and has been proposed as a mechanism for an accretionary model of stone (desert) pavement development (Brimhall et al.,

1991; McFadden, 2013). Several authors have noted that dust likely provides the majority of soil carbonate and clay minerals in many arid environments (Gerson and Amit, 1987; Reheis et al., 1995; Quade and Cerling, 1990; Dixon, 2009). Such a net gain in volume is not well accounted for in the usual soil mass balance equations.

Combining the mass conservation equations (14.1–14.3) and adding a term for volume per unit area dust input (α) yields the overall equation for soil thickness over time

$$h_t = \left(\frac{\rho_r}{\rho_s} \varepsilon \right) + \alpha - \nabla q_s \tag{14.4}$$

The *in situ* weathering of rock into regolith is a function of parent material, climate, slope, vegetation, and fauna activity. Whereas slope has been shown to effect erosion losses, recent studies have also demonstrated that the rate of regolith production (ε) is influenced by the relationship between slope and soil thickness. Although lower slopes allow for increased infiltration and chemical weathering (Laity, 2008), the thicker soil profiles that develop can slow the development of new regolith (Strudley and Murray, 2007; Strudley et al., 2006b). The rate of weathering declines exponentially as the soil depth increases and can be defined by:

$$\varepsilon = \frac{\varepsilon_o \; e^{-\mu h \cos\theta}}{\cos\theta} \tag{14.5}$$

where ε_o is the maximum rate of regolith production (when soil depth equals 0), μ is an exponential decay constant, h is soil profile thickness normal to the slope, and θ is local slope (Heimsath et al., 2001). The cosine terms are used to normalize soil depth to the topographic slope.

Under steady state conditions where erosion ($-\nabla q_s$) equals ε and ignoring external inputs like dust, Eq. 14.5 can be reworked to demonstrate the relationship between soil depth, slope, and weathering rates and shows that the depth of regolith varies nonlinearly with slope angle (Roering, 2008).

$$h = \frac{-\ln\left(\varepsilon \cos\theta / \varepsilon_o\right)}{\mu \cos\theta} \tag{14.6}$$

The relationship between desert regolith and weathering profiles has been proposed by Strudley et al. (2006a, 2006b) based on their analysis of pediment formation (a pediment is a gently sloping, low-relief erosional bedrock surface at the base of a weathering resistant, desert mountain) on granites in the Mojave Desert of California. Fig. 14.6 represents their model graphically by plotting a hypothesized regolith production rate, $W(h)$ (m/m.y.) as a function of regolith thickness, h. Strudley et al. (2006a) advocate that pediments are a function of a

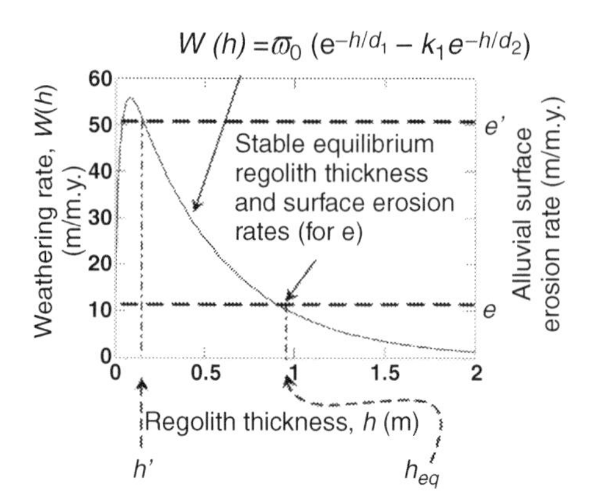

FIGURE 14.6 Weathering rate, *W*(*h*) (m/m.y.) as a function of regolith thickness, *h* (m). Dashed lines represent alluvial surface lowering rates (labeled *e* and *e′* [m/m.y.]. In this study, the bare-bedrock weathering rate is 14 m/m.y. and *h* and *h′* represent stable equilibrium regolith thicknesses. *(From Strudley et al. (2006b).)*

negative feedback between regolith thickness, *h*, and the rate of bedrock weathering, *W*(*h*) expressed as:

$$W(h) = \varpi(e^{-h/d_1} - K_1 e^{-h/d_2}) \quad (14.7)$$

where ϖ is a constant that provides a reference weathering rate given as 70 m/m.y., decay scaling factors to control the shape of the curve are given as $d_1 = 0.5$ m and $d_2 = 0.03$, and k_1 is a dimensionless coefficient of 0.8 that defines the rate of bare-rock weathering relative to the reference rate ϖ. The principal idea demonstrated in Eqs 14.5 and 14.7 is that, as the rate of regolith production is sensitive to the depth of the overburden which forms a protective mantle shielding the bedrock, resulting in decreased rates of bedrock weathering (Strudley et al., 2006b; Strudley and Murray, 2007). However, it has also been noted that a thin layer of regolith actually enhances weathering rates over bare rock due to enhanced chemical weathering resulting from water retention within the regolith cover (Strudley and Murray, 2007).

Strudley et al. (2006b) showed that the inverse relationship between regolith and weathering rates coupled with continuous lowering of the base level, would lead to a steepened piedmont, creating more competence for erosion of the surface and regolith thinning. When the weathering rate equals the erosion rate, regolith thickness equilibrium is reached h_{eq}. If $h > h_{eq}$, the regolith is thinning, thus increasing the bedrock weathering rate, *W*(*h*). If $h < h_{eq}$, the regolith is thickening, decreasing the bedrock weathering rate. For the CZ in deserts with pediments and other related granitic bedrock surfaces, aridity (and weathering resistant lithologies) is the key, thus limiting weathering rates and the above self-organized system is unlikely to produce deep weathering profiles and thus a very limited and weakly consolidated soil cover which is highly susceptible to erosion by fluvial and mass-movement processes. The previously discussed paradigm has been further developed by a more robust numerical model by Pelletier (2010), for pediments in southern Arizona. In addition to the variables given earlier, Pelletier (2010) adds a parameter for flexural-isostatic rebound as a result of erosional unloading (uplift?), and a maximum rate of bedrock weathering. For the purposes of our study with the CZ in drylands, the study confirms that the rate of soil production depends primarily on climate and lithology, with lower soil production rates associated with more arid (hyperarid) climates and more resistant lithologies (typically granite, granodiorite, or quartz monzonite, among others). All of the previous studies indicate that soil (and regolith) production tends to be limited in deserts characterized by granitic terrain and landforms, with typical soil thickness less than 1 m common, and that it requires significant geologic time, tectonic stability, and favorable lithologic units. For future CZ studies in the global arid zone, numerical modeling offers a unique opportunity to better understand the connections between soil, weathering, and regolith production on one hand; and climatic, lithologic, tectonic, and intrinsic sediment transport-weathering feedbacks, on the other.

14.5 LANDSCAPE DEVELOPMENT IN DESERTS – EXAMPLES FROM THE MOJAVE DESERT, CA

The varied landscapes of the Mojave Desert in southern California provide a wide range of examples of the processes involved in sediment production, transport, and deposition involved in the creation of the CZ. In this section we explore the role of dust as an allochthonous input into the CZ and the geomorphic environments of stone pavements and sand ramps as major sediment basins.

14.5.1 Aeolian Dust and the CZ in Deserts

Aeolian dust and mineral aerosols also influence the CZ in deserts. Figure 14.7 represents an idealized model of the potential aeolian inputs to mountain-piedmont-basin areas in the Mojave Desert, California (Hirmas et al., 2011). The aeolian component plays an important role in the development of soils on stone (desert) pavements, pediments, and on alluvial fans and playas, as well as on dune formation and deposition. Aeolian dust from playas and surrounding alluvial piedmonts affect development of the weathering profile on alluvial fans and stone pavements, as well as mountain pediments as represented by the cycle presented in Fig. 14.7. In Section 14.5.2, we will review some of the major roles and inputs of aeolian dust on desert geomorphic systems.

14.5.2 Soils and Landscape Development on Stone Pavements in the Mojave Desert

Aeolian dust deposition constitutes much of the pedogenic material in Pleistocene and Holocene soils of the Mojave Desert and other arid regions of the

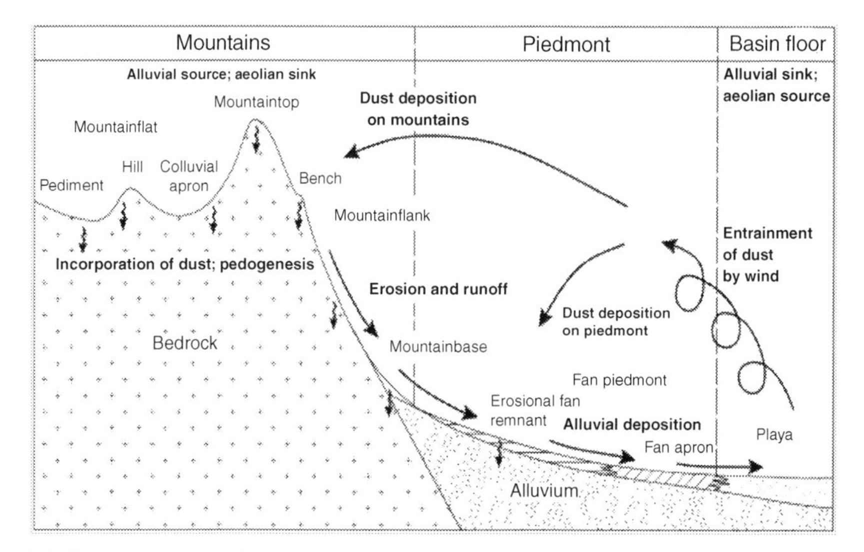

FIGURE 14.7 An idealized geomorphic model of a mountain-piedmont basin and coupled aeolian dust and mineral aerosol inputs and outputs, Mojave Desert, California. *(From Hirmas et al. (2011).)*

world (Reheis et al., 1995; Tchakerian, 1999). Incorporation of this dust into near surface soil horizons provides much of the water-holding capacity of desert soils. Dust accumulation and concomitant soil development are genetically linked to stone pavement formation (McFadden et al., 1987; 1998). Stone pavements on basaltic alluvial fans in the eastern Mojave Desert have been formed by the upward lifting of the surface clasts from the accumulation of aeolian fines (vesicular Av Horizon) below pavement clasts. In this way, Av horizons are formed as a result of aeolian addition of fine-grained materials (e.g., dust) and associated soluble salts, carbonates, and iron oxides. The vesicles form owing to entrapment of air by aeolian infall with subsequent expansion due to heating following rainfall events. Vesicular soil formation, along with clay and carbonate accumulation from aeolian input and localized weathering, accrete vertically over time and cause horizon thickening (Fig. 14.8).

Capo and Chadwick (1999) saw a 200% increase in soil profile thickness in calcretes near Las Cruces, New Mexico that they attributed to aeolian inputs of calcium. This vertical accretion model has supplanted earlier explanations for stone pavement formation such as wind deflation, water sorting, freeze and thaw, and wetting and drying. The sources of the aeolian dust for the development of Av horizons and stone pavements are the fine sediments from desiccating lake basins and nearby alluvial sources. Recently, Sweeney et al. (2013) using geochemical data, suggest that proximal alluvial sources, including fine sand and very fine-to-medium silt from ephemeral washes and distal alluvial fan deposits, are the primary contributors of aeolian sediment to Av

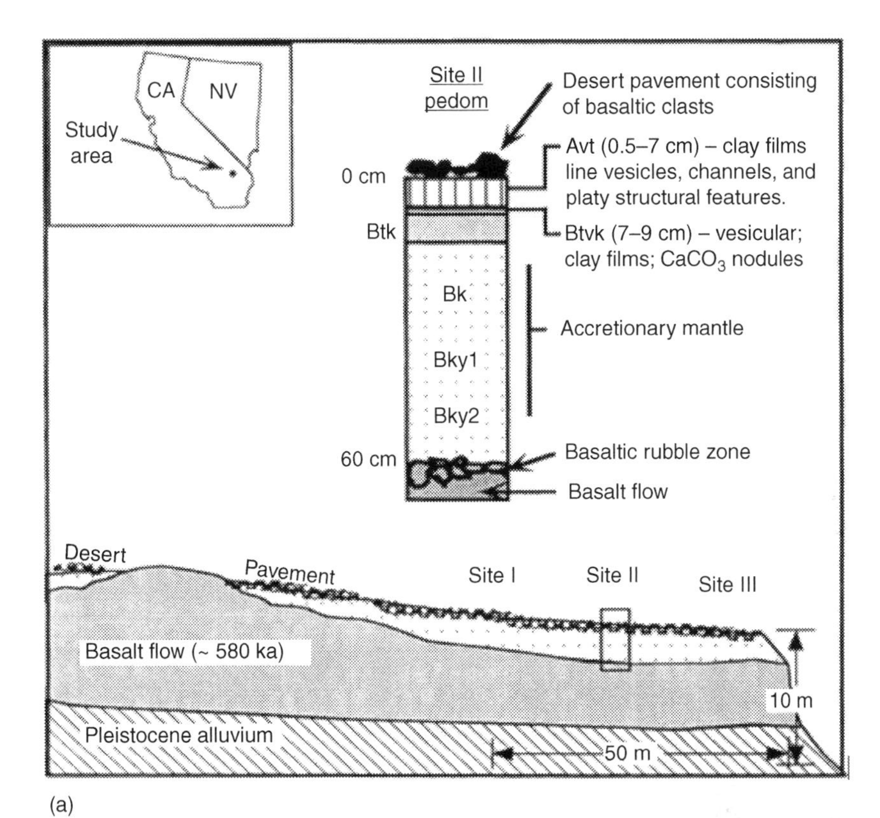

(a)

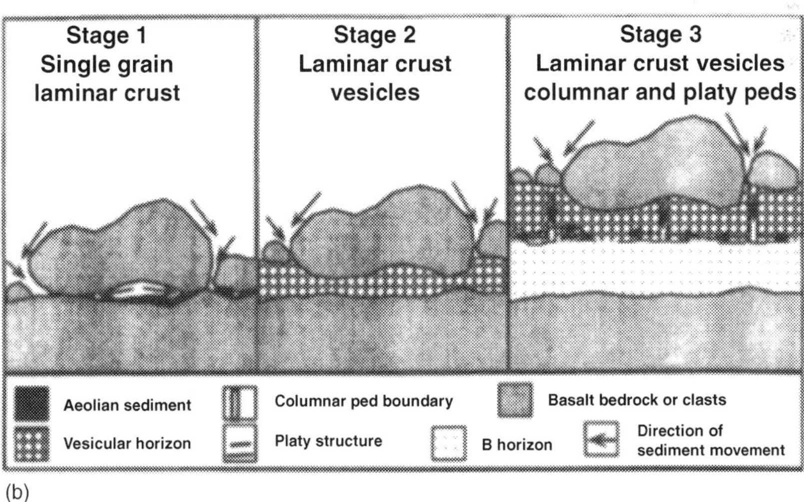

(b)

FIGURE 14.8 (a) Diagram showing the stratigraphy, soils, and aeolian accretionary mantles on basaltic stone (desert) pavements in the Cima Volcanic Field, Mojave Desert, CA (from Anderson et al., 2002); (b) close up view of idealized clay and carbonate accumulation, vesicular soil formation from aeolian sources, and vertical accretion as proposed in (a). *(Modified from Anderson et al. (1994) by Tchakerian (1999).)*

horizons accumulating beneath stone pavements in the southwestern deserts of the United States, and that playa sources are only secondary and contribute less than previously indicated.

In addition to providing significant mass to desert soils, aeolian dust has been shown to be a major contributor to soil mineralogy. A number of elements, including several important for plant growth, were found to have largely been derived from aeolian inputs in the Colorado Plateau (Reynolds et al., 2001). Using magnetite concentrations and trace elemental enrichments in desert soil, they determined that aeolian processes deposited significant amounts of mineral material. Many of the enriched elements were plant-essential nutrients meaning aeolian dust can have a profound impact on soil development and fertility in arid environments. In the Reynolds et al. (2001) study, enrichments of soil over bedrock included K (1.2X), Fe (1.6X), P (2X), Na (3.8X), Mg (4.4X), and Ca (10.5X).

Examining soil carbonate in B and K horizons of a calcrete in New Mexico, Capo and Chadwick (1999) found aeolian inputs of Ca at concentrations of 1.5 g Ca cm^{-3}. They used $^{87}Sr/^{86}Sr$ ratios to determine that between 94% and 98% of the Ca levels in the soils were derived from aeolian dust. Similarly, high levels of aeolian-derived salts have been found in soils in the Atacama Desert (Ewing et al., 2006). The region contains high levels of nitrate, iodate, and sulfate salts as well as other rarer compounds. Aeolian mass inputs reaching 830 kg/m^2 were noted in some areas, leading to a volumetric expansion of 120%. Ewing et al. (2006) further noted that soils in the dryer regions of the Atacama showed mass accumulation as solutes resided in the soil for extended time periods whereas wetter soils lost soluble material through leaching and biologic uptake and shrunk in volume.

14.5.3 Soils and Landscape Development on Sand Ramps in the Mojave Desert

In desert regions that have significant aeolian accumulations in the forms of dunes, sand sheets, and sand ramps, the latter offers the closest analog to a complete weathering or regolith profile such as those found in the CZ. Sand ramps consist mainly of aeolian depositional units, intermixed with fluvial and talus deposits, and lie against the flank of mountains fronts (Fig. 14.9).

Sand ramps tend to be located astride well-defined regional and local scale sand sources and transport corridors (Zimbelman et al., 1995). They typically contain multiple aeolian depositional units separated by paleosols with well-preserved Bk horizons and calcified root nodules, indicating the presence of past vegetation cover in the region. Most sand ramps are relict features and are not accumulating today; inactive surfaces are characterized with boulder-to-gravel size talus mantles and incised stream channels that have exposed the underlying sequence of sedimentary units (Lancaster and Tchakerian, 2003; Tchakerian and Lancaster, 2002; Tchakerian, 2009).

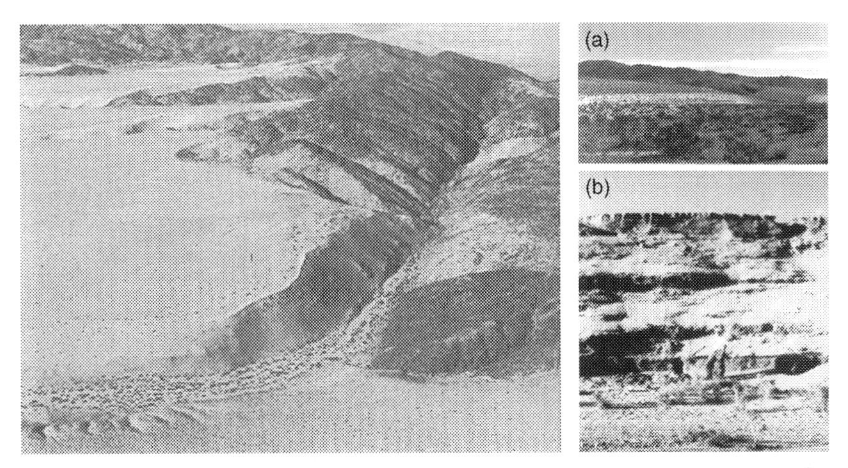

FIGURE 14.9 (a) The Dale Lake sand ramp viewed from above with the ephemeral wash dissecting the ramp (b) close up view of the sand ramp with exposed aeolian depositional units and paleosols. *(From Zimbelman et al. (1995) and Pease and Tchakerian (2003).)*

The mixed provenance of sediments including local weathering, aeolian, mass movement, and fluvial processes accompanied by vegetation dynamics, provides a unique laboratory for studying sediment dynamics of the CZ in a desert environment. In addition, sand ramps are useful indicators of palaeoenvironmental reconstruction as exemplified by extensive luminescence dating research in the region (Clarke and Rendell, 1998; Lancaster and Tchakerian, 2003; Bateman et al., 2012, among others). For CZ research, the provenance, transport, and deposition of sediments from desert basins and adjoining areas to the ramps, are an important component of desert sedimentary studies (i.e., sediment supply). These sand transport corridors have been the focus of numerous studies, principally through the use of geochemical analysis of sediments (e.g., Pease and Tchakerian, 2003). Three major wind/sand transport corridors have been identified (Fig. 14.10) for the eastern Mojave Desert.

14.5.3.1 The Mojave Wash to the Kelso Dunes

This corridor begins at the Mojave River Wash as it exits Afton Canyon, continues on towards the Devil's Playground sand sheet, and ends about 50km downwind in the Kelso Dunes.

Luminescence ages from the aeolian sediments at the Kelso Dunes indicate major dune constructional episodes between 26 ka and 15 ka, and from 12 ka to 7 ka (Lancaster and Tchakerian, 2003). The dune field has been substantially reactivated during the past 4000 years, and much of the earlier evidence of dune formation obscured. A recent re-examination of the Soldier Mountain Sand Ramp (as first discussed by Lancaster and Tchakerian, 1996) near the Mojave River Wash by Bateman et al. (2012), indicate a rather abrupt constructional period for the formation of the sand ramp with luminescence ages around 5000

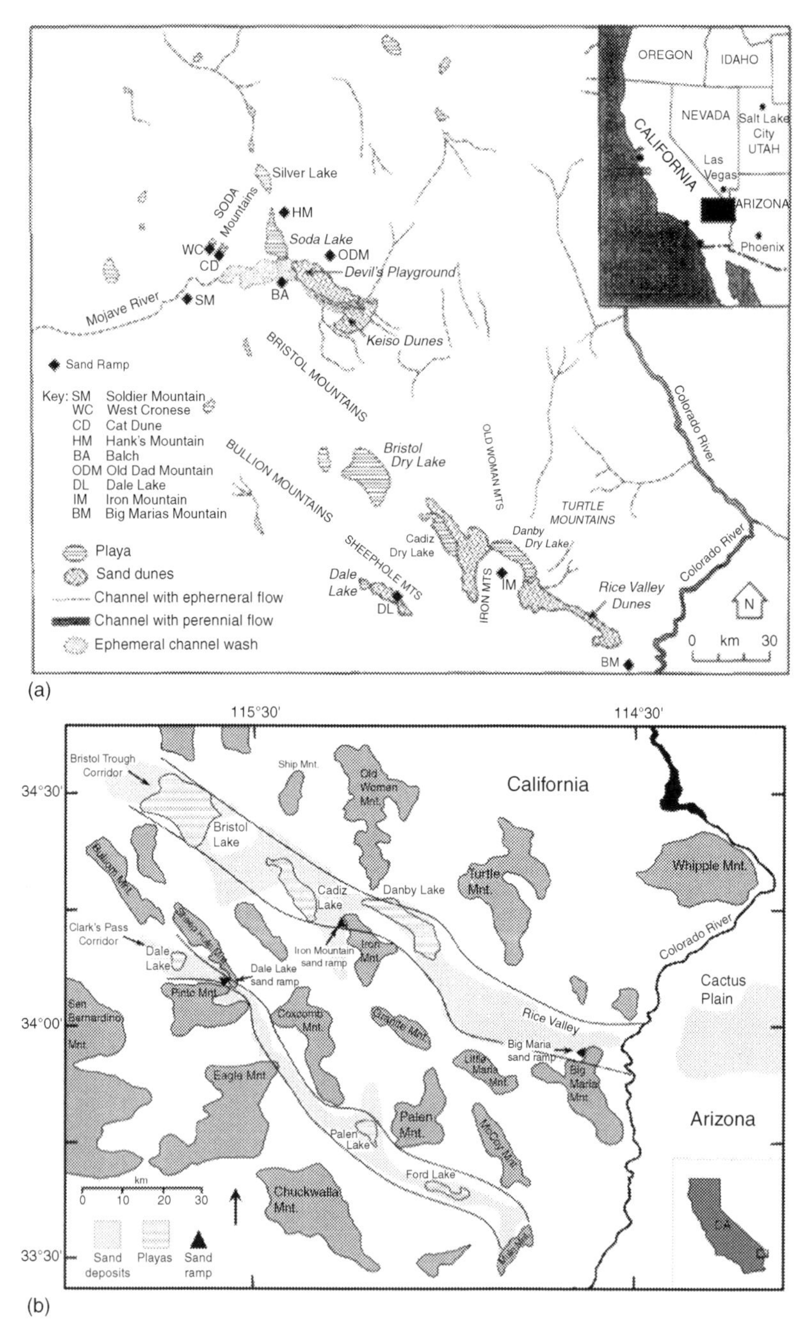

FIGURE 14.10 (a) Major sand transport corridors in the Mojave Desert (from Rendell and Sheffer, 1996); (b) close up view of the Dale Lake and Bristol Dry Lake sand transport corridors. *(From Pease and Tchakerian (2003).)*

years ago. They propose a model whereby some sand ramps can form during "a window of opportunity," when sediment supply is amply available for transport and deposition up a mountain, and cease when sediment supply and/or prevailing winds and deposition space is curtailed. This model of a high magnitude, low-frequency depositional event is in contrast to other sand ramps from the eastern Mojave Desert (such as the Kelso Dunes and Dale Lake), where sand construction proceeds at a more gradual cycle with deposition primarily controlled by longer-term climatic events and sediment availability, separated by periods of stability as indicated by the presence of many paleosols layers and carbonate horizons. Another major difference is their geographic locations. The Soldier Mountain Sand Ramp is located directly downwind from the Mojave River Wash and sand has only to traverse less than 2 km to reach the mountain and its final resting place. On the other hand, sediments forming the main units of the Dale Lake sand ramp (Section 14.5.3.2) had their origin at the currently dry Dale Lake Playa and surrounding lacustrine and alluvial piedmont areas, and had to traverse a distance of over 15 km up to Clark's Pass, with a net elevation change of 235 m (Dale Lake and its surrounding alluvial regions are at around 355 m in elevation, while Clark's Pass is at 590 m elevation). The time required for sediments to be mobilized, transported, and deposited could be one of the reasons why some of the sand ramps at Dale Lake and elsewhere, are significantly older than Soldier Mountain.

14.5.3.2 The Dale Lake to Palen and Ford Playas

Sometimes referred to as Clark's Pass system and first proposed by Zimbelman et al. (1995), the orientation of the Sheep Hole and Pinto Mountains have acted like a funnel enabling the moving aeolian sands to climb more than 250 m from the level of Dale Lake Playa to Clark's Pass and form numerous massive sand ramps (Fig. 14.10b). One of the best studied, the Dale Lake Sand Ramp, has been the focus of numerous studies (Tchakerian, 1991; Lancaster and Tchakerian, 1996, 2003). Multiple periods of aeolian accumulation, separated by paleosols formed in periods of geomorphic stability have been recognized. At present, the sand ramp is relict (stable) with boulder-to-gravel size talus mantling the surface. An incised channel has exposed the underlying sedimentary units. Luminescence ages from the sand ramp indicate two main periods of sand emplacement, between greater than 35–25 and 15–10 ka (Lancaster and Tchakerian, 2003). A few kilometers west from the main Dale Lake Sand Ramp, another (yet unnamed) significant sand ramp, with over 50 m of sedimentary units (including many aeolian units separated by paleosols, carbonate horizons, and talus), awaits further study (Fig. 14.11). Future studies in this region might decipher the complex interrelationships between the timing for the emplacement of the various depositional (and erosional) aeolian units, environmental changes, soil and weathering development, and the nature and dynamics of sediment delivery to these unique geomorphic systems.

FIGURE 14.11 Entrenched sand ramp a few kilometers west from the Dale Lake sand ramp as seen in Fig. 14.9. *(Photo by V.P. Tchakerian.)*

14.5.3.3 Bristol Dry Lake to the Big Maria Mountains

This corridor begins south of Bristol Dry Lake and proceeds over a number of basin and ranges, and terminates just before reaching the Colorado River, at the Big Maria Mountains. A number of sand ramps in this corridor have been studied using remote sensing, geomorphology, sedimentology, geochemistry, and luminescence dating (Zimbelman et al., 1995; Tchakerian, 1999; Lancaster and Tchakerian, 2003; Pease and Tchakerian, 2003). Although the corridor provides the geomorphic setting, geochemical analysis of sediments suggests that the sand in the sand ramps is often derived from discrete, local sources (e.g., dry lake bed sediments, alluvial fans, ephemeral streams, among others), and that there are also provenance variations between the west and east side of the corridor (Fig. 14.12). The sand ramp sediments from the west side of the transport corridor (Iron Mountain, Dale Lake) have similar compositions which are distinct from those on the east side (Big Maria Mountains, Arizona Sands). Pease and Tchakerian (2003) noted the similarity in geochemistry of western ramps and attributed it to locally derived sand from similar lithology originating from the Bullion Mountains. They further noted that Big Maria sand was chemically more similar to sand in the Bristol Trough corridor whereas the Arizona sands on the east side of the Colorado River were much more quartz-rich and not linked to the wind corridors. Internal variations in geochemistry lead Pease and Tchakerian (2003) to suggest that the local and regional sources of sediment in the wind corridors changed over time with climate fluctuations confirming models presented by Tchakerian and Lancaster (2002). These variations in

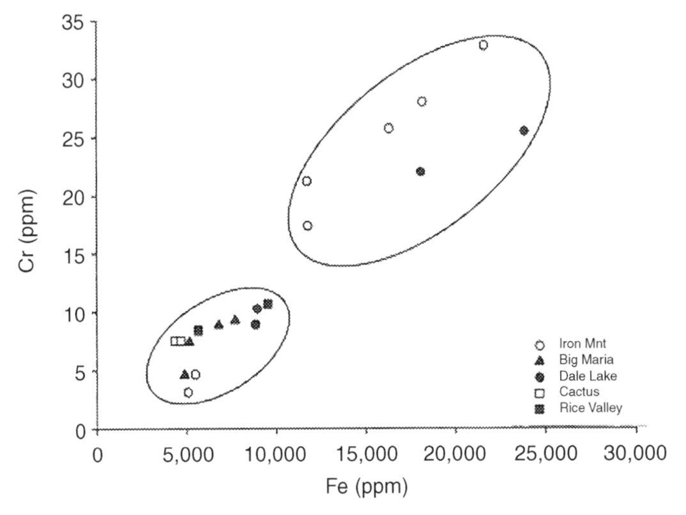

FIGURE 14.12 Chromium (Cr) and iron (Fe) elemental plot showing the relative concentration of mafic minerals in sediment samples from sand ramps along the transport corridors, as identified in Fig. 14.10. The samples from Cactus indicate dune sediments from across the Colorado River in Arizona. See text for further discussion. *(From Tchakerian (1999).)*

geochemistry, influenced primarily by differences in source mineralogy, weathering, and transport patterns, provide an understanding of the multiple and distinct sediment sources for these sand ramps (and other similar deposits). Such an understanding can better explicate the relations between weathering and sediment supply, transport, and deposition, as well as climate variability, arid/humid cycles in arid lands, and the formation of the CZ in desert environments.

14.6 CONCLUSIONS

Desert environments make up a significant portion of the Earth's surface and their CZ environments are an important component of the landscape. The unique environmental setting of the desert CZ, including rainfall scarcity, high temperatures and evapotranspiration, low humidity, and minimal vegetation cover leads to distinctive characteristics in the soil and regolith system. Modeling and quantifying the origin, transport, and deposition of weathering profiles/regolith/sediment is crucial for understanding and predicting the past, present, and future of the CZ in desert environments.

REFERENCES

Alvarez, R., Lavado, R.S., 1998. Climate, organic matter and clay content relationships in the Pampa and Chaco soils, Argentina. Geoderma 83, 127–141.

Anderson, K.C., Wells, S.G., Graham, R.C., McFadden, L.D., 1994. Processes of vertical accretion in the stone-free zone below desert pavements. In: GSA Abstracts with Programs 26, 7, Seattle, WA. Geological Society of America, Boulder, CO, p. 87.

Anderson, K.A., Wells, S.G., Graham, R.C., 2002. Pedogenesis of vesicular horizons, Cima volcanic field, Mojave Desert, California. Soil Sci. Am. J. 66 (3), 878–887.

Bateman, M.D., Bryant, R.G., Foster, I.D.L., Livingstone, I., Parsons, A.J., 2012. On the formation of sand ramps: a case study from the Mojave Desert. Geomorphology 161/162, 93–109.

Brimhall, B., Lewis, C., Ford, C., Bratt, J., Taylor, G., Warin, O., 1991. Quantitative geochemical approach to pedogenesis: importance of part material reduction, volumetric expansion, and eolian influx in lateritization. Geoderma 51, 51–91.

Büdel, B., Weber, B., Kühl, M., Pfanz, H., Sültemeyer, D., Wessels, D., 2004. Reshaping of sandstone surfaces by cryptoendolithic cyanobacteria: bioalkalization causes chemical weathering in arid landscapes. Geobiology 2 (4), 261–268.

Capo, R.C., Chadwick, O.A., 1999. Sources of strontium and calcium in desert soil and calcrete. Earth Planet. Sci. Lett. 170, 61–72.

Chen, J., Blume, H.-P., Beyer, L., 2000. Weathering of rocks induced by lichen colonization – a review. Catena 39 (2), 121–146.

Clarke, M.L., Rendell, H.M., 1998. Climate change impacts on sand supply and the formation of desert sand dunes in the south-west USA. J. Arid Environ. 39, 517–531.

Dixon, J., 2009. Aridic soils, patterned ground, and desert pavements. In: Parsons, A., Abrahams, A. (Eds.), Geomorphology of Desert Environments. Springer, Berlin, pp. 101–121.

Dregne, H.E., 1976. Soils of Arid Regions. Elsevier, New York, p. 250.

Egli, M., Dahms, D., Norton, K., 2014. Soil formation rates on silicate parent material in alpine environments: different approaches-different results? Geoderma 213, 320–333.

Ewing, S.A., Sutter, B., Owen, J., Nishiizumi, K., Sharp, W., Cliff, S.S., Perry, K., Deitrich, W., McKay, C.P., Amundson, R., 2006. A threshold in soil formation at Earth's arid–hyperarid transition. Geochim. Cosmochim. Acta 70, 5293–5322.

Gerson, R., Amit, R., 1987. Rates and modes of dust accretion and deposition in an arid region-The Negev, Israel. Geological Society [London] Special Publication 35, pp. 157–169.

Gómez-Heras, M., Smith, B.J., Fort, R., 2006. Surface temperature differences between minerals in crystalline rocks: implications for granular disaggregation of granites through thermal fatigue. Geomorphology 78, 236–249.

Goudie, A.S., 1986. Laboratory simulation of 'the wick effect' in salt weathering of rock. Earth Surf. Process. Land. 11 (3), 275–285.

Goudie, A.S., 2002. Great Warm Deserts of the World. Oxford University Press, Oxford, p. 444.

Gunzburger, Y., Merrien-Soukatchoff, V., 2011. Near-surface temperatures and heat balance of bare outcrops exposed to solar radiation. Earth Surf. Process. Land. 36, 1577–1589.

Hamerlynck, E.P., McAuliffe, J.R., McDonald, E.V., Smith, S.D., 2002. Ecological responses of two Mojave Desert shrubs to soil horizon development and soil water dynamics. Ecology 83 (3), 768–779.

Heimsath, A.M., Dietrich, W.E., Nishiizumi, K., Finkel, R.C., 1997. The soil production function and landscape equilibrium. Nature 388, 358–361.

Heimsath, A.M., Dietrich, W.E., Nishiizumi, K., Finkel, R.C., 2001. Stochastic processes of soil production and transport: erosion rates, topographic variation and cosmogenic nuclides in the Oregon coast range. Earth Surf. Process. Land. 26, 531–552.

Hirmas, D.R., Graham, R.C., Kendrick, K.J., 2011. Soil-geomorphic significance of land surface characteristics in an arid mountain range, Mojave Desert, USA. Catena 87 (3), 408–420.

Laity, J.E., 2008. Deserts and Desert Environments. Wiley-Blackwell Publishers, Oxford, p. 342.

Lancaster, N., Tchakerian, V.P., 1996. Geomorphology and sediments of sand ramps in the Mojave Desert. Geomorphology 17, 151–166.

Lancaster, N., Tchakerian, V.P., 2003. Later Quaternary eolian dynamics, Mojave Desert, California. In: Enzel, Y, Wells, S.G., Lancaster, N. (Eds.), Paleoenvironments and paleohydrology of the Mojave and southern Great Basin Deserts, Geological Society of America Special Paper 368, Boulder, CO, pp. 231–249.

Lin, H., 2010. Earth's Critical Zone and hydropedology. Hydrol. Earth Syst. Sci. 14, 25–45.

Luque, A., Ruiz-Aguda, E., Cultrone, G., Sebastián, E., Siegesmund, S., 2011. Direct observation of microcrack development in marble caused by thermal weathering. Environ. Earth Sci. 62, 1375–1386.

McFadden, L.D., 2013. Strongly dust-influenced soils and what they tell us about landscape dynamics in vegetated aridlands of the southwest United States. Geol. Soc. Am. Special Pap. 500, 501–532.

McFadden, L.D., Wells, S.G., Jercinovich, M.J., 1987. Influences of eolian and pedogenic processes on the origin and evolution of desert pavements. Geology 15, 504–508.

McFadden, L.D., McDonald, E.V., Wells, S.G., Anderson, K., Quade, J., Forman, S.L., 1998. The vesicular layer and carbonate collars of desert soils and pavements: formation age and relation to climate change. Geomorphology 24, 101–145.

NRC (National Research Council), 2001. Basic Research Opportunities in Earth Science. National Academy Press, Washington, p. 154.

Pease, P.P., Tchakerian, V.P., 2003. Geochemical analysis of sediments from Quaternary sand ramps, southern Mojave Desert, California. Quaternary Int. 104, 19–29.

Pelletier, J.D., 2010. How do pediments form? A numerical modeling investigation with comparison to pediments in southern Arizona, USA. Geol. Soc. Am. Bull. 122 (11/12), 1815–1829.

Phillips, J.D., Marion, D.A., Luckow, K., Adams, K.R., 2005. Nonequilibrium regolith thickness in the Ouachita Mountains. J. Geol. 113, 325–340.

Post, W.M., Emanuel, W.R., Zinke, P.J., Stangenberger, A.G., 1982. Soil carbon pools and world life zones. Nature 298, 156–159.

Quade, J., Cerling, T.E., 1990. Stable isotopic evidence for a pedogenic origin of carbonates in trench 14 near Yucca Mountain, Nevada. Science 250, 1549–1552.

Reheis, M., Goodmacher, J., Harden, J., McFadden, L., Rockwell, T., Shroba, R., Sowers, J., Taylor, E., 1995. Quaternary soils and dust deposition in southern Nevada and California. Geol. Soc. Am. Bull. 107, 1003–1022.

Rendell, H.M., Sheffer, N.L., 1996. Luminescence dating of sand ramps in the eastern Mojave Desert. Geomorphology 17, 187–197.

Reynolds, R., Belnap, J., Reheis, M., Lamothe, P., Luiszer, F., 2001. Aeolian dust in Colorado Plateau soils: nutrient inputs and recent change in source. Proc. Natl. Acad. Sci. 98 (13), 7123–7127.

Rodriguez-Navarro, C., Doehne, E., 1999. Salt weathering: influence of evaporation rate, supersaturation and crystallization pattern. Earth Surf. Process. Land. 24, 191–209.

Roering, J.J., 2008. How well can hillslope evolution models "explain" topography? Simulating soil transport and production with high-resolution topographic data. Geol. Soc. Am. Bull. 120 (9/10), 1248–1262.

Schlesinger, W.H., Pilmanis, A.M., 1998. Plant–soil interactions in deserts. Biogeochemistry 42, 169–187.

Strudley, M.W., Murray, A.B., 2007. Sensitivity analysis of pediment development through numerical simulation and selected geospatial query. Geomorphology 88, 329–351.

Strudley, M.W., Murray, A.B., Haff, P.K., 2006a. Regolith thickness instability and the formation of tors in arid environments. J. Geophys. Res. 111, 1–16.

Strudley, M.W., Murray, A.B., Haff, P.K., 2006b. Emergence of pediments, tors, and piedmont junctions from a bedrock weathering-regolith thickness feedback. Geology 34, 805–808.

Sweeney, M.R., McDonald, E.V., Markley, C.E., 2013. Alluvial sediment or playas: what is the dominant source of sand and silt in desert soil vesicular A horizons, southwest USA. J. Geophys. Res. Earth Surf. 118, 257–275.

Tchakerian, V.P., 1991. Late Quaternary aeolian geomorphology of the Dale Lake sand sheet, southern Mojave Desert, California. Phys. Geogr. 12, 347–369.

Tchakerian, V.P., 1999. Dune palaeoenvironments. In: Goudie, A.S., Livingstone, I., Stokes, S. (Eds.), Aeolian Environments, Sediments and Landforms. John Wiley & Sons, New York, pp. 261–292.

Tchakerian, V.P., 2009. Palaeoclimatic interpretations from desert dunes and sediments. In: Parsons, A.J., Abrahams, A.D. (Eds.), Geomorphology of Desert Environments. Springer-Verlag, New York, pp. 757–772.

Tchakerian, V.P., 2015. Deserts and desertification. In: North, G.R., , Pyle, J., Zhang, F. (Eds). Encyclopedia of Atmospheric Sciences, second ed., vol. 3. John Wiley & Sons, New York, pp. 185–192.

Tchakerian, V.P., Lancaster, N., 2002. Late Quaternary arid/humid cycles in the Mojave Desert and western Great Basin of North America. Quaternary Sci. Rev. 21, 799–810.

Tijani, M.N., Okunlola, O.A., Abimbola, A.F., 2006. Lithogenic concentrations of trace metals in soils and saprolites over crystalline basement rocks: a case study from SW Nigeria. J. Afr. Earth Sci. 46 (5), 427–436.

Viles, H., 1995. Ecological perspectives on rock surface weathering: towards a conceptual model. Geomorphology 13, 21–35.

Viles, H.A., Goudie, A.S., 2007. Rapid salt weathering in the coastal Namib Desert: implications for landscape development. Geomorphology 85 (1/2), 49–62.

Wells, S.G., McFadden, L.D., Dohrenwend, J.C., 1987. Influence of the late Quaternary climatic changes on geomorphic and pedogenic processes on a desert piedmont, eastern Mojave Desert, California. Quaternary Res. 27, 130–146.

Yaalon, D.H., Ganor, E., 1973. The influence of dust on soils during the Quaternary. Soil Sci. 116 (3), 146–155.

Young, M.H., McDonald, E.V., Caldwell, T.G., Benner, S.G., Meadows, D.G., 2004. Hydraulic properties of a desert soil chronosequence in the Mojave Desert, USA. Vadose Zone J. 3, 956–963.

Zimbelman, J.R., Williams, S.H., Tchakerian, V.P., 1995. Sand transport paths in the Mojave Desert, southwestern United States. In: Tchakerian, V.P. (Ed.), Desert Aeolian Processes. Chapman & Hall, London, pp. 101–130.

Chapter 15

The Critical Zone in Tropical Environments

Sara Mana*, Paulo Ruiz**, and Amalia Gutiérrez**
*Department of Earth and Environmental Sciences, University of Iowa, Iowa City, Iowa, USA;
**Laboratorio Nacional de Materiales y Modelos Estructurales, Universidad de Costa Rica, Ciudad de la Investigación, Costa Rica

15.1 INTRODUCTION

The Critical Zone is highly heterogeneous due to variations in physical, chemical, and mechanical weathering processes (Fig. 15.1). In tropical zones high microbiodiversity, high temperatures, and large amounts of rainfall can enhance weathering. Soil production can differ widely from temperate zones due to climatic conditions, seasonality, and differences in biota and microbiota. Studies regarding the CZ in tropical areas have not been numerous and known Critical Zone Observatories (CZOs) include only the Luquillo CZO in Puerto Rico and the Red Soils CZO in China (subtropical).

The coupling of chemical and physical weathering is responsible for the development of the landscape in which we live. Anderson et al. (2007) have analyzed the physical and chemical controls on CZ weathering by modeling the CZ as a feed-through reactor, whose size is controlled by the relative rates of downward progress of the weathering front and erosion from the top. Weathering rates are affected by many factors, recent work by Dupré et al. (2003) emphasizes how rock type, together with temperature and erosion are important factors controlling chemical weathering rates along with climate (Fig. 15.2). High precipitation and temperatures are commonly postulated in tropical areas to increase chemical weathering. Additionally, the amounts of dissolved gases (CO_2 and O_2) and the hydrogeology can control the rate of weathering flux (e.g., water table above or below the saprolite). The material type and its history control the rate of water percolation. Rock fracturing and mineral dissolution can increase permeability, whereas the alteration of minerals into clays can reduce it. The presence of fracturing is strictly related to the tectonic environment in which it is located, whereas mineral dissolution is controlled by temperature and climate. Therefore, a localized study of the processes affecting the CZ is necessary. This work aims at describing how the Critical Zone is modified by

Developments in Earth Surface Processes, Vol. 19. http://dx.doi.org/10.1016/B978-0-444-63369-9.00015-X

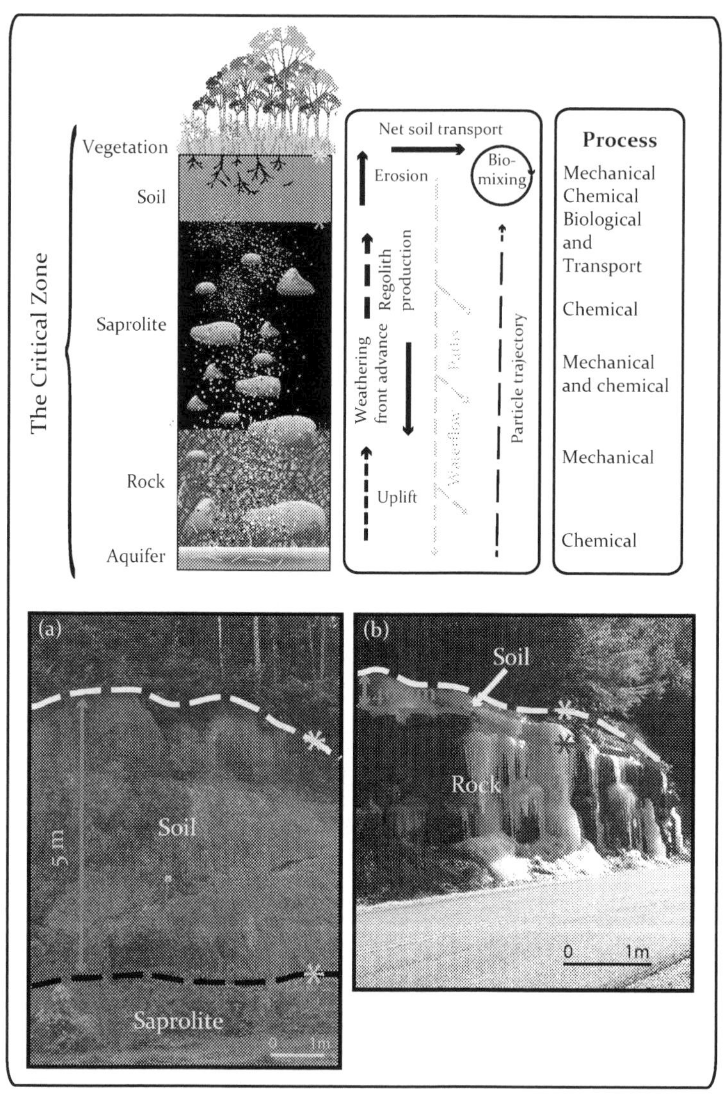

FIGURE 15.1 Vertical layering within the Critical Zone in Tropical Environments. Particle trajectories and water flow paths are highlighted. The section on the top left can be used as an inclusive example, although not all layers are always present. The photograph on the left shows an example in tropical climates and on the right an example for temperate zones is shown, asterisks indicate layer boundaries. In tropical climates, chemical and biological processes exacerbate weathering. Also, soil thickness in tropical climates exceeds mean thickness in temperate climates. On the other hand, rock layers tend to be thicker in temperate climates than in tropical climates due to less strong weathering forces. (a) In temperate zones: the image shows an outcrop related to the Rodinia rupture, a part of the Neoproterozoic rift (760–550 Myr) located near Floyd, Virginia. In spite of being a few hundred million years older than the rocks on the left, these rocks have only produced a thin layer of soil, due to physical (frost) weathering. (b) In the tropics: the image shows an outcrop of weathered volcanic rocks of the Miocene (23–5 Myr) Arc in Costa Rica, it is located 60 km SE of Los Chiles, Alajuela, facing the San Juan River. (Gazel et al., 2005).

FIGURE 15.2 The weathering process cycle for CZ in the tropics. Acting forces can be divided into climatic, anthropogenic, and tectonic origins. In tropical zones climate forcing is stronger than in temperate zones. The weathering processes include physical, chemical, and biological weathering, with intensified biological and chemical weathering in the tropics. In tropical climates intensified responses are visible in the soil, in the atmosphere, and in the hydrological cycle. Factors that are particularly enhanced in tropical as compared to temperate climates are framed in dark gray [green in the web version]. *(Image sources: NSF/Pacific Worlds, Telegraphy/ALAMY, Baum et al. (2001), Alvarado (2010), Dybas (2013), Telegraph/EPA, Jacobsen et al. (2012).)*

tropical processes. In tropical environments the CZ can be much thicker and rich in components and biodiversity than in nontropical regions (Fig. 15.3).

The tropics are formally defined as the region occurring between the Tropic of Cancer and the Tropic of Capricorn (Fig. 15.3). This definition can be broadened to enclose regions with high temperatures, large amounts of rainfall, and seasonality with alternating periods. Due to the vastness of this definition tropical climates may vary and are controlled principally by factors such as rainfall, topography, and winds, as well as incident radiation, and cloud coverage (Giller, 2001). Additionally, areas characterized by tropical climate can be subdivided into: (1) wet tropics, if there is at least 10 months of rainfall; (2) wet and dry tropics, if they are characterized by a dry season of at least 2 months; (3) dry tropics; and (4) cool tropics, in regions with higher altitudes (Norman et al.,

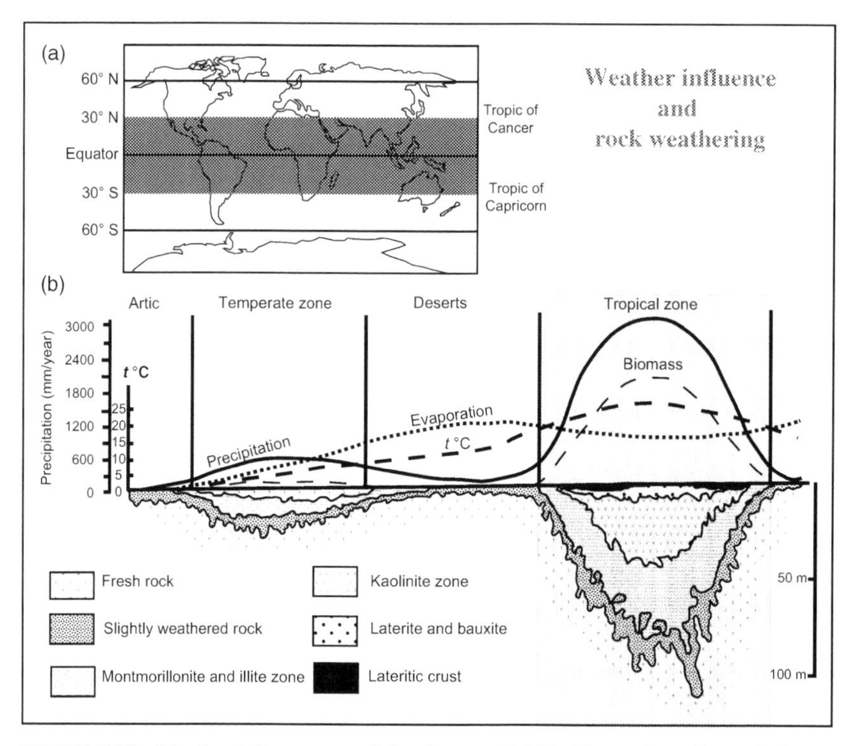

FIGURE 15.3 Weather influence on soil development. (a) World map shows the tropical zone, situated between the Tropic of Cancer and the Tropic of Capricorn. (b) The vertical profile shows the evaporation, precipitation, temperature, and biomass distribution for a range of climates. In tropical zones, the penetration depth of weathering is more extreme, producing a thicker section of alteration products. Additional layers of laterite and bauxite are present and the thickness of remaining layers is greater than in other zones. *(Modified from Smirnov (1982).)*

1995). Most tropical areas have alternate dry and rainy seasons, instead of the four seasons occurring in temperate zones. This means that frost or thawing never happen and no interruption of the biological cycles of fauna occurs due to low temperatures or snowfall. Also, mechanical weathering of rocks is related to roots or waves, but may not be due to frost weathering. Soil production is continuous due to biological and microbial activity, and both dry and rainy periods promote soil erosion to different extents. Excess water percolation is thought to accelerate processes such as chemical weathering and underground leaching.

The depth and degree of weathering are closely related to the high temperatures, precipitation, and biomass occurring in the tropics. The vertical section in Fig. 15.3 shows that weathering can reach depths up to 100 m, whereas weathering in temperate zones hardly reaches depths of 50 m. On the other hand, lateritic crusts and bauxite are present exclusively in tropical areas. Finally, thickness of kaolinite, montmorillonite, and illite zones are much larger than in temperate climates. This is a consequence of the weathering order of primary minerals (anorthite > albite > hypersthene > orthoclase > apatite) and formation of secondary minerals (kaolinite, allophane, gibbsite, Fe oxyhydroxide) (Sak et al., 2004).

The Critical Zone is evolving through time and humans are radically changing it (Vitousek et al., 1997; Brantley et al., 2006). The transformation of soils due to urbanization, agriculture, forestry, and mining change the global landscape and may affect our ability to feed human population (e.g., Stocking, 2003; Sanchez and Swaminathan, 2005). Concurrently, the worldwide population is increasing, resulting into a shrinking world, characterized by a reduced global diversity of natural environments.

15.2 IMPORTANCE OF TROPICAL ENVIRONMENT AND ITS RELATIONSHIP WITH CZ

Tropical environments and especially tropical rainforests are of great worldwide importance. They nourish extensive biodiversity, and through thermal and hydrological convection cycles are able to regulate regional and global climate systems. Critical Zones in the tropics host the densest life zones, hence the importance of studying them. An exceptional case is Costa Rica, with only 0.03% of the planet's surface area, its forests alone hold more than 4% of the Earth's biodiversity (Obando, 2002). Worldwide rainforests cover less than 7% of the Earth's surface area but harbor 50% or more of the plants and animals living on Earth (Bierregaard et al., 1992). On the other hand, heat and moisture rise from tropical latitudes to form part of atmospheric convection cells, the Hadley cells that dominate tropical and subtropical climates (Kennett, 1982), which then play an important part in the planet's climate. Equatorial latitudes receive the largest amount of light all year round, enhancing processes such as, photosynthesis and other chemical reactions vital to naturally occurring biochemical reactions in the soil (Paul, 2006).

Forests and their substrate play an important part in the transformation of atmospheric gases as carbon sinks, as well as triggering the nitrate cycle (Fig. 15.4). The dissimilatory nitrate reduction to ammonium (DNRA), an anaerobic microbial pathway of the N cycle that transforms NO_3 first to NO_2, and then on to NH_4, is enhanced in tropical soils (due to low redox potential, available NO_3, and labile C) thus decreasing the amount of nitrate susceptible to leaching into groundwater and denitrification into greenhouse gases (Silver et al., 2001). Harboring nutrients and nutrient pulses constitute one of the most important roles of soil in the tropical rainforests, with extensive impact on the ecosystem as well as plant productivity (Lodge et al., 1994).

The hydrological cycle presented by Giambelluca (2002) underlines the differences between cleared land and forests. In tropical forests evaporation of intercepted rainfall and transpiration is higher (Bruijnzeel, 2001) and infiltration is rapid. Thus, treetops and leaf coverage protect the soil from being washed away, and moisture contained in rainforests acts as a thermal regulator,

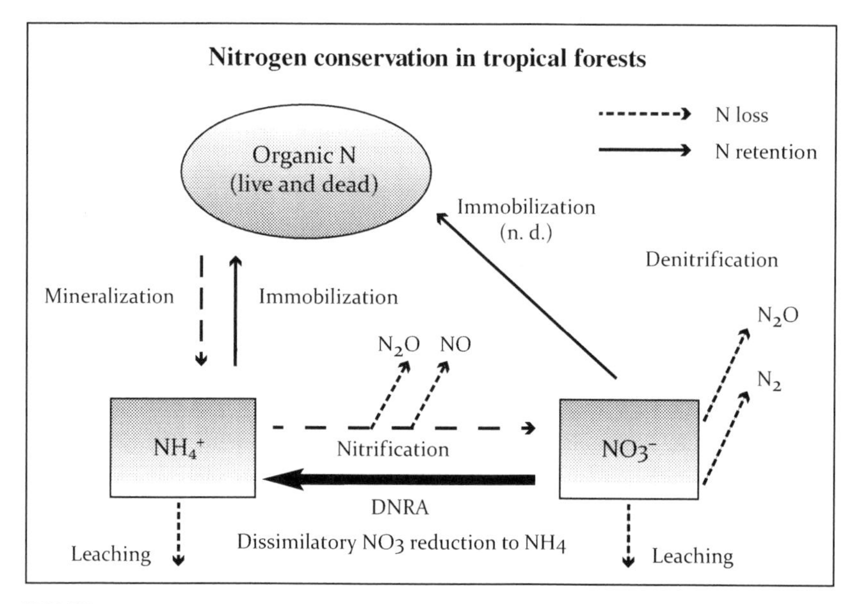

FIGURE 15.4 Nitrogen cycle in tropical environments. Nitrogen is converted through biological and physical processes. The dissimilatory nitrate reduction to ammonium (DNRA) is enhanced in tropical soils, decreasing the amount of nitrate susceptible to leaching into groundwater and denitrification into greenhouse gases. *(Modified from Silver et al. (2001).)*

preventing abrupt temperature changes and subsequent droughts. All these hydrological characteristics make tropical forests singular when compared to CZ in other zones.

Worldwide tropical environments are characterized by a great geological diversity with contributions from igneous, metamorphic, and sedimentary rocks. The rock-type distribution and emplacement respond to major events driven by continental drift, plate tectonics, and paleoclimates rather than geographical location and climate.

In our study we focus on observations derived from Costa Rica, which can be considered a case example for tropical environments. Costa Rica is a small tropical, mountainous country located on the Central American isthmus at 10° North from equator, with high mean annual temperatures (~25°C) and precipitation (~2926 mm/year). The region is formed by a segment of the North American craton, with contributions from island arc magmatism, radiolarite depositions, Galapagos hotspot activity, plate subduction, and accretionary terrain (e.g., Hauff et al., 2000; Gazel et al., 2009). In this area the weathering of rocks due to environmental conditions can be extreme (Laguna, 1984) to the point where it is difficult to find unweathered bedrock outcrops. The extreme weathering can result in the formation of bauxite-rich laterite deposits with particular economic and industrial applications. Consistently, Guinea, Vietnam, Jamaica, Guyana, and Brazil hold most of the world bauxite reserves (USGS, 2012).

15.3 BIOLOGY AND HOLDRIDGE LIFE ZONES

The variety of soils and the diversity in the Critical Zones occurring in tropical environments are large. A growing concern is focused on how to distinguish and organize different types of Critical Zones according to soil conditions, plants, animals, and other key factors. A valuable tool, introduced by L. R. Holdridge in 1967, is the classification of "life zones" that link the distribution of natural vegetation associations and undercanopy characteristics to climate variables. This classification is based on observations obtained in tropical climates, where ground-level environments in densely forested areas are hard to assess by techniques of remote sensing. The underlying postulate of this classification is that elements making up the climate have primary influence on the vegetation. Because the development of mature forests extends over many years, its vegetation reflects long-term properties of the climate (Holdridge et al., 1971). Combinations of three types of climatic data are used: potential evapotranspiration, biotemperature, and humidity provinces. Based on this dataset the life zones are organized into 30 groups; with each group being further subdivided into smaller, more specific zones increasing the number of total life zones present on the planet to 116. This system is commonly used due to its simplicity and overall accuracy (Enquist, 2002).

Holdridge (1967) stated that the soil is an integral part of every ecosystem. The soil is the base within, or upon which, every terrestrial community of living organisms has developed. It not only supports and provides food to animal and plant communities, but it also develops, thanks to these communities, with the participation of climatic factors. The soil holds abundant populations of protozoa, fungi, and other forms of life; plant roots traverse and feed it; and animals move on it and burrow in the soil. In this regards the aforementioned "live ecosystem" corresponds to the central element of a CZ. Therefore, the possibility of using the classification of Holdridge to define distinct types of Critical Zones should be considered. The Holdridge system recognizes that the tropics are complex, and at least 38 tropical life zones have been redefined. In Costa Rica (approx. 50,000 km^2), 12–24 life zones have been classified. Specifically, in Costa Rica the Tropical Science Center developed a system building on the fundaments of the Holdridge classification that includes 23 life zones and 11 transitional zones (Orozco, 2007). This is widely used because it is considered the most accurate classification for the particularities of this country.

15.4 ECONOMICAL IMPORTANCE OF THE CZ IN THE TROPICS

As the zone that sustains all life, and particularly human life, the CZ is essential to various aspects of economy, from a local to a global scale. In fact, the CZ includes not only the soil, but also incipient regolith, aquifers, microbiota, plants, trees, animals, and a thin layer of air and atmospheric gases. All of these factors play a role in our economy.

The soil is at the core of the agricultural system, which represents a considerable part in the economy of developing countries. The effects of loss of soil and declining soil fertility have been assessed and have led many policy makers to conclude that these events would threaten their sustainable development (Brantley et al., 2007). Nutrients are mainly stored in the above-ground vegetation and transferred together with enzymes and other compounds into the soils as litterfall and converted to available nutrients before being quickly reabsorbed by vegetation (Palm et al., 1996). The nutrients are available in slash-and-burn agriculture when land is left uncultivated 20 years or more, yet this is not the case in modern practices (Rumple et al., 2006). The absence of winter in tropical latitudes allows a continuous crop production, especially in humid tropics where incoming solar energy, adequate soils, and moisture balance render this feasible. It must be noted that a no-production lapse always occurs between a harvest and the next sowing. In some cases, for example, in sugar cane production, the dry season is necessary. The crops are slashed and burned in order to clear the land and finish the harvest before beginning to plant again.

Both maize and rice seed cultures seemingly have been introduced during prehistoric times (Harris, 1972) and have become dominant crops in tropical latitudes. Other economical activities include the harvest and export of tropical products such as banana, coffee, cacao, etc., which have become the principal source of income for some economies. For example, Guatemala, El Salvador, Nicaragua, Ethiopia, and Uganda rely heavily on coffee exports, whereas Argentina and Paraguay primarily export soybeans. Due to extremely rich and varied genetic crop resources, different varieties of plants from virgin rainforests are used to fortify already existing crops (Clement, 1999).

At the lower end of the CZ, rocks are a fundamental source of materials for economic purposes. For instance, limestone and gypsum are essential for construction, whereas mining is an important source of income for certain communities. Chile and Peru, Congo, South Africa, and other African countries depend primarily on the mining industry and oil is the first export product for Venezuela, the Middle East, and Indonesia. Deposits of bauxite and laterites develop in association with weathering, thus are principally found in wet hot tropical areas. These are exploited because of their enrichment in metals such as iron and aluminum becoming an integral part of the economy (e.g., Jamaica).

Aquifers are crucial in tropical environments. They store water used in dry seasons to sustain an all-year-round agriculture. Rivers and other forms of water-flow are used to produce renewable energy with hydroelectrical plants. Trees are another exploitable resource and many exotic species of wood come from tropical rainforests. Finally, microbiota including bacteria, fungi, and termites present in tropical forests are fundamental in the process of decomposing the decaying matter and rendering nutrients available in the ecosystem (Lal, 1988; Waldrop et al., 2000).

15.5 IMPORTANCE OF CZ STUDIES IN PREVENTING NATURAL DISASTERS

Natural hazards are a combination of geophysical events such as earthquakes, landslides, volcanic activity, and climate-induced hazards that threaten different social entities due to the process itself (natural vulnerability), but also due to vulnerability associated with human systems (human vulnerability) (Fig. 15.5). Natural disasters occur when natural hazards produce damage to physical and social goods with major impacts on society and/or infrastructures (Alcántara-Ayala, 2002). Hazards are the result of sudden changes in long-term behavior caused by minute changes in the initial conditions (Alcántara-Ayala, 2002). Natural disasters can have great impacts on developing countries located in tropical latitudes. This is not only due to the factors of human vulnerability but also to natural conditions such as the susceptibility of flooding caused by

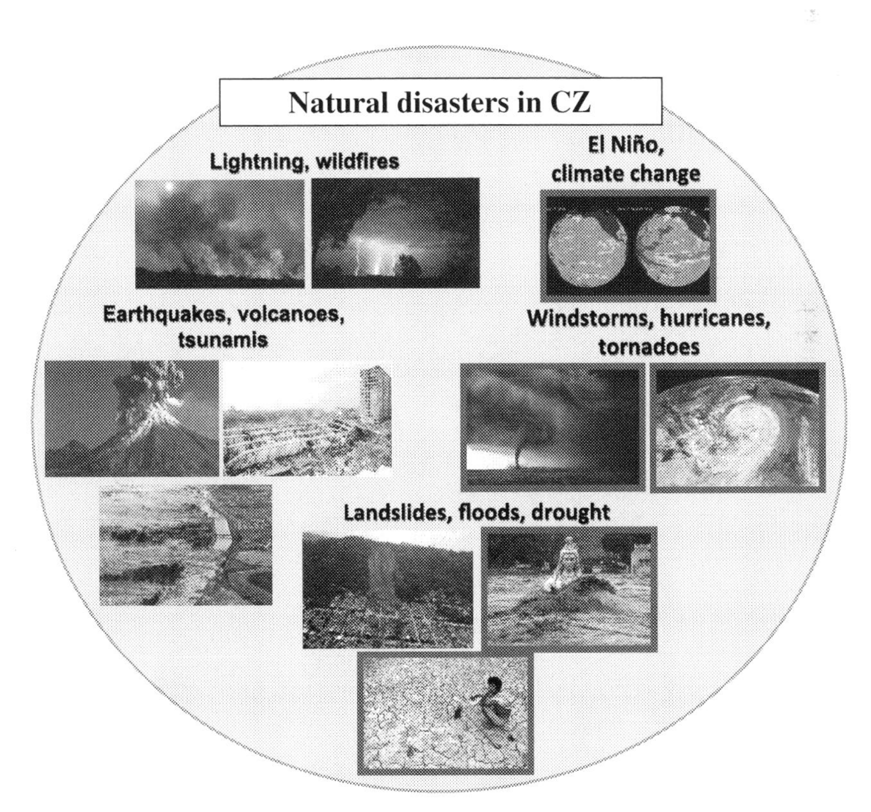

FIGURE 15.5 Natural disasters in the Critical Zone. Climate-induced hazards include ENSO, windstorms, hurricanes, tornadoes, flood, lightning, droughts, and wildfires. Tectonic hazards include earthquakes, volcanoes, and tsunamis. Anthropogenic activities can trigger any of these hazards and particularly accelerate climate change. Hazards enhanced in tropical climates are framed in dark gray [green in the web version].

tropical storms, monsoons, and hurricanes (Alcántara-Ayala, 2002). Natural hazards are inevitable processes, but they result in natural disasters only after we have decided to live or build infrastructure along their paths. Geomorphology and detailed CZ studies play an important role in preventing natural disasters. Mapping, modeling prediction, management proposals, effective volcanic-hazard zoning, and seismic-hazard assessment are essential (Alcántara-Ayala, 2002) and communities should focus on better understanding, analyzing, and forecasting hazards such as floods and mass movements.

Berz et al. (2001) classified natural hazards as follows: (1) earthquakes, volcanoes, and tsunamis; (2) windstorms including tropical cyclones, extra-tropical storms, regional storms, tornadoes, and hail; (3) lightning and wildfires; (4) floods; (5) effects of El Niño and climate change; (6) landslides; and (7) drought (Fig. 15.5). These can variably affect the CZ and the tropics.

Some natural hazards are linked to climate and are therefore more common in certain settings whereas others are independent of geographic location and can occur worldwide. The first are processes tied to climate settings. These, being dependent on heat and moisture that occur in tropical regions, are more frequent in said latitudes and include storms, tornadoes, hailstorms, ENSO, climate change, floods, and droughts. The latter category includes earthquakes, tsunamis, volcanic activity, lightning, landslides, and wildfires.

Earthquakes are regarded as the most destructive force of nature, even though storms and floods are responsible for the largest part in deaths and loss. Earthquakes can be either amplified (Mexico-city-effect) or attenuated by the composition and the thickness of sediment layers upon which human infrastructures are built (Berz et al., 2001). Depending on the magnitude and position, earthquakes can either severely disrupt the CZ and human activities, or pass unnoticed. *Tsunamis* can be generated by earthquakes (e.g., Japan, 2011) or underwater landslides. Even though they do not occur in the Critical Zone, in *sensu stricto* they modify and therefore affect ecosystems including water, plants, and soil in coastal areas (Berz et al., 2001). *Volcanic eruptions* represent a potential threat to all communities and ecosystems occurring in their vicinity. Damages can be caused due to ash deposits, tsunamis, lava and mudflows, avalanches, volcanic earthquakes, and gas release. These processes affect the CZ interrupting the weathering cycle, killing vegetation and wildlife, damaging the soil, and disrupting natural cycles of water and atmospheric gases.

Storms are a threat to the land, although predominantly influenced by sea winds and climate. Hazards include not only the direct impact of the storm but also high winds capable of bringing down trees and destroying infrastructure, storm surges, and pounding waves affecting both anthropic and natural habitats in coastal areas. Secondary effects include large amounts of rainfall inland resulting in severe floods (Berz et al., 2001). The result is an enormous catastrophic potential (e.g., Hurricane Katrina, 2005) and disturbance of the CZ. *Tornadoes* are not very frequent in tropical latitudes but can happen and significant losses have been recorded due to tornadoes in India, Japan, South Africa, and South America. *Hailstorms* are not recurrent in hot tropical areas, but when occurring they can cause extensive damage to agriculture, buildings, and vehicles (Berz et al., 2001). Due to the short time of occurrence their effects on vegetation, wildlife, and other components of the CZ are considered minimal in tropical areas.

Lightning can have massive impacts on human activities or no impact at all. Regarding human activities, thunderstorms can cause more fatalities than other natural hazards. They are dangerous for aircraft, electrical equipment, electronic functions, and control units resulting in the damage and interruption of electrical service (Berz et al., 2001). Lightning is the main cause of natural fires, with a direct impact on the CZ. *Wildfires* influence ecosystems directly by disturbing competition relationships between and within species. They reduce the potential coverage of forests and facilitate the expansion of fire-dependent grassland and shrubland. In tropical latitudes they have direct consequences on the hydrological cycle, they result in impoverishment and weathering of soil as well as degradation of ecosystems and modification of biotic relations (Jin, 2010). Remote sensing techniques such as LiDAR images are important to track the effects of wildfires; for example, during the Las Conchas wildfire aftermath in the Jemez Mountains (New Mexico), LiDAR was used to quantify postfire erosion (Orem and Pelletier, 2013).

Floods are common and have the highest damage rates among natural hazards. In tropical latitudes their frequency is particularly high due to high amounts of rainfall, tropical storms, monsoons, and other climate conditions. They are not limited to areas along rivers as the water can proceed for long distances. Effects of the excess moisture include washing away vegetation, rocks, soil, and human infrastructure, as well as disrupting natural habitats of many species.

The El Niño southern oscillation (ENSO) produces climate anomalies across the globe in interannual timescales, with impact in tropical eastern sea surface temperatures and rainfall (Hoerling et al., 2001). During *La Niña*, trade winds are intense and heavy rainfalls occur mainly over the far western tropical Pacific; during El Niño, the winds relax so that coastal zones of Ecuador and Peru have severe floods, whereas New Guinea and Indonesia experience relatively dry conditions (Fedorov and Philander, 2000; Fig. 15.6). Moreover, *global climate change* in general exacerbates extreme weather conditions through winds, snowfall, rain, or heat. In the tropics the most common setting is an increase in heat and rainfall, intensifying vulnerability to tropical vector-borne diseases (Adger et al., 2005).

Landslides can be caused by different events such as earthquakes, volcanic activity, and floods. Damages caused by landslides affect wildlife and people, disrupting the normal cycle of all the components of the CZ in the area. The following are three examples of landslides that happened in the tropics under different conditions, all with catastrophic consequences:

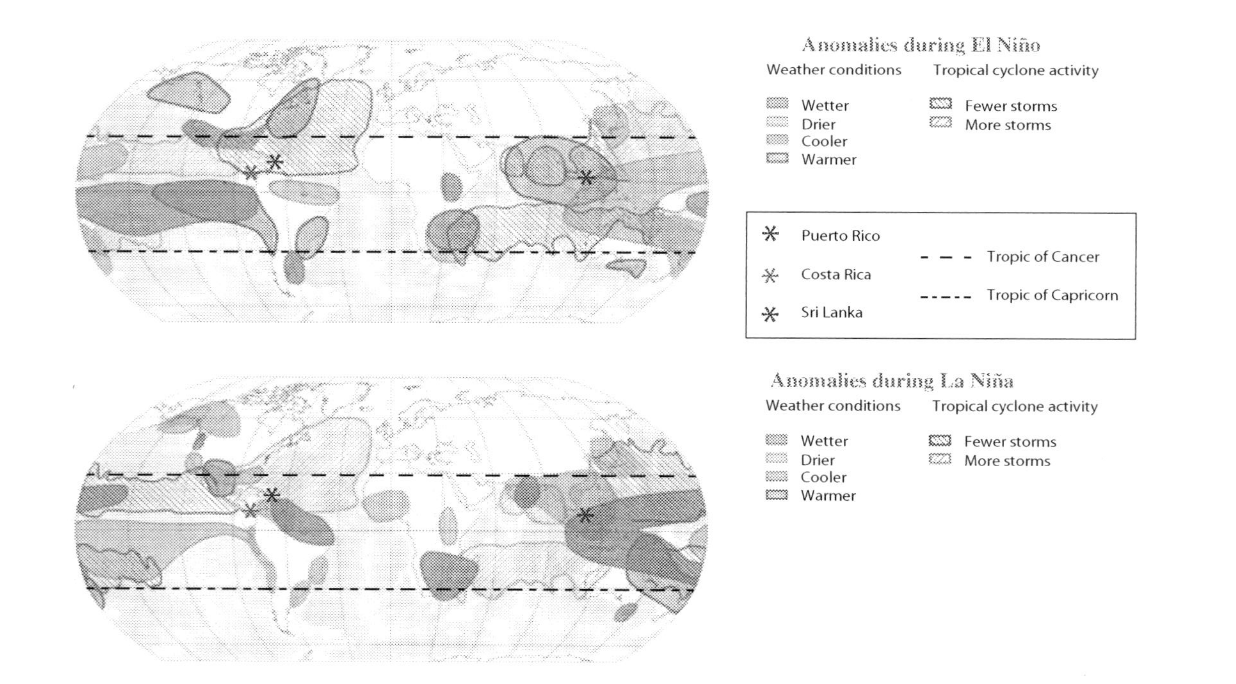

FIGURE 15.6 Changes in regional climate pattern during El Niño and La Niña anomalies. Changes affect mostly tropical latitudes and may result in wetter/drier and warmer/cooler climates with an incidence on the number of storms. Asterisks mark the location of Puerto Rico, Costa Rica, and Sri Lanka. *(Modified from Nathan (2011).)*

1. The volcanic eruption of the Nevado del Ruiz in Colombia interacted with snow and ice, producing meltwater, which triggered catastrophic lahars and killed more than 23,000 people (Pierson et al., 1990).
2. El Salvador earthquakes (2001), triggered hundreds of landslides killing almost 1200 people. Failures were independent of lithology and responded to steep slopes, although unconsolidated ash combined with groundwater resulted in flows, which traveled large distances and accounted for more than half of the deaths too. From this case study, what stands out is that vegetation coverage did not stop the landslides (most affected areas had good coverage) and that human activity can aggravate risks (urbanization on steep slopes and next to defined high hazard sites) (Bommer et al., 2002).
3. The Cinchona earthquake in 2009 shook the Central Valley of Costa Rica and was associated with a local fault (RSN, 2009). The seismic movement triggered thousands of landslides, causing casualties and extensive infrastructure damage (Ruiz et al., 2014). Research conducted by Ruiz (2012) and Ruiz et al. (2014) led to the following assessments: (a) Factors exerting an influence on landslide susceptibility include orographic regions, slope gradient, and lithology. The vegetation cover had no known influence. (b) Older geological units (>0.5 Ma) had thicker soil coverage and produced slides and slumps, whereas younger units (<0.06 Ma) had less soil coverage and produced debris flows.

Comparison of earthquake-triggered landslide volumes in tropical and temperate climates (Fig. 15.7) using events of the same magnitude and in the same conditions show that in most cases landslide volumes are greater in the tropics.

Droughts are natural hazards that can occur even in tropical climates. Droughts impact both surface and groundwater resources and affect all the components of the Critical Zone. Lack of moisture may stop the hydrological cycle, cause the soil to dry out, and microbial and microbacterial activity ceases temporarily, vegetation lacks nutrients, water dries up, and fauna may be threatened by the lack of food and suitable habitat, whereas trees can either endure this period or die (Condit, 1998).

15.6 A CASE STUDY IN SOIL WEATHERING IN TROPICAL ENVIRONMENTS: PUERTO RICO VERSUS SRI LANKA

The Critical Zone is highly heterogeneous due to a variety of physical, chemical, and mechanical weathering processes (Fig. 15.1). Anderson et al. (2007) modeled the CZ as a feed-through reactor in order to analyze physical and chemical controls in weathering. Chemical weathering is driven by thermodynamic disequilibrium relative to mineral formation conditions, whereas mechanical weathering and physical erosion are driven by stress gradients that break or move material. The process as a whole helps us understand how these affect atmospheric CO_2 and climate (see Dupré et al., 2003).

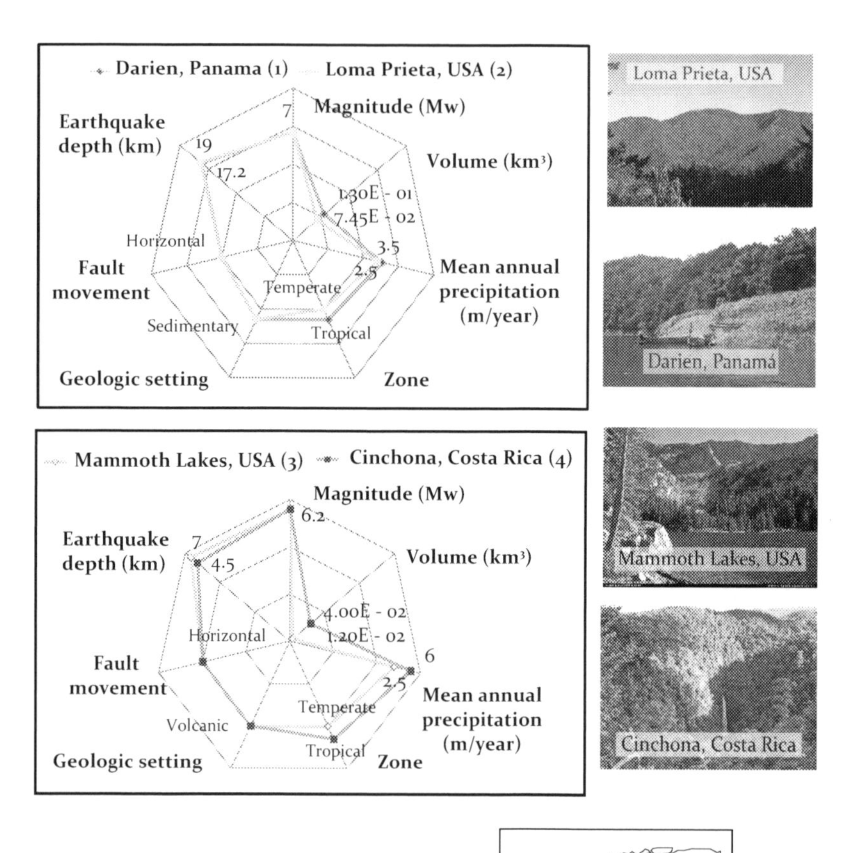

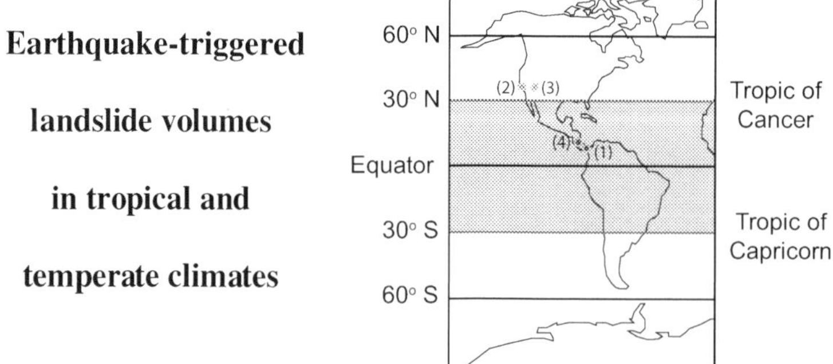

FIGURE 15.7 Landslide volume comparison in tropical and temperate climates. On top, magnitude 7 earthquakes are compared between the Darien, Panamá, 1976 earthquake and the Loma Prieta, USA, 1989 quake. Both events occurred in sedimentary rocks, with horizontal fault movements, with similar depths and mean annual precipitation. In the middle, magnitude 6.2 earthquakes are compared between the Mammoth Lakes, USA, 1980 earthquake and the Cinchona, Costa Rica, 2009 quake. Both events occurred in volcanic edifices, with horizontal fault movements, though depth and mean annual precipitation vary slightly. In both graphs landslide volume is greater in tropical climates than in temperate zones. Location of each event is shown on the map on the bottom. *(Based on Malamud et al. (2004) and Ruiz (2012).)*

In the feed-through reactor, in a region of net denudation (where uplift ≈ erosion), solid material enters through the lower boundary, sediments are skimmed off at the top by erosion, and dissolved products leak out through the porous sides (Anderson et al., 2007). A parcel of rock moves vertically through this system, fracturing and undergoing chemical weathering, until it reaches the rock–regolith boundary; then it moves randomly through the soil due to the effect of mechanical and biomechanical forces and with time it moves downslope. Finally, it undergoes abiotic weathering and biologically driven nutrient extraction (Fig. 15.1). Dupré et al. (2003) analyzed the weathering of the main types of continental silicate rocks, which is a process that is considered the main CO_2 sink in the geological timescale. Continental basalts have very high weatherability (eight times more than granite) and account for 30% of the total consumption of atmospheric CO_2. Weathering rates are controlled by surface area, regional continental runoff, and mean annual temperature (Dupré et al., 2003). In this perspective, formation of continental flood basalts is affected by the Earth's climate directly, over mega-year periods. Research conducted by White and Brantley (2003) considers that the time-dependent nature of silicate reaction rates reflects characteristics intrinsic to the mineral (surface area, depletion of energetically reactive surfaces, accumulation of leached layers, and secondary precipitates) and extrinsic to the weathering environment (low permeability, high mineral/fluid ratios, and increased solute concentration).

The following examples show us the variety of weathering rates in tropical climates (Fig. 15.6) and the main factors which control erosion in tropical environments. According to White et al. (1998), erosion rates for quartz diorites in Puerto Rico are extreme (5.8 mm/100 year), whereas basalts in Costa Rica have an erosion rate considered "normal" (2.9 mm/100 year) (Brantley et al., 2007). Granulites and amphibolites from Sri Lanka hold the lowest rate of weathering in tropical regions (silicate weathering rate of 0.2–0.7 mm/100 year) (von Blanckenburg et al., 2004). These can be considered extreme cases in comparison to forest soils that haven't been disturbed by human activity, which erode at rates of 0.76–1.5 mm/100 year (Dybas, 2013).

15.6.1 Puerto Rico

In the warm, wet, steep terrain of the Luquillo Experimental Forest (Puerto Rico), the silicate weathering flux is the highest recorded (5.8 mm/100 year). This is due to the fact that in Luquillo, the water table is below the saprolite and water flushes rapidly through the profile (Anderson et al., 2007). Studies conducted by White et al. (1998) in the Rio Icacos watershed in the Luquillo forest, characterized the weathering regimes within the regolith. This region is characterized by a steep topography, high denudation rates, extremely high rainfall (4200 mm/year), and tropical temperatures. The underground CZ vertical profile is composed of (from top to bottom): soil, oxidized saprolite (the association of both is called regolith), and a decomposed saprock, which grades to the quartz diorite bedrock. Weathering occurs in two distinct environments: dissolution of plagioclase and hornblende occurs at the saprock interface and weathering of biotite and quartz take place in the overlying thick saprolitic regolith (White et al., 1998). These produce two distinctive water chemistries: K, Mg, and Si are enriched in the regolith pore waters; whereas Ca, Na, and Si are more concentrated in steam waters (White et al., 1998). Porosities are high near the surface and infiltration models indicate that increasing capillary tension and gravimetric potential control pore water infiltration, with a rate of 1 m/year. Interestingly changes in patterns of precipitation, temperature, and vegetative cover have not significantly impacted weathering rates over the last several hundred thousand years (White et al., 1998). A notable exception is the promontory of El Yunque, a peak that presides over the island of Puerto Rico. Due to the intense rainfall, this area should be covered in vegetation and eroding rapidly. Nonetheless, this area is characterized by a slower erosion rate (0.39 mm/100 year). This is due to the texture and the composition of the rocks present at El Yunque that suffered a "cooking" process in the chamber of an ancient volcano.

15.6.2 Costa Rica

In Costa Rica, topography and microclimates are diverse. Weathering thickness calculated from alluvial fill terraces on the Pacific Coast yield weathering rates of 2.9 mm/100 year (Sak et al., 2004). This region is characterized by a flat topography (5–200 m asl), high rainfall (~3085 mm/year), and tropical temperatures (27.3°C). The calculated rate of rind advance is within an order of magnitude from the global average. In this case, Sak et al. (2004) observed that weathering thickness increase linearly as a function of time, and are proportional to mean annual temperature and precipitation increase.

15.6.3 Sri Lanka

In Sri Lanka, mountains are as high as 2500 m asl, slopes are steep (5–30°), monsoon rain is heavy (~5000 mm/year), and temperatures are tropical (~25°C). The island formed as a result of the continental breakup of Gondwana during the early Cretaceous. Uplift and tectonic forcing have not taken place ever since. Research conducted by von Blanckenburg et al. (2004) calculated the lowest denudation (0.2–1.1 mm/100 year) and silicate weathering (0.2–0.7 mm/100 year) rates on the planet. This contradicts the suggestion that high relief, temperature, and precipitation lead to high rates of weathering. In this area fresh mineral surfaces available for weathering are not supplied due to the absence of tectonic activity (von Blanckenburg et al., 2004). Soils develop into mature profiles resulting in thick clay layers, which constitute an obstacle for water filtration and slows down weathering and denudation.

15.7 CLIMATE CHANGE AND ITS EFFECT ON CRITICAL ZONES IN THE TROPICS

Prediction of global warming due to fuel emissions was made by Svante Arrhenius in 1896, but consensus on anthropogenic global warming was not reached until 1980, and widespread shift in public opinion did not happen until 2007 (Corlett, 2012). It was initially thought that climate change would be too slow to ever notice it, but in the last 10 years, evidences have shown that it is happening at a rate that is relevant for all life on the planet. Since the mid-1970s, the tropics have warmed at a mean rate of 0.26°C per decade (Corlett, 2012). Different models for greenhouse gas scenarios have been created, nonetheless little is known of what will really happen. What is certain is that the tropics will continue to get warmer, by at least 1–2°C by 2100, and possibly even more if fossil fuels emissions continue to increase and exceed expectations (Corlett, 2012).

Biological consequences of climate change and temperature increase are not well known, although recent evidence suggests that the tropics are only a few degrees away from major adverse impacts (Corlett, 2012). Among potential threats, droughts have received the most attention, although it is the increase in temperatures that will have a major impact on the tropical lowlands. Rainfall might either increase or decrease. Biological responses for species can occur through individual action and by reordering within a community and/or immigration from outside it (Corlett, 2012). Novel ecosystems are inevitable and future studies need to focus on how climate change will interact with other threats to biodiversity. For the moment, conservation management responses focus on minimizing nonclimate pressures on tropical rainforests, restoring connectivity to enable movement of genes and species between fragmented populations, and reconditioning canopy cover to reduce temperature maxima (Corlett, 2011).

Anthropogenic climate change has various effects besides global warming (or cooling in some places). Possible resulting scenarios include glacier melting and retreat of permafrost regions, rising of sea level, and changes in seasonal and regional precipitation patterns, resulting in more heavy rain, flash floods, landslides and tornadoes (Berz et al., 2001). All of these conditions affect the natural patterns and cycles in the Critical Zone, with variable implications, some conditions more easily withstood than others.

ENSO cycles (Fig. 15.6) have dramatic effects on tropical rainforests, plant and animal reproduction, as well as incidence of droughts and fires (Corlett, 2011). Permanent El Niño- or La Niña-like conditions could disrupt the synchronization between plants and animals across the tropics with potentially drastic consequences. Research shows a trend towards enhanced ENSO activity causing some concern (Corlett, 2011).

Evidence for climate change is also visible in the disappearance of *ice caps* and glaciers at high elevations in tropical latitudes (which are very sensitive to small changes in temperature), in the Andes and Africa (Thompson, 2000). Loss of natural glacier dams implies less water for agriculture and hydropower production (Thompson, 2000), and glacial melt water contribution to rivers is in danger. Jacobsen et al. (2012) showed that 11–38% of regional species pools, including endemic species, can be expected to be lost following complete disappearance of this water in glacier-fed communities.

Wildfires are principally driven by climate, increasing during the La Niña period of the ENSO in tropical latitudes (Jin, 2010). Areas with higher risk of wildfires occur in sub-Saharan and southeast Africa (dry tropics). They indirectly affect ecosystems by changing the climate, accelerating the natural carbon cycle, becoming sources of carbon emission, and emitting harmful trace gases (Jin, 2010). Goldammer and Price (1998) predicted changes in wildfire risk in the coming years: tropical forests will increase risk due to land-use changes, increasing fire sources, reduction of mean annual precipitation, and average prolongation of dry seasons. In contrast, Corlett (2012) has stated that undisturbed tropical rainforests do not generally burn because of high humidity and fuel moisture.

In recent years, the demand for water has increased, whereas floods and droughts have been experienced with higher severity levels. Mishra and Singh (2010) classify droughts into meteorological, hydrological, agricultural, socio-economic and groundwater droughts, all regarding a lack of moisture for a specific purpose. In tropical regions, a drought may correspond to a response to La Niña phenomena or simply be an exacerbation of normal "dry season" conditions, and are most common in the dry tropics. Droughts are slow developing events characterized by long duration, and have the least predictability (Mishra and Singh, 2010). Severe droughts can have large impacts on agriculture, domestic and industrial users, tourism, and ecosystems (Kahil et al., 2015). Prolonged lack of water directly affects soil conditions making it more vulnerable to dust storms during drought or to erosion due to rainfall after the drought has ended (Banwart, 2011).

According to Hunter (2003), changes in climate may affect public health through increased frequency of heavy rainfall events, associated flooding, and increased temperature. Many viral, bacterial, and parasitic diseases have been associated with waterborne transmission, and vector-borne diseases such as malaria may increase due to rising temperatures. Evidence exists of outbreaks following floods in developing nations, and malaria prevalence in South America and Asia varies in relation to ENSO events (Hunter, 2003).

15.8 COSTA RICA: SUGGESTING A NEW CZO

CZOs are crucial for understanding the physical, chemical, and biological processes that turn rock into soil and their reactions to climate and land use changes. Multidisciplinary teams study and analyze data collected in CZOs, and are discovering new interrelations. Tropical environments are extremely rich in natural species as well as processes, and due to climate change they are being threatened in ways in which we only begin to comprehend. It has come to our attention that the only CZOs in the tropics are those of Luquillo (Puerto Rico) and Red Soils (China), which give us only partial information on CZ in the tropics due to a lack of other sites for later comparison. Our focus on the importance of Critical Zones in tropical environments has led us to the conclusion that more CZOs are necessary in these latitudes and that Costa Rica is an ideal place. The country is not only extremely biodiverse but holds a large amount of different climates, at diverse altitudes and in a wide variety of settings, in only a few ten thousand square kilometers. Biological stations belonging to National System of Conservation Areas, the Organization for Tropical Studies (NGO) and the Soltis Center (TAM) already exist and would be a great base to build CZOs; also multidisciplinary groups of scientists could be easily formed. Recent studies include the characterization of two bioactive volcanic lakes with geogenic carbon dioxide reservoirs (Cabassi et al., 2014).

15.9 DISCUSSION

In tropical environments, Critical Zones are affected by a variety of phenomena similar to the ones conditioning the weathering of rocks. Acting forces include climatic, geologic, and anthropogenic activity (Fig. 15.2). These forces are directly or indirectly linked to natural processes occurring in the CZ and may in turn affect to a variable extent. Climatic factors include the presence or absence of heat, moisture, rainfall, and winds. In tropical climates heat and moisture dominate weather conditions and their variations have a direct repercussion on all of the processes that dominate in the CZ. Geological factors include tectonic and magmatic activity due to the geologic conditions present in tropical areas such as the ring of fire, divergent plate boundaries, hotspots, continental rift, and triple junctions. Natural hazards linked to geological activity constitute an important modification of CZ (earthquakes, landslides, volcanic eruptions). Finally anthropic activities in tropical latitudes have a direct impact on the CZ and all the factors that determine its functionality (presence of water and nutrients among others). Human activities with major impacts include logging and deforestation for land use, intensive farming and agriculture, emission of greenhouse gases, mining and oil extraction, as well as chemical and atmospheric pollution, all of which profoundly modify the CZ.

A variety of processes, including physical, chemical, and biological phenomenon interact in the CZ, leading to the weathering of rocks and the formation of soil. Physical processes include stresses due to temperature, pressure, weight, and abrasion exerted by water, ice, and wind. These act on basement rocks as well as soil and wildlife on a smaller scale. Chemical interactions are strictly related to the presence of water. Additionally the CZ is influenced by other mechanisms such as photosynthesis and nutrient cycles that involve a biological and a chemical interaction. Biological processes occur on microscopic to macroscopic scales. They include internal biochemical reactions as well as biological activity that result in the modification of external environments.

Combinations of the aforementioned processes occur simultaneously and interactively affect all the components of the CZ, from rocks to biota. In the tropics, these activities are enhanced due to the presence of heat, moisture, and abundant biota. This has consequences on the rate at which the soil is produced. In fact, according to Stanley (2010), fungi take longer to process organic material in cold temperatures, whereas the process is accelerated with warmer temperature. To study the CZ in detail it becomes necessary to categorize and quantify these interactions. Holdridge's classification of life zones (1967) is an adequate basis for further studies, principally of soil and biota.

Finally, modifications in the CZ have secondary effects on climate, biological, hydrological, and soil cycles. Absence of tree coverage has implications for moisture retention, temperature regulation and water evapotranspiration, condensation, evaporation and infiltration rates; modifying cloud formation and rainfall patterns. Loss of vegetation implies a loss of habitat and food sources for animals, and the degradation of soil quality is linked to absence of microorganisms and restrained organic matter production. Disruption of hydrological cycles has a direct impact both on biological and human activities, and along with soil impoverishment can hinder agricultural production to a great extent.

In order to promote further investigations regarding the CZ in tropical climates, we suggest establishing a new CZO in Costa Rica. The country not only already has existing biological observatories but it counts on experienced professionals in biology, geography, geology, engineering, and other sciences coming mostly from four major state universities that have the adequate equipment, laboratories, and experimental stations. These professionals would be able to assist already existing research groups or institutes, and/or form new ones with guidance from scientists already experienced in the study of Critical Zones. The region is very diverse. It has distinct geologic settings and it is influenced by conditions from both the Pacific Ocean and the Caribbean Sea. It is exposed to both El Niño and La Niña resulting in a large number of microclimates (Fig. 15.6). Moreover, forest regeneration is rapid due to soil, ecological, and climate conditions, thus in just a few decades the deforestation rate was reverted (Chazdon, 2008).

15.10 CONCLUSIONS

This work aims at describing how the Critical Zone is modified by tropical processes. In tropical environments, the CZ can be much thicker and more rich in components and biodiversity compared to nontropical regions. The tropics are regions with high temperatures, large amounts of rainfall, and alternating period seasonality. Tropical regions are of worldwide importance because of the extensive biodiversity, and their influence on thermal and hydrological convection cycles that have huge consequences on both regional and global climatic systems. Additionally, Critical Zones in the tropics host the densest life zones, hence the importance of studying them. Due to the profusion of biodiversity and microclimates, the opportunity to use the classification developed by Holdridge (1967) to define distinct types of Critical Zones should be considered.

As the zone that sustains all life, and particularly human life, the CZ is essential to various aspects of economies, from local to global scales. Variation in the CZ conditions may affect agricultural systems, harvest and export of tropical products, mining industry and oil, aquifers, and trees. Additionally, it's fundamental to study in more detail the CZ to prevent natural disasters. Natural hazards are inevitable processes, which result in natural disasters only when we live and/or build infrastructure along their paths. Geomorphology and detailed CZ studies play important roles in preventing natural disasters. Determining the conditions of soil, slope inclination, vegetation, rainfall patterns, and volcanic and seismic hazard enable the prediction of hazard risk and management of human activities in order to minimize potential threats.

Specific case studies such as the Puerto Rico, Costa Rica, and Sri Lanka example show us that soil erosion is determined by factors other than precipitation, temperature, and relief. This shows that thorough investigations of the CZ in the tropics are needed. Since global climate change is already modifying climates and natural habitats, it is important to know how Critical Zones work in the tropics in order to protect and preserve biodiversity and ecological cycles, as well as to prevent and mitigate natural hazards effectively.

ACKNOWLEDGMENT

We thank Chris Houser and Rick Giardino for the opportunity to contribute to this special publication on the Critical Zone.

REFERENCES

Adger, W.N., Hughes, T.P., Folke, C., Carpenter, S.R., Rockström, J., 2005. Social–ecological resilience to coastal disasters. Science 309, 1036–1039.

Alcántara-Ayala, I., 2002. Geomorphology, natural hazards, vulnerability and prevention of natural disasters in developing countries. Geomorphology 47, 107–124.

Alvarado, G.E., 2010. Hydrogeological and sedimentological aspects of the mud flows related to the Cinchona earthquake (Mw 6.2) of January 8, 2009, Costa Rica. Rev. Geol. Am. Centr. 46, 67–96.

Anderson, S.P., von Blanckenburg, F., White, A.F., 2007. Physical and chemical controls on the Critical Zone. Elements 3, 315–319.

Banwart, S., 2011. Save our soils. Nature 474, 151–152.

Baum, R.L., Crone, A.J., Escobar, D., Harp, E.L., Major, J.J., Matrinez, M., Pullinger, C., Smith, M.E., 2001. Assessment of landslide hazards resulting from the February 13, 2001, El Salvador earthquake. USGS open-file report 01-119.

Berz, G., Kron, W., Loster, T., Rauch, E., Schimetschek, J., Schmider, J., Siebert, A., Smolka, A., Wirtz, A., 2001. World map of natural hazards – a global view of distribution and intensity of significant exposures. Nat. Hazards 23, 443–465.

Bierregaard, Jr., R.O., Lovejoy, T.E., Kapos, V., dos Santos, A.A., Hutchings, R.W., 1992. The biological dynamics of tropical rainforest fragments. BioScience 42 (11), 859–866.

Bommer, J.J., Benito, M.B., Cuidad-Real, M., Lemoine, A., López-Menjívar, M.A., Madariaga, R., Mankelow, J., Méndez de Hasbun, P., Murphy, W., Nieto-Lovo, M., Rodríguez-Pineda, C.E., Rosa, H., 2002. The El Salvador Earthquakes of January and February 2001: context, characteristics and implications for seismic risk. Soil Dynam. Earthq. Eng. 22, 389–418.

Brantley, S.L., White, T.S., White, A.F., Sparks, D., Richter, D., Pregitzer, K., Derry, L., Chorover, J., Chadwick, O., April, R., Anderson, S., Amundson, R., 2006. Frontiers in exploration of the Critical Zone: report of a workshop sponsored by the National Science Foundation (NSF), 2005, Newark, DE, p. 30.

Brantley, S.L., Goldhaber, M.B., Ragnarsdottir, K.V., 2007. Crossing disciplines and scales to understand the Critical Zone. Elements 3, 307–314.

Bruijnzeel, L.A., 2001. Forest hydrology. In: Evans, J.C. (Ed.), The Forests Handbook. Blackwell Scientific, Oxford, UK, p. 12.

Cabassi, J., Tassi, F., Mapelli, F., Borin, S., Calabrese, S., Rouwet, D., Chiodini, G., Marasco, R., Chouaia, B., Avino, R., Vaselli, O., Pecoraino, G., Capecchiacci, F., Bicocchi, G., Caliro, S., Ramirez, C., Mora-Amador, R., 2014. Geosphere–biosphere interactions in bio-activity volcanic lakes: evidences from Hule and Río Cuarto (Costa Rica). PLoS ONE 9 (7), e102456.

Chazdon, R.L., 2008. Beyond deforestation: restoring forests and ecosystem services on degraded lands. Science 320, 1458–1460.

Clement, C.R., 1999. 1492 and the loss of Amazonian crop genetic resources. I. The relation between domestication and human population decline. Econ. Bot. 53 (2), 188–202.

Condit, R., 1998. Ecological implications of changes in drought patterns: shifts in forest composition in Panama. Climatic Change 39, 413–417.

Corlett, R.T., 2011. Impacts of warming on tropical lowland rainforests. Trends Ecol. Evol. 26 (11), 606–613.

Corlett, R.T., 2012. Climate change in the tropics: the end of the world as we know it? Biol. Conserv. 151, 22–25.

Dupré, B., Dessert, C., Oliva, P., Goddéris, Y., Viers, J., François, L., Millot, R., Gaillardet, J., 2003. Rivers, chemical weathering and Earth's climate. C. R. Geosci. 335, 1141–1160.

Dybas, C.L., 2013. Discoveries in the Critical Zone: where rock meets life. CZO, NSF, pp. 13–112.

Enquist, C.A.F., 2002. Predicted regional impacts of climate change on the geographical distribution and diversity of tropical forests in Costa Rica. J. Biogeogr. 29 (4), 519–534.

Fedorov, A.V., Philander, S.G., 2000. Is El Niño changing? Science 288, 1997–2002.

Gazel, E., Alvarado, G.E., Obando, J., Alfaro, A., 2005. Geología y evolución magmática del arco de Sarapiquí, Costa Rica. Rev. Geol. Am. Centr. 32, 13–31.

Gazel, E., Carr, M.J., Hoernle, K., Feigenson, M.D., Szymanski, D., Folkmar, H., van den Bogaard, P., 2009. Galapagos-OIB signature in southern Central America: mantle refertilization by arc-hot spot interaction. Geochem. Geophys. Geosyst. 10, Q02S11.

Giambelluca, T.W., 2002. Hydrology of altered tropical forest. Hydrol. Process 16, 1665–1669.

Giller, K.E., 2001. Nitrogen Fixation in Tropical Cropping Systems, second ed. CABI, Wallingford, UK, p. 423.

Goldammer, J.G., Price, C., 1998. Potential impacts of climate change on fire regimes in the tropics based on MAGICC and GISS GCM-derived lightning model. Climatic Change 39, 272–296.

Harris, D.R., 1972. The origins of agriculture in the tropics. Am. Scient. 60 (2), 180–193.

Hauff, F., Hoernle, K., Bogaard, P., 2000. Age and geochemistry of basaltic complexes in western Costa Rica: contributions to the geotectonic evolution of Central America. Geochem. Geophys. Geosyst. 1 (5), doi:10.1029/1999-GC000020.

Hoerling, M.P., Hurell, J.W., Xu, T., 2001. Tropical origins for recent North Atlantic climate change. Science 292, 90–92.

Holdridge, L.R., 1967. Life Zone Ecology. Tropical Science Center, San José, Costa Rica, p. 206.

Holdridge, L.R., Grenke, W.C., Hatheway, W.H., Liang, T., Tosi, Jr., J.A., 1971. Forest Environments in Tropical Life Zones: A Pilot Study. Pergamon Press, Oxford, UK, p. 747.

Hunter, P.R., 2003. Climate change and waterborne and vector-borne disease. J. Appl. Microbiol. 94, 37S–46S.

Jacobsen, D., Milner, A.M., Brown, L.E., Dangles, O., 2012. Biodiversity under threat in glacier-fed river systems. Nat. Climate Change 2, 361–364.

Jin, H., 2010. Drivers of global wildfires – statistical analyses. Master Thesis. Lund University, Sweden, p. 97.

Kahil, M.T., Dinar, A., Albiac, J., 2015. Modeling water scarcity and droughts for policy adaptation to climate change in arid and semiarid regions. J. Hydrol. 522, 95–109.

Kennett, J.P., 1982. Marine Geology. Prentice Hall, Englewood Cliffs, NJ, p. 813.

Laguna, J., 1984. Efectos de alteración hidrotermal y meteorización en vulcanitas del Grupo Aguacate, Costa Rica. Rev. Geol. Am. Centr. 1, 1–17.

Lal, R., 1988. Effects of macrofauna on soil properties in tropical ecosystems. Agric. Ecosyst. Environ. 24, 101–116.

Lodge, D.J., McDowell, C.P., MCSwiney, C.P., 1994. The importance of nutrient pulses in tropical forests. Tree 9 (10), 384–387.

Malamud, B.D., Turcotte, D.L., Guzzeti, F., Reichenbach, P., 2004. Landslides, earthquake, and erosion. Earth Planet. Sci. Lett. 229, 45–59.

Mishra, A.K., Singh, V.P., 2010. A review of drought concepts. J. Hydrol. 391, 202–216.

Nathan, 2011. World Map of Natural Hazards. Munich RE.

Norman, M.J.Th., Pearson, C.J., Searle, P.G.E., 1995. The Ecology of Tropical Food Crops, second ed. Cambridge University Press, Cambridge, UK, p. 430.

Obando, V., 2002. Biodiversidad en Costa Rica: Estado del conocimiento y gestión, first ed. National Institute for Biodiversity, Santo Domingo de Heredia, Costa Rica, p. 81.

Orem, C.A., Pelletier, J.D., 2013. Using airborne and terrestrial LiDAR to quantify and monitor post-fire erosion following the Las Conchas fire, Jemez Mountains, New Mexico. Abstract presented at Chapman Conference on Post-wildfire Runoff and Erosion Response. Estes Park, Colorado.

Orozco, E.G., 2007. Zonificación Climática de Costa Rica para la gestión de la infraestructura vial. Master Thesis. Univ. de Costa Rica, San José, Costa Rica. p. 369.

Palm, C.A., Swift, M.J., Woomer, P.L., 1996. Soil biological dynamics in slash-and-burn agriculture. Agric. Ecosyst. Environ. 58, 61–74.

Paul, E.A., 2006. Soil Microbiology, Ecology and Biochemistry, third ed. Elsevier Academic Press, USA, p. 552.

Pierson, T.C., Janda, R.J., Thouret, J.C., Borrero, C.A., 1990. Perturbation and melting of snow and ice by the 13 November 1985 eruption of Nevado del Ruiz, Colombia, and consequent mobilization, flow and deposition of lahars. J. Volcanol. Geotherm. Res. 41, 17–66.

Red Sismológica Nacional (RSN: ICE-UCR), 2009. El terremoto de Cinchona del jueves 8 de enero de 2009. Rev. Geol. Am. Centr. 40, 91–95.

Ruiz, P., 2012. Reconstruction of the paleo and neo stages of Poás and Turrialba volcanoes, Costa Rica: competing processes of growth and destruction. PhD Thesis. Rutgers University, New Jersey, p. 172.

Ruiz, P., Soto, G., Barrantes, R., 2014. Uso de imágenes Lidar en el estudio de la vulnerabilidad de la red vial nacional, caso de estudio ruta No. 126. Congreso de Ingeniería Civil, San José, Costa Rica. p. 11.

Rumple, C., Alexis, M., Chabbi, A., Chaplot, V., Rasse, D.P., Valentin, C., Mariotti, A., 2006. Black carbon contribution to soil organic matter composition in tropical sloping land under slash and burn agriculture. Geoderma 130, 35–46.

Sak, P.B., Fisher, D.M., Gardner, T.W., Murphy, K., Brantley, S.L., 2004. Rates of weathering rind formation on Costa Rican basalt. Geochim. Cosmochim. Acta 68, 1453–1472.

Sanchez, P., Swaminathan, M., 2005. Cutting world hunger in half. Science 307, 357–359.

Silver, W.L., Herman, D.J., Firestone, M.K., 2001. Dissimilatory nitrate reduction to ammonium in upland tropical soils. Ecology 82 (9), 2410–2416.

Smirnov, V., 1982. Geologia de yacimientos minerales. Editorial Mir Moscù, URSS.

Stanley, S.M., 2010. Relation of Phanerozoic stable isotope excursions to climate, bacterial metabolism, and major extinctions. PNAS 107, 19185–19189.

Stocking, M., 2003. Tropical soils and food security: the next 50 years. Science 302, 1356–1359.

Thompson, L.G., 2000. Ice core evidence for climate change in the tropics: implications for our future. Quaternary Sci. Rev. 19, 19–35.

United States Geological Survey (USGS), 2012. Mineral Commodity summaries, Bauxite and Alumina.

Vitousek, P.M., Mooney, H.A., Lubchenco, J., Melillo, J.M., 1997. Human domination of Earth's ecosystems. Science 277, 494–499.

von Blanckenburg, F., Hewawasam, T., Kubik, P.W., 2004. Cosmogenic nuclide evidence for low weathering and denudation in the wet, tropical highlands of Sri Lanka. J. Geophys. Res. 109, F03008.

Waldrop, M.P., Balser, T.C., Firestone, M.K., 2000. Linking microbial community composition to function in a tropical soil. Soil Biol. Biochem. 32, 1837–1846.

White, A.F., Brantley, S.L., 2003. The effect of the time on the weathering of silicate minerals, why do weathering rates differ in the laboratory and field? Chem. Geol. 202, 479–506.

White, A.F., Bluv, A.E., Schultz, M.S., Vivit, D.V., Stonestorm, D.A., Larsen, M., Murphy, S.F., Eberl, D., 1998. Chemical weathering in a tropical watershed, Luquillo Mountains, Puerto Rico: I. long term vs. short term weathering fluxes. Geochim. Cosmochim. Acta 62 (2), 209–226.

Chapter 16

The Critical Zone of Coastal Barrier Systems

Patrick Barrineau*, Phillip Wernette*, Bradley Weymer**, Sarah Trimble*, Brianna Hammond*, and Chris Houser*,**
*Department of Geography, Texas A&M University, College Station, Texas, USA; **Department of Geology and Geophysics, Texas A&M University, College Station, Texas, USA

16.1 INTRODUCTION

Over 2000 barrier islands occur worldwide, and they occur on every continent except Antarctica. Barrier islands represent over 10% of the world's open-ocean coastline and occur in inland seas and lakes (Otvos, 2012; Pilkey et al., 2009; Dean and Dalrymple, 2002) and are believed to provide important protection to the mainland during hurricanes and tropical storms. More elongated and thin barriers tend to occur in microtidal areas, whereas shorter barriers tend to occur in macrotidal environments and are associated with a greater number of inlets and tidal deltas. Barrier Islands are a significant part of the cultural and economic landscape throughout much of the world, particularly in the United States, where recreational and navigation development can dominate the Barrier Island landscape (Meyer-Arendt, 1990; Ehlers and Kunz, 1993; Nordstrom et al., 2000). Anthropogenic alterations to hydrological, physical, and other processes that rely on connectivity between seemingly disparate systems (i.e., beach–dune interaction, nutrient cycling in the near surface, etc.) can significantly degrade their functionality within the Critical Zone (CZ) of barrier islands and can play a significant role globally in the evolution of these environments.

Barriers range in length from >100 km (Padre Island, TX, USA) to just a few hundred meters, and they are highly heterogeneous, with a diverse range of evolutionary pathways as a result of complex feedbacks between sediment supply, hydrology, and nutrient exchange, the atmosphere, and ecological communities. In this respect, Barrier Islands represent an interesting but overlooked laboratory to explore CZ processes over modern and geological timescales. Traditionally seen as a result of variations in sea level, sediment supply, and accommodation space, barriers may be described in general using transgressive or regressive states to frame more detailed discussions of their various CZ dynamics. The shoreline of a Barrier Island moves landward as the island transgresses

Developments in Earth Surface Processes, Vol. 19. http://dx.doi.org/10.1016/B978-0-444-63369-9.00016-1

in response to relative sea-level rise and/or neutral or negative sediment supply (Morton and Sallenger, 2003). Regression occurs when the shoreline extends seaward in response to a positive supply of sediment and relatively lower sea level (Posamentier et al., 1992; Helland-Hansen and Martinsen, 1996). Such differences provide the foundation for studying Barrier Island CZ processes within a larger contextual framework. For instance, regressive barriers may have extensive backbarrier maritime forest and dune swale plant communities (i.e., Sapelo Island, Georgia; Tackett and Craft, 2010), but commonly lack well-developed transgressive dune sheets (see Hesp, 2013). Regressive (i.e., prograding) depositional sequences are the most likely coastal sedimentary structures to be preserved in the stratigraphic record, because transgressive sequences are characteristically reworked by wave action and falling sea level, and thus are subsequently eroded (Kraft and Brown, 1982).

Many barrier islands are in a state of transgression (landward retreat) in response to a combination of eustatic sea-level rise and local subsidence. Transgression is encouraged when the storm surge of tropical cyclones and other storms breaches coastal foredunes and redistributes sediment to the backbarrier as overwash deposits. The extent of sedimentation during an extreme storm and the rate of island transgression depends on the height and alongshore continuity of coastal foredunes (Thieler and Young, 1991; Sallenger, 2000; Morton, 2002; Nott, 2006; Houser et al., 2008), which over longer timescales is in turn dependent on the availability of sediment from alongshore and offshore sources (Hequette and Ruz, 1991; Psuty, 1992; Schwab et al., 2000; Hansom, 2001; Houser et al., 2008). Independent of sediment availability, the height and continuity of the dunes and the rate of island transgression depends on the time interval between storm landfall and the ability of the dunes to recover both vertically and horizontally (see Houser and Hamilton, 2009; Hesp, 2002). Barrier Islands impacted by infrequent storms tend to have larger dunes covered in vegetation that promotes dune growth, whereas islands impacted by frequent storms are characterized by low-profile dunes covered by burial-tolerant vegetation that do not promote foredune development (Stallins and Parker, 2003).

In this manner, Barrier Islands can be viewed through a CZ lens that accounts for differences in transgressive or regressive island state and explicitly recognizes complex feedbacks between sediment supply, aeolian transport, vegetation development, and hydrology. Alterations to these relationships can alter dune height and extent, which can reinforce and even amplify the vulnerability of a Barrier Island to future storms (see Houser et al., 2008), which in turn threatens the ecological function of the island and that puts commercial and recreational development at risk (see Nordstrom et al., 2000). This discussion first describes the external control represented by island transgression and regression. Then we focus on the vital role coastal dunes play in the Barrier Island CZ, as both an external control on multiple sets of processes, and indicators of past and present dynamic regimes. With attention paid to the central role of coastal dunes and island state, vegetation and soils are reviewed in the alongshore and across shore directions. Finally we use our discussion as a guide to suggest a number of areas for potentially valuable future research in Barrier Island environments using a CZ perspective.

16.2 TRANSGRESSION AND REGRESSION

Over geological timescales, sea-level change is a primary control on island evolution and migration, and the development of the Barrier Island CZ. Transgressive islands contain wider swaths of relatively unconsolidated overwash deposits, and are more susceptible to inundation during violent storms or higher spring tides. On the other hand, ridge-and-swale sequences, multiple foredune ridges, or other prograding sedimentary sequences generally dominate regressive barriers. These strata are more resistant to inundation via overwash and long-term sea-level rise, and as a result they typically host more mature ecosystems with decreased frequencies of storm-related disturbances.

16.2.1 Island Transgression

Typically, a rise in relative sea level results in a marine transgression. However, it is well documented that geological inheritance also plays a significant role in determining how Barrier Island systems evolve (Kraft, 1971; Kraft and John, 1979; Kraft and Brown, 1982; Riggs et al., 1995; Dillenburg et al., 2000; Harris et al., 2005; Short, 2010). The overall configuration of transgressive barriers (Fig. 16.1) is primarily controlled by antecedent geology, which in turn influences wave refraction and attenuation, beach location, shape, type, morphodynamics, and circulation. These processes within the CZ govern sediment transport across the dunes and transgressive barrier system (Short, 2010). Consequently, these inherited geological features also control contemporary response of the Barrier Island to extreme storms, and in turn the rate of island transgression to variations in sediment supply or an increase in mean sea level (Houser, 2012).

Modern Barrier Island transgression is accomplished primarily by relative sea-level rise and extreme storms that are capable of breaching the dunes and depositing sediment in the backbarrier in the form of blowouts, washover fans, and terraces (Houser, 2012; Morton and Sallenger, 2003; Stone et al., 2004). Houser (2012) suggests that the threshold storm surge required for foredunes to be overtopped or breached decreases as sea-level rises, and subsequently the probability of island overwash and island transgression increases. Dune scarping can induce blowouts that are part of the transgression and can potentially create washover channels during extreme storms (Houser, 2012). Breaching and overwash is focused in areas where dune height is low (i.e., overwash regime), creating the potential for rapid transgression and even overstepping (Sallenger, 2000). However, the two-dimensional storm impact model proposed by Sallenger (2000) does not account for alongshore variability in dune height. Along relatively short sections of the same beach, significant differences in foredune

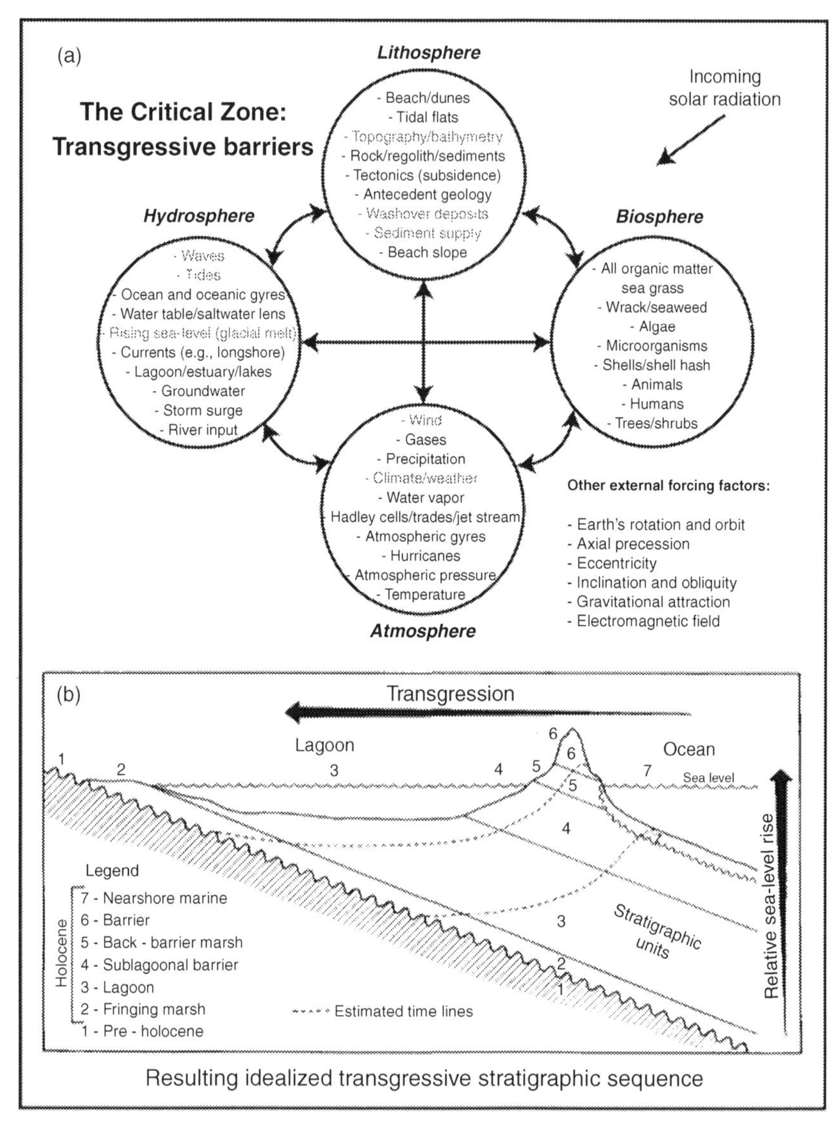

FIGURE 16.1 Simplified "open system" conceptual model of the transgressive barrier CZ and resulting idealized stratigraphic sequence. (a) The transgressive CZ system showing the 4 reservoirs (i.e., lithosphere, biosphere, atmosphere, and hydrosphere) of material with flows of matter and energy between them. Each system consists of feedback loops (process-form relationships) and subsystems; such as the ocean and groundwater (e.g., hydrosphere). The most significant processes and controlling factors for barrier transgression are highlighted in dark gray [red in the web version] and signify the main differences between transgressive and regressive barrier island systems. The combined processes, storages, flows of energy, and matter result in the idealized transgressive stratigraphic sequence. (b) Sequence model of transgressive barriers showing variants of depositional environmental sequences. Lithostratigraphic units cross time lines based on radiocarbon dates. *(Figure modified after Kraft and John (1979).)*

height can occur, which may lead to different transgression histories alongshore. In areas where the dune heights are low, lateral dune erosion through the expansion of washover conduits can develop, whereas in areas where the dunes are high only the base of the dune is scarped and sediment is transported seaward.

The rate of island transgression and the development of the CZ depends on the recovery time between extreme storm events, related to the height and extent of the foredune system (see Sallenger, 2000; Masetti et al., 2008). Dune scarping and washover depend on the elevation of the wave run-up relative to the elevation of the dune base and crest respectively. Where the elevation of the run-up is high, compared to the elevation of the dune base, blowouts may also develop from overwash hollows if vegetation is slow to recover (Leatherman, 1976; Ritchie and Penland, 1990; Saunders and Davidson-Arnott, 1990). The potential for scarping and washover is dependent on the elevation of the storm surge (R_{high}) relative to the elevation of the dune base (D_{low}) and crest (D_{high}). Dune scarping is initiated when and where $D_{low} < R_{high} < D_{high}$, while overwash develops when and where $R_{high} > D_{high}$ (Sallenger, 2000). Washover is a self-reinforcing process and once an overwash lobe develops it can expand quite rapidly alongshore through lateral erosion (Houser, 2013). Recent evidence suggests that driving on the beachface can reduce the elevation of both the dune base and crest (Houser et al., 2013), making the island susceptible to scarping and erosion by storm surge (Hesp, 2002; Davidson-Arnott, 2005) or direct washover and inundation (Sallenger, 2000).

The height of a dune when a storm makes landfall depends on the time since the previous storm, the level of erosion sustained in that storm and the rate of poststorm dune recovery (Houser et al., 2008). The rate of recovery, therefore, depends on the regional sediment budget (Psuty, 1992) and the ability of sediment to be transported from the beach and backshore to the dune (Sherman and Bauer, 1993; Short and Hesp, 1981). The preserved sedimentary structures (facies) of the beach and dune can provide a relative chronology of how the dunes respond to and recover from extreme storms and the implications for island response to relative sea-level rise (i.e., transgression). Many beaches along the passive continental margins of the US east and Gulf coasts commonly have barrier islands perched upon premodern (i.e., Pleistocene) stratigraphic units that occur beneath and seaward of the shoreface (Riggs et al., 1995). These stratigraphic units control the morphology of the shoreface and strongly influence modern beach–dune dynamics, sediment composition, and sediment fluxes, in turn affecting the islands' response to sea-level rise and extreme storms (Riggs and O'Connor, 1974; Riggs, 1979).

16.2.2 Island Regression

Regressive barrier islands are typically a result of a relative decrease in sea level (see Fig. 16.2). Over geological timescales, this primarily occurs during glacial periods, caused by changes in atmospheric–oceanic circulation and by

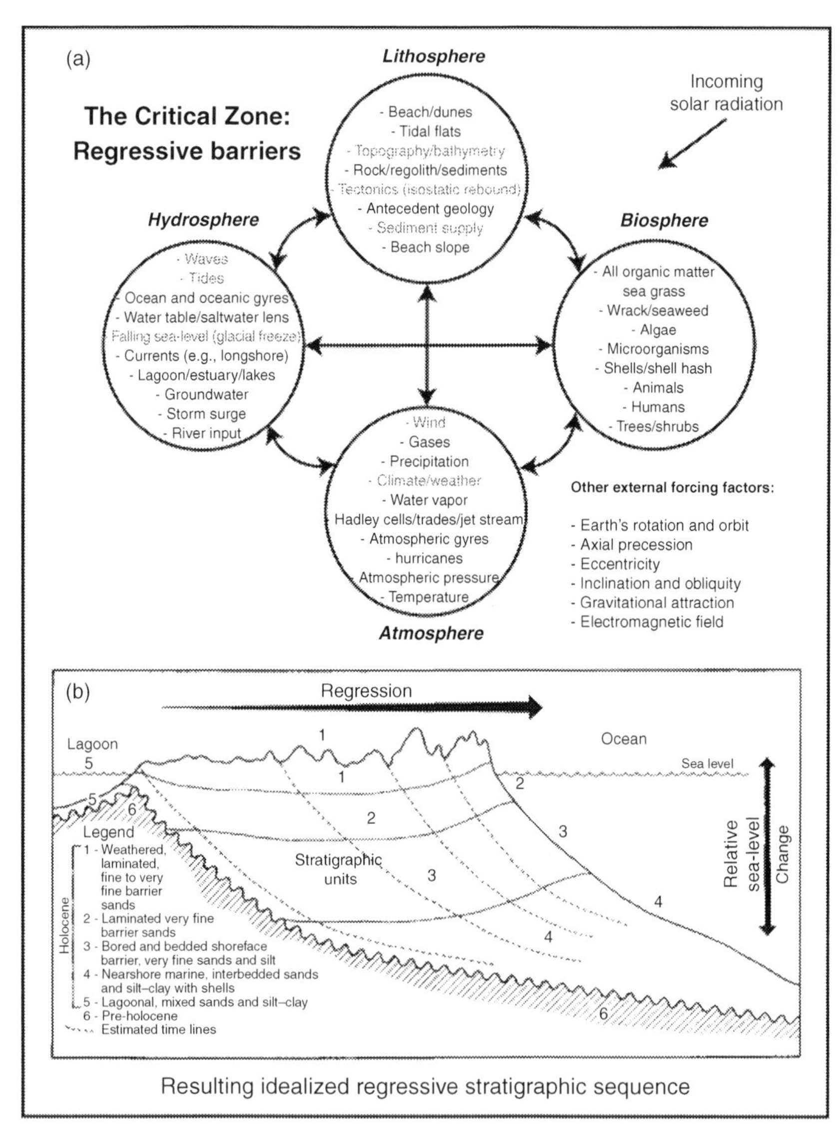

FIGURE 16.2 Simplified "open system" conceptual model of the regressive barrier CZ and resulting idealized stratigraphic sequence. (a) The regressive CZ system showing the 4 spheres (i.e., lithosphere, biosphere, atmosphere, and hydrosphere) of material with flows of matter and energy between them. Each system consists of feedback loops (process-form relationships) and subsystems; such as the ocean and groundwater (e.g., hydrosphere). The most significant processes and controlling factors for barrier regression are highlighted in dark gray [red in the web version] and signify the main differences between regressive and transgressive barrier island systems. The combined processes, storages, flows of energy, and matter result in the idealized regressive stratigraphic sequence. (b) Sequence model of regressive barriers showing variants of depositional environmental sequences. Lithostratigraphic units cross time lines based on radiocarbon dates. *(Figure modified after Kraft and John (1979).)*

external forcing factors such as Earth's axial precession, eccentricity, inclination, and obliquity (i.e., Milankovitch cycles). However, after the Last Glacial Maximum (Wisconsinan Glaciation), modern barrier regression can also be a result of isostatic rebound, especially in high-latitude regions where the rate of uplift exceeds the current rate of eustatic sea-level rise. Alternatively, barrier regression (progradation) can also occur within barriers that have an excess sediment supply (i.e., Galveston, TX). In other words, coastal progradation and the establishment of regressive sedimentary sequences occur when the shoreline translates seaward under a condition of falling sea level and/or excess sediment supply relative to the accommodation space on the shelf (Helland-Hansen and Martinsen, 1996). The translation can occur with stable or slowly rising sea levels ("normal" regression) or be strictly induced by a decrease in sea level ("forced" regression) (Posamentier et al., 1992). In both cases, given the wave dominance on coastal CZ processes, a prograded coastal barrier is formed. Progradation associated with slowly rising or stationary sea levels occurs mainly because of a positive longshore sediment imbalance (Lessa et al., 2000). Prograded barriers in wave-dominated settings are commonly characterized by a series of beach and foredune ridges aligned with the shoreline (strandplain). The resulting ridge-and-swale topography hosts sequential belts of vegetation assemblages arranged into a chronosequence of ecosystem development.

16.3 COASTAL DUNES AS A CENTRAL NODE IN THE BARRIER ISLAND SYSTEM

Coastal dunes are at a conceptual node between atmospheric, hydrological, biological, and physical processes and the physical interface between marine and terrestrial environments; they control the distribution of nearshore and beach sediment in the back barrier, indicate sediment availability (and therefore stability) for beach and dune environments, reflect prevailing atmospheric regimes, and even indicate former storm and overwash regimes. Dunes develop parallel to the shoreline where sediment supply and wind energy are sufficient to transport sediment to the backshore where it is captured by vegetation (Short and Hesp, 1981; Pye, 1983). The primary source of sediment to the dune is aeolian transport from the subaerial beachface, which is defined as the intertidal zone between the upper limit of the swash at high tide and the lower limit of swash at low tide. Aeolian transport rates are greatest on dissipative beaches (Short and Hesp, 1981) that are associated with relatively fine sediment and large wave heights, and are characterized by multiple offshore bars and a low-gradient and wide beach. During nonstorm conditions, the beachface accretes through the landward migration of the nearshore bars, leading to an intermediate beach state and ultimately to a relatively steep reflective beach if conditions persist (Thom and Hall, 1991). Reflective beaches are not conducive to aeolian transport, unless they are backed by an extensive and flat backshore, due to greater disruption to the airflow and limited fetch length (Bauer and Davidson-Arnott, 2003). The bars are re-established in response to storm waves that require a longer and flatter profile to dissipate the increased wave energy (see Wright and Short, 1984). Erosion of the beachface by storm waves can take several hours, but the landward migration of the bars and their welding to the beachface can take a significantly longer time. The disparate timescales of storm response and recovery create a significant control on beach recovery, dune recovery, and therefore the nature of the Barrier Island CZ (Aagaard et al., 2004).

The response of dunes and barrier islands to storms is complicated by the alongshore variation in dune height (see Fig. 16.3; Houser et al., 2013). Even large dunes can be eroded through lateral truncation and the extent of washover increases if adjacent smaller dunes are breached. Localized coastal response can be forced by wave refraction over the inner-shelf bathymetry (Demarest and Leatherman, 1985; Kraft et al., 1987; Pilkey et al., 1993; Riggs et al., 1995; Schwab et al., 2000; McNinch, 2004; Browder and McNinch, 2006; Schupp et al., 2006; Stockdon et al., 2007). For example, the shoreface of Santa Rosa Island in northwest Florida is characterized by a ridge-and-swale bathymetry that corresponds to the alongshore variation in beach- and dune-morphology. The

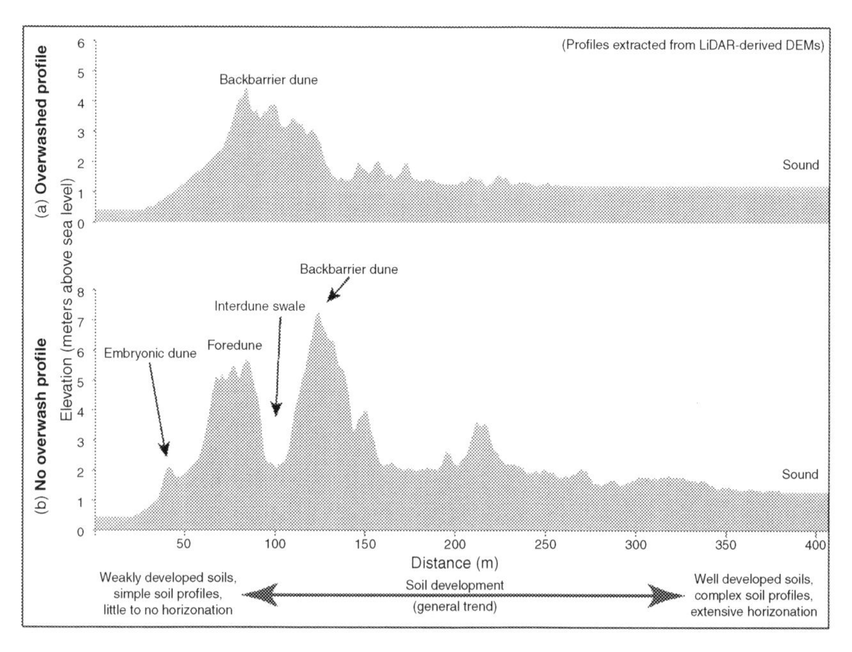

FIGURE 16.3 LiDAR-derived cross-shore profiles of Padre Island, Texas along different stretches of the barrier shoreface. Portions of Padre Island are transgressive, whereas others are regressive (the island is over 100-km long and hosts a wide variety of physical and biochemical environments). The variability in cross-shore morphology observed in relatively close portions of a single barrier highlights the importance of dune height and its relation to overwash potential; higher foredune ridges are typically associated with less-frequent storm overwash.

ridges and swales produce bathymetric highs and lows that force an alongshore variation in the surf similarity parameter and beach state (Houser et al., 2011). This in turn controls the aeolian transport potential and sediment supply to the foredunes. In general, the ridges are located seaward of the widest sections of the island that are backed by cuspate spits along the backbarrier shoreline, whereas the swales are seaward of narrow sections of the island that are susceptible to washover and breaching during storms (Houser et al., 2008). Houser (2012) presented evidence that the ridge-and-swale bathymetry is a transgressive surface and the remnant of cuspate spits that are present along the backbarrier shoreline. Specifically, the cuspate spits had to first develop along the backbarrier shoreline and eventually evolve into mud-cored ridges as the island transgressed with relative sea-level rise. Once the ridge-and-swale bathymetry emerged on the Gulf of Mexico shoreface it was able to reinforce the alongshore variation in dune height and storm response. In this respect, the alongshore variation in beach- and-dune morphology, which is a reflection of the CZ distribution, is a result of a large-scale feedback in which the barrier island CZ is dependent on the geological context that was in turn formed by the alongshore variation in the CZ early in the formation of the island. Similar results have been observed along Padre Island on the Texas coast (see Houser and Mathew, 2012).

Dune recovery requires the transfer of sediment from the nearshore to the beach and ultimately to the dune, assuming that dune-building vegetation is present. Recent evidence presented by Houser et al. (2015) suggests that the recovery of dunes along Galveston Island, Texas and Santa Rosa Island, Florida, follows a sigmoid curve consistent with the growth models used to quantify vegetation growth (see Hugenholtz and Wolfe, 2005a, 2005b). The first stage of recovery begins immediately after the storm with the development of a berm and the steepening of the beachface through the landward migration and welding of the innermost bar (see Morton et al., 1994). Development of a wide back-beach is dependent on deposition from swash excursions that exceed the elevation of the beach ridge in intermediate to reflective environments, or through the landward migration of the bars in dissipative environments. As wind speeds increase above threshold for aeolian transport, a narrow spatial or temporal window occurs in which sediment can be transported to the dune before the storm surge extends into the backshore and the transport system begins to shut down (see Houser and Hamilton, 2009), because storms winds capable of entraining sediment are often accompanied by elevated water levels. The final stages of dune recovery are vegetation recolonization, dune formation, and dune expansion. Initial vegetation colonization and growth can determine the form of the resultant foredune (discussed later in the chapter). The rate that vegetation is able to re-emerge depends on the depth of the washover and the species-dependent burial toleration. The faster the vegetation emerges, the greater the chance of that vegetation remaining viable (Maun, 2009). Some burial-tolerant species are adapted to washover and quickly colonize the surface through creeping, but are not effective at trapping sand and building dunes. Conversely, dune-builder species (i.e., *ammophilia sp.*) expend their energy growing upwards as sediment is trapped, but are slow to colonize following a storm.

If given sufficient time to develop, the primary limit on dune height and extent is the availability of sediment, which is in turn dependent on the regional sediment budget. On regressive barriers, dunes are commonly arranged in shore-parallel ridge-and-swale sequences that determine the accumulation of organic materials and drive biochemical cycling across the island. In contrast, transgressive barriers develop overwash deposits or extensive dune systems encompassing multiple phases of development (Hesp, 2012). Backbarrier physical processes are highly dependent on the prevailing morphodynamic forms of a particular island. In addition, barrier islands are aggregate entities generated from eroded mainland sediment and bedrock, and are thus affected by regional geological setting (e.g., mineralogy). Ridge-and-swale sequences or overwash deposits can provide the stratigraphic basis for the development of a backbarrier ecosystem, which can vary in form from maritime forests to active sand dunes. On transgressive barriers, increased overwash frequency results in widespread storm deposits, which are commonly unconsolidated and rapidly incorporated into the subaqueous backbarrier margin (Houser, 2012). In arid environments, the backbarrier may be completely devoid of vegetation and dominated by *nebkha* and active migrating sand dunes (Fryberger et al., 1988). In some cases these coastal dune systems grow large enough to be treated as discrete aeolian systems (Hesp, 2013; Hesp and Thom, 1990) and in sufficiently arid regions, the coastal dunes may extend seamlessly into a backbarrier aeolian landscape.

16.4 ALONGSHORE AND ACROSS-SHORE VARIATION IN THE BARRIER ISLAND CRITICAL ZONE

Barrier Island morphology can be described in terms of six depositional environments (see Figs 16.1 and 16.2) that represent different habitats and ecosystems that can be described with respect to CZ processes: (1) mainland coast that is protected by the barrier and may include bays and estuaries; (2) lagoon, bay, or extensive marsh that separates the barrier from the mainland; (3) subaerial barrier, including the beach, dunes, and backbarrier deposits; (4) subaqueous platform of sediments on which the subaerial barrier is built; (5) shoreface extending offshore from the exposed beach; and (6) inlets and associated tidal deltas (Reinson, 1979; Davidson-Arnott, 2010). As outlined in the previous sections, the cross-shore and alongshore variation in the CZ of a barrier island is primarily controlled by the height and extent of the foredune. The following sections describe how the foredune is involved in the distribution of vegetation, biogeochemical cycling, and the development of soils. The number of permanent inlets or storm-cut breaches is in part determined by the frequency of storm events and the ability of the dunes to recover in height and extent, and the tidal range.

16.4.1 Beachface

Beaches in the barrier island CZ tend to be sandy sediment with no persistent vegetation unless hardy pioneer plants are able to withstand frequent inundation by the storm tide and salt spray. Wave action and swash erodes deposit material from this part of the island on a regular basis, forming a zone of empty sediment at the water line that progresses to the "backbeach," an area where some rhizomatic plants may begin to take hold, provided enough time exists between disturbances such as large storms. These plant types include sparse forbs, prostrates, and annuals (see Fig. 16.4; Ehrenfeld, 1990). Higher moisture and salt contents in the air prevent some species from colonizing, and encourage others' growth. The predominance of on- and offshore wind regimes, largely determined by prevailing atmospheric conditions over a particular barrier island, can either enhance or suppress the relative importance of maritime air parcels in the development of vegetation at the land–water interface. In addition, the deposition of phytodetritus (more commonly referred to as "beach wrack") creates an "organic subsidy" to the nutrient-poor, beach environment. The rate of wrack deposition is dependent on the grain size of the beach, wave climate, and species of plant detritus being deposited (Orr et al., 2005). Across-shore variation in the deposition of beach wrack and alongshore variability in its distribution across a barrier island can cause significant differences in beach morphology and affect foredune initiation via alteration of vegetation-colonization patterns.

FIGURE 16.4 Backshore vegetation is limited to hardy creeping vines and low grasses and forbs, whereas dune grasses and sedges are capable of growing in the more sheltered portions of the dune toe and interdune areas. *(Photo Padre Island National Seashore, P. Barrineau, 08/13.)*

Soil development on the beach is also limited by the sediment porosity and frequent disturbances, although water percolated through the beach contains some soluble nutrients and sediments (Anschult et al., 2009). The greater porosity of most beach sands (as compared to a loamy soil occurring further inland, for example) enables water to freely percolate downward without nutrients accumulating enough to form a B horizon. As a result, beach sediments typically lack pedogenic horizonation. The lack of soil horizonation on the beach is not indicative of the complex internal stratigraphy. Water and wind redistribute and sort beach sediments into distinct stratigraphic structures. Differing sedimentary structures within the beach and nearshore subsurface create heterogeneous patterns of vulnerability to erosion (and overwash) similar to ridge-and-swale sequences informing island transgression (discussed earlier). In areas of significant wrack deposition, alternating layers of detritus and sand indicate that the beach is affected by both wind and waves. Stratified beach sediments are also evident in GPR data, where dipping planes are evident.

16.4.2 Foredunes

Coastal dunes vary in size from low, hummocky incipient foredunes to massive dune complexes more than 25-m tall (Hesp, 2002). Beach width and sediment supply, mentioned earlier, are primary controls on foredune development, along with wind velocity and direction. The genesis and evolution of coastal dune systems is highly situational, and a wide variety of morphologies and ecological communities result from particular conditions, including an on- or offshore wind regime and the frequency of disturbances like overwash.

Vegetation colonization in the foredune is commonly restricted by the nutrient availability, as well as external morphodynamic conditions (Martinez et al., 2004; Maun, 2009). Vertical zonation of foredune vegetation is related to the depth of the water table and persistence of salt- or sand-blasting. For example, at Padre Island, Texas the lowest portions of the dunes are generally dominated by species liked Marshhay cordgrass (*Spartina patens*) and morning glory (*Ipomoea* spp.), whereas the upper portions and landward side are dominated by grasses such as bitter panicum (*Panicum amarum*) and seacoast bluestem (*Andropogon scoparius littoralis*) (Weise and White, 1980). If the dunes are left undisturbed long enough – a window of time that varies by climate, location, and morphological history – then small shrubs may grow on the foredune and any backbarrier dunes. The species of vegetation growing in a particular area can dictate resultant foredune morphology, with lower creeping species providing less of a disturbance to wind flow than taller grass species or shrubs that drastically reduce sediment transport capacity higher above the surface (Hesp, 2002). Deposition of beach wrack near the high-tide line along the backshore may encourage vegetation colonization by providing sheltered microenvironments capable of hosting seedlings that would not survive otherwise, and can encourage the seaward advance of a foredune (Nordstrom et al., 2011; Hesp, 2002).

Dune soils are typically sandy Entisols (psamments), which means that they lack any diagnostic soil horizon, yet display some evidence of pedogenesis. A typical dune soil has one or more organic A horizons overlying one or more C horizons (see Fig. 16.5). The relatively thin organic A horizons are both formed by vegetation decomposition and nutrient cycling, and are typically differentiated by color or the degree of decomposition. The relatively thin A horizons

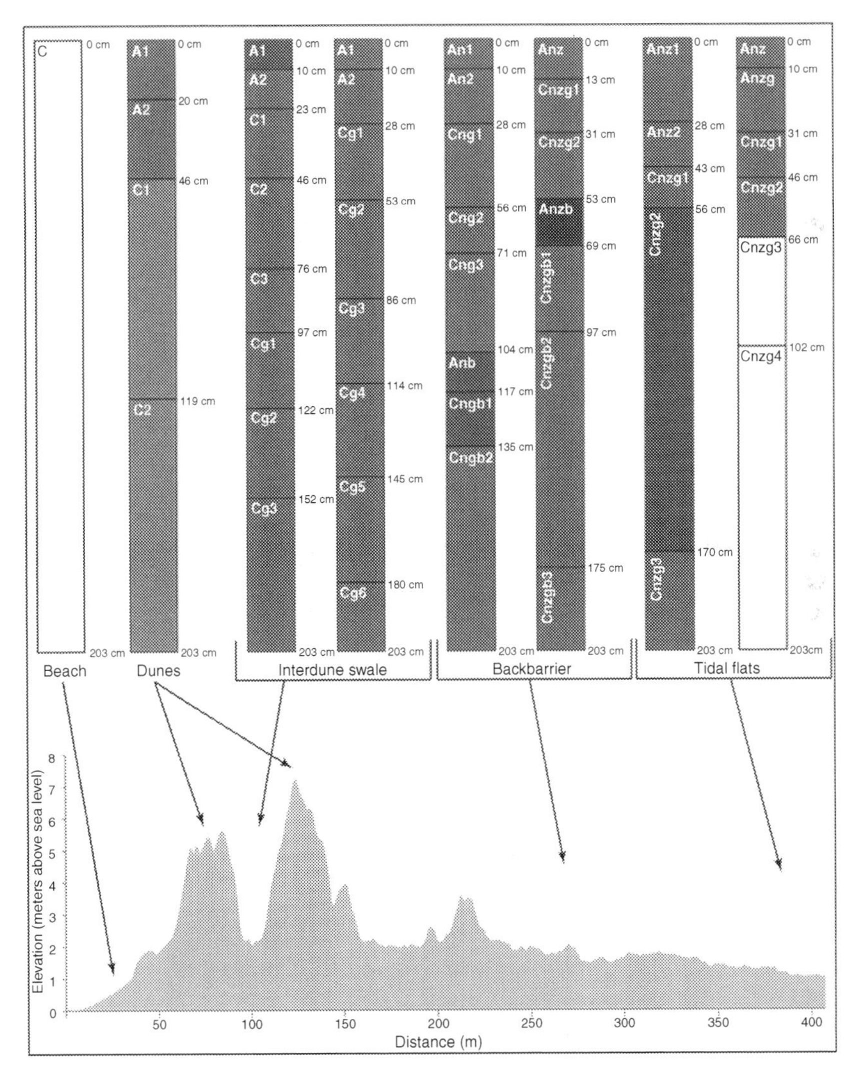

FIGURE 16.5 Representative "stacked" backbarrier and dune soil profiles collected at Padre Island, Texas. Buried "A" horizons in the foredune system and backbarrier indicate rapid deposition via overwash, whereas the tidal-flats landward of the barrier island host a greater degree of soil development due to enhanced vegetation productivity and a decreased overwash frequency.

occurring in dune soils are partially due to the slow rate of decomposition in organic materials, compared to areas farther from the shoreline. Grasses present on the dunes cycle some of the decomposed compounds in the upper solum. Slow decomposition, high sand content, and base cycling of grasses combine to limit A horizon development to a thin zone. In addition, dune soils on barrier islands typically lack B horizons, which are zones of illuviation where organic and inorganic materials accumulate. The Bw horizon, if present, indicates that the horizon exhibits weak evidence of pedogenic development. These Bw horizons are typically distinguished based on differences in hue, value, and/or chroma. One reason that B horizons are not common or, if present, are very weakly developed in barrier island dunes, is the combination of high porosity of the sandy parent material and frequent disturbance from storms. The high porosity enables water to infiltrate through the profile with relatively little resistance. This percolating water contains some soluble nutrients from the surface and A horizon, which are translocated deeper in the profile. However, since the parent material is very uniform, nutrients in the water are leached through the entire column and never accumulate to form a B horizon. Dune soils on barrier islands may contain redoximorphic features (i.e., gleying, mottling) if the water table is shallow enough to affect the described profiles. In the case of Padre Island, a Cg horizon indicates the presence of gleying at a depth of 97 cm (Weise and White, 1980).

16.4.3 Interdune

Topographic lows within foredune systems, including interdune swales and foredune blowouts, encourage patchy and diverse vegetation colonization by creating small sheltered microclimates between active foredunes. Within these microenvironments, enhanced nutrient exchange and moisture availability, along with decreased wave scarping and overwash, result in a higher vegetation density than on adjacent topographic highs and the backshore. A variety of plants populate swales including forbs and shrubs, but also hydrophytic grasses and sedges. Seasonality can play a significant role in the preferential colonization of one species over another, as many barriers experience wet and dry seasons associated with storm landfall or winter cold fronts. In more arid climates the growth of vegetation may be moisture-limited and foredunes may continue seamlessly into a backbarrier dune field (e.g., Padre Island, TX USA). The propagation of foredune sands into the backbarrier is often dictated by foredune blowouts, or "hollows formed by wind erosion" of an existing foredune (Hesp, 2002). Variations in vegetation density and species zonation can dictate the frequency and magnitude of foredune blowouts, which can complicate successional patterns of vegetation from the backshore to the backbarrier. In addition, long-term, sediment-supply regimes (discussed previously) can encourage the development of extensive foredune blowout systems and the translation of beach and dune sediments to the backbarrier as a response to regressive island states

(Psuty, 1992). Emerging research suggests the long-term stability of foredunes may be variable in the cross-shore direction as a result of inherited geological stability related to nearshore bathymetry; as a result the persistence of foredune blowouts may be dictated by the long-term island state (e.g., transgression or regression) as well as shorter-term parameters like overwash frequency and wave climate (see Weymer et al., 2015; Houser, 2012).

Soil development is enhanced in interdune swales due to the relatively shallow water table (i.e., increased moisture) and abundant vegetation, with redox features extending to a depth of ~28 cm on Padre Island (Weise and White, 1980). The high moisture, abundant vegetation, and rapid decomposition in these swales (Conn and Day, 1997) combine to increase the amount of organic material available for pedogenesis, and ultimately leading to greater horizonation and the development of Typic Psammaquents (typical sandy Entisols) or Aquic Psamments (wet sandy Entisols). With aeolian deflation, an interdune surface can be reduced to the level of the water table, leading to the development of isolated wetlands important for endemic and migratory species (Orme, 1992; Cooper, 1967). Depending on the depth of the water table, interdune swale soils may contain color patterns in the soil (redoximorphic features). Redox depletions (i.e., gray mottles) are areas where iron and manganese oxides have been reduced or removed, and redox concentrations (i.e., red mottles) are areas of the soil where iron and manganese have been oxidized, giving them a brown, orange, red, or yellow value and high chroma (Chorover et al., 2007).

16.4.4 Backbarrier and Washover Deposits

If the island is wide enough that the last dunes gently slope down to a backbarrier shoreline, maritime forests may develop on the open ground or among a mixture of smaller dunes and swales. Shrub thickets, grasses, and forbs may also be present or interspersed with a heathland. Heathland is an expanse of low-growing, woody vegetation that forms on well-drained acidic soils. Less disturbed portions of the backbarrier may strongly resemble the biochemical dynamics of the adjacent mainland, whereas more disturbed areas typically have less persistent vegetation and little to no soil development. A strong on- or offshore wind regime can dictate nutrient and moisture content – onshore winds will bring salt spray and moisture further into the backbarrier than consistent offshore flow. By creating heterogeneous patterns of nutrient distributions in the backbarrier, the persistence of salt spray and coastal fog can determine the resultant plant assemblages, varying from highly endemic collections of hardy salt- and moisture-tolerant species to those which resemble communities occurring on the adjacent mainland (Tackett and Craft, 2010; Levy, 1990). It should be noted that island *size* is not an absolute determinant of species diversity. Richness in species diversity is a function of island age, history of natural disturbance, history of use (by humans), topography, and soil quality, and varies widely for all island sizes. Total area is only a factor when islands are smaller than some threshold size (such as 1000 Ha for the eastern US).

Backbarrier soils commonly contained highly gleyed horizons due to frequent flooding and a shallow water table, so vegetation is typically not moisture-limited and is relatively robust compared to beach environments, for example. The high productivity of vegetation produces abundant organic matter that decomposes and releases important nutrients to the upper horizons. These nutrients are cycled in the upper solum by the dense grasses that persist in the swales, which reinforces development of the A horizon. The high primary productivity and periodic island washover result in stacked soils in the backbarrier (e.g., Fig. 16.5). Stacked soils are those that have multiple A horizons that are separated by some other horizon. These soils are typically composed of multiple sequences of A horizons over C horizons, such that a typical soil will contain an A–C–A–C sequence. The lower A horizon represents a period of stability when pedogenesis was able to stratify the initial sediments, which are represented by the lower C horizon(s). Washover sediments bury this initial soil, which separates the initial, A horizon from pedogenic modification and forms a C–A–C soil sequence. Following the burial event, the surface is stabilized again, which allows pedogenic processes to form another A horizon.

Backbarrier soils also have high salt concentrations, as indicated by the "sodic" modifier and the *n*, *y*, and/or *z* suffixes. The *n* and *z* suffixes indicate that the horizon contains significant amounts of exchangeable sodium (*n*) and other salts more soluble than gypsum (*z*). Saline horizons are defined by their value, chroma, and chemical properties (e.g., electrical conductivity). Although saline soils limit backbarrier vegetation to halophytes, soil development is promoted by the vegetation that can survive. The organic material from the vegetation decomposes and releases nutrients that are cycled by the grasses and form thick organic A horizons (see Figs 16.5 and 16.6).

16.4.5 Backbarrier Shorelines

Tidal-flat soils are the most saline in barrier island systems. The high salt content is derived from and varies with salt content of water in the backbarrier lagoon. Barrier Islands with a hypersaline lagoon, as is the case with Padre Island, have a particularly high salt content in backbarrier and tidal-flat soils. The high salt content retards soil development in tidal-flat soils, and is one reason that tidal-flat soils are not described as deep as soils in other locations on the barrier island. In addition to the high salt content, frequent saturation of tidal-flats retards soil development by limiting nutrient eluviation and illuviation.

Biochemical cycling that occurs in lagoons and marshes behind barrier islands is an important process within the barrier island CZ. Higher nutrient concentrations and biodiversity are frequent hallmarks of backbarrier lagoons and marshes, and studies of biochemical dynamics of these habitats reveal complex structures and processes dictating their evolution. Coastal marshes and lagoons

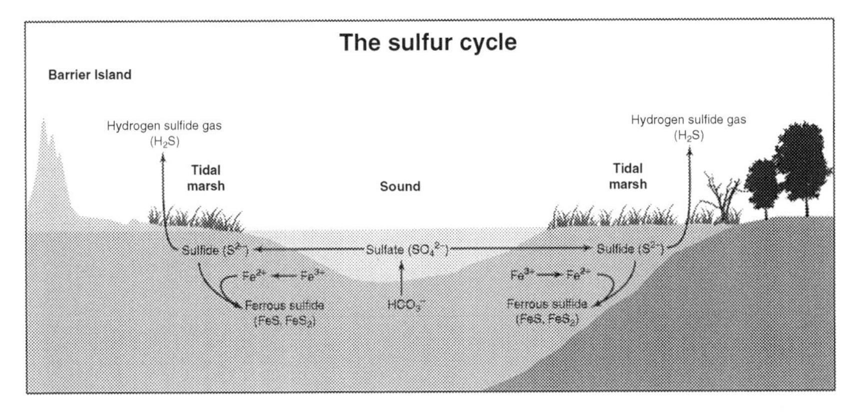

FIGURE 16.6 Nutrient cycles dictate the exchange of materials such as hydrogen, sulfur, and iron, and represent an external control on vegetation development and community succession, particularly in low-relief environments such as a backbarrier shoreline. The "rotten egg" smell associated with coastal marshes is resultant from a robust exchange of chemicals including sulfur in the lagoon and backbarrier environments, which in turn informs the evolution of plant communities within the physical framework.

are highly reductive (commonly anoxic) environments that may, under the right conditions, exhibit high- sulfur soils (see Fig. 16.6) with little to no horizonation because vegetation and microbiota are highly efficient in cycling the available nutrients in the sediments. As a result, these environments may have a "rotten egg" smell as a result of excess sulfur that accumulates in the soil from seawater incorporated into the soils by a process known as sulfidication (see Darmody et al., 1977). Because organic matter is a requirement of sulfidication, sulfur content typically increases with organic content.

16.4.6 Distribution of Vegetation

The species occurring on a barrier island are typically some subset of the nearby mainland, though some endemic species may survive only on the island itself. The harsh maritime conditions of barrier islands nurture a distinctive class of plants capable of surviving in spite of salt spray, salty groundwater, mobile sand, and poor soils. Barrier Islands are shaped by prevailing winds and nearshore currents into long, linear features, and with "steep gradients of environmental factors that occur across (the island) with increasing distance from the ocean (Ehrenfeld, 1990, p. 438)." Morphological variability creates microclimates capable of hosting endemic species and plant communities, whereas the directionality of prevailing winds dictates the persistence of maritime air parcels and the transport of salt spray inland. As with barrier morphology, the distribution and evolution of plant communities and biota is arguably most dependent on the interval between storm landfall and overwash events.

This complex suite of factors creates a distinctive heterogeneous zonation of species and ecological communities (Bockelmann et al., 2002). Plant communities are "usually more or less linearly arranged into belts parallel to the long dimension of the island" (Ehrenfeld, 1990, p. 438). Remnant morphodynamic regimes (i.e., transgressive or regressive island state) leave ridge-and-swale sequences as an external control on the development of this zonation, which is often associated with topographic highs and lows of dunes and swales. The exact species present in each "belt" or ecological structure vary with geographic location, but the general characteristics of the cross-shore variation remain consistent.

Human impact on barrier islands is a large control on alongshore variability in plant communities, "both through the influx of exotic species, and the loss of diversity through habitat destruction" (Ehrenfeld, 1990, p. 439). For beach and dune environments in particular, alteration of native communities creates vastly different morphodynamic regimes that can determine both short-term and longer-term stability. Although it is known that in some cases humans have introduced nonnative species, and in other locations have eradicated a species from an island, the exact holistic impact of human-induced habitat loss and alteration has not been quantified. Popular efforts such as dune grass planting programs and beach nourishment, though politically expedient and economically beneficial, may not represent adequately realistic changes needed to ensure the sustainability of the barrier island CZ.

16.4.7 Biogeochemical Cycling and Soil Development

The low relief, sandy sediments with little to no soil development commonly occurring on barrier islands belie a complex heterogeneous array of microenvironments created and informed by biochemical dynamics. Feedbacks between soils and microbiota determine initial community structure, which in turn affects soil development and stratigraphy to form discrete patches of subaerial and subaqueous habitats that diverge over time (Brantley et al., 2007). Species zonation and soil development are commonly linked to topographic controls like ridge-and-swale sequences, which in turn are affected by variety in biogeochemical dynamics. In this manner soils, nutrients, and morphology are related by a series of feedbacks dictating the framework within which ecosystem development and morphological change occur. Soil development and horizonation generally increase with distance from the shoreline, based on the frequency of dune breaching and washover. Beaches have very little to no soil development, though they typically have complex internal stratigraphies (discussed later in the chapter). Depending on the nature of the coastal dune system immediately landward of the beach, soil development may be extensive or nonexistent. The backbarrier is frequently flooded, which provides ample organic material for soil development. Frequent overwash can cause stacked soils in the backbarrier and even in the adjacent tidal flats, depending on the inland extent of overwash deposits. Tidal-flat soils are very poorly developed because the hypersaline environment limits vegetation and other pedogenic factors.

Alongshore variation in soil development is the result of irregular disturbance regimes, beach width, topographic relief, vegetation abundance, and subsurface conditions. Barrier Islands are subject to storms, which can produce waves capable of breaching the dunes and redistributing beach and dune sediment in the backbarrier. As noted earlier, offshore ridges-and-swales along Santa Rosa Island (Florida) determine the alongshore pattern of dune breaching and washover, which engages with the backbarrier in a positive feedback loop that reinforces the stability of nonoverwashed portions of the barrier island by encouraging the development of a mature maritime forest and dense thickets of shrubs and low trees. The lowering of dunes increases the likelihood of future breaching at that same location. Repeat breaching and sediment redistribution is a negative feedback loop that prohibits soil development in the dunes, but promotes soil development in the backbarrier.

16.5 DISCUSSION

As outlined in the previous sections, the interrelated nature of vegetation, soils, and landforms in a barrier island illustrates the connectivity of the CZ – an open system that exchanges matter and energy between the atmosphere, lithosphere, biosphere, and hydrosphere. Traditionally, barrier islands are studied by researchers in separate and characteristically isolated disciplines. For example, vegetation was studied by biologists, soils, the focus of soil scientists, groundwater, by hydrologists, and subsurface sediments and bedrock by geologists (Chorover et al., 2007). Discrete studies of these various systems and compartments are essential, however in recent years it has become clear that this somewhat reductive approach cannot be used to understand and predict the behavior of the barrier island CZ holistically. The functional, emergent properties of barrier islands are not only the result of its various environments, but also of the interactions among them (see Strahler, 1980; Huggett, 1985; Malanson et al., 1990; Phillips, 1999; NRC, 2001, 2010, 2012). Over time, physical and biochemical processes alter the character of the barrier's CZ, which can be observed and studied at a variety of spatiotemporal scales ranging from weathering of particles, to soil profiles and plant communities, and even to an entire coastal system (Chorover et al., 2007).

A process-form systems approach is a preferred conceptual framework because it accounts for feedbacks between seemingly disparate components of a barrier island CZ. For example, a barrier can be considered as a process-form system in which the morphological variables (i.e., grain size, beach slope, vegetation cover, soil) interact with flows of energy, mass, and momentum (i.e., wind, waves, tides). Moreover particular sections of a barrier island may represent discrete process-form systems that may only interact during extreme events (i.e., overwash deposits along the backbarrier margin). Within a process-form framework, models can be developed that use an iterative process to account for the complex feedbacks between physical and biogeochemical processes on landform evolution and in turn, how the landform influences these processes. We argue the most effective conceptual framework accounts for overarching physical processes that determine the system-scale state (i.e., transgressive or regressive) and biochemical processes that react to create heterogeneous patches of plant communities (i.e., beach dune ridge-and-swale sequences). Although the constant morphological change occurring along barrier island beaches and dunes, as well as backbarrier shorelines, dictates the progression of plant communities toward climax, the same biochemical processes at work within an island's physical setting cause the development of heterogeneous communities with differing rates of physical change and susceptibility to disturbances. As a result, alteration of the external state of a barrier island drives biochemical evolutionary pathways (i.e., storm overwash resulting in island denudation) and at a smaller scale, biochemical processes can influence physical dynamics (i.e., dune grass colonization).

It is clear that CZ research has made significant progress in understanding the complex interactions between the atmosphere, hydrosphere, biosphere and lithosphere. However, current CZ research is generally on individual attributes of the CZ and does not take full advantage of systems theory as proposed by Strahler (1952, 1980), Huggett (1985) and others. Researchers must improve the theoretical and conceptual ideas of how the CZ functions by investigating barrier islands as a series of subsystems within the process-form regime. Some of the most fundamental problems with understanding the CZ are related to the notion that it is extremely difficult to model both form and process.

Although the integration of a true systems approach into the current CZ conceptual framework may seem daunting, we argue that tools to address many problems are at hand. Improved technologies and methods including artificial neural networks, LiDAR, remote sensing, thermoluminescent dating, and near-surface geophysics will help enable researchers to develop more comprehensive deterministic models to understand and predict the evolution of the barrier CZ. Enhanced accuracy and resolution of elevation models enables more realistic projection of the evolution of a barrier island's physical state, while remote sensing and near-surface geophysics (i.e., ground-penetrating RADAR, airborne electromagnetic conductivity surveys) provide detailed characterization of the landscape at a larger scale than ever before. As a result researchers are now able to observe and quantify changes in the barrier CZ with greater accuracy, and the increasing occurrence of repeat surveys (i.e., government-sponsored annual LiDAR surveys and the ubiquity of satellite-based imagery via public and private efforts). These methods enable more consistent observation of changes in the system state of individual barriers (e.g., transgressive or regressive), while researchers are able to employ advanced dating and field survey methods to augment these data in order to paint a highly realistic picture using quantifiable parameters. Together, smaller-scale reductionist studies regarding individual processes or landforms encountered throughout the barrier CZ provide insight into the mechanisms that drive the evolution of these systems and inform the efforts of modelers employing analytical reasoning to predict an island's future behavior. Although many traditional studies had to rely on locally based expert knowledge of isolated forms and processes between individual barrier islands, robust methodologies are now available that can provide for both detailed characterization of endemic dynamics as well as standardized comparisons between far-flung barriers.

16.6 CONCLUSIONS

The barrier island CZ is a highly dynamic environment representing an ideal natural laboratory for investigating systems-scale dynamics that dictate the evolution of the CZ. Although the transgressive or regressive state of a barrier island sets the boundaries on development of certain plant communities and subsystems, a process-form approach accounts for both external controls, internal feedbacks, and changing roles of the systems' component parts. Coastal dunes occur at the physical interface of terrestrial and marine environments, and at a conceptual node between dynamics informed by the atmosphere, hydrosphere, biosphere, and lithosphere. As such they are an ideal platform for discussing many of the seemingly disparate processes evident within the barrier island CZ. For instance, foredunes connect nearshore bathymetry to backbarrier vegetation distributions via overwash frequency as described earlier. Atmospheric regimes (i.e., prevailing wind speed and direction) dictate the fetch length upwind from the dune toe and inform sediment delivery to the foredune system, which in turn determines the magnitude of storm surge needed to breach the dunes and inundate the backbarrier. Although across-shore variability in vegetation, soils, and morphology on barrier islands is somewhat obvious, alongshore changes are subtler but have (arguably) greater impact on the system state of the backbarrier and holistic evolution of the island.

These disconnected processes may be linked by conceptual nodes within a barrier island CZ system if a process-form, systems approach is used that allows for feedbacks between process and form, and changes in those relationships through time. The necessary tools for quantifying CZ variables in a variety of barrier-island environments now exist and are accessible for researchers, along with improved modeling capabilities that enable the detailed characterization of process and form, and their evolution through space and time. By using a dynamic process-form systems approach and employing the full arsenal of scientific methodologies now available, researchers concerned with various elements of the barrier island CZ should seek to illustrate connectivity between previously disparate processes and forms, in an effort to link concepts that may enable more accurate prediction of the barrier islands' response to storms and rising sea levels moving forward.

REFERENCES

Aagaard, T., Davidson-Arnott, R., Greenwood, B., Nielsen, J., 2004. Sediment supply from shoreface to dunes: linking sediment transport measurements and long-term morphological evolution. Geomorphology 60 (1–2), 205–224.

Anschult, P., Smith, T., Mouret, A., Deborde, J., Bujan, S., Poirier, D., Lecroart, P., 2009. Tidal sands as biogeochemical reactors. Estuar. Coast. Shelf Sci. 84, 84–90.

Bauer, B., Davidson-Arnott, R., 2003. A general framework for modeling sedimentary supply to coastal dunes including wind angle, beach geometry, and fetch effects. Geomorphology 49 (1–2), 89–108.

Bockelmann, A.C., Bakker, J.P., Neuhaus, R., Lage, J., 2002. The relation between vegetation zonation, elevation and inundation frequency in a Wadden Sea salt marsh. Aquat. Bot. 73 (3), 211–221.

Brantley, S.L., Goldhaber, M.B., Ragnarsdottir, V., 2007. Crossing disciplines and scales to understand the critical zone. Elements 3, 307–314.

Browder, A.G., McNinch, J.E., 2006. Linking framework geology and nearshore morphology: correlation of paleochannels with shore-oblique sandbars and gravel outcrops. Marine Geol. 231 (1–4), 141–162.

Chorover, J., Kretzschmar, R., Garcia-Pichel, F., Sparks, D.L., 2007. Soil biogeochemical processes within the critical zone. Elements 3, 321–326.

Conn, C.E., Day, F.P., 1997. Root decomposition across a barrier island chronosequence: litter quality and environmental controls. Plant Soil 195 (2), 351–364.

Cooper, W.S., 1967. Coastal sand dunes of California. Geological Society of America Memoirs 104, p. 131.

Darmody, R.G., Fanning, D.S., Drummond, W.J., 1977. Determination of total sulfur in tidal marsh soils by x-ray spectroscopy. Soil Sci. Soc. Am. J. 41 (4), 761–765.

Davidson-Arnott, R., 2005. Conceptual model of the effects of sea level rise on sandy coasts. J. Coast. Res. 21 (6), 1166–1172.

Davidson-Arnott, R., 2010. Introduction of Coastal Processes and Geomorphology. Cambridge University Press, Cambridge, UK, p. 442.

Dean, R.G., Dalrymple, R.A., 2002. Coastal Processes with Engineering Applications. Cambridge University Press, p. 475.

Demarest, J.M., Leatherman, S.P., 1985. Mainland influence on coastal transgression – Delmarva Peninsula. Marine Geol. 63 (1–4), 19–33.

Dillenburg, S.R., Roy, P.S., Cowell, P.J., Tomazelli, L.J., 2000. Influence of antecedent topography on coastal evolution as tested by shoreface translation-barrier model (STM). J. Coast. Res. 16 (1), 71–81.

Ehlers, J., Kunz, H., 1993. Morphology of the Wadden Sea, natural processes and human interference. In: Hillen, Verhagen (Eds.), Coastlines of the Southern North Sea. ASCE, pp. 65–84.

Ehrenfeld, J.G., 1990. Dynamics and processes of barrier island vegetation. Rev. Aquat. Sci. 2 (3–4), 437–480.

Fryberger, S., Schenk, C., Kyrstinik, L., 1988. Stokes surfaces and the effects of near surface groundwater table on aeolian deposition. Sedimentology 35 (1), 21–41.

Hansom, J.D., 2001. Coastal sensitivity to environmental change: a view from the beach. Royal Society of Edinburgh Conf. on Landscape Sensitivity, Stirling, Scotland, UK, pp. 291–305.

Harris, M.S., Gayes, P.T., Kindinger, J.L., Flocks, J.G., Krantz, D.E., Donovan, P., 2005. Quaternary geomorphology and modern coastal development in response to inherent geologic framework: an example from Charleston, South Carolina. J. Coast. Res. 21 (1), 49–64.

Helland-Hansen, W., Martinsen, O.J., 1996. Shoreline trajectories and sequences: description of variable depositional-dip scenarios. J. Sediment. Res. 66 (6), 670–688.

Hequette, A., Ruz, M.J., 1991. Spit and barrier-island migration in the southeastern Canadian Beaufort Sea. J. Coast. Res. 7 (3), 677–698.

Hesp, P.A., 2002. Foredunes and blowouts: initiation, geomorphology and dynamics. Geomorphology 48 (1–3), 245–268.

Hesp, P.A., 2012. Surfzone–beach–dune interactions. In: Kranenburg, Horstman, Wijnberg (Eds.), NCK-days 2012. Crossing Borders in Coastal Research. Jubilee Conference Proceedings 20th NCK-days. Netherlands Centre for Coastal Research, pp. 35–40.

Hesp, P.A., 2013. Conceptual models of the evolution of transgressive dune field systems. Geomorphology 199, 138–149.

Hesp, P.A., Thom, B.G., 1990. Geomorphology and evolution of active transgressive dunefields. In: Nordstrom, Psuty, Carter (Eds.), Coastal Dunes: Form and Process. John Wiley and Sons, Chichester, pp. 253–288.

Houser, C., 2012. Feedback between ridge and swale bathymetry and barrier island storm response and transgression. Geomorphology 173–174, 1–16.

Houser, C., 2013. Alongshore variation in the morphology of coastal dunes: implications for storm response. Geomorphology 199, 48–61.

Houser, C., Hamilton, S., 2009. Sensitivity of post-hurricane beach and dune recovery to event frequency. Earth Surf. Process. Land. 34, 613–628.

Houser, C., Mathew, S., 2012. Alongshore variation in foredune height in response to transport potential and sediment supply: South Padre Island, Texas. Geomorphology 125 (1), 62–72.

Houser, C., Barrett, G., Labude, D., 2011. Alongshore variation in the rip current hazard at Pensacola Beach, Florida. Nat. Hazards 57 (2), 501–523.

Houser, C., Hapke, C., Hamilton, S., 2008. Controls on coastal dune morphology, shoreline erosion and barrier island response to extreme storms. Geomorphology 100 (3–4), 223–240.

Houser, C., Labude, B., Haider, L., Weymer, B., 2013. Impacts of driving on the beach: case studies from Assateague Island and Padre Island National Seashores. Ocean Coast. Manag. 71, 33–45.

Houser, C., Wernette, P., Rentschler, T., Jones, H., Hammond, B., Trimble, S., 2015. Post-storm beach and dune recovery: implications for barrier island resilience. Geomorphology 234, 54–63.

Hugenholtz, C.H., Wolfe, S.A., 2005a. Biogeomorphic model of dunefield activation and stabilization on the northern Great Plains. Geomorphology 70 (1–2), 53–70.

Hugenholtz, C.H., Wolfe, S.A., 2005b. Recent stabilization of active sand dunes on the Canadian prairies and relation to recent climate variations. Geomorphology 68 (1–2), 131–147.

Huggett, R.J., 1985. Mathematical models in agriculture. Appl. Geogr. 5 (2), 172.

Kraft, J.C., 1971. Facies relations in Holocene-Pleistocene coastal sediments: model for interpretation of ancient transgressive-sequences. AAPG Bull. 56 (3), 634.

Kraft, J.C., Brown, H.F., 1982. Terrigenous and carbonate clastic facies in a transgressive sequence over volcanic terrain. AAPR Bull. 66 (5), 589.

Kraft, J.C., John, C.J., 1979. Lateral and vertical facies relations of transgressive barrier. AAPG Bull. 63 (12), 2145–2163.

Kraft, J.C., Chrzastowski, M.J., Belknap, D.F., Toscano, Fletcher, 1987. The transgressive barrier-lagoon coast of Delaware: morphostratigraphy, sedimentary sequences and responses to relative rise in sea level. In: Nummedal, Pilkey, Howard (Eds.), Sea-Level Fluctuation and Coastal Evolution. SEPM Spec. Pub. 41, pp. 129–143.

Leatherman, S.P., 1976. Barrier island dynamics: overwash processes and eolian transport. In: Proc. 15th Conf. Coastal Eng. 1958–1974. Honolulu, HI, USA.

Lessa, G.C., Angulo, R.J., Giannini, P.C., Araujo, A.D., 2000. Stratigraphy and Holocene evolution of a regressive barrier in south Brazil. Marine Geol. 165 (1–4), 87–108.

Levy, G.F., 1990. Vegetation dynamics on the Virginia barrier islands. Virginia J. Sci. 41 (4A), 300–306.

Malanson, G.P., Butler, D.R., Walsh, S.J., 1990. Chaos theory in physical geography. Phys. Geogr. 11 (4), 293–304.

Martinez, M.L., Maun, M.A., Psuty, N.P., 2004. The fragility and conservation of the world's coastal dunes: geomorphological, ecological and socioeconomic perspectives. In: Martinez, M.L., Psuty, N.P. (Eds.), Coastal Dunes: Ecology and Conservation. Springer-Verlag, New York.

Masetti, R., Fagherazzi, S., Montanari, A., 2008. Application of a barrier island translation model to the millennial-scale evolution of Sand Key, Florida. Continental Shelf Res. 28 (9), 1116–1126.

Maun, M.A., 2009. The Biology of Coastal Sand Dunes. Oxford University Press, Oxford, p. 288.

McNinch, J., 2004. Geologic Control in the Nearshore: Shore-Oblique Sandbars and Shoreline Erosional Hotspots. Mid-Atlantic Bight, USA.

Meyer-Arendt, K.J., 1990. Recreational business districts in Gulf of Mexico seaside resorts. J. Cultur. Geogr. 11, 39–55.

Morton, R.A., 2002. Factors controlling storm impacts on coastal barriers and beaches – a preliminary basis for near real-time forecasting. J. Coast. Res. 18, 486–501.

Morton, R.A., Sallenger, A.H., 2003. Morphological impacts of extreme storms on sandy beaches and barriers. J. Coast. Res. 19 (3), 560–573.

Morton, R.A., Paine, J.G., Gibeaut, J.C., 1994. Stages and durations of post-storm beach recovery, southeastern Texas coast. J. Coast. Res. 10 (4), 884–908.

Nordstrom, K.F., Lampe, R., Vandemark, L.M., 2000. Reestablishing naturally functioning dunes on developed coasts. Environ. Manag. 25 (1), 37–51.

Nordstrom, K.F., Jackson, N.L., Korotky, K.H., 2011. Aeolian sediment transport across beach wrack. J. Coast. Res. Special Issue 59, 211–217.

Nott, J., 2006. Tropical cyclones and the evolution of the sedimentary coast of northern Australia. J. Coast. Res. 221, 49–62.

National Research Council, 2001. Basic Research Opportunities in Earth Science. National Academy Press, Washington, DC, p. 168.

National Research Council, 2010. Landscapes on the Edge: New Horizons for Research on Earth's Surface. National Academy Press, Washington, DC, p. 180.

National Research Council, 2012. New Research Opportunities in the Earth Sciences. National Academy Press, Washington, DC, p. 132.

Orme, A., 1992. Late Quaternary deposits near Point Sal, south-central California: a time frame for coastal dune emplacement. In: Fletcher, Wehmiller (Eds.) Quaternary Coasts of the United States: Marine and Lacustrine Systems, IGCP Project 274: Society for Sedimentary Geology, Special Publication, vol. 48, pp. 310–315.

Orr, M., Zimmer, M., Jelinski, D.E., Mews, M., 2005. Wrack deposition on different beach types: spatial and temporal variation in the pattern of subsidy. Ecology 86 (6), 1496–1507.

Otvos, E.G., 2012. Coastal barriers – nomenclature, processes, and classification issues. Geomorphology 139, 39–52.

Phillips, J., 1999. Earth Surface Systems: Complexity, Order, and Scale. Basil Blackwell, Oxford, UK.

Pilkey, O.H., Young, R.S., Riggs, S.R., Smith, A.W.S., Wu, H., Pilkey, W.D., 1993. The concept of shoreface profile of equilibrium: a critical review. J. Coast. Res. 9, 255–278.

Pilkey, O.H., Cooper, J.A.G., Lewis, D.A., 2009. Global distribution of geomorphology of fetch-limited barrier islands. J. Coast. Res. 25 (4), 819–837.

Posamentier, H.W., Allen, G.P., James, D.P., Tesson, M., 1992. Forced regressions in a sequence stratigraphic framework – concepts, examples, and exploration significance. AAPG Bull. 76 (11), 1687–1709.

Psuty, N.P., 1992. Spatial variation in coastal foredune development. In: Carter, Curtis, Sheehy-Skeffington (Eds.), Coastal Dunes: Geomorphology, Ecology and Management for Conservation. Balkema, Rotterdam, Netherlands, pp. 3–13.

Pye, K., 1983. Coastal dunes. Prog. Phys. Geogr. 7 (4), 531–577.

Reinson, G.E., 1979. Facies models: barrier island systems. Geosci. Can. 6 (2), 51–68.

Riggs, S.R., 1979. A petrographic classification of sedimentary phosphorites. In: Cook, Shergold (Eds.), Proterozoic-Cambrian Phosphorites IGCP 156, vol. 1. Cambridge University Press, Canberra, Australia.

Riggs, S.R., O'Connor, M.P., 1974. Relict sediment deposits in a major transgressive coastal system. University of North Carolina Sea Grant Program, UNC-SG-74-04, p. 37.

Riggs, S.R., Cleary, W.J., Snyder, S.W., 1995. Influence of inherited geologic framework on barrier shoreface morphology and dynamics. Marine Geol. 126 (1–4), 213–234.

Ritchie, W., Penland, S., 1990. Aeolian sand bodies of the south Louisiana coast. In: Nordstrom, Psuty, Carter (Eds.), Coastal Dunes Forms and Process. Wiley and Sons, Chichester, West Sussex, UK, pp. 105–127.

Sallenger, A.H., 2000. Storm impact scale for barrier islands. J. Coast. Res. 16 (3), 890–895.

Saunders, K.E., Davidson-Arnott, R.G.D., 1990. Coastal dune response to natural disturbances. In: Davidson-Arnott, R.G.D. (Ed.), In: Proceedings of the Symposium on Coastal Sand Dunes. National Research Council of Canada, Ontario, pp. 321–346.

Schupp, C.A., McNinch, J.E., List, J.H., 2006. Nearshore shore-oblique bars, gravel outcrops, and their correlation to shoreline change. Marine Geol. 233, 63–79.

Schwab, W.G., Thieler, E.R., Allen, J.R., Foster, D.S., Swift, B.A., Denny, J.F., 2000. Influence of inner-continental shelf geologic framework on the evolution and behavior of the barrier-island system between Fire Island Inlet and Shinnecock Inlet, Long Island, New York. J. Coast. Res. 15, 408–422.

Sherman, D.J., Bauer, B.O., 1993. Dynamics of beach-dune systems. Prog. Phys. Geogr. 17 (4), 413–447.

Short, A.D., 2010. Sediment transport around Australia – sources, mechanisms, rates, and barrier form. J. Coast. Res. 26 (3), 395–402.

Short, A.D., Hesp, P.A., 1981. Wave, beach and dune interactions in southeastern Australia. Marine Geol. 48 (3–4), 259–284.

Stallins, J.A., Parker, A.J., 2003. The influence of complex systems interactions on barrier island dune vegetation pattern and process. Ann. Assoc.Am. Geogr. 93 (1), 13–29.

Stockdon, H.F., Sallenger, A.H., Holman, R.A., Howd, P.A., 2007. A simple model for the spatially-variable coastal response to hurricanes. Marine Geol. 238, 1–20.

Stone, G.W., Liu, B., Pepper, D.A., Wang, P., 2004. The importance of extratropical and tropical cyclones on the short-term evolution of barrier islands along the northern Gulf of Mexico, USA. Marine Geol. 210 (1–4), 63–78.

Strahler, A.N., 1952. Hypsometric analysis of erosional topology. Geol. Soc. Am. Bull. 63 (11), 1117–1142.

Strahler, A.N., 1980. Systems theory in physical geography. Phys. Geogr. 1, 1–27.

Tackett, N., Craft, C., 2010. Ecosystem development on a coastal barrier dune chronosequence. J. Coast. Res. 264, 736–742.

Thieler, E.R., Young, R.S., 1991. Quantitative evaluation of coastal geomorphological changes in south Carolina after Hurricane Hugo. J. Coast. Res. 8, 187–200.

Thom, B.G., Hall, W., 1991. Behavior of beach profiles during accretion and erosion dominated periods. Earth Surf. Process. Land. 16, 113–127.

Weise, B.R., White, W.A., 1980. Padre Island National Seashore: A Guide to the Geology, Natural Environments, and History of a Texas Barrier Island (Guidebook 17). Bureau of Economic Geology, University of Texas, Texas, US, p. 94.

Weymer, B., Everett, M.E., de Smet, T.S., Houser, C., 2015. Review of electromagnetic induction for mapping Barrier Island framework geology. Sediment. Geol. 321, 11–24.

Wright, L.D., Short, A.D., 1984. Morphodynamic variability of surf zones and beaches: a synthesis. Marine Geol. 56 (1–4), 93–118.

Chapter 17

Geospatial Science and Technology for Understanding the Complexities of the Critical Zone

Michael P. Bishop*,**, Iliyana D. Dobreva*,**, and Chris Houser*,**
*Department of Geography, Texas A&M University, College Station, Texas, USA;
**Center for Geospatial Sciences, Applications & Technology, Texas A&M University, College Station, Texas, USA

17.1 INTRODUCTION

The Critical Zone is undergoing constant change due to climatic, lithospheric, and anthropogenic forcing factors (Grujic et al., 2006; Stolar et al., 2006; Brantley et al., 2007; Harrison, 2009). Rapidly changing conditions are regulated by feedback mechanisms and highly coupled multiscale systems incorporating atmospheric, surface energy balance, biogeochemical, ecological, cryospheric, lithological, and cultural processes. Human and environmental relationships and dependencies associated with rapid change, raises serious questions about resource sustainability and the resilience of the Critical Zone (Smit and Wandel, 2006; Ernst, 2012). Climate change, soil erosion and sediment transfer, land cover dynamics, water quality and supply, ecosystem migration, biodiversity, and other issues need to be examined and understood from a spatio-temporal dynamic perspective, highlighting energy and mass transfer. Unfortunately, we have yet to fully understand complexities and scale dependencies of many processes and feedback mechanisms. Furthermore, it is necessary to understand how the Critical Zone is being transformed, and what the implications may be to an ever increasing human population that has an impact on resource availability and ecosystem sustainability (Corvalan et al., 2005; Lin et al., 2011).

Critical Zone systems are extremely complex and include coupled chemical, physical, biological, and geological processes that govern and modify the structural characteristics of ecosystems, topography, and the subsurface. Research on this complex topic is in its early infancy, as we attempt to identify key variables, characterize process mechanics, study scale dependencies, and model various

Developments in Earth Surface Processes, Vol. 19. http://dx.doi.org/10.1016/B978-0-444-63369-9.00017-3

system dynamics (Lin et al., 2011; Rasmussen et al., 2011). It is widely known that our knowledge and understanding of the Critical Zone is extremely limited (Brantley and Lebedeva, 2011), and this is especially reflected in a widely used definition of the Critical Zone, which commonly refers to the outer limits of vegetation down to the groundwater zone (e.g., Anderson et al., 2007, 2012; Völkel et al., 2011).

This reductionist approach to defining and understanding the Critical Zone does not formally account for the spatial extension of the Critical Zone into the atmosphere, due to the global weathering system that affects the sequestration and release of variable gases that influence global climate (Kump et al., 2000; Pielke et al., 2002), or the input of mass onto the land and oceans in the form of water, aerosols, and nutrients (Martin and Whitfield, 1983; Duce et al., 1991; Johnson et al., 2008). Furthermore, the Critical Zone extends to a much greater depth than near-surface groundwater zones, as climate forcing and high-magnitude erosion and denudation has been found to influence deep-seated tectonic processes that collectively govern uplift and the geological and topographic structure of the near surface environment, which controls relief production, and in turn climate, weathering, depth of regolith, and the depth to which water and organisms reside in the lithosphere (Gabet et al., 2004; Bishop et al., 2010).

Another dimension of understanding system complexity involves process–form dynamics that governs the physical structural characteristics of the Critical Zone. Globally, the Critical Zone can exhibit unique three-dimensional structural characteristics that are the result of, and evolve based upon, unique environmental conditions that are dependent upon geographic location, magnitude and type of processes, and/or system complexities. Tropical, temperate, high latitude, and high altitude Critical Zones exhibit very different vertical structural configurations based upon climate, ecological, geological, and anthropological influences. This includes variation in materials, processes, depths, and vertical layering. Assessing and characterizing these process–form spatial scale dependencies represents an important aspect of studying complex Critical Zones (Willett and Brandon, 2002; Bishop et al., 2003).

It is widely recognized that an integrated multidisciplinary approach to studying the Critical Zone is required (Lin et al., 2011). Fundamental research must include characterization and modeling of process and system dynamics, investigation of operational scale dependencies, spatial representation and mapping of the vertical structural, characterization of landscape-scale variability and complexity of phenomena, along with assessments on controlling factors and system resilience. Research will require attention to *in situ* data collection and observations from Critical Zone Observatories (Anderson et al., 2008; Kuntz et al., 2011). Geospatial technologies will also play an important role (Bishop et al., 2012), such that high-resolution multispectral/multitemporal imagery and subsurface data can be used to characterize the properties of the landscape and subsurface. For example, information regarding vegetation canopy structure (Lim et al., 2003; Disney et al., 2006; Schlerf and Atzberger, 2006), land cover and land use (Rozenstein and Karnieli, 2011; Verburg et al., 2011; Jin et al., 2013), topography (Giles, 1998; Hirano et al., 2003; Wechsler and Kroll, 2006), surface deformation (Berardino et al., 2002; Tralli et al., 2005), and surface biophysical properties of vegetation, soil, water, and topography can be extracted from remotely sensed data. Subsurface information about materials, density, moisture content, vertical zonation, weathering, stratigraphy, and geological structure can be obtained by using gravity, ground penetrating radar, electromagnetics, and seismic geophysical techniques (Njoku and Kong, 1977; Waters et al., 1990; Peters et al., 1994; Ouzounov and Freund, 2004).

Similarly, the field of geomorphometry will play a fundamental role in Critical Zone research, as the concept of *land surface* can serve as the foundation for quantitatively characterizing the properties of the topography and "surfaces" that vertically make up the vertical structure of the Critical Zone (e.g., vegetation, stratigraphic and geological surfaces). This is especially important as we need to represent and characterize the nature of the vertical zonation within the Critical Zone, study process–form relationships, and provide input data and constraints for numerical models. Geomorphometry represents the science of quantitatively characterizing the land-surface and its properties, including surface processes and the evolution of the landscape (Pike et al., 2009). Topographic parameters and topographic constraints govern many surface processes including, microclimate, evapotranspiration, erosion, weathering, sediment transport, and pedogenesis. Consequently, analysis of the topography will greatly assist scientists in a variety of academic disciplines, as they study various components of the Critical Zone that are inherently linked to the surface of the landscape.

Geocomputational technologies including artificial intelligence (AI), fuzzy systems (FS), and various forms of numerical modeling will also be required for geospatial information extraction and decision making. This includes progress on developing computational representation schemes and algorithms to characterize phenomena and the complexities of the Critical Zone to identify key variables and scale dependencies, formalize process mechanics, and model system dynamics. It also involves the development and testing of new computational approaches to investigate concepts, information, and relationships that have not been formalized, or are not yet mathematically tractable. Geocomputational approaches can also be used to address the issues of knowledge representation and data/information fusion; two essential topics when dealing with complex systems.

The purpose of this chapter is to demonstrate the significance of remote sensing, geomorphometry, and geocomputation in studying the complexities of the Critical Zone. We provide fundamental background information on the role of geospatial technology for data collection, information extraction, and numerical modeling of landscape conditions. This includes a treatment of remote sensing, geomorphometry, and geocomputation. We then provide an example of the integration of geospatial data and process modeling by describing a newly developed physics-based surface irradiance model that can be used to study alpine Critical Zones, as surface energy budget parameters govern meltwater production that influences natural hazards and water availability in lower-altitude Critical Zones. Finally, we highlight the use of a geocomputational approach that involves knowledge representation, integrates surface biophysical and topographic information, and provides a framework for studying landscape–subsurface relationships in the Critical Zone.

17.2 BACKGROUND

Critical Zone and systems research can be significantly facilitated through the use of geospatial technologies (Bishop et al., 2012; Bishop, 2013). There is a plethora of new sensor technologies and platforms specifically designed to capture atmospheric, landscape, and subsurface information that can be used to assess the physical, chemical, and process domains of the Critical Zone. Various remote sensing technologies enable spatio-temporal information to be extracted from geospatial data to support modeling efforts. Numerous conceptual/theoretical and methodological issues are at the heart of using geospatial data to improve our knowledge and facilitate the development of new geospatial technologies to address problems in the Earth sciences (Bishop et al., 2012; Bishop, 2013). Example issues include space and time representation schemes, process parameterization schemes, indeterminate boundaries, landscape biophysical assessment, spatial stratification of the landscape, integration of spectral and topographic information, integration of landscape and subsurface information, and geovisualization of spatial and temporal phenomena.

17.2.1 Remote Sensing

Remote sensing science and technology has rapidly evolved to enable the collection of geospatial data that cannot be obtained from other sources. The spatial and temporal dynamics of the atmosphere, lithosphere, ecosphere, and hydrosphere can be investigated due to new developments in sensors and platforms that provide high resolution spatial, spectral, radiometric, and temporal coverage. For example, sensors that sample different portions of the electromagnetic spectrum produce fundamentally different types of environmental information that are critical for understanding Critical Zone dynamics and system interactions. High resolution imagery and topographic information from drones can also produce data that permit change-detection studies (e.g., Berni et al., 2009; Rango et al., 2009; Niethammer et al., 2012) and direct assessment and estimation of fundamental lithospheric process rates (Wilson and Bishop, 2013). Geospatial robotic sensors, landscape sensor networks, and ubiquitous computing and telecommunications capabilities enable assessment of the Critical Zone like never before, enabling data collection, management, analysis and modeling, and geovisualization to facilitate the integration of data/information via geographic information system (GIS) technologies. Consequently, remote sensing and other geospatial technologies are essential for studying complex systems.

Remote sensing of the atmosphere is important, as atmospheric parameters dictate numerous processes, given temperature, and precipitation variations. Global and regional information on atmospheric conditions include atmospheric profiles for precipitable water vapor, aerosol particles, and cloud characteristics. Hyperspectral sensors can also be used to produce water vapor maps, and water vapor concentrations have been shown to vary with altitude (e.g., Green et al., 2006). The magnitude of precipitation is a critical parameter, and the Tropical Rainfall Measuring Mission (TRMM; Kummerow et al., 1998, 2000) has made significant contributions, as the microwave region of the spectrum can be used to estimate rates of precipitation. Various atmospheric parameters partially govern chemical weathering, evapotranspiration, regolith production, and surface erosion and deposition.

Remote sensing of the landscape is used to facilitate spatial stratification of the landscape, mapping of natural resources, assessment of land cover and land use dynamics, and the assessment of ecological, soil, geological, hydrological, and cryospheric systems. Our knowledge on how to apply advances in remote sensing for assessing such complex systems is still a work in progress (Mulder et al., 2011; Bishop, 2013; Bishop et al., 2014).

Spatial and temporal variations in vegetation and ecosystem structure are fundamentally linked with other systems (Wainwright et al., 2000; Rodriguez-Iturbe, 2000; Schmiedel et al., 2012). Remotely sensed data are routinely used to assess vegetation distribution and numerous biophysical properties including green biomass, leaf area index, chlorophyll concentration, leaf moisture, biochemicals, as well as canopy structure and parameters such as canopy height and basal area (Lefsky et al., 1999; Chen et al., 2003). Optical and microwave regions of the spectrum are required, such that there is a heavy reliance on using vegetation indices from reflectance data, and on LiDAR and microwave data for assessing canopy structural parameters. Assessment of biogeographical patterns with land cover and terrain information can greatly assist soil assessment and mapping (Moore et al., 1993).

Research has demonstrated that soil surface properties influence spectral reflectance curve form and the presence or absence of diagnostic absorption features (Clark et al., 1990; Cloutis, 1996; Serbin et al., 2009). Under laboratory conditions, spectral analysis can be used to assess several properties. Airborne or spaceborne spectroscopy permits soil assessment, although photosynthetic vegetation, nonphotosynthetic vegetation, and lichens complicate the purity of spectra, given spatial and inherent mixtures of matter. Nevertheless, great progress in assessing different aspects of soil, sediment, and rocks has occurred (Pour and Hashim, 2012; Zheng et al., 2014).

Mulder et al. (2011) provide a review of current soil and terrain assessment capabilities. They indicate that basic mineralogy can be determined if sensors sample the short-wave infrared (SWIR) and thermal infrared (TIR) regions of the spectrum. Rocks on the surface are best differentiated using the TIR region where carbonate, silicate, and mafic minerals can be discerned. Other researchers have demonstrated that clay and sulphate mineral species can be differentiated

(Hubbard et al., 2003; Hubbard and Crowley, 2005). The presence of vegetation complicates mineralogical mapping efforts.

Soil moisture, soil organic carbon, and iron content can also be assessed (Mulder et al., 2011). Soil moisture is predominately based upon microwave remote sensing and the contrast in the dielectric properties of surface materials. Soil organic carbon is related to soil color and the quantity of black humic acid, and the assessment of biochemicals including cellulose, starch and lignin. Finally, iron oxide and iron hydroxides have specific absorption features that can be assessed with hyperspectral data.

Terrain information is critical for landscape assessment, as the topography represents the integration of atmospheric, surface, and geological processes (Bishop et al., 2010; Wilson and Bishop, 2013). Remote sensing has long been used to generate topographic information via the collection of stereoscopic photography and imagery, radar imagery, and LiDAR data. Stereo-photogrammetric methods have been routinely used to generate digital elevation models (DEMs) from Satellite Pour l'Observation de la Terre (SPOT) and advanced spaceborne thermal emission and reflection radiometer (ASTER) data. Radar imagery such as the Shuttle Radar Mapping Mission (SRTM) and the TerraSAR-X can be used to assess global topographic conditions and assess deformation characteristics of the landscape, given earthquakes and human induced subsidence (Hensley and Farr, 2013). More recently, high-resolution LiDAR systems and terrestrial laser scanners now generate millions of 3D point measurements. The "point-clouds" must be processed to produce versions of the Earth surface that need to be semantically defined and mapped. Such high-resolution LiDAR-based DEMs enable geomorphometric analysis and modeling of the landscape and subsurface to support Critical Zone studies.

17.2.2 Geomorphometry

Geomorphometry plays a central role in a variety of academic disciplines, and is essential for studying surface processes and mapping and characterizing the landscape. Geomorphometry specifically addresses the issues of: (1) sampling attributes of the land surface; (2) geodesy and digital terrain modeling (DTM); (3) DEM preprocessing and error assessment; (4) generation of land-surface parameters, indices, and objects; and (5) terrain information production and problem-solving using parameters and objects. As Tomislav and Hannes (2009) and Bishop (2013) indicate, each aspect of geomorphometry represents a separate subdiscipline that contributes to the development of software tools and geospatial technology. This is especially important, as most GISs do not have adequate software for robust analysis of DEMs. Ultimately, geomorphometry is a rapidly evolving field that can contribute toward understanding surface and subsurface processes, process–form relationships, and the functionality of terrain units that govern matter cycling and resource availability (Tomislav and Hannes, 2009; Wilson and Bishop, 2013).

Remote sensing technologies serve as the foundation for collecting topographic information at a multitude of spatial and temporal scales. Change detections studies require temporal data that adequately sample the surface-given variations in process rates. Appropriate scale-dependent coverage is usually lacking, given that temporal operational scale dependencies can exist at a higher frequency than repeat coverage allows. Engineering related studies involving robotics, dynamic spatio-temporal coverage of the terrain, and unmanned aerial systems are rapidly changing our ability to better collect data from rapidly changing landscapes (e.g., Ramanathan et al., 2007; Firpo et al., 2011; Niethammer et al., 2012; Immerzeel et al., 2014).

The terrain and DTM represents a fundamental part of geomorphometry. Currently, most DEMs constitute a 2.5D representation of the terrain. True 3D analysis of the terrain will require accounting for variations in the Geoid. Nevertheless, DTM can be very complicated, and users must determine which concept of *land surface* they wish to represent. Error/uncertainty assessment is another important part of DTM, and great care must be exercised when comparing multitemporal DEMs generated from different data sources (Paul et al., 2004; Nuth and Kääb, 2011). Most DEMs are known to contain a variety of nonsystematic and systematic errors due to positional accuracy, sensor noise, and preprocessing and algorithm errors (Eckert et al., 2005; Höhle and Höhle, 2009; Nuth and Kääb, 2011).

The development and evaluation of geomorphometric parameters and objects to assess the landscape is an active research area (Tomislav and Hannes, 2009). First and second derivatives characterize the fundamental properties of the topography including slope gradient and angle, slope azimuth, and various curvature metrics. These parameters partially regulate many surface processes, and control the direction and velocities of mass fluxes over the landscape (i.e., water and sediment transport). These and many other scale-dependent parameters also govern surface energy fluxes, surface temperature, evapotranspiration, ablation, and meltwater production. Such parameters are used to characterize and map the functional and scale-dependent organization of the landscape that define surface topography and geological structure, constrain surface processes, and serve as a basis for mapping soils, landforms, and other landscape units.

Geomorphometric parameters and objects are also essential input into GIS-based simulations including surface energy budget, soil erosion, plant dispersion, and landscape evolution models (Pike, 2000; Haboudane et al., 2002; Dorsaz et al., 2013). Numerical simulations based upon topographic information can improve our understanding of complex systems related to geochronology, process mechanics, landscape mapping, resource assessment and inventory, and many other scientific and applied problems. Nevertheless, we must go beyond GIS-based empiricism (Bishop et al., 2012) and address the complex issues of landscape representation and data fusion, information integration, knowledge representation and discovery, conceptual testing and modeling, as well as intelligent decision making.

17.2.3 Geocomputation

Geocomputation can be considered a follow-on revolution to the rapid development and evolution of GIS technology (Openshaw and Abrahart, 2000). It is a part of the rapidly evolving field of geospatial sciences and technology, but focuses on: (1) exploiting various information technology developments, such as high performance computing, AI, and FS; (2) discovering and exploiting information and knowledge beyond data collection and management; and (3) spatio-temporal analysis and modeling to address real-world problems.

Geocomputation is based upon an integrative science framework that drives the use of geospatial data, information, and knowledge to address complex problems that can be investigated through the development of new concepts and geospatial technologies. Therefore, geocomputational investigations are not effectively accomplished using GIS because of the significant limitations associated with representational issues, the need for scale-dependent algorithms and modeling, and the computational speed requirements. The evolution of GIS technology, however, will most likely progress in this direction.

AI and FS are routinely used in most disciplines for mapping. Neural networks, fuzzy clustering, and many other techniques have been evaluated for soils, snow, land cover, vegetation, and geological mapping (e.g., Foody, 1996; Ahn et al., 1999; Dobreva and Klein, 2011). AI techniques including cellular automata and expert systems have also been used to investigate the fundamental rules, facts, and relationships that permit knowledge representation and the prediction of complex phenomena, including spatio-temporal dynamics of dune formation and migration (Nield and Baas, 2008), vegetation and ecosystem expansion and migration (Iverson et al., 2004), urban expansion and land cover dynamics (Liao et al., 2014), and image interpretation (Furfaro et al., 2010).

Fuzzy sets have also been investigated as part of landscape mapping efforts (e.g., Deng and Wilson, 2008), representation of indeterminant boundaries (e.g., Deng and Wilson, 2008), addressing issues of scale related to semantic modeling and mapping (e.g., Fallahi et al., 2008), and in the conceptual modeling and decision making for a variety of applications (e.g., Furfaro et al., 2010). Similarly, developments in agent-based modeling has permitted investigators to simulate the influence of individual agents or populations of agents to interact with each other and the environment based upon landscape information (e.g., Heckbert et al., 2010; Perez and Dragicevic, 2010). Collectively, geocomputation supports information integration, knowledge representation, pattern recognition and data mining, decision support, and model development and testing opportunities.

Geocomputational modeling can facilitate the study of the Critical Zone as we explore for patterns and assess conditions that may not be mathematically tractable at this time. This includes the fundamental issue of scale (Bishop et al., 2012). Given the scale disconnects between many environmental parameters, processes, and systems, and the need to drive and test computational models with scale-relevant geospatial data, there are numerous challenges in developing robust geocomputational models that realistically address a multitude of issues. Consequently, new approaches and models for studying the Critical Zone must be developed and evaluated. The next two sections demonstrate two geocomputational approaches to evaluating the landscape in two different environments.

17.3 SURFACE IRRADIANCE MODELING

Surface irradiance modeling is required to understand the spatio-temporal dynamics of energy input onto various components of the alpine Critical Zone (Kobierska et al., 2011; Meinzer et al., 2013). Surface energy-balance models account for the energy fluxes and include processes that are highly scale dependent. The components of the energy flux can be represented as:

$$Q_N + Q_H + Q_L + Q_G + Q_R + Q_M = 0, \tag{17.1}$$

where Q_N is the net radiation, Q_H is the sensible heat flux, Q_L is the latent heat flux (Q_H and Q_L are referred to as the turbulent heat fluxes), Q_G is the ground heat flux, Q_R is the sensible heat flux supplied by rain, and Q_M is the energy used for melting snow and ice at the appropriate time of year or in the cryosphere. Positive quantities represent energy gains while negative magnitudes indicate an energy loss at the surface.

The most significant energy input is the net radiation component that includes short- and long-wave radiation. The short-wave radiation component has the largest influence of affecting the landscape in the cryosphere, as snow and glacier melt governs a multitude of surface processes and hazard potential. Geographic location, atmospheric conditions, topographic characteristics, and rates of surface processes collectively determine the complexity and rate of change in the cryosphere. Yet the important concepts of landscape lability and cryospheric sensitivity to climate change is not clearly understood, and represents an important research topic, as alpine glacier downwasting and retreat will have a huge impact on human populations at lower altitudes, given water resource and hazard issues (Bishop et al., 1995, 2014; Kargel et al., 2005, 2014). Therefore, surface irradiance modeling must account for numerous issues that are not currently addressed in GIS-based solar radiation transfer models.

17.3.1 Orbital Parameters and Solar Geometry

The solar irradiance reaching the top of the Earth's atmosphere (E^0) is a function of the Sun–Earth orbital parameters, solar geometry, and latitude. The orbital parameters describe the Earth's orbital ecliptic, which include the semimajor axis (a_o), eccentricity (e), and obliquity (ε), and a parameter related to Earth's precession which is the longitude of perihelion (ϖ) measured from the moving vernal equinox of a date (Berger, 1978a, 1978b).

The long-term variations of the obliquity, eccentricity, and precession parameters may be calculated through trigonometric expansion of ε, $e \sin \varpi$, and e. The amplitudes, rates, and phases of the expansion are provided by Berger (1978a). This trigonometric expansion is used in our model and it is valid for reproducing the orbital parameters for 1.5 million years in the past and future (Berger and Loutre, 1991). For each year, ε, ϖ, and e are calculated. Inclusion of these orbital parameters is essential in the computation of solar geometry parameters and enables postdiction and prediction over geological timescales. This facilitates paleoclimatic investigations and future climate change scenario simulations. Existing GIS-based irradiance models do not have this capability.

We compute a variety of other orbital parameters including the Sun–Earth distance (D) and the eccentricity correction factor (F_{ec}) that is required for computing irradiance. The latter is computed as:

$$F_{ec} = \left(\frac{a_o}{D} \right)^2, \tag{17.2}$$

where the semimajor axis of the orbit is set to 150×10^6 km.

Computing the solar geometry requires calculations of solar declination (δ) and solar hour angle (H_{TS}) for a particular time. The solar declination is computed as:

$$\delta = \arcsin(\sin(\varepsilon)\sin(\lambda_{ts})), \tag{17.3}$$

where λ_{ts} is the longitude of the true Sun. Other critical solar geometry parameters include the solar zenith (θ_s) and azimuth angles (ϕ_s).

The cosine of the geocentric solar zenith angle (θ_s^g) is calculated as (Jacobson, 2005):

$$\cos \theta_s^g = \sin \varphi \sin \delta + \cos \varphi \cos \delta \cos H_{TS}, \tag{17.4}$$

where φ is latitude.

This parameter requires a parallax correction ($\delta\theta_p$) to compute the apparent solar zenith angle (θ_s^a) from the Earth's surface. The parallax correction is calculated as a function of Earth's radius relative to the ellipsoid (R) and distance from the Sun (Roderick, 1992; Blanco-Muriel et al., 2001):

$$\delta\theta_p = \frac{R(\varphi) + h}{D} \sin(\theta_s), \tag{17.5}$$

and added to the geocentric solar zenith angle:

$$\theta_s^a = \theta_s^g + \delta\theta_p. \tag{17.6}$$

Atmospheric refraction correction is also required for an accurate computation of apparent solar zenith angle, and we use the formulation in the Nautical

Almanac Office (2013) which is based upon atmospheric temperature and pressure. The atmospheric refraction correction term is subtracted from the apparent solar zenith angle to compute θ_s.

The noncorrected solar azimuth angle (ϕ_s^{nc}) is calculated as:

$$\phi_s^{nc} = \pi + a\ \tan2(Y_s, X_s), \tag{17.7a}$$

$$Y_s = (-\cos\delta \sin H_{TS}) / \cos\alpha_s, \tag{17.7b}$$

$$X_s = (\sin\alpha_s \sin\varphi - \sin\delta) / (\cos\alpha_s \cos\varphi), \tag{17.7c}$$

where α_s is the solar elevation angle $(90 - \theta_s)$.

Solar azimuth angle (ϕ_s^g) must be corrected for grid convergence (Roderick, 1992) when the grid north differs from true north. Grid convergence ($\delta\phi_{gc}$) is a function of the latitude and longitude (λ_l), and the longitude of the central meridian of the projection (λ_{cm}) such that:

$$\delta\phi_{gc} = -\sin(\varphi)\tan(\lambda_l - \lambda_{cm}). \tag{17.8}$$

True solar azimuth is calculated as:

$$\phi_s = \phi_s^{nc} - \delta\phi_{gc} \tag{17.9}$$

17.3.2 Atmospheric Transmittance

Short-wave irradiance (0.3–3.0 μm) is attenuated by the scattering and absorption of atmospheric constituents. Atmospheric extinction is wavelength (λ [μm]) dependent and the atmospheric optical mass (m) and transmittance is dependent upon solar geometry, altitude, and gravity and pressure variations.

The transmittance due to a single process i can be expressed as:

$$\mathbf{T}_i(\lambda) = \exp(-m_i \tau_i(\lambda)) \tag{17.10}$$

where m_i is the optical mass and $\tau_i(\lambda)$ is the optical thickness of an extinction process. A parametrization of the optical mass is provided by Gueymard (2005):

$$m_i = (\cos(\theta_s) + a_{i1}(\theta_{s,deg})^{a_{i2}}(a_{i3} - \theta_{s,deg})^{a_{i4}})^{-1}, \tag{17.11}$$

where the solar zenith angle is used with coefficients that are based upon rigorously calculated data (Miskolczi, 1990).

Ultimately, the total downward atmospheric transmittance ($\mathbf{T}^{\downarrow}$) is a function of the total optical depth of the atmosphere that can be represented as:

$$\mathbf{T}^{\downarrow}(\lambda) = \mathbf{T}_r(\lambda)\mathbf{T}_a(\lambda)\mathbf{T}_{O_3}(\lambda)\mathbf{T}_{gas}(\lambda)\mathbf{T}_{H_2O}(\lambda), \tag{17.12}$$

where $\mathbf{T}_r$ is Rayleigh transmittance, $\mathbf{T}_a$ is aerosol transmittance, $\mathbf{T}_{O_3}$ is ozone transmittance, $\mathbf{T}_{gas}$ is the transmittance for miscellaneous well-mixed gases, and $\mathbf{T}_{H_2O}$ is water-vapor transmittance. Atmospheric attenuation is highly variable with wavelength, with Rayleigh and aerosol scattering dominating at shorter wavelengths and water vapor at longer wavelengths. We exclude these parameterization schemes for the sake of brevity. We utilize standard atmospheric water vapor profiles from MOTRAN (Kniezys et al., 1996), and compute the latitude and altitudinal variations in gravitational acceleration and atmospheric pressure to compute wavelength dependent transmittance. Commercial GIS-based radiation models do not perform spectral computations.

17.3.3 Direct Irradiance

Standard Exoatmospheric spectra at 1 AU (Wehrli, 1985; Gueymard, 2005) providing top of the atmosphere solar irradiance (E_m^0) at mean Sun–Earth distance are used in our model. The E^0 spectra must account for variation in the Sun–Earth distance using the eccentricity correction factor and the solar zenith angle such that:

$$E^0(\lambda) = E_m^0(\lambda) F_{ec} \cos\theta_s. \quad (17.13)$$

The direct irradiance is also governed by multiscale topographic effects. Local or microscale topographic variation is represented by the incidence angle of illumination (i) between the Sun and the vector normal to the ground, such that

$$\cos i = \cos\theta_s \cos\theta_t + \sin\theta_s \sin\theta_t \cos(\phi_t - \phi_s), \quad (17.14)$$

where θ_t is the slope angle of the terrain and ϕ_t is the slope-azimuth angle of the terrain.

Estimation of i is possible with the use of a DEM, and uncertainty in the estimate is related to the measurement scale, as subpixel-scale topographic variation is not accounted for. Values of cos i can be ≤ 0.0, indicating no direct irradiance due to the orientation of the topography. It is important to note that the incident solar geometry varies across the landscape, although this is usually assumed to be constant when working with individual scenes (i.e., small-angle approximation). We account for changes in solar geometry.

The meso-scale topographic relief in the direction of ϕ_s determines if a location is in shadow (S). This parameter value will be 0.0 or 1.0 depending upon the presence or absence of a cast shadow, respectively. We use a 3D subpixel analysis to predict the altitude and determine if a pixel is in cast shadow. The local topography and the regional scale relief dictates the variability in the direct irradiance. Collectively, the aforementioned parameters define the direct surface irradiance component (E_b) as:

$$E_b(\lambda) = E^0(\lambda) \mathbf{T}^{\downarrow}(\lambda) \cos i S. \quad (17.15)$$

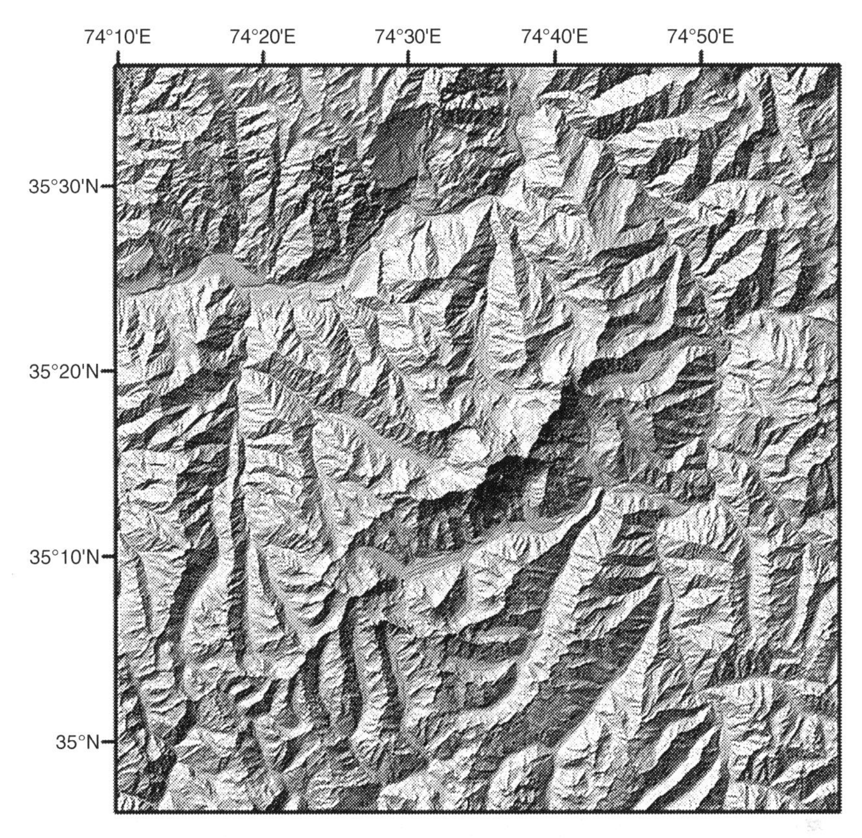

FIGURE 17.1 A shaded relief map of the Nanga Parbat Massif located in Northern Pakistan. The large Indus River can be seen flowing to the southeast in the northeastern region. Nanga Parbat is located in the central portion of this area.

We simulate direct irradiance for the Nanga Parbat Massif, which is the the ninth largest mountain in the world, located in northern Pakistan. A shaded relief map of the region (Fig. 17.1) depicts the extreme relief and knife-edge ridge of Nanga Parbat. Altitude (Fig. 17.2), slope angle (Fig. 17.3), and slope azimuth (Fig. 17.4) parameter maps depict altitude variations that influence atmospheric transmittance, and steep slopes and slope orientations that govern local topographic effects on the direct irradiance that reaches the Earth surface. Simulation results of daily direct irradiance on August 15, 2013 is depicted in Fig. 17.5. Temporal simulations depict variations in the direct irradiance as a function of regional relief structure (Fig. 17.6).

17.3.4 Diffuse-Skylight Irradiance

Atmospheric scattering also generates a hemispherical source of irradiance that represents an integration of the total sky irradiance. The diffuse-skylight

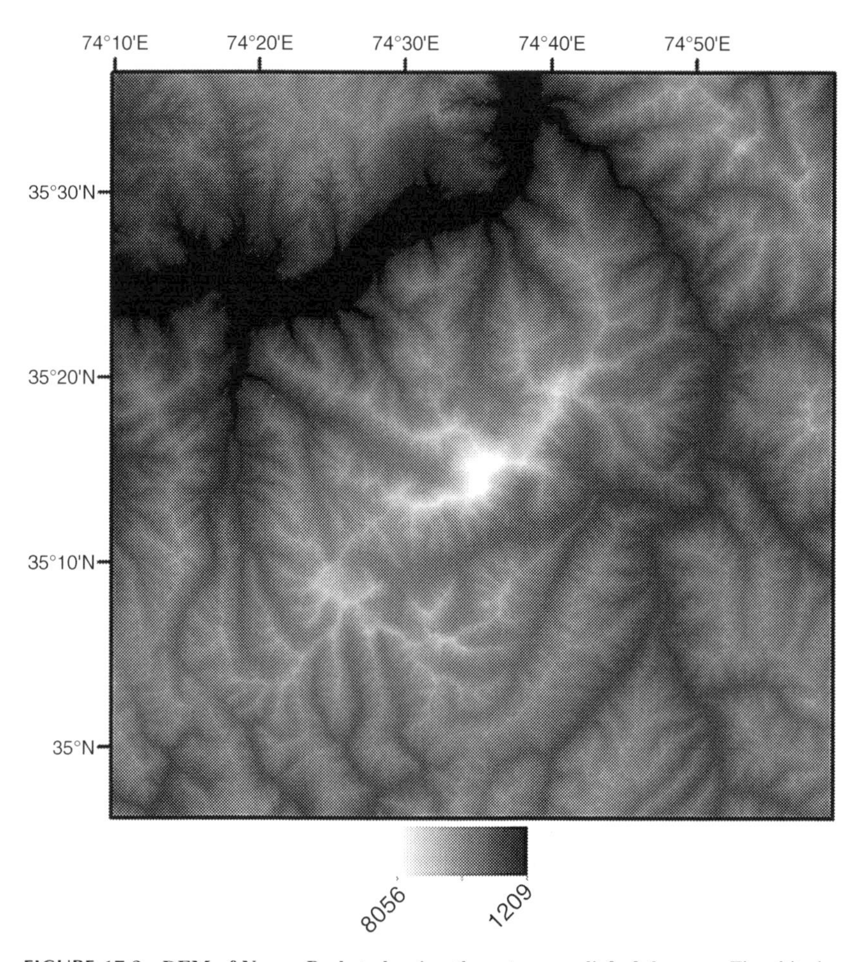

FIGURE 17.2 DEM of Nanga Parbat showing the extreme relief of the area. The altitudes within this region range from approximately 1100 m in the Indus valley to 8125 m at Nanga Parbat. This DEM was used for all simulation results. The extreme variations in altitude influences atmospheric attenuation.

irradiance (E_d) can be represented as a composite including a Rayleigh-scattered component (E_r), an aerosol-scattered component (E_a), and a ground backscattered component (E_g) that represents inter-reflections between the landscape surface and the atmosphere, such that:

$$E_d(\lambda) = E_r(\lambda) + E_a(\lambda) + E_g(\lambda). \tag{17.16}$$

Its accurate estimation is complicated by the fact that an anisotropic parameterization scheme is required. In general, the irradiance decreases with angular distance from the Sun. In addition, this irradiance component is also influenced by mesoscale hemispherical shielding of the topography. Consequently, only

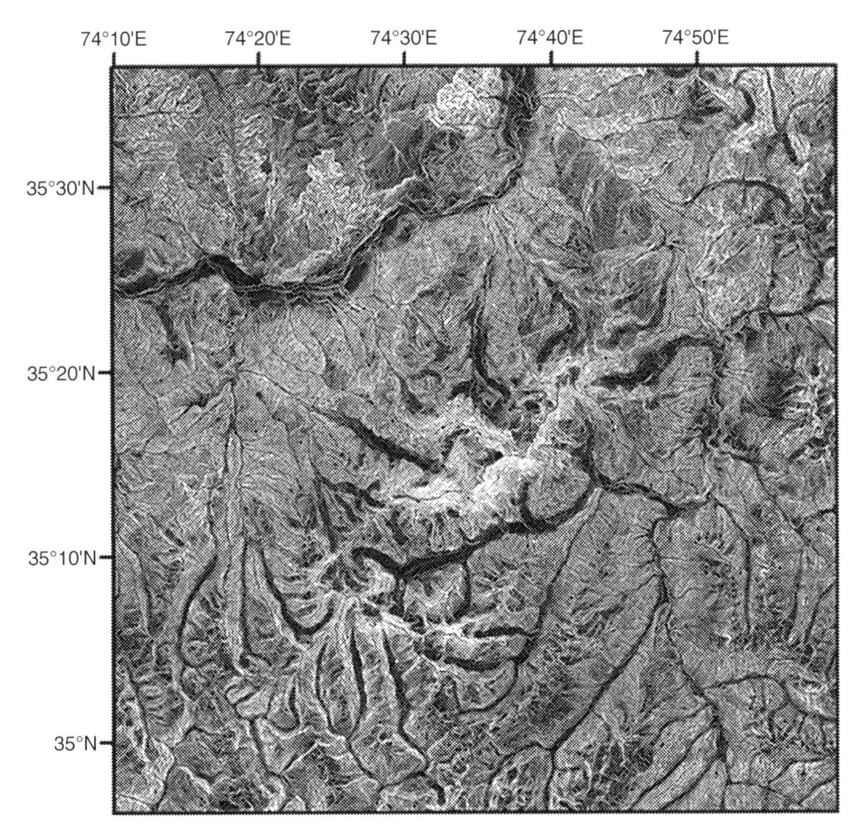

FIGURE 17.3 Terrain slope map of Nanga Parbat region. Relatively dark gray tones represent relatively low slope angles whereas lighter gray tones and white represent steep slope angles. Those areas exhibiting lower slope angles in general receive more direct irradiance.

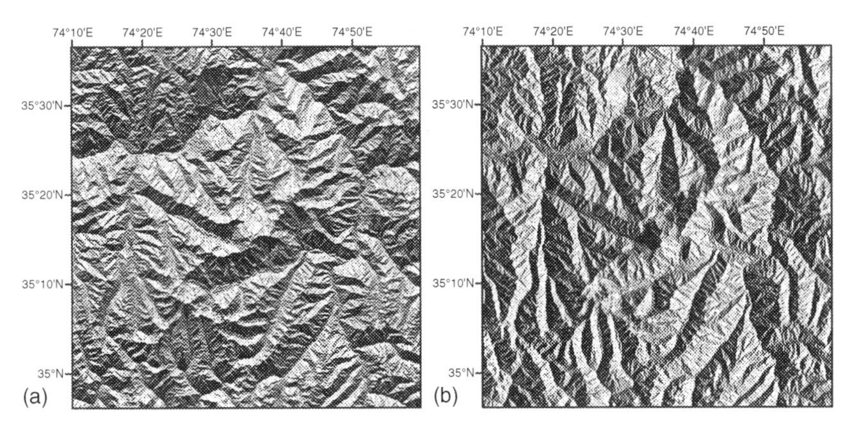

FIGURE 17.4 Transformed slope azimuth images of the Nanga Parbat region. (a) The cos ϕ_t image depicts the north-south variations in slope azimuth. Lighter gray tones represent northward facing slopes, while darker gray tones denote slopes oriented toward the south. (b) The sin ϕ_t image depicts the west-east variations in slope azimuth. Lighter gray tones represent east facing slopes, while darker gray tones denote slopes oriented toward the west.

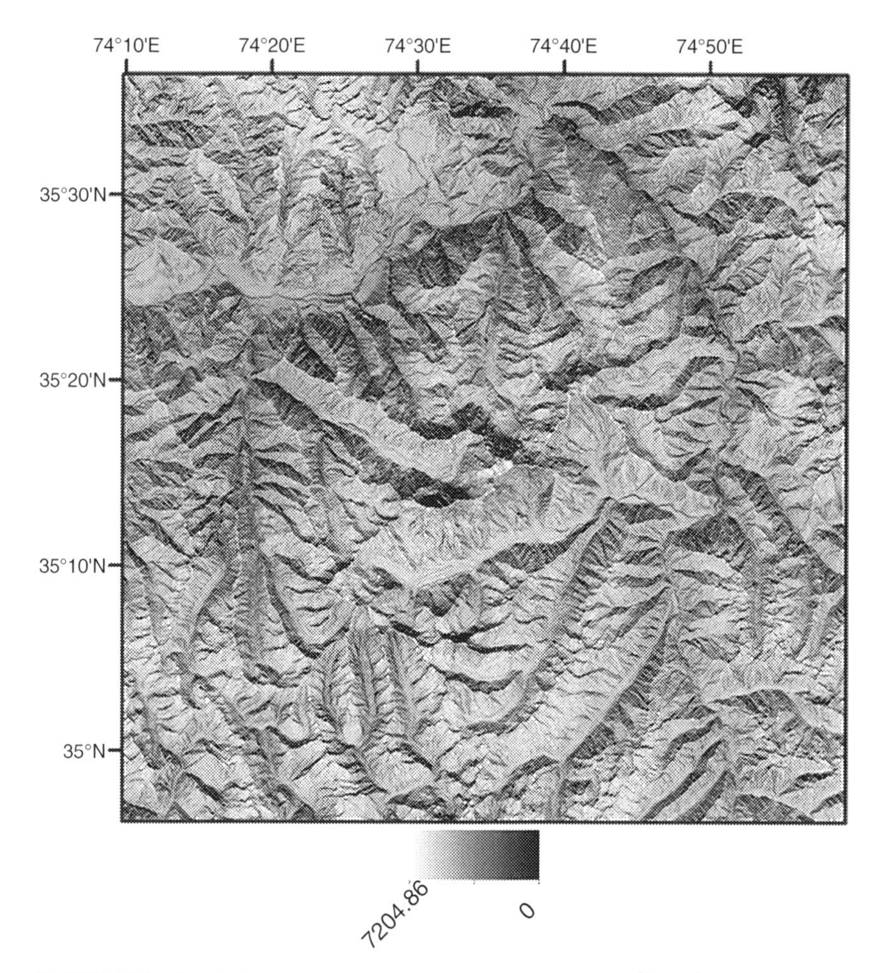

FIGURE 17.5 Simulation results for daily direct irradiance (W m^{-2} day^{-1}) over the Nanga Parbat region for August 15, 2013. Extreme altitude variation, slope angles and orientation, and regional relief cause significant spatial variation in direct irradiance over the landscape.

a solid angle of the sky will contribute to E_d, and this angle will change as a function of location. In general, the solid angle will increase with altitude. It is frequently referred to as the sky-view factor (V_f) in remote-sensing and energy-balance literature, and can be estimated using a DEM such that:

$$V_f = \sum_{\phi=0}^{360} \cos^2\theta_{\max}(\phi, d)\frac{\Delta\phi}{360}, \tag{17.17}$$

where $\theta_{\max}$ is the maximum local horizon angle at a given azimuth, ϕ, over a radial distance of d.

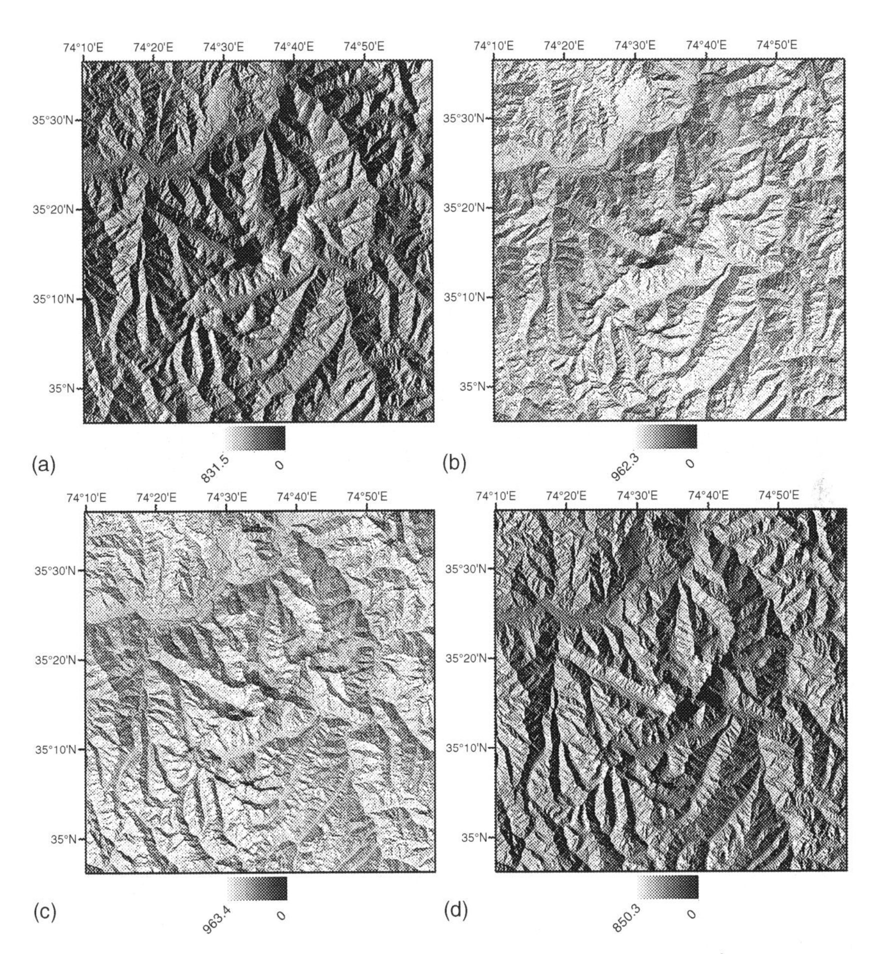

FIGURE 17.6 Simulation results for instantaneous direct solar radiation (W m^{-2}) on August 15, 2013 over the Nanga Parbat region at (a) 8 a.m., (b) 11 a.m., (c) 1 p.m., and (d) 4 p.m. Direct irradiance is in Watts per meter square. Solar geometry and terrain geometry variations dictate extreme space–time variability. Cast shadow can be clearly seen in early morning and late afternoon hours.

We use the parametrization of Bird and Riordan (1986) and have not accounted for the secondary influence by ground reflectance. The Rayleigh and aerosol components are computed as:

$$E_{\mathrm{r}}(\lambda) = E_{\mathrm{m}}^{0}(\lambda)F_{\mathrm{ec}}\cos\theta_{\mathrm{s}}\mathbf{T}_{\mathrm{O}_3}(\lambda)\mathbf{T}_{\mathrm{gas}}(\lambda)\mathbf{T}_{\mathrm{H}_2\mathrm{O}}(\lambda)\mathbf{T}_{\mathrm{aa}}(\lambda)\left(1-\mathbf{T}_{\mathrm{r}}(\lambda)\right)0.5, \quad (17.18\mathrm{a})$$

$$E_{\mathrm{a}}(\lambda) = E_{\mathrm{m}}^{0}(\lambda)F_{\mathrm{ec}}\cos\theta_{\mathrm{s}}\mathbf{T}_{\mathrm{O}_3}(\lambda)\mathbf{T}_{\mathrm{gas}}(\lambda)\mathbf{T}_{\mathrm{H}_2\mathrm{O}}(\lambda)\mathbf{T}_{\mathrm{aa}}(\lambda)\mathbf{T}_{\mathrm{r}}^{1.5}(\lambda)(1-\mathbf{T}_{\mathrm{as}}(\lambda))F_{\mathrm{s}}, \quad (17.18\mathrm{b})$$

$$\mathbf{T}_{\mathrm{as}}(\lambda) = \exp(-\omega(\lambda)\tau_{\mathrm{a}}(\lambda)m_{\mathrm{re}}), \quad (17.18\mathrm{c})$$

$$\mathbf{T}_{\mathrm{aa}}(\lambda) = \exp[-(1-\omega(\lambda))\tau_{\mathrm{a}}(\lambda)m_{\mathrm{re}}], \quad (17.18\mathrm{d})$$

where $\mathbf{T}_{aa}(\lambda)$ is the transmittance term for aerosol absorption, $\mathbf{T}_{as}(\lambda)$ is the transmittance term for aerosol scattering, F_s is the fraction of aerosol scatter that is downward, m_{re} is the relative air mass, and $\omega(\lambda)$ is the aerosol single scattering albedo defined by Gueymard (2005). The parameter $\tau_a(\lambda)$ is the aerosol optical depth and can be calculated using the Ångstrøm turbidity formula:

$$\tau_a(\lambda) = \beta\lambda^{-\alpha}, \tag{17.19}$$

where β is the turbidity coefficient and α is the aerosol size distribution parameter.

The fraction of aerosol scatter that is downward can be calculated as a function of the solar zenith angle:

$$F_s = 1 - 0.5\exp[(A_{fs} + B_{fs}\cos\theta_s)\cos\theta_s], \tag{17.20a}$$

$$A_{fs} = A_{lg}[1.459 + A_{lg}(0.1595 + A_{lg}0.4129)], \tag{17.20b}$$

$$B_{fs} = A_{lg}[0.0783 + A_{lg}(-0.3824 - A_{lg}0.5874), \tag{17.20c}$$

$$A_{lg} = \ln(1 - g_a). \tag{17.20d}$$

The diffuse irradiance can then be calculated as:

$$\begin{aligned} E_d = E_d^h \Big\langle \big\{ E_b / [E_m^0(\lambda) F_{ec} \cos(\theta_s)] \big\} + 0.5[1 + \cos(\theta_t)] \\ [1 - E_{bn} / (E_m^0(\lambda) F_{ec})] \Big\rangle + 0.5 E_t^h \alpha_g [1 - \cos(\theta_t)] V_f \end{aligned} \tag{17.21}$$

where E_t^h is the total surface irradiance on a horizontal surface and E_{bn} is the direct irradiance at a normal to the Sun surface.

Diffuse skylight variations for the Nanga Parbat Massif are depicted in Fig. 17.7. The magnitude of this irradiance component is strongly governed by hemispherical topographic shielding (Fig. 17.8).

17.4 GEOCOMPUTATIONAL MODELING

Given the rapid proliferation of geospatial technologies and high volumes of geospatial data, Earth scientists are faced with technical and scientific issues related to analysis and formalization of knowledge (Bishop et al., 2012). GIS-based empirical modeling is frequently dependent on simplifying assumptions and parameterization schemes that may not adequately characterize processes and dynamics, such that simulations do not correspond to objective measurements and observations obtained in the field. In contrast, many Earth scientists have historically been able to make relevant assessments of landscape change based on their conceptual knowledge developed through experience and

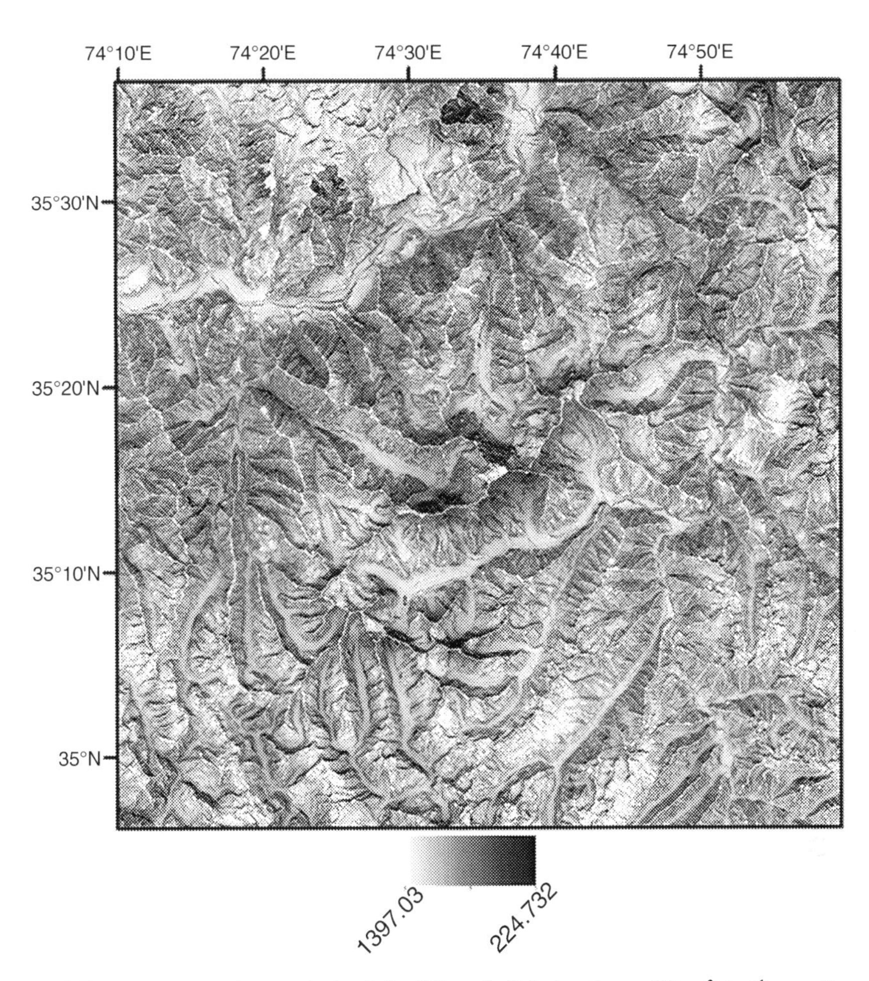

FIGURE 17.7 Simulation results for daily diffuse-skylight irradiance (W m^{-2} day^{-1}) over the Nanga Parbat region on August 15, 2013. The magnitude of irradiance over the landscape is significantly less than the direct irradiance (Fig. 17.5), and the upper portions of basins with steep relief exhibit relatively low diffuse irradiance. Ridge tops that delineate basins throughout the region exhibit relatively high diffuse irradiance due to the absence of hemispherical topographic shielding.

knowledge (Phillips, 1999), but do not have the ability to quantitatively formalize and evaluate their conceptual knowledge and predict dynamic patterns.

The Earth Sciences are rich in theory and concepts related to time, process, systems, and taxonomy, and there has been a recent attempt to formalize knowledge through cellular-automata (CA) modeling (e.g., Wootton, 2001; D'Ambrosio et al., 2001, 2003). It is a form of AI that iteratively makes use of deterministic, stochastic, or probabilistic rules in a map-like framework that has the potential to predict spatial and temporal patterns in various systems (Moody and Katz, 2004). The CA approach can be viewed as an expert system with

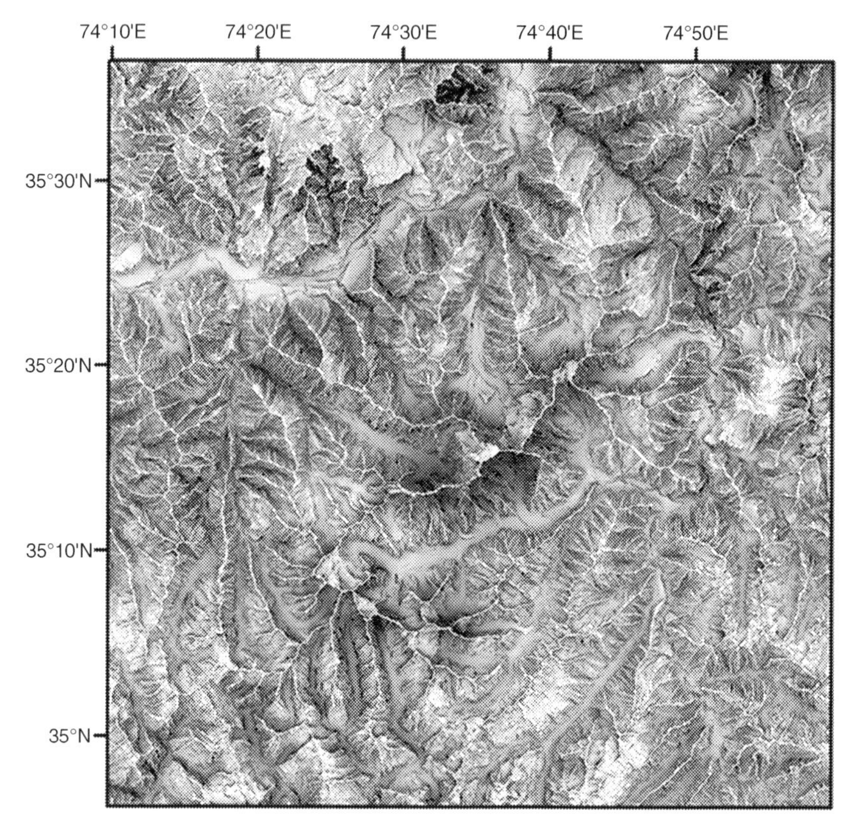

FIGURE 17.8 The Skyview coefficient map depicts hemispherical topographic shielding throughout the Nanga Parbat region. Dark gray tones depict areas with high shielding caused by steep slopes and extreme relief.

rules that are spatially and temporally dynamic, and can incorporate spatial and temporal interactions (Moody and Katz, 2004; Bishop et al., 2012). Despite the success of CA models in predicting realistic-looking patterns on the landscape, the results of CA models should be considered investigative hypotheses in exploring key variables and relationships, rather than quantitative predictions of parameter magnitudes or a formalization of process mechanics for feedback mechanisms.

As an alternative to CA, various methods of AI have been used to examine nonlinear dynamics and feedback processes to explore scenarios of landscape change, alternate futures, and divergent landscape patterns (Walsh et al., 2008). Specifically, analytical reasoning can be used to formalize process–form relationships and information integration analogous to human cognition, and therefore can be used to examine internal and external forcing factors that can be highly variable across a range of spatial and temporal scales (Bishop et al., 2012). This approach can be quite powerful when integrated with new geospatial data and geocomputational approaches that make it possible to quantify landscape morphology (Pike, 2000; Tomislav and Hannes, 2009), assess surface biophysical interactions (Liang, 2007; Smith and Pain, 2009; Tarolli et al., 2009), link process with form (Allen and Walsh, 1993; Montgomery et al., 2004), and characterize the scale-dependence and polygenetic nature of the landscape (Walsh et al., 1997; Tate and Wood, 2001; Bishop et al., 2003, 2012).

Analytical reasoning can significantly support Critical Zone research if our knowledge and conceptual understandings of system complexities can be formalized and tested using neural computing and FS approaches (Bishop et al., 2012). Geocomputational solutions warrant further investigation, as the issues of causation, knowledge representation, information integration, and scale can be addressed.

17.4.1 Fuzzy Cognitive Maps

Conceptual models have long-served as a means for Earth scientists to organize their conceptual understandings of feedback mechanisms and complex systems, and have served as a basis for human analytical reasoning and landscape interpretation. Scientists, however, have not mathematically formalized their conceptual understandings, so as to test the predictive capabilities of conceptual models for assessment and mapping. Simple conceptual models highlight variables, processes, forcing factors, relationships, and causal connections involving mass and energy transport. It is currently not possible to integrate and numerically model numerous simplistic conceptual models at scales relevant to scientists, planners, and managers (e.g., Werner, 1995; Thomas and Wiggs, 2008) or to include results of conceptual models to study new concepts that require investigation and inclusion into formal numerical modeling efforts. Essentially, there is a need to make better use of our existing knowledge and conceptual understandings.

AI and FS can be used to address these issues. Specifically, fuzzy cognitive maps (FCMs) represent an emerging geocomputational approach to formalize, implement, and test conceptual models in the physical sciences. FCMs are a knowledge-based AI technique that merges fuzzy logic and neural computing (Groumpos and Stylios, 2000). The technique was developed to deal with uncertain concept descriptions that were thought to be part of human reasoning (Papageorgiou et al., 2004). Knowledge or concepts are structured as a web of relationships that is similar to both human reasoning and the human decision-making process (Furfaro et al., 2010). FCMs have been successfully used in a variety of applications, mostly in the biomedical (Papageorgiou et al., 2003, 2008), industrial applications (Stylios and Groumpos, 2000), causal inference (Miao et al., 2007), and engineering areas (Espinosa-Paredes et al., 2008).

FCMs are diagraphs designed to mathematically formalize a conceptual model and capture the system-inherent cause/effect relationships. Designing an intelligent system that reasons using geospatial data is an attempt to utilize knowledge and inference to assess and map the landscape. The knowledge can

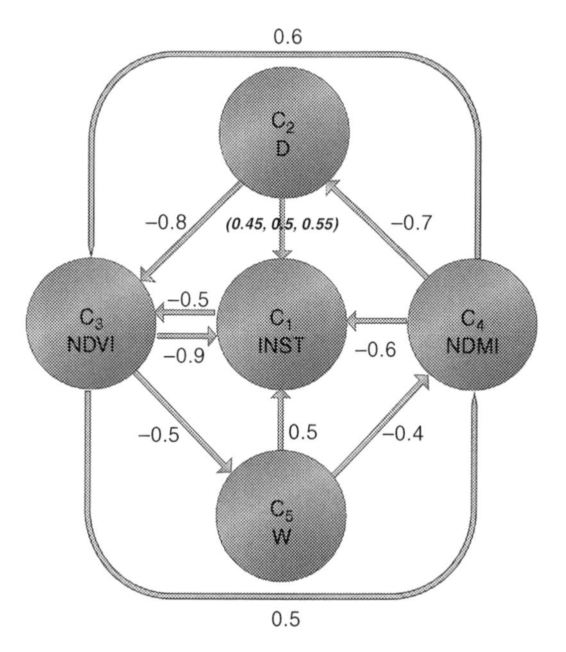

FIGURE 17.9 An FCM used to assess aeolian instability along the Texas coastline. This FCM has five concept nodes that are connected by arcs to reflect the causal connections between nodes. The concepts include: instability (C1), drought (C2), stability due to vegetation (C3), surface cohesion due to water (C4), and wind potential (C5). The magnitude and direction of the causal connections are also displayed.

be represented by concepts or variables from which it is possible to perform analytical plausible reasoning to obtain new facts and hypotheses. Knowledge must be coupled with an inference mechanism, defined as the process of integrating data and knowledge to infer new information. The way knowledge and inference are integrated into AI algorithms defines the difference between symbolic AI, fuzzy logic-based and neural-network computing approaches.

There are two basic elements for the structure of an FCM: concept nodes and arcs (Fig. 17.9). Concept nodes are the factors or attributes of the modeled system that can be inputs, outputs, process or system states, variables, or events. Arcs represent the causal relationships between concept nodes with an associated degree of causality. The weight of the interaction defines the perceived direction and degree of influence amongst the concepts with a numerical value ranging from −1 (inverse) to 1 (direct). The concept nodes and weighted arcs are initially based on knowledge and opinions of experts who describe each interconnection using a fuzzy rule, in which a linguistic variable is used to determine the degree of causality between concepts. The output of an FCM is computed by accounting for all possible influences derived from interconnected concepts (Furfaro et al., 2010). An iterative process is required to update concept node values, and full details of the methodology can be found in Furfaro et al. (2010).

The initial value (A_i) of a concept (C_i) is either defined by an expert or fuzzified and mapped onto the interval [0–1]. An iterative approach is used to update A_i by accounting for all possible influences derived from the free interaction between map concepts:

$$A_i(k+1) = f\left[A_i(k) + \sum_{i=1}^{n} A_j(k)\mathbf{W}_{ji}\right], \tag{17.22}$$

where $A(k) = [i = 1, \ldots, n]$ is a vector that represents the value of C_i at iteration k, $\mathbf{W}$ is a $n \times n$ matrix of causal connections, and f is a threshold function (we use a binary sigmoid function). In this respect, the FCM is a dynamic system that: (1) converges to an equilibrium state; (2) converges to a limit cycle; or (3) exhibits a chaotic behavior in which the concepts change their values in a nondeterministic fashion (Furfaro et al., 2010). The weights for the FCM can also be obtained by training using active Hebbian learning (Papageorgiou et al., 2004) in which the concepts are activated asynchronously and the weights are updated until a stable state is achieved according to the Hebbian rule (Hebb, 1949).

17.4.2 Aeolian Landscape Instability

We have developed and implemented a simplistic FCM conceptual model (Fig. 17.9) that is driven by satellite imagery and a geomorphometric parameter to address the difficult problem of aeolian landscape instability. Our contemporary understanding of instability in aeolian environments is largely the result of small-scale process studies in the field and in laboratory wind tunnels, but there is mounting evidence that this reductionist approach is not appropriate for landform- and landscape-scale assessment and mapping (Livingstone et al., 2007). The temporal and spatial limitations of many studies lead Walker (2010) to suggest that short-term process studies are necessity limited in scope and scale and can only be considered vignettes with regards to longer-term system dynamics. In this respect, there is a need to determine how experimental events fit into the broader frequency–magnitude regime of a system (Walker, 2010) and to determine the most appropriate scale-dependent variables to describe landscape instability and change. The inability to numerically up-scale small-scale sediment transport studies is partly responsible for the recent focus on CA models that use relatively simple rules to predict patterns of aeolian landscape change (e.g., Nield and Baas, 2008).

To our knowledge, this modeling approach has not been applied to aeolian environments where data is relatively limited. Advances in remote sensing technologies have the potential to alleviate our limited view of the landscape, but temporal data availability remains an issue to the timescales of landscape change that field scientists have described. Hugenholtz et al. (2012), however, argues that it will soon be possible to use remotely sensed data to develop models that

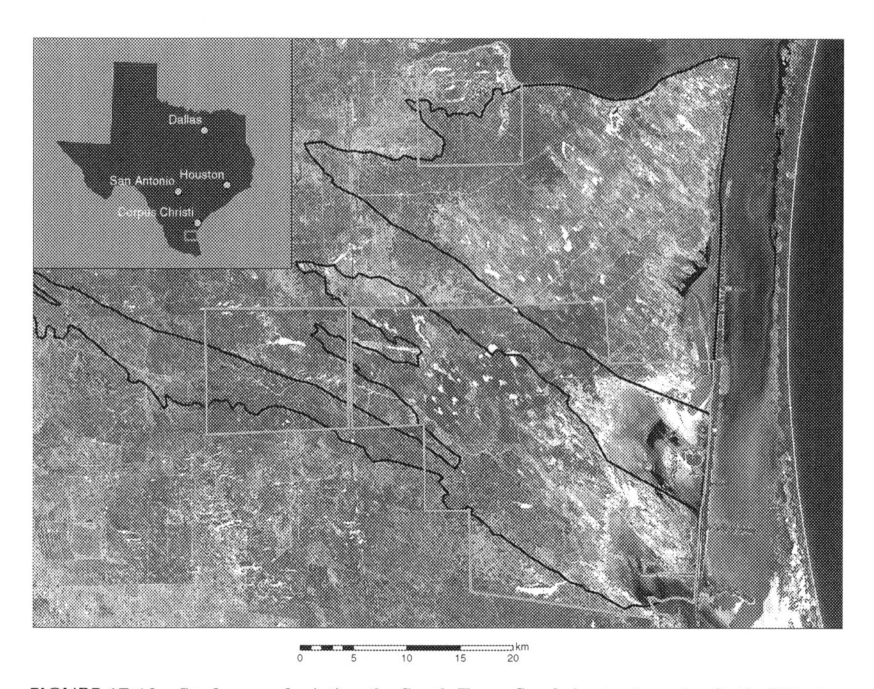

FIGURE 17.10 Study area depicting the South Texas Sand sheets along the Gulf of Mexico. The large sand sheets have been delineated in black and their orientation reflects the dominant prevailing wind direction.

integrate topography, climate, sediment, and ecosystem components to develop well-constrained estimates of landscape change in response to different climate change scenarios (e.g., Pelletier et al., 2009) and ultimately integrated into global climate models (Thomas et al., 2005), airflow models (Liu et al., 2011), and ecosystem models (Mangan et al., 2004).

We demonstrate this type of scale-dependent information integration over the South Texas sand sheet near Corpus Christi, Texas (Fig. 17.10). The South Texas sand sheet is a 2 million acre area between Corpus Christi and the Rio Grande Valley in South Texas that is part of the Kennedy and King Ranches, which have employed different land-use management strategies. The coastal sand plain is characterized by a grassland ecosystem with mottes of live oaks and mesquite along the margins of the sand sheet (Fulbright et al., 1990). Interspersed across the sand sheets are a combination of active parabolic dunes, trailing ridges, and banner dune complexes that are actively migrating to the northwest and were first described by Price (1958) and Russell (1981). The sand sheets have an average thickness of 3 m and overlie late Pleistocene fluvial-deltaic sands and strandplain sands (Bernard and LeBlanc, 1965; Russell, 1981) where aeolian activity has not been documented over the twentieth century or through the Holocene (Forman et al., 2008).

17.4.3 FCM Model

The output of our FCM model represents the instability concept (C1) that is governed by the concepts of climate influence or drought (C2), influence of vegetation on stability (C3), surface cohesion due to moisture (C4), and wind potential (C5). Expert knowledge and opinion was used to define the concept nodes, the linkage architecture, and the magnitude and direction of the causal connections. It encapsulates basic knowledge about the stabilizing influence of vegetation and surface moisture, while accounting for the influence of climate change and topography on the destabilizing influence of the wind.

We drive the FCM model using Landsat 8 imagery acquired on May 8, 2013. Spectral features were used as proxies for stability due to vegetation and moisture cohesion concepts. Specifically, we used the normalized difference vegetation index (NDVI) computed as:

$$\text{NDVI} = \frac{(L_{\text{NIR}} - L_{\text{red}})}{(L_{\text{NIR}} + L_{\text{red}})}, \tag{17.23}$$

where L represents the surface radiance in the near-infrared (NIR) and red (red) regions of the electromagnetic spectrum (Fig. 17.11). Surface moisture conditions

FIGURE 17.11 Normalized difference vegetation index image of the study area generated from Landsat 8 imagery acquired on May 8, 2013. Lighter gray tones and white signify relatively high photosynthetically active vegetation present on the surface. The landscape exhibits significant variation in terms of the influence of vegetation that protects the surface from wind erosion. Unique patterns also reflect the types of vegetation that colonize the coastal plain.

FIGURE 17.12 **Normalized difference moisture index image of the study area generated from Landsat 8 imagery acquired on May 8, 2013.** Lighter gray tones and white signify relatively high surface moisture conditions. The landscape exhibits significant variation in terms of the influence of land use and surface materials on surface moisture conditions.

related to surface materials and topographic position were accounted for using the normalized difference moisture index (NDMVI), which is computed as:

$$\mathrm{NDMVI} = \frac{(L_{\mathrm{NIR}} - L_{\mathrm{SWIR}})}{(L_{\mathrm{NIR}} + L_{\mathrm{SWIR}})}, \qquad (17.24)$$

where L represents the surface radiance in the NIR and SWIR regions of the spectrum (Fig. 17.12).

Airborne LiDAR (3-m resolution) data were obtained to compute the topographic shielding of the landscape using Eq. 17.17, as a proxy for topographically controlled wind potential (Fig. 17.13). This multiscale parameter identifies those locations more susceptible to wind and aeolian transport accounting for the hemispherical influence of the topography.

For low-to-moderate drought conditions ($D=0.45$), the model predicts instability across a large part of the southern lobe of the sand sheet (Fig. 17.14). This represents an area of the Kennedy Ranch that is leased to ranchers, and for which there is no common management plan. Therefore, instability varies by grazing plot. Grazing is no longer permitted in the northern lobe, such that aeolian instability is restricted to the currently active dunes. With increasing drought conditions ($D=0.5$), a larger area of the southern lobe becomes active and instability in the northern lobe extends to the southeast of the active

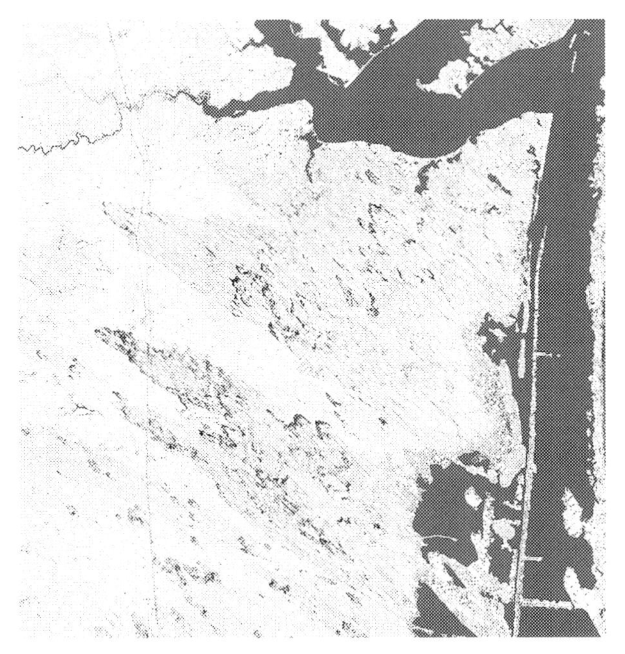

FIGURE 17.13 The Sky-view parameter image over the study area derived from 3 m airborne LiDAR data. Lighter gray tones and white signify less topographic shielding and greater potential for wind erosion, all other factors remaining constant. Darker gray tones signify greater hemispherical shielding due to active aeolian transport and deposition. Landscape variations in topographic shielding clearly reflect the leading edge of the sand sheets and localized aeolian instability.

dunes (Fig. 17.15). The instability that trails the active dunes is associated with a lagged vegetation recovery and the presence of sparse grasses and shrubs compared to adjacent areas where the dunes have not been present for decades or centuries. A moderate to high intensity drought ($D=0.55$) leads to instability across most of the southern lobe and greater instability in the areas trailing the active dunes in the northern lobe (Fig. 17.16). A preliminary comparison of the results with the extent of dune activity during and immediately following the prolonged drought from 1950 to 1956 suggests that the model accurately predicts dune reactivation in both the southern lobe and along the back-barrier shoreline of Padre Island National Seashore to the east. Given a relatively simple model of process–form relationships, the analytical reasoning model is able to map the influence of land management practices and the geomorphology of the inherited surface on aeolian instability.

17.5 DISCUSSION

Understanding the spatio-temporal dynamics of coupled systems of the Critical Zone will require *in situ* data collection, geospatial technology solutions, and numerical modeling (Rasmussen et al., 2011; Lin et al., 2011; Brantley and

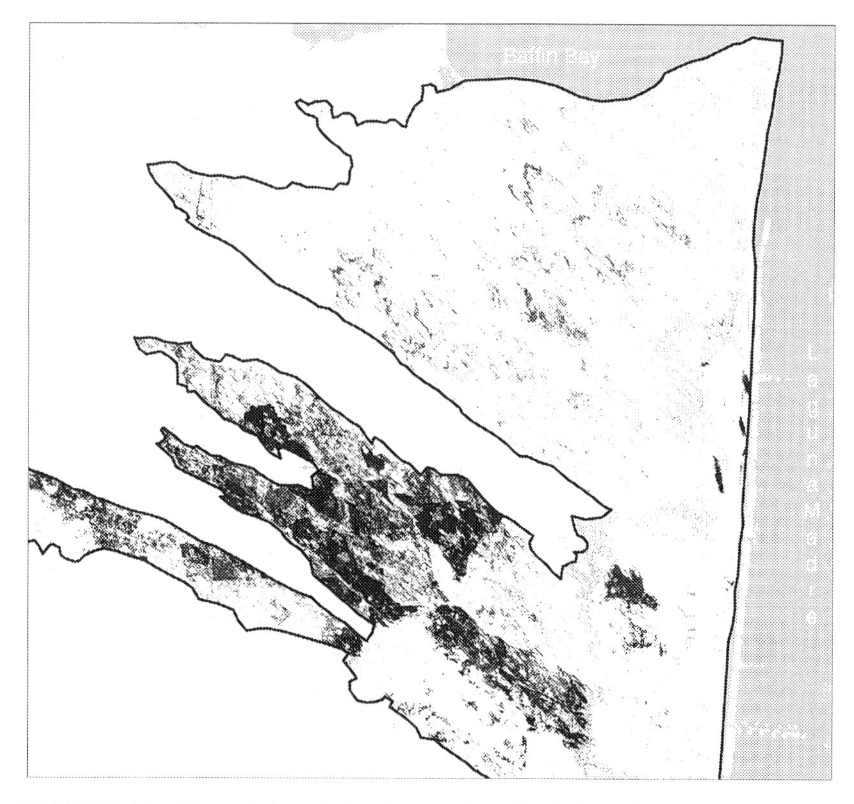

FIGURE 17.14 FCM model results for the South Texas Sand sheets along the Gulf of Mexico using a regional drought constant of 0.45.

Lebedeva, 2011). Accurately representing scale-dependent parameters and processes is a prerequisite for characterizing complex dynamics and environmental change. Numerous challenges include representational schemes, identification and characterization of parameters, and simplifying process equations to adequately account for topographic, climatic, tectonic, and anthropogenic forcings. Other challenges focus on distilling complex observations into simplified representational schemes that are compatible with geospatial data that serve as input into, and permit validation of simulation results. Our GIS-based surface irradiance model demonstrates multiscale characterization of external and internal forcing factors that govern the surface irradiance flux.

Simulation results clearly demonstrate significant spatial and temporal variation in the direct and diffuse-skylight irradiance components in mountain environments. Furthermore, these variations can be significant over geological timescales, as we seek to investigate the controversy associated with climate versus tectonic forcing in understanding mountain geodynamics. Surface irradiance variations influence weathering rates and meltwater production. This in turn influences the magnitude of mass movement, fluvial erosion, and glacier

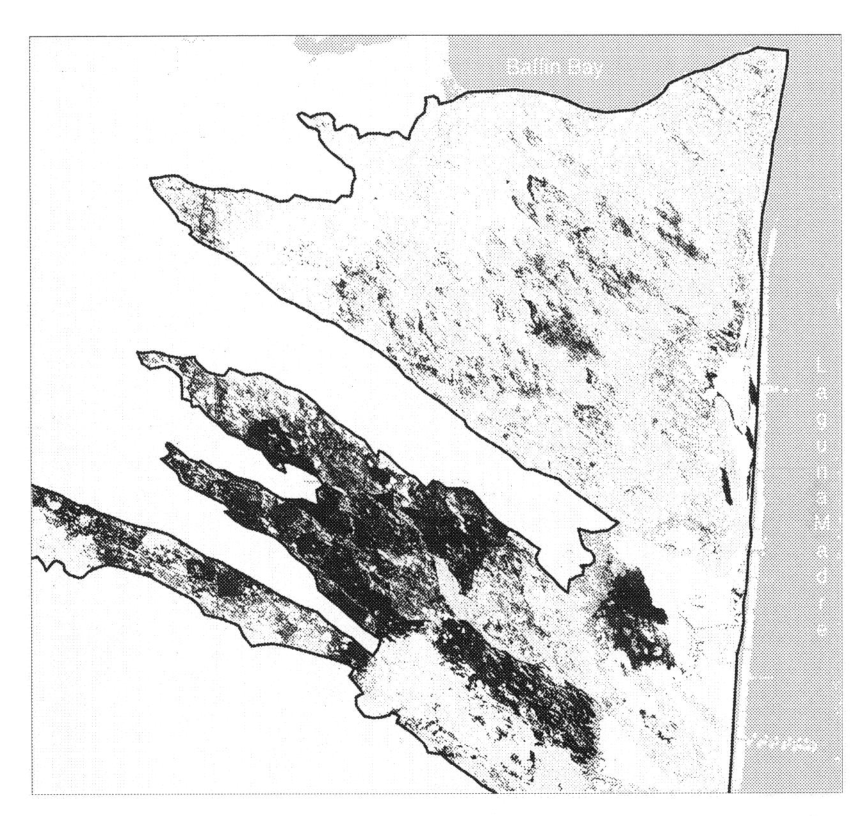

FIGURE 17.15 FCM model results for the South Texas Sand sheets along the Gulf of Mexico using a regional drought constant of 0.5.

erosion that collectively removes lithospheric mass. High-magnitude denudation is thought to cause decompression melting and localized tectonic uplift (Koons, 1995; Koons et al., 2002), while climate forcing directly affects glaciation and isostatic uplift (Whipple, 2009).

Such models are also required to assess climate–glacier dynamics, as glaciers are known to respond to short- and long-wave radiation (Kargel et al., 2005; Bishop et al., 2014). Other simulations conducted in the Karakoram and Nepalese Himalaya clearly reveal the role of surface irradiance in governing the altitudinal variation of glacial ice-flow velocity and surface melting, given variations in supra-glacial debris cover. Furthermore, we have discovered topographically controlled surface irradiance anomalies that are associated with large proglacial lakes, that represent a significant hazard to humans inhabiting valley bottoms at lower altitudes.

High-resolution surface irradiance models can be used to study irradiance partitioning and the nature of landscape resilience. Ultimately, they must be coupled with ecological, hydrological, and landscape evolution models. This is required as complex feedback mechanisms exist, such that climate and

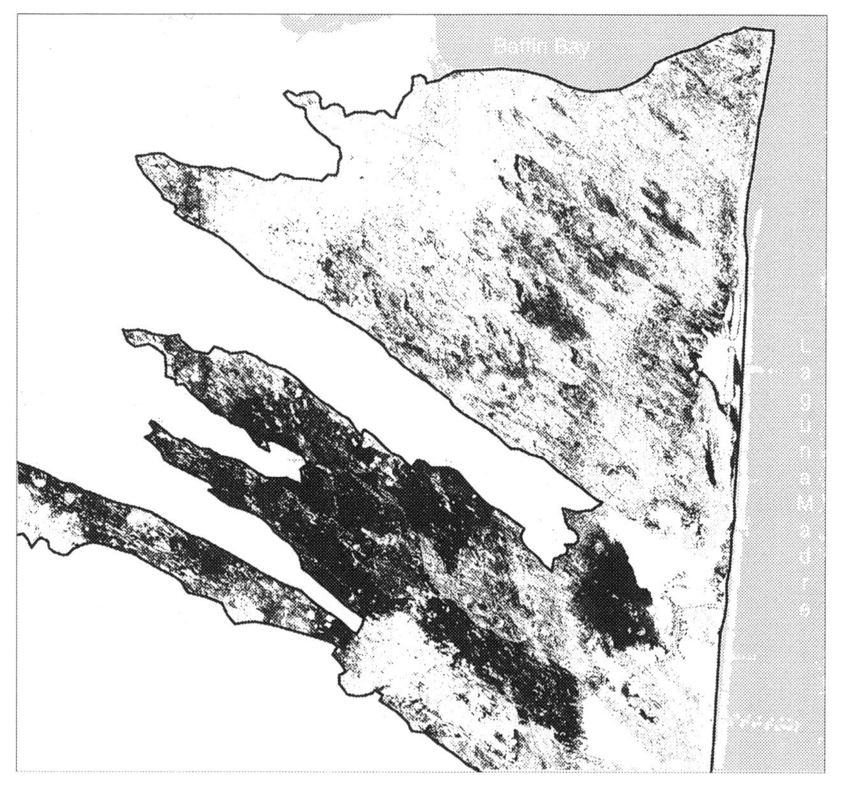

FIGURE 17.16 FCM model results for the South Texas Sand sheets along the Gulf of Mexico using a regional drought constant of 0.55.

topography govern radiation and precipitation forcing, that governs weathering, regolith production, mass movements, fluvial erosion, and glaciation. High-magnitude denudation influences tectonic processes which govern uplift and relief production, which alters the topography and climate.

Improvements to numerical models and complex relationships can be facilitated by conceptual modeling and analytical reasoning. Researchers can develop and test conceptual models based upon complex data acquired from the Critical Zone. New concepts can be hypothesized and tested to determine if geocomputational models can explain dynamic patterns in the Critical Zone.

This was demonstrated by modeling the degree of aeolian instability over the South Texas sand sheets, where our FCM model identified those locations of instability caused by anthropogenic forcing. The model also highlighted its ability to identify the role that the antecedent topography has on instability. Consequently, this approach can be used to identify key variables and forcing factors to improve mathematical formalization of processes and feedback mechanisms. Furthermore, expert knowledge of the Critical Zone can be tested through the use of different conceptual models.

Finally, Critical Zone studies must account for subsurface materials, properties, and structural characteristics. This is greatly facilitated by utilizing geophysical data to assess and map bulk density, magnetic, electrical, and structural properties. Geophysical and geospatial data integration, however, represents a significant problem, given inherent differences in sensor resolution, and limitations in data representation schemes (Wilkinson, 1996; Worboys, 1998). Kruse (2013) provides a review of geophysical techniques that are of value in geomorphological and Critical Zone studies. Ultimately, we need to integrate surface biophysical properties with subsurface biophysical properties in a three-dimensional framework to support integration science investigations. This can be challenging, although geocomputational modeling effectively permits such integration through the use of surface and subsurface concepts and causal connections. Therefore, FCM modeling can be used to study complex systems without a formal representation scheme that requires issues of scale and process mechanics to be mathematically tractable.

17.6 CONCLUSIONS

Geospatial technologies are indispensable for collecting geospatial data and assessing the complexities of the Critical Zone. Hyperspectral remote sensing and geophysical data provide a wealth of information about the biophysical properties of the landscape and near-surface environment. High-resolution DEMs and geomorphometric analysis provide unique information about multiscale properties of the topography that govern and constrain a multitude of atmospheric, surface, and tectonic processes. Therefore, geospatial technology solutions are required to develop new scale-dependent models and collect data that are needed to support numerical modeling efforts and validation. Unfortunately, there are a multitude of issues that need to be addressed involving data and information integration, knowledge representation, and the development and testing of our conceptual knowledge of the Critical Zone. Geocomputation modeling techniques can be extremely valuable to support integrative science investigations. Conceptual modeling and analytical reasoning support the identification of key variables and forcing factors to facilitate understanding and improve our attempts at mathematical formalization of processes and feedback mechanisms. Geocomputation also permits knowledge representation, and expert knowledge of the Critical Zone can be tested through the use of simplified to complex conceptual models. Collectively, geospatial technologies offer numerous opportunities to advance our understanding of the Critical Zone.

ACKNOWLEDGMENTS

We would like to acknowledge important discussions, contributions, and graphics from Roberto Furfaro from the University of Arizona, and Jeffrey Olsenholler and Patrick Barrineau from the Department of Geography at Texas A&M University.

REFERENCES

Ahn, C.W., Baumgardner, M.F., Biehl, L.L., 1999. Delineation of soil variability using geostatistics and fuzzy clustering analyses of hyperspectral data. Soil Sci. Soc. Am. J. 63, 142–150.

Allen, T.R., Walsh, S.J., 1993. Characterizing multitemporal alpine snowmelt patterns for ecological inferences. Photogrammetric Eng. Remote Sens. 59, 1521–1529.

Anderson, S.P., Anderson, R.S., Tucker, G.E., 2012. Landscape scale linkages in Critical Zone evolution. C. R. Geosci. 344, 586–596.

Anderson, S.P., Bales, R.C., Duffy, C.J., 2008. Critical Zone Observatories: building a network to advance interdisciplinary study of Earth surface processes. Mineral. Mag. 72, 7–10.

Anderson, S.P., von Blanckenburg, F., White, A.F., 2007. Physical and chemical controls on the Critical Zone. Elements 3, 315–319.

Berardino, P., Fornaro, G., Lanari, R., Sansosti, E., 2002. A new algorithm for surface deformation monitoring based on small baseline differential SAR interferograms. IEEE Trans. Geosci. Remote Sens. 40, 2375–2383.

Berger, A., Loutre, M.F., 1991. Insolation values for the climate of the last 10 million years. Quaternary Sci. Rev. 10, 297–317.

Berger, A.L., 1978a. Long-term variations of daily insolation and Quaternary climatic changes. J. Atmos. Sci. 35, 2362–2367.

Berger, A.L., 1978b. Long-term variations of caloric insolation resulting from the Earth's orbital elements. Quaternary Res. 9, 139–167.

Bernard, H.A., LeBlanc, R.J., 1965. Resume of Quaternary geology of the northwestern Gulf of Mexico province. The Quaternary of the United States: VII Congress of the International Association for Quaternary ResearchPrinceton University Press, Princeton, NJ.

Berni, J., Zarco-Tejada, P.J., Suarez, L., Fereres, E., 2009. Thermal and narrowband multispectral remote sensing for vegetation monitoring from an unmanned aerial vehicle. IEEE Trans. Geosci. Remote Sens. 47, 722–738.

Bird, R.E., Riordan, C., 1986. Simple solar spectral model for direct and diffuse irradiance on horizontal and tilted planes at the Earth's surface for cloudless atmospheres. J. Climate Appl. Meteorol. 25, 87–97.

Bishop, M.P., 2013. Remote sensing and GIScience in geomorphology: introduction and overview. In: Shroder, J.F. (Ed.), Treatise on Geomorphology. Academic Press, San Diego, pp. 1–24.

Bishop, M.P., Bush, A.B.G., Copland, L., Kamp, U., Owen, L.A., Seong, Y.B., Shroder, Jr., J.F., 2010. Climate change and mountain topographic evolution in the Central Karakoram, Pakistan. Ann. Assoc. Am. Geogr. 100, 772–793.

Bishop, M.P., Bush, A.B.G., Furfaro, R., Gillespie, A.R., Hall, D.K., Haritashya, U.K., Shroder, Jr., J.F., 2014. Theoretical foundations of remote sensing for glacier assessment and mapping. In: Kargel, J., Leonard, G., Bishop, M., Kaab, A., Raup, B. (Eds.), Global Land Ice Measurements from Space, Springer Praxis Books. Springer, Berlin, Heidelberg, pp. 23–52, Chapter 2.

Bishop, M.P., James, L.A., Shroder, Jr., J.F., Walsh, S.J., 2012. Geospatial technologies and digital geomorphological mapping: concepts, issues and research. Geomorphology 137, 5–26.

Bishop, M.P., Shroder, Jr., J.F., Colby, J.D., 2003. Remote sensing and geomorphometry for studying relief production in high mountains. Geomorphology 55, 345–361.

Bishop, M.P., Shroder, Jr., J.F., Ward, J.L., 1995. SPOT multispectral analysis for producing supraglacial debris-load estimates for Batura glacier, Pakistan. Geocarto Int. 10, 81–90.

Blanco-Muriel, M., Alarcón-Padilla, D.C., López-Moratalla, T., Lara-Coira, M., 2001. Computing the solar vector. Solar Energy 70, 431–441.

Brantley, S.L., Goldhaber, M.B., Ragnarsdottir, K.V., 2007. Crossing disciplines and scales to understand the Critical Zone. Elements 3, 307–314.

Brantley, S.L., Lebedeva, M., 2011. Learning to read the chemistry of regolith to understand the Critical Zone. Annu. Rev. Earth Planet. Sci. 39, 387–416.

Chen, J.M., Liu, J., Leblanc, S.G., Lacaze, R., Roujean, J.L., 2003. Multi-angular optical remote sensing for assessing vegetation structure and carbon absorption. Remote Sens. Environ. 84, 516–525.

Clark, R.N., King, T.V.V., Klejwa, M., Swayze, G.A., Vergo, N., 1990. High spectral resolution reflectance spectroscopy of minerals. J. Geophys. Res. Solid Earth 95, 12653–12680.

Cloutis, E.A., 1996. Review article: hyperspectral geological remote sensing: evaluation of analytical techniques. Int. J. Remote Sens. 17, 2215–2242.

Corvalan, C., Hales, S., McMichael, A., 2005. Ecosystems and Human Well-Being: Health Synthesis. Technical Report. World Health Organization.

D'Ambrosio, D., Di Gregorio, S., Gabriele, S., Gaudio, R., 2001. A cellular automata model for soil erosion by water. Phys. Chem. Earth B Hydrol. Oceans Atmos. 26, 33–39.

D'Ambrosio, D., Di Gregorio, S., Iovine, G., Lupiano, V., Rongo, R., Spataro, W., 2003. First simulations of the Sarno debris flows through cellular automata modelling. Geomorphology 54, 91–117.

Deng, Y., Wilson, J.P., 2008. Multi-scale and multi-criteria mapping of mountain peaks as fuzzy entities. Int. J. Geogr. Inform. Sci. 22, 205–218.

Disney, M., Lewis, P., Saich, P., 2006. 3D modelling of forest canopy structure for remote sensing simulations in the optical and microwave domains. Remote Sens. Environ. 100, 114–132.

Dobreva, I.D., Klein, A.G., 2011. Fractional snow cover mapping through artificial neural network analysis of MODIS surface reflectance. Remote Sens. Environ. 115, 3355–3366.

Dorsaz, J.M., Gironás, J., Escauriaza, C., Rinaldo, A., 2013. The geomorphometry of endorheic drainage basins: implications for interpreting and modelling their evolution. Earth Surf. Process. Land. 38, 1881–1896.

Duce, R.A., Liss, P.S., Merrill, J.T., Atlas, E.L., Buat-Menard, P., Hicks, B.B., Miller, J.M., Prospero, J.M., Arimoto, R., Church, T.M., Ellis, W., Galloway, J.N., Hansen, L., Jickells, T.D., Knap, A.H., Reinhardt, K.H., Schneider, B., Soudine, A., Tokos, J.J., Tsunogai, S., Wollast, R., Zhou, M., 1991. The atmospheric input of trace species to the world ocean. Global Biogeochem. Cycles 5, 193–259.

Eckert, S., Kellenberger, T., Itten, K., 2005. Accuracy assessment of automatically derived digital elevation models from ASTER data in mountainous terrain. Int. J. Remote Sens. 26, 1943–1957.

Ernst, W.G., 2012. Overview of naturally occurring Earth materials and human health concerns. J. Asian Earth Sci. 59, 108–126.

Espinosa-Paredes, G., Nuñez Carrera, A., Laureano-Cruces, A.L., Vázquez-Rodríguez, A., Espinosa-Martinez, E.G., 2008. Emergency management for a nuclear power plant using fuzzy cognitive maps. Ann. Nuclear Energy 35, 2387–2396.

Fallahi, G.R., Frank, A.U., Mesgari, M.S., Rajabifard, A., 2008. An ontological structure for semantic interoperability of GIS and environmental modeling. Int. J. Appl. Earth Observ. Geoinform. 10, 342–357.

Firpo, G., Salvini, R., Francioni, M., Ranjith, P.G., 2011. Use of digital terrestrial photogrammetry in rocky slope stability analysis by distinct elements numerical methods. Int. J. Rock Mech. Mining Sci. 48, 1045–1054.

Foody, G.M., 1996. Fuzzy modelling of vegetation from remotely sensed imagery. Ecol. Model. 85, 3–12.

Forman, S.L., Marín, L., Gomez, J., Pierson, J., 2008. Late Quaternary eolian sand depositional record for southwestern Kansas: landscape sensitivity to droughts. Palaeogeogr. Palaeoclimatol. Palaeoecol. 265, 107–120.

Fulbright, T., Diamond, D., Rappole, J., Norwine, J., 1990. The coastal sand plain of southern Texas. Rangelands 12, 337–340.

Furfaro, R., Kargel, J.S., Lunine, J.I., Fink, W., Bishop, M.P., 2010. Identification of cryovolcanism on Titan using fuzzy cognitive maps. Planet. Space Sci. 58, 761–779.

Gabet, E.J., Pratt-Sitaula, B.A., Burbank, D.W., 2004. Climatic controls on hillslope angle and relief in the Himalayas. Geology 32, 629–632.

Giles, P.T., 1998. Geomorphological signatures: classification of aggregated slope unit objects from digital elevation and remote sensing data. Earth Surf. Process. Land. 23, 581–594.

Green, R.O., Painter, T.H., Roberts, D.A., Dozier, J., 2006. Measuring the expressed abundance of the three phases of water with an imaging spectrometer over melting snow. Water Resour. Res. 42, W10402.

Groumpos, P.P., Stylios, C.D., 2000. Modelling supervisory control systems using fuzzy cognitive maps. Chaos Soliton. Fract. 11, 329–336.

Grujic, D., Coutand, I., Bookhagen, B., Bonnet, S., Blythe, A., Duncan, C., 2006. Climatic forcing of erosion, landscape, and tectonics in the Bhutan Himalayas. Geology 34, 801–804.

Gueymard, C., 2005. SMARTS2, A simple model of the atmospheric radiative transfer of sunshine: Algorithms and performance assessment. Technical Report FSEC-PF-270-95. Florida Solar Energy Center.

Haboudane, D., Bonn, F., Royer, A., Sommer, S., Mehl, W., 2002. Land degradation and erosion risk mapping by fusion of spectrally-based information and digital geomorphometric attributes. Int. J. Remote Sens. 23, 3795–3820.

Harrison, S., 2009. Climate sensitivity: implications for the response of geomorphological systems to future climate change. Geol. Soc. London Special Publ. 320, 257–265.

Hebb, D.O., 1949. The Organization of Behavior: A Neuropsychological Theory. John Wiley, New York.

Heckbert, S., Baynes, T., Reeson, A., 2010. Agent-based modeling in ecological economics. Ann. NY Acad. Sci. 1185, 39–53.

Hensley, S., Farr, T., 2013. Microwave remote sensing and surface characterization. In: Shroder, J.F. (Ed.), Treatise on Geomorphology. Academic Press, San Diego, pp. 43–79.

Hirano, A., Welch, R., Lang, H., 2003. Mapping from ASTER stereo image data: DEM validation and accuracy assessment. ISPRS J. Photogrammetry Remote Sens. 57, 356–370.

Höhle, J., Höhle, M., 2009. Accuracy assessment of digital elevation models by means of robust statistical methods. ISPRS J. Photogrammetry Remote Sens. 64, 398–406.

Hubbard, B.E., Crowley, J.K., 2005. Mineral mapping on the Chilean-Bolivian Altiplano using co-orbital ALI, ASTER and Hyperion imagery: Data dimensionality issues and solutions. Remote Sens. Environ. 99, 173–186.

Hubbard, B.E., Crowley, J.K., Zimbelman, D.R., 2003. Comparative alteration mineral mapping using visible to shortwave infrared (0.4–2.4 μm) Hyperion, ALI, and ASTER imagery. IEEE Trans. Geosci. Remote Sens. 41, 1401–1410.

Hugenholtz, C.H., Levin, N., Barchyn, T.E., Baddock, M.C., 2012. Remote sensing and spatial analysis of aeolian sand dunes: a review and outlook. Earth Sci. Rev. 111, 319–334.

Immerzeel, W.W., Kraaijenbrink, P.D.A., Shea, J.M., Shrestha, A.B., Pellicciotti, F., Bierkens, M.F.P., de Jong, S.M., 2014. High-resolution monitoring of Himalayan glacier dynamics using unmanned aerial vehicles. Remote Sens. Environ. 150, 93–103.

Iverson, L.R., Schwartz, M.W., Prasad, A.M., 2004. Potential colonization of newly available tree-species habitat under climate change: an analysis for five eastern US species. Landscape Ecol. 19, 787–799.

Jacobson, M.Z., 2005. Fundamentals of Atmospheric Modeling. Cambridge University Press, Cambridge, UK.

Jin, S., Yang, L., Danielson, P., Homer, C., Fry, J., Xian, G., 2013. A comprehensive change detection method for updating the National Land Cover Database to circa 2011. Remote Sens. Environ. 132, 159–175.

Johnson, A.G., Glenn, C.R., Burnett, W.C., Peterson, R.N., Lucey, P.G., 2008. Aerial infrared imaging reveals large nutrient-rich groundwater inputs to the ocean. Geophys. Res. Lett. 35, L15606.

Kargel, J., Leonard, G., Bishop, M., Kääb, A., Raup, B., 2014. Global Land Ice Measurements from Space. Springer, Berlin, Heidelberg.

Kargel, J.S., Abrams, M.J., Bishop, M.P., Bush, A., Hamilton, G., Jiskoot, H., Kääb, A., Kieffer, H.H., Lee, E.M., Paul, F., Rau, F., Raup, B., Shroder, J.F., Soltesz, D., Stainforth, D., Stearns, L., Wessels, R., 2005. Multispectral imaging contributions to global land ice measurements from space. Remote Sens. Environ. 99, 187–219.

Kniezys, F.X., Abreu, L.W., Anderson, G.P., Chetwynd, J.H., Shettle, E.P., Berk, A., Bernstein, L.S., Robertson, D.C., Acharya, P., Rothman, L.S., Selby, J.E.A., Gallery, W.O., Clough, S.A., 1996. The MODTRAN 2/3 Report and LOWTRAN 7 Model. Technical Report F19628-91-C-0132. Phillips Lab., Geophys. Directorate.

Kobierska, F., Jonas, T., Magnusson, J., Zappa, M., Bavay, M., Bosshard, T., Paul, F., Bernasconi, S.M., 2011. Climate change effects on snow melt and discharge of a partly glacierized watershed in Central Switzerland (SoilTrec Critical Zone Observatory). Appl. Geochem. 26 (Suppl.), S60–S62.

Koons, P.O., 1995. Modeling the topographic evolution of collisional belts. Annu. Rev. Earth Planet. Sci. 23, 375–408.

Koons, P.O., Zeitler, P.K., Chamberlain, C.P., Craw, D., Meltzer, A.S., 2002. Mechanical links between erosion and metamorphism in Nanga Parbat, Pakistan Himalaya. Am. J. Sci. 302, 749–773.

Kruse, S., 2013. Near-surface geophysics in geomorphology. In: Shroder, J.F. (Ed.), Treatise on Geomorphology. Academic Press, San Diego, pp. 103–129.

Kummerow, C., Barnes, W., Kozu, T., Shiue, J., Simpson, J., 1998. The Tropical Rainfall Measuring Mission (TRMM) sensor package. J. Atmos. Oceanic Technol. 15, 809–817.

Kummerow, C., Simpson, J., Thiele, O., Barnes, W., Chang, A.T.C., Stocker, E., Adler, R.F., Hou, A., Kakar, R., Wentz, F., Ashcroft, P., Kozu, T., Hong, Y., Okamoto, K., Iguchi, T., Kuroiwa, H., Im, E., Haddad, Z., Huffman, G., Ferrier, B., Olson, W.S., Zipser, E., Smith, E.A., Wilheit, T.T., North, G., Krishnamurti, T., Nakamura, K., 2000. The status of the Tropical Rainfall Measuring Mission (TRMM) after two years in orbit. J. Appl. Meteorol. 39, 1965–1982.

Kump, L.R., Brantley, S.L., Arthur, M.A., 2000. Chemical, weathering, atmospheric CO_2, and climate. Annu. Rev. Earth Planet. Sci. 28, 611–667.

Kuntz, B.W., Rubin, S., Berkowitz, B., Singha, K., 2011. Quantifying solute transport at the Shale Hills Critical Zone Observatory. Vadose Zone J. 10, 843–857.

Lefsky, M.A., Cohen, W.B., Acker, S.A., Parker, G.G., Spies, T.A., Harding, D., 1999. Lidar remote sensing of the canopy structure and biophysical properties of Douglas fir western hemlock forests. Remote Sens. Environ. 70, 339–361.

Liang, S., 2007. Recent developments in estimating land surface biogeophysical variables from optical remote sensing. Progr. Phys. Geogr. 31, 501–516.

Liao, J., Tang, L., Shao, G., Qiu, Q., Wang, C., Zheng, S., Su, X., 2014. A neighbour decay cellular automata approach for simulating urban expansion based on particle swarm intelligence. Int. J. Geogr. Inform. Sci. 28, 720–738.

Lim, K., Treitz, P., Wulder, M., St-Onge, B., Flood, M., 2003. LiDAR remote sensing of forest structure. Progr. Phys. Geogr. 27, 88–106.

Lin, H., Hopmans, J.W., Richter, D.D., 2011. Interdisciplinary sciences in a global network of Critical Zone Observatories. Vadose Zone J. 10, 781–785.

Liu, B., Qu, J., Zhang, W., Qian, G., 2011. Numerical simulation of wind flow over transverse and pyramid dunes. J. Wind Eng. Indus. Aerodyn. 99, 879–888.

Livingstone, I., Wiggs, G.F.S., Weaver, C.M., 2007. Geomorphology of desert sand dunes: a review of recent progress. Earth Sci. Rev. 80, 239–257.

Mangan, J.M., Overpeck, J.T., Webb, R.S., Wessman, C., Goetz, A.F.H., 2004. Response of Nebraska sand hills natural vegetation to drought, fire, grazing, and plant functional type shifts as simulated by the century model. Climatic Change 63, 49–90.

Martin, J.M., Whitfield, M., 1983. The significance of the river input of chemical elements to the ocean. Wong, C.S., Boyle, E., Bruland, K.W., Burton, J.D., Goldberg, E.D. (Eds.), Trace Metals in Sea Water, 9. Springer, US, pp. 265–296, Chapter 16.

Meinzer, F.C., Woodruff, D.R., Eissenstat, D.M., Lin, H.S., Adams, T.S., McCulloh, K.A., 2013. Above- and belowground controls on water use by trees of different wood types in an eastern US deciduous forest. Tree Physiol. 33, 345–356.

Miao, X., Mason, J.A., Swinehart, J.B., Loope, D.B., Hanson, P.R., Goble, R.J., Liu, X., 2007. A 10,000 year record of dune activity, dust storms, and severe drought in the central Great Plains. Geology 35, 119–122.

Miskolczi, F.M., 1990. High resolution atmospheric radiance-transmittance code (HARTCODE). Technical Report Version No. 01. Instituto per lo Studio delle Metodologie Geosiche Ambientali.

Montgomery, D.R., Hallet, B., Liu, Y.P., Finnegan, N., Anders, A., Gillespie, A., Greenberg, H.M., 2004. Evidence for Holocene megafloods down the Tsangpo River Gorge, southeastern Tibet. Quaternary Res. 62, 201–207.

Moody, A., Katz, D., 2004. Artificial intelligence in the study of mountain landscapes. In: Bishop, M., Shroder, Jr., J.F. (Eds.), Geographic Information Science and Mountain Geomorphology. Springer Praxis Publishing, Chichester, UK, pp. 219–251.

Moore, I.D., Gessler, P.E., Nielsen, G.A., Peterson, G.A., 1993. Soil attribute prediction using terrain analysis. Soil Sci. Soc. Am. J. 57, 443–452.

Mulder, V.L., de Bruin, S., Schaepman, M.E., Mayr, T.R., 2011. The use of remote sensing in soil and terrain mapping – a review. Geoderma 162, 1–19.

Nautical Almanac Office, 2013. The Astronomical Almanac for the Year 2013. United States Naval Observatory/Nautical Almanac Office.

Nield, J.M., Baas, A.C.W., 2008. The influence of different environmental and climatic conditions on vegetated aeolian dune landscape development and response. Global Planet. Change 64, 76–92.

Niethammer, U., James, M.R., Rothmund, S., Travelletti, J., Joswig, M., 2012. UAV-based remote sensing of the Super-Sauze landslide: evaluation and results. Eng. Geol. 128, 2–11.

Njoku, E.G., Kong, J.A., 1977. Theory for passive microwave remote sensing of near-surface soil moisture. J. Geophys. Res. 82, 3108–3118.

Nuth, C., Kääb, A., 2011. Co-registration and bias corrections of satellite elevation data sets for quantifying glacier thickness change. Cryosphere 5, 271–290.

Openshaw, S., Abrahart, R.J., 2000. Geocomputation. Taylor & Francis, London.

Ouzounov, D., Freund, F., 2004. Mid-infrared emission prior to strong earthquakes analyzed by remote sensing data. Adv. Space Res. 33, 268–273.

Papageorgiou, E.I., Spyridonos, P.P., Glotsos, D.T., Stylios, C.D., Ravazoula, P., Nikiforidis, G.N., Groumpos, P.P., 2008. Brain tumor characterization using the soft computing technique of fuzzy cognitive maps. Appl. Soft Comput. 8, 820–828.

Papageorgiou, E.I., Stylios, C.D., Groumpos, P.P., 2003. An integrated two-level hierarchical system for decision making in radiation therapy based on fuzzy cognitive maps. IEEE Trans. Biomed. Eng. 50, 1326–1339.

Papageorgiou, E.I., Stylios, C.D., Groumpos, P.P., 2004. Active Hebbian learning algorithm to train fuzzy cognitive maps. Int. J. Approximate Reason. 37, 219–249.

Paul, F., Huggel, C., Kääb, A., 2004. Combining satellite multispectral image data and a digital elevation model for mapping debris-covered glaciers. Remote Sens. Environ. 89, 510–518.

Pelletier, J.D., Mitasova, H., Harmon, R.S., Overton, M., 2009. The effects of interdune vegetation changes on eolian dune field evolution: a numerical-modeling case study at Jockey's Ridge, North Carolina, USA. Earth Surf. Process. Land. 34, 1245–1254.

Perez, L., Dragicevic, S., 2010. Modeling mountain pine beetle infestation with an agent-based approach at two spatial scales. Environ. Model. Softw. 25, 223–236.

Peters, L., Daniels, J.J., Young, J.D., 1994. Ground penetrating radar as a subsurface environmental sensing tool. Proc. IEEE 82, 1802–1822.

Phillips, J.D., 1999. Methodology, scale, and the field of dreams. Ann. Assoc. Am. Geogr. 89, 754–760.

Pielke, R.A., Marland, G., Betts, R.A., Chase, T.N., Eastman, J.L., Niles, J.O., Niyogi, D.D.S., Running, S.W., 2002. The influence of land-use change and landscape dynamics on the climate system: relevance to climate-change policy beyond the radiative effect of greenhouse gases. Philos. Trans. R. Soc. London 360, 1705–1719.

Pike, R.J., 2000. Geomorphometry - diversity in quantitative surface analysis. Progr. Phys. Geogr. 24, 1–20.

Pike, R.J., Evans, I.S., Hengl, T., 2009. Geomorphometry: a brief guide. In: Tomislav, H., Hannes, I.R. (Eds.), Geomorphometry: Concepts, Software, Applications, vol. 33. Elsevier, Amsterdam, The Netherlands, pp. 3–30, Chapter 1.

Pour, A.B., Hashim, M., 2012. The application of ASTER remote sensing data to porphyry copper and epithermal gold deposits. Ore Geol. Rev. 44, 1–9.

Price, W., 1958. Sedimentology and Quaternary geology of south Texas. Geol. Soc. Trans. 3, 4175.

Ramanathan, V., Ramana, M.V., Roberts, G., Kim, D., Corrigan, C., Chung, C., Winker, D., 2007. Warming trends in Asia amplified by brown cloud solar absorption. Nature 448, 575–578.

Rango, A., Laliberte, A., Herrick, J.E., Winters, C., Havstad, K., Steele, C., Browning, D., 2009. Unmanned aerial vehicle-based remote sensing for rangeland assessment, monitoring, and management. J. Appl. Remote Sens. 3, 033542.

Rasmussen, C., Troch, P., Chorover, J., Brooks, P., Pelletier, J., Huxman, T., 2011. An open system framework for integrating critical zone structure and function. Biogeochemistry 102, 15–29.

Roderick, M., 1992. Methods for calculating solar position and day length including computer programs and subroutines. Technical Report Resource Management, Technical Report No. 137. Land Management, Western Australian Department of Agriculture.

Rodriguez-Iturbe, I., 2000. Ecohydrology: a hydrologic perspective of climate-soil-vegetation dynamics. Water Resour. Res. 36, 3–9.

Rozenstein, O., Karnieli, A., 2011. Comparison of methods for land-use classification incorporating remote sensing and GIS inputs. Appl. Geogr. 31, 533–544.

Russell, J., 1981. The south Texas eolian sand sheet. In: Russell, J., Sterling, C. (Eds.), Modern Depositional Environments of Sands in South Texas. Gulf Coast Association of Geological Societies, Corpus Christi, Texas, p. 4346.

Schlerf, M., Atzberger, C., 2006. Inversion of a forest reflectance model to estimate structural canopy variables from hyperspectral remote sensing data. Remote Sens. Environ. 100, 281–294.

Schmiedel, U., Dengler, J., Etzold, S., 2012. Vegetation dynamics of endemic-rich quartz fields in the Succulent Karoo, South Africa, in response to recent climatic trends. J. Veg. Sci. 23, 292–303.

Serbin, G., Daughtry, C.S.T., Hunt, Jr., E.R., Reeves III, J.B., Brown, D.J., 2009. Effects of soil composition and mineralogy on remote sensing of crop residue cover. Remote Sens. Environ. 113, 224–238.

Smit, B., Wandel, J., 2006. Adaptation, adaptive capacity and vulnerability. Global Environ. Change 16, 282–292.

Smith, M., Pain, C., 2009. Applications of remote sensing in geomorphology. Progr. Phys. Geogr. 33, 568–582.

Stolar, D.B., Willett, S.D., Roe, G.H., 2006. Climatic and tectonic forcing of a critical orogen. Geol. Soc. Am. Special Pap. 398, 241–250.

Stylios, C.D., Groumpos, P.P., 2000. Fuzzy cognitive maps in modeling supervisory control systems. J. Intelligent Fuzzy Syst. 8, 83–98.

Tarolli, P., Arrowsmith, J.R., Vivoni, E.R., 2009. Understanding Earth surface processes from remotely sensed digital terrain models. Geomorphology 113, 1–3.

Tate, J., Wood, J., 2001. Fractals and scale dependencies in topography. In: Tate, N., Atkinson, P.M. (Eds.), Modelling Scale in Geographical Information Science. John Wiley and Sons, Chichester, UK, pp. 35–52.

Thomas, D.S.G., Knight, M., Wiggs, G.F.S., 2005. Remobilization of southern African desert dune systems by twenty-first century global warming. Nature 435, 1218–1221.

Thomas, D.S.G., Wiggs, G.F.S., 2008. Aeolian system responses to global change: challenges of scale, process and temporal integration. Earth Surf. Process. Land. 33, 1396–1418.

Tomislav, H., Hannes, I.R., 2009. Geomorphometry: Concepts, Software, Applications. Elsevier, Amsterdam, The Netherlands.

Tralli, D.M., Blom, R.G., Zlotnicki, V., Donnellan, A., Evans, D.L., 2005. Satellite remote sensing of earthquake, volcano, flood, landslide and coastal inundation hazards. ISPRS J. Photogrammetry Remote Sens. 59, 185–198.

Verburg, P.H., Neumann, K., Nol, L., 2011. Challenges in using land use and land cover data for global change studies. Global Change Biol. 17, 974–989.

Völkel, J., Huber, J., Leopold, M., 2011. Significance of slope sediments layering on physical characteristics and interflow within the Critical Zone examples from the Colorado Front Range, USA. Appl. Geochem. 26 (Suppl.), S143–S145.

Wainwright, J., Parsons, A.J., Abrahams, A.D., 2000. Plot-scale studies of vegetation, overland flow and erosion interactions: case studies from Arizona and New Mexico. Hydrol. Process. 14, 2921–2943.

Walker, I.J., 2010. Changing views in Canadian geomorphology: are we seeing the landscape for the processes? Can. Geogr. 54, 261–276.

Walsh, S., Moody, A., Allen, T., Brown, D., 1997. Scale dependence of NDVI and its relationship to mountainous terrain. In: Quattrochi, D.A., Goodchild, M.F. (Eds.), Scale in Remote Sensing and GIS. CRC Press, Boca Raton, FL, pp. 27–55.

Walsh, S.J., Messina, J.P., Mena, C.F., Malanson, G.P., Page, P.H., 2008. Complexity theory, spatial simulation models, and land use dynamics in the Northern Ecuadorian Amazon. Geoforum 39, 867–878.

Waters, P., Greenbaum, D., Smart, P.L., Osmaston, H., 1990. Applications of remote sensing to groundwater hydrology. Remote Sens. Rev. 4, 223–264.

Wechsler, S.P., Kroll, C.N., 2006. Quantifying DEM uncertainty and its effect on topographic parameters. Photogrammetric Eng. Remote Sens. 72, 1081–1090.

Wehrli, C., 1985. Extraterrestrial solar spectrum. Technical Report 615. World Radiation Center, Davos, Switzerland.

Werner, B.T., 1995. Eolian dunes: computer simulations and attractor interpretation. Geology 23, 1107–1110.

Whipple, K.X., 2009. The influence of climate on the tectonic evolution of mountain belts. Nat. Geosci. 2, 97–104.

Wilkinson, G.G., 1996. A review of current issues in the integration of GIS and remote sensing data. Int. J. Geogr. Inform. Syst. 10, 85–101.

Willett, S.D., Brandon, M.T., 2002. On steady states in mountain belts. Geology 30, 175–178.

Wilson, J.P., Bishop, M.P., 2013. Geomorphometry. In: Shroder, J.F. (Ed.), Treatise on Geomorphology. Academic Press, San Diego, pp. 162–186.

Wootton, J.T., 2001. Local interactions predict large-scale pattern in empirically derived cellular automata. Nature 413, 841–844.

Worboys, M., 1998. Computation with imprecise geospatial data. Comput. Environ. Urban Syst. 22, 85–106.

Zheng, B., Campbell, J.B., Serbin, G., Galbraith, J.M., 2014. Remote sensing of crop residue and tillage practices: present capabilities and future prospects. Soil Till. Res. 138, 26–34.

Chapter 18

The Built Environment in the Critical Zone: From Pre- to Postindustrial Cities

Nazgol Bagheri
Department of Political Science and Geography, University of Texas at San Antonio, San Antonio, Texas, USA

18.1 INTRODUCTION

This chapter examines the roles the human settlements and built environment play in the Earth's Critical Zone. Building upon the earlier definitions offered in earlier chapters of this volume – Critical Zone includes all the rock, soil, water, air, and living organisms between the top of the tree canopy and the base of groundwater aquifer. In other words, the Critical Zone, covering the entire surface of the Earth, makes life including human life possible. The future of human society and the Critical Zone are undoubtedly interconnected; however, the ways human settlements and the built environment affect and are affected by the Critical Zone and its life-sustaining resources remains understudied. Majority of the research on the Critical Zone, often done by hydrology, geology, biology, ecology, geochemistry, geomorphology scientists, explore the intertwined network of physical, chemical, and biological processes and study the interactive cycles of water, carbon, and energy. But, less is known about how our highly urbanized human settlements affect these cycles and the Critical Zone itself. To fill this gap in our knowledge, this chapter takes an unprecedented step toward integrating perspectives from architecture, urban planning, and urban geography disciplines and examines the relationship between the human settlements and the Critical Zone. This chapter is split into four sections. The first section offers a brief history of urbanization patterns as well as the changing concepts of urbanity and urban. Also, urban growth models from industrial city to postindustrial and later to postmodern city form along with their impacts on the Critical Zone will be discussed. The second section discusses how sustainable urban development models such as the smart growth, the green city, new urbanism and compact city emerged out of environmental sustainability efforts since the 1960s (Jenks and Burgess, 2000). The

Developments in Earth Surface Processes, Vol. 19. http://dx.doi.org/10.1016/B978-0-444-63369-9.00018-5

third section discusses the vernacular architecture and the traditional compact city model as a climatic response in desert cities of Iran. The harmonic relationship between the design of compact city, architectural elements, and the Critical Zone has also been illustrated in this section. The forth section outlines the numerous ways in which the contemporary rapid urbanization around the globe affects and is affected by the Critical Zone. Opportunities and challenges presented by the inevitable global urbanization will be contextualized in the framework of the Critical Zone. Recommendations address changes in research and practice to decrease the adverse impacts of urbanization on the Critical Zone and to make informed, proactive decisions in designing sustainable cities.

18.2 URBANIZATION AND THE CRITICAL ZONE

The contemporary world is witnessing the most rapid urbanization rate throughout the history. Beginning in 2008,[1] for the first time in human history, more than half of the world's population live in urban areas. There have been different theories about the origins of cities; however, access to water, or the hydraulic theory, has been one of the most accepted ones in the emergence of early cities (e.g., Nile Valley, Fertile Crescent, and Indus Valley). Urban scholars highlight the necessity of irrigation for urban development, particularly in the semi-arid climates of the Middle East where the first agricultural revolution occurred and resulted in the emergence of hydraulic societies, population agglomeration, and cities in the ancient world (Wittfogel, 1957). In the Critical Zone framework, water research scientists examine the ecology of streams, rivers, and their watersheds in both pristine and polluted forms and emphasize the role fresh water plays in sustaining life on the Earth. However, for urban scholars water is, in fact, the absolute precondition for urban development and growth of preindustrialized cities. Abbey (1968, p. 126) beautifully captures the relation between water and the city: "there is no lack of water here, unless you try to establish a city where no city should be." The Earth's Critical Zone, including fresh water, fertile soil, and mild climate, is an irreplaceable vital factor in the origin of cities and urbanity in the ancient world. The rise and fall of contemporary cities suggests that environmental conditions intertwined with the Earth's Critical Zone still play a significant role in whether or not a city develops.

While defining "urban" varies across the geographical and temporal boundaries, it can be considered as both an entity and a quality. Urban as an entity is often categorized based on certain quantitative characteristics such as population, population density, economic activity, legal and administrative boundaries. Urban as a quality deals with a more sociocultural and perceptional dimension

1. See the 2007 United Nations' State of World Population report at: http://www.unfpa.org/publications/state-world-population-2007

of the place and focuses on qualities associated with living in urban areas such as moving from *Gemeinschaft* (community) to *Gesellschaft* (society) lifestyle (Tönnies, 1963), living a fast-paced life, becoming a stranger in the city, and having more choices in different dimensions of life. Cities still remain the central gathering place for more employment, education, social, recreational, and political opportunities as well as distinct lifestyles, but they have dramatically changed in the recent decades.

If the urban studies in the last two centuries have had a master theoretical frame to organize their focus, problems, and explanations, it has been the key concept of industrialization. Urban theories have long been organized around the preindustrial, industrial, and postindustrial divide, seeking to understand the functional, environmental, and cultural transformations from one to the other. The preindustrial, traditional cities with their organic, piecemeal growth were replaced by early industrial cities immediately after the onset of the large-scale industrialization in the eighteenth century. Once the capitalist economies and industrialization came into play, the form and function of cities began changing dramatically due to new construction materials and methods (e.g., glass, steel, and concrete), shifting employment opportunities, rapid rural-to-urban migration, increased mobility, and new forms of transportations. Mass-transport system was a key player in this transition (Carmona et al., 2010). People were suddenly able to move around in cities with a speed never before experienced. The Industrial Revolution also led to increased efficiency in farming techniques, therefore a surplus of labor led to many flocking to city centers in search of new factory jobs and other opportunities. Traditional ecological models of the city by Chicago School scholars (see Park, 1925; Burgess, 1925; Park et al., 1925) described American industrial cities as concentric circles around the central business diacritic (CBD) where the urban growth resembled the ecological circles found in the nature. Industrial cities were often overcrowded – as best illustrated in Charles Dickens' novels about urban London during early industrialization. Industrial cities suffered from pollutants and pathogens, as a result of high air pollution, drinking water contamination, freshwater and marine pollution, poor housing qualities, other environmental hazards and related health effects. Different elements of the Earth's Critical Zone were contaminated: (1) Water in and around the cities was contaminated due to the lack of sewage connections and dumping of various industrial and new household waste products into water sources; (2) Air was unprecedentedly polluted due to the consumption of anthracite coal, heavy oil, and other fossil fuels; (3) Soil quality was deprived due to both increased solid waste and deforestation; (4) Living organisms between the top of the tree canopy and the base of groundwater aquifer became endangered and/or lost. Water, air, and soil contamination consequently disturbed the balance of the ecosystem inside and resulted in the death of various animal and plant species. Powered by fissile fuels, industrial cities introduced new job opportunities in urban areas but with high environmental costs.

During the early twentieth century, suburbanization or counterurbanization – a new form of urbanization as an escape from the industrial city – started in mature industrial and postindustrial cities. People began moving into suburban areas outside of the central business district and the urban core. Water and air pollution, crime, and overcrowding of the central city were among the many forces that led many to move to outlying regions. This marked the start of decentralization and urban sprawl, leading to the emergence of edge and satellite cities. New models of urban growth were introduced. Unlike industrial cities, the Los Angles (or perhaps Detroit) School's scholars (see Soja, 1989, 1992, 1996; Sorkin, 1992; Dear and Flusty, 1998; Dear, 2000, 2002) described the postindustrial cities as socially and spatially fragmented, exhibiting chaotic urban form, land use patterns, and geographies of social and environmental injustice. Information and communication, technological advancement, economic restructuring, globalization and environmental policies, and the emergence of shadow government (privatized proto-government) are the key forces behind the contemporary transformation of urban form in today's postmodern, neoliberal cities. Despite the resurgence of cities and culturally and physically restructuring the downtowns since the mid-1990s (Fishman, 2008), suburbanization is still an increasingly widespread urban phenomenon around the globe, resulting in salient sociospatial inequalities, fragmented land-use patterns, monopolizing resources, and intrusive environmental impacts. One of the major adverse impacts of urban sprawl comes from the fact that suburban development are often – if not always – auto-dependent (Sheller and Urry, 2000; Duany et al., 2000, 2009; Urry, 2004; Duany and Plater-Zyberk, 2011). Newman and Kenworthy (1999, 2000, p. 109) counts the following as the major problems of contemporary car-dependent, leapfrog suburban developments: oil vulnerability, petrochemical smog over urban and suburban areas, toxic emissions, high greenhouse gas contribution, and higher chance of stormwater problems from extra hard surfaces accommodating cars. Contextualizing these environmental impacts in the framework of the Critical Zone, one can easily conclude that suburbanization adversely affects all the elements of the Critical Zone: (1) Water quality is degraded due to the asphalt-covered surfaces and their effects on hydrology including an increase in the volume, rate, and pollutant content of stormwater runoff. Researchers have also found a connection between urban sprawl and flooding. New wasteful household consumption patterns (e.g., use of obsessive amounts of water for outdoor landscape irrigation) have caused overusage of finite, natural water resources and in some regions such as the US Southwest, resulted in seasonal draught and imbalance in groundwater aquifer. (2) Air quality has diminished due to the increased emission of toxic pollutants including carbon monoxide, carbon dioxide, nitrogen oxides, fine particulate matter, and volatile organic compounds from vehicles. Carbon dioxide (CO_2) – as one of the anthropogenic greenhouse gases with high-radiative forcing and relative longevity in the atmosphere – makes a strong connection between car-cased developments, greenhouse gas emissions, and the climate change. In addition, urban heat island increase average surface temperatures and consequently exacerbate air pollution and increase the energy consumption (higher demand for interior cooling). (3) Soil and rock: Farmland, open green space, forest, and natural habitats are lost or degraded due to excessive land consumption (10–20 times) underlying population growth, asphalt covering, and heat islands. (4) Besides the negative aspects of urbanization on living organisms as mentioned earlier, there is also significant loss of natural habitats and biodiversity, as well as the emergence of invasive plants (see McHarg, 1969; Fulton et al., 2002; Zeemerin, 2009; Carmona et al., 2010).

While the pace and spatial patterns of suburbanization vary around the world, the binary categorization of rural versus urban and urban versus suburban has been questioned, particularly in the global north where cities are still witnessing decentralization. With advancement in telecommunication in the information era, the rural areas often become urban as a quality and lifestyle despite them not being categorized as urban according to quantitative attributes commonly used to define urban as an entity. In urban geography subdiscipline, cities are ranked according to their size, functions, and the role they play in the global processes. Depending on the rank of a city in the urban hierarchy, cities may affect the Earth's Critical Zone in different dimensions and to different extents. Therefore, one must consider the variation of urbanization's impact on the Critical Zone. For instance, the impacts of rapid urban growth in China and India today are comparable to those of industrial cities of the nineteenth century.

It is also important to highlight that the city, from its preindustrial to its postindustrial urban form, has been dependent on city's hinterland – the surrounding nonurban areas extending out from the city borders – and resources found in the hinterland. This suggests that the impacts of urban growth are not spatially limited to the city's official boundaries but are important in the city's hinterland. White and Whitney (1992) suggest that majority of cities around the world go beyond their immediate, local hinterlands and depend on resources from elsewhere. To this end, Wackernagel and Rees (1998) have developed the concept of ecological footprint and warned us from the exponential increase of urban ecological footprint in cities. "Ecological footprint" (Rees and Wackernagel, 1998, p. 9) is defined as "an accounting tool that enables us to estimate the resource consumption and waste assimilation requirements of a defined human population or economy in terms of a corresponding productive land area."

According to the United Nation's report, more than two third of world population will live in urban areas. With its unrivaled speed and scale, global urban growth brings about critical questions of how long such growth – and in general, life on the Earth with its limited resources – will be sustained. In the next chapter, the concept of urban sustainability will be explained through a review of the early environmental sustainability movements. Multiple models of sustainable urban developments will be explained and evaluated.

18.3 ENVIRONMENTAL AND URBAN SUSTAINABILITY

The history of environmental sustainability may be reviewed from two viewpoints: (1) the perspectives of noted environmentalists, and (2) the responses of international organizations toward environmental issues. The beginning of environmental movements is marked by the publication of Rachel Carson's *Silent Spring* (1962). Carson's book deals with the devastating impacts of pesticides and weed killers on the environment and has influenced the formation of diverse groups such as Friends of the Earth, and the introduction of so-called green policies. Other significant works, which discussed the causes of environmental problems, and the principles of green alternatives, were E.F. Schumacher's *Small is Beautiful* (1974), McRobie's *Small is possible* (McRobie, 1981), and Meadows et al.'s *Limits to growth* (Meadows et al., 1972). Schumacher seriously challenges the proposition that humans can continue with their increasing rate of production and consumption on the Earth. He warns that our planet – the principal capital of humanity – is threatened by excessive production and consumption and that the human race is using up its capital at an alarming rate. Garrett Hardin's *Tragedy of the Commons* (1977) is another outstanding work dealing with environmental issues. Hardin argues that if every individual tried to maximize his consumption of resources held in common, this would lead to the destruction of these shared assets. In this view, endangered commons include all the elements of the Earth's Critical Zone: the air that we breathe, the water we drink, and the ecosystems that we have degraded. He accurately suggests that the prospect of an ever-growing population continuing to exist in a finite environment is impossible.

In the wake of the heated debates spurred by environmental literature on problems such as pollution, global climate change, and depletion of natural resources, the international organizations began to address environmental issues. Sustainable development became a topic of serious debate. In 1987, following the Bruntland report *Our Common Future* by the World Commission on Environment and Development (1987), environment and development were moved to the forefront of the political agenda. The report was representative of the growing global awareness in the second half of the century of the enormous environmental problems facing the planet, and of a growing shift toward global environment action. In Agenda 21 of the Earth Summit (Quarrie, 1992), the main policy document of the United Nations Conference on Environment and Development, held in Rio de Janeiro in 1992, 178 nations committed themselves to formulating national strategies and action plans to implement the guidelines of Agenda 21 in their respective countries. Within the provisions of the Local Agenda 21, urban planning and design has been viewed as a key instrument in pursuing sustainable development goals. In this framework, urban planners and designers have included environmentally sustainable architecture high on their agenda (Moughtin, 1996).

Although there is unanimity on the fact that the sole solution of global problems depends mostly on the adoption of policies that promote the sustainability of development, there seems to be no universal agreement on the general definition of sustainable development. Specialized literature defines the concept in a variety of ways. In certain cases, the term has even fallen victim to ambiguity or misuse. However, the most widely accepted definition is provided by the Brundtland Report: "sustainable development is the development that meets the needs of the present generation without compromising the ability of future generations to meet their own needs" (World Commission on Sustainable Development, 1987). This definition places particular stress on three key elements: development, needs, and future generations. In contrast to the term growth, which implies a physical and quantitative expansion, development is a qualitative state associated with exaltation and advancement, encompassing all cultural, social, economic, and physical dimensions of human life. The elements of needs and future generations are closely concerned with equity and equality in benefiting from opportunities, both at intergenerational and intragenerational levels.

With the sustainable development approach, humans' role in their relationship with the environment is one of a supervisor who constantly monitors losses and benefits arising from their interactions with the Critical Zone and the life-sustaining resources it offers. Humans are, indeed, the guardians and protectors of the Critical Zone. This new relationship between people and their environment is beautifully expressed by an Indian-American proverb: "we have not inherited the Earth from our parents; we have borrowed it from our children!"

While occupying only over two percent of the Earth's surface, cities consume more than three-fourth of the world's resources (Girardet, 1992). Haughton and Hunter (1994, p. 11) estimate on average, a city of one million people daily consumes 625,000 tons of water, 9,500 tons of fuel, and produces 500,000 tons of wastewater, 2,000 tons of waste solids, and 950 tons of air pollutants. These numbers suggest a linear nature of the metabolism in cities, which in its current form, is endangering the natures' cycles found in the Critical Zone. Since the early 1990s, architects and urban designers have presented new models to address these environmental problems. These efforts have manifested into a wide range of initiatives and innovative propositions; among them are the Green City movement, New Urbanism (NU), and the Smart Growth.

The Garden City Movement was originally introduced by Sir Ebenezer Howard in *Tomorrow: A peaceful path to real reform* (originally published in 1898; see Osborn, 1965) in which he promoted a detailed consideration of the environmental condition of an area before urban development. In this model, a sustainable environment was seen as something between urban and rural and was protected from the peripheral, industrial sections by a green belt. In other words, the Garden City model requires a comprehensive examination of the Earth's Critical Zone and its natural resources. The tradition of respecting the environment was continued by the New Urbanism and the Smart Growth models in land use planning and urban design practices. New Urbanism (NU) is an array of ideas and policies that appeared mostly in the United States during the early 1990s. Architects and urban planners associated with the New Urbanism are "...committed to re-establishing the art of building and the making of community, through citizen-based participatory planning and design." (Charter of New Urbanism, 2000). While NU was initially a social agenda to decrease the social injustice in fragmented, postmodern cities in the US, it also promoted environmental and urban sustainability due to its policies on compact and walkable neighborhood design (Katz, 1994; Charter of New Urbanism, 2000; Fishman, 2008). Smart growth emerged in the US during the mid-1990s as an anti-sprawl design movement to protect the open, green spaces around the cities. Smart growth is "a syncretic planning program of spatial regulation and public investment that in turn has informed, and been informed by, the more refined design aspirations of New Urbanism (NU)" (Dierwechter, 2014: pp. 691–692).

Reviewing these three key urban design movements/models, Elikin et al. (1991) suggest four principles for sustainable development: futurity and future generation, environment, equity, and participation. In general, sustainable development focuses on simultaneous protection for both the natural environment and the human-built environment. The principals of sustainable urban development may be summarized as follows: (1) to give priority to the recycling of existing buildings, places, and infrastructure. Indeed, sustainable urban design places particular emphasis on the maintenance and renovation of the existing city structure and fabric; (2) to protect natural resources, landscape, and wildlife. As such, any new construction materials should come from sustainable sources, such as wood and related materials from properly managed forests; (3) to reduce the consequences of energy in new urban development. In the next chapter, one of the popular models for sustainable urban development, the compact city model has been reviewed, by tracing its origins in the desert cities of the Middle East. The city of Yazd in Iran has been used to further explain the vernacular architecture and the traditional design details in line with local climate and construction materials availability in this model.

18.4 COMPACT CITY: A HARMONIC RELATIONSHIP BETWEEN THE CITY AND THE CRITICAL ZONE

Much of the planning literature from 1990 onwards focuses on the compact city model: a concept designed to implement sustainable development within the urban environment and to counteract the perceived negative social, economic, and environmental impacts of urban sprawl. This urban development model focuses on limiting the peripheral expansion of cities, and instead looks to direct development in the form of intensification, increasing the destinies of existing urban areas, and redeveloping underused or abandoned sites. The concepts of sustainable city, density, and compactness have a meaningful relationship, and the positive relationships between minimal population densities, physical compactness, reduction in energy consumption, and environmental sustainability are reflected in the compact city theory (Williams et al., 1996, 2000). Compact city model encourages design strategies such as infill development and higher-density mixed-use development.

A major concern plaguing urban planning and design today is the unbridled growth of cities. Throughout the twentieth century, cities have expanded in an unprecedented manner in all directions. The structure of urban development has also undergone substantial change, best reflected in the low-density sprawl in metropolitan areas. Affected by a combination of diverse factors such as sparse population, scattered workplaces, single-use zoning codes imposed by master plans, and the increased area allocated for motorized traffic and parking lots, urban districts have been exposed to mushrooming low-density expansions. By its very nature, density is both an urban and environmental issue. On the one hand, the cost of creating an urban infrastructure varies inversely with the density of development, but on the other hand, an excessively crowded urban population tends to increase the level of social stress. The compact city model reduces peripheral sprawl and contributes to the dynamism and liveliness of cities.

Table 18.1 compares the characteristics of compact city versus urban sprawl. Compact city is more energy efficient and less polluting because compact city dwellers can live closer to shops and work and can walk, bike, or take public transit. Low population density in a city is contradictory to the requirements of the "sustainable city". Low-density developments on the urban periphery have been heavily criticized by scholars from the viewpoint of social and environmental sustainability. Experts suggest that these type of fringe-settlements lead to an atomized way of life, void of a sense of belonging to the local society. According to researchers, low-density neighborhoods are socially neutral, and for the lack of socioeconomic diversity, life in these settlements can be monotonous and boring (Williams et al., 1996, 2000)

TABLE 18.1 Selective Characteristics of Compact City Versus Urban Sprawl

Compact city	Urban sprawl
• High residential densities • Mixture of land uses • Fine grain of land uses (proximity of varied uses and small relative size of land parcels) • Increased social and economic interaction • Contiguous development (some parcels or structure may be vacant or abounded or surface parking) • Multimodal transportation • High-degrees of accessibility • High-degrees of street connectivity: internal/external, including sidewalks and bicycle lanes • Low open-space ratio	• Low residential density • Unlimited outward extension of new development • Spatial segregation of different types of land uses through zoning • Leapfrog development • All transportation dominated by privately owned motor vehicles • Fragmentation of governance authority of land uses among many local governments • Great variance in the fiscal capacity of local governments • Widespread commercial strip development along major roadways • High indefensible places

Adapted from Ghobadian (1998), Neuman (2005), and Jungchan et al. (2015).

Many believe that the compact city promotes more community-based social patterns as well as social equity (Katz, 1994). Moreover, advocates of the compact city model voice their dissatisfaction with the ongoing sprawl of low-density cities, and with their implication for long-term environmental viability. Some of the unfavorable consequences of extended low-density settlements have been enumerated as follows (Unwin and Searle, 1991): (1) Wasting vast areas of quality agricultural lands for building housing and roads; (2) More surface runoff; (3) Increased consumption by open and single-household residential units, arising from their low rate of energy utility, as opposed to the more compact models of houses; and (4) Increased per capita consumption of gasoline leading to harmful emissions. Advocates of the concept of compact city argue that it imposes the least environmental burden by shortening the distance of personal trips, reducing land consumption, providing for public transport, and promoting the use of shared heating and cooling facilities.

The central part of Iran is covered by vast desert, with an altitude varying between 500 and 1500 m above sea level. The central desert is one of the most arid and hottest places on the planet. Despite this, through the ages, a considerable number of large and significant Iranian cities such as Sialk, Kerman, Yazd (see Fig. 18.1), and Kashan have developed and thrived on the fringes of this great desert (Beazley and Harverson, 1982). Human settlement of the desert could not have been made possible except through the intelligent exploitation of natural resources and through adjustments to the harsh climate of the region. Some of the problems that have induced local people in the hot and arid regions to seek inventive remedies are: intense sun and heat at daytime, wide temperature differences between day and night, extremely hot summers followed by rather cold winters, aridity of the air, scanty rainfall, shortage of water, hot dust, and sand storms.

FIGURE 18.1 Urban fabric of Yazd, Iran. *(Credit: Author.)*

Golany (1995) asserts the indigenous urban architecture in the Middle East has been centered on harmony and adaptation to the local environmental characteristics. The ultimate goal of urban design in hot and arid regions is to alleviate the impact of climatic stress on people who work, shop, exercise, or walk in open spaces. Furthermore urban design in these regions should seek to boost the capacity of buildings for accommodating interior spaces that provide climatic comfort while requiring minimal energy consumption. Four groups of stresses experienced by people in the open spaces of the desert may be classified as follows: (1) intense summer heat and solar radiation; (2) glaring light caused by direct sunshine; (3) desert storms, mostly in afternoons; and (4) cold winds in winter (Ghobadian, 1998).

Modern urban design knowledge suggests that in order to reduce heat stress, residential neighborhoods should be designed in a manner that shortens walking distances and protects playgrounds. Footpaths should, if possible, have plentiful shady spaces, provided by adjacent buildings or trees. To provide shade, especially where people and children gather during daytime, is of utmost significant. In order to adjust and control the urban microclimate, the details of buildings in hot and arid region need to be worked out in a way that provides shady footpaths and playgrounds as well as adequate ventilation, shelter against dust, and protection from glaring light. Examination of Iranian desert cities including Yazd, Kashan, and Kerman indicates that physical features of the traditional districts of these cities largely conform to modern design recommendations for the compact city. It appears that environmental adaptation in such desert communities is the result of a lengthy process of trial and error, which has spanned in view of the evolutionary course of designing and building urban fabrics.

Overall, in view of the common physical features of traditional Iranian desert cities, a model entitled the Iranian compact city can be identified. The proposed model is compatible with the principles contained in the compact city theory on the one hand, and conforms to the principles of sustainable urban design, on the other hand. As already noted, a sustainable urban design is one that can be responsive to three basic requirements: energy efficiency, protection of natural resources, and recycling of the existing building, fabrics, and infrastructure. In the Iranian compact city model, energy is saved through the application of various methods. These may be classified into saving energy through the city form, city structure, and passive systems. In desert cities, the city form, whether in terms of morphology or the composition and distribution of diverse functions, provides for reduced energy consumption. For example, in traditional neighborhoods, the center of the neighborhood embodies a combination of residential, commercial, health, and educational functions. As such, in contrast to new cities based on zoning regulations, the inhabitants are not obliged to travel long distances by car to access municipal services. Also, building forms in desert cities have been adapted to climatic characteristics, implying a reduction in energy consumption for heating and cooling architectural spaces.

To provide climatic comfort in the arid and hot regions, three methods have been applied in Iranian desert cities. These include the use of shade and wind, use of water, and minimizing the impact of solar radiation. Examinations of Iranian desert cities reveal that all three methods have been used. Reducing the exposure of building and urban fabrics to solar radiation is another way of providing climatic comfort. This can be achieved through the concept of compactness. The term compactness may be explained in a variety of ways; however, the main indicator of compactness is the ratio of the building surface exposed to natural impacts to its covered surface (the built-up area). In other words, a one-storied villa-type building erected in the middle of plot of land has minimal compactness. However, if another floor were added on the top, the structure would be more compact than before. This is because while the built-up space is doubled, the surface exposed to the sun or other natural agents is increased less than twofold. This principle has been duly observed in the physical organization of Iranian desert cities through the intertwined compactness of the residential units, so that in most cases dwellings adjoin other units on all four sides. Closely integrated and coherent roofs of urban blocks are common sights in all desert cities.

Traditional desert architecture uses passive systems to contribute to the efficient consumption of energy. Elements such as wind towers, water reservoirs, windmills, and watermills function at costs far lower that the modern-day energy consumption appliances such as air conditioners, water-coolers, etc.

In his studies on the technical and service-providing structures of Iranian desert architecture, Tavassoli (1983) asserts functional elements such as badgir (windcatcher) that can also serve as an architectural detail (see Fig. 18.2). Badgir is an architectural element that generates natural ventilation throughout the buildings. Figure 18.2 illustrates how the ghanat system – the underground tunnels that provide irrigation in Persian cities – is connected to windcatcher elements to create a cool breeze in buildings located in the desert fringe cities of central Iran, where there is large diurnal temperature variation with an arid climate.

Buildings and urban fabrics within traditional desert cities have developed in a manner as to impose the least damage on valuable natural resources, such as

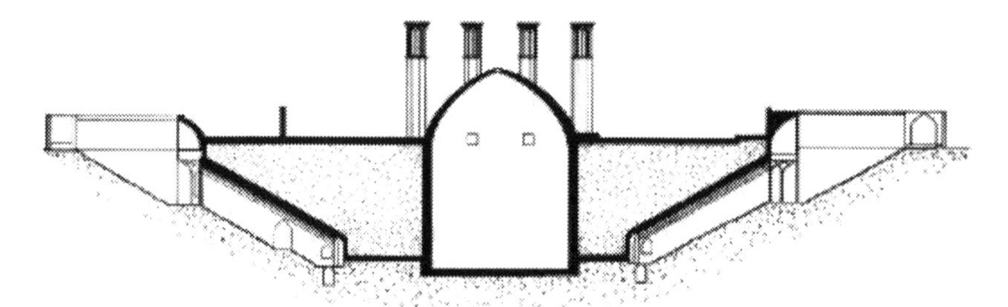

FIGURE 18.2 Sesh Badgireh Ghanat (water reservoir with six windcatchers), Yazd, Iran. *(Source: Ghobadian (1998, p. 43).)*

land, vegetation cover, and water. In contrast to modern cities, the compactness of the urban fabric in desert cities has enabled settlements to absorb large populations with relatively small surface areas. This has spared the need for encroachment on the periphery which contains quality agricultural land; building materials used are mostly adobe and brick, made of local clays and mixed with other items such as straw, which is a by-product of local farming. The thick layer provides high-insulation values.

These local-based materials are recommended as a basic component of traditional desert architecture.

Recycling buildings is another important feature of traditional Iranian architecture and urban design. It has been practiced throughout history, even at times of profound social and cultural transitions. For instance, after Islam came to Iran, indigenous religious buildings, such as fire-temples were not torn down, rather, they were maintained and modified for use as mosques. Even in recent history, we have witnessed numerous examples of how dwellings have been repeatedly renovated and handed down from one generation to another, sometimes allowing successive generations of a family to live in the same house. Indeed, one of the virtuous traditions of traditional architecture is that of preserving existing buildings to be used again and again. Such a culture is in stark contrast to the prevailing quasi-modern culture one that tears down even usable buildings to erect new ones in their stead, just to keep pace with changing trends and fashions in the popular culture.

The detailed review of vernacular architecture and organic urban fabric of the city of Yazd assure us that urban sustainability's goals can be achieved through the application of appropriate urban forms, appropriate types of construction and material, proper spatial distribution of land uses, and optimal densities. In other words, energy can be saved by creating closer links among diverse urban functions and limiting energy consumption. Within the dynamic context of sustainable urban development, new buildings need to be designed in a flexible way to allow for future changes to accommodate various functions over their life span. The transportation system selected for the new urban structure should be a type capable of creating a harmonious balance among the imperatives of economic development, environmental protection of the Critical Zone and the natural resources found in it, and quality of society's future life.

18.5 CONCLUSIONS

This chapter has examined the correlation between the Earth's Critical Zone and the built environment, particularly by focusing on the urbanization and the recent urban policies and design codes. A review of the hydraulic theory in urban studies has emphasized the vital role water (and other natural resources found in the Critical Zone) play in the fall or rise of cities throughout the urbanization history. Setting industrialization as shifting point in the history of cities, the impacts of urban form the preindustrial, industrial, to the postindustrial cities have been evaluated on the key elements of the Critical Zone including water, air, land (soil and rock), and the living organisms. These impacts, often negative, have warned architects, city planners, and urban policy makers about the limited natural resources and have questioned current, human-urban survival and development on the planet Earth. In response to these environmental crisis, new urban development models such as the smart growth, the green city, new urbanism and compact city model have been proposed and practiced since the 1960s. What these different models under the umbrella of green or sustainable label have is their commitment to ensure availability and sustainable management of the Earth's Critical Zone and its natural resources to the future generations as well as to establish sustainable consumption and production patterns. The compact city model was evaluated through an example of the city of Yazd in Iran. Traditional desert cities have been examined and it was found that certain traditions of constitution that had evolved over time in consonance with the environment. Through a combination of theoretical findings, principles of sustainable urban design, and the rich traditional architectural heritage, strategic guiding principles can be recommended. These principles will replace low-density sprawl with a more harmonious relationship between natural environment and human-built environment within the sustainable desert cities. The principles are: (1) Reconstruction, renovation, and regeneration of the existing fabrics in the central and old neighborhoods;(2) Improvement and restoration of the existing fabrics located in peripheral neighborhoods; (3) Grasping the opportunities which may arise for the renovation of old neighborhoods or new developments, as an occasion to apply better design standard within the context of sustainable environment planning. The future of human existence and progress on the planet Earth is undoubtedly, critically intertwined to the future of the Critical Zones and its life-sustaining resources. Sustainable urban growth is not a recommended approach, but rather it is the only practice to sustain life on the planet Earth. We must change our consumption and production patterns in order to solve the most urgent problems of global change and sustainability – including climate change, biodiversity loss, water and food security, energy consumption, and social and environmental injustice.

REFERENCES

Abbey, E., 1968. Desert Solitaire: A Season in the Wilderness. Touchstone, New York.

Beazley, E., Harverson, M., 1982. Living with Desert: Working Buildings of the Iranian Plateau. Aris & Philips, Warminster, Wiltshire, UK.

Burgess, E.W., 1925. The growth of the city. In: Park, R.E., Burgess, E.W., McKenzie, R. (Eds.), The Suggestions of Investigation of Human Behavior in Urban Environment. The University of Chicago Press, Chicago, IL, pp. 47–62.

Carmona, M., Heath, T., Oc, T., Tiesdell, S., 2010. Public Places, Urban Spaces: The Dimensions of Urban Design. Architectural Press, Oxford, second ed. (first ed.: 2003).

Charter of New Urbanism, 2000. http://www.cnu.org/charter (accessed March 21, 2015).

Dear, M., 2000. The Postmodern Urban Condition. Blackwell Publishers, Oxford.

Dear, M., 2002. Los Angeles and the Chicago school: invitation to a debate. City Commun. 1 (1), 5–13.
Dear, M., Flusty, S., 1998. Postmodern urbanism. Ann. Assoc. Am. Geogr. 88 (1), 50–72.
Dierwechter, Y., 2014. The spaces that smart growth makes: sustainability, segregation, and residential change across Greater Seattle. Urban Geogr. 35 (5), 691–714.
Duany, A., Plater-Zyberk, E., 2011. Garden Cities: Theory and Practice of Agrarian Urbanism. The Prince's Foundation for the Built Environment, London.
Duany, A, Plater-Zyberk, E., Speck, J., 2000. Suburban Nation. North Point Press, New York.
Duany, A., Lydon, M., Speck, J., 2009. The Smart Growth Manual. McGraw-Hill, New York.
Elikin, T., McLaren, M., Hillman, M., 1991. Reviving the City: Towards Sustainable Urban Development. Friends of the Earth with Policy Studies Institute, London.
Fishman, R., 2008. New Urbanism in the age of re-urbanization. In: Hass, T. (Ed.), New Urbanism and Beyond: Designing Cities for the Future. Rizzoli International, New York, pp. 296–298.
Fulton, W., Pendall, R., Nguyen, M., Harrison, A., 2002. Who Sprawls Most? How Growth Patterns Differ Across the U.S. Brookings, Washington, DC.
Ghobadian, V., 1998. Climatic Analysis of the Iranian Traditional Buildings. Tehran University Press, Tehran.
Girardet, H., 1992. Cities: New Directions for Sustainable Urban Living. Gaia Books, London.
Golany, G.S., 1995. Ethics and Urban Design. John & Sons, New York.
Haughton, G., Hunter, C., 1994. Sustainable Cities. J. Kingsley Publishers, London.
Jenks, M., Burgess, R. (Eds.), 2000. Compact Cities: Sustainable Urban Forms for Developing Countries. E & FN Spon, London.
Jungchan, L., Kurisu, K., An, K., Hanaki, K., 2015. Development of the compact city index and its application to Japanese cities. Urban Stud. 52, 1054–1070.
Katz, P., 1994. The New Urbanism: Toward an Architecture of Community. McGraw-Hill, New York.
McHarg, I.L., 1969. Design with Nature. Natural History Press, New York.
McRobie, G., 1981. Small is Possible. Harper and Row, New York.
Meadows, D.H., Meadows, D.L., Randers, J., Behrens, W., 1972. Limits to Growth. Universe Books, New York.
Moughtin, C., 1996. Urban Design: Green Dimensions. Butterworth Architecture, Oxford.
Neuman, M., 2005. The Compact City fallacy. J. Plan. Educ. Res. 25 (1), 11–26.
Newman, P., Kenworthy, J., 1999. Sustainability and Cities: Overcoming Automobile Dependence. Island Press, Washington, DC.
Newman, P., Kenworthy, J., 2000. The ten myths of automobile dependence. World Transport Policy Pract. 6 (1), 15–25.
Osborn, F.J., 1965. Garden Cities of Tomorrow. The MIT Press, Boston, MA.
Park, R., 1925. The City. The University of Chicago Press, Chicago, IL.
Park, R.E., Burgess, E.W., McKenzie, R.D., 1925. The City. The University of Chicago Press, Chicago, IL.
Quarrie, J., 1992. Earth Summit 92: The United Nations Conference on Environment and Development. Regency Press, London.
Sheller, M., Urry, J., 2000. The city and the car. Int. J. Urban Reg. Res. 24 (4), 737–757.
Soja, E., 1989. Postmodern Geographies: The Reassertion of Space in Critical Social Theory. Verso, New York.
Soja, E., 1992. Inside exopolis: scenes from Orange County. In: Sorkin, M. (Ed.), Variations on Theme Park: The New American City and the End of Public Space. Hills & Wang, New York.

Soja, E., 1996. Los Angles 1965–1992: the six geographies of urban restricting. In: Scott, A.J., Soja, E. (Eds.), The City: Los Angeles and Urban Theory at the End of the Twentieth Century. University of California Press, Los Angeles, CA, pp. 426–462.

Sorkin, M. (Ed.), 1992. Variations on a Theme Park: The New American City and the End of Public Space. Hills & Wang, New York.

Tavassoli, M., 1983. City Planning in the Hot and Dry Climate of Iran. Tehran University Publication, Tehran.

Tönnies, F., 1963. Community and Society. Harper & Row, New York, NY.

Unwin, N., Searle, G., 1991. Ecologically sustainable development and urban development. Urban Future 4, 1–12, Special Issue.

Urry, J., 2004. The 'System' of automobility. Theory Cult. Soc. 21 (4/5), 25–39.

Wackernagel, M., Rees, W., 1998. Our Ecological Footprint: Reducing Human Impact on the Earth. New Society Publishers, Gabriola Island, Canada.

White, R., Whitney, J., 1992. Cities and the environment: an overview. In: Stern, R., White, R., Whitney, J. (Eds.), Sustainable Cities. Westview Press, Oxford, pp. 8–51.

Williams, K., Burton, E., Jenks, M., 1996. Achieving the compact city through intensification: an acceptable option? In: Jenks, M., Burton, E., Williams, K. (Eds.), The Compact City: A Sustainable Urban Form?. E & F N Spon, London, pp. 83–96.

Williams, K., Burton, E., Jenks, M., 2000. Achieving Sustainable Urban Form. E & FN Spon, London.

Wittfogel, K., 1957. Oriental Despotism: A Comparative Study of Tribal Power. Yale University Press, New Haven, CT.

World Commission on Environment and Development, 1987. Our Common Future. Oxford University Press, Oxford.

Zeemerin, E., 2009. What does sustainability mean to city officials? Urban Aff. Rev. 45 (2), 247–273.

Chapter 19

Natural and Anthropogenic Factors Affecting Groundwater in the Critical Zone of the Texas Triangle Megaregion

Kevin R. Gamache* and John R. Giardino**

*Water Management and Hydrological Sciences Graduate Program and High Alpine and Arctic Research Program, and Office of the Vice President for Research, Texas A&M University, College Station, Texas, USA; **High Alpine and Arctic Research Program, Department of Geology and Geophysics, Texas A&M University; Water Management and Hydrological Sciences Graduate Program, Texas A&M University, College Station, Texas, USA

19.1 INTRODUCTION

Rapid changes occurring on Earth since the beginning of the Industrial Revolution are leading to a new geological epoch referred to as the Anthropocene (Amundson et al., 2007; Crutzen and Steffen, 2003). The Anthropocene (~250 years BP to present) encompasses some of the most pronounced changes in the history of Earth by any measurement: rates of erosion, deforestation, extinction, extent of climate change, and so forth (Amundson et al., 2007).

In less than three centuries, 1.86 million hectares (46 million acres) of the virgin landscape in the United States have been converted to urban uses; in the next 25 years that converted area will more than double to 45.32 million hectares (112 million acres; Carbonell and Yaro, 2005). During this time period, more than half of the land surface has been "plowed, pastured, fertilized, irrigated, drained, fumigated, bulldozed, compacted, eroded, reconstructed, manured, mined, logged, or converted to new uses" (Richter and Mobley, 2009, p. 1067). Activities like these have far-reaching impacts on life-sustaining processes of the near-surface environment, recently termed the "Critical Zone" (Richter and Mobley, 2009).

Numerous modifications to the definition of the Critical Zone have emerged since the original statement issued by the National Research Council (National Research and Council, 2001). The Critical Zone is the vertical and spatial zone of the surface and near surface systems that extends from the bottom of an

Developments in Earth Surface Processes, Vol. 19. http://dx.doi.org/10.1016/B978-0-444-63369-9.00019-7

aquifer to the atmosphere boundary layer (Anderson et al., 2010; National Research and Council, 2001). The NRC actually defined the upper extent of the Critical Zone as the top of the canopy. However, for discussion of water resources in the Critical Zone, we have redefined the upper layer to extend into the atmosphere, as it serves as a source and pathway for precipitation in the Critical Zone.

The Critical Zone plays an extremely important role in sustaining life on Earth because it lies at the interface of the lithosphere, atmosphere, and hydrosphere (Amundson et al., 2007) and encompasses soils and terrestrial ecosystems. Although not usually recognized in definitions of the Critical Zone (Anderson et al., 2010; National Research and Council, 2001), this zone also includes human systems (see Chapter 18). Thus, the Critical Zone is a complex mixture of air, water, biota, organic matter, earth materials, energy, human capital, and associated infrastructure and alterations, both natural and human (Brantley et al., 2007).

The Critical Zone has been defined as "the heterogeneous, near-surface environment in which complex interactions involving rock, soil, water, air, and living organisms regulate the natural habitat and determine the availability of life-sustaining resources" (Lin, 2010, p. 25). It has evolved as a dynamic and generally self-sustaining system (Amundson et al., 2007). This thin, fragile envelope that includes the land surface and its canopy of vegetation, rivers, lakes, and shallow seas (Wilding and Lin, 2006) is critical from a human perspective because it is the environment in which most people live and work (Graf, 2008).

Future global change has implications for the Critical Zone because of changes in phenomena such as rates of evapotranspiration, precipitation characteristics, plant distributions, and human responses (Goudie, 2006). Global climate models predict a warmer planet (Bradley et al., 2003). For Texas, this could mean changes to its climate – specifically temperature, evaporation, rainfall, and drought (Mace and Wade, 2008). At the same time, rapidly growing demand for water in urban areas is already straining local and regional water supplies. Concerns about the scarcity of urban water in the United States are becoming more prominent (Levin et al., 2002; Padowski and Jawitz, 2012). Water shortages in Atlanta, Georgia, in 2008, and San Francisco, California, in 2006–2007 (Dorfman, 2011, p. 1470; Padowski and Jawitz, 2012) are illustrative of the potential impacts of climate change on population growth, environmental regulation, and water supplies.

19.2 GOAL OF THE CHAPTER

The goal of this chapter is to focus on water resources as an important component of the Critical Zone. To illustrate the important roles that the environment and humans play in using and managing water resources, we focus on an aquifer that is currently experiencing rapid depletion, as a result of rapid human population growth. Our focus is on the Trinity, Carrizo–Wilcox, Gulf Coast, and Edwards aquifers, which are located in one of the fastest growing areas of Texas. Thus, our goal is to examine the potential impact of anthropogenic changes on the Critical Zone in this area. Our point is illustrated by focusing in one area of Texas that is experiencing exceptionally dynamic alterations to the Critical Zone.

19.3 IMPACT OF POPULATION GROWTH

In the United States, more than 80% of the population now lives in urban areas, compared to 64% in 1950 (Padowski and Jawitz, 2012). Further, population in the United States will likely increase by 40% by 2050 with the growth concentrated in 8–10 megaregions (Dewar and Epstein, 2007). A *megaregion* consists of two or more metropolitan areas linked with interdependent environmental systems, a multimodal transportation infrastructure, and complementary economies (Butler et al., 2009; Zhang et al., 2007). Ensuring that cities have an adequate supply of water will become increasingly important as human populations continue to concentrate in these highly urbanized megaregions. Thus, interaction between humans and the other natural systems in the Critical Zone will become more complex.

As populations continue to increase during a period of rapid global change, far-reaching impacts will occur within the Critical Zone. It will be increasingly important to understand these changes to the Critical Zone to mitigate them effectively. The highly urbanized megaregions that have been identified in Texas are the metropolitan areas of Dallas/Fort Worth, Austin, San Antonio, and Houston.

19.4 DESCRIPTION OF THE STUDY AREA

The Texas-Triangle Megaregion (TTMR) is one of the emerging megaregions initially identified by the University of Pennsylvania with the Regional Plan Association and the Lincoln Institute (Zhang et al., 2007). The region is spatially delineated by the metropolitan areas of Dallas/Fort Worth, Austin, San Antonio, and Houston (Fig. 19.1), with a total land size of approximately 155,000 km^2 (59,900 mi^2) encompassing 65 of the 254 counties (Fig. 19.2) in the state (Butler et al., 2009, p. 1439; Neuman et al., 2010; Zhang et al., 2007). The metro areas of Dallas/Fort Worth, Houston, and San Antonio form the vertices of the TTMR, which measure 701, 531, and 624 km (436, 319, and 388 mi), respectively (Butler et al., 2009).

The Texas Triangle is a singular, new, complex, and important urban phenomenon (Neuman et al., 2010). One of the most dynamic urban regions in the nation, with a present population of more than 17 million, the TTMR represents a new urban phenomenon: a "triangular megalopolis whose development is not linear and contiguous, like prior megalopolises" (Neuman and Bright, 2008, p. 9). This region has been characterized as the "core area of Texas," a single megacity forming the nucleus of Texas and rivaling New York and Los Angeles (Neuman and Bright, 2008, p. 9).

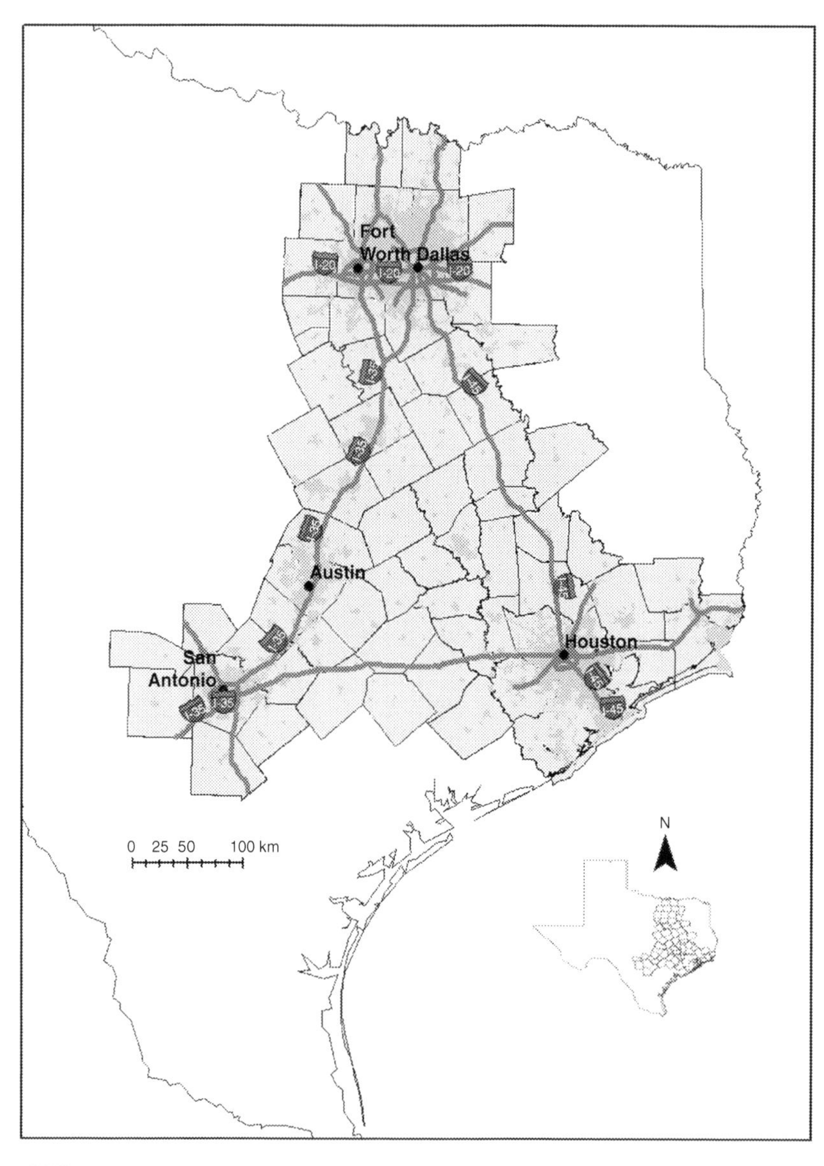

FIGURE 19.1 The TTMR.

The Triangle includes 70% of the population of the state, 80% of the employment, and 85% of the wages (Neuman et al., 2010). Based on the 2010 US Census Bureau TIGER Data, the region contains 109 urbanized area clusters and 17 urbanized areas totaling 16,312 km^2. The region is emerging as a new urban megaregion in its own right (Neuman et al., 2010).

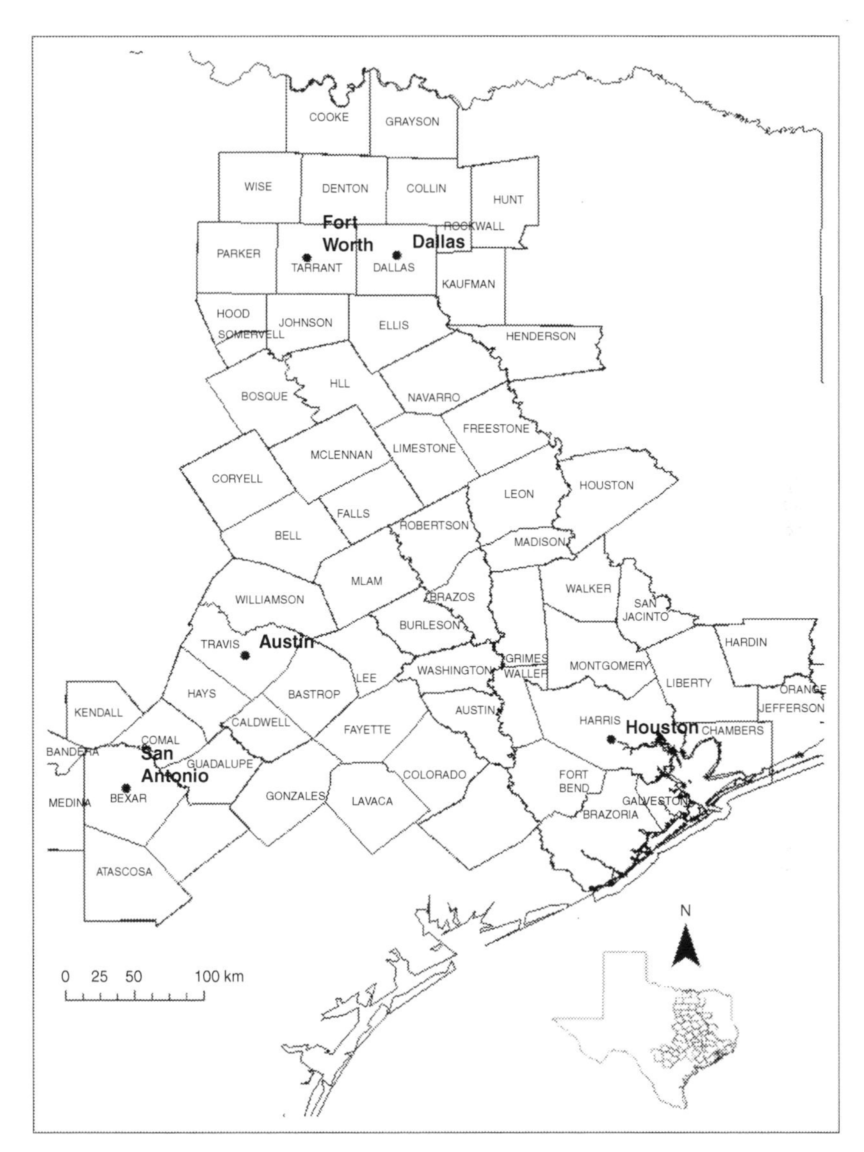

FIGURE 19.2 Counties in the TTMR.

19.5 PHYSICAL DIVISIONS AND ECOREGIONS

The TTMR contains a diverse landscape including portions of four of Bailey's Ecoregions (Bailey and Bailey, 1981; Bailey et al., 1976, 1978): from west to east, the Southwest Plateau and Plains Dry Steppe and Shrub Province, the Prairie Parkland (Subtropical) Province, the Southeastern Mixed Forest Province, and the Outer Coastal Plain Mixed Forest Province (Fig. 19.3).

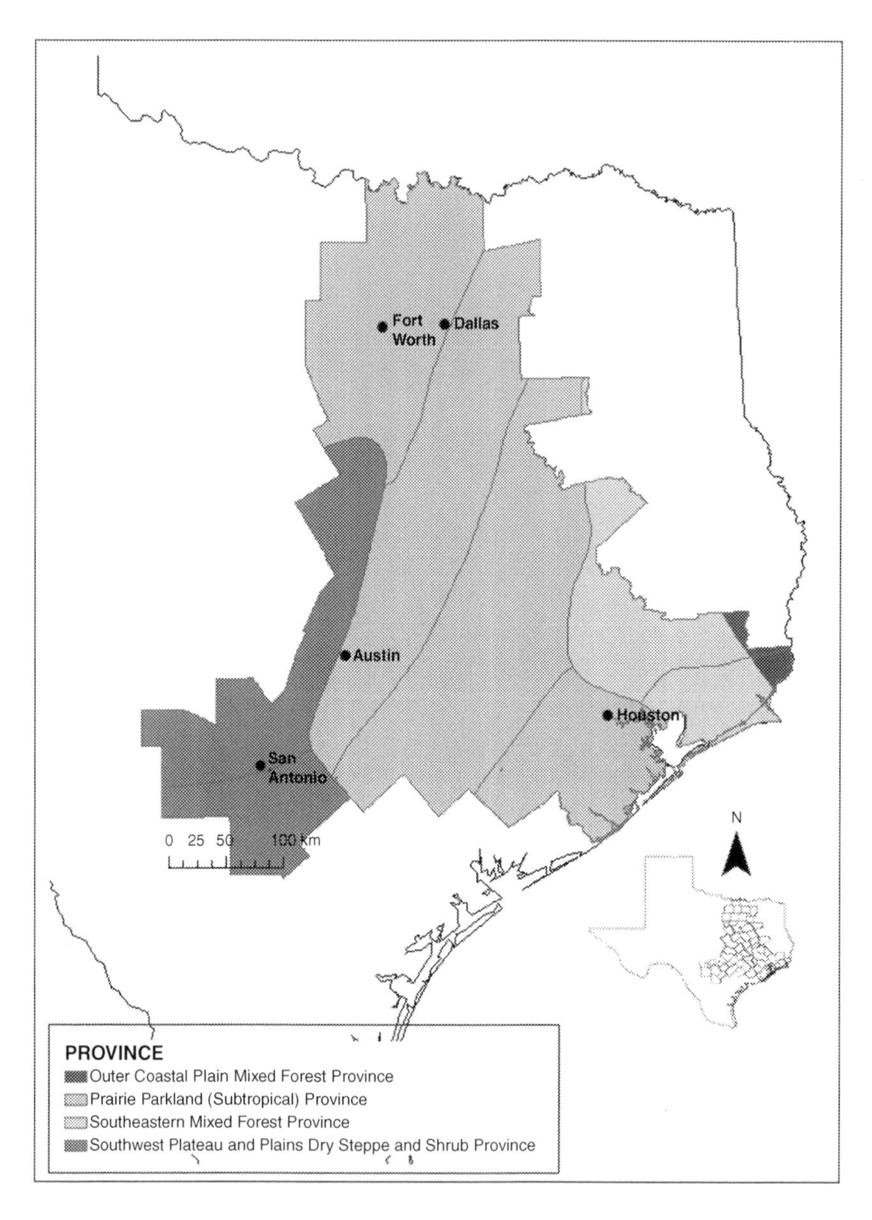

FIGURE 19.3 Bailey's Ecoregion Provinces in the TTMR.

19.6 SOUTHWEST PLATEAU AND PLAINS DRY STEPPE AND SHRUB PROVINCE

This region of flat to rolling plains and plateaus is occasionally dissected by canyons at the western end of the Gulf Coastal Plain and the southern end of the Great Plains (Bailey and Bailey, 1981; Bailey et al., 1976, 1978). The Balcones

FIGURE 19.4 **Edwards Plateau near Enchanted Rock, north of Austin, Texas.** *(Photo by the author.)*

Fault Zone and Escarpment sharply delineates the Southwest Plateau ecoregion from the prairie lands to the east. This area is characterized by hilly limestone terrain dissected by spring-fed streams of tremendous ecological, recreational, and aesthetic importance (Butler et al., 2009; Zhang et al., 2007). Within the TTMR, elevations range from sea level to 1100 m (3600 ft.) on the Edwards Plateau (Fig. 19.4).

The native vegetative cover is diverse and largely evergreen, dominated by juniper and live oak (Butler et al., 2009; Zhang et al., 2007). Live Oak-Ashe Juniper-Mesquite parks are the predominant (27%) vegetation cover in this province. Oak-Mesquite-Juniper park/woods occupy about 13% of the province within the megaregion. Thicker stands of Live Oak-Ash Juniper woods occupy approximately 12% of the area. Lesser coverage of Post-Oak Woods, Forest and Grassland Mosaic (7%), Bluestem Grassland (6%), Mesquite-Blackbush Brush (5%), Mesquite-Granero Woods (5%), and Mesquite-Live Oak-Bluewood Parks (3%) are also present. Cropland occupies about 21% of the area and surface water accounts for only 1% of the surface area in this province. The landcover of Southwest Plateau and Plains Dry Steppe and Shrub Province is summarized in Table 19.1.

19.6.1 Prairie Parkland (Subtropical) Province

The Prairie Parkland Province is the predominate ecoregion within the TTMR, occupying approximately 69% (104,152 km^2) of the total area. An extensive

TABLE 19.1 Southwest Plateau and Plains Dry Steppe and Shrub Province Vegetation Cover

Cover	Area (km^2)
Bluestem Grassland	1,260
Silver Bluestem-Texas Wintergrass Grassland	6,439
Oak-Mesquite-Juniper Parks/Woods	2,453
Live Oak-Mesquite-Ashe-Juniper Parks	288
Live Oak-Ashe Juniper Parks	318
Post Oak Parks/Woods Forest and Grassland Mosaic	33,011
Willow Oak-Water Oak-Blackgum Forest	60
Elm-Hackberry Parks/Woods	2,734
Water Oak-Elm-Hackberry Forest	1,246
Cottonwood-Hackberry-Saltcedar Brush/Woods	118
Pecan Elm	1,764
Young Forest/Grassland	97
Pine Hardwood	2,652
Marsh Barrier Island	552
Crops	11,187
Urban	3,852
Surface water	1,728
Total area	69,759

border of marshes stretches inland 8–16 km (5–10 mi), sometimes farther, from the Gulf Coast of Mexico (Bailey and Bailey, 1981; Bailey et al., 1976, 1978) in this province, encompassing about 552 km^2.

Vegetation in the Prairie Parkland Province is comprised mainly of a Post-Oak Parks/Woods Forest (Fig. 19.5) and Grassland Mosaic (47%). Silver Bluestem-Texas Wintergrass Grassland makes up approximately 11% of the province within the TTMR. Oak-Mesquite-Juniper Parks/Woods and Water Oak-Elm-Hackberry Forests/Woods make up lesser percentages of the vegetation cover in this area. Surface water occupies approximately 2% of the area.

The urban areas of Austin, Fort Worth, Dallas, and Houston are contained within this province of the TTMR and account for approximately 6% of the total land area of the province. The metropolitan areas of Dallas and Austin are located in and along the interface between the Blackland Prairie and Edwards Plateau (Butler et al., 2009, p. 1439; Zhang et al., 2007). These ecoregions are generally perpendicular to the Gulf Coast margin and to the major watersheds and river corridors, as they extend to the coast (Zhang et al., 2007). The Houston

FIGURE 19.5 Example of Post Oak Park/Woods. *(Photo courtesy of Tarleton State University.)*

metropolitan area and its associated communities, closer to the Gulf Coast, are located on terrain that is very flat and predominately covered in grassland, with forest or savannah-type vegetation in areas further inland (Zhang et al., 2007).

The Prairie Parkland ecoregion is highly fertile and agriculturally productive, comprised of fine-textured clay soils and only small remnants of a formerly extensive natural prairie (Butler et al., 2009, p. 1439; Zhang et al., 2007). A considerable portion of agricultural land (about 16%) still exists, although urban and industrial growth and development is a persistent challenge to the preservation of the intrinsic resources in the region (Zhang et al., 2007). The landcover of the Prairie-Parkland (Subtropical) Province is summarized in Table 19.2.

19.6.2 Southeastern Mixed Forest Province

The Southeastern Mixed Forest Province occupies 12% or 17,730 km^2 (6,846 mi^2) of the TTMR. Local relief is 30–180 m (100–600 ft.) on the Gulf Coastal Plains. The flat Coastal Plains have gentle slopes and local relief of less than 30 m (100 ft.). Most of the numerous streams move slowly; marshes, lakes, and swamps are numerous (Fig. 19.6).

Approximately half (48%) of the vegetation in this province is Pine Hardwood (Fig. 19.7). Lesser amounts (~11% total) of Young Forest/Grassland, Willow Oak-Water Oak-Blackgum Forest and Bald Cypress-Water Tupelo Swamp make up the remainder of the native vegetation in this province in the TTMR. Marsh Barrier Islands make up nearly 7% of the area. Cropland occupies about 20% of the land-surface here. Urban areas comprise only 2% of the land area. Surface water accounts for ~2.5% of the area. The Southeastern Mixed Forest Province vegetation cover is summarized in Table 19.3.

TABLE 19.2 Prairie Parkland (Subtropical) Province Vegetation Cover

Cover	Area (km^2)
Bluestem Grassland	1,645
Mesquite-Blackbrush Brush	1,184
Mesquite-Granjeno Woods	1,371
Mesquite-Live Oak-Bluewood Parks	767
Oak-Mesquite-Juniper Park/Woods	3,284
Live-Oak-Ashe Juniper Parks	7,155
Live-Oak-Mesquite-Ashe Juniper Parks	3,255
Post-Oak Woods, Forest and Grassland Mosaic	1,927
Pecan Elm	46
Crops	5,529
Other	17
Urban	73
Surface water	189
Total area	26,442

FIGURE 19.6 Mixed hardwood forest and swamp typical of the southwestern Mixed Forest Province. *(Photo courtesy of the US National Parks Service.)*

FIGURE 19.7 Pine hardwood forest. *(Photo courtesy of Texas Parks and Wildlife Department.)*

TABLE 19.3 Southeastern Mixed Forest Province Vegetation Cover

Cover	Area (km^2)
Bluestem Grassland	192
Post-Oak Woods, Forest, Grassland Mosaic	63
Willow Oak-Water Oak-Blackgum Forest	807
Bald Cypress-Water Tupelo Swamp	97
Young Forest/Grassland	783
Pine Hardwood	8,529
Marsh Barrier Island	1,157
Crops	3,532
Other	1,811
Urban	296
Surface water	441
Total area	17,708

19.6.3 Outer Coastal Plain Mixed Forest Province

Representing only 12% (1259 km^2) of the TTMR, the Outer Coastal Plain Mixed Forest Province is restricted to the flat and irregular southern Gulf Coastal Plains and is located in the far southeastern corner of the TTMR. The area is gently sloping, with relief typically less than 90 m (300 ft.; Bailey et al., 1980). Most of its numerous streams are sluggish; marshes, swamps, and lakes are numerous (Bailey and Bailey, 1981; Bailey et al., 1976, 1978).

Soils in this region tend to be wet, acidic, and low in major plant nutrients, having been derived mainly from Coastal Plain sediments, ranging from heavy clay to gravel (Bailey and Bailey, 1981; Bailey et al., 1976, 1978). Soil is comprised predominantly of sandy materials, with silty soils occurring mainly on expansive, level areas (Bailey and Bailey, 1981; Bailey et al., 1976, 1978). Vegetation is predominately pine hardwood (51%), with the other native vegetation comprised mainly of Willow Oak-Water, Oak-Blackgum Forest, Bald Cypress-Water, Tupelo Swamp, and Young Forest Grassland (17% combined). Approximately 15% of the province is cropland. Marsh Barrier Islands make up approximately 13% of the area. Urban areas occupy 4% of the area. Surface water, as a percentage of total land cover, is negligible in this province. The landcover of the Outer Coastal Plain Mixed Forest Province is summarized in Table 19.4.

19.7 AQUIFER STRUCTURE AND STRATIGRAPHY

According to the 2012 Texas State Water Plan, published by the TWDB, groundwater represents 60% of the total water used statewide (Vaughn et al., 2012). The amount of water used in the TTMR roughly mirrors the statewide water use data. In the TTMR, groundwater is supplied by numerous aquifers

TABLE 19.4 Outer Coastal Plain Mixed Forest Province Vegetation Cover

Cover	Area (km^2)
Willow Oak-Water Oak-Blackgum Forest	83
Bald Cypress-Water Tupelo Swamp	25
Young Forest/Grassland	96
Pine Hardwood	636
Marsh Barrier Island	163
Crops	189
Urban	57
Surface water	2
Total area	1251

capable of providing groundwater in quantities sufficient to support household, industrial, municipal, and irrigation needs throughout the region (Kelley et al., 2009; Vaughn et al., 2012, p.1361)

Thirty aquifers have been identified in Texas (Fig. 19.8), 21 of which are defined as "minor" and 9 of which are defined as "major," based on production (Brown and Farrar, 2008). Portions of four major aquifers underlie the TTMR: Trinity, Carrizo–Wilcox, Gulf Coast, and Edwards.

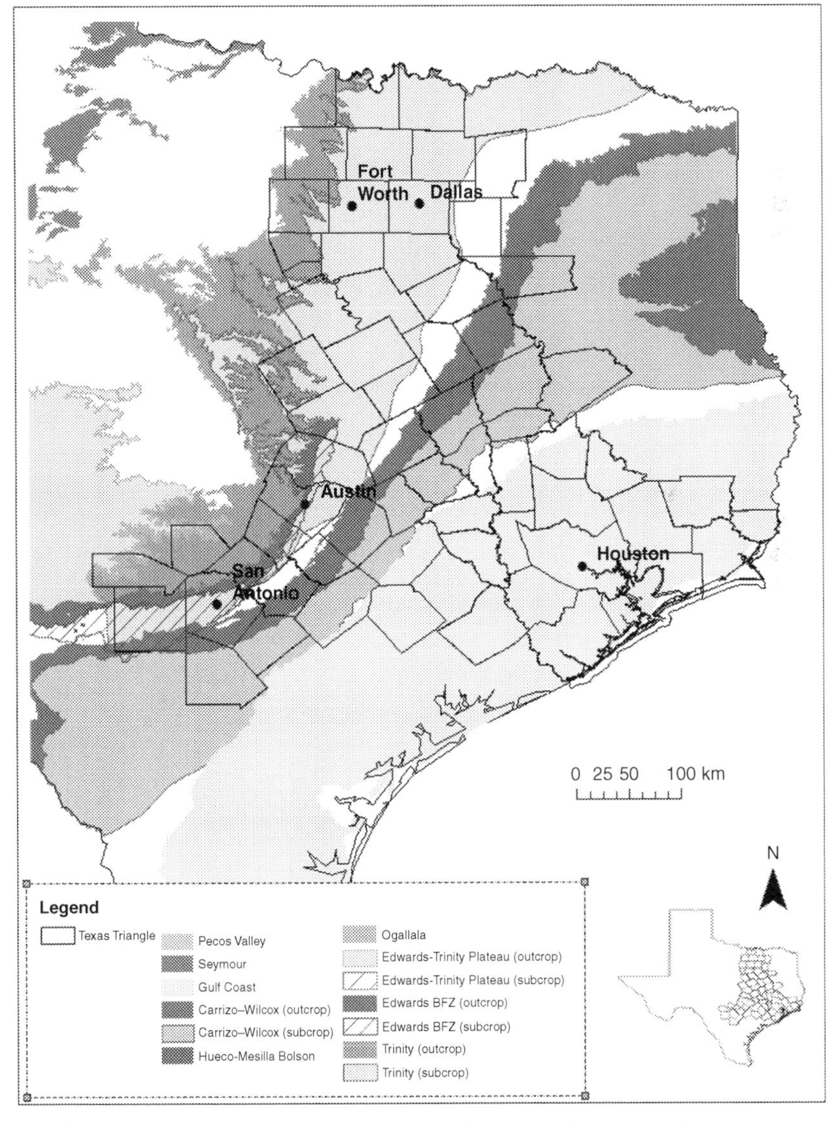

FIGURE 19.8 Major aquifers of Texas. *(Source: Texas Water Development Board.)*

The characteristic lithology of the relatively flat-lying karstic Edwards Aquifer and the Trinity Aquifer north of Austin are primarily massive limestones, sands, clays, gravels, and conglomerates. They are also unconfined aquifers (George et al., 2011). The Carrizo–Wilcox and Gulf Coast aquifers are principally confined clastic aquifers (Mace and Wade, 2008; Pearson and White, 1967). The age of groundwater in downdip areas of the Carrizo–Wilcox Aquifer can be more than 30,000 years (Mace and Wade, 2008; Pearson and White, 1967).

19.7.1 Carrizo–Wilcox Aquifer

The Carrizo–Wilcox Aquifer ranks third in the state for water use of 555 million m^3 (450,000 acre ft.) per year in 2003 behind the Gulf Coast Aquifer and the Ogallala Aquifer (Kelley et al., 2009). The Carrizo–Wilcox has been identified as a potential groundwater source to serve growing demands along the IH 35 corridor. The aquifer extends from the Rio Grande in south Texas northeastward into Arkansas and Louisiana, generally parallel to and east of IH 35.

The Carrizo–Wilcox is a hydrologically-connected system consisting of the Wilcox Group and the overlying Carrizo Formation of the Claiborne Group of fluvio-deltaic origin (Ashworth and Hopkins, 1995). The Wilcox Group contains a complex distribution of shale and sand facies that were deposited by ancient river systems (Fisher and McGowen, 1969; Thorkildsen et al., 1989). Because the sands of the Wilcox Group are locally hydraulically connected with the Carrizo Sand, both aquifers are jointly referred to as the Carrizo–Wilcox Aquifer (Green et al., 2011; Klemt et al., 1976).

Underlying the Carrizo–Wilcox Aquifer, the Paleocene Midway Formation acts as a regional confining unit (Kelley et al., 2009). Deposits of the Claiborne Group overlying the Carrizo–Wilcox Group include the fluvio-deltaic Queen City and Sparta formations separated from the Carrizo–Wilcox Aquifer by the Reklaw Formation, a marine shale unit (Kelley et al., 2009), as shown in Fig. 19.9.

The Carrizo–Wilcox Aquifer is predominantly composed of sand locally interbedded with gravels, silts, clays, and lignite deposited during the Tertiary. South of the Trinity River and north of the Colorado River, the Wilcox Group is divided into three distinct formations: Hooper, Simsboro, and Calvert Bluff. Of the three formations, the Simsboro typically contains the most massive water-bearing sands. Aquifer thickness in the downdip portion ranges from less than 61 m (200 ft.) to more than 914 m (3000 ft.) (Ashworth and Hopkins, 1995). Although the Carrizo–Wilcox Aquifer extends 914 m (3000 ft.) in thickness, the freshwater saturated-thickness of the sands averages 204 m (670 ft.; George et al., 2011).

The Carrizo Sand and Wilcox Group outcrop along a narrow band that parallels the Gulf Coast and dips beneath the land surface toward the coast. The Carrizo–Wilcox Aquifer has three principal sources of recharge: (1) subsurface

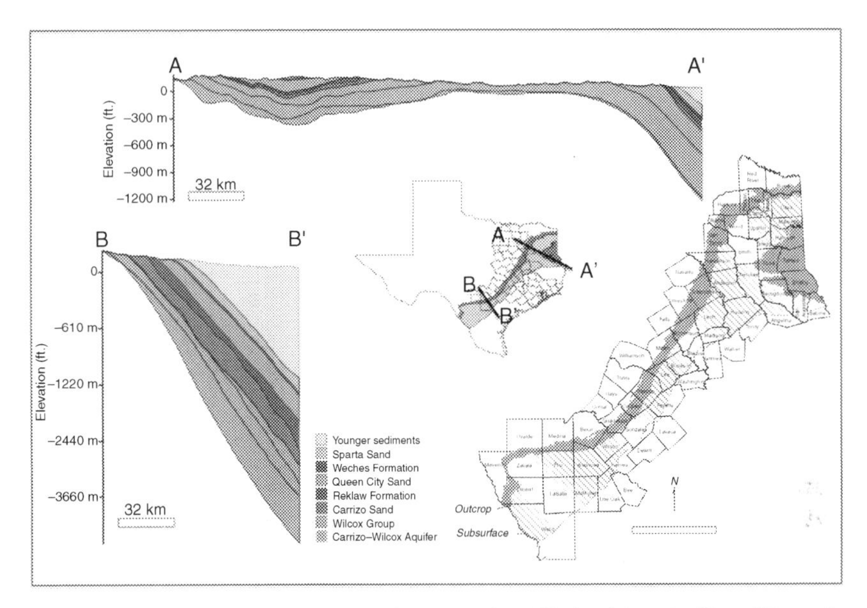

FIGURE 19.9 Carrizo–Wilcox structural cross section. *(Modified from* Aquifers of Texas, *by George et al. (2011). Because the horizontal scale is so much greater than that of the vertical dimension, the considerable vertical exaggeration steepens the dips in the cross sections at far greater angles than in reality.)*

interformational flow from other stratigraphic units; (2) distributed recharge from precipitation over the Carrizo–Wilcox Aquifer recharge zone; and (3) focused recharge in stream and river channels (Green et al., 2011). Irrigation pumpage during the recent drought has increased substantially in the Wintergarden area of southwest Texas, as has pumping of groundwater to support oil and gas exploration and production activities associated with the Eagle Ford Shale (Neffendorf and Hopkins, 2013). Water-level changes in the 11 Carrizo–Wilcox Aquifer recorder wells, managed by the TWDB, ranged from +2.62 m (8.6 ft.) in the Bastrop County well to −22 m (−72.2 ft.) in the LaSalle County well during the period 2011–2012 (Neffendorf and Hopkins, 2013). The median water-level change was −0.27 m (−0.9 ft.) and the average change was −2.74 m (−9.0 ft.; (Neffendorf and Hopkins, 2013, p.1351).

19.7.2 Gulf Coast Aquifer

The Gulf Coast Aquifer forms a wide belt paralleling the Gulf of Mexico from the Louisiana border to the border of Mexico (George et al., 2011). In Texas, the aquifer provides water to all or parts of 54 counties and extends from the Rio Grande northeastward to the Louisiana–Texas border. Municipal and irrigation uses account for 90% of the total pumpage from the aquifer. The greater Houston metropolitan area is the largest municipal user, where well yields average about 6056 L (1600 gal) per minute (Ashworth and Hopkins, 1995).

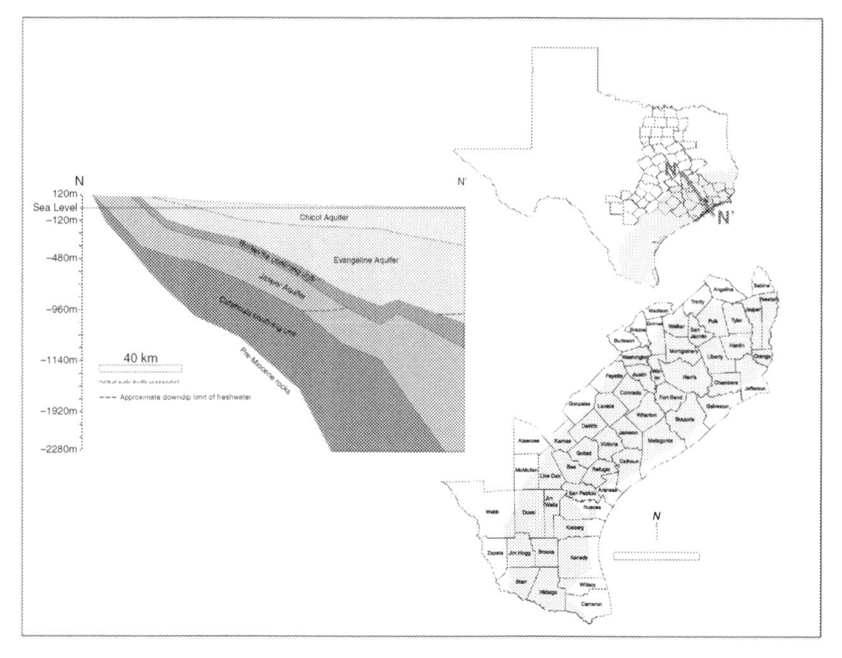

FIGURE 19.10 Gulf Coast Aquifer location and cross structure.

The Gulf Coast Aquifer (Fig. 19.10) consists of several aquifers, including the Chicot, Evangeline, and Jasper Aquifers (George et al., 2011). The aquifer consists of a complex of interbedded clays, silts, sands, and gravels of Cenozoic age, which are hydrologically connected to form a large, leaky artesian aquifer system (Ashworth and Hopkins, 1995). This system comprises four major components, consisting of the following generally recognized water-producing formations. The deepest is the Catahoula, which contains groundwater near the outcrop in relatively restricted sand layers. Above the Catahoula is the Jasper Aquifer, primarily contained within the Oakville Sandstone (Ashworth and Hopkins, 1995). The Burkeville confining layer separates the Jasper from the overlying Evangeline Aquifer, which is contained within the Fleming and Goliad sands. The Chicot Aquifer, or upper component of the Gulf Coast Aquifer System, consists of the Lissie, Willis, Bentley, Montgomery, and Beaumont formations, and overlying alluvial deposits. Not all formations are present throughout the system, and nomenclature commonly differs from one end of the system to the other (Ashworth and Hopkins, 1995). Maximum total sand thickness ranges from 213 m (700 ft.) in the south to 396 m (1300 ft.) in the northern extent (Ashworth and Hopkins, 1995; George et al., 2011). Freshwater saturated thickness averages about 304 m (1000 ft.; (George et al., 2011, p. 1335).

The aquifer is used for municipal, industrial, and irrigation purposes (George et al., 2011). Years of heavy pumpage for municipal and manufacturing use in portions of the aquifer have resulted in areas of significant water-level decline (Ashworth and Hopkins, 1995).

Declines of 61 m (200 ft.) to 91 m (300 ft.) have been measured in some areas of eastern and southeastern Harris and northern Galveston counties (Ashworth and Hopkins, 1995). In Harris, Galveston, Fort Bend, Jasper, and Wharton counties, water level declines of as much as 107 m (350 ft.) have led to land subsidence (George et al., 2011). From 2011 to 2012, water level changes in the 11 Gulf Coast Aquifer recorder wells ranged from 6 m (+19.7 ft.) in the Karnes County well to −5.1 m (16.7 ft.) in the northernmost Wharton County well, with a median change of 15.24 cm (0.5 ft.) and an average change of 27.4 cm (0.9 ft.; (Neffendorf and Hopkins, 2013, p. 1351).

19.7.3 Edwards Aquifer

The Edwards (Balcones Fault Zone, or BFZ) Aquifer covers approximately 11,266 km^2 (4350 mi^2) in parts of 11 counties (Ashworth and Hopkins, 1995). The aquifer forms a narrow belt extending from a groundwater divide in Kinney County through the San Antonio area northeastward to the Leon River in Bell County.

A poorly defined groundwater divide near Kyle in Hays County hydrologically separates the aquifer into the San Antonio and Austin regions. The name Edwards (BFZ) distinguishes this aquifer from the Edwards-Trinity (Plateau) and the Edwards-Trinity (High Plains) aquifers (Ashworth and Hopkins, 1995).

Consisting of partially dissolved limestone formed during early Cretaceous, the highly permeable aquifer exists under water-table conditions in the outcrop and under artesian conditions where it is confined below the overlying Del Rio Clay (Ashworth and Hopkins, 1995; George et al., 2011). The Edwards Aquifer consists of the Georgetown Limestone, formations of the Edwards Group (the primary water-bearing unit) and the equivalents, and the Comanche Peak Limestone where it exists (Ashworth and Hopkins, 1995). Thickness ranges from 61 m (200 ft.) to 183 m (600 ft.; (Ashworth and Hopkins, 1995, p. 1352). The extent of the Edwards Aquifer is delineated in Fig. 19.11.

The Edwards Aquifer responds rapidly to rainfall events and periods of drought (Mace and Wade, 2008; Pearson and White, 1967). Recharge to the aquifer occurs primarily by the downward percolation of surface water from streams draining off the Edwards Plateau to the north and west and by direct infiltration of precipitation on the outcrop (Ashworth and Hopkins, 1995). This recharge reaches the aquifer through fractures, faults, and sinkholes in the unsaturated zone. Unknown amounts of groundwater enter the aquifer as lateral underflow from the Glen Rose Formation (Ashworth and Hopkins, 1995).

Water from the aquifer is used primarily for municipal, irrigation, and recreational purposes. San Antonio obtains almost all of its water supply from the Edwards (BFZ) Aquifer (George et al., 2011). Water is also discharged artificially from hundreds of pumping wells, particularly municipal-supply wells in the San

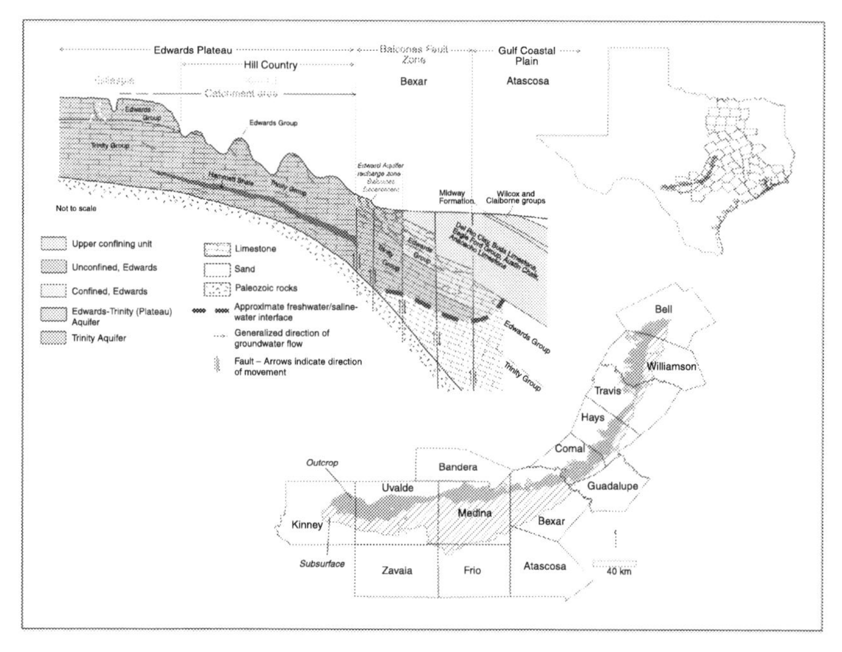

FIGURE 19.11 The Edwards Aquifer (Balcones Fault Zone). *(Modified from* Aquifers of Texas, *by George et al. (2011). Dips in the bedrock and topographic variation are severely distorted because of contrasts between horizontal and vertical scales.)*

Antonio region and irrigation wells in the western extent. In the four Edwards Aquifer (BFZ) recorder wells monitored by the TWDB, changes from 2011 to 2012 ranged from +0.76 m to −3.26 m (+2.5 ft. to −10.7 ft.) with a median change of +27.4 cm (+0.9 ft.) and an average change of −1.0 m (−3.3 ft.). From 2010 to 2011, changes ranged from + 3.04 m to −6.52 m (+10.4 to −21.4 ft.) with a median change of −1.07 m (−3.5 ft.) and an average change of −1.37 m (−4.5 ft.; (Neffendorf and Hopkins, 2013, p. 1351).

19.7.4 Trinity Aquifer

Extending across much of the western portion of the TTMR, the Trinity Aquifer consists of early Cretaceous formations of the Trinity Group where they occur in a band extending through the central part of the state in all or parts of 55 counties, from the Red River in north Texas to the Hill Country of south central Texas (Ashworth and Hopkins, 1995). The extent of this aquifer is delineated in Fig. 19.12. Trinity Group deposits also occur in the Panhandle and Edwards Plateau regions where they are included as part of the Edwards-Trinity (High Plains and Plateau) Aquifers (Ashworth and Hopkins, 1995). This aquifer is one of the most extensive and highly used groundwater resources in Texas (George et al., 2011).

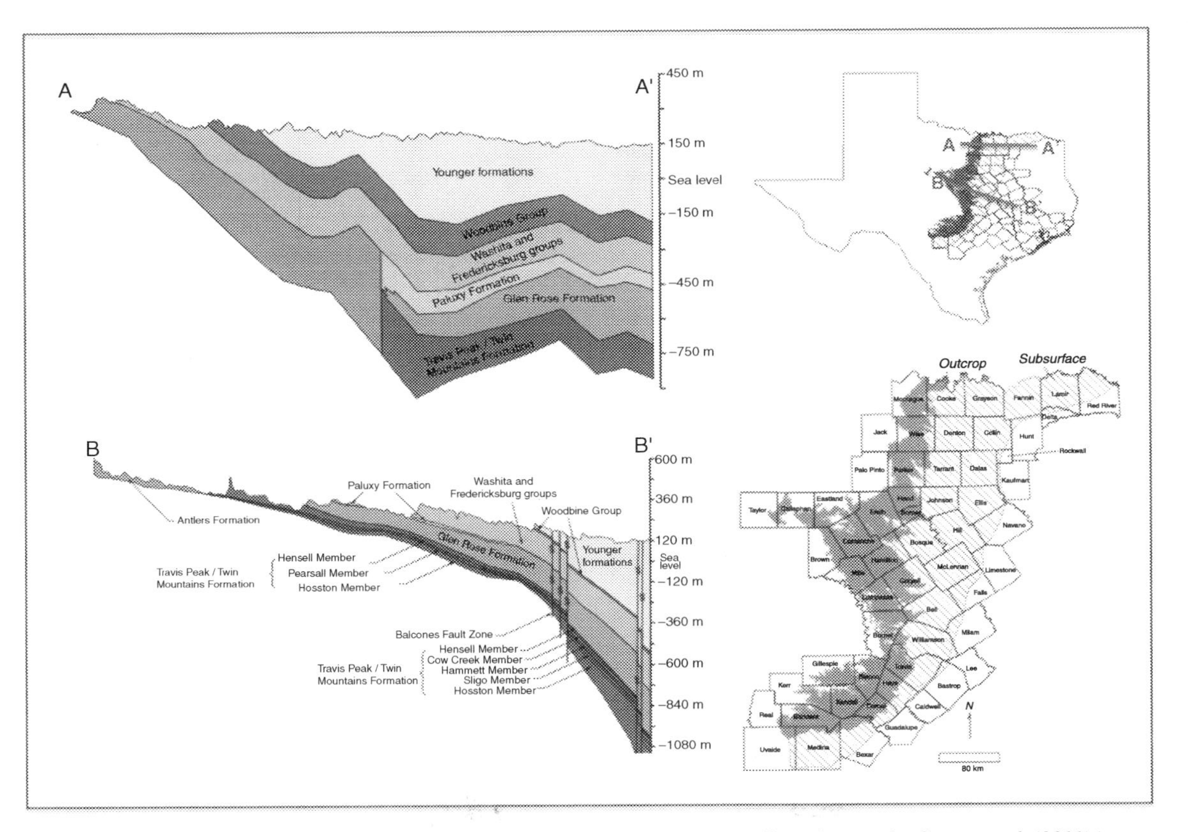

FIGURE 19.12 Trinity Aquifer location and cross structure. *(Adapted from* Aquifers of Texas, *by George et al. (2011).)*

The Trinity Group is comprised of (from youngest to oldest) the Antlers, Glen Rose, Paluxy, Twin Mountains, Travis Peak, Hensell, and Hosston Aquifers (George et al., 2011). Updip, where the Glen Rose thins or is missing, the Paluxy and Twin Mountains coalesce to form the Antlers Formation (Ashworth and Hopkins, 1995). Forming the upper unit of the Trinity Group, the Paluxy Formation consists of up to 122 m (400 ft.) of predominantly fine- to coarse-grained sands interbedded with clays and shales (George et al., 2011). These aquifers consist of limestones, sands, clays, gravels, and conglomerates with a combined freshwater saturated thickness averaging ~183 m (600 ft.) in north Texas and about 579 m (1900 ft.) in central Texas (George et al., 2011). The Antlers consists of up to 274 m (900 ft.) of sands and gravels, with clay-beds in the middle section (Ashworth and Hopkins, 1995).

The Trinity Aquifer is most extensively developed from the Hensell and Hosston Members in the Waco area, where the water level has declined by as much as 122 m (400 ft.; Ashworth and Hopkins, 1995). Water from the Antlers is used mainly for irrigation in the outcrop area of north and central Texas (George et al., 2011). Some of the largest declines in water levels range from 107 m (350 ft.) to more than 305 m (1000 ft.) and have occurred in counties along the IH 35 corridor from McLennan County to Grayson County (George et al., 2011). Extensive development of the Trinity Aquifer has occurred in the Fort Worth-Dallas region, where water levels have historically dropped as much as 168 m (550 ft.; (George, 2011, p. 1335). These declines are primarily attributed to municipal pumping, but they have slowed over the past decade as a result of increasing reliance on surface water (George et al., 2011).

Since the mid-1970s, many public supply wells have been abandoned in favor of a surface-water supply, and water levels have responded with slight rises (Ashworth and Hopkins, 1995). Water level declines of as much as 30.5 m (100 ft.) are still occurring in Denton and Johnson counties (Ashworth and Hopkins, 1995).

19.8 SURFACE WATER RESOURCES

Texas has significant surface water resources, as well as groundwater resources. The surface waters of the Texas Triangle include a vast array of streams, lakes, and reservoirs occupying more than 2450 km^2 (946 mi^2) of surface area (Fig. 19.13). Over 11,811 km (7,339 mi) of major streams and rivers occur in the region, including portions of 10 major river basins, including Brazos, Colorado, Guadalupe, Neches, Nueces, Red, Sabine, and San Trinity.

19.9 LIVING IN THE CRITICAL ZONE: ANTHROPOGENIC FACTORS AFFECTING GROUNDWATER IN THE CRITICAL ZONE OF THE TEXAS-TRIANGLE MEGAREGION

Anthropogenic forces, ranging from population growth to physical alteration of the landscape, are driving significant changes to the Critical Zone in the TTMR. Population in the Texas Triangle increased more rapidly than in any other region

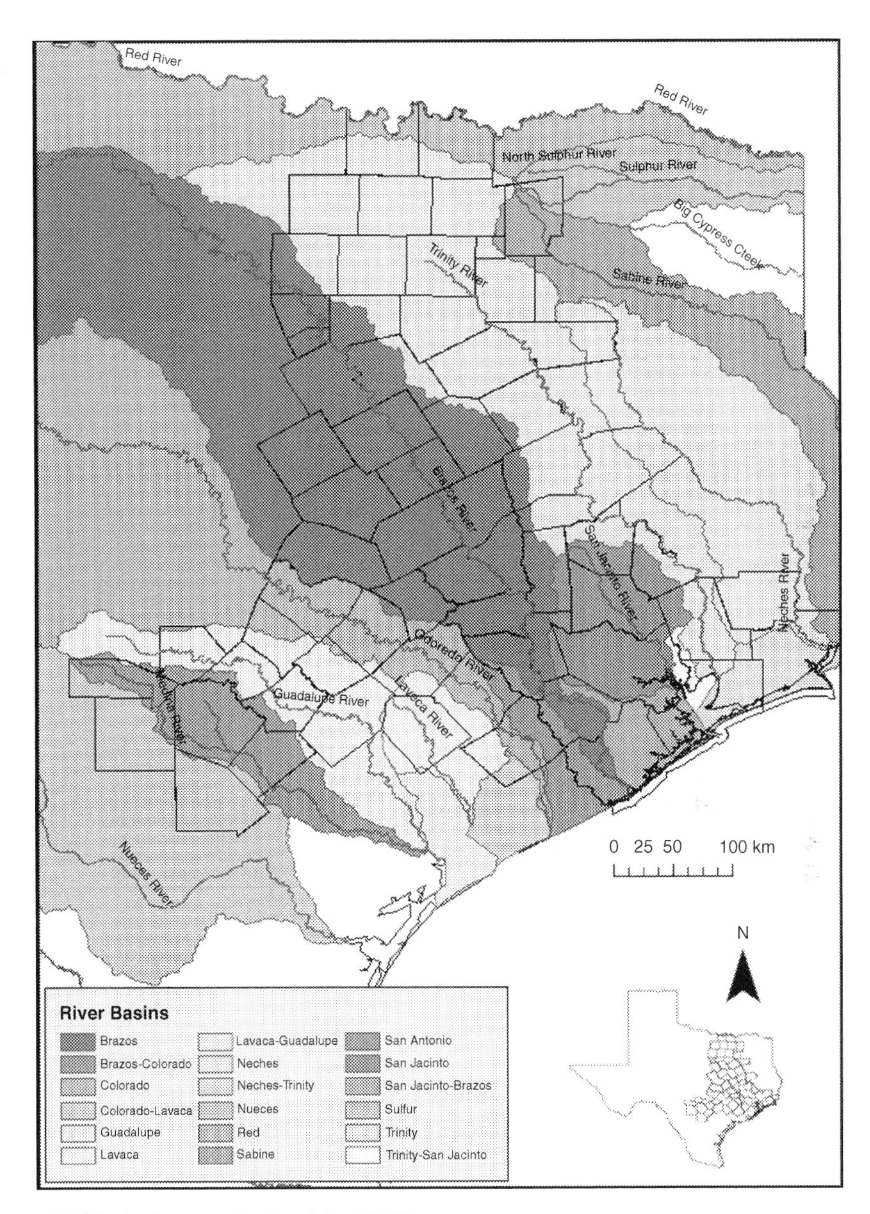

FIGURE 19.13 River basins of the TTMR.

in Texas during the last half of the twentieth century (Economics and Statistics Administration, 2014) and that growth is projected to continue at an accelerated rate during the first half of the twenty-first century (Potter and Hogue, 2011). The expanding population will result in accelerated urbanization and other land use changes that will impact the availability and quality of water resources in the region. With urbanization come drastic changes in landcovers. Many of these landcovers (i.e., concrete, asphalt, etc.) result in drastic changes in rates of runoff and evaporation, as well as permeability of the surface.

19.9.1 Population Growth

The US Census Bureau has forecast that the population of the nation will grow by 40% to 430 million by 2050 (Carbonell and Yaro, 2005). Whereas the population of the United States is projected to double between 2000 and 2100, the population of Texas is projected to increase by about 2.5 times (Butler et al., 2009). As Texas continues to grow steadily, growth in the TTMR is expected to be even faster (Neuman and Bright, 2008).

Population in the TTMR is projected to increase by 57% between 2000 and 2030, above the 42% increase for the rest of the state (Neuman et al., 2010). Projections indicate that, over the next 20 years, population in the area will account for more than 80% of the total population in the state (Neuman et al., 2010). By 2070, the 65 counties of the TTMR will have a projected population of 38.5 million people.

19.9.2 Urbanization

This expanding Texas population will be accommodated primarily within the Texas Triangle. This has been the fastest-growing region of the state for decades (Neuman et al., 2010). The size of the city of Houston today is equal to the land area of the cities of Boston, Denver, Las Vegas, Orlando, and Philadelphia combined (Butler et al., 2009). The axis from San Antonio to Dallas is on its way to becoming fully urbanized because of the proximity of the string of cities along IH 35: New Braunfels, San Marcos, Austin, Georgetown, Temple, Killeen, and Waco (Neuman et al., 2010). In contrast, along Interstate 45 between Dallas and Houston, and along Interstate 10 between Houston and San Antonio, only small villages and towns exist (Neuman et al., 2010).

According to the 2011 National Land Cover Database (Jin et al., 2013), approximately 14% of the total surface area of the TTMR is developed to some degree, with approximately 4% of that area being medium- or high-intensity development. The area of impervious surface in the TTMR is shown in Fig. 19.14. Table 19.5 depicts the major land cover categories in the TTMR.

According to Riebsame et al. (Riebsame et al., 1994), land-use and land-cover changes are gaining recognition as key drivers of environmental change. Urbanization and industrial development in the megaregion are the primary agents of change in land use (Butler et al., 2009). In the TTMR, the most rapid urban growth and land consumption in the state is in the fringes of the Triangle cities (Neuman et al., 2010). Carbonell and Yaro (Carbonell and Yaro, 2005) estimated the need to build 50% as much housing and 100% of the commercial and retail space as were built over the past 200 years to support the growth anticipated in the next five decades in this area.

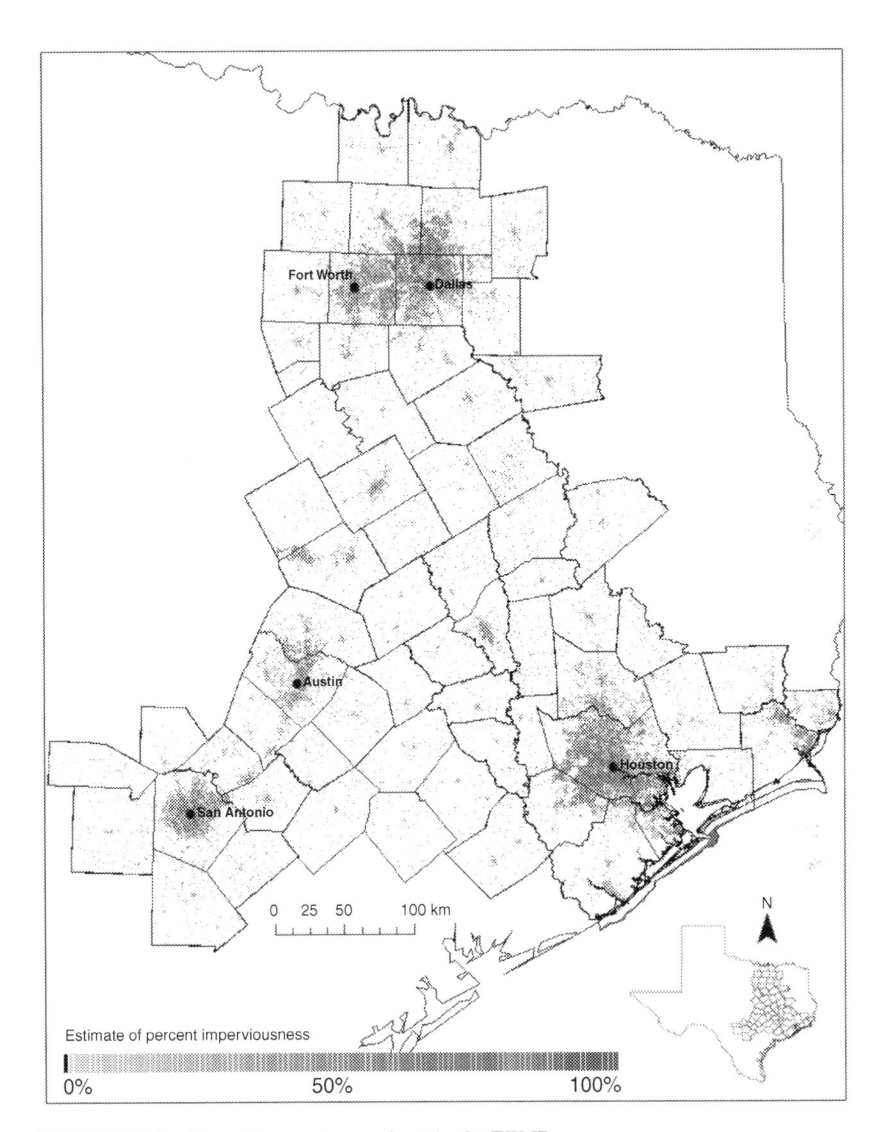

FIGURE 19.14 **Map of impervious surfaces in the TTMR.**

19.10 ANTHROPOGENIC HYDROLOGICAL ALTERATIONS IN THE CRITICAL ZONE

Supporting the modern industrial infrastructure of a major metropolitan megaregion has required extensive modifications to the various components of the Critical Zone. These modifications come in the form of an extensive network of dams and reservoirs; a high-density matrix of wells for extracting water, oil, and gas from the Critical Zone; significant landcover alterations; and interbasin transfer of ground and surface water.

TABLE 19.5 TTMR Land Cover by Category

Land use category	Area (sq km)	% of total
Open water	5,996	4
Developed, open space	9,558	6
Developed, low intensity	521	4
Developed, medium intensity	3,929	3
Developed, high intensity	1,622	1
Barren land (rock/sand/clay)	731	0.1
Deciduous forest	13,085	9
Evergreen forest	11,436	7
Mixed forest	3,963	3
Scrub/shrub	15,574	10
Grassland herbaceous	25,421	17
Pasture/hay	31,871	21
Cultivated crops	13,803	9
Woody wetlands	8,387	5
Emergent herbaceous wetlands	2,557	2
Total area	153,554	100

19.10.1 Dams and Reservoirs

More than 76 major dams and reservoirs provide a maximum storage capacity in excess of 39.47 km^3 (32 million acre ft.) of water for the TTMR (Fig. 19.15). Normal storage capacity for these reservoirs is approximately 16.40 km^3 (13.3 million acre ft.). The largest of these reservoirs, Medina Lake in Medina County, has a maximum storage capacity of 403 million m^3 (327,250 acre ft.). Lake McQueeny, built on the Guadalupe River in Guadalupe County, is the smallest of the reservoirs, with a maximum capacity of 6.2 million m^3 (5050 acre ft.).

The first of these dams and reservoirs was constructed on Shawnee Creek to form Randell Lake, which provides water for the city of Denton. The last major construction effort was completed in 1987 on the Elm Fork of the Trinity River to form Lake Ray Roberts, which provides water to Cooke, Denton, and Grayson counties.

Most of the dams and reservoirs within the TTMR serve multiple purposes, including public water supplies, recreation, flood control, hydroelectric power generation, and irrigation. The primary purpose (34%) of these dams and reservoirs is to provide a public water supply for municipalities within the TTMR.

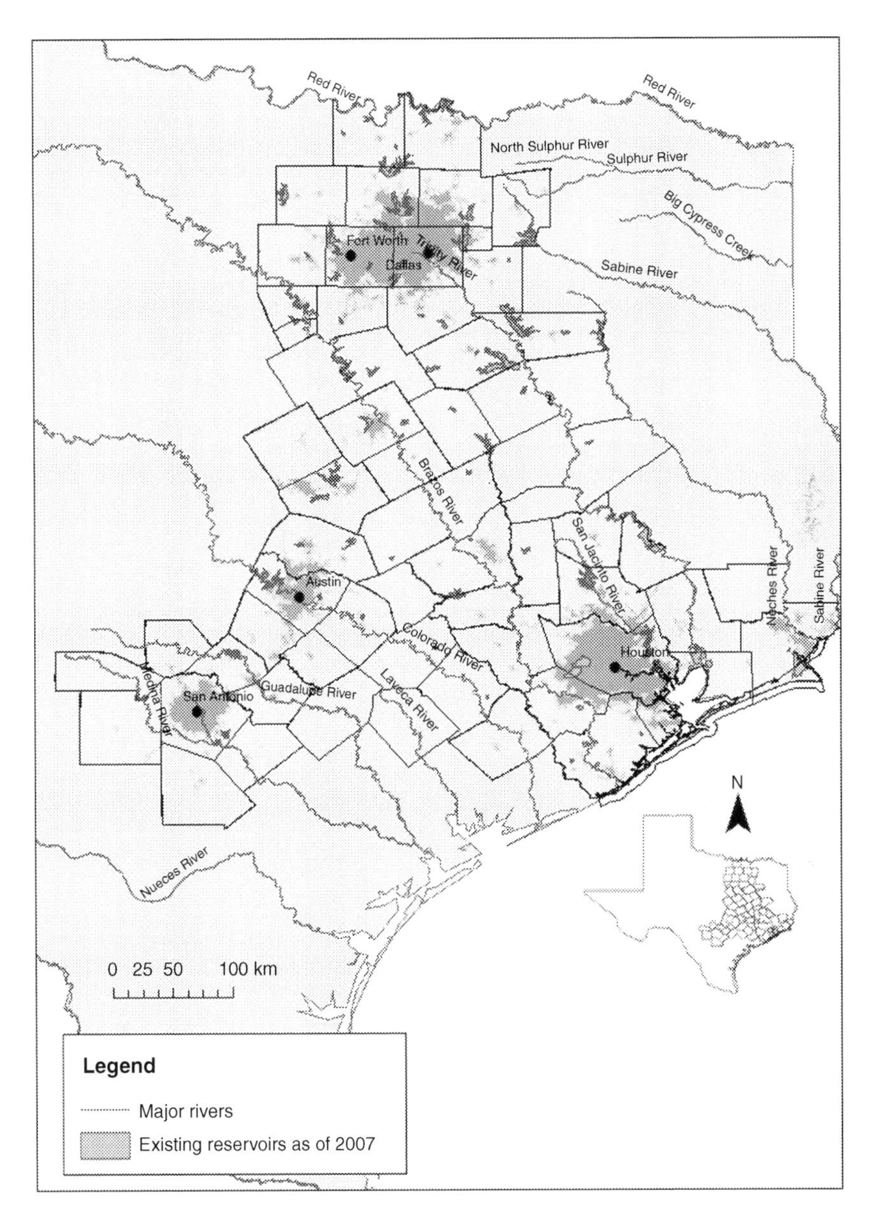

FIGURE 19.15 Existing Reservoirs in the TTMR.

Approximately one-quarter of the reservoirs also provide a source of recreation for the region. Fifteen percent of the dams and associated reservoirs were constructed to serve a flood control function. A small percentage of the reservoirs (9%) provide a source of water for irrigation. Only 4% of the dams provide a source of hydroelectric power (Fig. 19.16).

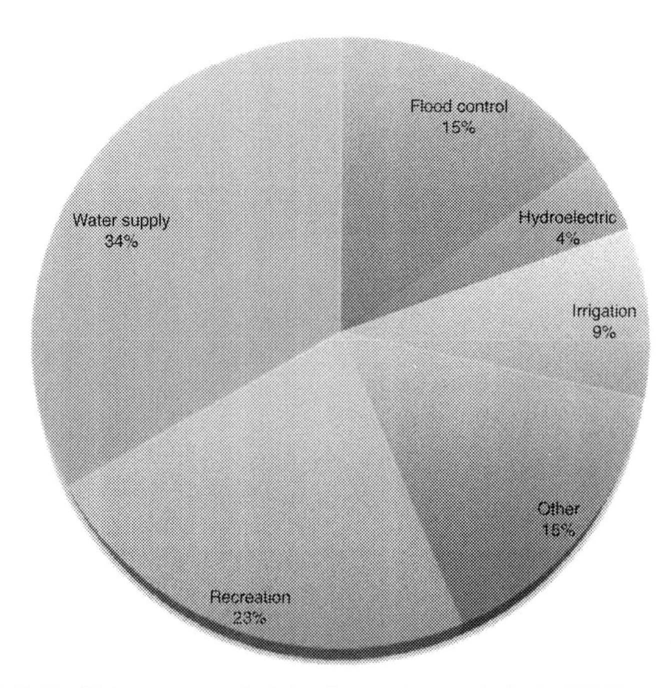

FIGURE 19.16 **Major purposes of existing dams and reservoirs in the TTMR.**

19.10.2 Water Use

Many communities in central Texas along the IH 35 corridor have experienced double-digit growth rates over the past 10 years, and this rate of growth is expected to continue in many communities of the region (Vaughn et al., 2012). The ability of the region to sustain this growth is largely dependent on the ability to provide adequate water supplies. According to the TWDB (Vaughn et al., 2012), municipal, industrial, and other uses for water will increase 22% by 2060 and failure to meet demand could cost businesses and workers in Texas approximately $11.9 billion per year (Combs, 2009; Vaughn et al., 2012).

Politicians and managers of water resources are increasingly recognizing the important role of groundwater resources in meeting the demands for drinking water, agricultural and industrial activities, sustaining ecosystems, and adaptation to, and mitigation of, the impacts of climate change and coupled human activities (Green et al., 2011, p. 429; Kaiser and Skillern, 2000). In the next 25 years, the fastest-growing categories of use are projected to be in municipal and manufacturing use and, by the 2040s, municipal and industrial uses of water are expected to exceed agricultural use of water (Kaiser and Skillern, 2000).

Throughout Texas, landowners and municipalities have depended on groundwater as a primary water resource because of local availability and quality (Vaughn et al., 2012). Groundwater provides about 60% of the 19.86 km^3 (16.1 million acre ft.) of water used in the state each year (Fipps, 2002, p. 1419;

Levasseur, 2012; Vaughn et al., 2012). The TWDB (Vaughn et al., 2012) predicts that, over the next 50 years, agricultural use of groundwater will experience a dramatic decline because of aquifer depletion and rising energy costs. At the same time, municipal share of groundwater use will double (Kaiser and Skillern, 2000).

The increasing demand for groundwater in the region has resulted in a decrease in aquifer levels of more than 244 m (800 ft.) in the Dallas area and 122 m (400 ft.) in the Houston area in less than a century (Neuman and Bright, 2008). In the Gulf Coast Aquifer in Houston and the Carrizo–Wilcox Aquifer close to Tyler, Lufkin and College Station-Bryan water levels have dropped by more than 91.4 m (300 ft.; (Neuman and Bright, 2008, p. 1223). It is projected that two of the five largest aquifers in the region will have less than 45% of the reservoir storage capacity remaining by 2050 (Butler et al., 2009). A recent estimate indicated a potential reduction of about 31% in the total groundwater supply in the state by the year 2060 (Chaudhuri and Ale, 2013; Vaughn et al., 2012).

19.10.3 Municipal Use

The TWDB (Vaughn et al., 2012) estimated that the rapidly growing population in the state will spur changes in the demand for and use of water. In 2010, irrigation was projected to account for 56% of the water use in Texas, followed by municipal use at 27% (Combs, 2014). By 2060, municipal water use is expected to become the largest category, at 38.3% of all water use, followed closely by irrigation at 38.1% (Combs, 2014). Bryan-College Station, Lufkin-Nacogdoches, Bastrop, and Tyler are the major municipalities that rely on groundwater from the Carrizo–Wilcox Aquifer (Boghici, 2008). Whereas San Antonio currently gets most of its water for municipal use from the Edwards Aquifer, it has entered into contractual negotiations with Alcoa Corporation for the rights to purchase groundwater originating from a lignite mining operation in the Carrizo–Wilcox Aquifer more than 161 km (100 mi) from the city (Neuman and Bright, 2008).

The city of San Antonio has also begun to incorporate aquifer storage and recovery as a key component of its attempt to achieve water supply diversity. The Twin Oaks Aquifer Storage and Recovery System stores up to 148 million m^3 (120,000 acre ft.) of Edwards Aquifer water that becomes available during wet periods in the Carrizo Aquifer for later use.

19.10.4 Irrigation

As is the case in the United States as a whole, irrigation represents the highest use of Texas groundwater (Kelley et al., 2009). Agricultural irrigation consumes about 80% of all groundwater pumped annually in Texas (Kaiser and Skillern, 2000).

19.10.5 Groundwater Quality

The principal type of aquifer in the TTMR is unconsolidated sand and gravel. This makes the aquifers susceptible to contamination because of the high permeability and hydraulic conductivity (Neuman et al., 2010). More than 1250 active and former municipal solid waste sites exist within the Texas Triangle (Fig. 19.17), with the potential to impact groundwater quality negatively. In addition, 139 permitted Industrial and Hazardous Waste Sites are present. The region is also home to 74 Environmental Protection Area Superfund sites. The TWDB (Vaughn et al., 2012) has logged 38,581 wells, each of which has the potential to impact groundwater quality negatively if not properly maintained.

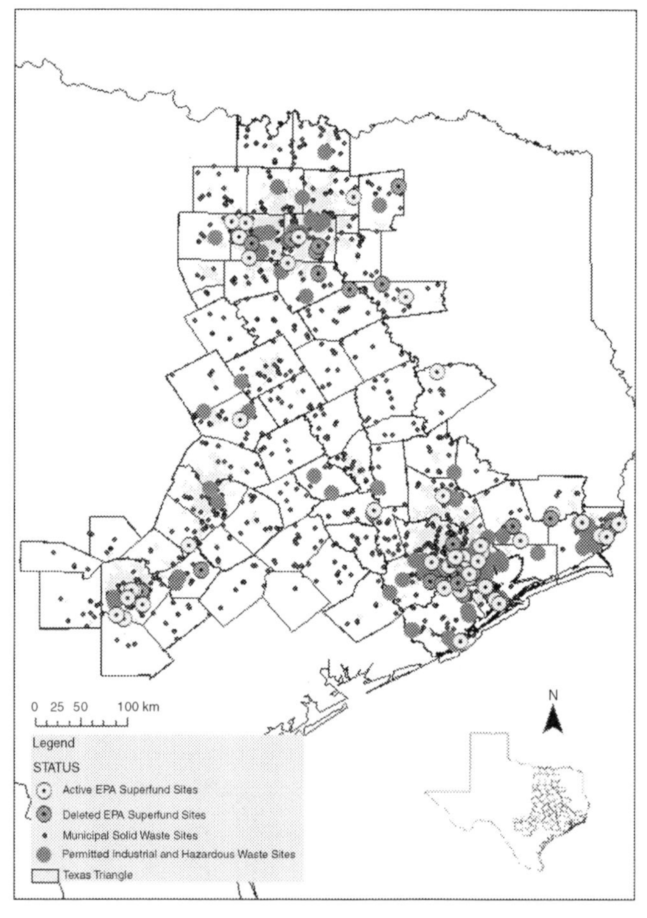

FIGURE 19.17 Municipal solid waste, industrial waste, and Environmental Protection Agency Superfund Sites in the TTMR. *(Source data from the Texas Commission on Environmental Quality retrieved from http://www.tceq.texas.gov/gis/download-tceq-gis-data.)*

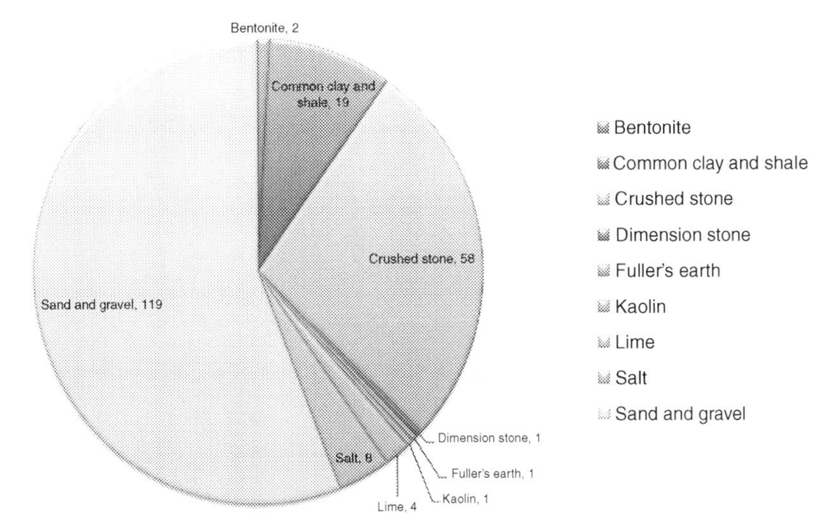

FIGURE 19.18 **Mining activities in the TTMR by commodity.**

19.10.6 Mining Operations

The TTMR supports significant mining operations of nine major commodities, ranging from bentonite to sand and gravel, as shown in Fig. 19.18. More than 235 active mines exist in the TTMR (Fig. 19.19), supporting mining operations of these commodities. The vast majority of the operations mine sand and gravel (119) and crushed stone (58; USGS, 2003, p. 1477).

19.10.7 Oil and Gas Development

Renewed interest in the Carrizo–Wilcox Aquifer and the Trinity Aquifer comes from the energy sector, particularly from companies engaged in development of oil and gas resources, such as the Eagle Ford Shale and Barnett Shale (Levasseur, 2012; Nicot and Scanlon, 2012). Investment in the unconventional reservoirs by the petroleum industry has exceeded \$1 billion to date (Carlos Dengo, Director, Texas A&M Berg-Hughes Center for Petroleum and Sedimentary Systems, July 8, 2014, personal communication). Productive Mississippian Barnett Shale is found at depths of 2,000–2,600 m (6,561–8,530 ft.) near the Dallas–Fort Worth metroplex and in the Eagle Ford Shale play area extending over portions of 24 counties ($\sim$50,000 km^2 or 19,305 mi^2) in south Texas. Figure 19.20 shows the extent of the Eagle Ford.

This interest is placing new demands on groundwater in areas where the resource is already becoming constrained, particularly in the Wintergarden and Dallas–Fort Worth areas. On an average, oil and gas companies utilize 5.7–11.3 million liters (1.5–3.0 million gallons) of water per horizontal well drilled and hydraulically fractured, but these estimates are largely dependent

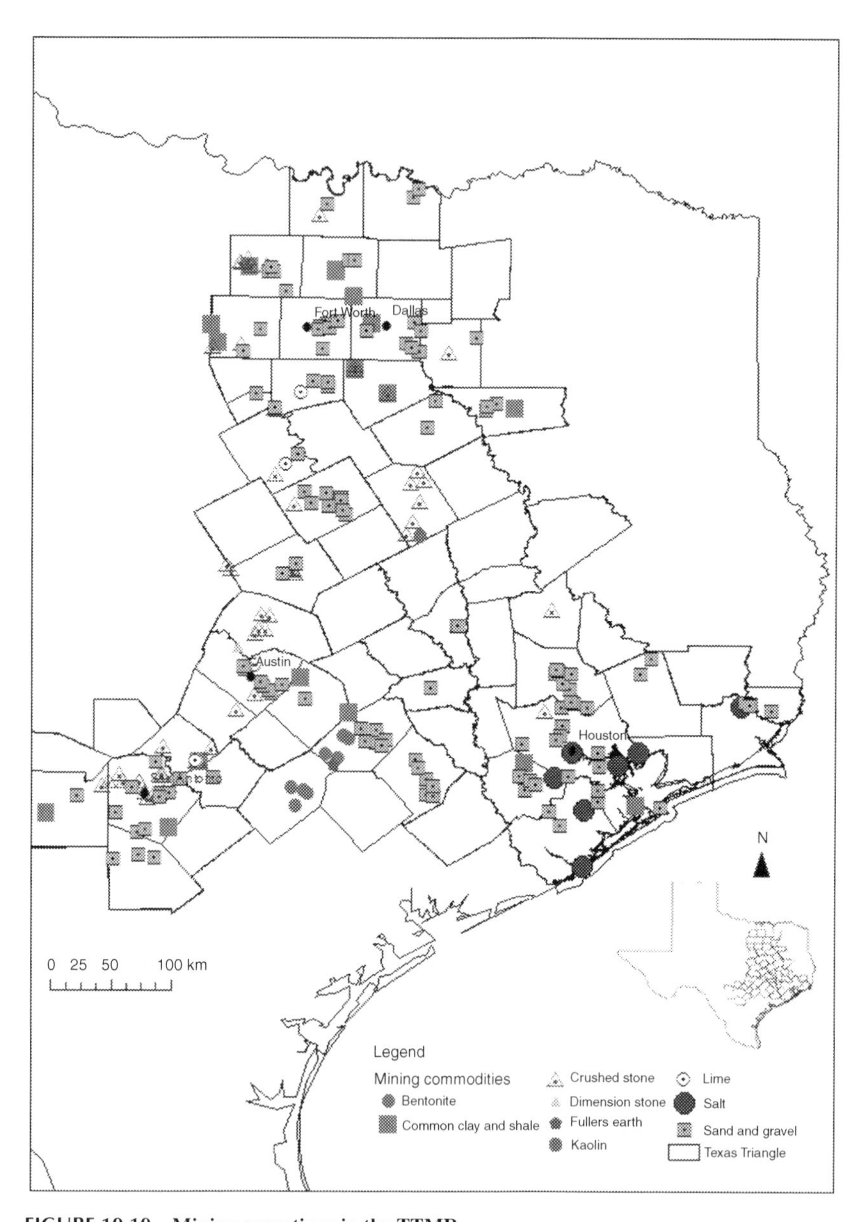

FIGURE 19.19 Mining operations in the TTMR.

on the length and number of stages in the well design for a given formation (Nicot and Scanlon, 2012). The TWDB estimated that in 2008 about 44.16 million m^3 (35.8 thousand acre ft.) of water was used for hydraulic fracturing ((Levasseur, 2012, p. 1297; Vaughn et al., 2012). Fracking water use in the Barnett Shale (Fig. 19.20) in 2010 represented approximately 9% of the

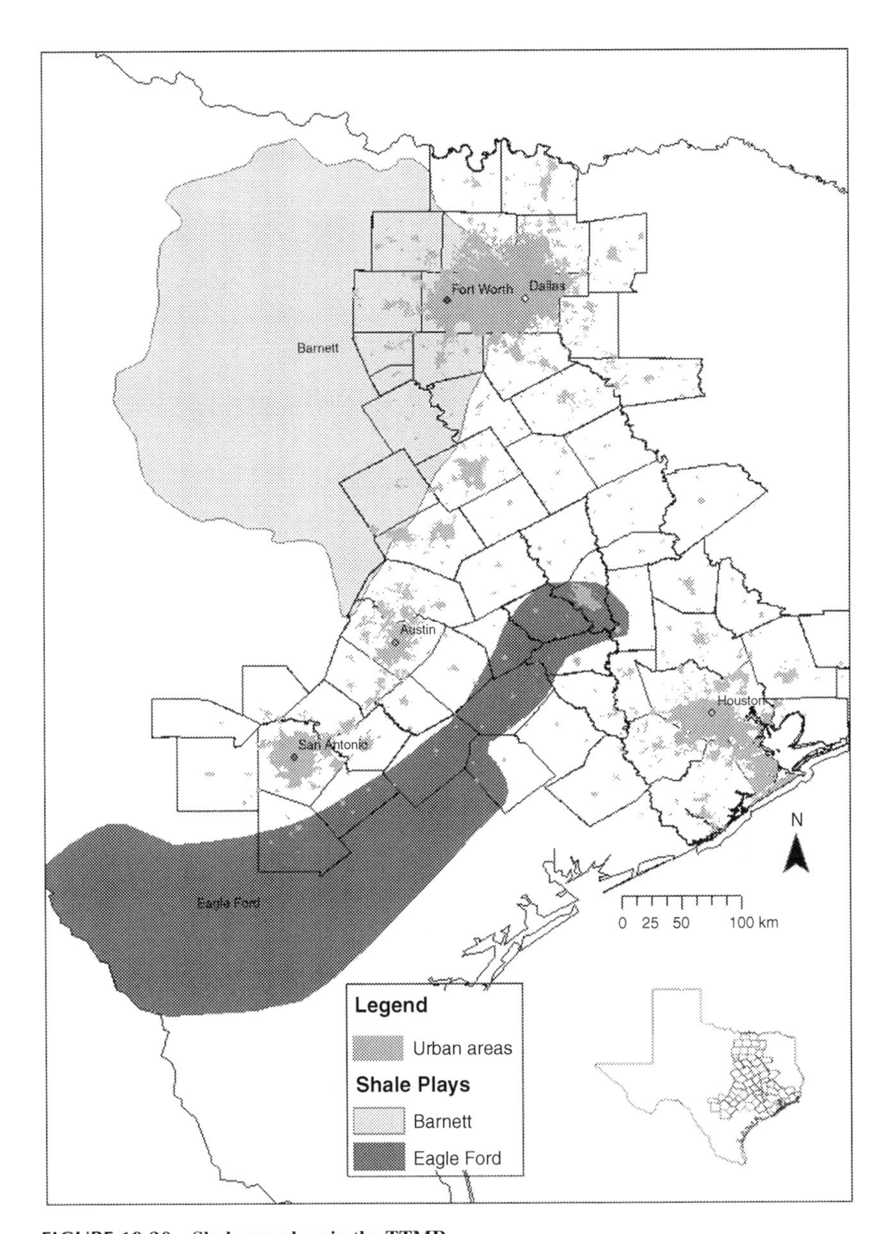

FIGURE 19.20 Shale gas plays in the TTMR.

308.37 million m^3 (250,000 acre ft.) of water used by the city of Dallas, the ninth-largest city in the United States (Nicot and Scanlon, 2012). Fracking for shale-gas production in the Eagle Ford Shale began in 2008, and wells drilled in the Eagle Ford Shale totaled 1040, with cumulative water use of 1.8 million m^3 (14,600 acre ft.) by mid-2011 (Nicot and Scanlon, 2012).

While surface water is available in the Barnett Shale from the Trinity and Brazos rivers and reservoirs, it is not as readily available in the Eagle Ford Shale region (Nicot and Scanlon, 2012). Groundwater resources are generally available in each of the shale gas plays, and, unlike surface water, groundwater is ubiquitous and generally available close to production wells. However, in the Eagle Ford Shale region groundwater has already been significantly depleted for irrigation in the Winter Garden region of South Texas, resulting in water-level declines >60 m (197 ft.) over a 6500 km^2 (2510 mi^2) area, disappearance of several large springs, and transition from predominantly gaining to mostly losing streams (Nicot and Scanlon, 2012). Population growth will also increase demand for this resource and possibly compound stress on the aquifer in which water levels have significantly declined in past decades (Nicot and Scanlon, 2012).

19.11 WATER MANAGEMENT POLICY IN THE TEXAS TRIANGLE MEGAREGION: IMPLICATIONS FOR THE CRITICAL ZONE

Unlike scientists who recognize that all water is interconnected, Texas law distinguishes between surface water and groundwater for the purpose of regulation with different rules governing each class (Kaiser, 1988, p. 525; Vaughn et al., 2012). The state recognizes that a landowner owns groundwater (fresh and brackish) underlying his or her land as real property (Combs, 2014). In contrast, with the exception of diffused water, such as storm water runoff, all surface water, including streams, rivers, and lakes, is "held in trust" by the state and appropriated to users through permits or "water rights" (Fipps, 2002). The complicated system in Texas arose from Spanish and English common law, the laws of other Western states, and state and federal case law and legislation (Vaughn et al., 2012).

Commonly known as the "Rule of Capture," groundwater law in Texas is based on the English common law doctrine, which says that the landowner may withdraw groundwater without limitations and without liability for losses to neighbors' wells as long as water is not wasted or taken maliciously (Combs, 2014; Fipps, 2002; Kaiser, 2005, p. 739). The Texas Supreme Court in its 1904 decision *Houston and T.C. Railway Co. v. East* adopted the "rule of capture" doctrine in part because the science of quantifying and tracking the movement of groundwater was so poorly developed at the time that it would be practically impossible to administer any set of legal rules to govern its use (Kaiser, 1988; Vaughn et al., 2012).

The right of landowners to capture and make "nonwasteful" use of groundwater has been upheld by Texas courts over the years, with only a few exceptions: drilling a well on someone else's property or drilling a "slant" well on adjoining property that crosses the property line (trespass), pumping water for the sole purpose of injuring an adjoining landowner (malicious or wanton conduct), or causing land subsidence on adjoining land from negligent overpumping (Fipps, 2002). Texas groundwater law has often been called the "law of the biggest pump" because the deepest well and most powerful pump get the water (Fipps, 2002; Kaiser, 1987).

In Texas, all surface water is held in trust by the state, which grants permission to groups and individuals to use the water (Vaughn et al., 2012). The state owns all waters flowing on the surface of Texas (Combs, 2014; Kaiser, 2005). The TCEQ issues and manages permits based on a "first in time, first in right" principle, meaning that those holding the oldest permits have first access to available water (Combs, 2014).

Texas recognizes two basic doctrines of surface water rights: the riparian doctrine and the prior appropriation doctrine. Introduced more than 200 years ago when Spanish settlers first arrived in Texas, the riparian doctrine permits landowners whose property is adjacent to a river or stream to make reasonable use of the water (Kaiser, 1988). First adopted in Texas in 1895, the prior appropriation system has evolved into the modern system used today (Vaughn et al., 2012). Under prior appropriation, landowners who live near many of the water bodies in the state are allowed to divert and to use water for domestic and livestock purposes, not to exceed 247,000 m^3 (200 acre ft.) per year (Kaiser, 1988; Vaughn et al., 2012).

19.12 MANAGING THE WATER RESOURCES OF THE CRITICAL ZONE IN TEXAS

Four agencies hold the primary responsibility for managing and enforcing water planning, water quality, and water quantity in Texas: TWDB, Texas Parks and Wildlife Department, TCEQ, and Texas Soil and Water Conservation Board.

The TWDB was created in 1959 as the primary water supply planning and financing agency. The TWDB supports the development of 16 regional water plans and is responsible for developing the state water plan every 5 years.

The Texas Parks and Wildlife Department works with regional and state water planning stakeholders and regulatory agencies to protect and enhance water quality and to ensure adequate environmental flows for rivers, bays, and estuaries. It also provides technical support to the environmental flows process and is a member of the Texas Water Conservation Advisory Council.

The TCEQ is the environmental regulatory agency for the state, focusing on water quality and quantity through various state and federal programs. The agency issues permits for the treatment and discharge of industrial and domestic wastewater and storm water, reviews plans and specifications for public water systems, and conducts assessments of surface water and groundwater quality. The TCEQ regulates retail water and sewer utilities, review rates increase by investor-owned water and wastewater utilities, and administers a portion of the Nonpoint Source Management Program. In addition, TCEQ administers the surface water rights permitting program and a dam safety program, designates Priority Groundwater Management Areas, creates some GCDs, and enforces requirements of groundwater management planning.

The Texas State Soil and Water Conservation Board administers soil and water conservation law in Texas and coordinates conservation and nonpoint source pollution abatement programs. The agency also administers water quality and water supply enhancement programs.

19.12.1 Managing Water Use Through Water Planning

In response to the most severe drought of record in Texas in the 1950s, the Texas Water Planning Act of 1957 created the TWDB and gave it authority to develop a State Water Plan (Brown, 1997). Although the state had legislated the water planning process in 1957, it took little or no action on the first two water plans developed in 1961, 1968, 1987, 1990, and 1992. In 1997, once again acting in response to a drought (1995–1996), the TWDB, in conjunction with the Texas Natural Resources Conservation Commission and the Texas Parks and Wildlife Department, developed the first "consensus-based" plan (Brown, 1997).

Recognizing that water is the single most important factor for the future economic viability of Texas, the legislature passed Senate Bill 1, the Comprehensive Water Management Bill, which was signed into law on June 19, 1997 (Brown, 1997). Senate Bill 1 put in place the "bottom-up" approach to water planning rooted in local, consensus-based decision making that Texas uses for water planning today (Combs, 2009). Senate Bill 1 resulted in designation of water planning regions based on geographical, hydrological, and political boundaries, water utility development patterns, and socioeconomic characteristics (Brown, 1997).

19.12.2 Managing Groundwater Through Conservation Districts

State policy in Texas dictates that groundwater management is best accomplished through locally elected, locally controlled GCDs, suggesting that any modification or limitation on the rule of capture will be made by local groundwater districts (Brown, 1997). In 1949, the Texas legislature first provided for voluntary creation of GCDs over any groundwater reservoir designated by the state (Fipps, 2002). While continuing to acknowledge the "rule of capture" of groundwater by landowners, the Texas legislature passed additional legislation in 1985 and 1997 to encourage establishment of GCDs and, in limited cases, to allow for the creation of districts by state initiative (Fipps, 2002).

As of April 2014, a total of 101 GCDs had been created in the state. The total includes 98 established (i.e., confirmed) districts and 3 unconfirmed districts. The 98 established districts cover all or part of 179 of the 254 counties in the state. The TTMR has 50 GCDs in place (TWDB, 2014). GCDs are charged to manage groundwater by providing for the conservation, preservation, protection, recharging, and prevention of waste of the groundwater resources within their jurisdictions (Fipps, 2002).

GCDs can be created by one of four procedures (Lesikar et al., 2002): (1) GCDs can be established through the action of the legislature; (2) GCDs can be created through a landowner petition procedure based on state law in Subchapter B, Chapter 36 of the Texas Water Code; (3) GCDs can be created by the TCEQ on its own motion in a designated Priority Groundwater Management Area (PGMA) through a procedure similar in principle to procedure (2) above but in which action is initiated by the TCEQ rather than by petition; or (4) an alternative to creating a new GCD is to add territory to an existing district if the existing district is willing to accept the new territory.

GCDs are authorized by the state with powers and duties that enable them to manage groundwater resources. The three primary GCD legislatively mandated duties are permitting water wells, developing a comprehensive management plan, and adopting the necessary rules to implement the management plan (Lesikar et al., 2002).

The principle power of a GCD to prevent waste of groundwater is to require that all wells, with certain exceptions, be registered and permitted. Wells with permits are subject to GCD rules governing spacing, production, drilling, equipping, and completion or alteration. Even exempt registered wells are subject to GCD rules governing spacing, tract size, and well construction standards to prevent unnecessary discharge of groundwater or pollution of the aquifer. Permits may be required by a GCD for all wells except for wells specifically exempted by a GCD and statutorily exempt wells (i.e., wells used solely for domestic use or for providing water for livestock or poultry purposes; the drilling of a water well used solely to supply water for a rig actively engaged in drilling or exploration operations for an oil or gas well permitted by the RRC; or the drilling of a water well authorized by the RRC for mining activities; Lesikar et al., 2002).

In 1985 the Texas legislature passed House Bill 2, containing provisions for the Texas Water Commission (a predecessor to the TWDB) to identify areas of the state that have critical groundwater problems, such as aquifer depletion, water quality contamination, land subsidence, or shortage of water supply. Accordingly, beginning in 1986, the TWC and the TWDB identified possible critical areas and conducted further studies (Fipps, 2002). Portions of 11 Groundwater Management Agencies are located within the TTMR (TWDB, 2014).

Groundwater Management Areas were created "to provide for the conservation, preservation, protection, recharging, and prevention of waste of the groundwater, and of groundwater reservoirs or their subdivisions, and to control subsidence caused by withdrawal of water from those groundwater reservoirs or their subdivisions" (Texas Water Code §35.001, Added by Acts 1995, 74th Leg., ch. 933, §2, eff. Sept. 1, 1995).

Beginning in 2005, Texas required, through legislation, that GCDs meet regularly and define the "desired future conditions" of the groundwater resources within designated management areas (Vaughn et al., 2012). Based on these desired future conditions, TWDB delivers modeled available groundwater values to GCDs and regional water planning groups for inclusion in their plans.

The previous discussion illustrates how controlled surface water and groundwater are in the Critical Zone in Texas, especially in the TTMR. Despite this control regime, the waters in the Critical Zone are being depleted and degraded at an alarming rate.

19.13 SUMMARY AND CONCLUSIONS

The TTMR has become the "core area of Texas," creating an urbanized area rivaling New York and Los Angeles and expected to accommodate more than 80% of the population of the state by 2030. By 2070, the 65 counties of the TTMR will support a projected population of almost 40 million people. This rapid growth will put significant impacts on the water resources of this part of the Critical Zone.

The TTMR has historically derived approximately 60% of its water for all major uses from groundwater from the four primary aquifers located in the Critical Zone of this megaregion. Despite the significant sources of surface water within the TTMR, as the population of the region continues to grow, water managers will increasingly rely on groundwater as a reliable water source to sustain this growth.

The increasing pressure to use groundwater will result in significant alterations to the Critical Zone, such as those already being used by the city of San Antonio. The transfer of water from the Carrizo–Wilcox Aquifer to service the city of San Antonio, which overlies the Edwards Aquifer, as well as the storage of significant volumes of water from the Edwards Aquifer in the Carrizo–Wilcox for later use represent significant anthropogenic alterations to the Critical Zone.

Land-use and landcover changes will continue to significantly affect the Critical Zone in this region as well. As more land-surface area is converted from a natural state or agricultural use to urban development, the increased impervious surface will impact the Critical Zone in a variety of ways. Increased runoff and evaporation rates will occur. Expanded urban infrastructure will result in mixing of chemicals and petroleum products with surface water, resulting in degraded water quality. Much of this degraded water will eventually make its way into the aquifer system underlying the area. The IH 35 corridor, which serves as the lifeblood of the TTMR, overlaps a significant portion of the groundwater recharge zones for the region. As the area becomes more congested and the impervious surface increases even more, the ability to recharge the aquifers in this area naturally will decrease. The enhanced runoff from the impervious surface will also pose a significant water quality hazard to surrounding surface water and groundwater sources.

Development of shale plays for hydrological fracturing of hydrocarbons in the Eagle Ford and Barnett shale regions will continue to have increasing significant impacts on the Critical Zone of the TTMR. As these areas are further developed in the future, that development will place additional demands on groundwater resources even as these resources are receiving increased demand to supply a growing municipal population. Use of these groundwater resources to support hydrofracturing operations introduces the added anthropogenic changes to the Critical Zone associated with disposal of hazardous waste generated from the fracturing process.

The increasing urbanization of the TTMR will continue to affect surface and groundwater quality in the region. In addition to water-quality issues generated by increased impervious surfaces, the potential for surface-water and groundwater contamination resulting from municipal solid-waste sites and permitted industrial-waste sites will be ever present.

The significant anthropogenic alterations to the Critical Zone in the TTMR will require robust, forward-thinking laws, policies, and management structures to mitigate the negative impact on this Critical Zone. This is particularly true regarding groundwater management because groundwater law in Texas has changed very little since the beginning of the twentieth century.

Life on Earth depends on the uninterrupted provision of "Critical Zone services," ranging from the provision of water of a quality and in a quantity that will support human activities and ecosystems to the production of food and fiber for a growing global population (Anderson et al., 2010). Providing these Critical Zone services will become increasingly difficult in the TTMR, as the population approaches nearly 40 million people – more than a fivefold increase from the 7.1 million people who occupied the area in 1970.

Supporting the modern industrial infrastructure of the TTMR has required extensive modifications to the Critical Zone in the form of an extensive network of dams and reservoirs; a high-density matrix of wells for extracting water, oil, and gas from the Critical Zone; significant landcover alterations; and interbasin transfer of ground and surface waters. Progressive depletion of Critical Zone reserves threatens sustainable development in the heavily groundwater-dependent Texas Triangle and requires robust and effective water resource policy for the megaregion to remain economically viable. According to the TWDB, demand for water will increase by 22% by 2060 and failure to meet the demand could cost businesses and workers in Texas approximately $11.9 annually (Vaughn et al., 2012).

Progressive depletion of freshwater reserves threatens sustainable development in many parts of the United States, including Texas, that are heavily reliant on groundwater resources (Chaudhuri and Ale, 2013). In a state where more than 95% of land is privately owned, the emphasis on private-property rights relating to groundwater has resulted in the rule of capture being held by many to be sacrosanct (Brown, 1997). Any full-scale revision of the rule of capture in Texas will most likely arise from attitudinal changes that evolve with the growth of free-market forces on precious water resources of Texas (Brown, 1997).

REFERENCES

Amundson, R., Richter, D.D., Humphreys, G.S., Jobbágy, E.G., Gaillardet, J., 2007. Coupling between biota and Earth materials in the Critical Zone. Elements 3 (5), 327–332.

Anderson, R.S., Anderson, S., Aufdenkampe, A.K., Bales, R., Brantley, S., Chorover, J., Duffy, C.J., Scatena, F.N., Sparks, D.L., Troch, P.A., 2010. Future Directions for Critical Zone Observatory (CZO) Science, National Science Foundation, Washington, DC.

Ashworth, J.B., Hopkins, J., 1995. Aquifers of Texas: Texas Water Development Board Report 345. Texas Water Development Board, Austin, TX.

Bailey, R.G., Bailey, R.G., 1981. Ecoregions of the United States: 1976. The Service, Washington, DC.

Bailey, R.G., Bailey, R.G., United States Forest Service., US Fish and Wildlife Service., 1976. Ecoregions of the United States. US Forest Service, Washington, DC.

Bailey, R.G., Davis, G.D., United States Forest Service., 1978. Ecosystems of the United States: US Dept. of Agriculture, Forest Service, Washington, DC.

Bailey, R.G., U.S. Fish and Wildlife Service, United States Forest Service, 1980. Description of the ecoregions of the United States. Miscellaneous publication. U.S. Dept. of Agriculture, Forest Service, Ogden, Utah.

Boghici, R., 2008. Changes in Water Levels in Texas, 1990–2000. Texas Water Development Board, Austin, Texas.

Bradley, R.S., Alverson, K., Pedersen, T.F., 2003. Challenges of a Changing Earth: Past Perspectives, Future Concerns. Paleoclimate, Global Change and the Future. Springer Verlag, Berlin, pp. 163–167.

Brantley, S.L., Goldhaber, M.B., Ragnarsdottir, K.V., 2007. Crossing disciplines and scales to understand the critical zone. Elements 3 (5), 307–314.

Brown, C.R., Farrar, B., 2008. A Hole in the bucket: Aspermont's impact on groundwater districts and what it says about Texas groundwater policy. Texas Environ. Law J. 39, 1.

Brown, J.E., 1997. Senate bill 1: we've never changed Texas water law this way before. B. Texas Environ. Law J. 28, 152.

Butler, K., Hammerschmidt, S., Steiner, F., Zhang, M., 2009. Reinventing the Texas Triangle: solutions for growing challenges. Center for Sustainable Development, School of Architecture, University of Texas at Austin, p. 29.

Carbonell, A., Yaro, R.D., 2005. American spatial development and the new megalopolis. Land Lines 17 (2), 1–4.

Chaudhuri, S., Ale, S., 2013. Characterization of groundwater resources in the Trinity and Woodbine aquifers in Texas. Sci. Total Environ. 452, 333–348.

Combs, S., 2009. Liquid Assets: The State of Texas' Water Resources, Data Services Division, State of Texas, Austin, TX.

Combs, S., 2014. Texas Water Report: Going Deeper for the Solution. Data Services Division, State of Texas, Austin, TX.

Crutzen, P.J., Steffen, W., 2003. How long have we been in the anthropocene era? Climat. Change 61 (3), 251–257.

Dewar, M., Epstein, D., 2007. Planning for "Megaregions" in the United States. J. Plann. Lit. 22 (2), 108–124.

Dorfman, M., Mehta, M., Chou, B., Fleischli, S., Rosselot, K., 2011. Thirsty for Answers: Preparing for the Water-Related Impacts of Climate Change in American Cities. National Resources Defense Council, Washington, DC.

Economics and Statistics Administration, U.S.C.B., 2014. Average annual population growth 1930–2000. US Department of Commerce Washington, DC. Average annual growth rate by county.

Fipps, G., 2002. Managing Texas Groundwater Resources Through Groundwater Conservation Districts. Texas Agricultural Extension Service, The Texas A&M University System, College Station, TX, p. 11.

Fisher, W.L., McGowen, J.H., 1969. Depositional systems in Wilcox Group (Eocene) of Texas and their relation to occurrence of oil and gas. AAPG Bull. 53 (1), 30–54.

George, P.G., Mace, R.E., Petrossian, R., 2011. Aquifers of Texas. Texas Water Development Board, Austin, Texas.

Goudie, A.S., 2006. Global warming and fluvial geomorphology. Geomorphology 79 (3–4), 384–394.

Graf, W.L., 2008. In the Critical Zone: geography at the U.S. geological survey. Prof. Geogr. 56 (1), 100–108.

Green, T.R., Taniguchi, M., Kooi, H., Gurdak, J.J., Allen, D.M., Hiscock, K.M., Treidel, H., Aureli, A., 2011. Beneath the surface of global change: impacts of climate change on groundwater. J. Hydrol. 405 (3–4), 532–560.

Jin, S., Yang, L., Danielson, P., Homer, C., Fry, J., Xian, G., 2013. A comprehensive change detection method for updating the national land cover database to circa 2011. Remote Sens. Environ. 132, 159–175.

Kaiser, R., 1987. Hand Book of Texas Water Law: Problems and Needs. Texas Water Resources Institute, Texas Agricultural Experiment Station, Texas A&M University.

Kaiser, R., Skillern, F.F., 2000. Deep trouble: options for managing the hidden threat of aquifer depletion in Texas. Texas Tech. Law Rev. 32, 249.

Kaiser, R.A., 1988. Handbook of Texas Water Law: Problems and Needs. Texas A&M University, Texas Water Resources Institute, College Station, Texas.

Kaiser, R., 2005. Groundwater management in Texas: evolution or intelligent design. Kan. JL & Pub. Pol'y, vol. 15, p. 467.

Kelley, V., Deeds, N., Fryar, D., Senger, R., 2009. Development of regional groundwater availability models of the Carrizo–Wilcox aquifer in Texas. Gulf Coast Assoc. Geol. Soc. Trans. 59, 401–410.

Klemt, W.B., Duffin, G.L., Elder, G.R., 1976. Groundwater resources of the Carrizo aquifer in the Winter Garden area of Texas. Texas Water Development Board Report, p. 210.

Lesikar, B., Kaiser, R., Silvy, V., 2002. Questions About Groundwater Conservation Districts in Texas. Texas Water Resources Institute, College Station, TX.

Levasseur, P.G., 2012. Current regulations, scientific research, and district rulemaking processes to protect and conserve the Carrizo–Wilcox Aquifer in Texas by groundwater conservation districts. Master of Arts, The University of Texas at Austin, Austin, TX, p. 201.

Levin, R.B., Epstein, P.R., Ford, T.E., Harrington, W., Olson, E., Reichard, E.G., 2002. US drinking water challenges in the twenty-first century. Environ. Health Perspect. 110 (Suppl 1), 43–52.

Lin, H., 2010. Earth's critical zone and hydropedology: concepts, characteristics, and advances. Hydrol. Earth Syst. Sci. 14, 25–45.

Mace, R.E., Wade, S.C., 2008. In hot water? How climate change may (or may not) affect the groundwater resources of Texas, 2008, Joint Meeting of the Geological Society of America, Soil Science Society of America, American Society of Agronomy, Crop Science Society of America, Gulf Coast Association of Geological Societies with the Gulf Coast Section of SEPM. GCAGS Transactions, pp. 655–668.

National Research Council. Committee on Basic Research Opportunities in the Earth, S., 2001. Basic Research Opportunities in Earth Science. 0-309-07133-X, National Research Council, Washington, DC.

Neffendorf, B., Hopkins, J., 2013. Summary of Groundwater Conditions in Texas: Recent (2011–2012) and Historical Water-Level Changes in the TWDB Recorder Network. 13-02, Texas Water Development Board, Austin, TX.

Neuman, M., Bright, E., Morgan, C., 2010. Texas Urban Triangle: Creating a Spatial Decision Support System for Mobility Policy and Investments that Shape the Sustainable Growth of Texas. Report UTCM 09-30-10, Texas Transportation Institute, The Texas A&M University System, College Station, TX.

Neuman, M., Bright, E.M., 2008. Texas Urban Triangle: Framework for Future Growth. SWUTC/08/167166-1, Texas Transportation Institute, The Texas A&M University System, College Station, TX.

Nicot, J.-P., Scanlon, B.R., 2012. Water use for shale-gas production in Texas. US Environ. Sci. Technol. 46 (6), 3580–3586.

Padowski, J.C., Jawitz, J.W., 2012. Water availability and vulnerability of 225 large cities in the United States. Water Resour. Res. 48 (12), 16.

Pearson, F.F., White, D.E., 1967. Carbon 14 ages and flow rates of water in Carrizo sand Atascosa county Texas. Water Resour. Res. 3 (1), 251–261.

Potter, L.B., Hogue, N., 2011. Texas Population Projections, 2010–2050, Office of the State Demographer, State of Texas, Austin, TX.

Richter, D.d., Mobley, M.L., 2009. Environment. Monitoring Earth's critical zone. Science 326 (5956), 1067–1068.

Riebsame, W.E., Parton, W.J., Galvin, K.A., Burke, I.C., Bohren, L., Young, R., Knop, E., 1994. Integrated modeling of land use and cover change. Bioscience 44 (5), 350–356.

Thorkildsen, D., Quincy, R., Preston, R., 1989. A digital model of the Carrizo–Wilcox Aquifer within the Colorado River basin of Texas. LP-208, Texas Water Development Board, State of Texas, Austin, TX.

TWDB, 2014. Groundwater Management Areas. Texas Water Development Board, Austin, TX.

USGS, 2003. Active mines and mineral plants in the US. In: U.G. Survey (Ed.), USGS Minerals Yearbook – 2003. United States Geological Survey, Reston, VA.

Vaughn, E.G., Crutcher, J.M., Labatt, T.W., McMahan, L.H., Bradford, Jr., B.R., Gruver, M.C., Gruver, M.C., 2012. Water for Texas 2012. Texas Water Development Board, Austin, TX.

Wilding, L.P., Lin, H., 2006. Advancing the frontiers of soil science towards a geoscience. Geoderma 131 (3–4), 257–274.

Zhang, M., Steiner, F., Butler, K., 2007. Connecting the Texas Triangle: economic integration and transportation coordination, The Healdsburg Research Seminar on Megaregions, pp. 21–36.

Chapter 20

A Summary and Future Direction of the Principles and Dynamics of the Critical Zone

John R. Giardino* and Chris Houser**

*High Alpine and Arctic Research Program, Department of Geology and Geophysics and the Water Management and Hydrological Science Graduate Program, Texas A&M University, College Station, Texas, USA; **Department of Geography, Department of Geology and Geophysics, Office of the Dean of Geosciences, Texas A&M University, College Station, Texas, USA

The distinction between the past, present and future is only a stubbornly persistent illusion.

Albert Einstein

20.1 INTRODUCTION

As we immersed ourselves in all the literature relating to Critical Zone research, as mentioned in our introduction, we were both excited and discouraged. Excited to see that researchers from various disciplines are focusing on specific locations, creating a warehouse of data and beginning to engage in what we would describe as true interdisciplinary work. But, we were also discouraged to see that a lot of the research being produced at the various CZOs is still discipline-focused and published in discipline-specific journals. We think the collection of essays in the *Dynamics and Principles of the Critical Zone* brings together scientists from various disciplines to illustrate the exact point the Critical Zone Observations Network is trying to achieve. Numerous points of view are expressed in the various chapters, and they all focus on the dynamics and principles of the Critical Zone.

We, the editors (Giardino and Houser) used the introduction to set the stage for understanding the dynamics and principles of the Critical Zone as established by the National Research Council (NRC, 2001). We provided a brief history of the various committees and workshops that were held in formulating the structure and focus of Critical Zone research. We acknowledged that the term Critical Zone was first used for this zone by Ashley (1998). We also compiled a

Developments in Earth Surface Processes, Vol. 19. http://dx.doi.org/10.1016/B978-0-444-63369-9.00020-3

FIGURE 20.1 The diagram shows the Critical Zone with its various landscapes. The missing puzzle part represents the various components of the Critical Zone. The puzzle parts around the 3D diagram represent the chapters in this book.

map of 64 locations around the world that are focusing on Critical Zone issues. We hope this chapter helped frame the structure and content of all the chapters in this volume.

Figure 20.1 is our way of summarizing the contribution of this book in setting a stage for expanding the study, examination, and appreciation of the importance of the Critical Zone. The cover of the book shows a painting of the Critical Zone by Jeff Wheatcraft after we described our views to him that the Critical Zone extended from the top of the canopy or interface between the atmosphere and the solid Earth to the bottom of the aquifer. We also discussed that Earth consists of various environments. Figure 20.1 summarizes the contributions of this book as puzzle parts that when all put together provides an explanation of the Critical Zone of Earth. The Critical Zone is introduced from the perspective of Systems' Theory. It is noted that Earth scientists have come to rely on various ways to study complex systems. Understanding the flows of energy and mass from one subsystem to another will be fundamental in refining the accuracy of the predictive models developed from Critical Zone research. The focus will be on the identification of thresholds and linked systems and components.

The contributions of the chapters in the book are briefly summarized in the following paragraphs.

Timothy White, Susan Brantley, Steve Banwart, Jon Chorover, Bill Dietrich, Lou Derry, Kathleen Lohse, Suzanne Anderson, Anthony Aufdendkampe, Roger Bales, Praveen Kumar, Dan Richter, Bill McDowell, all of whom are active researchers at various CZOs around the US used *The Role of Critical Zone Observatories (CZO) in Critical Zone Science* to point out that the National Science Foundation (NSF) has created a nontraditional approach to investing considerable financial and human resources into the funding of CZOs around the US. It is their vision that the network of CZOs will empower an array of new approaches to scientific investigations that were not considered by the traditional NSF funding mechanism. They suggest that this new approach will lead to creative and stimulating efforts in understanding the fundamental operations of the important systems of Earth. As mentioned many times throughout the book, the US NSF initiated an integrated approach to the study of Critical Zone of Earth by establishing a series of CZOs. The important point is, these CZOs are selected locations where intense monitoring and data collection sites are focusing on a range of Critical Zone processes. The sites are at various locations, which represent various environments.

Steven M. Quiring, Trenton W. Ford, and Shanshui Yuan use *Climate of the Critical Zone* to show that the Critical Zone, being an open and interconnected system, is affected by climate in both time and space. One of the most significant impacts of climate variability, which manages to affect every other Critical Zone system, is soil moisture. Soil moisture has an impact on vegetation growth and viability, streamflow, evapotranspiration, and weathering processes, which in turn control biotic and abiotic processes throughout the Critical Zone. Understanding the interactions of the Critical Zone with local, regional, and global climate regimes will allow us to better understand not only how climate varies with time and space, but also the other systems it influences.

Gregory A. Pope discusses *Regolith and Weathering (Rock Decay) in the Critical Zone* in his chapter to tie together several processes in the Critical Zone. Weathering and rock-decay processes are central within the Critical Zone theme. Weathering is the process by which the crust of Earth adapts to atmospheric, hydrologic, and biotic influences, making it key in many fields of Critical Zone research. Regolith is the decayed body of rock and sediment that makes up much of the Critical Zone. The process of weathering is complex, including both mechanical and chemical deterioration of solid Earth materials. Most weathering processes take place at the scale of nanometers, involving the fundamental building blocks of the surface of Earth: minerals. Because each system composition is unique, the rates, methods, and efficiency of decay mechanisms vary greatly from location to location. These varying rates of decay allow for varied regolith thickness and age throughout the Critical Zone. In most environments, regolith may be tens of meters thick and may be older than one million years.

John Dixon in his chapter on *Soil Morphology in the Critical Zone: The Role of Climate, Geology, and Vegetation in Soil Formation in the Critical Zone* describes that the formation and nature of soils reflects a complex feedback among climate, geology, and vegetation. Water is the primary mechanism of exchange between the atmosphere, biosphere, hydrosphere, and lithosphere, and the local topography and vegetation cover control effective-water availability. The author argues that the climate and geology are largely independent controls, but that vegetation responds relatively quickly to changes in climate and can have a profound impact on the evolution of the Critical Zone through the recent history of Earth and in the future.

Julia Perdrial, Aaron Thompson, and Jon Chorover in their chapter on *Soil Geochemistry in the Critical Zone: Influence on Atmosphere, Surface- and Groundwater Composition* show that soil is fundamentally the loose material sitting atop bedrock. Because the Critical Zone is an open system, the soil is understandably influenced by many pathways of matter and energy flux, with water being the most important. These fluxes are the drivers of all biotic and abiotic interactions within the soil column. Just as weathering occurs at the nano-particle scale, so do biogeochemical reactions. Many climatic and land-use changes have occurred in what is now referred to as the Anthropocene, which have far reaching effects on the biogeochemical reactions taking place within soil systems. These changes naturally affect the health and viability of the Critical Zone, influencing its ability to sustain life as we are accustomed to it as well as impacting other Critical Zone systems: surface-water chemistry, groundwater composition, and atmospheric quality.

Aniela Chamorro, John R Giardino, Raquel Granados-Aguilar, and Amy E. Price in their chapter on *A Terrestrial Landscape Ecology Approach to the Critical Zone* use landscape ecology to illustrate the need for a transdisciplinary approach to Critical Zone studies. Landscape ecology a discipline in its own right, provides a fundamental framework for studying and understanding the Critical Zone. Landscape ecology and the Critical Zone incorporate humans and their structures and actions as a part of the system to be understood. Landscape ecology strives to examine the systems of Earth as cohesive units by incorporating composition, structure, and function in a horizontal perspective. Critical Zone research has a similar goal of using an interdisciplinary approach to understand landscapes, but in the vertical direction. Because these disciplines are so complimentary, employing a combination of both landscape ecology and Critical Zone research may provide understanding beyond traditional interdisciplinarity, by utilizing a transdisciplinary approach.

G.W. Moore, K. McGuire, P.A. Troch, and G. Barron-Gafford use their chapter on *Ecohydrology and the Critical Zone: Processes and Patterns Across Scales* to demonstrate that ecohydrology investigates the effects of the hydrologic cycle on ecosystem distribution, structure, and function, as well as the effects of biologic processes on the hydrologic cycle. The hydrologic cycle is connected to the biota of Earth from the root–soil–rock interfaces to the vegetation–atmosphere interactions. These interactions can be considered as feedback loops between vegetation and water in environments ranging from water-limited to water-rich and at scales from micropores to catchments.

Ellen Wohl in her chapter, *Rivers in the Critical Zone*, demonstrates that the phrase "river systems" is a broad terminology, which encompasses all streams/channels, their floodplains, and hyporheic zones. By definition, river systems are complex near-surface, heterogeneous Critical Zone regimes, which, along with soil moisture, have far reaching effects on habitat, availability of resources, and the transference of matter and energy among systems within the Critical Zone. River systems are ever-changing interconnected networks that transport material such as sediment and organic carbon. Anthropogenic activity and need has altered the connectivity and stability of many river systems through channelization, construction of dams and levees, as well as through usage for transportation and drinking or irrigation. The goal of most river restoration projects therefore is to restore, as close as possible, natural connectivity conditions. Even though rivers only account for a small portion of the fresh-water supply of Earth, because of their connectivity and locations, they are exceptionally important to both biotic and abiotic functioning within the landscape.

Quanrong Wang and Hongbin Zhan in their chapter, *Characteristic and Role of Groundwater in the Critical Zone*, argue that groundwater is a key element of the Critical Zone. Fortunately, existing flows and transport models that are already commonly used in a number of applications within the Critical Zone. Unfortunately, like all models, these are far from perfect, and we have a long way to go toward understanding the dynamic nature of groundwater and its interactions with other systems within the Critical Zone.

Netra R. Regmi, John R. Giardino, Eric V. McDonald, and John D. Vitek, examine the role of mass movement in their chapter *A Review of Mass Movement Processes and Risk in the Critical Zone of Earth.* They argue that the Critical Zone is a fragile network of interconnectedness between the vegetation atop the soil all the way to the bottom of the aquifer within the regolith and bedrock layers. Mass movement is one of the most violent ways in which the Critical Zone, its life-sustaining resources and life itself, can be harmed. Disaster management is the attempt to understand the driving mechanisms (spatial, temporal, and mechanical) of natural disasters, such as various types of mass movement. The goal of disaster-management research and implementation is to preserve human life, infrastructure, and promote safe human occupation and environmental protection in regions and specific locations prone to mass movement events.

Kevin Gamache, John R. Giardino, Netra Regmi, and John D. Vitek, in their chapter, *The Impact of Glacial Geomorphology on Critical Zone Processes*, discuss the important role glacial landscapes play in the Critical Zone. Globally, Antarctica has the largest concentration of glaciers than any other continent; 98% of Antarctica is covered by ice averaging 1.6 km thick. In the northern hemisphere, Greenland is the leader in ice coverage with over 1 million km^2 and thicknesses as great as 3.2 km. Understandably, in environments such as these, individual large glaciers may play significant roles in their local environment. Glaciers play many roles, including acting as a sink for fresh water, within the Critical Zone. The definition of the Critical Zone (top of canopy to bottom of aquifer) can still apply in these unique systems, but with the sparse vegetation associated with these environments, the "canopy" is the tops of the trees, shrubs, bushes, grasses, mosses, etc., which grow on glacial deposits. In the Polar Regions, an assumption is made that the top of the canopy includes the lower portion of the stratosphere. Glaciers have powerful erosional capabilities; the processes by which glaciers and ice-sheets erode landscapes influence the continuity of the deformed regions and resulting landforms. The authors argue that by studying the landscapes formed by glaciation, we have a window to examine past environments and climates.

Taylor Rowley, John R. Giardino, Raquel Granados-Aguilar, and John D. Vitek present a commonly ignored landscape in the Critical Zone in their chapter, *Periglacial Processes and Landforms in the Critical Zone*. Like glaciated environments, the periglacial environment is dynamic and may have far reaching effects on the Critical Zone. The presence of frost within the periglacial environment influences entities such as hydrologic activity and mass movement, which alter the shape of this unique landscape. Inclusion of periglacial-environment research is absolutely necessary for understanding the Critical Zone of Earth. Human activity and global climate change are altering all systems within the Critical Zone, including the periglacial landscape. Therefore, if we have any intention of preserving this remarkable landscape, we must first understand it and how it interacts with the rest of the Critical Zone.

V.P. Tchakerian and P. Pease in their chapter, *The Critical Zone in Desert Environments*, provide a perspective that deserts or arid landscapes, make up a large part of the surface of Earth. Because deserts are defined by low moisture and sparse vegetation, they commonly have poorly developed soils with low-organic content and high concentrations of salt and carbonates in some cases. Many geologic and geomorphic factors influence the development of desert soils. Formation of regolith, is no different, and develops at a different rate and through altered processes than temperate or tropical soils. Models can be employed to estimate soil and regolith development and hillslope movement potential using a variety of input factors.

Sara Mana, Pablo Ruiz, and Amalia Gutiérrez in *The Critical Zone in Tropical Environments* illustrate that these landscapes are of particular interest in the Critical Zone. The tropical ecosystem has tremendous biodiversity and is of great importance for regional and global-climate regulation. Weathering processes in tropical environments are accelerated by the high temperatures, increased precipitation, and amount of biomass compared to other biomes on Earth, or, regions of the Critical Zone. The tropical environment not only affects the climate system and the rest of the Critical Zone, but is greatly affected by it: increased rain and temperatures act as accelerators on an already fast-paced weathering and biologic activity system and human activity (landuse changes, chemical additives, etc.) have far reaching impacts on not only the immediate tropical system, but also the rest of the Critical Zone.

P. Barrineau, P. Wernette, B. Weymer, S. Trimble, B. Hammond, and C. Houser, explain in their chapter, *Critical Zone of Coastal Barrier Systems* that some of the most important, dynamic, and complex coastal systems within the Critical Zone are barrier islands. Contrary to previous hypotheses, barrier islands have developed and evolved to be complex elements based on a variety of feedback loops. Previously, Barrier Islands were thought to be regressional or transgressional, controlled by variations in sea level, sediment supply, and ideal location conditions. The components of these feedback loops affecting barrier-island formation, maintenance, and destruction include sediment, surface water, precipitation, and ecology. This is the first time that a coastal environment has been considered from a Critical Zone perspective.

Michael P. Bishop, Iliyana D. Dobreva, and Chris Houser in their chapter, *Geospatial Science and Technology for Understanding the Complexities of the Critical Zone*, consider the Critical Zone as a dynamic, ever-changing system from a variety of outside forcing agents (climate change, lithospheric evolution, human activity, etc.). We know a multidisciplinary approach to Critical Zone research is necessary to understand the complexity of the overall system. Geospatial technologies (remote sensing, geomorphometry, and geocomputation) used for different approaches to study large areas with ease, compared to prior methods that relied solely on fieldwork. Geospatial technology allows for data collection, information extraction, and numerical modeling of Critical Zone landscape conditions.

Nazgol Bagheri, in her chapter on *The Built Environment in the Critical Zone: From Pre- to Postindustrial Cities* adds a unique view to the study of the Critical Zone by examining the Critical Zone within the built environment. Whereas most Critical Zone research up to this point has taken place at CZOs, the rest of the Critical Zone is also of great importance and is being impacted by human activity. Anthropogenic activity such as industry, creation of large cities, and other factors have had immense, often adverse, impacts on the Critical Zone. To this point, little research has been conducted to examine the relationship and interactions of urbanization and the Critical Zone. By examining the ways in which Critical Zone resources, particularly water, have influenced decision making regarding urban location and growth in the past, it is possible to model urban growth possibilities of the future. Combining these theoretical models with environmental-sustainability research, it may be possible to design less harmful cities and practices for the future. By implementing vernacular architecture and organic urban design practices, we can hopefully make proactive decisions and changes for designing sustainable cities. Kevin R. Gamache and John R. Giardino use *Natural and Anthropogenic Factors Affecting Groundwater in the Critical Zone of the Texas Triangle Megaregion* to discuss the impact of rapid, current growth in urban areas today and in the future on water resources in the Critical Zone. This type of growth will continue throughout the world. To support the contemporary industrial infrastructure of megaregions requires comprehensive water-related alterations of the Critical Zone. In this region of Texas, humans have constructed an extensive network of dams and reservoirs to capture surface water. In addition to this network, an extensive number of wells for extracting water, oil, and gas from the Critical Zone are also having an historic impact on the Critical Zone. Also, significant, rapid alterations in landcover and interbasin transfer of ground and surface water are exacerbating the situation. Continuing depletion of the water resources of the Critical Zone threatens sustainable development in this heavily groundwater-dependent megaregion of Texas. To remain economically viable in the future will necessitate vigorous, effective, water-resource policies.

20.2 FROM PRESENT TO THE FUTURE

An article in *The Economist* quite pointedly summed up changes we can expect in the future, such that "*we will experience huge pressures to grow food and provide clean water to growing numbers of humans*" (Anon., 2011, p. 11). Thus, increased pressure will be brought to bear on the importance of the Critical Zone for the production of food as well as providing water resources. Critical Zone is a term that will become a common word in both the scientific and the layperson communities. All will come to understand and appreciate the imperative, critical role this relatively thin zone performs in the existence of life on Earth, as we know it today.

We believe two things are in the process of occurring today: an explosion and a revolution. We began this book by talking about what one sees on television, reads in a newspaper or magazine and hears on the radio. Unfortunately, one of the things that seem to have become commonplace are explosions, be they roadside, dynamite, body-carried, gas leaks, or other means. Nature is also responsible for explosions varying from internal pressure in an individual rock, excessive hydrostatic head behind a dam, either natural or human-made, all the way to a supernova. An explosion is simply the quick intensification of the volume and an abrupt liberation of energy. We are not suggesting that the creation of CZOs is an explosion equal to a supernova, but we do think the CZOs are resulting in an explosion of data focused on specific locations.

The second event that we believe is happening via the creation of the Critical Zone concept is a revolution. This revolution involves a structural change in the view organized scientific-funding organizations are taking, which is occurring rather rapidly. We believe that this revolution, is by no means equivalent to the "scientific revolution", which saw the rise of modern science. Nevertheless, a revolution in the way Earth is viewed and data are collected is occurring through integrating a true interdisciplinary approach that expands the spatial extent and spatial connections of Critical Zone research. The spatial connection is very important, as it is formalizing geographical-zone linkages between and among all the Critical Zone Observation sites across the planet.

Three very important mission objectives for the CZOs are to identify and measure fluxes that are occurring, determine the cumulative impact of these fluxes, and then use all these data to develop models that will be accurate in both short- and long-term forecasts of change (Banwart et al., 2011). In geomorphology, the concept of uniformitarianism was applied to the interpretation of

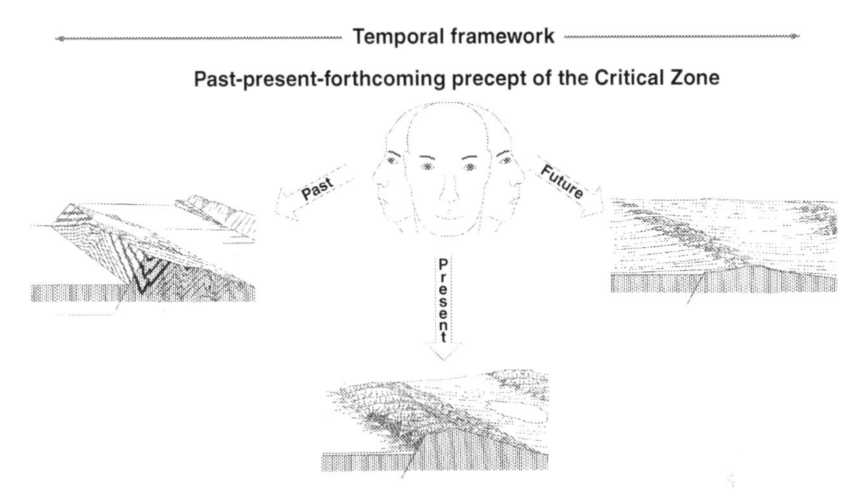

FIGURE 20.2 The diagram illustrates the past, present, and forthcoming precept. This precept is fundamental to applying a temporal framework to the data being collected at the various Critical Zone observation sites around Earth.

landscapes by Hutton (1795) and Playfair (1802), which can be simply stated as the "*present is the key to the past.*" Mathewson (1981) expanded on this original vision for application in engineering geology where the past plus the present can be used to predict the future (Fig. 20.2). We believe the data collected by the expanding network of CZOs along with accurate models will facilitate a type of uniformitarianism for the Critical Zone that we describe as *the past, present and forthcoming precept.* (Fig. 20.2).

We also believe that the definition of the Critical Zone will be modified in the future. The depth will not be confined by the bottom of the aquifer, but will be extended to include the depth of human impact. As we search for resources, the depths that humans mine or drill will continue to increase. Thus, the impacts from human society will reach further into Earth.

A thread that wove through every chapter was water. The thread that connects the various components and dynamics in the Critical Zone is water. As we mentioned in our introduction, we concur that soil plays an important role, but we do not think it is the major connection between all the systems. Thus, we strongly argue that the link between all these components or systems is water and that the research concentration on fluxes will begin to show that water is the true thread that stiches these systems together. NRC (2005) suggested that focus should be on water as the unifying theme in the study of complex environmental systems. Thus, we see a slow migration to water as the major linkage in the Critical Zone in the future.

Lastly, we believe that the geosciences will play a major role in the future study and management of the Critical Zone. As we mentioned, an increased population across the planet will place more and more pressures on the Critical

Zone. In fact, it is becoming rather obvious that "*humans have become a force of nature reshaping the planet on a geological scale – but at a far – faster-than-geological speed*" (Anon., 2011, p. 11). We believe this speed of change will drive a re-engineering of the global environment. This last aspect is an opportunity for the Earth, biological and physical sciences to work with the social sciences to develop models of circumstances and consequences based on human behavior.

ACKNOWLEDGMENTS

We would be remiss if we did not acknowledge the help of Ms. Raquel Granados-Aguilar, Ms. Amy Price, and Ms. Taylor Rowley for all the time and effort they put forth in searching and collecting references, drafting diagrams, and helping us not to lose our sense of humor. We also thank the numerous authors for their contributions to putting a wide-ranging focus on the Critical Zone in a single volume. Finally, we thank Mr. Jeff Wheatcraft for taking our description of the Critical Zone and turning it into a piece of art.

REFERENCES

Anonymous. 2011. The Economist, vol. 399, number 8735. pp. 81–83.

Ashley, G.M., 1998. Where are we headed? "Soft rock research into the new millennium". Geol. Soc. Am. Abstract/Program 30, A-148.

Banwart, S., Bernasconi, S.M., Bloem, J., Blum, W., Brandao, M., Brantley, S., Chabaux, F., Duffy, C., Kram, P., Lair, G., Lundin, L., Nikolaidis, N., Novak, M., Panagos, P., Ragnarsdottir, K.V., Reynolds, B., Rousseva, S., de Ruiter, P., van Gaans, P., van Riemsdijk, W., White, T., Zhang, B., 2011. Soil Processes and functions in Critical Zone Observatories: hypotheses and experimental design. Vadose Zone J. 10, 974–987.

Hutton, J., 1795. Theory of the Earth with Proofs and Illustrations. Edinburgh.

Mathewson, C.C., 1981. Engineering Geology. Bell and Howell Co, North Carolina, p. 3.

National Research Council Committee on Basic Research Opportunities in the Earth Sciences, 2001. Basic Research Opportunities in the Earth Sciences. National Academies Press, Washington, DC.

NRC, 2005. Review of the GAPP Science and Implementation Plan. Committee to Review the GAPP Science and Implementation Plan, National Research Council. Washington, DC.

Playfair, J., 1802. Illustrations of the Huttonian Theory of the Earth. Coldell and Davies, London.

Subject Index

C

D

E

F

G

H

I

K

L

M

O

P

Q

R

S

T

U

X

Z

Printed in the United States
By Bookmasters